Amazonian Deforestation and Climate

Amazonian Deforestation and Climate

Edited by

J.H.C. GASH
Institute of Hydrology, UK

C.A. NOBRE
Centro de Provisão de Tempo e Estudos Climáticos, Brazil

J.M. ROBERTS
Institute of Hydrology, UK

R.L. VICTORIA
Centro de Energia Nuclear na Agricultura, Brazil

JOHN WILEY & SONS
Chichester • New York • Brisbane • Toronto • Singapore

Published by John Wiley & Sons Ltd,
Baffins Lane, Chichester,
West Sussex PO19 1UD, England

National 01243 779777
International (+44) 1243 779777

Reprinted March 1997

Other Wiley Editorial Offices

John Wiley & Sons, Inc., 605 Third Avenue,
New York, NY 10158-0012, USA

Jacaranda Wiley Ltd, 33 Park Road, Milton,
Queensland 4064, Australia

John Wiley & Sons (Canada) Ltd, 22 Worcester Road,
Rexdale, Ontario M9W 1L1, Canada

John Wiley & Sons (Asia) Pte Ltd, 2 Clementi Loop #02-01,
Jin Xing Distripark, Singapore 0512

British Library Cataloguing in Publication Data

A catalogue record for this book is available from the British Library

ISBN 0-471-96734-3

Produced from camera-ready copy supplied by the Institute of Hydrology
Printed and bound in Great Britain by Bookcraft (Bath) Ltd
This book is printed on acid-free paper responsibly manufactured from sustainable forestation,
for which at least two trees are planted for each one used for paper production.

CONTENTS

List of Contributors

I. do Amaral
Instituto Nacional de Pesquisas da Amazônia, CP 478, 69.011, Manaus, AM, Brazil

S. Amaral
Instituto Nacional de Pesquisas Espaciais, Av. dos Astronautas, 1758 CP 515-12201, São José dos Campos, SP, Brazil

R.C.S. Alvalá
Instituto Nacional de Pesquisas Espaciais, Av. dos Astronautas, 1758 CP 515-12201, São José dos Campos, SP, Brazil

J. Brass
NASA Ames Research Centre, Moffet Field, Ca 94035, USA

L.A. Bruijnzeel
Faculty of Earth Sciences, Free University, De Boelelaan 1085, 1081 HV Amsterdam, The Netherlands

C.B. Bunton
Hadley Centre for Climate Prediction and Research, Meteorological Office, London Road, Bracknell, RG12 2SZ, United Kingdom

O.M.R. Cabral
CNPMA-EMBRAPA, CP 69, Jaguariùna, CEP 13820-000, SP, Brazil

G.J. Collatz
NASA/Goddard Space Flight Center, Greenbelt, MD 20771, USA

E. Cuevas
Centro de Ecologia y Ciencias Ambientales, Instituto Venezuelano de Investigaciones Cientificas, Aptdo 21827 Caracas, Venezuela

A.D. Culf
Institute of Hydrology, Crowmarsh Gifford, Wallingford OX10 8BB, United Kingdom

P.J. Curran
University of Southampton, Highfield, Southampton SO9 5NH, United Kingdom

A.C.L. da Costa
Universidade Federal do Pará, Belém, CEP 66075-900, PA, Brazil

J. de P.R. da Costa
Universidade Federal do Pará, Belém, CEP 66075-900, PA, Brazil

H.R. da Rocha
Universidade de São Paulo, Rua do Matão, 1226, 05508-900 São Paulo SP, Brazil

E.P. da Rocha
Universidade de São Paulo, Rua do Matão, 1226, 05508-900 São Paulo SP, Brazil

R.L.C. de Araujo
Federal University of Amazonas, 69072-970, PO Box 2032, Manaus, AM, Brazil

J.R.S. de Souza
Federal University of Pará, 66075-900, PO Box 1611, Belém, PA, Brazil

M.A.F.S. Dias
DCA IAG/USP, Cidade Universitaria, Rua do Matão, 1226, 05508-900 São Paulo SP, Brazil

R.E. Dickinson
University of Arizona, Institute of Atmospheric Physics, PAS Bldg 81, Room 542, Tucson AZ 85721, USA

J.L. Esteves
INCRA, CP 113, 78934 Ji-Paraná RO, Brazil

G.D. Farquhar
Research School of Biological Sciences, Australian National University, Canberra, Australia

G. Fisch
Centro Técnico Aeroespacial/Instituto de Aeronautica e Espaço (CTA/IAE), Av Nelson D'Avila s/n, CP 6019, 12 228-904 São José dos Campos SP, Brazil

G.M. Foody
University of Wales, Singleton Park, Swansea, West Glamorgan SA2 8PP, United Kingdom

J.H.C. Gash
Institute of Hydrology, Crowmarsh Gifford, Wallingford OX10 8BB, United Kingdom

R. Gielow
Instituto Nacional de Pesquisas Espaciais, São José dos Campos, SP, Brazil

B.M. Gomes
Universidade Federal de Rondônia, Ji-Paraná, RO, Brazil

J. Grace
Institute of Ecology & Resource Management, University of Edinburgh, The King's Buildings, Mayfield Road, Edinburgh EH9 3JU, United Kingdom

M.G. Hodnett
Institute of Hydrology, Crowmarsh Gifford, Wallingford OX10 8BB, United Kingdom

M. Honzák
University of Wales, Singleton Park, Swansea, West Glamorgan SA2 8PP, United Kingdom

B. Kruijt
Institute of Ecology & Resource Management, University of Edinburgh, The King's Buildings, Mayfield Road, Edinburgh EH9 3JU, United Kingdom

J. Lean
Hadley Centre for Climate Prediction and Research, Meteorological Office, London Road, Bracknell, RG12 2SZ, United Kingdom

J. Lloyd
Research School of Biological Sciences, Institute of Advanced Studies, Australian National University, GPO Box 475 Canberra, Australia

R. Lockwood
USDA Forest Service, Pacific Southwest Research Station, 4955 Canyon Crescent Drive, Riverside CA 92507, USA

R.M. Lucas
University of Southampton, Highfield, Southampton S017 1BJ, United Kingdom

R.F. da F. Lyra
Universidade Federal de Alagoas, Maceió, CEP 57072-970, AL, Brazil

G. Maitelli
Universidade Federal de Mato Grosso, Av. Feranando Corrêa, s/n, Cuiabá, CEP 78090, MT, Brazil

O. Manzi
Instituto Nacional de Pesquisas Espaciais, São José dos Campos, SP, Brazil

A. de O. Marques Filho
Instituto de Pesquisas da Amazônia, CP478, 69011 Manaus, AM, Brazil

L.A. Martinelli
Centro de Energia Nuclear na Agricultura, USP, Piracicaba, SP, Brazil

P. McCracken
Institute of Ecology & Resource Management, University of Edinburgh, The King's Buildings, Mayfield Road, Edinburgh EH9 3JU, United Kingdom

J.A. McIntyre
Institute of Ecology & Resource Management, University of Edinburgh, The King's Buildings, Mayfield Road, Edinburgh EH9 3JU, United Kingdom

A-L.C. McWilliam
Institute of Hydrology, Crowmarsh Gifford, Wallingford OX10 8BB, United Kingdom

E. Medina
Centro de Ecologia y Ciencias Ambientales, Instituto Venezuelano de Investigaciones Cientificas, Aptdo 21827 Caracas, Venezuela

P. Meir
Institute of Ecology & Resource Management, University of Edinburgh, The King's Buildings, Mayfield Road, Edinburgh EH9 3JU, United Kingdom

A.C. Miranda
Departamento de Ecologia, Universidade Nacional de Brasilia, 70910-900 Brasilia DF, Brazil

H.S. Miranda
Departamento de Ecologia, Universidade Nacional de Brasilia, 70910-900 Brasilia DF, Brazil

M.Z. Moreira
Centro de Energia Nuclear na Agricultura, USP, Piracicaba, SP, Brazil

C.A. Nobre
Centro de Previsao de Tempo e Estudos Climaticos, Instituto Nacional de Pesquisas Espaciais, Rodovia Presidente Dutra km 40, C. Postal 1, 12630-000 Cachoeira Paulista, SP, Brazil

M.D. Oyama
Centro de Previsao de Tempo e Estudos Climaticos, Instituto Nacional de Pesquisas Espaciais, Rodovia Presidente Dutra km 40, C. Postal 1, 12630-000 Cachoeira Paulista, SP, Brazil

H.S. Pinheiro
Centro de Geociencias da Universidade Federal do Pará, Belém, CEP 66075-900, PA, Brazil

F.M.A. Pinheiro
Universidade Federal do Pará, Belém, CEP 66075-900, PA, Brazil

S. Planton
Météo-France, Centre National de Recherche Météorologique, 21057 Toulouse, France

P. Regnier
Laboratoire de Météorologie Dynamique, Ecole Normale Supérieure, 75231 Paris, France

A. Ribeiro
Centro de Energia Nuclear na Agricultura, Universidade de São Paulo, 13400 Piracicaba SP, Brazil

P. Riggan
USDA Forest Service, Pacific Southwest Reserch Station, 4955 Canyon Crescent Drive, Riverside CA 92507, USA

J.M. Roberts
Institute of Hydrology, Crowmarsh Gifford, Wallingford OX10 8BB, United Kingdom

P.R. Rowntree
Hadley Centre for Climate Prediction and Research, Meteorological Office, London Road, Bracknell RG12 2SZ, United Kingdom

T.D. de A. Sá
EMBRAPA-CPATU, Tv Enéas Pinheiro s/n, CEP 66095 Belém, PA, Brazil

E. Salati
FBDS, São Conrado, 22 610-040 Rio de Janeiro, RJ, Brazil

P.J. Sellers
NASA/Goddard Space Flight Center, Greenbelt, MD 20771, USA

J. Tomasella
Instituto de Pesquisas Hidraulicas, CP 15029, 91501-970 Porto Alegre, RS, Brazil

V.N. Ubarana
Fundação Cearense de Meteorologia e Recursos Hidricos (FUNCEME), CP 3221, Av Bezerra de Menezes, 1900, 60325-002 Fortaleza CE, Brazil

E. Vertamatti
Centro Técnico Aeroespacial/Instituto de Aeronautica e Espaço (CTA/IAE), Av Nelson D'Avila s/n, CP 6019, 12 228-904 São José dos Campos SP, Brazil

R.L. Victoria
Centro de Energia Nuclear na Agricultura, Universidade de São Paulo, 13400 Piracicaba SP, Brazil

I.R. Wright
Institute of Hydrology, Crowmarsh Gifford, Wallingford OX10 8BB, United Kingdom

Foreword

E. SALATI
FBDS, 22 610-040 Rio de Janeiro, Brazil

The ABRACOS project was initiated at the end of the nineteen eighties when attitudes to Amazonian deforestation were more polarised than they are today. At one extreme conservationists saw Amazonia as a "green paradise" to be conserved in its entirety at all costs; while at the other extreme the government agencies and the settlers who were carrying out the deforestation saw themselves as pioneers putting unproductive land to good use. They referred to the forest as the "inferno verde" or "green hell". Today attitudes among environmentalists are more pragmatic and it is recognised that, although large reserves of forest must be kept untouched, it is not realistic to keep the whole of Amazonia as a living museum. Conservation must be combined with sustainable development. Equally, from the development point of view, it is now acknowledged that the settlement of Amazonia over the past 30 years has not been successful — people and farming methods cannot simply be transferred from relatively temperate areas to the humid tropics. The climate, rainfall distribution and intensity, and the nature and structure of the soils, as well as crop pests and diseases, all combine to create a very different environment. Development strategies must exploit these differences by adapting or developing technologies appropriate for the environmental conditions of the tropical rainforests. This can only be achieved by research, such as that described in this book, to establish the processes controlling both the forest and the agriculture environment of Amazonia.

The ABRACOS project had the ambitious objective of discovering how the soil and vegetation of Amazonia interact with the atmosphere to affect climate. Implementing a project in the humid tropics is a challenge for experimentalists: conditions in Amazonia are difficult and unforgiving, and the ABRACOS team are to be congratulated on their success. They have collected a wealth of data, and their analyses, which are described in this book, have resulted in a step increase in our knowledge of the movement of water through the soil, vegetation and atmosphere of Amazonia and how that may change with deforestation.

Preface

This book describes the results of the Anglo-Brazilian Amazonian Climate Observation Study (ABRACOS), which were presented at the end-of-project symposium held in Brasilia on 19 to 21 September 1994. The primary objective of ABRACOS was to improve climate model predictions of how Amazonian deforestation might affect climate. This was to be achieved by first collecting new field data and then using those data to calibrate the land surface sub-models of the General Circulation Models which are used in climate prediction. The papers presented here describe how those soil, vegetation and atmospheric data were collected and how they have been analysed in pursuit of the general objective. However as the project progressed it was possible to widen the scope of the study using the ABRACOS framework to include studies of the carbon cycle and the atmospheric boundary layer. The data have also been used in modelling studies using one-dimensional, meso-scale and General Circulation Models. The results reported here cover the whole spectrum of work carried out within the ABRACOS framework.

The papers are arranged essentially in the order in which they were presented in Brasilia: hydrology, soils and microclimate; plant physiology and vegetation; followed by atmospheric boundary layer, and meso- and large scale meteorology. Each of these groups is introduced by a review paper and we are grateful to L.A. Bruijnzeel, E. Medina and R.E. Dickinson for making these presentations.

ACKNOWLEDGEMENTS

ABRACOS was funded by the UK Overseas Development Administration (ODA) in collaboration with the Agência Brasileira de Cooperação (ABC). We are grateful for this financial support, and we would also like to thank all those at ODA and ABC, as well as the British Embassy and the British Council in Brasilia who have helped to make ABRACOS such a smoothly run project. We also thank the UK NERC TIGER programme, who provided additional finance and funded the British universities' contribution; and the Conselho Nacional de Desenvolvimento Científico e Tecnológico (CNPq), the Fundação de Amparo à Pesquisa do Estado de São Paulo (FAPESP) and the Coordenadoria de Aperfeiçoamento de Pessoal de Nível Superior (CAPES), for their help with funding the Brazilian contributions, particularly the Rondônian boundary layer measurements. We are grateful to the Instituto Brasileiro de Meio Ambiente e Recursos Renováveis (IBAMA) and to the Companhia de Vale do Rio Doce for allowing us to use their forest reserves and facilities, and for all the help they have given us.

The initial collaborating organisations in ABRACOS were the UK Institute of Hydrology and the Brazilian Instituto Nacional de Pesquisas Espaciais (INPE), Instituto Nacional de Pesquisas da Amazônia (INPA), Centro de Energia Nuclear na Agricultura (CENA) and Centro de Pesquisa Agroflorestal da Amazônia Occidental (CPAA/EMBRAPA), but as can be seen from the list of contributors to this volume many other British and Brazilian institutions were involved. We are grateful to all the contributors and also to all those who have worked on the project but whose names do not appear as authors — their contribution has been equally important.

We are grateful to Marcos D. Oyama for translating the abstracts into Portuguese, and to Celia Kirby, John Griffin, Charlotte Allen and Nick Fey of the Institute of Hydrology Publications Section for copy-editing, attention to diagrams and preparation of final camera-ready copy.

We would also like to acknowledge a debt to W.J. Shuttleworth who was responsible for the design and initiation of ABRACOS, and who also made a very substantial contribution to the early stages of its implementation.

J.H.C. Gash
C.A. Nobre
J.M. Roberts
R.L. Victoria

February 1996

1 An overview of ABRACOS

J.H.C. GASH[1], C.A. NOBRE[2], J.M. ROBERTS[1] and R.L.VICTORIA[3]
[1] *Institute of Hydrology, Wallingford, Oxon, UK*
[2] *Instituto Nacional de Pesquisas Espaciais, São José dos Campos, Brazil*
[3] *Centro de Energia Nuclear na Agricultura, Universidade de São Paulo, Brazil*

INTRODUCTION

For more than ten years evidence has been accumulating from experiments with General Circulation Models (GCMs) that the replacement of Amazonian rainforest by pasture will produce significant changes in climate: recent studies include those by Nobre *et al.* (1991), and Lean & Rowntree (1993). These experiments are essentially sensitivity analyses to the land surface parameterization, and the credibility of the results therefore depends on how well the land surface sub-models represent the vegetation and soil. Field measurements are a critical component in the process of improving the land surface representation, and of crucial importance to the further understanding of global model experiments and hence of the likely impact of tropical deforestation (Henderson-Sellers, 1991).

Over the past ten years there has been continuing progress in improving land surface sub-models in GCMs. Schemes such as those developed by Dickinson *et al.* (1986), Sellers *et al.* (1986), and Noilhan & Planton (1989) incorporate realistic descriptions of the plant physiological controls on transpiration, and, increasingly, improved descriptions of rainfall interception (e.g. Dolman & Gregory, 1992; Taylor, 1994). However, as models become more realistic, so the need increases for field data against which to calibrate the parameters and validate predictions.

From 1983 to 1985 the UK Institute of Hydrology and the Brazilian National Institutes of Amazonian and Space Research (INPA and INPE respectively) collaborated in the Amazon Region Micrometeorological Experiment (ARME) in the primary rainforest in Reserva Ducke near Manaus. This study produced a dataset (Shuttleworth *et al.*, 1984; Lloyd *et al.*, 1988; Shuttleworth, 1988) which has since been used to calibrate all the GCMs currently being used for Amazonian deforestation experiments, either directly, by optimisation against the actual data (e.g. Sellers *et al.*, 1989), or indirectly by application of the values of the key parameters – albedo, surface roughness, surface conductance and canopy interception capacity – derived during that study. The data have also been valuable for validating model predictions of the current climate (e.g. Shuttleworth & Dickinson, 1989, Lean & Rowntree, 1993).

Using data from this single site study to represent all the rainforest in the world

Amazonian deforestation and climate. Edited by J.H.C. Gash, C.A. Nobre, J.M. Roberts and R.L. Victoria.

clearly involves the pragmatic, simplifying assumption that rainforest is a uniform vegetation type. In reality this is not the case. Even the term rainforest as used in this context is vague and, for example, includes seasonally dry tropical forest, which may behave quite differently to true rainforest growing in areas with no dry season. Nevertheless the data from the forest in Reserva Ducke are probably representative of at least a substantial proportion of Amazonia, and compared to the modelled scenario of a deforested Amazonia, the forested control runs of GCM deforestation experiments were relatively well calibrated. Model experiments had used parameters derived from savannah or temperate grass, neither of which should necessarily be a good representation of the actual pasture. The most urgent requirement was thus for data to calibrate the pasture of the post-deforestation scenario. It was this need which provided the primary motivation for the Anglo-BRazilian Amazonian Climate Observation Study (ABRACOS).

EXPERIMENTAL DESIGN

ABRACOS was designed to run on two time scales (Shuttleworth *et al.*, 1991). On a climatological or hydrological time scale, continuous measurements were made, these being designed to study long-term trends and differences between vegetation types. Meteorological and plant physiological processes were then also studied in detail during intensive measurement campaigns, or missions, typically lasting several weeks or months.

SITES

Three pairs of climatological sites were established in eastern, central and south-western Amazonia. Measurements were made at these sites to compare the long term differences in climate (Culf *et al.*, 1996) and soil moisture (Hodnett *et al.*, 1996) between undisturbed forest and the pasture that has replaced it. The location of these sites was chosen to be broadly representative of different climatic zones in Amazonia, and of the different types of clearance which have been carried out. The location of the sites is shown in Figure 1. The position of each of the sub-sites in relation to the major surrounding areas of forest and pasture is shown in Figure 2.

Table 1 gives the coordinates of each site and their names, which follow those used by Culf *et al.* (1995). The mean monthly rainfall for the standard rainfall stations closest to the ABRACOS sites are shown in Figure 3. The sites near Ji-Paraná in Rondônia are close to the south western edge of the forest zone and have a pronounced dry season between June and August, often with periods of several weeks without rain. The wettest period is December to April. The sites near Manaus, Amazonas, in central Amazonia are subject to a dry period between June and September, but even during these months periods of more than a week without rain are rare. In this case the wettest months are December to April. The sites near Marabá in the state of Pará, in eastern Amazonia, are near to the edge of the zone

Figure 1 The location of the ABRACOS sites

Figure 2 Satellite images showing the areas around the sites of Ji-Paraná (above), Manaus and Marabá (overleaf), showing the location of the sites and the major areas of forest and clearing. The light colour represents cleared forest.

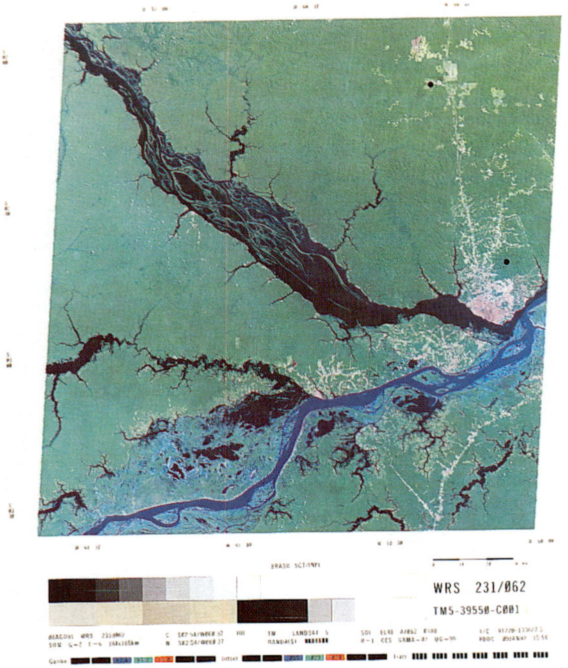

Figure 2 (continued) Satellite images showing the areas around the sites of Manaus (above) and Marabá (below), showing the location of the sites and the major areas of forest and clearing. The light colour represents cleared forest.

Table 1 The principal ABRACOS sites

Site	Location	Vegetation	Established	Coordinates
Reserva Vale do Rio Doce	Marabá, Pará	Forest	July 1991	5° 45' S, 49°10' W
Fazenda Boa Sorte	Marabá, Pará	Grass	July 1991	5° 10' S, 48° 45' W
Reserva Ducke	Manaus, Amazonas	Forest	Sept. 1990	2° 57' S, 59° 57' W
Fazenda Dimona	Manaus, Amazonas	Grass	Sept. 1990	2° 19' S, 60° 19' W
Reserva Jaru	Ji-Paraná, Rondônia	Forest	Oct. 1991	10° 05' S, 61° 55' W
Fazenda Nossa Senhora	Ji-Paraná, Rondônia	Grass	Oct 1991	10° 45' S, 62° 22' W

which is naturally forest, also with a seasonally dry period between June and August. The wettest months again occur between December and April. An analysis of the climate data collected during the project is given by Culf *et al.* (1995b). The soils of Manaus and Ji-Paraná are discussed by Wright *et al.* (1996), who point out that the soils at these sites are close to the textural extremes of Amazonian soils and the sites are therefore well-positioned to capture the range of variability in soil properties.

RONDÔNIA

During the last two decades the forest in Rondônia has been progressively cleared by means of organised colonisation schemes, which involves laying out a road

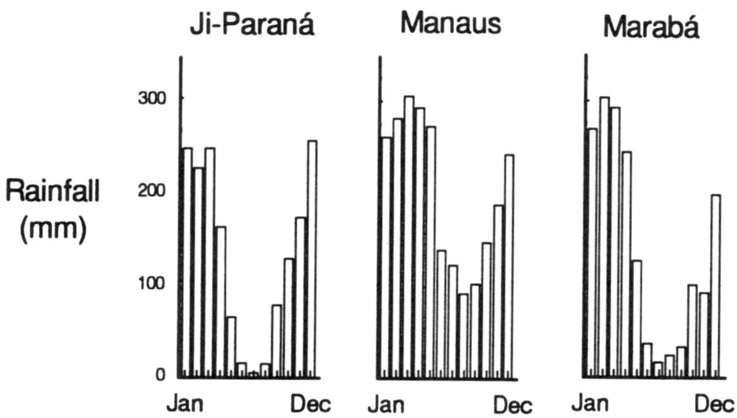

Figure 3 The mean monthly rainfall for the standard climate stations nearest to Ji-Paraná, Manaus and Marabá.

network and then allocating 2 km × 2km square plots to settlers. The settlers clear progressively backwards from the roads, creating the characteristic fishbone pattern which can be seen on satellite images of this area (Figure 2). The area immediately around Ji-Paraná is therefore a mixture of cleared grassland interspersed with strips of forest. The relative sizes of these components depends on when the colonization scheme was initiated; the older the scheme the less forest remains.

To the north east of Ji-Paraná there is a large Indian reserve and beyond that the Reserva Jaru, which is an ecological reserve owned by the Brazilian Environmental Protection Agency (Instituto Brasileiro de Meio Ambiente e Recursos Renováveis, IBAMA). At this site the forest climate measurements were made from the top of a 52 m high tower, while long term measurements of soil moisture (Hodnett *et al.*, 1996), rainfall interception (Ubarana, 1996) and leaf litter fall (Roberts *et al.*, 1996) were made in the immediate vicinity. Climate and soil moisture measurements were also made at the pasture site on the Fazenda Nossa Senhora (da Aparecida), situated north of the town of Ouro Preto d'Oeste, about 80 km south west of the forest site.

The data collection at both these sites is the responsibility of the National Institute of Colonization and Agrarian Reform (Instituto Nacional de Colonização e Reforma Agrária, INCRA) in collaboration with the Federal University of Rondônia and IBAMA. Details of the vegetation at both these sites are given by McWilliam *et al.* (1996). With a view to the use of this pair of sites for atmospheric boundary layer studies they were selected as representing a large area of forest, with an adjacent large area of substantial clearance — large enough for the atmospheric boundary layer to be in equilibrium with the composite surface of cleared areas and forest strips. An upwind fetch of least 30 km was thought necessary for this to occur.

AMAZONAS

There is very little forest clearance in the state of Amazonas, which has poor soils and a lack of roads. There is however a ribbon of clearance along the road which runs north from Manaus to Boa Vista. A site was established on one of these farms, Fazenda Dimona, with parallel climate measurements being made some 60 km away in the Reserva Ducke, using the 45 m tower originally set up for the ARME project in 1983. Reserva Ducke is an area of protected primary forest belonging to the Brazilian National Institute for Amazonian Research (INPA).

Although the expanding city of Manaus is now only some 15 km from the tower site, the reserve is downwind of an extensive area of forest. Soil moisture measurements were made at the Fazenda Dimona clearing and also in forest adjacent to the clearing (Hodnett *et al.*, 1996, 1995). The data collection at these sites is the responsibility of INPA, in collaboration with the Western Amazonian Agroforestry Research Centre (Centro de Pesquisa Agroflorestal da Amazonia Ocidental, EMBRAPA-CPAA). Details of the vegetation around the tower in the Reserva Ducke are given by Roberts *et al.* (1990), and for the Fazenda Dimona site by McWilliam *et al.* (1993).

Together the pair of sites represent a small clearing surrounded by a large area of virtually undisturbed forest. A third climate station was deployed in Manaus city to enable the effect of urbanisation to be assessed (Maitelli & Wright, 1996).

PARÁ

The oldest Amazonian forest clearance is in the state of Pará and is typified by large cattle ranches in which almost all the forest has been cleared. The remaining areas of forest in this region are mainly either Indian reservations, ecological reserves, or large areas still used for Brazil nut collection. A forest site was established in the 17 000 ha ecological reserve belonging to the mining company, Companhia Vale do Rio Doce. The reserve is some 50 km south of Marabá. A pasture site was established some 50 km north east of Marabá, on the Fazenda Boa Sorte, close to the village of São Raimundo. Forest climate measurements were made on top of a 52 m high tower reaching above the canopy; soil moisture measurements were made close to the base of the tower. Data collection for both the pasture and forest sites is the responsibility of the Federal University of Pará. Other long term measurements at the forest site were rainfall interception (Ubarana, 1995) and leaf litter fall (Roberts *et al.*, 1995). Together these sites represent a relatively small forest surrounded by a large clearing. The vegetation at this pair of sites is described by Sá *et al.* (1996).

Table 2 The timing and location of the main experimental campaigns. The micrometeorological campaigns are designated M 1 to 6; the boundary layer campaigns of RBLE are designated 1 to 3.

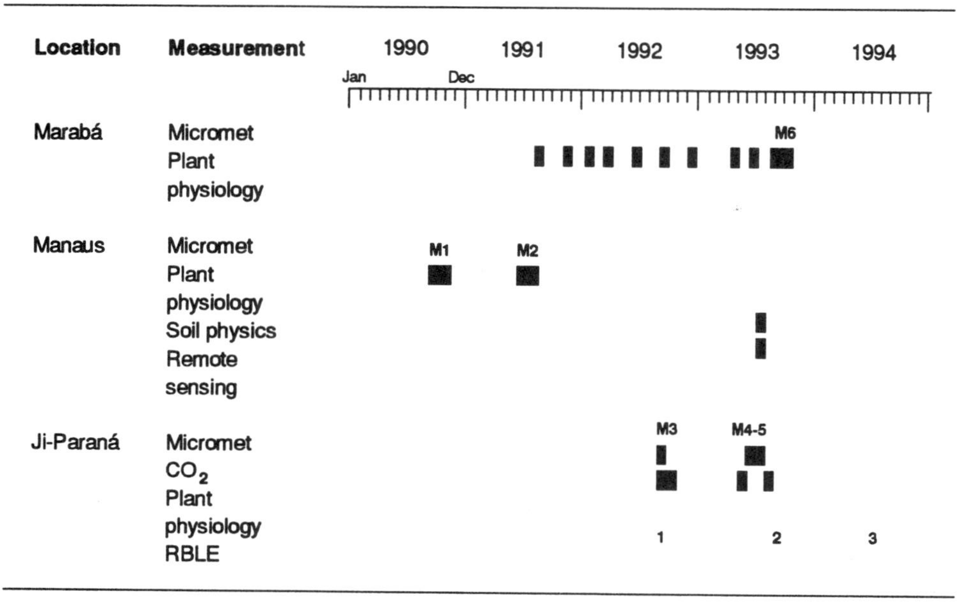

INTENSIVE FIELD CAMPAIGNS

The intensive field campaigns took place during the period 1990 to 1994, with the timing and location of the principal campaigns as summarised in Table 2. Six intensive micrometeorological missions were mounted, all with plant physiological measurements being made in parallel, but additional plant physiological campaigns were also run separately, most notably at Marabá. Studies of the isotopic composition of the water — in both vapour and the liquid phase — were also carried out concurrently with the micrometeorological campaigns (Ribeiro *et al.*, 1996), with the objective of identifying the provenance of the water being evaporated. Soil thermal properties were studied both during the main campaigns (Alvalá *et al.*, 1996), and over longer time periods (de Souza*et al.*, 1996), but a special soil physics campaign was held at Fazenda Dimona during 1993, aimed at deriving soil hydraulic properties, (Tomasella & Hodnett, 1996).

Exploiting the infrastructure put in place for the originally planned measurements, the overall scope of the project was expanded to incorporate atmospheric boundary layer, carbon dioxide and remote sensing studies. The Rondônia Boundary Layer Experiment (RBLE) comprised three campaigns measuring the development of the atmospheric boundary layer over forest and pasture (Nobre*et al.*, 1996). The RBLE campaigns were held in parallel with the micrometeorological campaigns Mission 3 (M3) and Missions 4 and 5 (M4-5), and, in 1994, with limited micrometeorological measurements being made in parallel. Micrometeorological carbon dioxide flux measurements (Grace *et al.*, 1996, 1995a, 1995b), and measurements of stem and soil respiration (Meir *et al.*, 1996) were made in parallel with these campaigns, as well as in the dry savannah, or cerrado, forest near Brasilia (Miranda *et al.*, 1996). Carbon isotope studies (Kruijt*et al.*, 1996) investigated the sources and sinks of CO_2 in the forest canopy and the extent of refixation of respired CO_2. During an intensive campaign in 1993 remote sensing ground truth data were collected in previously cleared, but now abandoned farms, north of Manaus (Honzak *et al.*, 1996; Lucas *et al.*, 1996).

SUMMARY OF PROJECT ACTIVITY AND PROGRESS

1990 AND 1991

ABRACOS started on 1st January 1990. The first measurements were made in one of the cattle ranches in the clearings north of Manaus. Parallel climate measurements were made from the Reserva Ducke tower. During March 1990 the Ducke tower was refurbished, and Fazenda Dimona was selected as the clearing site. The measurements of Mission 1 (M1) were made during August to October of 1990. Although this campaign was originally intended as a trial, in fact a great deal of useful micrometeorological data were obtained from Fazenda Dimona (Wright *et al.*, 1992). The aerodynamic roughness parameters were derived, and a rapid response of the grass to water stress was observed. The evaporation fell from 3.8 mm per day to 2.1 mm per day during an unusual three week period without rain. Comparison of the first two months of climate data by Bastable *et al.* (1993) showed significant differences in the local climate resulting directly from the clearance. Bastable *et al.* (1993) also reported the first measurements of albedo from Amazonian grassland. At the same time as the micrometeorological measurements were being made a major effort was directed to measuring the total above-ground biomass and leaf area of a 20 m × 20 m sample plot of undisturbed forest and comparing the results with those from the pasture (McWilliam *et al.*, 1993). This comparison showed that the total above-ground dry biomass for the forest was the equivalent of 265 tonnes per hectare, while the pasture dry biomass was equivalent to only 6 tonnes per hectare.

Satellite imagery had been used to define the general area for the other two sites around the towns of Ji-Paraná and Marabá. However the scale of the clearance in these areas is such that a field visit revealed only three possible sites near Marabá, and because of the additional requirement for the sites in Rondônia to have sufficiently large up wind fetch for boundary layer studies, only two possible sites could be located, both near Ji-Paraná. The sites of the Reserva Vale do Rio Doce, near Marabá, and Reserva Jaru, near Ji-Paraná, were chosen for logistical reasons. A second micrometeorological mission (M2) was carried out at Fazenda Dimona during July to August 1991. This was earlier in the dry season than M1 and the soils were moister than previously encountered during M1. The consolidated dataset therefore covered a wide range of soil moisture values: this range of conditions has been useful in parameterizing the surface conductance (Wright *et al.*, 1995; da Rocha *et al.*, 1995). Plant physiological measurements were made during this period, both at Fazenda Dimona and in Reserva Ducke. At Reserva Ducke parallel measurements of in-canopy climate were made to allow the leaf-scale measurements to be scaled up to derive canopy fluxes (Cabral *et al.*, 1996). At the same time as M2 was being held near Manaus, a tower was erected and climate and soil moisture stations established near Marabá in the forest at Reserva Vale do Rio Doce and the pasture at Fazenda Boa Sorte. This was followed in October 1991 by the erection of the tower and establishment of long term measurements at Reserva Jaru and Fazenda Nossa Senhora. The Reserva Jaru tower is shown in Figure 4.

1992 TO 1994

After two successful intensive micrometeorological campaigns at Fazenda Dimona, the primary activity during Missions 3, 4 and 5 was moved from Manaus to Ji-Paraná. This allowed an assessment to be made of how representative the Manaus site was, both in terms of differences in climate and soil-type, and by working at a site closer to the forest margin, to give data with more extreme soil moisture deficits. The large cleared area around Ji-Paraná also allowed the atmospheric boundary layer studies to observe the difference in boundary layer growth over the forest and cleared areas.

Mission 3 was held in Fazenda Nossa Senhora, near Ji-Paraná, in August and

Figure 4 The tower in Reserva Jaru

September 1992, and it was also possible to make measurements of evaporation and CO_2 flux, in-canopy climate, and undertake plant physiological and isotope studies from the Jaru tower. The first boundary layer sounding campaign of RBLE was also held at this time.

Missions 4 and 5, originally planned as separate wet and dry season campaigns, were run together to give a continuous set of data from April to July 1993, over both forest and pasture.

Although throughout the project there was an active programme of plant physiological measurements at Marabá, no micrometeorological measurements were possible until Mission 6 was held there in September and October 1993. This was aimed at calibrating the model parameters, such as surface roughness, which are required for deriving evaporation from plant physiological measurements of stomatal conductance.

The evaporation of intercepted rainfall is an important component of the forest water balance. Previously Lloyd & Marques (1988) and Lloyd *et al.* (1988) had reported interception measurements from Reserva Ducke, but there had been no measurements of interception loss from seasonally dry Amazonian forest, where the different leaf area index and pattern of storms might be expected to result in a different fraction of the rainfall being evaporated. To remedy this lack of knowledge measurements of interception loss were made at the Reserva Vale do Rio Doce and Reserva Jaru forest sites to give two years of data between 1992 and 1994. Preliminary results are given by Ubarana (1995).

The only intensive campaign which took place during 1994 was RBLE3. During this campaign, held in August, measurements of surface fluxes were made to coincide with boundary layer measurements made simultaneously over pasture and forest.

CONCLUDING REMARKS

The present paper is intended as an introduction to the thinking behind ABRACOS, and a summary and overview of the observational programme. The results from the component studies of ABRACOS are presented in the separate papers of this volume, and in other papers in the open literature, while the immediate conclusions that can be drawn from the project are summarised here by Nobre *et al.* (1995). The design of ABRACOS was ambitious — it involved building new experimental sites in remote and difficult locations, and running a challenging programme of routine and intensive measurements. The primary aim of the project was to collect data for calibrating climate models, and, as is shown in this and the following papers in this volume, a large amount of data has indeed been collected. Some of this material has already been used for calibrating climate models, but the objective is that the data should be as widely used as possible and for this reason will be assembled into a database and made available in the public domain.

ACKNOWLEDGEMENT

The design and initiation of ABRACOS is a result of the foresight of W.J. Shuttleworth (Shuttleworth, 1989; Shuttleworth *et al.*, 1991), who also made a very substantial contribution to the early stages of its implementation. We are happy to acknowledge that debt.

REFERENCES

Alvalá, R.C., Gielow, R., Wright, I.R. and Hodnett, M.G. 1996. Thermal diffusivities of Amazonian soils. This volume.

Bastable, H.G., Shuttleworth, W.J. Dallarosa, R.L.G., Fisch, G. and Nobre, C.A., 1993. Observations of climate, albedo and surface radiation over cleared and undisturbed Amazonian forest. *Int. J. Climatology* **13**, 783-796.

Cabral, O.M.R., McWilliam, A.L.C. and Roberts, J.M., 1996. In-canopy microclimate of Amazonian forest and estimates of transpiration. This volume.

Culf, A.D., Fisch, G. and Hodnett, M.G. 1995. The albedo of Amazonian forest and ranchland. *J. Climate*, in press.

Culf, A.D., Esteves, J.L., Marques, A.de O. and da Rocha, H.R., 1996. Radiation, temperature and humidity over forest and pasture in Amazônia. This volume.

Dickinson, R.E., Henderson-Sellers, A., Kennedy, P.J and Wilson, M.F., 1986. Biosphere- Atmosphere transfer scheme (BATS) for the NCAR community climate model. NCAR Technical Note, NCAR/ TN-275+STR, 69.

Dolman, A.J. and Gregory, D., 1992. The parameterization of rainfall interception in GCMs. *Q. J. R. Meteorol. Soc.* **118**, 455-467.

Grace, J, Lloyd, J., McIntyre, J., Miranda, A., Meir, P. and Miranda, H., 1996. Carbon flux to Amazon forest in Rondônia. This volume.

Grace, J, Lloyd, J., McIntyre, J., Miranda, A., Meir, P., Miranda, H., Moncrieff, J., Massheder, J. Wright, I and Gash, J, 1995a. Fluxes of carbon dioxide and water vapour over an undisturbed tropical forest in south-west Amazonia. *Global Change Biology* **1**, 1-12..

Grace, J., Lloyd, J., McIntyre, J., Miranda, A., Meir, P., Miranda, H., Nobre, C., Wright, I. and Gash, J., 1995b. Net carbon dioxide uptake by undisturbed tropical rainforest in 1992/93. Submitted to Nature.

Henderson-Sellers, A, 1991. A commentary on: Tropical deforestation: albedo and the surface energy balance. *Climatic Change* **19**, 135-137.

Hodnett, M.G., Pimental da Silva, L., da Rocha, H.R. and Senna, R.C., 1995. Seasonal soil water storage changes beneath central Amazonian rainforest and pasture. *J. Hydrol.*, in press.

Hodnett, M.G., Oyama ,M. and Tomasella, J., 1996. Comparisons of long-term soil water storage behaviour under pasture and forest in three areas of Amazonia. This volume.

Honzak, M., Foody, G.M., Lucas, R.M., Curran, P.J. Amaral, I. and Amaral, S. 1996. Estimation of the leaf area index and total biomass of tropical regenerating forest: A comparison of methodologies. This volume.

Kruijt, B., Lloyd, J., Grace, J., McIntyre, J., Raupach, M., Farquhar, G., Miranda, A.C. and Cracken, P. 1996. Sources and sinks of CO_2 in Rondonian tropical rainforest, inferred from concentrations and turbulence along a vertical gradient. This volume.

Lean, J. and Rowntree, P., 1993. A GCM simulation of the impact of Amazonian deforestation on climate using an improved canopy representation. *Q. J. R. Meteorol. Soc.* **119**, 509-530.

Lloyd, C.R., Gash, J.H.C., Shuttleworth, W.J. and Marques, A. de O. 1988. The measurement and modelling of rainfall interception by Amazonian rain forest. *Agric. and Forest Meteorol.* **43**, 277-

294.
Lloyd, C.R. and Marques, A.de O. 1988. Spatial variability of throughfall and stemflow measurements in Amazonian rainforest. *Agric. For. Meteorol.* **42**, 63-73.
Lucas, R.M., Curran, P.J., Honzak, M., Foody, G.M., do Amaral, I. and Amaral, S. 1996. Disturbance and recovery of tropical forests: balancing the carbon account. This volume.
Maitelli, G.T. and Wright, I.R. 1996. The climate of a riverside city in the Amazon Basin: urban-rural differences in temperature and humidity. This volume.
McWilliam, A.L.C., Roberts, J.M., Cabral, O.M.R., Leitão, M.V.B.R., de Costa, A.C.L., Maitelli, G.T. and Zamporoni, C.A.G.P. 1993. Leaf area index and above-ground biomass of terra firme rain forest and adjacent clearings in Amazonia. *Functional Ecol.* **7**, 310-317.
McWilliam, A.L.C., Cabral, O.M.R., Gomes, B.M., Esteves, J.L. and Roberts, J. 1996. Forest and pasture leaf-gas exchange in south-west Amazonia. This volume.
Meir, P., Grace, J and Lloyd, J. and Miranda, A. 1996. Soil respiration in a rainforest in Amazonia and in cerrado in central Brazil. This volume.
Miranda, A.C., Miranda, H.S., Lloyd, J. Grace, J., McIntyre, J.A., Meir, P., Riggan, P., Lockwood, R. and Brass, J. 1996. Carbon dioxide fluxes over a cerrado *sensu stricto* in Central Brazil. This volume.
Nobre, C.A., Sellers, P.J. and Shukla, J. 1991. Amazonian deforestation and regional climate change. *J. Climate* **4**, 957-988.
Nobre, C.A., Gash, J.H.C., Roberts, J.M. and Victoria, R.L. 1996. The conclusions from ABRACOS. This volume.
Nobre, C.A., Fisch, G., da Rocha, H.R., Lyra, R.F.F., da Rocha, E.P. and Ubarana, V.N. 1996. Observations of the atmospheric boundary layer in Rondônia. This volume.
Noilhan, J. and Planton, S. 1989. A simple parameterization of land surface processes for meteorological models. *Mon. Weath. Rev.* **117**, 536-549.
Ribeiro, A., Victoria, R.L., Martinelli, L.A. Moreira, M.Z. and Roberts, J.M. 1996. The isotopic composition of atmospheric water vapour inside a canopy in the Amazon forest: vertical and diurnal variation. This volume.
Roberts, J., Cabral, O.M.R. and de Aguiar, L.F. 1990. Stomatal and boundary-layer conductances measured in a terra firme rain forest, Manaus, Amazonas, Brazil. *J. Appl. Ecol.* **27**, 336-353.
Roberts, J., Cabral, O.M.R, de Costa, J.P., McWilliam, A.L.C. and Sá, T.D. de A. 1996. An overview of the leaf area index and physiological measurements during ABRACOS. This volume.
da Rocha, H.R., Nobre, C.A., Bonatti, J.P., Wright, I.R. and Sellers, P.J. 1995. A vegetation-atmosphere interaction study for Amazonian deforestation using field data and a single-column model. *Q. J. R. Meteorol. Soc.*, in press.
Sá, T.D. de A., da Costa, J. de P.R. and Roberts, J.M. 1996. Forest and pasture conductances in southern Pará, Amazonia. This volume.
Sellers, P.J., Mintz, Y., Sud, Y.C. and Dalcher, A.,1986. A simple biosphere model (SiB) for use with general circulation models. *J. Atmos. Sci.* **43**, 505-531.
Sellers, P.J., Shuttleworth, W.J., Dorman, J.L., Dalcher, A. and Roberts, J. 1989. Calibration, the simple biosphere model (SiB) for Amazonian tropical forest using field and remote sensing data: Part 1, average calibration with field data. *J. Appl. Meteorol.* **28**, 727-759.
Shuttleworth, W.J. 1988. Evaporation from Amazonian rain forest. *Proc. Roy. Soc.(Lond.) B*, **233**, 321-346.
Shuttleworth, W.J. 1989. Priorities in climate-related hydrological research in Amazonia. Institute of Hydrology report (unpublished).
Shuttleworth, W.J., Gash, J.H.C., Lloyd, C.R., Moore, C.J., Roberts, J., Marques, A. de O., Fisch, G., Silva, V. de P., Ribeiro, M.N.G., Molion, L.C.B., de Abreu Sá, L.D., Nobre, C., Cabral, O.M.R., Patel, S.R., and de Moraes, J.C. 1984. Eddy correlation measurements of energy partition for Amazonian forest. *Q. J. R. Meteorol. Soc.* **110**, 1143-1162.
Shuttleworth, W.J. and Dickinson, R.E. 1989. Tropical deforestation: comment on the discrepancy between recent GCM and observational results. *Q. J. R. Meteorol. Soc.* **115**, 1177-1179.

Shuttleworth, W.J., Gash, J.H.C., Roberts, J.M., Nobre, C.A., Molion, L.C.B. and Ribeiro, N.M.G. 1991. Post-deforestation Amazonian climate: Anglo-Brazilian research to improve prediction. *J. Hydrol.* **129**, 71-86.

de Souza, J.R.S., Pinheiro, F.M.A., de Araujo, R.L.C., Pinheiro, H.S. and Hodnett, M.G. 1996. Temperature and moisture profiles in soil beneath forest and pasture areas in eastern Amazonia. This volume.

Taylor, C.M. 1995. Aggregation of wet and dry land surfaces in interception schemes for General Circulation Models. *J.Climate.* **8**, 441-448.

Tomasella, J. and Hodnett, M.G. 1996. Soil hydraulic properties and van Genuchten parameters for an oxisol under pasture in central Amazonia. This volume.

Ubarana, V.N. 1995. Observation and modelling of rainfall interception loss in two experimental sites in Amazonian forest. This volume.

Wright, I.R., Gash, J.H.C., da Rocha, H.R., Shuttleworth, W.J., Nobre, C.A., Maitelli, G.T.M., Zamporoni, C.A.G.P. and Carvalho, P.R.A. 1992. Dry season micrometeorology of central Amazonian ranchland. *Q. J. R. Meteorol. Soc.* **118**, 1083-1099.

Wright, I.R., Manzi, A.O. and da Rocha, H.R., 1995. Canopy surface conductance of Amazonian pasture: model application and calibration for canopy climate. *Agric. For. Meteorol.* **75**, 51-70.

Wright, I.R., Nobre, C.A., Tomasella, J., da Rocha, H.R., Roberts, J.M., Vertamatti, E., Culf, A.D., Alvalá, R.C., Hodnett, M.G. and Ubarana, V.N. 1996. Towards a GCM surface parameterisation for Amazonia. This volume.

RESUMO

O objetivo principal do Projeto ABRACOS (Anglo-Brazilian Amazonian Climate Observation Study) consistiu em fornecer dados para melhorar as previsões dos impactos climáticos causados pelo desmatamento da Amazônia. Esse objetivo pode ser alcançado melhorando-se a representação, nos Modelos de Circulação Global (MCGs), da vegetação de pastagem, que geralmente substitui a floresta, e coletando-se dados climáticos a fim de validar algumas das previsões desses modelos. Dados foram coletados e utilizados para calibrar modelos de parametrização da superfície, os quais foram então usados em experimentos mais realísticos com MCGs.

O Projeto ABRACOS foi realizado em três áreas experimentais da Amazônia brasileira, entre 1990 e 1994. A estratégia consistiu na coleta contínua de dados climáticos e de água no solo nas áreas experimentais (floresta e pastagem) dos três sítios. Estas medições foram complementadas por uma série de seis missões experimentais, nas quais medidas intensivas micrometeorológicas, de solo e de fisiologia vegetal foram obtidas. Durante o projeto, foi possível expandir o escopo dos experimentos para incluir: medidas da camada limite atmosférica acima da floresta e pastagem, medidas de fluxos de dióxido de carbono e também o uso de sensoriamento remoto para estimar o crescimento de florestas secundárias.

2 Predicting the hydrological impacts of land cover transformation in the humid tropics: the need for integrated research

L.A. BRUIJNZEEL
Vrije Universiteit, Amsterdam, the Netherlands

INTRODUCTION

Despite widespread concern and increased international efforts at conservation, the world's tropical forests continue to disappear at an increasing rate. Land-use practices that proved stable for centuries under low to moderate population pressures are breaking down; large numbers of landless people invade logged-over forests to practice slash-and-burn agriculture; vast tracts of tropical forest are also being converted to other uses, such as livestock ranches, extractive tree plantations (rubber, oil palm, cocoa, etc.) and, to a lesser extent, forest plantations, settlements, and mining. Estimates of tropical forest alteration rates during the period 1981-1990 converge around a figure of about 17 million ha a^{-1} (Lanly *et al.*, 1991), which is considerably higher than the 11.5 million ha of forest that were estimated to have disappeared annually during the second half of the previous decade (Lanly, 1982).

Moreover, the latest figures on tropical forest alteration suggest a further increase to about 20 million ha a^{-1} for the early 1990s (Lal, 1995). While one could argue about the precision of these values, the wealth of individual publications presenting evidence of the rapidly diminishing forest cover in a large number of tropical countries (*cf.* Proctor, 1989) is such that the problem has forced itself onto the agendas of economic summit conferences. Vivid descriptions of environmental havoc wrought by 'deforestation' have led many to believe that tropical forest alteration or removal will *automatically* cause: major floods; the choking of rivers, reservoirs, irrigation canals and coral reefs with sediment; lower water yields of springs and streams; lower and more erratic rainfall and hence 'desertification', etc. Hamilton and King (1983) and Bruijnzeel (1986; 1990) have shown that apparently conflicting views of the role of various land-use types with regard to hydrology and sedimentation in the humid tropics can be reconciled by describing the impacts of *specific activities* and types of conversion, each of which may be characterized by a particular level of disturbance, rather than by lumping them under such meaningless

Amazonian deforestation and climate. Edited by J.H.C. Gash, C.A. Nobre, J.M. Roberts and R.L. Victoria.

general terms as 'deforestation'. As such, forests subject to occasional fire ('natural' or in the context of shifting cultivation), cyclones or (selective) logging may recover more or less to their previous state if left alone long (Saldarriaga, 1987; Whitmore, 1989; Horne and Hickey, 1991). Such disturbances may therefore be ranked as of intermediate intensity. The environmental impacts of various logging regimes have been reviewed in detail by Bruijnzeel (1992a). This contribution will focus on the hydrological impacts of more intense perturbations such as the conversion of tropical forest to some form of agriculture or plantation. The paper will summarize the available evidence with respect to the effects of forest conversion on (meso)-climate, total water yield, dry season flow and catchment response to rainfall. On the basis of this review, issues requiring specific research efforts will be identified and suggestions made as to how these could be effected through the establishment of a pantropical network of research sites that capture the environmental variability encountered in the humid tropics (Bruijnzeel and Abdul Rahim, 1992). Although based to a large extent on two earlier reviews of the literature (Bruijnzeel, 1990; 1993) an effort has been made to update them to the autumn of 1994.

HYDROLOGICAL EFFECTS OF LAND COVER TRANSFORMATION IN THE HUMID TROPICS

Forest land can be cleared in a number of ways, each of them characterized by a specific degree of soil disturbance. This affects soil compaction, infiltration, water retention capacity and hence the growth (and water use) of the vegetation replacing the forest. Therefore, before reviewing the respective hydrological impacts of forest conversion, the direct effects of some of the most common methods of forest clearing are discussed briefly.

METHODS OF LAND CLEARING

Couper *et al.* (1981) compared several methods of forest clearing under Nigerian conditions, ranging from traditional slash-and-burn through modernised manual clearing (chainsaw) to highly mechanized techniques, in terms of man hours and energy expenditure, whereas Lal (1981) reported on the associated surface runoff percentages and erosion rates for no-tillage agriculture during the first year after clearing. Manual clearing was the slowest and most expensive method of forest clearing, yet subsequent soil erosion was as low as 0.4 t ha^{-1} a^{-1} compared to about 4 t ha^{-1} a^{-1} after clearing by crawler tractor with a shear blade attached, and 15 t ha^{-1} a^{-1} in the case of crawler tractors with tree pusher/root rake attachments. Corresponding values of surface runoff were 1%, 6.5% and 12% of incident rainfall, respectively. Similarly, Dias and Nortcliff (1985b) reported negligible change in (physical) topsoil properties in the case of traditional slash-and-burn clearance but considerable deterioration after clearing by bulldozers, both because of compaction and removal of topsoil, thereby exposing material with less favourable properties

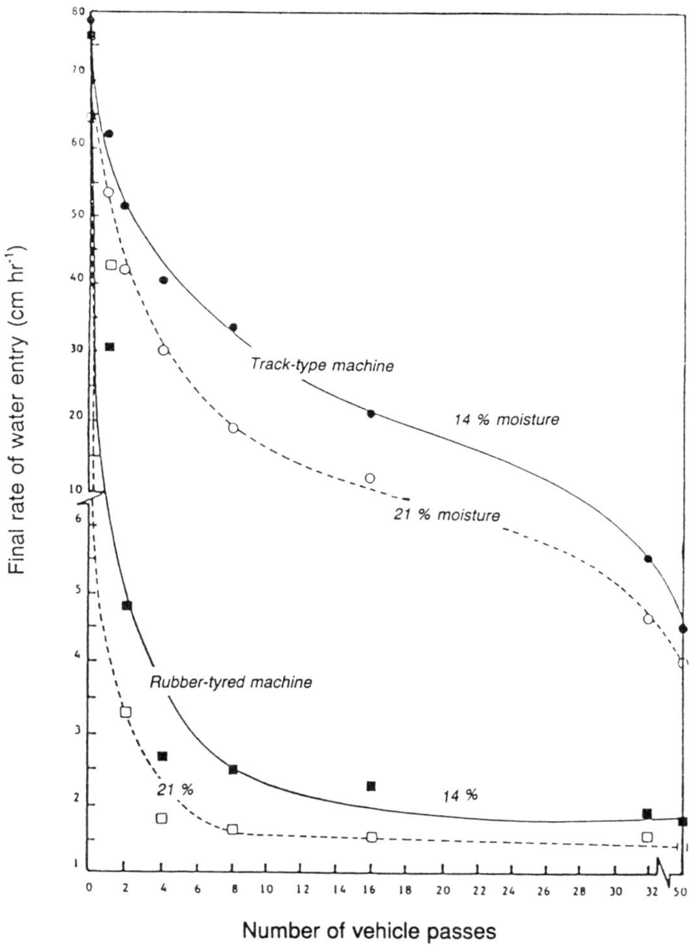

Figure 1 The impact of rubber-tyred and tracked skidding machinery on topsoil infiltration capacity (cm h^{-1})as a function of the number of vehicle passes and soil wetness for an Ultisol in Peninsular Malaysia (adapted from Kamaruzaman Jusoff,1991).

(*cf.* Malmer and Grips 1990). The experiments of Dias and Nortcliff (1985a) and Kamaruzaman Jusoff (1991) in central Amazonia and Peninsular Malaysia, respectively, with various types of tractor equipment revealed a close relationship between the number of tractor passes and the resulting degree of soil compaction. In both cases, soil deterioration increased with soil moisture content. Interestingly, the impact of wheeled machinery was considerably greater than that of tracked vehicles (Figure 1), particularly after manoeuvres involving wheel slippage (Dias and Nortcliff, 1985a). The effects usually extend over the first 15-20 cm of soil and, in the case of former log haulage tracks, have been reported to show very little improvement with time, even after many years of non-usage (Hendrison, 1990;

Malmer and Grips, 1990; van der Plas and Bruijnzeel, 1993). The examples cited thus far all pertained to loamy or clayey soils. Sandy soils, on the other hand, may be expected to show a somewhat different response to disturbance because of their skeletal (as opposed to aggregated) structure. Mechanised clearing of forest on podzols in Sabah did not result in a significant increase in topsoil bulk density although steady-state infiltration capacity (but not sorptivity) was reduced significantly ($P<0.1$ only; Malmer and Grips, 1990). A similar result was reported recently by Jetten (1994) for white sand soils in Guyana. However, taking only such basic soil properties as bulk density, porosity and saturated hydraulic conductivity into consideration may give a false impression. With the exception of valley bottoms, saturated conditions will only rarely if ever occur in sandy soils. It is important, therefore, to examine the effects of disturbance over the entire range of soil water suctions as represented by soil water retention and unsaturated hydraulic conductivity curves. Jetten (1994) compared the values of the so-called van Genuchten parameters of fitted water retention and unsaturated conductivity curves for undisturbed and disturbed sandy soils in Guyana. White sand soils were affected most by the repeated passage of wheeled skidders. The corresponding pF-curves became more gradual, i.e. suction values for a given moisture content increased because of a shift in pore distribution towards finer pores (Figure 2, top). Unsaturated hydraulic conductivities decreased accordingly, particularly at the wetter end of the suction range (Figure 2, bottom). The explanation for this phenomenon may lie in compaction of the originally loosely packed sand but possibly also in the mixing of fine organic matter with mineral soil particles. The effect was less pronounced for brown sand soils but the changes occurred over the entire suction range (Jetten, 1994).

Apart from the 'direct' consequences of forest disturbance such as conversion, there are also important 'indirect' changes in soil physical characteristics, particularly where deforestation is followed by agricultural cropping. For example, the exposure of mineral soil to the elements often results in a deterioration of the soil's rainfall absorbing capacity due to a combination of heating and drying out of the topsoil, the disappearance of organic matter and the faunal communities that used to thrive there, splash erosion, surface sealing, etc. Such effects are well documented (Lal, 1987). Also, as the organic matter content of a soil decreases after clearance, organic acids released in the process cause the partial destruction of the clay-humus complex. This, in turn, leads to a greater dispersion and downward migration of clay particles which tends to obstruct macropores and reduce the hydraulic conductivity (Martins *et al.*, 1991).

Although manual methods of forest clearing are superior to most mechanical equipment in terms of damage to the forest floor, there will often be no choice but to resort to mechanical means. However, when bringing new land into cultivation has the objective of carrying out as much as possible as 'cheaply' as possible, short-term gains may well be offset by the costs of restoring or maintaining soil productivity in the long-term. For instance, van der Weert (1974) and Seubert *et al.* (1977) drew attention to the negative effects of mechanised clearing on root development and therefore on agricultural production. Similarly, soil compaction resulted in the poor

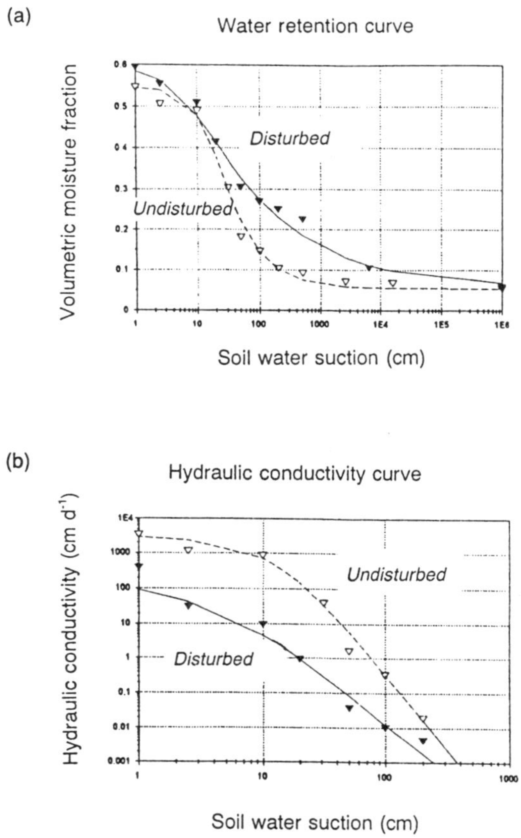

Figure 2 Changes in the water retention curve (top) and the hydraulic conductivity curve (bottom) for a 'Giant Podzol' in Guyana after mechanised logging (after Jetten, 1994).

growth of newly planted eucalypts and *Acacia mangium* trees following forest clearance using heavy equipment, 'windrowing' and burning of logging debris in French Guyana (Fritsch 1992) and Sabah, Malaysia (Malmer 1993), respectively. Interestingly, in the Sabahan study the manual extraction of timber and leaving slash on-site, although initially more expensive than the use of crawler tractors and burning of slash, proved to be more economical soon after planting of the new trees because the former treatment required far less weeding than the latter; tree growth was also double (Sim and Nykvist, 1991).

In view of the adverse effects of mechanized clearing and burning of felling debris, there are strong arguments for the use of alternative timber extraction methods, such as skyline yarding (Ludwig, 1992), elephants, where the size of the logs allow (Whitmore, 1990), or even helicopters in the case of well-stocked forests in steep terrain (Blakeney, 1992) and to leave slash on site to rot rather than burning it. Sim and Nykvist (1991) discussed the considerable problems faced by workers on the ground when felling debris is not burned and proposed various compromises (see also Martin 1970; Van der Weert 1974; Couper *et al.* 1981).

EFFECT ON MESOCLIMATES

Rainfall in the tropics is notoriously variable, both in space and in time (Nieuwolt 1977). In addition, irregular cyclic patterns occur, such as the Quasi-Biennial Oscillation (QBO) and the El Niño Southern Oscillation (ENSO), which exhibit cycles of 2-2.5 years (Partharasathy and Dhar, 1976; Lhomme, 1981) and 2.8-8 years (WMO, 1988), respectively. Moreover, at a longer time scale, it has been suggested that cyclic patterns in rainfall activity of about 10, 21 and 32 years are related to variations in single, double and triple sunspot activity, respectively (Vines, 1986). The sound establishment of any trends in rainfall in a particular region is difficult. Even more difficult, if not intractable in most cases, is the determination of a direct cause and effect relationship between 'deforestation' and diminished or otherwise altered rainfall patterns. There are two basic approaches: (i) the (statistical) analysis of long-term rainfall records in combination with concurrent information on changes in land use; and (ii) computer simulation of regional climate. As will be clear from the examples discussed below, each of the two methods has its limitations.

Time series analysis

Circumstantial evidence for (at least temporarily) decreased rainfall at individual or groups of rainfall stations in the tropics abounds in the literature (see Meher-Homji (1989) for a recent review). Although such reports often blame 'deforestation', they have rarely taken into account large-scale weather patterns (including ENSO events) or cyclical fluctuations, nor have they applied rigorous statistical techniques to their data. Those who did, usually found trends to be either non-significant or only weakly significant. For example, Mooley and Parthasarathy (1983) examined the occurrence of above- and below-average annual rainfall totals for 306 stations all over the Indian subcontinent between 1871 and 1980 and were unable to detect any statistically significant trends. Fleming (1986) analysed time trends in annual rainfalls for ten stations in Costa Rica having records ranging between 28 and 95 years and found that lowland sites all showed a decrease with time, while virtually all hill stations exhibited an increase. Although these trends were rarely statistically significant, Fleming (1986) suggested that they could have been caused by the conversion of (semi-evergreen) forest to pastures and annual cropping in the lowlands. Tangtham and Sutthipibul (1989) reported a significant negative correlation between 10-year-moving averages of annual rainfall and remaining forest area in Northern Thailand for the period 1951-1984 whilst a positive correlation was found between forest area and the number of rainy days. In other words, showers tended to become more frequent and smaller, although the authors pointed out that the effect of deforestation, if any, was still within one standard error of the means of the respective time series. Similarly, Ojo (1987) concluded from his analysis of long-term (1901-1985) rainfall data for 60 stations in West Africa that no regular pattern could be observed in trends, periodicities and persistence for decadal, quinquennial or annual rainfall totals. Nevertheless, he noted that drought conditions had been particularly persistent since the mid 1960's and called for follow-up studies to confirm his findings.

Despite these difficulties, evidence of persistent trends in either the total amount of rainfall or the intensity of the dry season, or both, is accumulating for various parts of the world. Wasser and Harger (1992) reported that the average length of the dry season for five stations in Java had increased from 4.4 months at the beginning of the century to 5.4 in 1991. The slope of this trend differed from zero at the 1% significance level. Interestingly, drought years in Java before 1970 were not always related to ENSO events while the seven droughts occurring after 1970 were (Wasser and Harger, 1992). Wisely, the authors did not touch upon the question as to whether the inferred tendency towards increased aridity was related to changes in land use or not. However, the wealth of climatic information for the Indonesian archipelago (more than 4600 rainfall stations at present, with many records spanning more than a century; Berlage, 1949; Wasser and Harger, 1992) certainly invites further in-depth analyses. Another long-term negative trend in rainfall (of about 5.5 mm a^{-1} on average) has been reported for upland Sri Lanka (by Bandara and Kuruppuarrachchi, 1988). Although the area in question experienced substantial conversion of (montane) forest to tea plantations, it is difficult to ascertain whether this could have affected regional rainfall. Overall water use of tea plantations growing on deep volcanic soils in Kenya has been shown not to differ dramatically from that of montane forests (Blackie, 1979a; Cooper, 1979). However, the situation in upland Sri Lanka differs in that not only are the soils much shallower but also the dry season is much more severe (Werner, 1988). As such, there may have been a substantial reduction in evapotranspiration following the conversion of forest. Alternatively, the observed drop in rainfall may be related to a shift in the location of the ITCZ/equatorial trough over Sri Lanka reflecting larger-scale changes in atmospheric circulation (Arulanantham, 1982).

A particularly interesting case is presented by the rainfall records for Côte d'Ivoire. Early analyses (Snoeck, 1975) did not reveal any anomalies but later work (Quencez, 1989; Olivry *et al.*, 1993) strongly suggested a 'rupture' in rainfall patterns that took place in the late 1960s to early 1970s. A similar tendency has been observed for Nigeria (Adejuwon *et al.*, 1990). The latter study revealed not only a general decrease in annual rainfall totals but also noted a delay in the onset of the rainy season and a more rapid retreat of the rains. These findings are in close agreement with an earlier analysis of long-term rainfall records for the somewhat drier zone to the north by Hubert *et al.* (1989). Despite all this, rainfall in Côte d'Ivoire has been on the rise again since 1990 (R. van Rompaey, *personal communication*), conforming to a postulated 32-year cyclic pattern (Figure 3).

In view of the potentially devastating consequences of a negative long-term trend in rainfall for the availability of water for agricultural and other purposes (Olivry *et al.*, 1993; Servat and Sakho, 1993) a major research programme (ICCARE) has

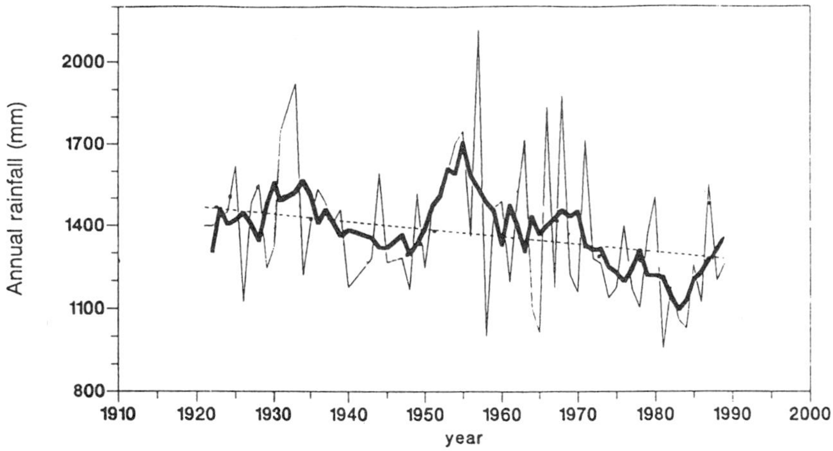

Figure 3 Variations in annual rainfall (mm) at Daloa, Côte d'Ivoire, between 1922 and 1990 (adapted from Van Rompaey, 1993): the solid line represents the five-year running mean.

been launched recently to identify the real extent of the problem and to assess the necessity of adjustments of current engineering hydrological practices in West Africa (Servat, 1994). Although the causes of the apparent change in regional climate will not be addressed by the ICCARE project (Servat, 1994), it is suggested here, as a first step, that the project's comprehensive climatic and hydrologic data base be combined with information on concurrent changes in regional vegetation cover and land use over the past 45 years (*cf.* Bertrand, 1983; Sayer *et al.*, 1992; Gautier, 1994). Together with an improved hydrological parametrisation of the respective post-forest vegetation types, this information could be used within the framework of a regional-scale modelling exercise (*cf.* McGuffie *et al.*, 1994). Ideally, such an exercise would also have to take into account atmospheric oceanic interactions (Cadet and Nnoli, 1987; Leroux, 1988). This approach might also reveal to what extent changes in land use in the more humid zones of West Africa have affect rainfall patterns in the drier belts to the north (Monteny and Cassenave, 1989; Savenije and Hall, 1994).

Simulation studies

Since the early 1960s the rate of forest clearance in Amazonia has increased considerably, largely for the creation of pastures (Malingreau and Tucker, 1988; Houghton *et al.*, 1991). In view of the perceived role of the Amazon forest in the regulation of the regional climate (Salati and Vose, 1984) a number of increasingly sophisticated computer simulations have been carried out to assess the climatic consequences of large-scale forest conversion in the region. At least 15 such modelling studies have been published for the Amazonian case, some of which have addressed the corresponding changes caused by forest clearing in the African and

South-east Asian rain forest regions as well. The results of the respective modelling efforts have been reviewed recently by Henderson-Sellers *et al.* (1993) and McGuffie *et al.* (1994). Model predictions vary considerably depending, amongst others, on the land surface parametrisation scheme used. However, whilst the magnitude of the predicted changes in surface temperature, evaporation and precipitation may differ between simulations of forest conversion, there is a growing consensus that temperatures will increase whereas both evaporation and rainfall will be reduced (Henderson-Sellers *et al.*, 1993; McGuffie *et al.*, 1994).

Interestingly, the magnitude of the predicted changes seems to become smaller as the models become more refined and the parametrisation of the land surface improved (*e.g.* Lean and Warrilow, 1989; Shukla *et al.*, 1990; Dolman and Gregory, 1992; Lean and Rowntree, 1993; Lean *et al.*, 1996; see also the comments of Bruijnzeel, 1990, p. 78). However, as pointed out by Eltahir and Bras (1993), there is less agreement for the predicted change in runoff (derived as the difference between the change in rainfall minus change in evaporation). Some models have predicted an increase in runoff and others a decrease. We will come back to this important point later. More recently, the same authors also challenged the widely accepted notion that about 50 per cent of the rainfall over Amazonia's forest is generated by the forest itself (Salati and Vose 1984) and instead suggested a moisture recycling ratio of 25-35 per cent. The difference between the two estimates was explained in terms of a different perception of the degree to which the Amazon Basin represents a closed system (Eltahir and Bras 1994).

Tropical montane 'cloud forests'

Although forests may not directly determine the amounts of precipitation they receive when occurring as scattered stands of limited areal extent, there are specific locations, such as coastal and montane fog or cloud belts, where the presence of tall vegetation does increase the amount of water reaching the forest floor as canopy drip. This is effected via the process of fog or cloud 'stripping', *i.e.* the capturing of atmospheric moisture by the canopy of such 'cloud forests' when subjected to persistent wind driven fog or clouds (Zadroga, 1981). Contributions by cloud stripping (also called 'occult' precipitation; Zadroga 1981) generally lie within the range of 5-20% of ordinary rainfall in wet tropical locations (Bruijnzeel and Proctor, 1995) but can be much higher at particularly exposed locations (Stadtmüller and Agudelo, 1990) or during the 'dry' season in more seasonal climates (Vogelmann, 1973; Cavelier and Goldstein, 1989).

Cloud forests may be particularly vulnerable to global warming as they often occur on exposed ridges with shallow soils (Van Steenis, 1972; Stadtmüller, 1987; Werner, 1988). For example, the average cloud base rose temporarily by about 100 m after Hurricane 'Hugo' destroyed large tracts of forest in the Luquillo Mountains of Puerto Rico, thereby exposing stunted ridge top vegetation normally enshrouded in cloud (F.N. Scatena, *pers. comm.*; *cf.* Walker *et al.*, 1991). As will be discussed in the next section, the hydrological consequences of converting montane cloud forests are largely unknown. There is an urgent need to increase hydrological and plant physiological research efforts in cloud forest (*cf.* Bruijnzeel and Proctor, 1995; Bruijnzeel and Veneklaas, 1995).

EFFECTS ON WATER YIELD

'Contradictory' results

A common notion about the hydrological role of forests is that the complex of forest soil, roots and litter acts as a sponge soaking up water during rainy spells and releasing it evenly during dry periods. Indeed, accounts of springs and streams in the tropics drying up after forest removal are numerous (Hamilton and King, 1983; Bartarya, 1989; Pereira, 1989). On the other hand, the number of reports of the reverse (*i.e.* streams drying up after reforestation of degraded land) is increasing as well. When trying to reconcile these apparent contradictions, it is helpful to distinguish between effects of clearing on *total* water yield and the *seasonal distribution* of flows (Bruijnzeel, 1989). Before addressing the issue, a few methodological comments are in order. Firstly, simply comparing streamflow totals for basins with contrasting land-use types may produce erroneous results because of the possibility of geologically determined differences in basin leakage. Catchments underlain by sandstones, limestones, basalts and volcanic ashes are particularly notorious in this respect (Gonggrijp, 1941; Davis and de Wiest, 1966). Also, the strong spatial and year-to-year variability of tropical rainfall may compound direct comparisons between basins or years (Blackie, 1979a,b; Malmer, 1993). Rigorous experimental designs (*e.g.* the 'paired catchment' or 'single basin' techniques; Hewlett and Fortson, 1983; Ibrahim and Chang, 1989) are needed to overcome such problems. However, the experimental error of the paired catchment method is such that reductions in forest cover of, say, less than 20 per cent failed to produce a detectable change in streamflow, be it under warm-temperate (Hibbert, 1967), subtropical (Subba Rao *et al.*, 1985) or humid tropical (Gilmour, 1977; Jetten, 1994) conditions. In addition, such controlled experiments are time-consuming and expensive and, consequently, relatively few of them have been conducted in the tropics (Bruijnzeel, 1990). Figure 4 summarises the evidence available to date.

In all cases, the removal of at least 33 per cent of forest cover resulted in significant increases in annual streamflow during the first three years, whereas initial gains in water yield after complete forest clearance ranged between 125 and 820 mm a^{-1} (Figure 4a). In addition, increases in water yield proved to be roughly proportional to the fraction of biomass removed (Figure 4a). There can be little doubt that these changes in water yield mainly reflect the different evaporative characteristics of mature tropical forest and young secondary or planted vegetation and, therefore, differences in soil water status (Figure 5). Under mature tropical forest much of the water infiltrating into the soil (typically 80-95 per cent of incident rainfall) is transpired again by the trees (generally at a rate of about 1000 mm a^{-1} where soil moisture is not limiting) whereas the remainder is used to sustain streamflow (Bruijnzeel 1990). The observed variation in initial response to clearing (Figure 4a) is considerable and can be explained only partially by differences in rainfall between

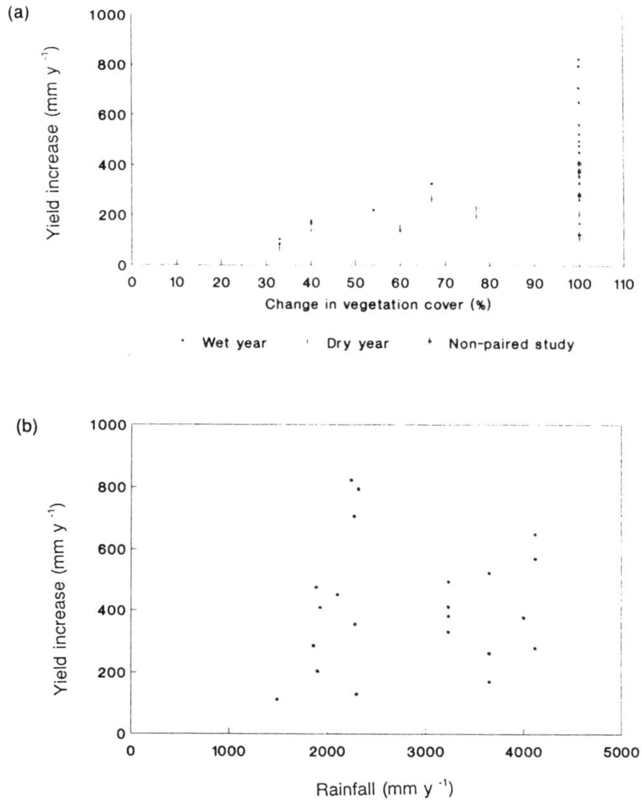

Figure 4 (a) Increases in water yield (mm a^{-1}) during the first three years after clearing tropical forest *vs.* percentage of area cleared; (b) *Idem vs.* corresponding amounts of annual rainfall (based on Table 4 in Bruijnzeel (1990), supplemented with data from Malmer (1993)).

locations (Figure 4b; Bosch and Hewlett, 1982). Other factors include differences in elevation (affecting evaporation), catchment steepness and soil depth (governing the residence time of the water in the catchment and baseflow recession; Ward and Robinson, 1990), the degree of disturbance of undergrowth and soil by machinery or fire (Swindel *et al.*, 1982; Malmer, 1993), and the fertility status of the soil (influencing post-clearing plant productivity and water uptake; Monteith, 1988; Uhl *et al.*, 1988; Denich, 1989; Brown and Lugo, 1990). Naturally, the relative importance of the respective factors will vary between sites and additional process studies are usually required if the results of paired basin experiments (which essentially constitute a black box approach) are to be fully understood (Bruijnzeel, 1990; Bonell and Balek 1993).

Changes in water yield during forest regeneration

Generally, there is a more or less irregular decline to pre-clearing levels of streamflow with time that is associated with the development of the new vegetation

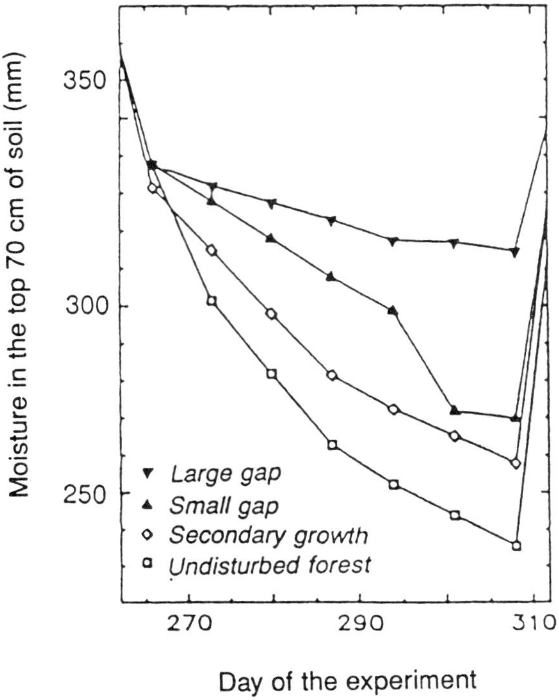

Figure 5 Soil water contents (mm $70cm^{-1}$) in undisturbed rain forest, 6-year-old regrowth as well as in narrow (10×50 m) and large (50×50 m) clearings during the dry season, Costa Rica (after Parker, 1985).

and year-to-year variability in rainfall. For warm-temperate conditions this may take as long as 35 years (Kovner, 1956; Hibbert, 1967). A return to predisturbance levels of streamflow during forest regeneration in the humid tropics may be expected to proceed much faster than that in view of the rapid growth of young tropical secondary vegetation (Brown and Lugo, 1990). However, published hydrological information in this respect is scarce and contradictory and pertains mostly to selectively logged terrain. Kuraji and Paul (1994) presented results from a water balance study involving two small catchments in Sabah, East Malaysia. The basins had been subjected to different degrees of forest exploitation about two and a half years prior to the start of the streamflow observations. One catchment had been selectively logged and the other had been clearfelled and burned. Estimates of forest evapotranspiration (ET) during the third and fourth year after forest disturbance were about 1450 mm a^{-1} for the logged-over catchment. The corresponding figures for the clearfelled basin (whose vegetation was dominated by *Macaranga spp.*) were close to 1200 mm a^{-1}. Malmer (1993), also working in Sabah, conducted streamflow measurements on two small catchments which had lost about one third of their overstorey biomass during logging five years previously. His results did not reveal any trend in annual ET over the next five years, except for higher values derived by Kuraji and Paul (1994) for more recently logged-over forest are close to the overall mean of 1465 mm yr^{-1} derived for undisturbed rain forests in the region by Bruijnzeel (1990). As such, it could be argued that the Sabahan forests had reattained pre-logging levels of ET after three to five years. However, because basin leakage via the sandy valley fills of the catchments studied by Malmer (1993) cannot be excluded (A. Malmer, *personal communication*), the actual ET may well have been lower. Conversely, Abdul Rahim and Zulkifli (1994) did not observe any decline in yield increase during the first seven years after harvesting 40 per cent of the commercial stocking in a lowland rain forest catchment in Peninsular Malaysia, suggesting that water use by the regrowth in the gaps created by logging still remained below that of the original stock after seven years. Unfortunately, observations had to be terminated after the seventh year when the area became inundated during the creation of a reservoir (N. Abdul Rahim, *personal communication*).

As for the water use by young secondary vegetation, the results obtained by Kuraji and Paul (1994) for three-to-four-year-old regrowth in Sabah suggest that streamflow from such areas may be about 250 mm a^{-1} higher than that for catchments with mature forest. Concurrent measurements of throughfall in the respective forests revealed that the difference in overall ET could be accounted for entirely by differences in rainfall interception (Paul and Kuraji, 1993). On the other hand, recent work in two-year-old secondary vegetation in the Zona Bragantina near Belém, eastern Amazonia, using the energy balance Bowen ratio technique, did not suggest ET to be dramatically low (D. Hölscher, *pers. comm.*). Interestingly, the partitioning of rainfall by a nearby 10-year-old stand dominated by a banana-like palm species differed considerably from that usually observed in mature lowland forest in the Amazon (*e.g.* Lloyd and Marques, 1988) in that the stemflow ratio was much higher (Sá *et al.* 1993). Clearly, hydrological research efforts in tropical secondary vegetation need to be increased considerably, also because of the fact that, with few exceptions, most tropical countries now have larger areas under secondary forests than under primary ones (Brown and Lugo, 1990). Such work might also lead to a better quantitative explanation of the apparent lack of hydrological response to forest alteration in the case of large catchment areas having vegetation in various stages of regeneration (Qian, 1983; Dyhr-Nielsen, 1986).

Forest conversion and the reforestation of degraded land

In the case of a true forest conversion (*i.e.* no regeneration), water yields after maturation of the new crop may remain above or return to their original values. Permanent *increases in water yield* have been observed in the case of conversions to (i) annual cropping (typically by 300-450 mm a^{-1}, depending on rainfall; Edwards, 1979; Lawson *et al.*, 1981; Lal, 1983); (ii) grassland (200-400 mm a^{1}; Mumeka 1986; Fritsch, 1992; Waterloo, 1994); (iii) tea plantations (*idem*; Blackie, 1979a; Callander and Woodhead, 1979); and (iv) rubber or cocoa plantations (200-300 mm yr^{-1}; Monteny *et al.*, 1985; Imbach *et al.*, 1989). On the other hand, water yields may be expected to eventually *return to their original levels* after full canopy closure in the case of fast-growing plantation forests (*e.g.* conifers; Blackie 1979b; Bruijnzeel, yields may be expected to eventually *return to their original levels* after full canopy closure in the case of fast-growing plantation forests (*e.g.* conifers; Blackie 1979b; Bruijnzeel, 1988; Waterloo, 1994) or certain extractive tree crops (*e.g.* palm-oil; Foong *et al.*, 1983). The hydrology of even-aged plantation forests in the humid tropics has recently been reviewed in some detail by Bruijnzeel (1995). He drew attention to the fact that very little is known about the hydrological consequences of planting such widely used hardwood species as *Acacia mangium*, *Gmelina arborea*, *Tectona grandis*, and, to a lesser extent, *Eucalyptus spp.* In view of the very high rainfall interception fractions reported for *Acacia mangium* (Bruijnzeel, 1995) and the exceptionally rapid initial growth of this species (Lim,, 1988), it is not inconceivable that the total ET of vigorously growing *Acacia* stands might exceed that of the original forest. Further work is needed to confirm this.

The differences in water yield quoted in the previous paragraph are generally thought to reflect the contrast in ET between tall, aerodynamically rough, deep-rooted vegetation types with relatively low reflection coefficients and high leaf biomass on the one hand and short, smoother and more shallow-rooted vegetation types with higher reflection coefficients and lower leaf biomass on the other (Calder, 1990; Bruijnzeel, 1990). During rainy spells the difference manifests itself mainly as a result of enhanced evaporation from the wetted forest canopy (Shuttleworth, 1988a; Calder, 1990). In the case of prolonged dry spells the generally less extensive root network of short crops is unable to exploit subsoil moisture reserves as efficiently as the root system of mature trees (Eeles, 1979; Waterloo, 1994; *cf.* Figure 6). It could be argued that many natural grasslands in the tropics grow several metres tall and this may have a bearing on rainfall interception. Waterloo (1994) conducted a preliminary investigation of rainfall interception by a 1.75 m tall *Pennisetum polystachyon* grassland in Fiji. He derived the low value of 4% of incident rainfall for the living portion of the grass, whereas a further 7% were estimated to have been

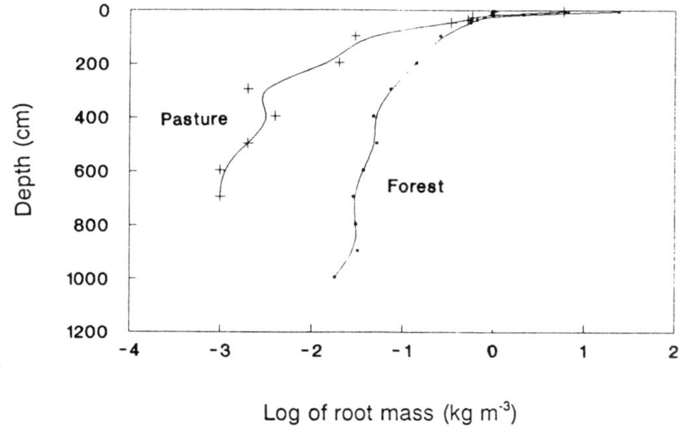

Figure 6 The distribution of the logarithm of root mass (kg m^{-3}) with depth for intact forest and abandoned pasture land in Eastern Amazonia (based on data in Nepstad 1989).

intercepted by the litter layer. The former estimate is remarkably close to the value reported by Leopoldo *et al.* (1981) for sugar cane in Brazil, which lends some confidence to the result. As such, the idea that rainfall interception totals in tall tropical grasslands might approach those for tropical forest would seem to be ungrounded. As there are many kinds of tropical grasslands (Walter and Breckle, 1986), one could question to what extent their hydrological behaviour can be expected to be similar. However, the observation of Waterloo (1994) that the wet-season relationship between net radiant energy and global radiation above his *Pennisetum* grassland in Fiji was very similar to those derived for well-watered *Panicum maximum* in Côte d'Ivoire (Monteny and Gosse, 1976) and a *Brachiaria* pasture in central Amazonia (wet season conditions; Bastable *et al.*, 1993) stems hopeful in this respect. Nevertheless, differences in biomass and leaf area index (LAI) between man-made pastures and natural grasslands may be large, depending on grazing intensity and climatic and edaphic factors. For example, the ungrazed Fijian grassland showed LAI values ranging from 0.2 at the end of a severe dry season to 2.0 at the height of the rainy season (Waterloo, 1994) whereas that for the Amazonian site was only 0.8 (measured at the end of a moderately developed dry season; McWilliam *et al.*, 1993). There may be equally important differences in below-ground biomass between sites (*cf.* Nepstad, 1989).

As shown in Figure 6, rainforest roots may extend to considerable depths where local geology allows it. Similarly, roots have been observed to extract groundwater from a depth of 5-6 m in a rainforest in Surinam not unlike that found in Central Amazonia (Poels, 1987). Also, Eernisse (1993) observed an increase in the number of live roots just above the weathering front at 4 m depth in a 'brown sand' soil as well as close to an iron pan at 3 m depth in a 'white sand' soil in Guyana. Although the significance of deep roots for the uptake of water and nutrients in different rainforest settings remains to be established, these observations suggest that the overriding importance usually attached to the surface root mat in Amazonian forests growing on infertile substrates (Jordan, 1985) may require modification (Nepstad *et al.*, 1994). In addition, these findings have consequences for the experimental design of soil water depletion studies in tropical forests (*cf.* Huttel, 1975; Eeles, 1979; Hodnett *et al.*, *this volume*). On a different note, Bonell (1993) has drawn attention to the fact that subsurface stormflow in northern Queensland travelled preferentially along channels that were occupied by large (living) roots. Much more work is needed on root distribution and functioning in tropical forest ecosystems, including secondary vegeation and plantations at various stages of growth (Berish and Ewel, 1988; Brown and Lugo, 1990; Evans, 1992).

One of the most important consequences of the contrast in rooting depth between forests and most other vegetation types is that the reforestation of degraded scrub or grassland with fast-growing exotic tree species will result in a *decrease* in streamflow totals, usually shortly after canopy closure. Values reported for seasonal tropical uplands in Madagascar and India (1500-2000 m a.s.l.) are in the order of 100-200 m a^{-1} (Bailly *et al.*, 1974; Samraj *et al.* 1988), but stronger reductions have been observed in certain lowland situations (up to 400 mm a^{-1} after reforestation with vigorously growing *Pinus caribaea*; Waterloo, 1994). Interestingly, the latter figure exceeds the predicted reductions in total evaporation after conversion of Amazon rain forest to pasture by some of the more recent simulations (Lean and Rowntree, 1993; Lean *et al.*, *this volume*). Much more work needs to be done in this respect (see also the final sections).

The replacement of natural grass or scrub vegetation by one particular exotic species, *Eucalyptus*, has been severely criticised for a variety of reasons, one of them being that eucalypts are allegedly voracious consumers of water (Vandana Shiva and Bandyopadhyay, 1983). The observations of Smith and Scott (1992), who reported streamflow from reforested areas in upland South Africa to decline faster and earlier in the case of *Eucalyptus grandis* as compared to *Pinus patula*, would seem to confirm such fears. Because rainfall interception by tropical conifers is usually higher than that for eucalypts (Bruijnzeel, 1995), such differences in overall ET must reflect different rates of water uptake. Recent work on young eucalypts in subhumid South India (annual rainfall *c*. 800 mm) suggested that transpiration can be severely reduced under such conditions as a result of soil moisture stress but also that daily uptake rates can be as high as 5-7 mm under conditions of ample soil water supply (Calder *et al.*, 1992; Roberts and Rosier, 1993). Although it would take a long time before the tree roots would have ready access to the groundwater table (found at 30-100 m depth in this particular case) and thus be able to transpire continuously at a potential rate (Calder, 1986), there was growing evidence that the trees were 'mining' soil water reserves built up during previous years as the root network developed (Calder *et al.*, 1992). Soil water contents under (dry) indigenous forest were similar to those recorded under *Eucalyptus*, however, and subsequent recharge to the groundwater table was presumed to be equally small (Harding *et al.*, 1992).

However, severe reductions in streamflow have been reported elsewhere in India after coppicing mature *E. grandis* and *E. camaldulensis* trees whose root systems remained intact (up to 285 mm a^{-1}; Vishwanatham *et al.*, 1980; 1982). In conclusion it can be stated that most eucalypts are subject to normal stomatal control under conditions of soil water stress. If excessive water use is to be avoided, they should not be planted in depressions or along water courses nor be coppiced after maturation. Alternatively, it has been suggested that the trees be grown in patches in the landscape and in rotation with rainfed agriculture to minimize any adverse effects on groundwater recharge (Calder *et al.*, 1993). Further work, covering more humid climatic conditions and other widely planted fast-growing species (such as *Acacia mangium*) or eucalypt clones with contrasting water use efficiencies (Olbrich *et al.*, 1993; Calder *et al.*, 1993) is desirable. The same, incidently, holds for a variety of agroforestry systems. Agroforestry is often thought of as a panacea for many environmental problems in the tropics but reliable information on the water (and nutrient) dynamics of such systems is still woefully inadequate (Young, 1989; Bruijnzeel, 1990; Ong *et al.*, 1991; Ong and Khan, 1993).

Converting tropical montane cloud forest

Again, cloud forests may represent a specific case. Evapotranspiration in these forests is known to be low to very low (Bruijnzeel and Proctor, 1995). This, together with the extra inputs of moisture via cloud stripping discussed earlier, causes runoff ratio's for such areas to be very high (Herrmann, 1970; Zadroga, 1981; Lugo, 1986). Interestingly, the low rates of water use of tropical montane cloud forests are not so much a direct consequence of low radiation inputs and high humidities (Weaver *et al.*, 1973; Ash, 1987) or the frequent wetting of the canopy by passing clouds suppressing transpiration (Calder, 1982). Recent observations in Malaysian (Bruijnzeel *et al.*, 1993) and Indonesian (Hafkenscheid, 1994) cloud forests have demonstrated that water uptake also remained strongly suppressed during periods of bright sunshine under conditions of ample soil water and low to moderate vapour pressure deficits. The causes of the low transpiration and productivity of these often stunted forests are still unknown although a wealth of hypotheses have been advanced over the years (see Bruijnzeel and Proctor (1995) and Bruijnzeel and Veneklaas (1995) for recent reviews of the hydrological and physiological aspects, respectively).

Despite the demonstrated importance of tropical montane cloud forests for the continuing supply of water to adjacent lowlands, they are being converted to agricultural uses in many places in Latin America (Hamilton *et al.*, 1995). The associated effect on water yield is as yet unknown, but will presumably reflect a trade-off between the loss of the extra water formerly gained via cloud stripping and the difference in water use between the old and the new vegetation (Bruijnzeel and Proctor, 1995). There is circumstantial evidence that cloud forest clearance has led to decreased flow in some cases (Stadtmüller and Agudelo, 1990). Process-based research is urgently needed.

Effect of scale

The results presented thus far mostly pertained to small catchment areas (less than 1 km^2) involving a unidirectional change in cover. Although these provide a clear and consistent picture of increased water yield following replacement of tall vegetation by a shorter one and *vice versa*, effects may be more difficult to discern in larger basins (>1000 km^2) having a variety of land-use types and natural vegetation in various stages of regeneration. In addition, the strong spatial and temporal variation in tropical rainfall (Nieuwolt, 1977) tends to obscure any effects associated with modifications of the surface (Qian, 1983; Dyhr-Nielsen, 1986).

Nevertheless, Bandara and Kuruppuarachchi (1988) reported a highly significant increase in annual runoff coefficients for the 1108 km^2 Mahaweli basin in Sri Lanka over the period 1944-1981, despite a concurrent trend towards diminished rainfall. This increased hydrological response was accompanied by the widespread conversion of tea plantations (not forest!) to annual cropping and home gardens without appropriate soil conservation measures. In the first two examples (Qian, 1983; Dyhr-Nielsen, 1986), there was a fair degree of vegetation regrowth, possibly in the form of resprouting from stumps which would tend to increase transpiration; Vishwanathan *et al.*, 1980; *cf.* Denich, 1989 in areas that had been subjected previously to slash-and-burn agriculture, whereas in the Sri Lankan case the soils remained more exposed to the elements. Quantifying the hydrological behaviour of large basins is difficult in any situation but particularly so in the humid tropics where variability in space and time tends to be large whilst data availability is generally poor (Bonell *et al.*, 1993).

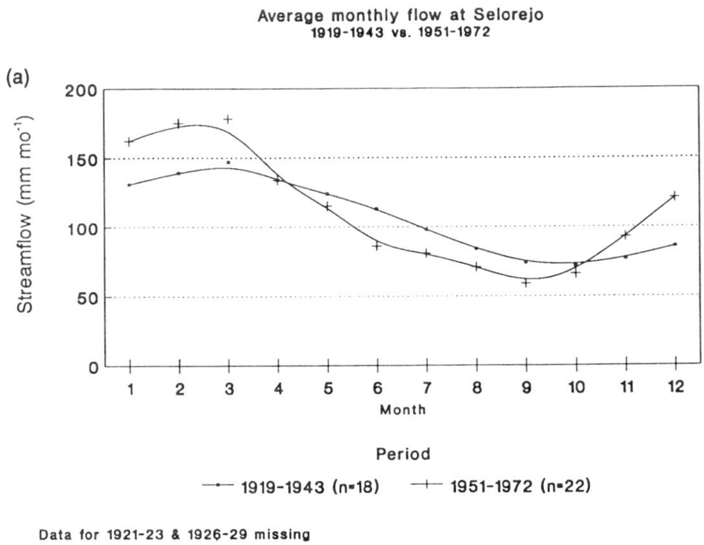

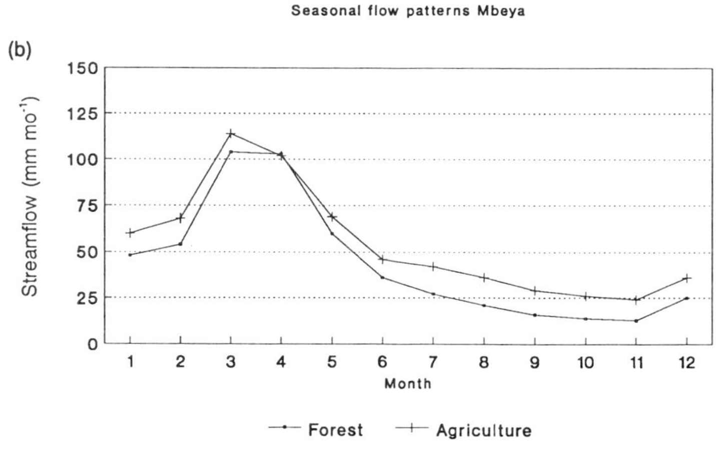

Figure 7 Changes in seasonal distribution of streamflow following changes in land use. (a) Konto area, East Java (after Rijsdijk and Bruijnzeel 1991); (b) Mbeya, Tanzania (based on data in Edwards 1979).

EFFECTS ON DRY SEASON FLOW

Reports of greatly diminished streamflow during the dry season after forest removal abound in the literature (Hamilton and King, 1983; Hardjono, 1980; Bartarya, 1989). At first sight, this might seem contradictory to the evidence presented in Figure 4 of increased flows following the clearing of forested catchments, particularly in view of the fact that, generally, the bulk of these increases has been observed during baseflow or dry season conditions (Bruijnzeel, 1990). This apparent conflict can be resolved by taking into account the net effect of the changes in ET and infiltration capacity that are brought about by the land use change (Bruijnzeel, 1989). If infiltration opportunities after forest removal have deteriorated to the extent that increases in stormflow during rainfall exceed the gains in baseflow associated with the lower ET of a cleared surface, then diminished dry season flow is the result (Figure 7a). The converse applies as well (Figure 7b).

It is well-known that a catchment's infiltration capacity may be reduced considerably after forest exploitation or conversion due to compaction of the soil by heavy machinery, overgrazing or prolonged exposure to raindrop impact, by an increase in the area occupied by impervious surfaces such as roads, tracks and residential areas, or by open cast mining, etc. (Lal, 1987; Bandyopadhyay and Vandana Shiva, 1987; Bons, 1990; Rijsdijk and Bruijnzeel, 1990; Bruijnzeel, 1992a). Variations on this theme are, of course, widespread in the tropics and must be held responsible for the deterioration in streamflow regimes that is so often observed (*cf.* Figure 7a; see also the next section). If, on the other hand, surface infiltration characteristics are maintained over most of the cleared catchment, either because of careful extraction of timber along a well-planned and maintained road system (Abdul Rahim, 1990; Malmer, 1993), or subsequent soil conservation practices such as controlled grazing (Pereira, 1989), or simply because of a non-erosive rainfall regime (Edwards, 1979), then the effect of reduced ET after clearing will show up as increased baseflow during dry spells (Figure 7b).

As pointed out by Bonell and Balek (1993), this hypothesis still requires experimental verification. However, the observation that dry season water yields from scrub or grassland are usually reduced considerably following forestation, already indicates that the evapotranspiration factor may easily override the infiltration factor, particularly in areas with shallow soils and a well-developed dry season (Mathur *et al.*, 1976; Gilmour *et al.*, 1987; Smith and Scott, 1992; Waterloo, 1994). Again, there is scope for much more work in this respect which will need to cover different combinations of environmental factors (climate, soils) and species.

It could be argued that the controlled conditions of catchment experiments bear little relevance to the real world situation. However, the lesson that can be learned from such experiments is a most important one: the frequently observed deterioration in tropical river regimes following forest removal is not so much the result of clearing *per se* but rather due to a lack of good land husbandry during and after the operation (Bruijnzeel, 1986; Pereira, 1989).

EFFECTS ON FLOODING

Local effects

The hydrological response of small catchment areas to rainfall depends on the interplay between climatic, geological and land-use variables. Key parameters in this respect include the hydraulic conductivity of the soil at different depths, rainfall intensity and duration, and slope morphology (Dunne, 1978). Generally speaking, infiltration capacities of undisturbed forest soils are such that they are easily capable of absorbing most, if not all, rainfall intensities (see Bonell (1993) for a discussion of a few exceptional cases). Under such conditions, a catchment's response to rainfall usually represents a mixture of contributions by 'saturation' overland flow from valley bottoms and other depressions and subsurface stormflow from the sideslopes. The relative magnitude of the respective components will vary, both between catchments as a result of differences in topography and soils (possibly also climate) and between events as a result of differences in antecedent moisture status and storm characteristics (see Bonell and Balek (1993) for a comprehensive review). Normally, peaks that are produced by some form of overland flow are more pronounced than those generated by subsurface types of stormflow (Dunne, 1978; Ward, 1984). As such, it is especially the shift from subsurface flow to overland flow dominated stormflow patterns, which so often accompanies land clearing due to effects of soil compaction and burning, that produces the frequently reported dramatic increases in peakflows (Bailly *et al.*, 1974; Fritsch, 1992; Malmer, 1993; van der Plas and Bruijnzeel, 1993). However, in catchments where overland flow (usually of the 'saturation' type) is already rampant under undisturbed conditions (mainly due to the presence of an impeding layer at shallow depth; Bonell and Gilmour, 1978; Elsenbeer and Cassell, 1990), the response to rainfall after forest removal is hardly changed (Gilmour, 1977). Whilst storm runoff patterns may not be affected very much under such conditions, this does not imply that forest removal would not have adverse consequences. On the contrary, surface erosion and basin sediment yield normally show dramatic increases in such cases (Gilmour, 1977; Fritsch, 1992).

Carefully planned and conducted conversion operations will be able to keep soil compaction and disturbance, and hence occurrence of infiltration-excess overland flow, to a minimum, particularly if there is no burning (Hsia, 1987; Malmer, 1993). However, even with minimum soil disturbance, there will be increases in peakflows after forest removal, because the associated reduction in ET will cause soils to be wetter (*cf.* Figure 5) and therefore more responsive (Pearce *et al.*, 1980a; Hsia, 1987; Fritsch, 1992). Relative increases in response tend to be largest for small events (roughly 100%), declining to 10% or less for large events (Gilmour, 1977; Pearce *et al.*, 1980a; Hewlett and Doss, 1984). Although such increases can be expected to diminish as a new cover establishes itself within one to two years (Hsia, 1987; Fritsch, 1992; Malmer, 1993), they may become 'structural' because of contributions from roads and residential areas where there were none before (Rijsdijk and Bruijnzeel, 1990; Douglas *et al.*, 1993) or because soils remain wetter (*e.g.* in the case of a conversion to grassland; Fritsch, 1992). Whilst the general trends in catchment response to rainfall after forest conversion may be known, there is much scope for further work.

The paired catchment approach is of limited value in this respect if not accompanied by sufficiently detailed hydro-pedological studies (Fritsch, 1992) and process studies (Bonell, 1993). For example, Malmer (1993) examined the effect of mechanized timber removal followed by burning on catchment stormflow response in East Malaysia. Despite a severe reduction in final topsoil infiltration capacity over almost one quarter of the catchment (Malmer and Grip, 1990), there was no significant change in stormflow response during the actual conversion period of eight months. In addition, catchment response was even significantly reduced during the first year after conversion and increased only slightly (but not significantly so) during the second year. Peakflows showed a similar pattern (Malmer, 1993). Such findings are hard to explain in the absence of detailed observations. Malmer (1993) explained his results in terms of temporary detention storage of overland flow in the ruts of old tractor tracks which only started to erode at a later stage..

Off-site effects

Whilst it is beyond doubt that adverse land-use practices after forest clearance generally cause serious local increases in stormflow volumes and peakflows, one has to be careful to extrapolate such effects over larger scales. High stormflows generated by heavy rain on a misused part of a river basin may be 'diluted' by more modest flows from other parts receiving less or no rainfall at the time or with better land-use practices (Hewlett, 1982; Qian, 1983; Dyhr-Nielsen, 1986), whereas within-channel storage can be an important dampening factor on very large river systems (Richey *et al.*, 1989; Zhang, 1990). Also, one has to be careful when ascribing observed trends in the frequency of occurrence of peak discharges or floods to upstream changes in land use. Gentry and Lopez-Parodi (1980), for example, examined the 1962-1978 water level records for the Amazon River at Iquitos, Peru, and found a distinct increase in the height of the annual flood crest that they linked to large-scale deforestation in the Andes. A subsequent analysis by Richey *et al.* (1989) of a much longer time series (1903-1983) for a gauging station further downstream (Manaus) revealed that the increase in flow during the 1962-1978 period was within the range of longer term cycles which, in turn, were largely influenced by large-scale fluctuations in climate.

Truly devastating and large-scale (thousands of km^2) flooding is the result of an equally large field of extreme rainfall (Raghavendra, 1982; Mooley and Parthasarathy, 1983), particularly when this occurs at the end of a rainy season when soils have become wetted up thoroughly by antecedent rains. Under such extreme conditions, basin response will be governed mainly by soil water storage opportunities rather than topsoil infiltration capacity (Hamilton, 1987; Bruijnzeel and Bremmer, 1989; *cf.* Gilmour, 1977). Nevertheless, it cannot be excluded that *widespread* forest removal, followed by poor cultivation practices, may have a cumulative effect and there are signs (*e.g.* the widening of river beds) that the latter may indeed be happening in various parts of the tropics, *e.g.* in Nigeria (Odemerho, 1984) and the Himalayas (Bruijnzeel and Bremmer, 1989). Similar reports of deteriorating flow regimes are available for large basins (>10,000 km2) in southern China (Chen, 1987). However, other factors, which often tend to be overlooked, include torrential rains on the flood (*sic*) plains themselves during times of high water; backwater effects; raised river beds due to high sedimentation; and the breaking up of temporary barriers (such as huge landslides) to river flow (Bruijnzeel and Bremmer, 1989; Zhang, 1990). Finally, Hamilton (1987) made the important point that equating economic losses associated with a flood with the severity of that flood may introduce a certain bias. Increased economic losses may give an impression of more frequent and damaging events while in reality the former may reflect economic growth and increased floodplain occupancy.

Effects of land-use changes on the magnitude of flood peaks in large rivers are very difficult to evaluate because such changes are rarely very fast and consistent and normally compounded by climatic variation (Richey *et al.*, 1989). Moreover, such an analysis requires long-term and high quality data on land use, streamflow and rainfall, which are rarely available for humid tropical regions (particularly with respect to spatial distribution of rainfall; Medina *et al.* 1987; Bonell *et al.*, 1993). Although attempts at routing flood peaks from headwater areas through river systems in warm humid regions are becoming increasingly successful (Dubreuil, 1986; Zhang, 1990; Sambou and Thirriot, 1993), the evaluation of land-use effects on such peaks still constitutes one of the greatest challenges in the field of environmental hydrology. However, rates of forest conversion are faster than ever in many areas and the availability of remotely sensed information on land use, cloud cover and rain fields (Stewart and Finch, 1993) as well as macro-scale hydrological models (Vörösmarty and Moore, 1991; Vörösmarty *et al.*, 1991) is improving rapidly. As such, the question of 'highland-lowland interaction' may be coming closer to an answer in the not too distant future.

ISSUES REQUIRING FURTHER RESEARCH

Having broadly reviewed the various hydrological impacts of forest conversion, what then are the chief gaps in our knowledge hindering improved management of soil and water resources in the humid tropics? This question is perhaps not as straightforward as it would seem at first sight and answers coming from different interest groups are likely to differ. There are those who believe that 'greater emphasis on biophysical research strategies (including hillslope hydrology) may hold the key to (solving) land management problems in the humid tropics' (Bonell and Balek, 1993). Others (Hamilton and King, 1983; Bruijnzeel, 1986; Pereira, 1989) have emphasized the need to put into practice what is known already. The answer lies probably somewhere in the middle. In the following, a selection is presented of what this author perceives as the most important gaps in hydrological knowledge with respect to the proper understanding and prediction of the hydrological consequences of land cover transformations in the tropics.

Effects of forest conversion on local to regional rainfall patterns

The trends towards increased aridity observed in different parts of the tropics are a matter of the utmost concern. *Rigorous analyses of long-term rainfall records* for the regions concerned (*e.g.* the countries around the Gulf of Guinea in West Africa; parts of Southern India and Sri Lanka; possibly Java and Sumatra) are needed which will have to take into account the various fluctuations referred to earlier. Such analyses will also need to establish the link between seasonal rainfall totals and the prevailing synoptic situations.

Key rainfall stations may be identified which are located in areas that have remained under forest throughout the observation period and others which have experienced large-scale forest removal. The respective time series for such stations could be analysed in a manner similar to the 'paired catchment' method, *i.e.* with a 'calibration' period and a 'treatment' period (*cf.* Zhang, 1986). In addition, *modelling studies of the meso-climatic effects of specific land cover transformations will need to be carried out for regions other than Amazonia*, particularly for south-east Asia as well as for west- and central Africa (*cf.* Monteny and Casenave. 1989).

Due to the fact that its climate is influenced so strongly by the surrounding seas the response of the south-east Asian rainforest block to conversion can be expected to be quite different from that obtained for the more continental situation of central Amazonia (McGuffie *et al.*, 1994). However, before meaningful results can be obtained with the modelling approach, the parametrizations of the rainforests and grasslands of south-east Asia need to be improved. For instance, there are indications that rainfall interception percentages under tropical maritime conditions (Gilmour, 1975; Bruijnzeel, 1988; Scatena, 1990; Burghouts, 1993; Waterloo, 1994) may be up to twice as high as those reported for tropical continental conditions (Jordan and Heuveldop, 1981; Lloyd and Marques, 1988; *cf.* Ubarana, *this volume*).

It has been suggested (Shuttleworth, 1989) that such differences reflect different measuring techniques but most of the quoted 'maritime' studies (at least the more recent ones) employed numerous gauges in a roving arrangement (Lloyd and Marques, 1988). As such, the difference may be real and due to advected energy from the surrounding seas (*cf.* Pearce *et al.*, 1980b; Bruijnzeel and Wiersum, 1987). Similarly, the extensive *Imperata* and *Themeda* grasslands of south-east Asia still await hydrological description. With the planned reforestation of several million hectares of these grasslands with fast-growing tree species like *Acacia mangium* and various eucalypt hybrids (C. Cossalter, *pers. comm.*) a unique opportunity may present itself to study the effects (if any) of large-scale reforestation on regional rainfall patterns and so verify the predictions of model simulations.

Effects of forest conversion on catchment water yield

Whilst the trend of the change in water yield following tropical forest clearing is well-established, the observed variations are such that no real quantitative predictions can be made (Figure 4). As such, there is a distinct need to supplement the more traditional catchment studies (which remain essential when studying effects on stormflow and water quality; see below) with process orientated work and physically based model applications (Bonell and Balek, 1993; Vertessy *et al.*, 1993). Such work would seem particularly urgent with respect to (i) the *effects of reforesting degraded scrub- and grasslands with fast-growing exotics*; and (ii) the clarification of the *effects of converting tropical montane cloud forests to agricultural cropping* (potatoes, cabbages and other 'temperate' vegetables) *and pasture.*

Arguably, the most urgent area in which a major research effort is needed concerns the determination of the hydrological characteristics of a number of fast-growing tree species (*e.g. Acacia* and *Albizzia spp.*, various eucalypts — including clones — and pines) which are becoming widely used for the reforestation of degraded land all over the tropics (Evans, 1992; Bruijnzeel, 1995). Ideally, observations should cover a range of edaphic and climatic conditions, including shallow and/or infertile soils. In addition, measurements will need to be made during the respective stages of the plantations if representative results are to be obtained. Running a long-term study that spans a full rotation period can be costly and becomes impractical where a species is grown over a period of, say, 15-40 years (*cf.* Blackie, 1979b). Observations in stands of different age classes in the same region coupled with robust models of rainfall interception (*e.g.* Gash; Rutter) and stand water use (Penman-Monteith) are probably the best solution to this problem although there may be complications because of differences in tree spacing and/or soil type (Rogerson, 1967; Waterloo, 1994).

The research methodology for this type of study is well-established and could broadly follow the lines of the Institute of Hydrology's work in India (Calder *et al.*, 1993) and central Amazonia (Shuttleworth, 1988a) wherever topographic conditions and considerations of 'fetch' permit. A particularly promising micrometeorological technique in this respect is the so-called temperature fluctuation method (Tillman, 1972) which only requires a simple fast-responding dry bulb thermocouple for the determination of the sensible heat flux density. In combination with measurements of available energy the latent heat flux density can be derived.

The method has been applied successfully under both dry and wet tropical conditions (Lloyd *et al.* (1991) and Vugts *et al.* (1993), respectively). The latter authors also employed wet-bulb thermocouples in an attempt to estimate the latent heat flux density above a pine forest in Fiji in a more direct manner and obtained generally good agreement with the more traditional energy balance Bowen ratio approach (Angus and Watts, 1984). Also, there should be considerable scope for the use of plant physiological approaches (Cermak and Ku era, 1987; Hatton and Vertessy, 1990; Calder, 1991), especially in relation to the evaluation of tree productivity at different positions in the terrain (dry ridges *vs.* moist footslopes; areas with poor growth due to soil compaction during the clearing phase, etc.) and where shallow soils are underlain by fractured rocks. It should be stressed that, wherever possible, such studies should be carried out within a catchment hydrological framework, as this would help to maintain the link with the study of other, equally important, aspects, such as changes in basin response to rainfall (see below), site fertility (and thus plantation sustainability; Bruijnzeel, 1992b; Nykvist *et al.*, 1994), and sediment production (Malmer, 1993; Bruijnzeel, 1995).

As for the hydrological effects of cloud forest clearance, any changes in water yield will presumably reflect the trade-off between the loss of the 'occult' precipitation component and the net change in water use by the old and the new vegetation. Studies are needed of the water use and degree of 'cloud stripping' along altitudinal gradients, both under maritime conditions (where cloud forests are found at relatively low elevations) and further inland (where they usually occur above 1500-2000 m; Stadtmüller, 1987). Bruijnzeel and Proctor, 1995) recently discussed the problems of determining rates of transpiration and cloud stripping under the conditions prevailing in most tropical montane cloud forests.

Needless to say, the combination of the above observations with a physically based distributed catchment hydrological model capable of simulating not only water yield but also the spatial distribution of soil moisture with time, would be especially powerful. However, as pointed out by Bonell (1993), the transfer of physically based models to the humid tropics may not be without complications. For instance, the surface of the bedrock may not be parallel to that of the topography in deeply weathered tropical terrain. In such cases the commonly made assumption that the hydraulic gradients of subsurface flow will run parallel to the soil surface (O'Loughlin, 1990) may not be valid (*cf.* Quinn *et al.*, 1991).

Arguably, the most promising catchment hydrological model package currently available is the topography-driven TOPOG modelling framework developed by O'Loughlin and co-workers (O'Loughlin, 1986; O'Loughlin *et al.*, 1989; Vertessy *et al.*, 1993). Although physically based distributed modelling has come under close scrutiny in recent years (Beven, 1989; Grayson *et al.*, 1992; Jensen and Mantoglou, 1992; and others) the TOPOG modelling framework is not as data demanding as some other models of the same class (*cf.* Bathurst and O'Connell, 1992). Yet it has proved remarkably versatile in its applications. These have ranged from the prediction of saturated zones in the landscape (O'Loughlin, 1990); steady-state catchment moisture distributions during wet and dry conditions and their relationship with storm runoff (Moore *et al.*, 1988; O'Loughlin *et al.* 1989); hazard predictions for occurrence of sheet erosion (Vertessy *et al.*, 1990), gully erosion (Dietrich *et al.*, 1992; Constantini *et al.*, 1993), and mass wasting (Dietrich *et al.*, 1992); hydro-ecological response to elevated CO_2 at the catchment scale (Hatton *et al.*, 1992); and the long-term water yields associated with mature aggrading *Eucalyptus regnans* forests (Vertessy *et al.*, 1993, 1995). Additional applications of this type of models include the study of soil water dynamics in relation to occurrence of ecological habitats, forest growth and resilience to perturbation, and forest plantation establishment and yield (Moore *et al.*, 1991).

Thus far, experience with the TOPOG package under humid tropical conditions has been limited but work is underway for the data-rich Babinda and 'La Cuenca' catchments in North Queensland and Peru, respectively (R.A. Vertessy, *personal communication*). Both sites experience rampant overland flow (Bonell and Gilmour, 1978; Elsenbeer and Cassel, 1990). A third study which employs the TOPOG package (this time to evaluate the changes in water yield associated with forest recovery after destruction by hurricane) has been initiated recently in the Bisley experimental catchments in Puerto Rico (Bruijnzeel and Scatena, 1994).

Much of what has been discussed in the previous paragraphs also pertains to studies aiming to *evaluate the water relations and hydrological behaviour of various agro-forestry systems* (*cf.* Imbach *et al.*, 1988; Young 1989; Ong *et al.*, 1991; Ong and Khan, 1993). Finally, it should be mentioned that *detailed studies of variations in soil hydraulic conductivity in the respective land-use systems (such as vegetation types) are a sine qua non when addressing the effects of land-use change on dry season water yield* (Bruijnzeel, 1989; Bonell, 1993). Again, such observations should be made at sites representing the various stages of vegetation development and on different soil types (and depths), possibly using the soil catena concept as the framework for measurement (Chappell and Ternan 1992). The obtained conductivities may then be compared with prevailing short-term rainfall intensities as a first step towards predicting any changes in the frequency of occurrence of overland flow associated with the change in land use (Gilmour *et al.*, 1987). The scope for close collaboration with pedologists in this regard is evident.

Effects of forest conversion on catchment response to rainfall

This aspect of tropical forest conversion is probably the least researched of the various hydrological consequences discussed so far. This led Bonell (1993) to call for a major increase in field and modelling studies, notably those related to hillslope hydrology, which he believed to 'hold the key to land management problems in the humid tropics'. Whilst such a statement may be valid from the standpoint of the research scientist interested in runoff generation, it remains to be seen whether it is true from a resource management point of view. Most experimental evidence

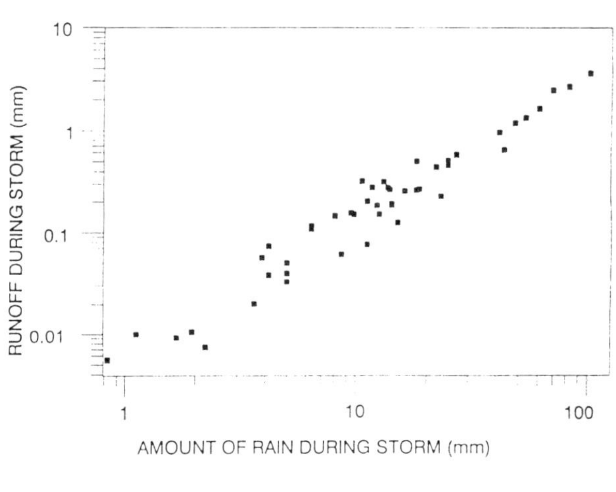

Figure 8 Stormflow volume as a function of amount of rainfall in a small forested catchment in central Amazonia (adapted from Lesack, 1993).

available to date (Bruijnzeel, 1990) suggests that no major increases in stormflow volume occur after *controlled* deforestation. Therefore, the adverse effects of forest removal can be kept to a minimum by adhering to a number of well-documented guidelines in most cases (*e.g.* Pearce and Hamilton, 1986). As such, the widely observed contrast in this respect between theory and practice is more a socio-economic rather than a technical problem (Hamilton and King, 1983; Bruijnzeel 1986; Pereira, 1989).

Nevertheless, there is a lack of studies dealing with the effects on stormflow (and basin sediment yield) associated with the rehabilitation of degraded land, be it by soil conservation practices, the introduction of agro-forestry systems or full-scale reforestation (Bruijnzeel, 1990). *The data base is particularly weak on the time lag between land rehabilitation and any subsequent reductions in stormflow and sediment production.* This holds for catchments of all sizes but especially for (very) large ones (Goswami, 1985; Hamilton, 1987; Bruijnzeel and Bremmer 1989). The general problem of scaling up basin hydrological response to rainfall continues to be the subject of intense research (van de Ven *et al.*, 1991; Bolle *et al.*, 1993) but its discussion falls largely outside the scope of the present paper.

With respect to the Amazonian context, the finding of Nortcliff and Thornes (1984), Fritsch (1992) and Lesack (1993) that the bulk of storm runoff in their respective forested catchments was generated mostly by a well-defined saturated valley bottom (*cf.* Figure 8) could be combined with the observation of Kahn (1987) that the vegetation in these poorly drained riparian zones is strongly dominated by palms (Figure 9). Although this configuration should not be overgeneralised (Elsenbeer *et al.*, 1992) it might open up an avenue for the prediction of the magnitude of storm-

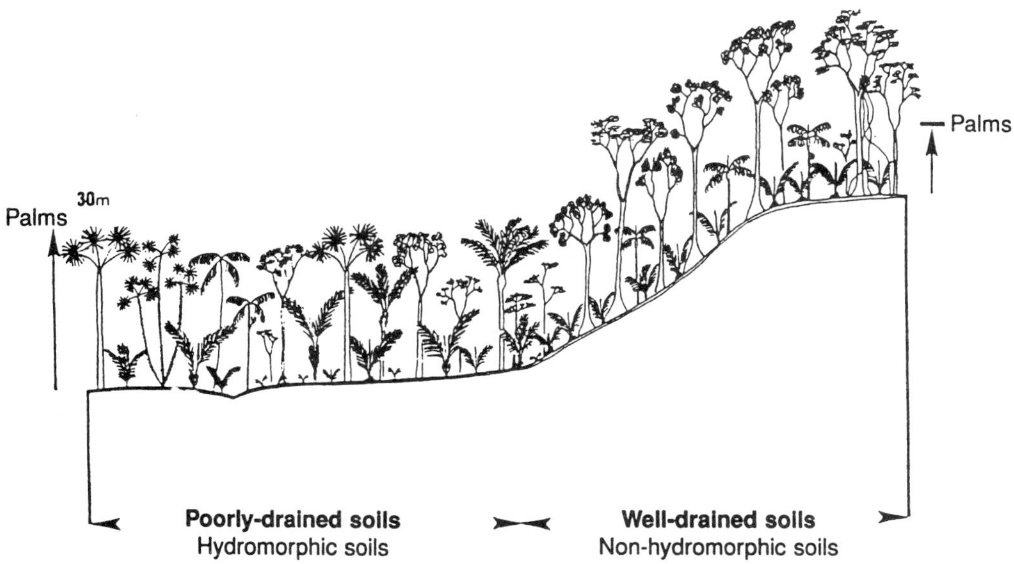

Figure 9 Distribution of palms along the catena in an undisturbed forest near Manaus (after Kahn, 1987).

and peakflows over a range of scales, provided that there is sufficient spatial information on the associated rainfall. This could be effected by a judicious combination of remotely sensed information on topography (Ritchie *et al.*, 1993) and vegetation (*cf.* Lucas *et al., this volume*) with, for instance, the TOPOG model for the prediction of the extent of saturated zones (O'Loughlin, 1990) and empirical relationships between the latter and stormflow volumes or peakflows (Moore *et al.*,1988; Lesack, 1993).

A similar lack of reliable information exists with respect to patterns of sediment delivery in relation to land-use change in the humid tropics. Suffice to say here that the study of sediment delivery in tropical catchments has only rarely been undertaken outside the context of small research basins (see the reviews by Walling, 1983; Bruijnzeel, 1990; 1995). There is a clear need for much more work which could usefully be combined with the equally much needed improvement of the description of the terrestrial hydrology component in global circulation models. As indicated earlier, large-scale simulations of the changes in rainfall and evaporation associated with a massive change in land cover have produced contrasting results for the prediction of the corresponding changes in runoff (Dickinson, 1991). Indeed, the perceived lack of collaboration between climatologists and hydrologists is a matter of concern (N mec, 1994) and needs to be overcome if true progress with the modelling of basin response at the macroscale is to be made (*cf.* Vörösmarty *et al.*, 1991; Vörösmarty and Moore, 1991).

CONCLUDING REMARKS

The response of the hydrological scientific community to the daunting environmental problems presented by the mass destruction of tropical forests has been diverse. On the one hand, there have been those that considered the incorporation of hydrological expertise into global circulation models (GCMs) in conjunction with remote sensing techniques to be the most promising way forward (*e.g.* Shuttleworth, 1988b; Shuttleworth *et al.*, 1991). Others reckoned that this 'macrohydrological' approach could cover only a relatively narrow part of the overall spectrum of water management issues (*e.g.* Bonell and Balek, 1993; Bruijnzeel, 1993). These hydrologists called for the establishment of a network of 'reference catchments' in which process studies and (physically based) hydrological models could be used to evaluate human- and climate-induced changes at various levels of scale.

One conclusion that has hopefully emerged from the present review is that the two approaches are complementary rather than contradictory and that their integration would be a major step forward. It is proposed that future research efforts be concentrated on a relatively small number of well-researched key locations with long-term records which are to be linked together in a pan-tropical network that captures the main features of the environmental variability encountered in the humid tropics (Bruijnzeel and Abdul Rahim, 1992; *cf.* Pearce, 1990).

A survey of the main hydrological research sites in the humid tropics (Bruijnzeel,

1990; Bonell and Balek, 1993) suggests several important deficiencies in data coverage. First, with the exception of the well-known East African experiments that were terminated long ago and did not include much process orientated work (Blackie *et al.*, 1979), there are no rigorous experiments in volcanic uplands. In view of the fact that such areas are usually densely populated and essentially stripped of their protective cover to make way for agricultural cropping in many places (*e.g.* Java, the Philippines, the Andes), many of them are eroding severely and in need of rehabilitation (El-Swaify *et al.*, 1982). Secondly, the virtual absence of paired studies to evaluate the hydrological effects of reforesting degraded areas is striking. Finally, there do not seem to be any paired catchment experiments in areas having 'cloud forests'. An attempt in the Philippines had to be terminated for socio-economic reasons (S. Peñafiel, *personal communication*).

As indicated repeatedly in recent review papers (Bonel 1993; Bruijnzee 1993), the time seems ripe for a concerted effort to step up our efforts in the field of environmental hydrology in the tropics. The establishment of a pan-tropical network consisting of the various well-known data-rich sites (Bruijnzee 1990, 1993), supplemented with a few carefully chosen additional locations, might be a first step towards reaching improved exchange of information (*cf.* N mec, 1994) and standardization of measurements. It is proposed that the coordination of such a (possibly rather informal) network be handled through such international bodies like IHP-UNESCO, the BAHC programme, or the newly established Center for International Forestry Research (CIFOR) in Bogor. The continued rapid changes in land use in the humid tropics, coupled with steady improvements in equipment, data availability (including remotely sensed data) and computational facilities, guarantee that there are exciting times ahead for tropical hydrologists.

ACKNOWLEDGEMENTS

I am grateful to Dr John Gash and Dr Carlos Nobre for the invitation to compile this paper and for the opportunity to participate in the Brasilia meeting. Dr Renaat van Rompaey (formerly Wageningen Agricultural University) is thanked for introducing me to lots of literature on rainfall in West Africa and for stimulating discussions. Dirk Hölscher (University of Göttingen) generously shared (partially unpublished) information on the ecology and hydrology of young secondary vegetation in eastern Amazônia.

Dr Rob Vertessy (Cooperative Research Centre for Catchment Hydrology, Canberra) is thanked for exposing me to the TOPOG modelling framework and for convincing me that there is a future for physically based distributed parameter hydrological models in tropical land-use hydrology. The manuscript benefited from the comments of Dr John Roberts and two anonymous reviewers. Finally, thanks are due to Irene Sieverding whose sustained support was essential to the completion of the manuscript.

REFERENCES

Abdul Rahim, N. 1990. *The effects of Selective Logging Methods on Hydrological Parameters in Peninsular Malaysia*. PhD Thesis, University of Wales, Bangor, UK.

Abdul Rahim, N. & Zulkifli, Y. 1994. Hydrological response to selective logging in Peninsular Malaysia and its implications for watershed management. In: *Proc. Int. Symp. Forest Hydrology, Tokyo, Japan*, 263-274.

Adejuwon, J.O., Balogun, E.E. and Adejuwon, S.A. 1990. On the annual and seasonal patterns of rainfall fluctuations in sub-Saharan West Africa. *Int. J. Climatol.*, **10**, 839-848.

Angus, D.E. & Watts, P.J. 1984. Evapotranspiration – How good is the Bowen ratio method? *Agric. Wat. Manage.*, **8**, 133-150.

Arulanantham, J.T. 1982. The effects, if any, on rainfall due to the deforestation of Sinharaja forest. In: *Tropical Environments*, M.M. Yoshino, I. Kayane and C.M. Bandara (Eds.), Institute of Geoscience, University of Tsukuba, Tsukuba, Japan, 169-173.

Ash, J. 1987. Stunted cloud forest in Taveuni, Fiji. *Pacific Science*, **41**, 191-199.

Bailly, C., Benoit de Cognac, G., Malvos, C., Ningre, J.M. and Sarrailh, J.M. 1974. Etude de l'influence du couvert naturel et de ses modifications à Madagascar; expérimentations en bassins versants élémentaires. *Cahiers Scientifiques du Centre Technique Forestier Tropical*, **4**, 1 - 114.

Bandara, C.M. & Kuruppuarachchi, T.A. 1988. Land-use change and hydrological trends in the upper Mahaweli basin. Paper presented at the Workshop on Hydrology of Natural and Man-made Forests in the Hill Country of Sri Lanka, Kandy, October 1988, 18 pp.

Bandyopadhyay, J. & Vandana Shiva. 1987. Devastating the Doon valley. *India Today*, **June 1987**, 67-75.

Bartarya, S.K., 1989. Hydrogeology, geo-environmental problems and watershed management strategies in a central Himalayan river basin, Kumaun, India. In: *Headwater Control*, J. Krecek *et al.* (Eds.), IUFRO/WASWC/CSVIS, Plzen, Czechoslovakia, 308-318.

Bastable, H.G., Shuttleworth, W.J., Dallarosa, R.L.G., Fisch, G. and Nobre, C.A. 1993. Observations of climate, albedo and surface radiation over cleared and undisturbed Amazonian forest. *Int. J. Climatol.*, **13**, 783-796.

Bathurst, J.C. & O'Connell, P.E. 1992. Future of distributed modelling: the Système Hydrologique Européen. *Hydrol. Proc.*, **6**, 265-277.

Berish, C.W. & Ewel, J.J. 1988. Root development in simple and complex tropical successional ecosystems. *Plant and Soil*, **106**, 73-84.

Berlage, H.P. 1949. Rainfall in Indonesia. Mean rainfall figures for 4399 rainfall stations in Indonesia, 1879-1941. *Verhandelingen van het Koninklijk Magnetisch en Meteorologisch Observatorium te Batavia*, **34**, 212 pp.

Bertrand, A.1983. La déforestation en zône de forêt en Côte d'Ivoire. *Revue des Bois et Forêts des Tropiques*, **202**, 3-17.

Beven, K. 1988. Changing ideas in hydrology - the case of physically-based models. *J. Hydrol.*, **105**, 157-172.

Blackie, J.R. 1979a. The water balance of the Kericho catchments. *E. African Agric. For. J.*, **43**, 55-84.

Blackie, J.R. 1979b. The water balance of the Kimakia catchments. *E. African Agric. For. J.*, **43**, 155-174.

Blackie, J.R., Edwards, K.A. and Clarke, R.T. (Eds.), 1979. *Hydrological Research in East Africa. E. African Agric. For. J.*, **43** (special issue).

Blakeney, K.J. 1992. Cable and helicopter logging for reduced damage. Paper presented at the International Symposium on Harvesting and Silviculture for Sustainable Forestry in the Tropics (Kuala Lumpur, 5-9 October 1992).

Bolle, H.J., Feddes, R.A. and Kalma, J. (Eds.), 1993. Exchange Processes at the Land Surface for a Range of Space and Time Scales. *IAHS Publ. No.* **212**.

Bonell, M. 1993. Progress in the understanding of runoff generation dynamics in forests. *J.Hydrol.*, **150**, 217-275.

Bonell, M. & Gilmour, D.A. 1978. The development of overland flow in a tropical rainforest catchment. *J. Hydrol.*, **39**, 365-382.

Bonell, M. & Balek, J. 1993. Recent scientific developments and research needs in hydrological processes of the humid tropics. In: *Hydrology and Water Management in the Humid Tropics. Hydrological Research Issues and Strategies for Water Management*, Bonell, M., Hufschmidt, M.M. and Gladwell, J.S. (Eds.), Cambridge University Press, Cambridge. 167-260.

Bonell, M. Hufschmidt, M.M. and Gladwell, J.S. (Eds.), 1993. *Hydrology and Water Management in the Humid Tropics. Hydrological Research Issues and Strategies for Water Management. UNESCO*, Paris, and Cambridge University Press, Cambridge.

Bons, C.A. 1990. Accelerated erosion due to clearcutting of plantation forest and subsequent "Taungya" cultivation in upland West-Java, Indonesia. In: *Research Needs and Applications to Reduce Erosion and Sedimentation in Tropical Steeplands*, R.R. Ziemer, C.L. O'Loughlin and L.S. Hamilton (Eds.), *IAHS Publ.* No. **192**, 279-288.

Bosch, J.M. & Hewlett, J.D. 1982. A review of catchment experiments to determine the effect of vegetation changes on water yield and evapotranspiration. *J. Hydrol.*, **55**, 3-23.

Brown, S. & Lugo, A.E., 1990. Tropical secondary forests. *J. .Trop. Ecol.*, **6**, 1-32.

Bruijnzeel, L.A. 1986. Environmental impacts of (de)forestation in the humid tropics: a watershed perspective. *Wallaceana*, **46**, 3-13.

Bruijnzeel, L.A. 1988. Estimates of evaporation in plantations of *Agathis dammara* Warb. in south-central Java, Indonesia. *J. Trop. For. Sci.*, **1**, 145-161.

Bruijnzeel, L.A., 1989. (De)forestation and dry season flow in the tropics: a closer look. *J. Trop. For. Sci.*, **1**, 229-243.

Bruijnzeel, L.A. 1990. *Hydrology of Moist Tropical Forest and Effects of Conversion: a State of Knowledge Review*. UNESCO, Paris, and Free University, Amsterdam, The Netherlands.

Bruijnzeel, L.A. 1992a. Managing tropical forest watersheds for production: where contradictory theory and practice co-exist. In: *Wise Management of Tropical Forests 1992*, F.R. Miller & K.L. Adam (Eds.), Oxford Forestry Institute, Oxford, 37-75.

Bruijnzeel, L.A. 1992b. Sustainability of fast-growing plantation forests in the humid tropics, with special reference to nutrition. In: *Taungya: Forest Plantations with Agriculture in South-east Asia*. C.F. Jordan and Gajaseni, J. (Eds.), CAB International, Wallingford, 51-67.

Bruijnzeel, L.A. 1993. Land-use and hydrology in warm humid regions: where do we stand? In: *Hydrology of Warm Humid Regions*, J.S. Gladwell (Ed.), *IAHS Publ., no. 216*, 1-34.

Bruijnzeel, L.A. 1995. Hydrology of forest plantations in the humid tropics. In: *Management of Soil, Nutrients, and Water in Tropical Plantation Forests*, E.K.S. Nambiar & A.G. Brown (Eds.), ACIAR, Canberra, and CIFOR, Bogor (in press).

Bruijnzeel, L.A. & Wiersum, K.F. 1987. Rainfall interception by a young *Acacia auriculiformis* A. Cunn. plantation forest in West Java, Indonesia: application of Gash's analytical model. *Hydrol. Proc.*, **1**, 309-319.

Bruijnzeel, L.A. & Bremmer, C.N. 1989. Highland-lowland interactions in the Ganges Brahmaputra River Basin: a review of published literature. *ICIMOD Occasional Paper*, **11**, International Centre for Integrated Mountain Development, Kathmandu, Nepal.

Bruijnzeel, L.A. & Abdul Rahim, N. 1992. Ecological and environmental services provided by tropical trees and forests. In: *Center for International Forestry Research (CIFOR). Strategic Planning Thematic Papers*. ACIAR, Canberra, 23-26.

Bruijnzeel, L.A., Waterloo, M.J., Proctor, J., Kuiters, A.T. and Kotterink, B. 1993. Hydrological observations in montane rain forests on Gunung Silam, Sabah, Malaysia, with special reference to the *'Massenerhebung'* effect. *J. Ecol.*, **81**, 145-167.

Bruijnzeel, L.A. & Scatena, F.N. 1994. *Hydrological modelling in a humid tropical island setting: with special reference to The Luquillo Experimental Forest, Puerto Rico*. Project description, Vrije Universiteit Amsterdam.

Bruijnzeel, L.A. & Proctor, J. 1995. Hydrology and biogeochemistry of tropical montane cloud forests: what do we really know? In: *Tropical Montane Cloud Forests*, L.S. Hamilton, J.O. Juvik and F.N. Scatena (Eds.), *Ecological Studies*, **110**, Springer, New York, 38-78.

Bruijnzeel, L.A. & Veneklaas, E. 1995. Climatic conditions and tropical montane forest productivity: the fog has not lifted yet. *Ecology* (in press).

Burghouts, Th.B.A. 1993. *Spatial Variability of Nutrient Cycling in Bornean Rain Forest.* PhD Thesis, Vrije Universiteit, Amsterdam.

Cadet, D.L. & Nnoli, N.O. 1987. Water vapour transport over Africa and the Atlantic Ocean during summer 1979. *Q. J. Roy. Meteorol. Soc.*, **113**, 581-604.

Calder, I.R. 1982. Forest evaporation. In: *Hydrological Processes of Forested Areas*, Anonymous, *National Research Council of Canada Publication* no. 20548, NRCC, Ottawa, 173-193.

Calder, I.R. 1986. Water use of eucalypts: a review with special reference to South India. *Agric.Wat. Manage.*, **11**, 333-342.

Calder, I.R. 1990. *Evaporation in the Uplands*. J. Wiley, New York.

Calder, I.R. 1991. Deuterium tracing for the estimation of transpiration from trees. Part 2. Estimation of transpiration rates and transpiration parameters using a time-averaged deuterium tracing method. *J. Hydrol.*, **130**, 27-35.

Cermak, J. & Ku era, J. 1987. Transpiration of mature stands of spruce (*Picea abies* (L.) Karst.) as estimated by the tree trunk heat balance method. *IAHS Publ. no. 167*, 311-317.

Calder, I.R., Swaminath, M.H., Kariyappa, G.S., Srinivasalu, N.V., Srinivasa Murty, K.V. and Mumtaz, J. 1992. Deuterium tracing for the estimation of transpiration from trees. Part 3. Measurements of transpiration from *Eucalyptus* plantation, India. *J. Hydrol.*, **130**, 37-48.

Calder, I.R., Hall, R.L. and Prasanna, K.T. 1993. Hydrological impact of *Eucalyptus* plantation in India. *J. Hydrol.*, **150**, 635-648.

Callander, B.A. & Woodhead, T. 1979. Eddy correlation measurements of convective heat flux and estimation of evaporative heat flux over growing tea. *E. African Agric. For. J.*, **43**, 85-101.

Cavelier, J. & Goldstein, G. 1989. Mist and fog interception in elfin cloud forests in Colombia and Venezuela. *J. Trop. Ecol.*, **5**, 309-322.

Cermak, J. & Ku era, J. 1987. Transpiration of mature stands of spruce (*Picea abies* (L.) Karst.) as estimated by the tree trunk heat balance method. *IAHS Publ. no. 167*, 311-317.

Chappell, N. & Ternan, L., 1992. Flow path dimensionality and hydrological modelling. *Hydrol. Proc.*, **6**, 327-345.

Chen, C.Y. 1987 Hydrological effects of forests in the Hengduan Mountains, Southwest China. In: *Forest Hydrological Resources in China. An Analytical Asessment*, P.F. Ffolliot & D.P. Guertin (Eds.), U.S. MAB Program, Washington D.C., 57-60.

Cooper, J.D. 1979. Water use of a tea estate from soil moisture measurements. *East African Agric. For. J.*, **43**, 102-121.

Constantini, A., Dawes, W., O'Loughlin, E. and Vertessy, R. 1993. Hoop pine plantation management in Queensland: I. Gully erosion hazard prediction and watercourse classification. *Australian J. Soil & Wat. Conserv.*, **6**, 35-39.

Couper, D.C., Lal, R. and Claassen, S.L, 1981. Land clearing and development for agricultural purposes. In: *Tropical Agricultural Hydrology*, R. Lal & E.W. Russell (Eds.), J. Wiley, New York, 119-130.

Davis, S.N. & De Wiest, R.J.M. 1966. *Hydrogeology*. J. Wiley, New York.

Denich, M., 1989. Untersuchungen zur Bedeutung junger Sekundärvegetation für die Nutzungssystemproduktivität im östlichen Amazonasgebiet, Brasilien. *Göttinger Beiträge zur Land- und Forstwirtschaft in den Tropen und Subtropen*, **46**, 1-265.

Dias, A.C.C.P. & Nortcliff, S. 1985a. Effects of tractor passes on the physical properties of an Oxisol in the Brazilian Amazon. *Trop. Agric. (Trinidad)*, **62**, 137-141.

Dias, A.C.C.P. & Nortcliff, S. 1985b. Effects of two land clearing methods on the physical properties of an Oxisol in the Brazilian Amazon. *Trop. Agric. (Trinidad)*, **62**, 207-212.

Dickinson, R.E. 1991. Global change and terrestrial hydrology - a review. *Tellus*, **43AB**, 176-181.

Dietrich, W.E., Wilson, C.J., Montgomery, D.R., McKean, J. and Bauer, R. 1992. Erosion thresholds and land surface morphology. *Geology*, **20**, 675-679.

Dolman, A.J. & Gregory, D. 1992. The parameterization of rainfall interception in GCM's. *Q. J. Roy. Meteorol. Soc.*, **118**, 455-467.

Douglas, I., Greer, A.J., Kawi Bidin and Sinun, W. 1993. Impacts of roads and compacted ground on post-logging sediment yield in a small drainage basin, Sabah, Malaysia. In: *Hydrology of Warm Humid Regions*, J.S. Gladwell (Ed.), *IAHS Publ. no. 216*, 213-218.

Dubreuil, P. 1986. Review of relationships between geophysical factors and hydrological characteristics in the tropics. *J. Hydrol.*, **87**, 201-222.

Dunne, T. (1978). Field studies of hillslope flow processes. In: *Hillslope Hydrology*, M.J. Kirkby (Ed.), J. Wiley, New York, 227-293.

Dyhr-Nielsen, M. 1986. Hydrological effect of deforestation in the Chao Phraya basin in Thailand. Paper presented at the International Symposium on Tropical Forest Hydrology and Application, Chiangmai, Thailand, 11-14 June, 1986.

Edwards, K.A. 1979. The water balance of the Mbeya experimental catchments. *E. African Agric. For. J.*, **43**, 231-247.

Eeles, C.W.O. 1979. Soil moisture deficits under montane rain forest and tea. *E. African Agric. For. J.*, **43**, 128-138.

Eernisse, N. 1993. *Relative Depths of the Weathering Front and the Fine Root Network: a Pilot Study, Mabura Hill, Guyana*. Report, Tropenbos Programme, Guyana.

Elsenbeer, H. & Cassel, D.K. 1990. Surficial processes in the rainforest of western Amazonia. In: *Research Needs and Applications to Reduce Erosion and Sedimentation in Tropical Steeplands*, R.R. Ziemer, C.L. O'Loughlin and L.S. Hamilton (Eds.), *IAHS Publi. no. 192*, 289-297.

Elsenbeer, H., Cassel, K. and Castro, J. 1992. Spatial analysis of soil hydraulic conductivity in a tropical rain forest catchment. *Wat. Resour. Res.*, **28**, 3201-3214.

El-Swaify, S.A., Dangler, E.W. and Armstrong, C.L. 1982. *Soil Erosion by Water in the Tropics. University of Hawaii Research and Extension Series*, **24**, University of Hawaii, Honolulu.

Eltahir, E.A.B. & Bras, R.L. 1993. On the response of the tropical atmosphere to large-scale deforestation. *Q. J. Roy. Meteorol. Soc.*, **119**, 779-793.

Eltahir, E.A.B. & Bras, R.L. 1994. Precipitation recycling in the Amazon basin. *Q. J. Roy. Meteorol. Soc.*, **120**, 861-880.

Evans, J. 1992. *Plantation Forestry in the Tropics*. Clarendon Press, Oxford.

Fleming, T.H. 1986. Secular changes in Costa Rican rainfall: correlation with elevation. *Journal of Trop. Ecol.*, **2**, 87-91.

Foong, S.F., Syed Sofi, S.O. and Tan, P.Y. 1983. A lysimetric simulation of leaching losses from an oil palm field. In: *Proceedings of the Seminar on Fertilizers in Malaysian Agriculture*, Malaysian Society of Soil Science, Kuala Lumpur, 45-68.

Fritsch, J.M. 1992. *Les Effets du Défrichement de la Forêt Amazonienne et de la Mise en Culture sur l'Hydrologie des Petits Bassins Versants*. ORSTOM Editions, Paris.

Gautier, L 1994. Deforestation in southern Côte d'Ivoire: a high-resolution remote sensing approach. *The Biodiversity of African Plants*, L.J.G. van der Maesen (Ed.), WICC-IAC, Wageningen, The Netherlands, 56-57.

Gentry, A.H. & Lopez-Parodi, J. 1980. Deforestation and increased flooding of the upper Amazon. *Science*, **210**, 1354-1356.

Gilmour, D.A. 1975. *Catchment Water Balance Studies on the Wet Tropical Coast of North Queensland*. PhD Thesis, James Cook University, Townsville, Australia.

Gilmour, D.A. 1977. Effect of rainforest logging and clearing on water yield and quality in a high rainfall zone of north-east Queensland. In: *Proceedings of the Brisbane Hydrology Symposium 1977*, Institution of Engineers Australia, Canberra, 156-160.

Gilmour, D.A., Bonell, M. and Cassells, D.S. 1987. The effects of forestation on soil hydraulic properties in the Middle Hills of Nepal: a preliminary assessment. *Mountain Res. & Develop.*, **7**, 239-249.

Gonggrijp, L. 1941. Evaporation from montane forest in West Java at an altitude of 1750-2000 m a.s.l. *Tectona*, **34**, 437-447.

Goswami, D.C. 1985. Brahmaputra River, Assam, India: physiography, basin denudation and channel aggradation. *Wat. Resour. Res.*, **21**, 959-978.

Grayson, R.B., Moore, I.D. & McMahon, T.A. 1992. Physically based hydrologic modeling - 2. Is the concept realistic? *Wat. Resour. Res.*, **28**, 2659-2666.

Hamilton, L.S. 1987. What are the impacts of deforestation in the Himalaya on the Ganges-Brahmaputra lowlands and delta? Relations between assumptions and facts. *Mountain Res. & Develop.*, **7**, 256-263.

Hamilton, L.S. & King, P.N. 1983. *Tropical Forested Watersheds. Hydrologic and Soils Response to Major Uses or Conversions*. Westview Press, Boulder, Colorado.

Hamilton, L.S., J.O. Juvik and F.N. Scatena. (Eds.) 1993. *Tropical Montane Cloud Forests. Proceedings of an International Symposium*, East-West Center, Honolulu.

Harding, R.J., Hall, R.L., Swaminath, M.H. and Srinivasa Murty, K.V. 1992. The soil moisture regimes beneath forest and agricultural crop in southern India - measurements and modelling. In: *Growth and Water Use of Forest Plantations*, I.R. Calder, R.L. Hall & P.G. Adlard (Eds.), J. Wiley, New York, 244-269.

Hardjono, H.W. 1980. Influence of a permanent vegetation cover on streamflow. In: *Proceedings of the Seminar on Watershed Management, Development and Hydrology*, Surakarta, Indonesia, 3-5 June 1980, 280-297 (in Indonesian).

Hatton, T.J. & Vertessy, R.A. 1990. Transpiration of plantation *Pinus radiata* by the heat pulse method and the Bowen ratio. *Hydrol. Proc.*, **4**, 289-298.

Hatton, T.J., Walker, J., Dawes, W. and Dunin, F.X. 1992. Simulations of hydro-ecological responses to elevated CO_2 at the catchment scale. *Austral. J. Botany*, **40**, 679-696.

Henderson-Sellers, A., Dickinson, R.E., Durbidge, T.B., Kennedy, P.J., McGuffie, K. and Pitman, A.J. 1993. Tropical deforestation: modelling local to regional-scale climatic change. *J. Geophys. Res.*, **98(D4)**, 7289-7315.

Hendrison, J. 1990. *Damage-Controlled Logging in Managed Tropical Rain Forest in Suriname*. Pudoc, Wageningen, The Netherlands.

Herrmann R. 1970. Vertically differentiated water balance in tropical high mountains - with special reference to the Sierra Nevada de Sa. Marta, Colombia. *International Association of Hydrological S Publ. no. 93*, 262-273.

Hewlett, J.D. 1982. Forests and floods in the light of recent investigation. In: *Hydrological Processes of Forested Areas*. National Research Council of Canada Publication no. 20548, NRCC, Ottawa, 543-559.

Hewlett, J.D. & Fortson, J.C. 1983. The paired catchment experiment. In: *Forest Water Quality*, J.D. Hewlett (Ed.), School of Forest Resources, University of Georgia, Athens, Ga., 11-14.

Hewlett, J.D. & Doss, R. 1984. Forests, floods, and erosion: a watershed experiment in the southeastern Piedmont. *For. Sci.*, **30**, 424-434.

Hibbert, A.R. 1967. Forest treatment effects on water yield. In: *International Symposium on Forest Hydrology*, W.E. Sopper & H.W. Lull (Eds.), Pergamon Press, Oxford, 527-543.

Horne, R. & Hickey, J. 1991. Review. Ecological sensitivity of Australian rain forests to selective logging. *Austral. J. Ecol.*, **16**, 119-129.

Houghton, R.A., Lefkowitz, D.S. and Skole, D.L. 1991. Changes in the landscape of Latin America between 1850 and 1985. I. Progressive loss of forests. *For. Ecol. & Manag.*, **38**, 143-172.

Hsia, Y.J., 1987. Changes in storm hydrographs after clearcutting a small hardwood forested watershed in central Taiwan. *For. Ecol. & Manag.*, **20**, 117-134.

Hubert, P., Carbonnel, J.P. and Chaouche, A. 1989. Segmentation des séries hydrométriques. Application à des séries de précipitations et de débits de l'Afrique de l'Ouest. *J. Hydrol.*, **110**, 349-367.

Huttel, Ch. 1975. Recherches sur l'écosystème de la forêt subéquatoriale de basse Côte d'Ivoire. IV Estimation du bilan hydrique. *La Terre et la Vie*, **29**, 192-202.

Ibrahim, M.A. & Chang, M.T. 1989. On the calibration of forested watersheds: single vs. paired watershed approaches. Paper presented at the FRIM-IHP-UNESCO Regional Seminar on Tropical Forest Hydrology, Kuala Lumpur, 4-9 September 1989.

Imbach, A.C., Fassbender, H.W., Borel, R., Beer, J. and Bonnemann, A. 1989. Modelling agroforestry systems of cacao (*Theobroma cacao*) with laurel (*Cordia alliodora*) and poro (*Erythrina poeppigiana*) in Costa Rica. IV. Water balances, nutrient inputs and leaching. *Agrofor. Systems*, **8**, 267-287.

Jensen, K.H. & Mantoglou, A. 1992. Future of distributed modelling. *Hydrol. Proc.*, **6**, 255-264.

Jetten, V.G. 1994. *Modelling the Effects of Logging on the Water Balance of a Tropical Rain Forest. A Study in Guyana. Tropenbos Series*, **6**, Tropenbos Foundation, Wageningen, The Netherlands.

Jordan, C.F. 1985. *Nutrient Cycling in Tropical Forest Ecosystems*, J.Wiley, New York.

Jordan, C.F. & Heuveldop, J. 1981. The water budget of an Amazonian rainforest. *Acta Amazonica*, **11**, 87-92.

Kahn, F. 1987. The distribution of palms as a function of local topography in Amazonian terra-firme forest. *Experientia*, **43**, 251-259.

Kamaruzaman Jusoff. 1991. Effect of tracked and rubber-tyred logging machines on soil physical properties of the Berkelah Forest Reserve, Malaysia. *Pertanika*, **5**, 265-276.

Kovner, J.L. 1956. Evapotranspiration and water yields following forest cutting and natural regrowth. *Soc. Amer. For. Proc.s*, 106-110.

Kuraji, K. & Paul, L.L. 1994. Effects of rainfall interception on water balance in two tropical rainforest catchments, Sabah, Malaysia. In: *Proceedings of the International Symposium on Forest Hydrology, Tokyo, Japan*, 291-298.

Lal, R. 1981. Deforestation of tropical rainforest and hydrological problems. In: *Tropical Agricultural Hydrology*, R. Lal & E.W. Russell (Eds.), J. Wiley, New York, 131-140.

Lal, R. 1983. Soil erosion in the humid tropics with particular reference to agricultural land development and soil management. In: *Hydrology of Humid Tropical Regions*, R. Keller (Ed.), *IAHS Publ. no. 140*, 221-239.

Lal, R. 1987. *Tropical Ecology and Physical Edaphology*. J. Wiley, New York.

Lal, R. 1995. Soil resources for plantation forestry and their characteristics. In: *Management of Soil, Nutrients, and Water in Tropical Plantation Forests*, E.K.S. Nambiar & A.G. Brown (Eds.), ACIAR, Canberra, and CIFOR, Bogor (in press).

Lanly, J.P. 1982.) Tropical Forest Resources. *FAO Forestry Paper*, **30**. FAO, Rome

Lanly, J.P., Singh, K.D. and Janz, K. 1991. FAO's 1990 reassessment of tropical forest cover. *Nature & Resources*, **27**, 21-26.

Lawson, T.L., Lal, R. and Oduro-Afriyie, K. 1981. Rainfall redistribution and microclimatic changes over a cleared watershed. In: *Tropical Agricultural Hydrology*, R. Lal & E.W. Russell (Eds.), J. Wiley, New York, 141-151.

Lean, J. & Warrilow, D.A. (1989) Simulation of the regional climatic impact of Amazonian deforestation. *Nature* **342**, 411-413.

Lean, J. & Rowntree, P.R. 1993. A GCM simulation of the impact of Amazonian deforestation on climate using an improved canopy representation. *Q. J. Roy. Meteorol. Soc.*, **119**, 509-530.

Leopoldo, P.R., De Padua Sousa, A. and Filho, S.T. 1981, Interceptação da água de chuva em cultura de cana-de-açúcar. *Brasil Açucareira*, **6** , 9-16.

Leroux, M., 1988. La variabilité des précipitations en Afrique Occidentale. Les composantes aérologiques du problème. *Veille Climatique Satellitaire*, **22**, 26-46.

Lesack, L.F.W. 1993. Water balance and hydrologic characteristics of a rain forest catchment in the Central Amazon basin. *Wat. Resour. Res.*, **29**, 759-773.

Lhomme, J.P. 1981. L'évolution de la pluviosité annuelle en Côte d'Ivoire au cours des soixante dernières années. *La Météorologie VIe Série*, **25**, 135-140.

Lim, M.T. 1988. Studies on *Acacia mangium* in Kemasul Forest, Malaysia. I. Biomass and productivity. *J. Trop. Ecol.*, **4**, 293-302.

Lloyd, C.R. & Marques, A. de O. 1988. Spatial variability of throughfall and stemflow measurements in Amazonian rain forest. *For. Agric. Meteorol.*, **42**, 63-73.

Lloyd, C.R., Culf, A.D., Dolman, A.J. & Gash, J.H.C. 1991. Estimates of sensible heat flux from observations of temperature fluctuations. *Bound. Layer Meteorol.*, **57**, 311-322.

Ludwig, R. 1992. Cable crane yarding: an economical and ecologically suitable system for commercial timber harvesting in logged-over rainforests of the Philippines. Paper presented at the International Symposium on Harvesting and Silviculture for Sustainable Forestry in the Tropics (Kuala Lumpur, 5-9 October 1992).

Lugo, A.E. 1986. Water and the ecosystems of the Luquillo experimental forest. *General Technical Report*, **SO-63**, USDA Forest Service Southern Forest Experiment Station/Institute of Tropical Forestry, Rio Piedras, Puerto Rico.

Malingreau, J.P. & Tucker, C.J. 1988. Large-scale deforestation in the southeastern Amazon Basin of Brazil. *Ambio*, **17**, 49-55.

Malmer, A. 1993. *Dynamics of Hydrology and Nutrient Losses as Response to Establishment of Forest Plantation. A Case Study on Tropical Rainforest Land in Sabah, Malaysia.* PhD Thesis, Swedish University of Agricultural Sciences, Umeå, Sweden.

Malmer, A. & Grip, H. 1990. Soil disturbance and loss of infiltrability caused by mechanized and manual extraction of tropical rainforest in Sabah, Malaysia. *For. Ecol. Manage.*, **38**, 1-12.

Martin, G. 1970. Le défrichement mécanique pour la création de palmeraies industrielles. *Oléagineux*, **25**, 575-580.

Martins, P.F. da S., Cerri, C.C., Volkoff, B., Andreux, F. and Chauvel, A. 1991. Consequences of clearing and tillage on the soil of a natural Amazonian ecosystem. *For. Ecol. Manage.*, **38**, 273-282.

Mathur, H.N., Ram Babu, Joshi, P. and Singh, B. 1976. Effect of clearfelling and reforestation on runoff and peak rates in small watersheds. *Indian Forester*, **102**, 219-226.

McGuffie, K., Henderson-Sellers, A., Zhang, H., Durbidge, T.B. and Pitman, A.J. 1994. Global climate sensitivity to tropical deforestation. *Global and Planetary Change* (in press).

McWilliam, A-L.C., Roberts, J.M., Cabral, O.M.R., Leitao, M.V.B.R., de Costa, A.C.L., Maitelli, G.T. and Zamparoni, C.A.G.P. 1993. Leaf area index and above-ground biomass of *terra firme* rain forest and adjacent clearings in Amazonia. *Funct. Ecol.*, **7**, 310-317.

Medina, R.R., Jinno, K., Ueda, T. and Kawamura, A. 1987. The identification of recent rainfall fluctuations in the Philippines. In: *The Influences of Climate Change and Climate Variability on the Hydrologic Regime and Water Resources*, S.I. Solomon, M. Beran and W. Hogg (Eds.), *IAHS Publ. no. 168*, 233-244.

Meher-Homji, V.M. 1989. Deforestation and probable climatic change: a review. In: *Global Warming and Climatic Change. Perspectives from Developing Countries*, S. Gupta & R.K. Pachauri (Eds.), Tata Energy Research Institute, New Delhi, India, 115-129.

Monteith, J.L. 1988. Does transpiration limit the growth of vegetation or *vice versa*? *J. Hydrol.*, **100**, 57-68.

Monteny, B. and Gosse, G. 1976. Analyse et estimation du rayonnement net d'une culture de *Panicum maximum* en zone tropicale humide. *Oecologia Plantarum*, **11**, 173-191.

Monteny, B.A. Barbier, J.M. and Bernos, C.M. 1985. Determination of the energy exchanges of a forest type culture: *Hevea brasiliensis*. In: *The Forest-Atmosphere Interaction*, B.A. Hutchinson & B.B. Hicks (Eds.), D. Reidel, Dordrecht, 211-233.

Monteny, B.A. & Casenave, A. 1989. The forest contribution to the hydrological budget in tropical West Africa. *Annales Geophysicae*, **7**, 427-436.

Mooley, D.A. & Parthasarathy, B. 1983. Droughts and floods over India in summer monsoon seasons 1871-1980. In: *Variations in the Global Water Budget*, A. Street-Perrott *et al (Eds.)*, D. Reidel, Dordrecht, The Netherlands, 239-252.

Moore, I.D., O'Loughlin, E. and Burch, G.J. 1988. A contour-based topographic model for hydrological and ecological applications. *Earth Surf. Proc. & Landforms*, **13**, 305-320.

Moore, I.D., Grayson, R.B. and Ladson, A.R. 1991. Digital terrain modelling: a review of hydrological, geomorphological and biological applications. *Hydrol. Proc.*, **5**, 3-30.

Mumeka, A., 1986. Effect of deforestation and subsistence agriculture on runoff of the Kafue river headwaters, Zambia. *Hydrol. Sci. J.*, **31**, 543-554.

N mec, J., 1994. Climate variability, hydrology and water resources: do we communicate in the field? *Hydrol. Sci. J.*, **39**, 193-197.

Nepstad, D.C. 1989. *Forest Regrowth in Abandoned Pastures of Eastern Amazonia: Limitations to Tree Seedling Survival and Growth*, PhD thesis, Yale University, New Haven, Connecticut.

Nepstad, D.C., Carvalho, C.R. de, Davidson, E.A., Jipp, P.H., Lefebvre, P.A., Negreiros, G.H., da Silva, E.D., Stone, T.A., Trumbore, S.E. & Vieira, S. 1994. The role of deep roots in the hydrological and carbon cycles of Amazonian forests and pastures. *Nature*, **372**, 666-669.

Nieuwolt, S. 1977. *Tropical Climatology*. J. Wiley, New York.

Nortcliff, S. & Thornes, J.B. 1984. Flodplain response of a small tropical stream. In: *Catchment Experiments in Fluvial Hydrology*. T.P. Burt & D.E. Walling (Eds.), Geo-Books, Norwich, 73-85.

Nykvist, N., Grip, H., Sim, B.L., Malmer, A. and Wong, F.K. 1994. Nutrient losses in forest plantations in Sabah, Malaysia. *Ambio*, **23**, 210-215.

Odemerho, F.O. 1984. The effects of shifting cultivation on stream-channel size and hydraulic geometry in small headwater basins of southwestern Nigeria. *Geografiska Annaler*, **66A**, 327-340.

Ojo, O. 1987. Rainfall trends in West Africa, 1901-1985. In: *The Influences of Climate Change and Climate Variability on the Hydrologic Regime and Water Resources*, S.I. Solomon, M. Beran and W. Hogg (Eds.), *IAHS Publ.* no. *168*, 37-43.

Olbrich, B.W., Le Roux, D., Poulter, A.G., Bond, W.J. & Stock, W.D., 1993. Variation in water use efficiency and ^{13}C levels in *Eucalyptus grandis* clones. *J. Hydrol.*, **150**, 615-633.

O'Loughlin, E.M. 1986. Prediction of surface saturation zones in natural catchments by topographic analysis. *Wat. Resour. Res.*, **22**, 794-804.

O'Loughlin, E.M., Short, D.L. and Dawes, W.R.,. 1989. Modelling the hydrologic response of catchments to land use change.*Institute of Engineers in Australia National Conference Publication*, **89/19,** 335-340.

O'Loughlin, E.M. 1990. Modelling soil water status in complex terrain. *Agric. For. Meteorol.*, **50**, 23-38.

Ong, C.K., Odongo, J.C.W., Marshall, F. & Black, C.R. 1991. Water use by trees and crops. Five hypotheses. *Agroforestry Today*, **3(2)**, 7-10.

Ong, C.K. & Khan, A.H.H. 1993. The direct measurement of water uptake by individual tree roots. *Agroforestry Today*, **5(4)**, 2-5.

Olivry, J.C., Bricquet, J.P. and Mahe, G. 1993. Vers un appauvrissement durable des ressources en eau de l'Afrique humide? In:*Hydrology of Warm Humid Regions, J.S. Gladwell (Ed.), International Association of Hydrological Sciences Publication*, **216**, 67-78.

Parker, G.G. 1985. *The Effect of Disturbance on Water and Solute Budgets of Hillslope Tropical Rainforest in Northeastern Costa Rica*. PhD Thesis, University of Georgia, Athens, Ga., USA.

Parthasarathy, B. & Dhar, O.N. 1976. A study of trends and periodicities in the seasonal and annual rainfall of India. *J. Meterol., Hydrol. & Geophys.*, **27**, 23-28.

Paul, L.L. & Kuraji, K. 1993. *Transition Report of Hydrological Study on Forested Catchments in Sabah, 1990-1992*. Research & Development Division, Sabah Forestry Department, Sandakan, Sabah, Malaysia.

Pearce, A.J. 1990. Streamflow generation processes: an Austral view. *Wat. Resour. Res.*, **26**, 3037-3047.

Pearce, A.J., Rowe, L.K. and O'Loughlin, C.L. 1980a. Effects of clearfelling and slashburning on water yields and storm hydrographs in evergreen mixed forests, western New Zealand.*IAHS Publ. 130*, 119-127.

Pearce, A.J., Rowe, L.K. and Stewart, J.B. 1980b. Nighttime, wet canopy evaporation rates and the water balance of an evergreen mixed forest. *Wat. Resour. Res.*, **16**, 955-959.

Pearce, A.J. & Hamilton, L.S. 1986. *Water and Soil Conservation Guidelines for Land-Use Planning*. Summary of Seminar - Workshop on Watershed Land-use Planning, Gympie, May 1985. Environment & Policy Institute, Hawaii.

Pereira, H.C. 1989. *Policy and Practice of Water Management in Tropical Areas*. Westview Press, Boulder, Colorado.

Poels, R. 1987. *Soils, Water and Nutrients in a Forest Ecosystem in Suriname*. PUDOC, Wageningen, The Netherlands.

Proctor, J. 1989. Progress report on tropical rain forests. *Prog. in Phys. Geogr.*, **13**, 409-430.

Quencez, P. 1989. Evolution de la pluviosité en Afrique de l'Ouest. *Oléagineux*, **44(4)**, 39-41.

Quinn, P.F., Beven, K.J., Chevallier, P. and Planchon, O 1991. The prediction of hillslope flow paths for distributed hydrological modelling using digital terrain models. *Hydrol. Proc.*, **5**, 59-79.

Qian, W.C. 1983. Effects of deforestation on flood characteristics with particular reference to Hainan island, China. In: *Hydrology of Humid Tropical Regions*, R. Keller (Ed.), *IAHS Publ. no. 140*, 249-258.

Raghavendra, V.K. 1982. A case study of rain storm in the eastern mountainous catchments of Brahmaputra leading to highest ever recorded flood level at Dibrugarh. In: *Proceedings of the International Symposium on Hydrological Aspects of Mountainous Watersheds*, Roorkee, India. Manglik Prakashan, Saharanpur, India, I-39 - I-44.

Richey, J.E., Nobre, C. and Deser, C. 1989. Amazon river discharge and climate variability: 1903-1985. *Science*, **246**, 101-103.

Rijsdijk, A. & Bruijnzeel, L.A. 1990. Erosion, sediment yield and land use patterns in the upper Konto watershed, East Java, Indonesia. *Konto River Project Communication*, **18**, 3 Volumes. Konto River Project, Malang, Indonesia.

Ritchie, J.C., Murphey, J.B., Grissinger, E.H. and Garbrecht, J.D. 1993. Monitoring streambank and gully erosion by airborne laser. In: *Sediment Problems: Strategies for Monitoring, Prediction ad Control*, R.F. Hadley & T. Mizuyama (Eds.), *IAHS Publ. no. 217*, 161-166.

Roberts, J. & Rosier, P.T.W. 1993. Physiological studies in young *Eucalyptus* stands in southern India and derived estimates of forest transpiration. *Agric. Wat. Manage.*, **24**, 103-118.

Rogerson, T.L. 1967. Throughfall in pole-sized lloblolly pine as affected by stand density. In: *International Symposium on Forest Hydrology*, W.E. Sopper & H.W. Lull (Eds.), Pergamon Press,Oxford, 187-190.

Sá, T.D. de A., Hölscher, D., Bastos, T.X. and Pacheco, N.A. 1993. Partição da àgua da chuva em vegetação secundária do Nordeste Paraense. Poster presentation at the Seventh Brazilian Congress on Agrometeorology, July 1993, Rio Grande do Sul, Brazil.

Salati, E. & Vose, P.B. 1984. Amazon Basin: a system in equilibrium. *Science*, **225**, 129-138.

Saldarriaga, J.G. 1987. Recovery following shifting cultivation. In: *Amazonian Rain Forests. Ecosystem Disturbance and Recovery*, C.F. Jordan (Ed.), Springer, New York, 24-33.

Sambou, S. & Thirriot, C. 1993. Algorithme non-récursif et récursif: application á la prévision des crues du haut bassin du Fleuve Sénégal. In: *Hydrology of Warm Humid Regions*, J.S. Gladwell (Ed.), *IAHS Publ. no. 216*, 139-148.

Samraj, P., Sharda, V.N., Chinnamani, S., Lakshmanan, V. and Haldorai, B. 1988. Hydrological behaviour of the Nilgiri sub-watersheds as affected by bluegum plantations, Part I. The annual water balance. *J. Hydrol.*, **103**, 335-345.

Savenije, H.H.G. & Hall, M.J. 1993. Climate and land use: a feedback mechanism? In: *Water and Environment: Key to Africa's Development. IHE Report Series*, **29**, IHE, Delft, 93-108.

Sayer, J.A., Harcourt, C.J. and Collins, M.N. 1992. *The Conservation Atlas of Tropical Forests. Africa*. MacMillan Publishers, Basingstoke.

Scatena, F.N. 1990. Watershed scale rainfall interception on two forested watersheds in the Luquillo Mountains of Puerto Rico. *Journal of Hydrology*, **113**, 89-102.

Servat, E. 1994. *ICCARE. Identification et Conséquences d'une Variabilité du Climat en Afrique de l'Ouest non-Sahélienne. Présentation du Programme*. Antenne Hydrologique, ORSTOM - Côte d'Ivoire, Abidjan, Côte d'Ivoire.

Servat, E. & Sakho, M. 1993. Incidences de l'instabilité des ressources en eau sur la gestion d'un système d'eau aménagé. Cas du Sassandra en Côte d'Ivoire. In: *Hydrology of Warm Humid Regions*, J.S. Gladwell (Ed.), *IAHS Publ. no. 216*, 107-116.

Seubert, C.E., Sanchez, P.A. and Valverde, C. 1977. Effects of land clearing methods on soil properties of an ultisol and crop performance in the Amazon jungle of Peru. *Trop. Agric. (Trinidad)*, **54**, 307-321.

Shukla, J., Nobre, C. and Sellers, P.J. 1990. Amazon deforestation and climatic change. *Science*, **247**, 1322-1325.

Shuttleworth, W.J, 1988a. Evaporation from Amazonan rain forest. *Phil. Trans. Roy. Soc. (London), Ser. B*, **233**, 321-346.

Shuttleworth, W. J. 1988b. Macrohydrology - the new challenge for process hydrology. *J. Hydrol.*, **100**, 31-56.

Shuttleworth, W.J. 1989. Micrometeorology of temperate and tropical forest. *Phil.Trans. Roy. Soc. (London), Ser. B* **324**, 299-334.

Shuttleworth, W.J., Gash, J.H.C., Roberts, J., Nobre, C.A., Molion, L.C.B. and Ribeiro, M. de N.G. 1991. Post-deforestation Amazonian climate: Anglo-Brazilian research to improve prediction. *J. Hydrol.*, **129**, 71-85.

Sim, B.L. & Nykvist, N. 1991. Impact of forest harvesting and replanting. *Journal of Tropical Forest Science*, **3**, 251-284.

Smith, R.E. & Scott, D.F. 1992. The effects of afforestation on low flows in various regions of South Africa. *Water SA*, **18**, 185-194.

Snoeck, J. 1975. Variations de la pluviosité en zône forestière Ivoirienne. *Café, Cacao, Thé*, **19**, 165-175.

Stadtmüller, T. 1987. *Cloud Forests in the Humid Tropics. A Bibliographic Review*. The United Nations University, Tokyo, and Centro Agronomico Tropical de Investigacion y Ensenanza, Turrialba, Costa Rica.

Stadtmüller, T. & Agudelo, N. 1990. Amounts and variability of cloud moisture input in a tropical cloud forest. In: *Hydrology of Mountainous Regions*, H. Lang & A. Musy (Eds.), *International Association of Hydrological Sciences Publication*, **193**, 25-32.

Stewart, J.B. & Finch, J.W., 1993. Application of remote sensing to forest hydrology. *Journal of Hydrology*, **150**, 701-716.

Subba Rao, B.K., Ramola, B.C. and Sharda, V.N., 1985. Hydrologic response of a forested mountain watershed to thinning. *Indian Forester*, **111**, 681-690.

Swindel, B.F., Lassiter, C.J. and Riekerk, H., 1982. Effects of clearcutting and site preparation on water yield from slash pine forests. *Forest Ecology and Management*, **4**, 101-113.

Tangtham, N. & Sutthipibul, V., 1989. Effects of diminishing forest area on rainfall amount and distribution in north-eastern Thailand. Paper presented at the FRIM-IHP-UNESCO Regional Seminar on Tropical Forest Hydrology, Kuala Lumpur, 4-9 September 1989.

Tillman, J.E., 1972. The indirect determination of stability, heat and momentum fluxes in the atmosphere boundary layer from simple scalar variables during dry unstable conditions. *Journal of Applied Meteorology*, **11**, 783-792.

Uhl, C., Buschbacher, R. and Serrão, E.A.S., 1988. Abandoned pastures in eastern Amazonia. I. Patterns of plant succession. *Journal of Ecology*, **76**, 663-681.

Vandana Shiva & Bandyopadhyay, J, 1983. *Eucalyptus* - a disastrous tree for India. *The Ecologist*, **13**, 184-187.

Van de Ven, F.H.M., Gutnecht, D., Loucks, D.P. and Salewicz, K.A. (Eds.), 1991. *Hydrology for the Water Management of Large River Basins. International Association of Hydrological Sciences*, **201**.

Van der Plas, M.C. & Bruijnzeel, L.A., 1993. Impact of mechanized selective logging of rainforest on topsoil infiltrability in the Upper Segama area, Sabah, Malaysia. In: *Hydrology of Warm Humid Regions*, J.S. Gladwell (Ed.), *International Association of Hydrological Sciences Publication*,

216, 203-211.

Van der Weert, R., 1974. The influence of mechanical forest clearing on soil conditions and resulting effects on root growth. *Tropical Agriculture (Trinidad)*, **51**, 325-331.

Van Rompaey, R.S.A.R., 1993. *Forest Gradients in West Africa, a Spatial Gradient Analysis*. PhD Thesis, Department of Forestry, Wageningen Agricultural University, Wageningen, The Netherlands.

Van Steenis, C.G.J., 1972. *The Mountain Flora of Java*. Brill, Leiden, The Netherlands.

Vertessy, R.A., Wilson, C.J., Silburn, D.M. Connolly, R.D. and Ciesiolka, C.A., 1990. Predicting erosion hazard areas using digital terrain analysis. In: *Research Needs and Applications to Reduce Erosion and Sedimentation in Tropical Steeplands*, R.R. Ziemer, C.L. O'Loughlin and L.S. Hamilton (Eds.), *IAHS Publication* no. 192: 298-308.

Vertessy, R.A., Hatton, T.J., O'Shaughnessy, P.J. and Jayasurya, M.D.A., 1993. Predicting water yield from a mountain ash forest using a terrain analysis based catchment model. *Journal of Hydrology*, **150**, 665-700.

Vertessy, R.A., Hatton, T.J., Benyon, R.J. & Dawes, W.R., 1995. Long-term growth and water balance predictions for a mountain ash (*Eucalyptus regnans*) forest catchment subject to clearfelling and regeneration. *Tree Physiology*, **15** (in press).

Vines, R.G., 1986. Rainfall patterns in India. *Journal of Climatology*, **6**, 135-148.

Viswanatham, N.K., Joshie, P. and Ram Babu, 1980. Influence of forest on soil erosion control - Dehradun. In: *Annual Report 1980*,. Central Soil and Water Conservation Research and Training Institute, Dehradun, India, 100-104.

Viswanatham, N.K., Joshie, P. and Ram Babu, 1982. Influence of forest on soil erosion control - Dehradun. In: *Annual Report 1982*, Central Soil and Water Conservation Research and Training Institute, Dehradun, India, 40-43.

Vogelmann, H.W., 1973. Fog precipitation in the cloud forests of Eastern Mexico. *BioScience*, **23**, 96-100.

Vörösmarty, C.J & Moore, B., 1991. Modelling basin-scale hydrology in support of physical climate and global biogeochemical studies: an example using the Zambezi River. *Surveys in Geophysics*, **12**, 271-311.

Vörösmarty, C.J., Moore, B., Grace, A., Peterson, B.J., Rastetter, E.B. and Melillo, J. 1991. Distributed parameter models to analyse the impact of human disturbance on the surface hydrology of a large tropical river basin in southern Africa. In: *Hydrology for the Water Management of Large River Basins*, F.H.M. van de Ven, D. Gutnecht, D.P. Loucks and K.A. Salewicz (Eds.), *IAHS Publ. no. 201*, 233-244.

Vugts, H.F., Waterloo, M.J., Beekman, F.J., Frumau, K.F.A. and Bruijnzeel, L.A. 1993. The temperature variance method: a powerful tool in the estimation of actual evaporation rates. In: *Hydrology of Warm Humid Regions*, J.S. Gladwell (Ed.), *IAHS Publ. no. 216*, 251-260.

Walker, L.R., Brokaw, N.L., Lodge, D.J. and Waide, R.B. 1991. *Ecosystem, Plant, and Animal Response to Hurricanes in the Caribbean. Biotropica*, **23(4)** (special issue).

Walling, D.E. 1983. The sediment delivery problem. *J. Hydrol.*, **65**, 209-237.

Walter, H. & Breckle, S.W. 1986. *Ecological Systems of the Geobiosphere. 2. Tropical and Subtropical Zonobiomes*. Springer, New York.

Ward, R.C. 1984. On the response to precipitation of headwater streams in humid areas. *J. Hydrol.*, **74**, 171-189.

Ward, R.C. & Robinson, M. 1990. *Principles of Hydrology*. McGraw Hill, London.

Wasser, H.J. & Harger, J.R.E. 1992. *Several Environmental Factors Affecting the Rainfall in Indonesia*. ROSTSEA/UNESCO, Jakarta, Indonesia.

Waterloo, M.J. 1994. *Water and Nutrient Dynamics of Pinus Caribaea Plantation Forests on Former Grassland Soils in Southwest Viti Levu, Fiji*. PhD Thesis, Faculty of Earth Sciences, Vrije Universiteit, Amsterdam.

Weaver, P.L., Byer, M.D. & Bruck, D.L. 1973. Transpiration rates in the Luquillo Mountains of Puerto Rico. *Biotropica*, **5**, 123-133.

Werner, W.L. 1988. Canopy dieback in the Upper Montane rain forests of Sri Lanka. *GeoJournal*, **17**, 245-248.

Whitmore, T.C. 1989. Changes over twenty-one years in the Kolombangara rain forests. *J. Ecol.*, **77**, 469-483.

Whitmore, T.C. (1990) *Introduction to Tropical Rain Forests*. Clarendon Press, Oxford.

World Meteorological Organization. 1988. The tropical ocean and atmosphere project. *WMO Bulletin*, **37**, 99-104.

Young, A. 1989. *Agroforestry for Soil Conservation*. CAB International, Wallingford.

Zadroga, F. 1981. The hydrological importance of a montane cloud forest area of Costa Rica. In: *Tropical Agricultural Hydrology*, R. Lal & E.W. Russell (Eds.), J. Wiley, New York, 59-73.

Zhang, K.Y. 1986. The influence of deforestation of tropical rainforest on local climate and disaster in the Xishuangbanna region of China. *Climatological Notes*, **35**, University of Tsukuba, Tsukuba, Japan, 223-236.

Zhang, R.F. 1990. The flood forecasting system for the Three Gorges of the Yangtze River. *IAHS Publ.*, **197**, 485-495.

RESUMO

Este artigo faz uma revisão da literatura sobre os possíveis impactos da conversão de florestas tropicais para outros tipos de uso do solo (pastagens, plantações), com relação ao regime pluviométrico, balanço hídrico total e sazonal e resposta da bacia de drenagem à precipitação. Questões que requerem pesquisas são identificadas dentro de cada categoria e sugestões metodológicas de como lidar com problemas específicos são oferecidas. Ênfase é colocada na integração dos trabalhos de pesquisa e das simulações com modelos de base física, dentro do contexto de bacia hidrográfica, para facilitar a integração das diversas áreas. Um apelo é feito para concentrar os esforços de pesquisas (integradas) futuras em poucos e bem conhecidos locais, unidos numa rede que represente a variabilidade ambiental encontrada nos trópicos.

3 Comparisons of long-term soil water storage behaviour under pasture and forest in three areas of Amazonia

M.G. HODNETT[1], M.D. OYAMA[2], J. TOMASELLA[3] and A. de O. MARQUES FILHO[4]
[1]*Institute of Hydrology, Wallingford, UK*
[2]*CPTEC, INPE, SP, Brazil*
[3]*Instituto de Pesquisas Hidraulicas, Porto Alegre, RS, Brazil*
[4]*Instituto Nacional de Persquisas da Amazonia, Manaus, Brazil*

INTRODUCTION

In simple terms, transpiration can be regarded as being controlled only by atmospheric factors while the vegetation is "well supplied with water" (Penman, 1956). When the supply of soil water becomes limiting, the vegetation suffers "soil water stress" and transpiration decreases, altering the partitioning of incoming solar radiation into sensible and latent heat. Less water (and therefore, latent heat) is returned to the atmosphere and the proportion returning as sensible heat increases. The supply of water to the vegetation depends on the soil properties, and on the type of vegetation and its ability to abstract water from the soil (in particular, its rooting depth). Land-use change, particularly deforestation, on a large enough scale could potentially modify the water and energy balance sufficiently to modify climate. The influence of land surface evapotranspiration on the Earth's climate has been discussed by Shukla and Mintz (1982).

General Circulation Models (GCMs) are being used to predict the effects of large scale forest clearance on climate (e.g. Lean and Warrilow, 1989; Shukla *et al.*, 1990; Nobre *et al.*, 1991). However, their validity and accuracy are dependent on representative parameterisation of the soil and of the vegetative surface before and after deforestation (Henderson-Sellers, 1987). More sophisticated Soil-Vegetation-Atmosphere Transfer (SVAT) models are being developed, for example SiB (Sellers *et al.*, 1986), BATS (Dickinson *et al.*, 1986) and the parameterisation of Noilhan and Planton (1989). The input parameter requirements for these models are increasing, particularly for soil data. Long-term, and, ideally, large scale, soil moisture data are required for the verification of SVAT models, but as Owe and Van de Griend (1990) have noted, the lack of reliable large scale soil moisture data has been a long-standing problem.

As far as is known, the only published medium-term records of soil water content

Amazonian deforestation and climate. Edited by J.H.C. Gash, C.A. Nobre, J.M. Roberts and R.L. Victoria.

for forested or cleared forest sites in Amazonia are those of Nepstad *et al.* (1994), which show nine months of data from sites in Pará, in eastern Amazonia. One of the main aims of the ABRACOS soil water studies was to obtain a minimum of two or three years of weekly soil water content and soil profile storage measurements to enable the calibration of SVAT and soil water uptake models.

The continuous weekly soil water record was particularly important to allow the detailed study missions (micro-meteorology and plant physiology) to be located within the seasonal cycle and within the interannual variability of profile storage change. Periods of very low soil profile storage are of particular interest because of the effects of soil water stress on the energy balance. The longer term context of the soil water data-set from Manaus is discussed in a later paper (Hodnett *et al.*, 1995c).

This paper presents and compares the soil profile storage data for the three paired (forest and pasture) ABRACOS sites. Comparisons are made between sites and between years, and the physical processes influencing the soil water storage behaviour at the different sites are discussed.

MATERIALS AND METHODS

LOCATION

A general outline of the ABRACOS project, and the location of the three main study sites has been presented by Gash *et al.* (1996). Further details can be found in Wright *et al.* (1992), Bastable *et al.* (1993), McWilliam *et al.* (1993) and Hodnett *et al.* (1995).

It should be noted that the forest soil water monitoring site near Manaus was in an area of forest adjacent to the pasture site at Fazenda Dimona, and not in the Reserva Ducke, where the tower was located. This was to ensure that the soil type and rainfall were, as far as possible, identical and to allow simultaneous observations to be made at both sites.

SOILS AND VEGETATION

Site details are summarised in Table 1. N.B.: the "English" soil type names given first are translations from the Brazilian classification.

MANAUS

At the pasture and adjacent forest sites, the soil was a clayey yellow latosol (Latossolo amarelo, álico, textura argilosa - Brazil, Haplic Acrorthox - Soil Taxonomy, or xanthic ferrasol - FAO), very similar to those described by Ranzani (1980) and Correa (1984). The soils have a very high clay content (>75%) and a low dry bulk density of between 0.93 and 1.15 Mg m^{-3}, giving a very high porosity of between

56% and 64%. The lowest bulk density was in the top 0.1 m in the forest; in the pasture the surface density was 1.06 Mg m^{-3}. However, much of the porosity is concentrated in the macropores and large mesopores, which drain rapidly, and the very fine pores containing water that is inaccessible to plants. This means that the available water capacity is low, only about 70 mm m^{-1} in the upper metre of the profile (Correa, 1984). More details of the soil at this site and its hydraulic properties are given by Hodnett *et al.* (1995) and Tomasella and Hodnett (1996).

During most of the period of study reported here, the pasture was in good condition, with little invasion and degradation by woody shrubs (<2% cover). There were two main grass species, *Brachiaria humidicola* and *Brachiaria decumbens*. By July 1993, the shrub cover had increased to about 30%; most of this increase occurred after late 1992.

Table 1 Summary of site details

LOCATION		WEST		CENTRAL		EAST	
SITE NAME		JI-PARANÁ		MANAUS		MARABÁ	
DRY SEASON[1] Months with <150 mm		May - September		June - Sept ember		May - October	
< 50 mm		June - August		----		June - August	
		Pasture	Forest	Pasture	Forest	Pasture	Forest
Soil Type (FAO)		Orthic Acrisol	Orthic Acrisol	Xanthic Ferralsol	Xanthic Ferralsol	Orthic Acrisol	Humic Cambisol
Surface Texture		Loamy Sand	Sand	Clay	Clay	Sandy Clay Loam	Loam
Depth (m)							
0-0.05	Clay (%)	9	4	86	83	20	18
	Sand (%)	85	88	8	16	74	45
	Bulk density (Mgm m^{-3})	1.50	1.38	1.06	0.93	1.40 (0.15 m)	1.24 (0.15 m)
	Porosity (%)	42	47	59	64	46	52
0.5-1.0	Clay (%)	33	24	75	77	40	12
	Sand (%)	58	67	88	80	53	13
	Bulk density (Mgm m^{-3})	1.30	1.38	1.14	-	1.42	1.33
	Porosity (%)	50	47	56	-	45	49
Access Tube Depth (m)		3.6	2-3.6	2.0	1-3.6	3.6	3.6
No of tubes		6	8	15	15	6	6
Start of observations		11/91	11/91	9/90	9/90	8/91	8/91

[1]Mean monthly rainfall data.
Ji-Paraná, Ouro Preto, 1982 - 1992.
Manaus, Reserva Ducke, 1966 - 1992
Marabá, Marabá, 1973 - 1993

MARABÁ

The soil at the pasture site (Fazenda Boa Sorte) was a medium to clayey textured red-yellow podzolic soil (podzólico vermelho-amarelo Tb álico a moderado textura média/argilosa - Brazil, typic haplustult - Soil Taxonomy, or orthic acrisol - FAO). The texture in the upper 0.5 m of the profile was sandy loam to sandy clay loam.

Below 0.5 m the clay content increased to give a sandy clay texture. The bulk density in the top 0.5 m was 1.52 Mg m^{-3} (42%) and between 1.32 Mg m^{-3} (49%) and 1.49 Mg m^{-3} (43%) in the profile to 2.0 m depth. The values in parentheses are porosities based on a particle density of 2.6 Mg m^{-3}. The pasture was mainly grass, but it was becoming degraded with woody shrubs and small palms. The main grass species was *Panicum maximum*.

At the forest site (Reserva Companhia do Vale do Rio Doce), the soil has been classified as a medium textured yellowish cambisol (Cambissolo Tb álico amarelado textura média, probably a typic dystrochrept - Soil Taxonomy, or humic cambisol - FAO). The texture of the profile to about 0.6 m depth was sandy loam, but it was very stony (gravel/stone content, 36% w/w). The stone content decreased abruptly below this depth, where the texture was silt loam. Below about one metre depth, the profile consisted of saprolite (heavily weathered rock, retaining some of the original rock

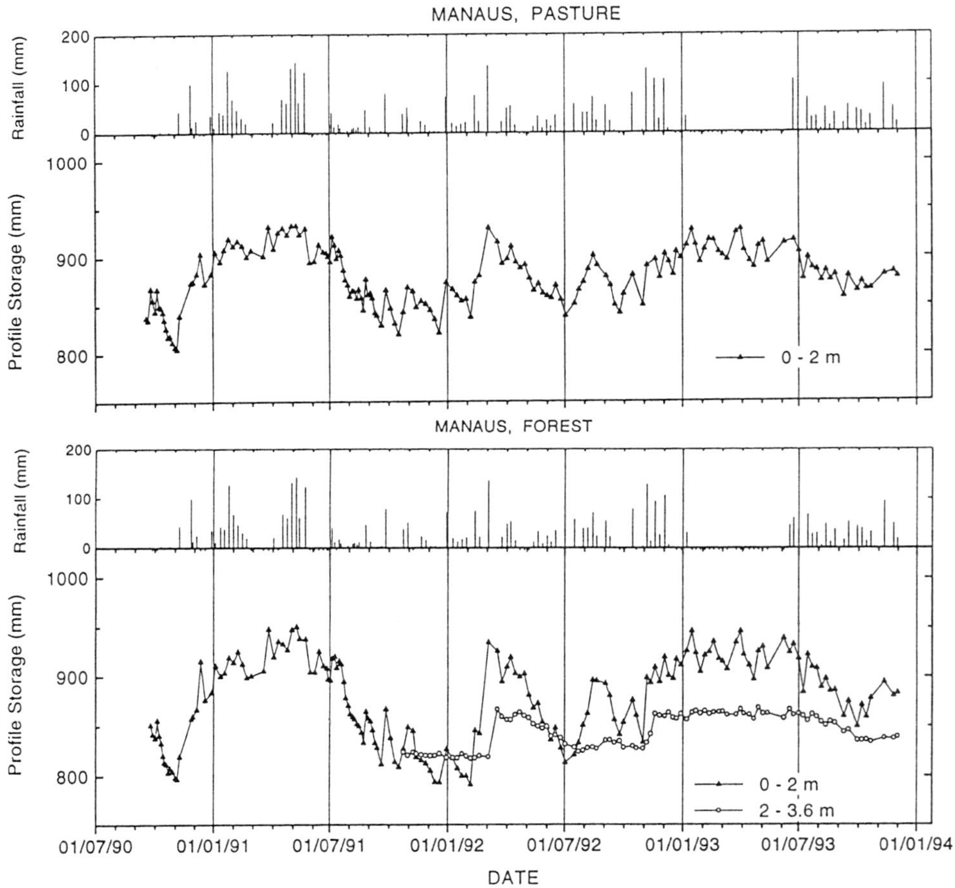

Figure 1 Manaus, Fazenda Dimona, pasture (top): profile storage (0-2 m) and rainfall between reading dates. Note: in all graphs in the following figures, negative values indicate unreliable rainfall data. Manaus Fazenda Dimona, forest (bottom): profile storage in the 0-2 m and 2-3.6 m layers, and rainfall between reading dates.

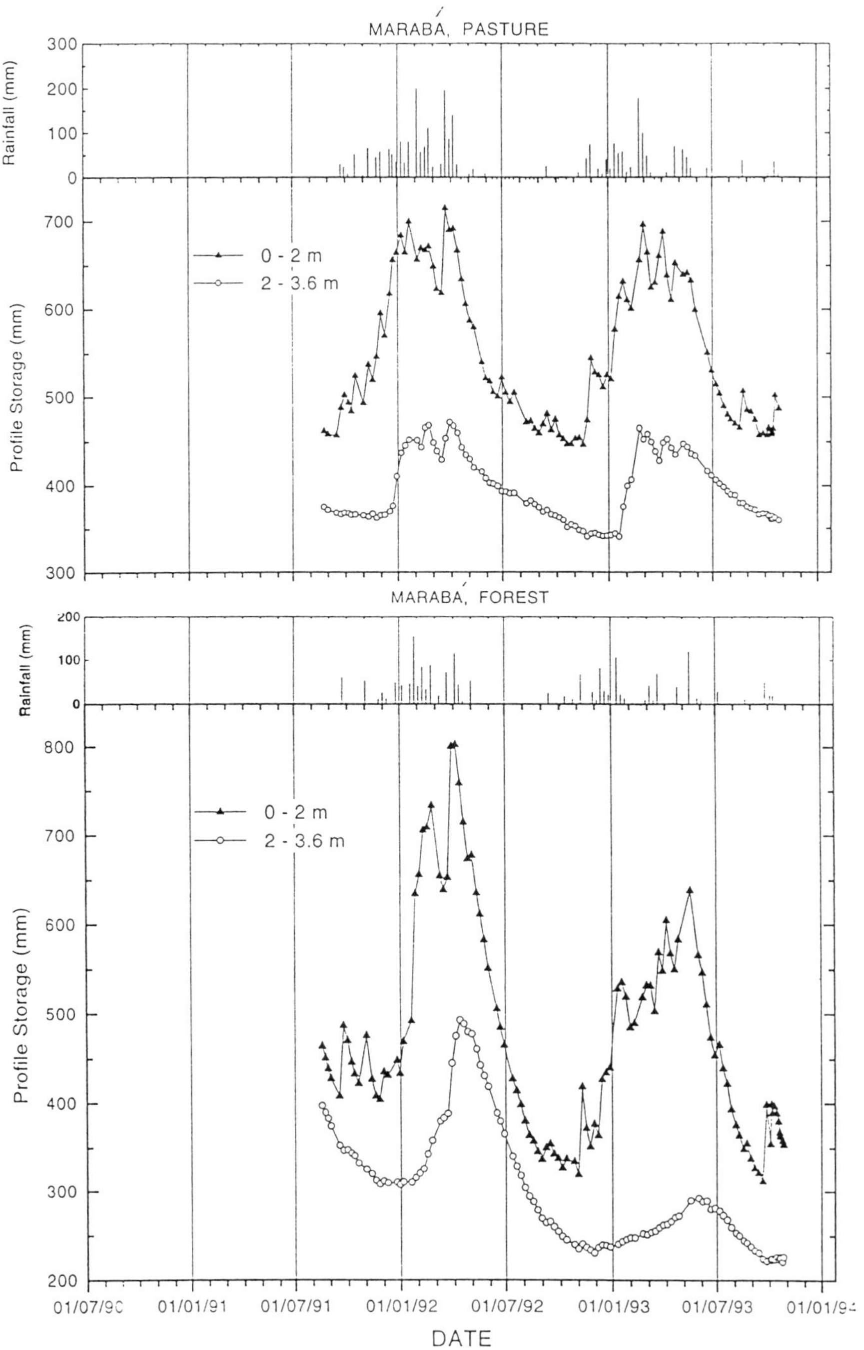

Figure 2 Marabá, pasture (top): profile storage in the 0-2 m and 2-3.6 m layers, and rainfall between reading dates. Marabá forest (bottom): profile storage in the 0-2 m and 2-3.6m layers, and rainfall between reading dates.

structure) which was texturally a silt loam or a silt. The bulk density in the top 0.5 m was 1.19 Mg m^{-3} (54%) and increased gradually to 1.49 Mg m^{-3} (43%) at a depth of 2.3 m.

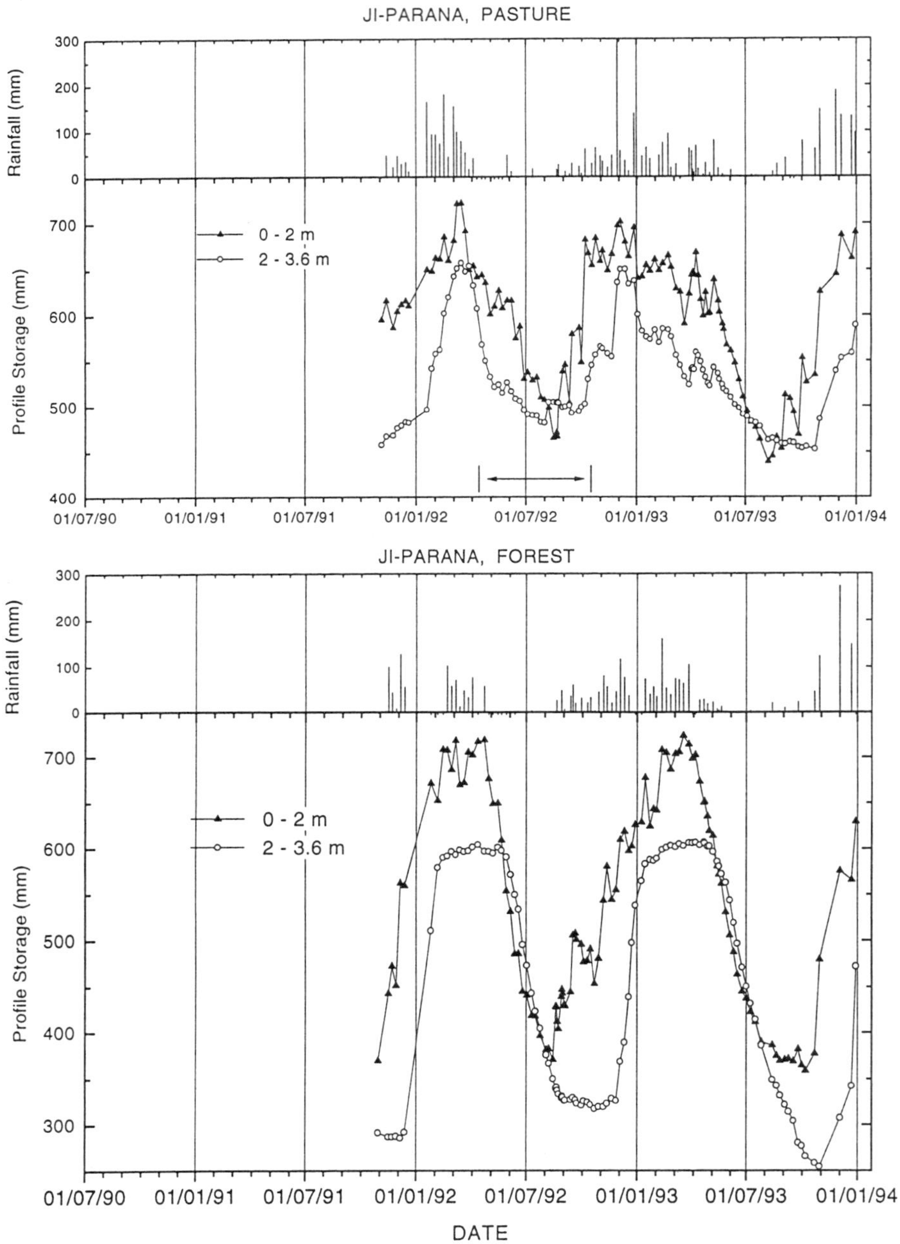

Figure 3 Ji-Parana, pasture (top): profile storage in the 0-2m and 2-3.6m layers and rainfall between reading dates. Ji-Paraná, forest (bottom) Profile storage in the 0-2m and 2-3.6m layers and rainfall between reading dates

JI-PARANÁ

At the pasture site (Fazenda Nossa Senhora da Aparecida) we have provisionally classified the soil as a medium textured red-yellow podzol (Podzólico vermelho amarelo A moderado textura média - Brazil, typic paleudult - Soil Taxonomy, or orthic acrisol - FAO). The surface soil was a sand/loamy sand (85% sand). The clay content increased gradually with depth from 7% in the top 0.1 m to 35% between 1 m and 1.5 m, where the texture was sandy clay loam. Between 1 m and 2 m, the soil was gravelly (28% w/w), merging into saprolite below 2 m. Bedrock was not encountered within the upper 3.8 m of the profile. Roots were observed throughout the profile examined in a pit to a depth of 1.65 m. Roots were sometimes found concentrated in the channels left by rotted tree roots. The bulk density decreased from 1.50 Mg m^{-3} (42%) at the surface to 1.24 Mg m^{-3} (52%) at 1.35 m.

The pasture was in good condition with little invasion by shrubs. There were a few scattered large palms, but these were not near to the soil water monitoring sites. The grass species was exclusively *Brachiaria brizantha.*

We have provisionally classified the soil at the Ji Paraná forest site (Reserva Biológica do Jaru) as a medium textured red-yellow podzol (Podzólico vermelho-amarelo A moderado textura média - Brazil, typic paleudult - Soil Taxonomy, or orthic acrisol - FAO). The surface soil was a sand (88%) and the clay and silt content increased with depth. At 0.6 m the texture was a sandy loam. Below this depth, the profiles differed because of the large range in the depth to the granite bedrock which varied from 1.6 m in one soil sampling pit to > 3.8 m at two of the measurement sites. In one of the deeper profiles sampled, the clay content reached 34% between 1 m and 1.5 m (sandy clay loam) and then decreased to around 17% at 2.7 m. In most places the soil merged downwards into saprolite, then into fairly hard weathered granite. The bulk density at the surface was 1.38 Mg m^{-3} (47%), which was less than in the pasture. The density at 0.2 m was 1.55 Mg m^{-3} (40%) and decreased gradually with depth to 1.49 Mg m^{-3} (43%) at 0.6 m.

INSTRUMENTATION

Soil water content was measured using a neutron probe. The Fazenda Dimona sites near Manaus were established in September 1990. To represent the "mean" soil water behaviour in the area, the neutron probe access tubes were installed to represent three different landscape elements: plateau, slope and valley floor. Soil water storage behaviour was expected to differ in each element because of the presence of a shallow water table in the valley, and the greater likelihood of surface runoff on the slopes. Access tubes were installed to a depth of 2.0 m in the plateau and slope elements and to 1.0 - 1.8 m in the valley floor. In October 1991, the five forest plateau tubes were replaced by tubes with a maximum reading depth of 3.6 m.

The soil water measurement sites near Marabá and Ji-Paraná were established in August and October 1991 respectively. There were six access tubes at each site, each with a maximum reading depth of 3.6 m, except at the Ji-Paraná forest site. Here, the depth of six of the eight access tubes was limited by the depth to the granite bedrock. Maximum reading depths ranged from 2.0 to 3.6 m. The two deepest tubes did not reach bedrock. At the Marabá and Ji-Paraná sites, only the plateau landscape element was represented. Water content was measured at depths of 0.1 m and 0.2 m, and then at 0.2 m intervals for the remainder of the depth of the tube. At the Marabá sites the measurement depth interval was 0.3 m below 1.2 m depth. Details of the numbers and depths of the neutron probe access tubes at each of the sites are given in Table 1. Observations were made on a weekly basis with twice weekly observations being made during the intensive field missions. At the Manaus sites, simultaneous observations were made in the forest and pasture but at the Marabá and Ji-Paraná sites observations were made at the forest and pasture sites on separate days.

The neutron probe count rates were converted to volumetric water content θ using a soil calibration of the form:

$$\theta = m\,(R/R_w) + n$$

where R is the count rate at a given depth in the soil, R_w is the count rate in a water standard, and m and n are the slope and intercept of the calibration. These were determined by the method of Couchat *et al.* (1975). Bulk density profiles were determined to allow the calibration to be adjusted accordingly. For the Manaus sites, gravimetric calibration was carried out for the readings made at 0.1 m depth. This was not done for the other sites and may have led to an under-estimation of the maximum profile water storage changes of 10 - 20 mm.

The profile storage (S) to depth (D) was determined using:

$$S = \sum_{i=1}^{D} \theta_i \, \Delta_i z_i$$

where θ_i and z_i are the water content and thickness respectively of layer i. For each reading date, the profile storage to the maximum depth in each tube, and the mean storage for each site were calculated.

The soil water balance was used to determine evaporation or deep drainage for selected periods:

$$P = R + E + D + \Delta S$$

where P is the net rainfall, R is the surface runoff, E is the evapotranspiration and ΔS is the change in profile water storage to the depth of interest, all during the same time period.

RESULTS

SEASONAL PROFILE STORAGE CHANGES

Figures 1, 2 and 3 show the mean profile water storage (a) for pasture and (b) for forest for the layers from 0 to 2 m and from 2 m to 3.6 m for the Manaus, Marabá and Ji-Paraná sites respectively. The data are shown from the start of observations until the end of December 1993 (October 1993 for Marabá). All graphs are on the same scale. The data are summarised in Table 2, which shows the maximum recorded seasonal water content variation in the 0-2 m and 2-3.6 m layers, and in the full 3.6 m profile. It should be noted that the maximum change in the full 3.6 m profile is not the sum of the change in two layers. This is because the timing of the minimum water content is not the same in the two layers, mainly because drainage may be reducing the storage in the lower layer while the upper layer is wetting up at the start of the wet season.

Unlike the Manaus sites, the Marabá and Ji-Paraná "paired" sites were far enough apart for there to have been significant differences in rainfall on a weekly basis. Observations were also made on separate (usually successive) days. The Manaus pasture and forest data might therefore appear more similar than those from the other paired sites.

Table 2 Maximum recorded soil water storage changes at each site

Layer	**0–2 m**		**2–3.6 m**		**0–3.6 m**	
Site	Pasture	Forest	Pasture	Forest	Pasture	Forest
Manaus	132	154	-	48	-	200
Marabá	263	483	131	263	376	724
Ji-Paraná	262	365	165	353	450	701

MANAUS

The forest and pasture profile water storage changes in the upper 2 m were very small compared to the Marabá and Ji-Paraná sites. There was also only a relatively small difference between the forest and pasture changes in the 2 m profile. The maximum seasonal variation at the Manaus sites was 154 mm for the forest and 132 mm for the pasture.

The minimum storage recorded for both forest and pasture was on 5 November 1990, at the end of the dry season. In the pasture, the storage never fell this low again over the next three years, but in the forest the same minimum was recorded on three subsequent occasions, in December 1991 (twice), and February 1992 (once). It is notable that February is (on average) one of the wettest months.

During each wet season, the same narrow range of maxima was seen in the 2 m profile, which wetted up to the same extent (a storage of 900 mm - 920 mm) each year. Large peaks in storage were never seen in the weekly data because drainage from the profile is rapid. The pattern of wet season storage behaviour was very similar in the forest and pasture.

There were considerable differences in the annual cycle of storage change between years as a result of inter-annual rainfall variability. The 1991 dry season extended into 1992, which was very different to the mean pattern, and the 1991-1992 wet season was very weak; storage only remained above 900 mm for about two months, compared to seven months in 1990-91 and nine months in 1992-93. The 1993 dry season was the wettest observed. This three year run of data has been examined in the context of a 27 year rainfall record from Reserva Ducke in a subsequent paper (Hodnett *et al.*, 1996).

Deeper (3.6 m) access tubes were installed in the forest in October 1991 because there was evidence from the water balance for October 1990 that water uptake was taking place from below 2 m depth. After the installation of these tubes, the storage in the 2 - 3.6 m layer decreased only 5 mm, reaching the minimum recorded during the study in January and February 1992. It is clear from these data that most of the storage in this layer had already been utilised by October 1991. Hodnett *et al.* (1995), using a water balance based on these data, have shown that the forest was almost certainly using water from depths greater than 3.6 m during this period. Rainfall and uptake from storage in the upper 2 m of the profile only contributed between 1 and 2 mm day^{-1}, when typical dry season evaporation rates are assumed to be about 3.5 mm day^{-1} (Shuttleworth, 1988).

The maximum observed storage change in the 2 m to 3.6 m layer was 48 mm, equivalent to a mean water content change of only 0.03 moisture volume fraction (MVF). In 1992 and 1993, the maximum depletions in this layer were 40 mm and 31 m respectively. With the data available, it is not possible to define with any certainty how much of the observed change was due to drainage, but the water availability is clearly below 2 m and does not exceed 0.03-0.04 MVF, or 30-40 mm m^{-1}. These are exceptionally low values for any soils. They are about half of the values quoted by Correa (1984) for the upper profile.

MARABÁ – PASTURE

The largest seasonal water storage change in the Marabá pasture 2 m profile (wet season peak - dry season minimum) was 269 mm, which was 114 mm more than for the forest in Manaus. In the two wet seasons shown, the maximum storage recorded was very similar. The minimum storages recorded in all three dry seasons were also similar.

The very low rates of depletion at the end of the dry season in 1992 and 1993 indicate that the abstraction limit was being reached. Between 20 October and 17 November 1992 the depletion in the 0 - 2 m and 2 - 3.6 m layers was 0.5 mm and 4.8 mm respectively. There were 11.7 mm of rain. Applying the water balance and assuming negligible drainage, the evapotranspiration for the period was 17.0 mm, or an average rate of 0.61 mm day^{-1}. Most of this was the result of the evaporation of the rainfall input.

In the 2 - 3.6 m layer, the maximum storage variation was 131 mm. This layer wetted up to the same extent in both wet seasons but the minimum storage was 23 mm lower in the 1992 dry season, which continued longer than in 1991. From May until November 1992, depletion in this layer continued at an almost constant rate of 0.4 mm day^{-1} ceasing only when the upper profile began to wet up. This implies that the depletion below 2 m at this time was the result of water uptake by the roots of the grasses and shrubs and was not due to drainage. Wetting of the profile below 2 m was rapid, but only occurred after the profile above had wetted up by almost 200 mm to within 60 mm of the maximum observed.

MARABÁ – FOREST

The seasonal changes in the 0 - 2 m layer in the forest were far larger than in the pasture. The maximum changes in 1992 and 1993 were 483 mm and 327 mm respectively. A very important observation is that, during the study period, this was the only site where the profile did not rewet to the same condition in both wet seasons. In the 0 - 2 m layer, the 1993 dry season began with 164 mm less water in storage than in 1992. Over the full 3.6 m measured profile, the 1993 dry season began with 349 mm less water in storage.

The rainfall in the 1992/3 season was almost certainly less than in 1991/2 (the rainfall data for the site were not reliable in this period). In addition, the storage at the start of the 1991/2 wet season was 89 mm higher than at the start of the 1992/3 season.

In the 2 - 3.6 m layer, the difference in storage behaviour between years was very pronounced. In the 1992/3 wet season, this layer wetted up only 62 mm, compared to 185 mm in the previous wet season. The minimum storage also varied from year to year; in 1992 it was 77 mm lower than in 1991. The maximum soil water depletion in a single season was 263 mm during 1992, compared to only 72 mm during 1993. The maximum seasonal change in storage over the full 3.6 m profile was 724 mm between 6 April and 2 November 1992.

Rates of soil water uptake from the 3.6 m profile were still substantial even at the end of the 1992 and 1993 dry seasons when the profile storage was very low (606 and 535 mm respectively). Over a 21 day period in August 1992 (rainfall 0.4 mm), the mean evapotranspiration rate determined from the water balance was 2.5 mm day^{-1}, of which 1.17 mm day^{-1} was contributed from the 2 - 3.6 m layer. Over a similar period in September 1993, with 5.4 mm of rain, the mean evapotranspiration rate was 2.16 mm day^{-1}, but in this case, the contribution from the 2 - 3.6 m layer was only 0.66 mm day^{-1}. In contrast, at the end of the 1991 dry season, when the 3.6 m profile storage was 226 mm more than in 1993, the mean evaporation rate from the profile over a 31 day dry period was 3.25 mm day^{-1}, of which 1.43 mm day^{-1}

was from the 2 - 3.6 m layer.

It is notable that in the dry season, depletion in the 2 - 3.6 m layer slowed or stopped almost as soon as the top 2 m began to rewet. Unlike in the pasture, the profile below 2 m began to wet up gradually, well before the profile above had been fully wetted. It is unlikely that there was any significant deep drainage in the 1993 wet season.

JI-PARANÁ – PASTURE

At this site, the maximum storage change in the 0 - 2 m layer in 1993 was 262 mm, which was similar to the change at the Marabá pasture site. This layer wetted up to the same extent in the two wet seasons studied. The minimum storage was not recorded reliably in 1992 because of inconsistencies in the depth location of the neutron probe between April and October 1992.

Over a two week period in late July and early August 1993, at the end of the dry season, the rate of uptake from this layer was about 1.8 mm day^{-1}, indicating that the pasture was still able to continue abstracting water at a significant rate. This is in strong contrast to the end of the dry season uptake rate at the Marabá pasture site.

The 2 - 3.6 m layer wetted up to the same extent in the 1991/2 and 1992/3 wet seasons. In both seasons there was a very steep increase in storage during very wet periods, followed by a rapid decrease. This was caused by the rise of the water table which saturated the profile from below, indicating that there is a low permeability layer below 3.6 m. The total depletion during the 1993 dry season was 165 mm.

In the 2 - 3.6 m layer, the mean rate of depletion over the last 56 days of the 1993 dry season was 0.60 mm day^{-1}. After the 0 - 2 m layer began to rewet, the depletion rate in the 2 - 3.6 m layer decreased to 0.17 mm day^{-1} and remained almost constant for 70 days until the wetting front arrived. The sudden decrease in depletion rate indicates that root uptake from this layer had been occurring but stopped when water became readily available in the upper profile. The rate of 0.17 mm day^{-1} is probably a reasonable estimate of the underlying drainage rate, implying an average rate of root uptake from below 2 m (over the 56 days) of about 0.4 mm day^{-1}. Over the last two weeks of the 1993 dry season, the evapotranspiration rate estimated from the water content change in the 3.6 m pasture profile was 2.5 mm day^{-1} even though the 0 - 2 m profile storage had decreased by 262 mm from the maximum observed.

JI-PARANÁ – FOREST

In 1993, the change in the 0 - 2 m layer was 365 mm. Peak storage was very similar in 1992 and 1993. The Ji-Paraná forest site differs from all of the other sites in that the water table rises to within 1.2 m of the ground surface at the peak of the wet season. The root zone and the zone of seasonal water table fluctuation overlap and the observed storage changes therefore also reflect the rise and fall of this water table.

Mean rates of water loss from the 2 m profile (evapotranspiration and drainage) determined from the water balance exceeded 8 mm day^{-1} in the early dry season, confirming that the fall of the water table and subsequent unsaturated drainage play an important role in the storage behaviour. The depletion rates decreased gradually, and almost levelled out in August/September 1993. Applying the water balance to the dry period between 13 and 26 August 1993, gave a mean depletion/evapotranspiration rate of 1.28 mm day^{-1}. Between 3 and 17 September, when 7.2 mm of rain fell, the depletion was only 1.8 mm, giving a net uptake rate of 0.64 mm day^{-1}. The very low depletion rate shows that the abstraction limit for the 2 m profile was being reached. Although similar minimum storages were seen at the end of the 1991 and 1992 dry seasons, the levelling out of the storage, as seen in 1993, was not observed because of the timing of the start of the wet season in those years.

The storage response of the 2 - 3.6 m layer was very different to that at the other sites because of the influence of the water table. The flat top to the storage curve marks the period when the entire profile below 2 m depth was saturated. This lasted about three months in both 1992 and 1993. The seasonal change in water content was large, and varied from 287 mm in the 1992 dry season to 353 mm in 1993. In 1992, depletion in the 2 - 3.6 m layer ceased abruptly as soon as the upper profile began to wet up, as observed at other sites. In 1993, there was 63 mm more depletion in this layer than in 1992, most of which occurred after the reserves 0 - 2 m layer had been largely exhausted.

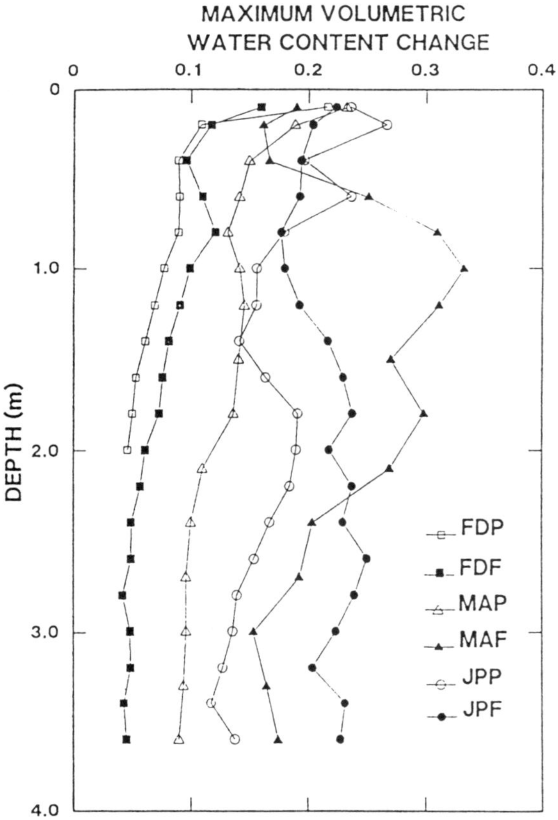

Figure 4 Profiles of maximum observed water content change at each depth for all sites: FDP – Manaus pasture, FDF – Manaus forest, MAP – Maraba pasture, MAF – Maraba forest, JPP – Ji-Paraná pasture, JPF – Ji-Paraná forest

Between 8 and 26 July 1993, the mean depletion rate, determined from the water balance of the entire 3.6 m profile, was 4.5 mm day^{-1} compared to the mean measured evapotranspiration rate of 3.77 mm day^{-1} (Wright, 1994, pers. comm.). The difference is probably caused by drainage from the lower part of the profile. It would appear that, at this stage, the water supply to the forest was not limiting. However, only five weeks later, at the end of the dry season, the depletion rate had decreased to 1.7 mm day^{-1}. Unfortunately, no directly measured evapotranspiration data are available for this period for comparison. Total profile storage on 17 September was 93 mm lower than at the end of July. If evapotranspiration was being sustained at a rate of about 3.6 mm day^{-1}, the remainder must have been supplied from the profile below 3.6 m.

DISTRIBUTION OF WATER CONTENT CHANGES WITH DEPTH

Figure 4 shows the difference between the wettest and driest water contents measured at each depth for all of the sites. Between sites, there was a large range of water content change at all depths.

Apart from in the top 0.2 m, water content changes at all depths were consistently much greater in the forest than in the pasture. At the Fazenda Dimona sites (where the soil type was identical), the differences were less than at the other sites at all depths. The change only exceeded 0.1 in the upper metre of the profile and below 3.0 m the change was only 0.03 - 0.04 MVF. The lowest change at any of the other sites was 0.09 MVF below 3 m beneath the Marabá pasture.

The very large variations between sites and with depth are of note. At 0.4 m, for example, differences ranged from 0.09 MVF in the Fazenda Dimona pasture to 0.19 in the Ji-Paraná pasture. At 1.0 m depth, differences ranged from 0.08 in the Manaus forest to 0.33 in the Marabá forest, and at 2.0 m, from 0.05 in the Manaus forest to 0.28 MVF in the Marabá forest. Below 2.2 m the largest changes were in the Ji-Paraná forest. At all except the Manaus sites, there were significant changes of water content below a depth of 2.0 m, ranging from 0.09 to 0.22 MVF.

DISCUSSION

In the upper part of the observed profiles, the same minimum water content was seen each dry season. Here, the vegetation had the opportunity to extract all of the available water before the profile was rewetted and the lowest water content marks the abstraction limit. The highest measured water contents are likely to have been between "field capacity" (when drainage becomes negligible, about -10 kPa matric potential) and saturation at all of the sites except at Ji-Paraná, where the highest water contents marked saturation below 1.2 m in the forest and below about 2.4 m in the pasture. The differences between the sites seen in the profiles of maximum water content change (Figure 4) therefore mainly reflect the differences in soil properties between the sites.

Of particular note are the very small seasonal changes of water content at the Manaus sites. It might be considered that this would have been the result of the shorter dry periods between rainfall events compared to the other sites, giving the vegetation less opportunity to take up stored water before the profile was rewetted. Although this determines the total change of storage within the profile, the very small changes of water content are due to the very low available water capacity (AWC) of this type of soil, despite its very high clay content. The low AWC has been shown by Correa (1984) and can be seen from the water release data presented by Ranzani (1980) and the mercury porosimetry curves of Chauvel *et al.* (1991). The low AWC has been discussed in more detail by Hodnett *et al.* (1995) who suggested that it is not advisable to predict the water content of these soils (e.g. at field capacity or wilting point) from their clay content using relationships derived originally from data for temperate soils.

The depth to which the extraction limit was reached depended on the vegetation and also on the rainfall amounts in, and duration of, the dry season. Inter-annual rainfall variability is high, and it is not known for the Marabá and Ji-Paraná sites how representative the seasonal profile storage changes observed during this study were of the normal, or whether abnormally wet or dry conditions were observed. Hodnett *et al.* (1996) examined the longer term context of the Manaus soil water record, addressing in particular the amount of water that might be taken up from below a depth of 2 m in drier years.

Deeper in the profile, the extraction limit was not reached, particularly at the pasture sites. At the forest sites, the fact that significant changes of water content occurred at 3.6 m late in the dry season indicates that root uptake, and not drainage, was mainly responsible. This, and the low rates of depletion in the forest profiles (Manaus, in particular) at this time, implies that uptake almost certainly occurs to greater depth. Several studies in Brazil have shown that the roots of rainforest trees can extend down to considerable depths.

Chauvel *et al.* (1992) found more roots at 6.0 m depth than at 3.0 m beneath forest near Manaus on almost identical soils to those at the ABRACOS Manaus site. Nepstad (1989) reported that mature forest on an oxisol at Paragominas (Pará, eastern Amazonia), can access soil moisture from at least 6 m depth and Nepstad *et al.* (1991 and 1994) found forest roots to 18 m depth in the same area. It is of note that all of this evidence of very deep rooting is from sites with oxisols, which have a very low water availability. Nepstad *et al.* (1994) found that in the severe 5.5 month dry season in 1992, more than 75% of the water for transpiration was supplied by the profile below 2 m depth. The mean evapotranspiration rate for this dry season determined from the water balance was 3.6 mm day^{-1}, which is very similar to the dry season values modelled by Shuttleworth (1988) for Reserva Ducke, near Manaus. These data appear to indicate that the forest at the Paragominas site was not suffering from soil water stress.

It has become clear from studies such as these that the roots of rainforest trees do penetrate and abstract water from deep in the soil profile. However, it is not known how much, or how often, deep abstraction is possible, nor whether the forest is stressed and transpiration rates are restricted when it occurs. This is extremely important because the reliable prediction of climate change depends strongly on this knowledge, particularly if longer dry seasons are likely in future. At present, there is no evidence from plant physiological or micro-meteorological measurements in ABRACOS that the forest transpiration rates decrease in response to lowered soil moisture (Roberts *et al.* 1996; Wright *et al.*, 1996). However, although the detailed plant physiological and micro-meteorological measurements were made under a wide range of soil water storage conditions, the study missions during which they were carried out did not coincide with the driest conditions observed during the weekly soilwater monitoring regime.

Increased leaf fall in the dry season may be an indication of soil water stress. However, although some forest species lose their leaves in the dry season, there is likely to be a compensating increase in transpiration from the lower canopy, which was previously shaded. This may result in only minor reductions in the transpiration rate from the forest as a whole.

An indication of the amount of deep uptake can be obtained from the data presented here. At the Marabá and Ji-Paraná forest sites, the mean uptake rates from the 3.6 m measured profiles at the end of the 1993 dry season, were 2.16 mm day^{-1} and 1.70 mm day^{-1} respectively. This was when the profile storage at both sites was close to the minimum measured during the study, and there was no readily available water in the upper profile following rainfall events. (Note that when the 3.6 m profile is treated as a single layer, a 30 mm rainfall input near the end of the dry season appears to have an insignificant effect on the profile storage. However, this input is in the zone with the highest concentration of roots and is therefore readily available. Stress conditions may be temporarily relieved despite a very large deficit remaining.)

If it is assumed that the forest was not stressed, and a dry season transpiration rate of 3.5 mm day^{-1} is taken (data from Shuttleworth, 1988), 38% and 51% respectively of the water requirements of the forest at the Marabá and Ji-Paraná sites, were being taken up from below 3.6 m depth at the end of the 1993 dry season. At the Marabá forest site, the proportion of water taken up from below 2 m depth at the end of the dry season varied from 48% in 1991 to 62% in 1992. It is not known how extreme these conditions were in the context of the longer rainfall record. These values are much lower than the 75% of uptake from below 2 m (for the entire dry season) found by Nepstad *et al.* (1994). This difference is almost certainly the result of the much lower water availability in the oxisol at the Paragominas site, where the average water content change over the 2 m - 8 m layer was 63 mm m^{-1}, compared to at least 150 mm m^{-1} at the Marabá site.

At the Ji-Paraná site, 6 of the 8 access tubes reached solid bedrock at depths of between 2.0 and 3.2 m. The unweathered rock probably has little ability to store water, and it is clear that the soil water reservoir at this site is probably limited compared to at the other sites. However, deep roots may find their way via cracks and fissures in the bedrock to reach the declining dry season water table.

At the pasture sites, there were significant reductions in transpiration rate as soil water deficits increased, but there were marked variations between the sites.

Hodnett *et al.* (1995) showed that the mean pasture evaporation rate at the Manaus (Fazenda Dimona) fell to 1.2 mm day^{-1} after the 2 m profile storage had decreased 130 mm from the maximum observed. There was little uptake below a depth of 1.5 m. At the Ji-Paraná site, the lowest evaporation rate was 2.50 mm day^{-1}, but this occurred after the 2 m profile storage had decreased 262 mm from the maximum observed. Uptake took place from at least 2.5 m depth. The Marabá site showed a very strong contrast; the evaporation rate fell to 0.61 mm day^{-1} after storage in the 2 m profile decreased 269 mm from the maximum. At this site the grass senesced almost completely and only a few scattered shrubs and stunted palms remained green. The very marked difference between the Marabá and Ji-Paraná pasture sites is partly due to the difference in water availability, but may also reflect the difference in response to water stress between the grass species. Nepstad *et al.* (1991) observed that *Panicum maximum* (the grass species at Marabá) is a nutrient demanding forage species and that in older pasture it suffers competition from less demanding shrub and grass species as nutrient stocks decline after clearance. The Marabá pasture is about 20 years old, and it is likely that nutrient levels are low, which may affect the stress response of the grass. Skerman and Riveros (1990) report that *Panicum maximum* does not tolerate severe drought and is the least drought tolerant of the species at the study sites.

The very large differences in soil water storage behaviour between sites clearly demonstrate the need to quantify and map the distribution of the water retention and hydraulic properties of Amazonian soils. In view of the depth to which abstraction can occur, these properties also need to be determined for the subsoil and soil parent materials, particularly in areas with a long dry season. Very few studies of the hydraulic properties of Amazonian soils have been made, and many more will be required to allow the SVAT sub-models of GCMs to be applied with confidence.

Extensive soil survey data (Projeto Radambrasil) are available, often to a depth of 3 m, covering texture (% sand, silt and clay), and in some cases bulk density, which allows an estimate of porosity to be obtained. Unfortunately, soils are not mapped on the basis of their hydraulic properties, but on a pedological basis, which includes soil-forming processes, texture, colour, chemical properties and parent material, among others. There are fairly well established means of estimating soil hydraulic properties from other, more commonly measured soil parameters, but all have been designed for temperate soils. There are strong indications that the oxisols studied so far have very different properties, which cannot be reliably predicted using these relationships. The derivation of similar relationships specifically for Amazonian soils is a priority, requiring widespread measurement of hydraulic properties on a range of soil types. These relationships can then be extrapolated to larger areas using the survey data.

In general, the soil classification is based mainly on the upper metre of the profile, with some account being taken of the properties to 2 m, or at most, 3 m. This means that later in the dry season, when stress is most likely, water is being taken up from layers which are hardly regarded as soil (parent material) and which are not, or only barely, included in the soil classification. In the case of Paragominas, more than 75% of the soil water deficit developed in the dry season was in material not regarded as soil. The hydraulic and water retention properties of these materials, usually saprolite (very weathered rock, or less weathered rock) are unknown as they are almost never sampled for soil mapping. It is not known how widespread saprolite is in the subsoils of Amazonia, but it is likely to occur commonly in the "shield" areas. It is unlikely to occur in areas with deep, unconsolidated sediments such as those of the Barreiras formation which lies to the north and south of the Amazon in the central part of the basin. The ABRACOS data provide some indications as to the properties to a depth of 3.6 m at the project sites, but otherwise they are virtually unknown.

SUMMARY AND CONCLUSIONS

There was a large difference in the seasonal variation of soil water storage between sites, determined by soil properties, vegetation cover, length of dry season and water table response. At all sites, there was a consistently larger seasonal variation beneath forest, compared to pasture.

The smallest seasonal storage changes occurred, as expected, at the Manaus sites, which have the shortest dry season. Here, the maximum seasonal change recorded to 2 m depth was 154 mm in the forest and 132 mm in the pasture. For the 3.6 m profile, the maximum change in the forest was 200 mm.

At the Manaus site, where data are available for 4 dry and 3 wet seasons, the driest conditions were observed at the end of the 1990 dry season for forest and pasture. Peak storage was similar in all wet seasons, but the duration of the very wet conditions was 2 months in the delayed and weak 1991/2 wet season, compared to 7 - 9 months in the other years studied.

The largest seasonal change occurred at the Marabá forest site where the change in the 3.6 m measured profile from the 1991/2 wet season to the driest observed, in the 1993 dry season, was 724 mm. The pasture change was 376 mm. The Ji-Paraná forest and pasture sites showed maximum changes of 701 mm and 450 mm respectively, but there was a major contribution from saturated drainage in these figures. The water table rose to within 1.2 m of the soil surface in the forest, and 2.4 m in the pasture. There appeared to be no water table influence at the Manaus plateau or Marabá sites. Water table behaviour must be taken into account in modelling data from sites such as Ji-Paraná.

The forest took up water from beyond 3.6 m at all sites, although in Manaus, this appears to have been necessary only because of the extremely low water availability in the oxisol at that site. On the other soils studied, 200 mm of water could have been readily supplied by less than 2 m depth of profile. At the end of the 1993 dry season, the profile below 3.6 m was estimated to have been supplying 1.34 mm day^{-1} (38%) and 1.80 mm day^{-1} (51%) respectively of the water for transpiration at the Marabá and Ji-Paraná sites. This was based on an assumed unstressed actual transpiration rate of 3.5 mm day^{-1}. Whether or not the transpiration from the forest as a whole is significantly reduced through soil water stress remains an open question.

Since there is significant uptake from the profile below 2 m, it is important to have soil hydraulic properties data from these layers for use in SVAT models. Very little, if any, data is available. There is an urgent need to obtain more data on these properties for all of the main soil types in Amazonia and in particular for their parent materials.

Soil water uptake in the pasture was strongly affected by the increasing soil water deficit. The least affected site was Ji-Paraná, where the end of dry season uptake rate was 2.5 mm day^{-1}, compared to 1.2 mm day^{-1} in Manaus and only 0.61 mm day^{-1} in Marabá. The grass species were different at all of the sites and it is not clear whether the differences were caused by this, or by differences in water availability.

The Marabá forest site was the only site where the profile did not wet up to the same extent each year during the period of study. The 1993 dry season began with 350 mm less water in storage then the 1992 dry season. There was little wetting below a depth of 2.5 m and a very significant deficit (in the measured profile, and probably also below) was carried forward into the next year. It is not known, at this stage of analysis, how frequent an occurrence this is.

REFERENCES

Bastable, H.G., Shuttleworth, W.J., Dallarosa, R.L.G., Fisch, G. and Nobre, C.A. 1993. Observations of climate, albedo, and surface radiation over cleared and undisturbed Amazonian rainforest. *Int. J. Climatol.* **13**, 783-796.

Chauvel, A., Vital, A.R.T, Lucas, Y., Desjardins, T., Franken, W.K., Luizão, F.J., Araguás, L.A., Rozanski, K. and Bedmar, A.P. 1992. O papel das raízes no ciclo hidrológico da floresta Amazônica. Anais do VII Congresso Brasileiro de Meteorologia, São Paulo, SP, Brazil. 298-302.

Chauvel, A., Grimaldi, M. and Tessier, D. 1991. Changes in soil pore space distribution following deforestation and revegetation: an example from the Central Amazon Basin, Brazil. *For. Ecol. and Manage.* **38**, 259-271.

Correa, J.C. 1984. Características físico hídricas dos solos latossolo amarelo, podzolico vermelho amarelo e podzol hidromórfico do estado do Amazonas. *Pesquisa Agropecuaria Brasileira* **20**, 1317-1322.

Couchat, P., Carré, C., Marcesse, J. and Le Ho, J. 1975. The measurement of thermal neutron constants of the soil: Application to the calibration of neutron moisture gauges and to the pedological study of the soil. Proc. Conf. Nuclear Data Cross Sections on Technology, Washington DC, USA.

Dickinson, R.E., Henderson-Sellers, A., Kennedy, P. and Wilson, M. 1986. Biosphere/Atmosphere Transfer Scheme for NCAR Community Climate Model. NCAR Tech. Note TN275+STR, National Centre for Atmospheric Research, 69pp.

Gash, J.H.C., Nobre, C.A., Roberts, J.M. and Victoria, R.L. An overview of ABRACOS. This volume.

Lean, J. and Warrilow, D.A. 1989. Simulation of the regional climatic impact of Amazonian deforestation. *Nature* **342**, 411- 413.

Henderson-Sellers, A. 1987. Effects of land use on climate in the humid tropics. In: R.E. Dickinson (Ed.) *The geophysiology of Amazonia*. 463-496. John Wiley and Sons Ltd., New York.

Hodnett, M.G., da Silva, L.P., da Rocha, H.R. and Senna, R.C. 1995. Seasonal soil water storage changes beneath central Amazonian rainforest and pasture. *J. Hydrol.* **170**, 233-254.

Hodnett, M.G., Tomasella, J. and Oyama, M.D. 1996. Deep soil water uptake by forest in central Amazonia: predictions from long-term daily rainfall data using a simple water balance model. This volume.

McWilliam, A.-L. C., Roberts, J.M., Cabral, O.M.R., Leitao, M.V.B.R., de Costa, A.C.L., Maitelli, G.T. and Zamparoni, C.A.G.P. 1993. Leaf area index and above-ground biomass of *terra firme* rain forest and adjacent clearings in Amazonia. *Functional Ecology* **7**, 310-317.
Nepstad, D.C. 1989. Forest regrowth in abandoned amazonian pastures of eastern Amazonia: limitations to tree seedling survival and growth. PhD Thesis, Yale University, New Haven, USA.
Nepstad, D.C., Uhl, C. and Serrão, E.A.S. 1991. Recuperation of a degraded Amazonian landscape: forest recovery and agricultural restoration. *Ambio* **20**, 248-255.
Nepstad, D.C., de Carvalho, C.R., Davidson, E.A., Jipp, P.H., Lefebvre, P.A., Negreiros, G.H., da Silva, E.D., Stone, T.A., Trumbore, S.E. and Vieira, S. 1994. *Nature* **372**, 666-669.
Nobre, C.A., Sellers, P.J., Shukla, J. 1991. Amazonian deforestation and regional climate change. *J. Climate* **4**, 957-988.
Noilhan, J. and Planton, S. 1989. A simple parameterisation of land surface processes in meteorological models. *Mon. Weath. Rev.* **117**, 536-549.
Owe, M. and Van de Griend, A.A. 1990. Daily surface moisture model for large area semi-arid land application with limited climate data. *J. Hydrol.* **121**, 119-132.
Penman, H.L. 1956. Evaporation: an introductory survey. *Neth. J. Agric. Sci.* **4**, 9-22.
Ranzani, G. 1980. Identifição e caracterização de alguns solos da Estação Experimental de Silvicultura Tropical do INPA. *Acta Amazonica*, **10**, 7-41.
Roberts, J.M., Cabral, O.M.R., da Costa, J.P., McWilliam, A.-L.M. and Sá, T.D de A. 1996. Leaf area index and leaf gas exchange at the ABRACOS sites. An overview. This volume.
Sellers, P.J., Mintz, Y., Sud, Y. and Dalcher, A., 1986. A simple biosphere model (SiB) for use within general circulation models. *J. Atmos. Sci.*, **43**, 506-531.
Shukla, J. and Mintz, Y., 1982. Influence of land surface evapotranspiration on the Earth's climate. *Science* **215**, 1498-1501.
Shukla, J., Nobre, C. and Sellers, P. 1990. Amazon deforestation and climate change. *Science* **247**, 1322-1325.
Shuttleworth, W.J. 1988. Evaporation from Amazonian rainforest. *Proc. R. Soc. Lond.* **B233**, 321-346.
Skerman, P.J. and Riveros, F. 1990. Tropical grasses. *FAO Plant Production and Protection Series*, **No 23**. FAO, Rome. 832pp.
Tomasella, J. and Hodnett, M.G. 1996. Hydraulic parameters of an oxisol under pasture in Central Amazonia: measurement and derivation. This volume.
Wright, I.R., Gash, J.H.C., da Rocha, H.R., Shuttleworth, W.J., Nobre, C.A., Carvalho, P.R.A., Leitao, M.V.B.R., Maitelli, G.T. and Zamparoni, C.A.G.P., 1992. Dry season micrometeorology of Amazonian ranchland. *Q. J. R. Meteorol. Soc.* **118**, 1083-1099.
Wright, I.R., Nobre, C.A., Tomasella, J., Da Rocha, H.R., Roberts, J.M., Vertamatti, E., Culf, A.D., Alvalá, R.C., Hodnett, M.G. and Ubarana, V. 1996. Towards a GCM surface parameterisation for Amazonia. This volume.

RESUMO

Em áreas experimentais (pastagem e floresta) de Marabá (Pará), Manaus (Amazonas) e Jiparaná (Rondônia), a umidade volumétrica do solo foi medida em várias profundidades até 3,6 m. São apresentados três anos de dados para os sítios de Manaus e dois anos para os demais.

Entre os sítios, devido a diferenças nas propriedades do solo, cobertura vegetal, duração da estação seca e comportamento do lençol freático, houve grande diferença na variação sazonal do armazenamento de água no solo. A variação sazonal foi consistentemente maior na floresta, em comparação à pastagem. A menor variação ocorreu nos sítios de Manaus, os quais tiveram estação seca mais curta. Neste caso, a máxima variação registrada no perfil de 2 m foi de 154 mm para a floresta e de 132 mm para a pastagem. No perfil de 3,6 m, a máxima variação na floresta foi de 200 mm. A máxima variação na floresta de Marabá foi de 724 mm, o qual foi a maior variação registrada entre os sítios, e na pastagem, foi de 349 mm. A floresta e pastagem de Ji-Paraná apresentaram, como máxima variação, 701 mm e 511 mm, respectivamente, mas houve importante contribuição da drenagem saturada nesses valores. O lençol freático elevou-se até 1,2 m da superfície na floresta, e até 2,4 m na pastagem, o que indica a necessidade de se levar em conta o comportamento do lençol freático e o movimento lateral de água no solo nos resultados apresentados para Ji-Paraná. Nos sítios de Manaus e Marabá, parece não ter havido influência do lençol freático.

Houve extração de água abaixo de 3,6 m em todos os sítios de floresta. Apesar da variação sazonal do perfil ter sido pequena em Manaus, a retirada de água a maiores profundidades foi necessária devido à baixa disponibilidade do oxisol nesse sítio. Ao final da estação seca de 1993, estima-se que cerca 38% e 51% da água usada para transpiração, na floresta de Marabá e Ji-Paraná, respectivamente, provém do perfil abaixo de 3,6 m, assumindo-se a taxa de transpiração sem estresse de 3,5 mm d^{-1}.

A extração de água na pastagem foi fortemente afetada pelo déficit hídrico do solo. A pastagem de Ji-Paraná foi a menos afetada, com a taxa de extração de água, ao final da estação seca, de 2,5 mm d^{-1}, enquanto em Manaus foi de 1,2 mm d^{-1} e em Marabá, somente 0,61 mm d^{-1}.

A floresta de Marabá foi o único sítio onde não houve recarga completa do perfil na estação úmida. A estação seca de 1993 iniciou-se com um armazenamento 400 mm menor que a estação seca de 1992. Houve pequena recarga abaixo de 2,5 m e a estação seca subsequente iniciou-se com um déficit substancial (no perfil medido, e provavelmente abaixo). Não é sabido, até o presente, a frequência de ocorrência desse tipo de evento.

4 Deep soil water uptake by forest and pasture in central Amazonia: predictions from long-term daily rainfall data using a simple water balance model

M.G. HODNETT[1], J. TOMASELLA[2], A. de O. MARQUES FILHO[3] and M.D. OYAMA[4]
[1] *Institute of Hydrology, Wallingford, UK*
[2] *Instituto de Pesquisas Hidraulicas, Porto Alegre, Brazil*
[3] *Instituto de Pesquisas da Amazônia, Manaus, Brazil.*
[4] *CPTEC, INPE, Cachoeira Paulista, Brazil*

INTRODUCTION

Details of the ABRACOS soil water data set and results for the three forest and pasture sites have been presented by Hodnett *et al.* (1996). For each of the ABRACOS sites, the two or three year record of weekly data provided the "setting" for detailed studies, by identifying where in the seasonal cycle of soil water storage change they were carried out. During ABRACOS considerable variation in the behaviour of soil water storage was observed between years. The purpose of this paper is to determine, for the three-year record from the Manaus (Fazenda Dimona) site:

(a) The context of the soil water conditions that were observed (particularly during the detailed studies) in relation to the longer-term record, with particular reference to the occurrence of conditions which might result in soil water stress.

(b) How often the uptake of deep soil water is required and how much may be taken up

If measurements of micro-meteorological variables and plant physiological responses have been made under extreme conditions of soil water stress during the ABRACOS studies, parameters derived during these periods may be used in SVAT models with confidence and without undue extrapolation. However, if extreme conditions have not been observed during these studies, or if Global Circulation Models (GCMs) predict longer dry seasons and more severe stress conditions than those under which

Amazonian deforestation and climate. Edited by J.H.C. Gash, C.A. Nobre, J.M. Roberts and R.L. Victoria.

the parameters were derived, the reliability of the predicted climatic changes will be open to question.

Chauvel *et al.* (1991), working in an area with similar soils about 30 km to the south of Fazenda Dimona study site, found forest roots to a depth of 6 m with more roots at this depth than at 4 m and 5 m. Nepstad *et al.* (1994) also observed deep roots under forest near Manaus (to at least 8 m depth), but there is some discussion as to the role of these roots because the same article showed that uptake of soil water from below a depth of 1 m is not required in this area. Objective (b) addresses this apparent paradox.

In the absence of long-term soil water datasets for comparison, soil water storage data for forest and pasture were estimated from long-term daily rainfall data using simple soil water models. These were derived and calibrated using the ABRACOS Fazenda Dimona soil water dataset. Detailed long-term meteorological data were not required.

SITE AND INSTRUMENTATION

The Fazenda Dimona pasture site is located on a cattle ranch about 100 km to the north of Manaus, in the State of Amazonas. The pasture under study had a grass cover of about 75%, the remainder being bare soil, old stumps and fallen trunks. Further details are given by Wright *et al.* (1992) and Gash *et al.* (1996). The forest site, where soil water monitoring was carried out, is in an area of forest adjacent to the Fazenda Dimona ranch.

The soils under both the forest and pasture sites are clayey yellow latosols (Soil taxonomy: Haplic acrorthox, FAO: xanthic ferralsol), very similar to those described by Ranzani (1980), and Correa (1984). The soils have a very high clay content (>75%), and a dry bulk density between 0.95 and 1.15 gm cm^{-3}, giving a very high porosity. However, most of the porosity is concentrated in the macro and meso-pores (which drain rapidly) and in the very fine pores (in which the water is inaccessible to plants). This leads to a low available water capacity of only about 70 mm m^{-1} in the upper metre of the profile (Correa, 1984). The hydraulic properties of the soils at the site have been described by Tomasella and Hodnett (1996).

Soil water content was measured under forest and pasture at weekly intervals using a neutron probe. Access tubes for the neutron probe were installed to a depth of 2 m in September 1990, with groups of tubes representing plateau, slope and valley floor areas. Deeper tubes (3.6 m) were installed at the forest plateau sites in October 1991. The soil water monitoring instrumentation has been described in more detail by Hodnett *et al.* (1995). The data collected from the plateau areas (eight tubes in the pasture and five in the forest) up to December 1993 have been used in the analysis presented here.

At the pasture site, an automatic weather station was operated to take hourly averages of various meteorological variables throughout the study period (Bastable *et al.*, 1993, Culf *et* al., 1996). In addition, intensive measurements of energy partition, necessary for the micrometeorological studies, were made between September and November 1990, and between June and September 1991 (Wright *et al.*, 1992; Wright *et al.*, 1995).

THE WATER BALANCE

The soil water storage, to any given depth in the soil profile can be determined on a daily basis using the water balance equation:

$$S_{t+1}=S_t+P_n-Q-E-Dr \tag{1}$$

where S is the storage (mm), t is the time, and P_n, Q, E and Dr are the cumulative net rainfall, surface runoff, evapotranspiration and drainage (expressed in mm), between times t and t+1.

The profile storage, S, was determined from the profile of n watercontent measurements using:

$$S=\sum_{i=1}^{n}\theta_i\Delta z_i \tag{2}$$

where θ_i is the soil water content of the i th layer of thickness Δz_i.

Rainfall was measured at the pasture site, but not at the forest site, which was only 1 km distant. Interception was taken to be negligible at the pasture site, so that net rainfall was equal to rainfall. For the forest site, the interception was estimated using the analytical model of Gash (1979) using the parameters determined for the Reserva Ducke site by Lloyd *et al.* (1988).

Surface runoff was taken to be zero for both forest and pasture. The surface soil in the forest is highly permeable and runoff is not observed on undisturbed soils even after very heavy rain (Nortcliff and Thornes, 1981; Chauvel *et al.*, 1991). In the pasture, local surface redistribution and ponding were seen during and soon after rainfall, implying lower surface hydraulic conductivity in some places. The water, which redistributes, soon infiltrates and it appears that significant surface runoff is rare. This is shown by the profile storage data presented by Hodnett *et al.* (1995) which shows little difference in the mean wetting up response beneath forest and pasture at this site.

Evapotranspiration was measured or estimated from meteorological data. Since measurements of evapotranspiration were only available for the pasture for short periods covered by the intensive field missions the pasture evapotranspiration series for the rest of the period was filled using the Penman-Monteith (Monteith, 1965) equation, using parameters derived by Wright *et al.*(1992):

$$\lambda E=\frac{\Delta A+\rho c_p SHD g_{av}}{\Delta+\gamma(1+g_{av}/g_{sv})} \tag{3}$$

where E is the water vapour flux, λ is the latent heat of evaporation, A is the available energy, Δ is the rate of change of saturated specific humidity with temperature, c_p is the specific heat of air, ρ is the density of the air, SHD is the specific humidity deficit, γ is the psychrometric constant, g_{av} is the aerodynamic conductance for water vapour transport between the canopy height and the reference level and g_{sv} is the bulk stomatal conductance. The stomatal conductance was estimated using the Jarvis (1976) model:

$$g_{sv}=a_1 LAI f(SHD)\, f(R_s)\, f(\theta) \tag{4}$$

where a_1 is the maximum stomatal conductance, LAI is the leaf area index and R_s is the solar radiation. The control functions for f(SHD), f(R_s) and f(θ), and the associated parameters derived by Wright *et al.*(1995) for the Fazenda Dimona pasture were used. This allowed the determination of evaporation on an hourly basis. The Penman-Monteith formula strictly only estimates transpiration, but for the Fazenda Dimona pasture, it has been found that after rainfall the contribution from soil evaporation is very small and of short duration and may be ignored (Wright, I.R. *pers. comm.* 1994).

Since there were no meteorological data available for the forest site where the neutron probe access tubes were installed, forest transpiration was estimated using an expression based on the Priestley-Taylor (1972) formula, assuming no stress condition during the dry season. The net all-wave radiation, R_N, was estimated on an hourly basis from the solar radiation measured at the adjacent pasture site using the expressions adjusted by Bastable *et al.* (1993) for the forest site at Reserva Ducke. Shuttleworth *et al.* (1984) found that on dry days the mean daily evaporative fraction (the ratio between latent heat flux and available energy) for Reserva Ducke was about 0.7. Roberts *et al.* (1993), applying a five layer canopy conductance model, found that, over daylight periods, an average of 74.5% of the net radiation was required for transpiration. For this study (neglecting soil heat flux) the latent heat flux was estimated using:

$$\lambda E=0.7R_n \tag{5}$$

The estimated hourly transpiration was accumulated during the day to generate total daily transpiration, which was used later to calibrate the model.

The deep drainage, Dr, below the maximum depth of soil water measurement (2.0 m) was determined using Equation 1, for the periods between soil moisture measurements, accumulating daily rainfall and transpiration over the same periods. Figures 1a and b show, for pasture and forest respectively, transpiration and drainage (average daily rate between two subsequent neutron probe measurements) plotted against storage in the 2 m profile. Figure 2 shows the mean profile storage for the 0-2 m and 2-3.6 m layers in the forest.

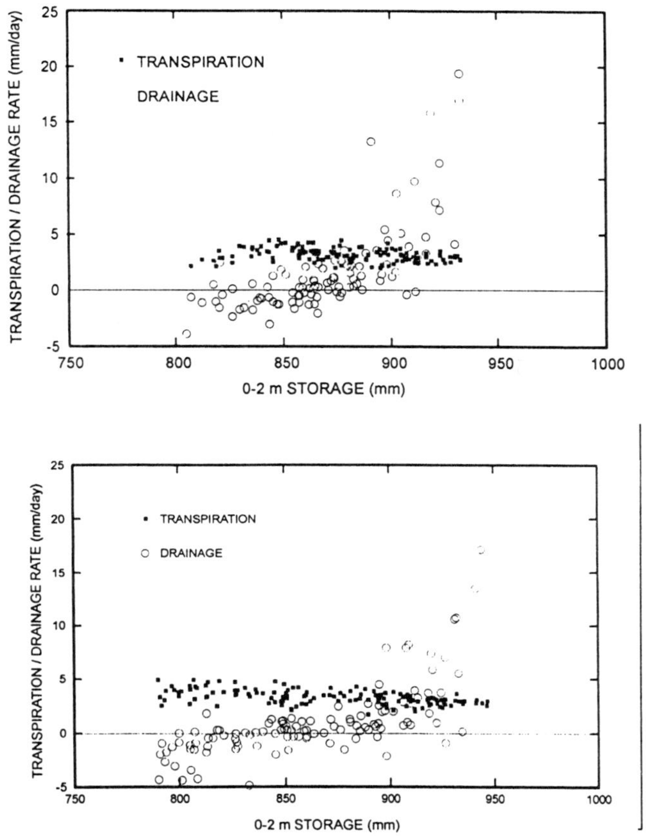

Figure 1 Fazenda Dimona. Estimated evapotranspiration and drainage below 2 m depth (from the water balance), plotted as a function of storage in the top 2 m of the soil profile for pasture (top) and for forest (bottom)

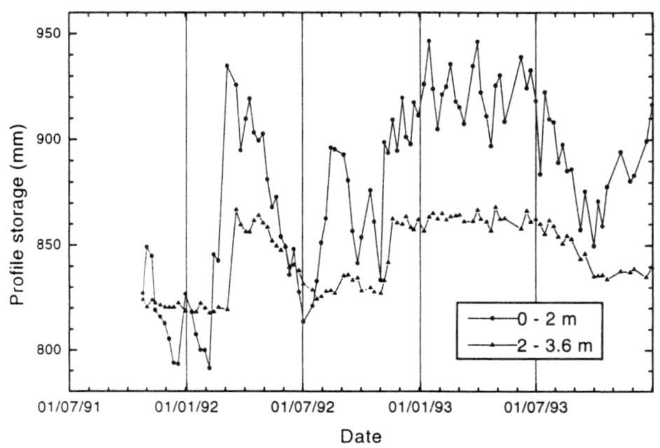

Figure 2 Fazenda Dimona, forest. Measured soil water storage in the 0-2 m and 2-3.6 m layers between October 1991 and December 1993.

PASTURE

When the profile storage was less than 850 mm, there was a significant reduction of transpiration rate as storage decreased, associated with the increasingly difficult water extraction by the roots at lower soil water contents. When storage exceeded 850 mm (generally in the wet season), the transpiration rate showed a gradual decrease as profile storage increased. This indicates a reduction in energy availability associated with increased cloudiness in very wet periods, in agreement with Wright *et al.* (1992). The very cloudy and rainy periods correspond with the highest profile water storage values, resulting in a fairly close relationship between transpiration rate and profile water storage. This relationship forms part of the basis of the model used to predict profile storage from rainfall data alone.

The curve relating the rate of deep drainage to the profile storage for the pasture showed that drainage was negligible at a storage of about 870 mm, and appeared to increase asymptotically, becoming very rapid when storage exceeded 900 mm. The shape of the curve reflects gravity drainage. This relationship shows much scatter, but this is the result of relating the mean drainage rate for a period (typically one week) to the profile storage at the end of that period. This is clearly a simplification because the drainage is averaged over the period, regardless of when it actually occurred. The storage value also takes no account of the distribution of the water within the soil profile.

FOREST

The calculated transpiration rates (assuming no influence of stress in the dry season) shown in Figure 2 are on average constant when the profile storage is less than 850 mm. As for the pasture, when the storage exceeds 850 mm there is a gradual decrease as a result of lower energy availability. The drainage curve shows a similar behaviour to that of the pasture, except when the storage decreases below 810 mm, when the drainage apparently becomes strongly negative. This implies that there is upward movement of water, or abstraction, from below a depth of 2 m during these periods because the storage changes in the 2 m profile are too small to account for the evaporative demand. Hodnett *et al.* (1995), using the same data set, showed that uptake rates from the 2 m profile fell to as low as 1.2 mm day^{-1} when storage was below 800 mm. This uptake rate is clearly too low to support a forest showing no obvious signs of stress, such as a marked increase in leaf fall. The transpiration data from Reserva Ducke presented by Shuttleworth *et al.* (1984), when examined together with the profile storage data from that site, suggest very little reduction in transpiration during dry periods. There are also few plant physiological indications of stress in these periods (Roberts, J.M. *pers. comm.*, 1993). These data strongly imply that the forests in the Manaus area are able to take up water from below 2 m depth, when necessary. The assumption that the forest never suffers from soil water stress is questionable, but the data that exist (which may not have been for periods of extreme stress) appear to imply that forest transpiration rates are not significantly affected.

The minimum storage for the 2 m profile, observed beneath the forest on several occasions during the study period, was 790 mm. This value appears to be the storage beyond which the forest does not (or cannot) extract any more water from the top 2 m and was adopted as the minimum possible profile storage for both the forest and the pasture. This apparently very high value for the minimum possible storage is the result of the very unusual water release properties of this soil. This water is held in such fine pores that plants are unable to extract it.

Figure 2 shows the storage in the upper 2 m and in the layer from 2 to 3.6 m for the period from October 1991 until December 1993. Changes in the 2 - 3.6 m layer are limited to about 40 mm and an abstraction limit is apparent between October 1991 and February 1992 which was abnormally dry. This period has been discussed by Hodnett *et al.* (1995). These data also show that, in the dry season, the storage in the layer below 2 m decreases at exactly the same time as in the upper 2 m. This appears to indicate that although water content changes are larger in the upper layer, uptake takes place concurrently from all layers and is not a sequential, from the top downwards, process.

MODEL FORMULATION

Simple models were derived to describe the relationships between transpiration and profile storage, and drainage and profile storage shown in Figures 1 and 2. The equations used to estimate the pasture transpiration rate from the profile storage were:

$$E_t = a_P \frac{S_t}{D} + b_P \tag{6}$$

for $S_t < 850$ mm, and:

$$E_t = c_P \frac{S_t}{D} + d_P \tag{7}$$

for $S_t \geq 850$ mm, where E_t is the transpiration rate (mm day^{-1}), D is the soil depth (2.0 m) and a_P, b_P, c_P, d_P are parameters for the pasture. Equation 6 simulates the effects of soil water stress on transpiration and Equation 7 takes account of the lower wet season transpiration rate due to cloudiness.

For the forest, the equations were:

$$E_t = c_P \frac{S_t}{D} + d_P \tag{8}$$

for $E_t \geq E_{VM}$, and:

$$E_t = E_{VM} \tag{9}$$

at other times, where c_F, d_F and E_{VM} are parameters: E_{VM} is the average dry season transpiration rate.

The drainage for forest and pasture was modelled using an exponential function:

$$Dr_t = \exp\left(e_{P,F}\left(\frac{(S_t - f_{P,F})}{D} \right) \right) - 1 \tag{10}$$

where Dr is the deep drainage rate (mm day^{-1}), $e_{P,F}$ is the exponent and $f_{P,F}$ is the storage at which drainage is zero.

The water balance (Equation 1) was solved on daily basis in the following way:

(i) Calculate interception for day d (I_d) from the rainfall for day d, using the method of Gash (1979).
(ii) Calculate the net rainfall for day d.
(iii) Calculate the transpiration and drainage from the storage for day d (S_d), using Equations 6 - 10.
(iv) Calculate the storage for the next day (S_{d+1}), using the water balance (Equation 1).

To model soil water uptake by the forest from below 2 m, the drainage function (Equation 10) was permitted to allow drainage to become negative, simulating upward water movement from below 2 m. When the storage in the top 2 m had been completely depleted, all of the forest water requirement was taken from below 2 m. A storage of 790 mm in the top 2 m of the profile (as discussed earlier) was taken as the threshold and the profile below 2 m was treated as an unlimited reservoir in which the accumulated deficit showed the amount of water supplied to sustain the transpiration. The layer below 2 m was replenished by drainage from the layer above.

OPTIMISATION

The parameters of Equations 6 - 10 were optimised using the Rosenbrock (1960) procedure. The objective function chosen was:

$$\sum_{i=1}^{m} \left(\frac{S_{t+1,c} - S_{t+1,o}}{S_{t+1,o}} \right)^2 + \left(\frac{CE_{t,c} - CE_{t,o}}{CE_{t,o}} \right)^2 \tag{11}$$

where S is the storage in the top 2 m, CE is the cumulative transpiration, the subscripts t, c and o indicate the time, and calculated and observed values, respectively.

The calculated values were derived on a daily basis using Equation 1, starting on a date when storage had been measured. Since there were some gaps in the rainfall data, the series was divided into several periods for which complete information was available. The optimisation was performed through these periods using a measured storage as the initial value, and the calculation was continued until the next gap, without applying any correction or updating of the predicted storage. After the parameters had been optimised using the measured soil water data-sets, this simple water balance approach allowed the profile storage response to be predicted from rainfall data alone.

RESULTS

The set of parameters resulting from the optimisation procedure is shown in Table 1. The parameters optimised for the forest and the pasture are very similar. In the case of the parameters which control the transpiration in wet periods (c & d), the agreement is remarkable, which suggests that the transpiration during the wet season should be very similar in the forest and in the pasture. The main difference is limited to the parameters of the drainage function (e & f).

In the case of (f), the value of profile storage when drainage is zero, there is a difference of 8 mm between forest and pasture, but the shape of the curves is such that this has a negligible impact. The (e) parameter is higher for the forest, resulting in a higher drainage rate when the profile storage is large, but differences in drainage rate are only significant above about 940 mm of storage. For a very high storage value, Equation 10 will predict an excessively high drainage rate, which will, in turn, result in an unrealistically low profile storage. In practice, the drainage rate will have a maximum rate determined by the saturated hydraulic conductivity. For this reason, the maximum drainage rate was limited to that corresponding to a storage of 960 mm in the top 2 m of the profile.

Table 1 Parameter set fitted for the forest and the pasture at Fazenda Dimona site

Parameter	a	b	c	d	e	f	E_{VM}	
							mm	mm day^{-1}
Pasture		90.81	-34.42	-28.22	15.75	82.27	869.0	
Forest				-28.55	15.74	93.6	877.5	3.89

If (f), 877.5 mm for the forest, is taken to represent field capacity, the apparent water availability in the 2 m profile is only 87.5 mm (877.5 - 790), but it should be noted that significant water uptake can occur when the profile storage is above 'field capacity'. As an example, for a 2 m profile storage of 930 mm and a wet season transpiration rate of 3.3 mm day^{-1}, 26.4 mm of water may be taken up before the combined processes of water uptake and drainage reduce the profile storage to 'field capacity'. For this site, this results in a maximum possible water uptake from the 2 m profile of about 114 mm.

Figure 3 shows the measured and predicted soil profile storage for the pasture and the forest respectively. For the forest, the estimated deficit in the profile below 2 m is also shown. The measured and predicted values are in very good agreement over the full range of the measured data. In the forest, the most severe drought conditions observed in the 2 m profile during the study period occurred at the end of the 1990 dry season and the 1991 dry season (which extended into February 1992). In the pasture, the driest conditions were observed at the end of the 1990 dry season, but similarly dry conditions did not occur in subsequent years. These results have been discussed in more detail by Hodnett *et al.* (1995). During the extended dry season from

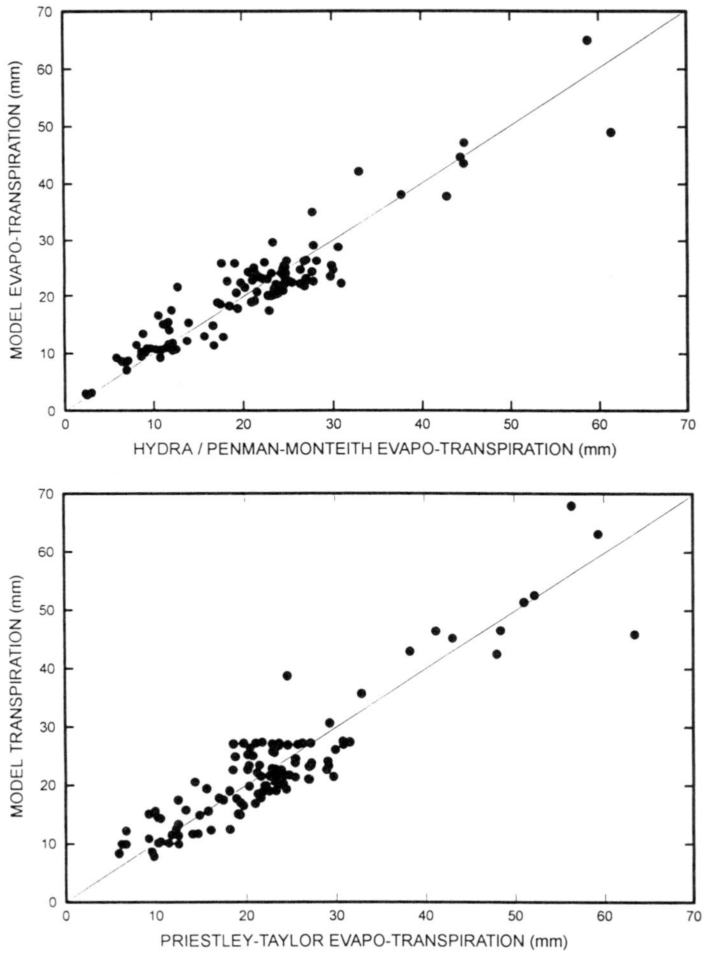

Figure 3 Pasture: measured (Hydra) and estimated (Penman-Monteith) evapo-transpiration, compared with that predicted from soil water storage (top). Forest: estimated (Priestley-Taylor) evapo-transpiration, compared with that predicted from soil water storage (bottom).

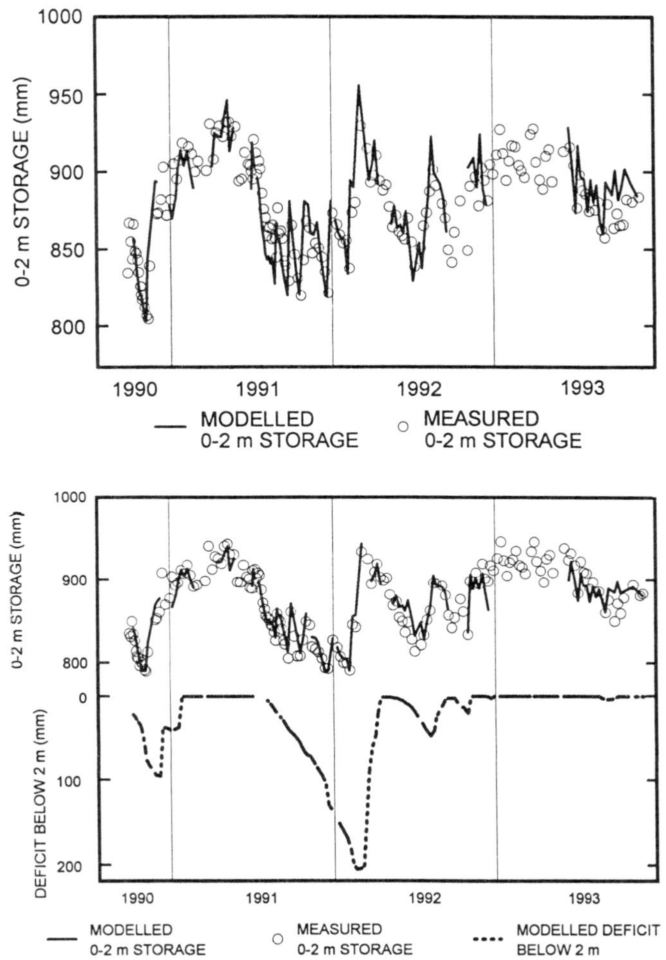

Figure 4 Measured and modelled storage in the top 2 m of the soil profile for pasture (top) and forest (bottom). Also shown is the predicted deficit in the profile below 2 m depth.

July 1991 to mid-February 1992, there was some evidence of local differences in rainfall between forest and pasture sites, but the observed difference in response was mainly a consequence of the differences in transpiration and interception behaviour between pasture and forest.

Figure 4 compares the estimated transpiration rates with those calculated by the model, for the pasture and the forest respectively. The transpiration predicted by the model shows a good correlation with the values measured with the "Hydra" and estimated by the Penman-Monteith formula (Monteith, 1965) for the pasture and by the Priestley-Taylor (1972) formula for the forest. It is clear that the model is predicting the soil profile storage well (Figure 3) and is doing so by predicting the transpiration (and therefore drainage) correctly.

LONG-TERM ESTIMATION OF SOIL MOISTURE SERIES

To assess the representativeness of the ABRACOS soil data-set, with particular reference to the frequency and intensity of severe drought periods, the adjusted model was run using a 27-year rainfall record from the Reserva Ducke climate station which is approximately 70 km south of the study site. Figure 5 shows the average monthly precipitation from Reserva Ducke, with the associated standard deviation. The average monthly rainfall for three years from the ABRACOS Fazenda Dimona site is also shown. Since there were some gaps in the ABRACOS rainfall record, all the months with more than five days of rainfall data missing were discarded, and the average monthly precipitation is shown only for those months where data for at least two years are available. Comparisons indicate that the rainfall regime at both sites is similar, allowing the Reserva Ducke rainfall record to be used to assess the frequency of droughts using the soil water model calibrated using the Fazenda Dimona soil water data. Also shown in Figure 5 are the mean monthly rainfall for Manaus for the period 1945-1975 (Anon, 1978) and the mean monthly total forest evaporation (transpiration + interception) estimated by Shuttleworth (1988).

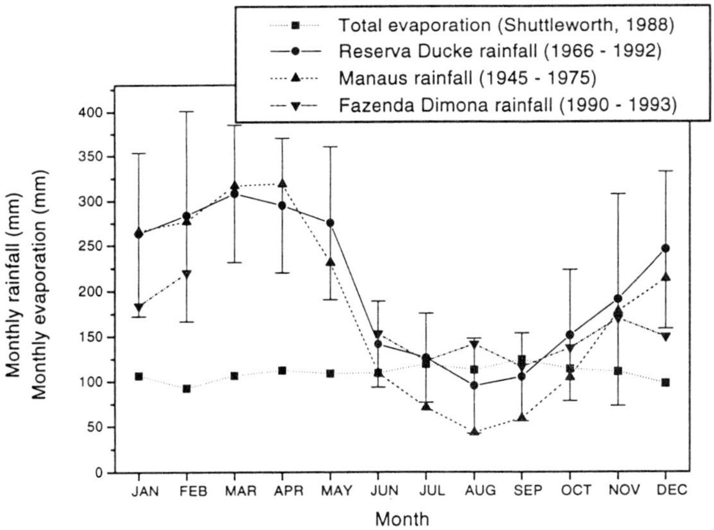

Figure 5 Mean monthly rainfall from Reserva Ducke, Manaus and Fazenda Dimona. Mean monthly forest evapo-transpiration is also shown.

PASTURE

Figure 6 shows the soil profile storage for the pasture generated by the model using the Reserva Ducke rainfall data for the period 1966-1992. The model was run on a daily basis, but data have been plotted weekly for clarity. In general, the behaviour

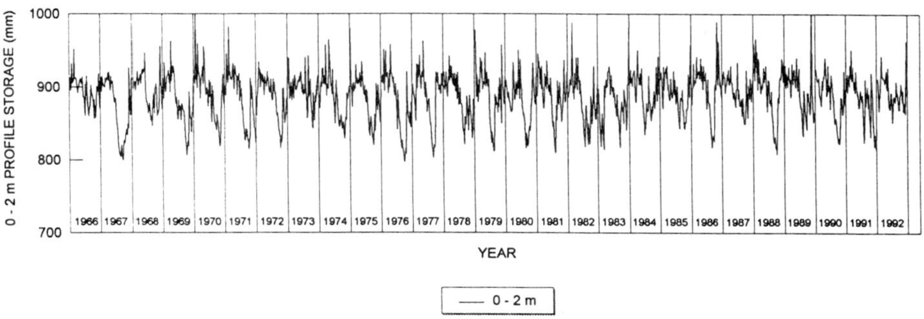

Figure 6 Pasture: profile storage (0–2 m) predicted from the Reserva Ducke rainfall record (1966–1992) using a daily water balance model. Data are plotted at weekly intervals.

through the year follows the rainfall pattern shown in Figure 5. The minimum storage usually occurred in September or October, but in some years occurred in December–January, indicating that low profile water storage conditions can occur even in periods normally regarded as being in the wet season (as indicated by mean rainfall data).

Figure 7a shows the number of years out of the 27 year record in which the estimated minimum storage (or annual maximum stress) fell into different classes. The minimum storage measured in each dry season during the ABRACOS study period is also shown. The estimated minimum covered a very limited range from 794 mm and 850 mm. There were only four years out of the 27-year record when the storage fell below the ABRACOS measured minimum of 804 mm. Although this is a fairly rare event, it is of note that in 13 of the 27 years (48%) the storage falls to within 16 mm of this value. For the three other ABRACOS measured years, 1991 was normal, 1992 was wetter than normal and 1993 was at the wettest extreme of the estimated data.

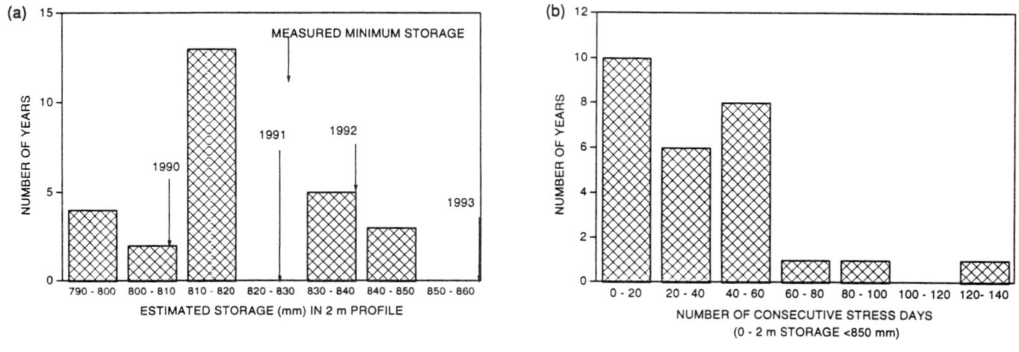

Figure 7 Pasture: (a) number of years out of the 27 year record when different classes of minimum storage occurred. The minimum storages measured in the 4 dry seasons studied are also shown. (b) Consecutive stress days: number of years in which different classes occurred.

For the pasture, the threshold for soil water stress (i.e. causing a reduction in transpiration below the potential rate) is a storage of 850 mm. Although the minimum storage indicates the maximum level of stress in a given season, the duration of periods when the storage is < 850 mm is also important. Figure 7b shows the frequency of occurrence of the number of consecutive days in which the storage remains below 850 mm. For the 27-year record, there were less than 61 consecutive stress days in 24 of the years (89%). The maximum was 130 days, in 1967. For the ABRACOS soil moisture series, the maximum number of consecutive days was 42, which indicates a longer than average dry period.

FOREST

Figure 8 shows the estimated profile storage (0–2 m) and the estimated deficit in the profile below 2 m for the period from 1966 to 1992. In general, the profile storage during dry periods was lower than in the pasture, and remained low for longer, as a consequence of the difference in transpiration demand in dry periods. There were eight years in which there was still a deficit in the profile below 2 m at the start of January, and in three of these years this 'carried over' deficit was larger than that of the subsequent dry season. These deficits have been counted as belonging to the previous year's dry season. The profile below 2 m depth rewetted completely in all years, with no occasions when a deficit was carried over from one dry season to the next.

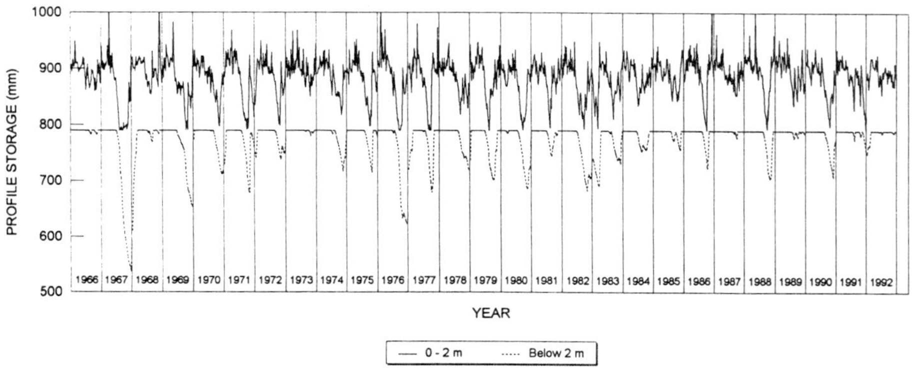

Figure 8 Forest: profile storage (0 - 2 m) and deficit in the profile below 2 m depth predicted from the Reserva Ducke rainfall record (1966 - 1992) using a daily water balance model. Data are plotted at weekly intervals.

Figure 9 shows the number of years out of the 27-year record in which different classes of soil water uptake occurred from below 2 m depth. There was less than 30 mm of uptake in six years out of the 27 (22%), and in 14 years (52%) the uptake was between 30 and 90 mm. Uptake exceeded 90 mm in seven of the years (26%). The

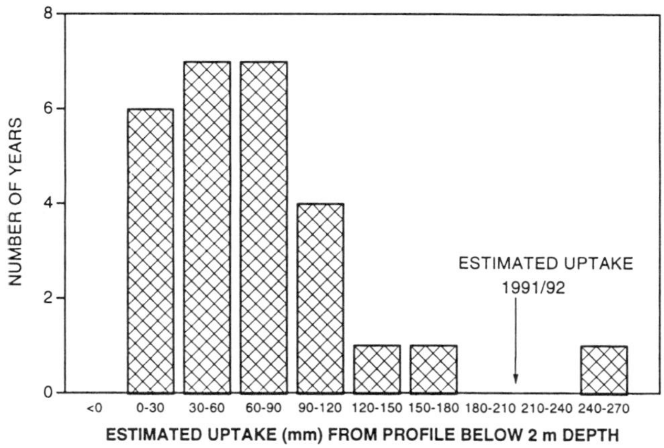

Figure 9 Forest: number of years out of the 27 year record when different classes of deficit occurred in the profile below 2 m depth

overall mean uptake from below 2 m depth was 72 mm, with a maximum of 254 mm, in 1967. The three years with deficits exceeding 120 mm in the profile below 2 m were 1967, 1969 and 1976.

The deficit estimated for the forest at Fazenda Dimona using the ABRACOS data for 1991 was 210 mm, which places it as the second largest in the Reserva Ducke estimated record. However, it is of note that the predicted deficit at Reserva Ducke, in 1991, was only 38 mm, indicating a very significant variability in the dry season rainfall between Reserva Ducke and Fazenda Dimona for that year. It is clear that the rainfall distribution in time and space is crucial in assessing deficits. This is of considerable importance to the energy partition process and the possible interaction on a greater scale.

It is also of note that the transpiration measurements described by Shuttleworth *et al.* (1984), and Shuttleworth (1988) were carried out in 1983, 1984 and 1985, and appear from this analysis, to have been carried out in years when large soil water deficits did not occur. The opportunity to observe transpiration rates and plant physiological responses under conditions of potentially much greater stress has not arisen.

DISCUSSION

One of the main aims of this study was to place the ABRACOS data-set in the context of the longer-term soil moisture record to identify any years when more severe drought conditions than those observed during ABRACOS may have occurred. Although the models are site specific, both should apply to areas with similar soils. The pasture model is both soil and grass species specific. The pasture at the other

ABRACOS sites responded differently to soil water stress (Hodnett*et al.*, 1996). The models were not validated, but because they were calibrated successfully for a period with a wide range of observed profile water storage conditions, it was considered that they should perform well.

FOREST

In the absence of any evidence from the literature that central Amazonian forest (as a whole) suffers from moisture stress, the model was run on the basis that forest transpiration is not limited by stress. The results indicated that uptake from below a depth of 2 m occurred in all years out of the 27 and soil water deficits (in the whole profile) varied from about 50 mm in wet dry-seasons, to almost 341 mm in 1967. They also showed that, at the ABRACOS forest site at Fazenda Dimona, the dry season of 1991/2 resulted in the second driest soil conditions when compared to those predicted from the Reserva Ducke rainfall record.

The mean 'dry season' transpiration rate, $E_{VM,}$ optimised for the forest and used in the prediction of soil water storage, was 3.89 mm day^{-1}. This is slightly higher than the mean value of 3.73 mm day^{-1}, estimated for the wet month of September 1983 by Shuttleworth *et al.* (1984), but lower than the peak measured daily values of 4.18 and 4.22 mm day^{-1} quoted in the same paper. In eastern Amazonia, Nepstad *et al.* (1994) obtained a mean dry season transpiration rate of 3.67 mm day^{-1}, over a 5.5 month dry season using a soil water balance approach. The value used in this study appears, from these comparisons, to be a reasonable estimate of the transpiration rate for rain-free periods in the dry season.

These results showing deep soil water uptake appear to conflict with the observation from the climatic mean monthly rainfall and transpiration data for Reserva Ducke (Figure 5) that transpiration only exceeds rainfall in August and September and that large deficits are unlikely. The large error bars on the rainfall data (± 1 SD) show that, in broad terms, transpiration may exceed rainfall in about one year in three in any of the months between June and November. The mean monthly rainfall data cannot be used to determine the soil water deficit resulting from a sequence of dry months. This can only be obtained reliably through the use of a daily water balance model.

Ranzani (1980) and Correa (1984) have presented water release data for similar soils which show that the water availability is low, and decreases with depth. These results have been confirmed by the *in situ* measurements of Hodnett *et al.* (1995, 1996) who observed a maximum seasonal water content change averaged over the layer from 2 m to 3.6 m of 0.037 MVF. The proportion of this change which resulted from drainage is not known, but the available water capacity is unlikely to exceed 30 mm m^{-1} of profile. Assuming a constant water availability with depth, the provision of 210 mm of water from below 2 m depth (as predicted for the forest site at Fazenda Dimona in 1991/2) would require uptake from 7 m of profile, i.e. to a total depth of 9 m. These results indicate that in this season, deep rooting would have been essential to maintain evapotranspiration at the potential level.

In a recent paper, Nepstad*et al.*(1994) reported roots to a depth of at least 8 m depth at Reserva Ducke. They also modelled the soil water uptake in Amazonia to produce a map indicating the zone where the forest does not need deep soil water (defined as from below 1 m depth). The ABRACOS study area is found in this zone, but the predicted lack of a need for deep soil water is contrary to the findings presented here, and by Hodnett *et al.* (1995, 1996). It is possible that the soils at the ABRACOS site are not representative, but this is unlikely because of their close similarity to those described by Anon.(1978), Ranzani (1980), Chauvel (1982) and Correa (1984) in the same area. The difference may be the result of the use of monthly rainfall data, as described above. It may also be due to over-estimation of the soil water availability through the use of methods to predict this from soil texture. Hodnett et al (1995) and Tomasella and Hodnett (1996) have cautioned against using relationships derived for temperate soils to estimate the water availability of tropical soils from their texture. Saxton *et al.* (1986) have derived such relationships but noted that for very high clay contents, the calculated values did not match reported values as well as desired. As a result, the textural range to which their method could be applied was limited to a maximum of 60% clay. The soils studied, which are found widely in the region, particularly on plateaux and upper slopes, are typically 75–85% clay.

PASTURE

The model used for the pasture did not allow for uptake of water from below a depth of 2 m. The minimum storage in the 2 m profile was taken to be 790 mm, but there were no years in which all of the water was taken up. The lowest estimated storage was 794 mm, in 1976. In contrast, the lowest storage in the forest occurred in 1967. This lack of coincidence between the forest and pasture minimum storage was also seen in the field measured data (Hodnett *et al.*, 1995). At the time of the study, the pasture had a grass cover of 75%, with the remainder being bare soil, logs, stumps and shrubs (Wright *et al.*, 1992). The shrub cover was about 1% for most of the study, so the data and model apply to a pasture without deep rooted plants. More degraded pastures, with a higher proportion of woody species and palms, will have deeper root systems and will have access to soil water from greater depth. Larger soil water deficits, as observed by Nepstad et al. (1994) will occur under such pastures. The one-layer model used, which incorporated a very simple stress function, appeared to represent the data well (Figure 3a). It was expected that it might have performed less well following rainfall events when profile storage was low (i.e. <850 mm) after a long dry period, because the stress function takes no account of the distribution of water in the profile.

RAINFALL VARIABILITY

The results presented are based on the 27-year daily rainfall record from a single raingauge site, at Reserva Ducke. There are two sources of rainfall variability which will result in larger deficits than those shown in the data presented and a different year to year distribution of maximum deficits. These are the lower rainfall close to major rivers in the Amazon basin, shown by Molion and Dallarosa (1990) and attributed to a river breeze effect by Oliveira and Fitzjarrald (1992) and the large spatial variability of dry season rainfall. In Figure 5, the monthly mean data for the Manaus station (nearer to the Rio Negro) can be seen to be consistently lower than at Reserva Ducke, particularly in the dry season months from June to October. If these data are representative, it is clear that much larger deficits will occur close to major rivers. The spatial variability of dry season rainfall is illustrated by the large difference in the maximum deficit predicted for Reserva Ducke and for the Fazenda Dimona forest site in 1991. It appears likely that, away from major rivers, larger deficits than those estimated in this study may occur at times, but in isolated patches. Both the 'random' rainfall variation caused by the predominantly convective events in the dry season, and the apparently more consistent variation caused by the proximity to major rivers must be taken into account for water balance studies which cover large areas of Amazonia.

CONCLUSIONS

The simple daily soil water balance models which were developed for forest and pasture were calibrated successfully using the ABRACOS Fazenda Dimona soil water data-sets. For the pasture, a simple linear function relating transpiration to soil water storage was used to model the impact of stress. Currently, there is little direct evidence from central Amazonia to indicate that forest transpiration is significantly affected by soil water stress and for the forest model, no stress function was included. To predict long-term seasonal variations in soil water uptake, the 27-year rainfall record from the Reserva Ducke site was used as the model input. If forest transpiration is reduced by soil water stress in severe drought conditions, then the model will over-estimate soil water uptake, but should nonetheless provide a good estimate of the maximum expected uptake, and how frequently such periods may occur. Both models are site specific but, the forest model should apply to areas with similar soils.

For the forest, the maximum possible uptake from the upper 2 m of the profile was 114 mm, including 26 mm which can be taken up when the soil storage is above field capacity (defined as when drainage from the upper 2 m is zero). The model indicated that storage uptake from below 2 m occurred in all years. The mean dry season uptake from below 2 m depth was 72 mm (over the 27 years). The maximum uptake from below 2 m was 254 mm in 1967, with a total soil water uptake in that year of 341 mm. At the Fazenda Dimona site, where the water availability at depth is about 30 mm m^{-1}, an additional 9 m depth of profile below 2 m would be required to provide 254 mm of water. It is clear that the forest requires a deep rooting strategy to survive on these soils, particularly in the drier dry seasons. The use of a daily water balance model revealed that the soil water deficits which could develop under forest were much larger than those implied by the use of long term monthly mean rainfall data.

For the ABRACOS forest data, the largest uptake from below 2 m (210 mm) occurred in December 1991. This was larger than in all years predicted from the Reserva Ducke rainfall data, except for 1967. The deficit predicted for Reserva Ducke in 1991 was only 38 mm, showing that the spatial variability of the rainfall in Central Amazonia has a major impact on the incidence of severe deficits. Other rainfall data from the area confirm that this spatial variability is not unusual.

For the pasture, the most severe observed deficit occurred in 1990, not in 1991, as in the forest. The maximum observed deficit was only 12 mm less than the maximum deficit predicted from the Reserva Ducke rainfall data and was only surpassed in four years out of the 27-year record. In the most severe dry season, the forest used 234 mm more soil water than the pasture.

The long-term Reserva Ducke rainfall record revealed that although more extreme dry seasons than those observed during ABRACOS have occurred, the ABRACOS soil water data series covers all but the most extreme conditions. Detailed micro-meteorological and plant physiological studies were carried out in the pasture during the dry season of 1990, but no studies were carried out at the forest site in the extreme dry season of 1991/2, when the forest may have been suffering from soil water stress. The key question that remains concerns the response of the forest to soil water stress. This is crucial to GCM predictions, particularly if longer dry seasons are likely. Measurements of actual transpiration and stomatal response under conditions of severe soil water deficit are essential and remain to be carried out.

ACKNOWLEDGEMENTS

The authors would like to acknowledge I. R. Wright for all the helpful suggestions and contributions which made possible the advance of this work.

REFERENCES

Anon. 1978. Projeto Radambrasil, Levantamento de Recursos Naturais. Vol 18, Folha SA.20 Manaus. Ministério das Minas e Energia, Departamento Nacional da Produção Mineral, Rio de Janeiro, Brasil, 623pp.

Bastable, H.G., Shuttleworth, W.J., Dallarosa, R.L.G., Fisch G. and Nobre C.A. 1993. Observation of climate, albedo, and surface radiation over cleared and undisturbed Amazonian forest. *Int. J. Climatol.*, **13**, 783-796.

Chauvel, A. 1982. Os latossolos amarelos, álicos, argilosos dentro dos ecossistemas das bacias experimentais do INPA e da região vizinha. *Acta Amazonica*, **12**, 47-60.

Chauvel, A., Grimaldi, M. and Tessier, D. 1991. Changes in soil pore space distribution following deforestation and revegetation: an example from the Central Amazon Basin, Brazil. *Forest Ecol. Manage.*, **38**, 259-271.

Chauvel, A., Vital, A.R.T, Lucas, Y., Desjardins, T., Franken, W.K., Luizão, F.J., Araguás, L.A., Rozanski, K. and Bedmar, A.P. 1992. O papel das raízes no ciclo hidrológico da floresta Amazônica. Anais do VII Congresso Brasileiro de Meteorologia, São Paulo, SP, Brazil. pp298-302.

Correa, J.C. 1984. Características físico hídricas dos solos latossolo amarelo, podzolico vermelho amarelo e podzol hidromórfico do estado do Amazonas. *Pesquisa Agropecuaria Brasileira*, **20**,1317-1322.

Culf, A.D., Esteves, J.L., Marques, A.de O. and da Rocha, H.R. 1996. Radiation, temperature and humidity over forest and pasture in Amazonia. This volume.

Gash, J.H.C. 1979. An analytical model of rainfall interception by forests. *Q.J.R. Meteorol. Soc.*, **105**, 43-55.

Gash, J.H.C., Nobre, C.A., Roberts, J.M. and Victoria, R.L. 1996. An overview of ABRACOS. This volume.

Hodnett, M.G., Pimentel da Silva, L., da Rocha, H.R. and Cruz Senna, R. 1995. Seasonal soil water storage changes beneath Central Amazonian rainforest and pasture. *J. Hydrol.*, **170**, 233-254.

Hodnett, M.G., Oyama, M.D. and Tomasella, J. 1996. Comparisons of long-term soil water storage behaviour under pasture and forest in three areas of Amazonia. This volume.

Jarvis, P.G. 1976. The interpretation of the variation in leaf water potential and stomatal conductances found in canopies in the field. *Phil. Trans. Roy. Soc., London, Ser. B*, **273**, 593-610.

Lloyd, C.R., Gash, J.H.C., Shuttleworth, W.J. and Marques, A de O. 1988. The measurement and modelling of rainfall interception by Amazonian rain forest. *Ag. and For. Met.*, **43**, 277-294.

Molion, L.C.B. and Dallarosa, R.L.G. 1990. Pluviometria da Amazonia:são os dados confiaveis? *Climanalise*, **5**, 40-42.

Monteith, J.L. 1965, Evaporation and the environment. *Symp. Soc. Exp. Biol.* **19**, 205-234.

Nepstad, D.C., de Carvalho, C.R., Davidson, E.A., Jipp, P.H., Lefebvre, P.A., Negreiros, G.H., da Silva, E.D., Stone, T.A., Trumbore, S.E. and Vieira, S. 1994. The role of deep roots in the hydrological and carbon cycles of Amazonian forests and pastures. *Nature*, **372**, 666-669.

Nortcliff, S. and Thornes, J.B. 1981. Seasonal variations in the hydrology of a small forested catchment near Manaus, Amazonas, and its implications for management. In:*Tropical Agricultural Hydrology*. (Eds. R. Lal and E.W. Russell). John Wiley, Chichester and New York, 482 pp.

Oliveira, A.P. and Fitzjarrald, D.R. 1992. The Amazon river breeze and the local boundary layer: I, Observations. *Bound. Layer Meteorol.*, **63**, 141-162.

Priestley, C.H.B. and Taylor, R.J. 1972. On the assesment of surface heat flux and evaporation using large scale parameters. *Mon. Weather Review*, **100**, 81-92.

Ranzani, G. 1980. Identificação e caracterização de alguns solos da Estação Experimental de Silvicultura Tropical do INPA. *Acta Amazonica*, **10**, 7-41.

Roberts, J.M., Cabral, O.M.R., Fisch, G., Molion, L.C.B., Moore, C.J. and Shuttleworth, W.J. 1993. Transpiration from an Amazonian rainforest calculated from stomatal conductance measurements. *Ag. and For. Met.*, **65**, 175-196.

Rosenbrock, H.H. 1960. An automatic method for finding the greatest or least value of a function. *Computer J.*, **3**, 175-184.

Saxton, K.E., Rawls, W.J., Romberger, J.S. and Papendick, R.I. 1986. Estimating generalised soil-water characteristics from texture. *Soil. Soc. Am. J.*, **50**, 1031-1036.

Shuttleworth, W.J., Gash J.H.C., Lloyd, C.R., Moore, C.J., Roberts, J.M., Marques Filho, A. de O., Fisch, G., Silva Filho, V. de P., Ribeiro, M. de N.G., Molion, L.C.B., de Abreu Sá, L.D., Nobre, C.A., Cabral, O.M.R., Patel, S.R. and de Moraes, J.C. 1984. Eddy correlation measurements of energy partition for Amazonian forest. *Quart.J.R. Met. Soc.*, **110**, 1143-1162.

Shuttleworth, W.J. 1988. Evaporation from Amazonian rainforest. *Proc. R. Soc. Lond.*, **B233**, 321-346.

Tomasella, J. and Hodnett M.G. 1996. Soil hydraulic properties and van Genuchten parameters for an oxisol under pasture in central Amazonia. This volume.

Wright I.R., Gash J.H.C., da Rocha, H.R., Shuttleworth, W.J., Nobre, C.A., Maitelli, G.T., Zamparoni, C.A.G.P., and Carvalho, P.R.A., 1992. Dry season micrometeorology of central Amazonian ranchland. *Quart. J. R. Meteorol. Soc.*, **118**, 1083-1099.

Wright I.R., Manzi A. and da Rocha, H.R. 1995. Surface conductance of Amazonian pasture: model application and calibration for canopy climate. *Agric. For. Meteorol.*, **75**, 51-70.

RESUMO

Foram desenvolvidos modelos simples de balanço hídrico visando prever o armazenamento de água nos primeiros 2 m do perfil de solo sob floresta e pastagem, utilizando somente a precipitação diária como entrada. Para a floresta, a extração de água abaixo de 2 m foi também prevista. O modelo foi calibrado usando dados de precipitação, micrometeorológicos e de água no solo obtidos no Projeto ABRACOS. A evaporação e a drenagem profunda foram relacionados com o armazenamento de água por meio de funções lineares e exponenciais, respectivamente, permitindo a determinação do armazenamento a partir de um simples balanço hídrico. Embora simples, os modelos forneceram um valioso "insight" no entendimento da variabilidade anual da extração de água no solo.

O armazenamento foi calculado a partir de uma série histórica de precipitação de 27 anos da Reserva Ducke, próximo à Manaus, assumindo-se que a evaporação na floresta não é afetada pelo estresse hídrico. Para a floresta, o modelo previu uma extração abaixo de 2 m em todos os anos, com uma extração média de 72 mm e uma máxima de 254 mm (em 1967). A disponibilidade de água nos solos próximos à Manaus é muito baixa, portanto, a água retirada deve ser proveniente de um perfil bastante profundo. A floresta necessita de raízes profundas para sobreviver, particularmente nas estações secas mais intensas. O uso do modelo diário de balanço hídrico revelou que, sob a floresta, o déficit hídrico pode ser muito maior que o inferido com o uso de dados mensais de precipitação de longo termo.

Os dados para floresta do Projeto ABRACOS mostraram que a maior extração de água abaixo de 2 m (210 mm) ocorreu em dezembro de 1991. Esta extração foi maior que a prevista, a partir dos dados de precipitação da Reseva Ducke, em todos os anos, exceto 1967. O déficit previsto para a Reserva Ducke em 1991 foi somente de 38 mm, o que mostra o importante papel da variabilidade espacial da precipitação na Amazônia Central na incidência de grandes déficits. Para a pastagem, o déficit mais severo observado ocorreu em 1990, não em 1991, como na floresta. Os déficits na pastagem calculados pelo modelo mostraram que o déficit observado foi superado em somente 4 dos 27 anos.

Uma questão chave que ainda permanece pendente refere-se à resposta da floresta sob condições de estresse hídrico. Esta é crucial nas previsões dos GCMs, particularmente se longas estações secas são prováveis. Medições de transpiração e resposta estomática sob condições severas de déficit são essenciais e necessitam ser realizadas.

5 Soil hydraulic properties and van Genuchten parameters for an oxisol under pasture in central Amazonia

J. TOMASELLA[1] **and M.G. HODNETT**[2]
[1]*Instituto de Pesquisas Hidraulicas, Porto Alegre, Brazil*
[2]*Institute of Hydrology, Wallingford, OX10 8BB, UK*

INTRODUCTION

Surface parameterisation of General Circulation Models (GCM) requires information on soil hydraulic properties to provide an accurate quantitative description of soil water storage behaviour. This is important because of the control that soil water availability exerts on evaporation. Changes in water availability on a large enough scale could affect climate. A sound knowledge of the soil physical properties in the unsaturated zone is also vital for hydrological studies, since the processes of infiltration, drainage and runoff generation are largely controlled by the soil properties.

Recent effort to improve the representation of surface processes in GCM has resulted in the introduction of multi-layer soil schemes (Sellers *et al.*, 1986; Verseghy, 1991) to improve the simulation of infiltration and root uptake. Although several semi-empirical descriptions of soil hydraulic properties are available (Brooks and Corey, 1964; Mualem, 1976; van Genuchten, 1980) the most sophisticated, and widely adopted for GCM is that of van Genuchten.

The use of more sophisticated subsurface models requires spatially representative parameters describing the soil physical properties at several depths. However, although more sophisticated models may be developed relatively easily, the problems of obtaining suitable data should not be overlooked. Indeed, the demand for sound parameters and data has far outstripped the methodologies and resources available to provide them. The lack of data is a particular problem for Amazonia because very little is known about the hydraulic properties of the soils, and even less about how they change as a result of deforestation. One of the aims of the ABRACOS soil water studies was to add to this knowledge base. This paper describes the measurement of the soil hydraulic properties of the oxisol at the ABRACOS pasture site near Manaus, and derivation of the van Genuchten parameters required to describe both the water release and hydraulic conductivity curves of oxisols.

Amazonian deforestation and climate. Edited by J.H.C. Gash, C.A. Nobre, J.M. Roberts and R.L. Victoria.

MATERIALS AND METHODS

Site, vegetation and soils

The site is a typical cattle ranch, Fazenda Dimona (2° 19' S, 60° 03' W), about 100 km north of Manaus, Amazonas. The pasture was created in about 1980 by cutting and burning the forest and sowing tropical grasses. A detailed description of the site is given by Wright *et al.* (1992). At the time of the study (July 1993), the pasture had become degraded (invaded by shrubs), grass cover was reduced to 50% and shrubs covered 35% of the area. The remainder was unvegetated, either bare soil (cattle tracks and trampled termite mounds), old stumps from the original forest, or fallen trunks.

This study was carried out in a largely flat plateau area where the soil was a clayey yellow latosol (Latossolo amarelo, álico, textura argilosa - Brazil, Haplic Acrorthox - Soil Taxonomy, or xanthic ferralsol - FAO). The soil was very similar to those described by Ranzani (1980), Chauvel (1982) and Correa (1984) within 30 km of this site. The soils have a high clay content (>75%), and a low dry bulk density of between 0.93 and 1.15 Mg m^{-3}. The porosity is between 56% and 64%, but is largely concentrated in the finest pore size classes containing water which is held at such low potentials that it is inaccessible to plants. This means that the available water capacity is low, only about 70 mm m^{-1} in the upper metre of the profile (Correa, 1985). More details of the soil at this site are given by Hodnett *et al.*(1995; 1996).

Ring permeameter measurements

Among the available methods for obtaining the saturated hydraulic conductivity value, the CSIRO ring permeameter (Perroux and White, 1988) provides rapid and simple measurements. The method for calculating the saturated conductivity is based on the analysis of Wooding (1968) for three dimensional flow from a ponded circular surface. The saturated hydraulic conductivity is calculated from:

$$K_s = \frac{q}{\pi r_0^2} - \frac{4bS_0^2}{\pi r_0(\theta_s - \theta_i)} \quad (1)$$

where K_s is the saturated hydraulic conductivity (mm h^{-1}), q is the steady-state volumetric flow rate (mm h^{-1}), S_0 is the sorptivity (mm $h^{-1/2}$), θ_i is the initial water content (m^3m^{-3}), θ_s is the saturated water content (m^3m^{-3}), b is a parameter considered to have a value of 0.55 and r_0 is the radius of the permeameter ring.

The sorptivity can be estimated using the data from the early stages of infiltration, where the cumulative infiltration Q (mm) can be approximated by (Philip, 1969):

$$\frac{Q}{\pi r_0^2} = S_0 t^{\frac{1}{2}} \quad (2)$$

The macroscopic capillary length λ_c, which is the reciprocal of the α_c parameter of the quasi-linear exponential model, can be expressed as (White and Sully, 1987):

$$\lambda_c=\alpha_c^{-1}=\frac{bS_0^2}{(\theta_0-\theta_i)K_0} \tag{3}$$

Using the ring permeameter (diameter 200 mm), a total of 28 *in situ* ponded measurements were made at the surface and at depths of 0.3 m and 1.05 m using a head of approximately +10 mm. To make measurements at these depths, pits were excavated to 0.25 m and 1.0 m. The final 0.05 m was trimmed carefully to minimise disturbance to the infiltration surface before setting the permeameter ring in place. Attempts to make measurements in the forest were not successful because of the presence of large macropores and roots, the uneven soil surface, which made ring installation and sealing difficult, and because infiltration rates were also so high that it was not possible to make reliable readings of the water reservoir.

Instantaneous Profile Method (IPM)
The IPM (Watson, 1966) uses simultaneous measurements of water content and matric potential during transient drainage on an initially saturated soil, to calculate the instantaneous soil water flux q (mm) using Equation 4.

$$q(z,t)=q(0,t)-\frac{\partial}{\partial t}\int_0^z \theta dz \tag{4}$$

The hydraulic conductivity value for a given time and depth are obtained from Equation 5, where K(z,t) can be expressed as a function of ψ or θ.

$$K(z,t)=\frac{-q(z,t)}{\frac{\partial\psi(z,t)}{\partial t}-1} \tag{5}$$

For the IPM, two field plots were selected randomly. Each plot consisted of an undisturbed soil column 1.5 m in diameter, which was laterally isolated from the surrounding soil by carefully excavating a trench to a depth of 1.3 m around the perimeter of a 1.5 m diameter plot to create a cylindrical column of undisturbed soil. A plastic sheet was wrapped around the column, leaving the top open. The overlap in the sheet was sealed with tape and the trench was then backfilled to hold the sheet tightly in place against the walls of the column. The plastic sheet extended 0.1 m above the soil surface level to allow water to be ponded on the plot.
In general, when the IPM is used, an area of 2 – 3 m radius is saturated by irrigation to permit the assumption that flow in the instrumented centre of the plot is vertical. However, this assumption may not be valid when there are impeding layers in the profile which can cause saturation and lateral flow. In this study, the aim of isolating the column was to eliminate the lateral flow and to reduce the amount of water

required for irrigation. In each plot, two neutron probe access tubes were installed to measure water content at depths of 0.1, 0.2, 0.3, 0.4, 0.6, 0.8, 1.0, 1.2 and 1.4 m. Mercury manometer tensiometers were installed at depths of 0.2, 0.4, 0.6, 0.9, 1.2 and 1.5 m. The plot surface was covered by a plastic sheet which was sealed around the instruments to prevent evaporation from the soil surface and to prevent rainwater from entering. The plots were sheltered to keep the sun off the tensiometers and to stop rainwater accumulating on top of the plastic sheet cover.

Water was ponded on the soil surface in each plot until the potential profile stabilised. The initial water content was measured at this stage. The starting time for drainage was taken as when free water disappeared from the surface, after irrigation had ceased. As internal drainage proceeded, periodic measurements of soil water potential and soil water content were carried out. The interval between measurements increased from 10 minutes during the rapid initial stages of drainage to 48 h after eight days. Measurements continued until the drainage rate had become too slow to measure reliably. This took ten days by which time the matric potentials had decreased to about -9 kPa. Before the hydraulic conductivity was calculated, the moisture content and matric potential versus time curves for each depth were smoothed by eye.

In-situ *water release curves*

In addition to the data from the IPM, *in-situ* water release curves covering range from –3 kPa to –80 kPa were obtained for forest and pasture using water content and potential data from weekly monitoring of neutron probe access tubes and tensiometers. Details of these installations are given by Hodnett *et al.* (1996). These data also recorded the seasonal variation in soil water content and allowed the abstraction limit to be determined. In the dry season, the water contents measured in the upper metre of the profile showed a limit beyond which the water content did not fall, although water contents continued to decrease at greater depths. This field measured minimum water content, θ_{fmin}, (from three years of field data, including a very dry, dry season, in 1990) was taken to be the abstraction limit after examination of published water release data for very similar soils published by Ranzani (1980) and Correa (1984) and mercury porosimetry data (Chauvel *et al.*, 1991; Hodnett, unpublished data).

Theory

The van Genuchten (1980) model is widely used and provides a continuous and smooth description of the unsaturated hydraulic conductivity and water release curve, with no sharp transition at the air entry value, as is produced by the Brooks-Corey model (1964). It is also versatile, and sufficiently accurate for practical purposes (Stephens and Rehfeldt, 1985). Van Genuchten (1980) proposed the following theoretical expression for the soil water release curve:

$$S_e = \frac{\theta - \theta_r}{\theta_s - \theta_r} = [1 + (\alpha\psi)^n]^{-1+1/n} \qquad (6)$$

where ψ is the soil matric potential (which, following Van Genuchten (1980) is assumed to be positive, to simplify notation), θ_s is the saturated volumetric water content, θ_r is the residual volumetric water content, S_e is the effective saturation, and α and n are soil parameters. These parameters are inversely related to the air entry value and width of the pore size distribution respectively (Russo, 1988). Using Equation (6) with the Mualem (1976) model, van Genuchten (1980) derived the following equation for unsaturated hydraulic conductivity:

$$K(\psi)=K_s\frac{(1-(\alpha\psi)^{n-1}[1+(\alpha\psi)^n]^{-(1-1/n)})^2}{[1+(\alpha\psi)^n]^{\ell(1-1/n)}} \qquad (7)$$

where $K(\psi)$ (mm h^{-1}) is the unsaturated hydraulic conductivity, and ℓ is an additional parameter which accounts for the tortuosity and the correlation factors on the water content. Mualem (1976) suggested an average value of $\ell = 0.5$, and van Genuchten (1980) adopted this value when Equation (7) was used as a predictive model. However, Russo (1988) showed that the use of as an additional parameter improved the performance of van Genuchten's model in many soils. Wösten and van Genuchten (1988) found that fixing ℓ at 0.5 provided a reasonable approximation for coarse-textured soils but for medium and fine textured soils, it did not result in acceptable fits. Consequently, ℓ was included in the optimisation process.

Optimisation method

The parameters of the van Genuchten model were estimated simultaneously from the measured conductivity and water release data, using a similar procedure to that of Wösten and van Genuchten (1988). The optimisation was carried out using the Rosenbrock (1960) algorithm, minimising the following objective function:

$$\sum_{i-1}^{m}\left(\frac{K(S_e)_c-K(S_e)_o}{K(S_e)_o}\right)^2+\left(\frac{\psi(S_e)_c-\psi(S_e)_o}{\psi(S_e)_o}\right)^2 \qquad (8)$$

where the subscript o indicates a measured value and the subscript c an estimated value. Estimated values were obtained using Equation (7) for the hydraulic conductivity and the inverse function of Equation (6) for the water release curve.

Equation (8) guarantees equal weight for all measurements. It was chosen because the objective of the problem also requires a good approximation for dry conditions. Kool *et al.* (1985) suggested the use of the logarithm of the conductivity data and a data dependent parameter to give the same weight to the water content data. In order to avoid the use of weighting parameters, the relative error of ψ estimated upon θ was used and provided a good fit to both curves. For the van Genuchten (1980) model, the values of n, α, θ_s, θ_r, ℓ and K_s were optimised.

The water release and hydraulic conductivity data from the IPM are limited by the potential at which drainage becomes too slow to measure. Fitting a curve to the IPM data alone may lead to substantial errors at the dry end of the water release curve (e.g. at wilting point, conventionally, but arbitrarily taken as -1500 kPa). To make the fitted curves as accurate as possible over the entire field water content/potential range, *in-situ* measured water release data from a routinely monitored site in the pasture were used to extend the range of the observed potential data to -80 kPa. In addition, the value of the residual water content, θ_r, derived during the optimisation procedure, was constrained to be less than the field measured minimum water content, θ_{fmin}.

To link the data from the two conductivity measurement plots and the regular field observation site, the values of soil moisture in Equation (10) were non-dimensionalised using the effective saturation, S_e, preserving a constant difference between the θ_s and θ_r. Thus, using different θ_s and θ_r for the three sites, but with equal effective porosity (θ_s - θ_r), it was possible to estimate a single relationship between ø and the effective saturation θ_{se}. This procedure ensured a single set of shape-parameters (n, α, ζ, ψ_e) at each depth. The assumption of equal effective porosity (for a particular depth) was supported by the low variability of the difference between the field measured maximum and minimum water contents at all of the tubes. It was also supported by the fact that the water release curves from the various plots were almost identical in form but were offset from each other on the water content scale. Initial estimates of the effective porosity were obtained from θ_{fmin}, for θ_r, and from the regularly monitored field site data for θ_s.

In this way, the water release curve was extended beyond the range of the data obtained from the IPM method. Although this methodology has obvious limitations, the values of θ_r obtained by including the extraction limit water content will clearly lead to physically more realistic estimates of water availability. Stephens and Rehfeldt (1985) concluded that the reliability of the van Genuchten (1980) method may depend on physically reasonable estimates of θ_r.

RESULTS

Ring permeameter data

Table 1 shows the arithmetic mean value of the different parameters for each layer, with the standard deviation (SD) given in parentheses. Some data have been discarded since the sorptivity value obtained from Equation 2 was high, resulting in a negative value of saturated hydraulic conductivity in Equation 1. The reasons for this are not clear but it is probable that, in these soils with unusual macropore properties, some of the suppositions inherent in Equation 1 were not satisfied. If this is the case, it is possible that the mean value of K_s is biased downwards because the method failed for the locations with a very high macroporosity (hence, high K_s).

The saturated hydraulic conductivity varied significantly with depth, the mean conductivity ranging from 97 mm h^{-1} at 0.3 m depth, down to 17 mm h^{-1} at 1.05 m. There was a large variability at the surface, with values ranging from 18 mm h^{-1} to 220 mm h^{-1}. The variability at 0.3 m depth was also very large, particularly compared with that at 1.05 m depth. The high saturated hydraulic conductivity is primarily due to the macropores and structure and this is clearly variable in the upper profile. The lower conductivity and variability at 1.05 m indicates that the soil is less structured and more uniform; this was apparent in the soil profile.

The bulk density value and the saturated water content are slightly less than the bulk density and porosity respectively measured by Correa (1984). In the case of the saturated water content, this is usually less than the porosity as a result of the effect of entrapped air.

Table 1 Summary of field results from CSIRO ring permeameter measurements at Fazenda Dimona (n is the number of determinations)

Depth (m)	0	0.30	1.05
θ_i (m^3m^{-3})	0.391 (0.046)	0.430 (0.012)	0.450 (0.008)
θ_s (m^3m^{-3})	0.483 0.049)	0.520 (0.011)	0.516 (0.010)
S_0 (mm $h^{-1/2}$)	26.82 (18.05)	66.47 (46.05)	15.45 (9.80)
K_s (mm h^{-1})	66.4 (55.8)	96.6 (45.9)	16.8 (7.2)
α_c (mm^{-1})	0.036 0.048)	0.017 (0.010)	0.034 (0.039)
Bulk density (Mgm m^{-3})	1.059 (0.104)	1.070 (0.037)	1.118 (0.020)
n	10	6	7

Water release curves

Figure 1 shows the water release curves for the two conductivity plots. Both showed the same sequence of change of water release characteristics with depth, indicating that the soil properties beneath both plots were similar. Above 1.2 m depth water contant decreased as the matric potentials fell from 0 (saturation) to –2 kPa. Below this potential the water content decrease was much more gradual with a volumetric water content change of between 0.02 and 0.05 moisture volume fraction (MVF) between –2 kPa and –8 kPa.

Under the definitions suggested by Luxmoore (1981) the pores which empty between 0 and –2 kPa are macropores (emptying between 0 and–0.3 kPa) and mesopores (between 0.3 and –30 kPa). Under these definitions, most of the porosity

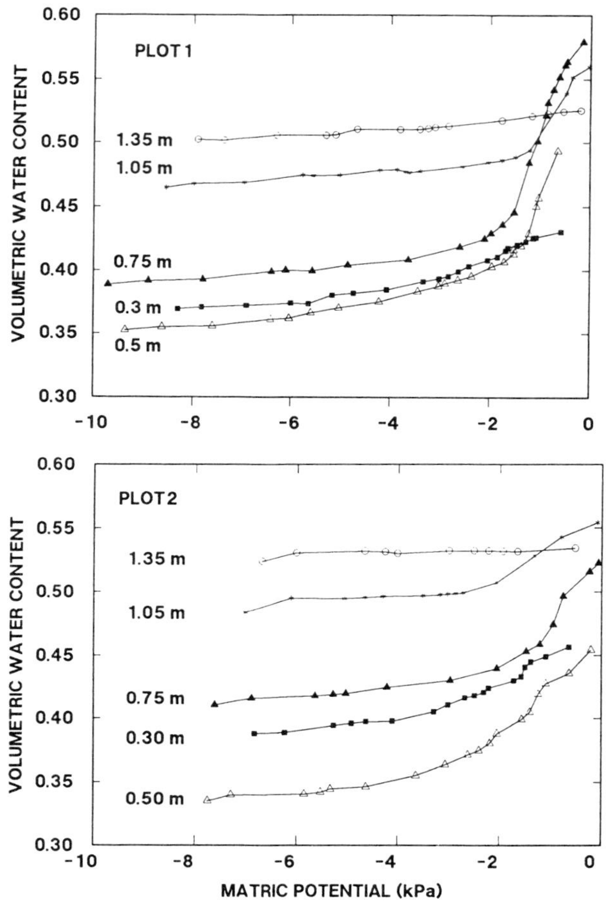

Figure 1 Water release curves derived using the instantaneous profile method (IPM) for plots 1 (top) and 2 (bottom).

appears to be concentrated at the larger end of the mesopore range. From the water release curve, the volume of the pores which emptied between 0 and –2 kPa was typically 0.05–0.10 MVF (moisture volume fraction) and was largest at the 0.5 and 0.75 m depths in both plots, but notably in plot 1 (0.14 and 0.21 MVF respectively). These results are confirmed by those of Chauvel *et al.* (1987b) who found that the macroporosity reached a maximum of almost 24% in this layer. It should be noted that at some depths, although the potential was close to saturation (–0.2 kPa) and it is likely that the values above underestimate the macro/meso-porosity because the macropores were not filled.

The curves for the 1.35 m depth in both plots showed only a very small water content change (0.01 MVF) between 0 and –2 kPa, indicating that the volume of macro/meso-pores at this depth was small. This is more typical of the behaviour of clay soils. The water release curves presented here (for pasture) confirm the observations of Chauvel (1982), who stated that the forest soils were "characterised by a porous,

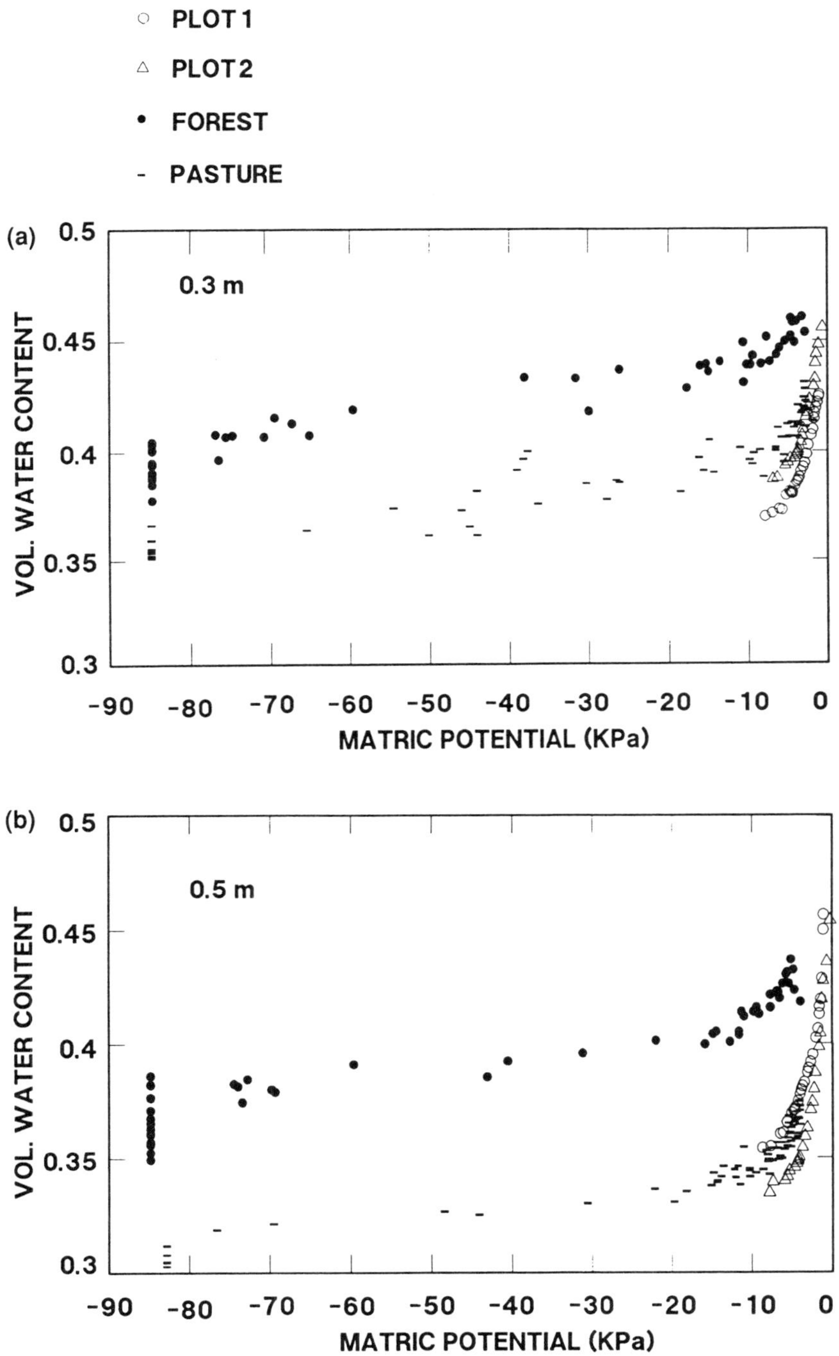

Figure 2 (a-b) *In situ* derived water release curves from long term monitored field and IPM, plots. (a) 0.3 m, (b) 0.5 m.

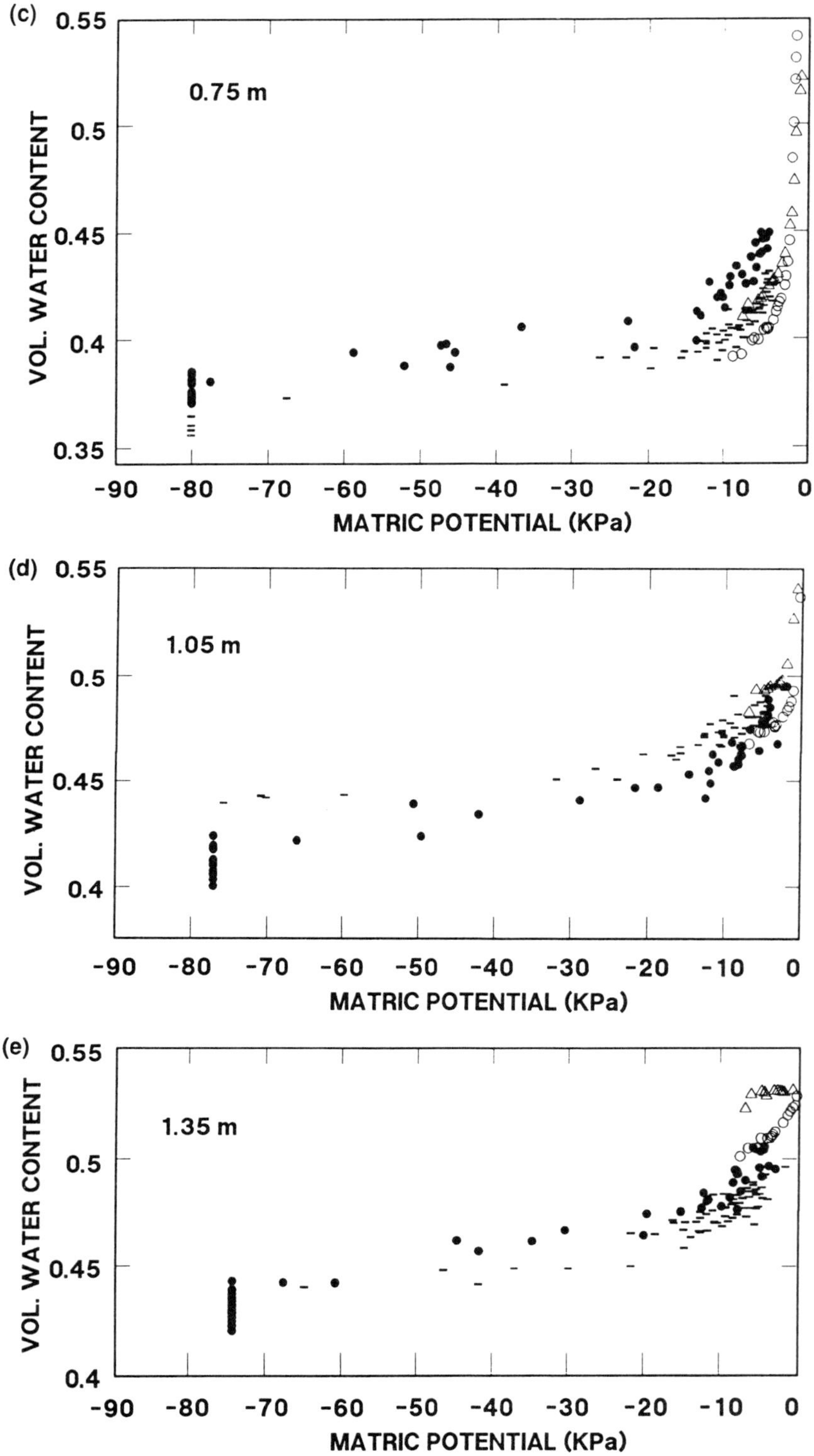

Figure 2 (c-e) *In situ* derived water release curves from long term monitored field and IPM, plots. (c) 0.75 m, (d) 1.05 m, (e) 1.35 m.

strongly micro-aggregated, middle horizon situated between two less porous horizons".

Figure 2 shows the water release curves for five depths at the weekly monitored field sites in the pasture and in the forest between saturation and –80 kPa. The curves for the two IPM plots are also shown for comparison. It is of note that the steep, wet end of the curve is not shown in the weekly data, because drainage is so rapid between saturation and –4 kPa that these conditions were never observed. Also shown in Figure 2 are water content values for dry periods when the tensiometers were off-scale (–80 kPa), plotted corresponding to the minimum matric potential recorded at that depth. These data have been shown to illustrate that the changes of water content that occurred below –80 kPa were between 0.01 - 0.03 MVF, even in the forest, where there was more abstraction at depth compared to in the pasture. These data, indicating a very low water availability in this type of soil, are confirmed by those of Ranzani (1980) and Correa (1984). Ranzani (1980) found water content changes of between 0.03 and 0.07 MVF between –33 kPa and –1500 kPa. Chauvel *et al.* (1991) and Hodnett (1994, unpublished data) using mercury intrusion porosimetry showed that the amount of water released between –100 kPa and –500 kPa is about 0.025 MVF and the soil only begins to free significant quantities of water beyond –4000 kPa. The pores which hold water at these very negative potentials are less than 0.1 μm in diameter and are known as "cryptopores" under the porosity definitions suggested by Brewer (1964).

The forest and pasture curves are almost identical in form at any given depth, but at 0.3 m and 0.5 m (Figure 2), they are displaced from each other on the water content axis. The data from the regularly monitored tubes showed that although the maximum (and minimum) water contents showed a considerable variation from tube to tube, the range of water content from maximum to minimum showed a much lower variability, reflecting the uniformity of the water holding capacity of the soils. Over the range where the curves are parallel, the water release characteristics and pore size distributions at the different sites are identical. The offset between the curves may be due to differences between the sites in the macro/meso-porosity (which, at this site, empties of water between 0 and –3 kPa).

Conductivity curves

Figure 3 shows the hydraulic conductivity as a function of matric potential for the two plots. The variation of the conductivity characteristic with depth caused by the macro/meso-porosity is clearly seen. The highest conductivities, and the most rapid rates of decrease in conductivity with potential, were in the layers with the largest volume of macro/meso-pores, notably at 0.5 and 0.75 m. The lowest conductivities were at 1.35 m where the macro/meso-pore volume was least. Between potentials of –2 kPa and –4 kPa the curves for the different depths converge and cross, so that at a potential of –6 kPa, the conductivity is higher at 1.05 m and 1.35 m (0.1 - 0.2 mm h^{-1}) than at 0.3 m (0.01 - 0.02 mm h^{-1}). The curve for 0.75 m shows transitional behaviour. The general form of the curves was similar in both plots, illustrating the uniformity of conditions at the site.

Figure 4 shows the $K(\psi)$ curves for the 0.3 m and 0.75 m depths. The $K(\theta)$ curves

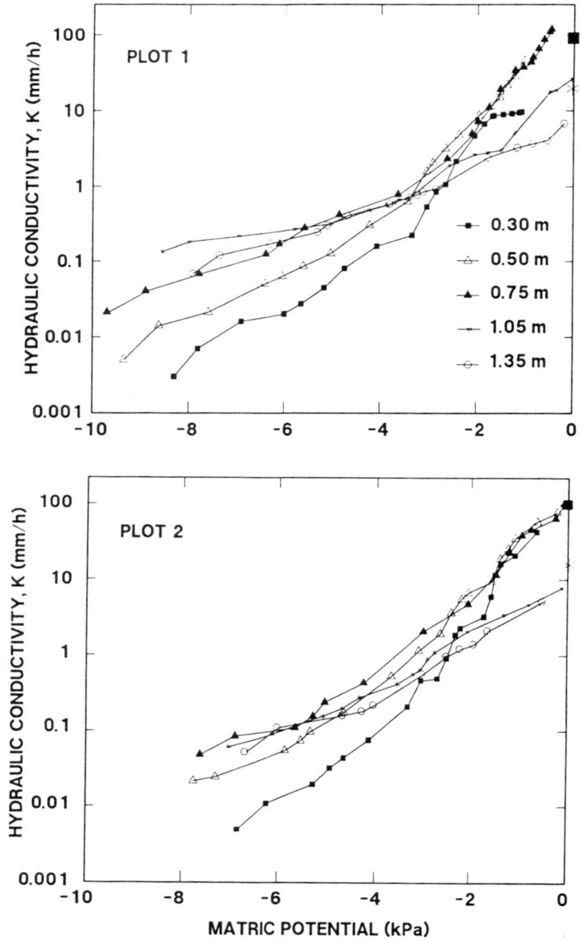

Figure 3 Conductivity curves (K – ψ), derived using the instantaneous profile method (IPM) for plots 1 (top) and 2 (bottom) respectively. The mean saturated conductivity determined using the ring permeameter is also shown for the 0.3 m (■) and 1.05 m (✱) depths

for the same two depths are shown in Figure 5. The close similarity of the K(ψ) curves from the two plots in Figure 4 suggests that, at each depth, a single relationship between K and ψ exists for both plots. This was observed for all depths. However, in Figure 5, an almost parallel behaviour is seen in the K(θ) relationships for the two depths shown. Like the water release data this was seen at all depths and may be caused by spatial variability in the macro/meso-porosity.

The optimised set of parameters for the van Genuchten (1980) model are presented in Table 2, together with the associated standard errors and the mean relative errors for the hydraulic conductivity and water release curves. Also shown, for comparison, are the parameters for sand and clay texture soils taken from van Genuchten *et al.*(1991), summarised from the data of Carsel and Parrish (1988).

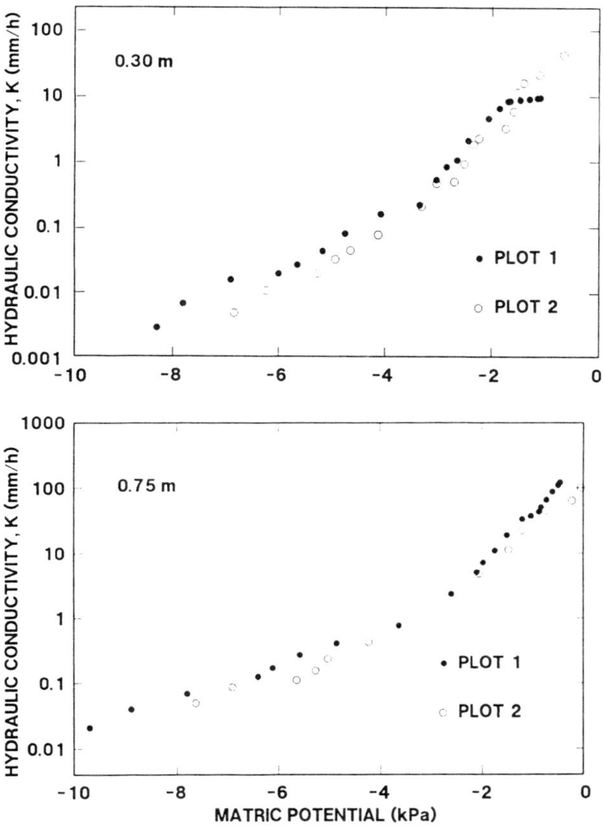

Figure 4 Conductivity curves (K – ψ) from plots 1 and 2, for 0.3 m (top) and 0.75 m (bottom)

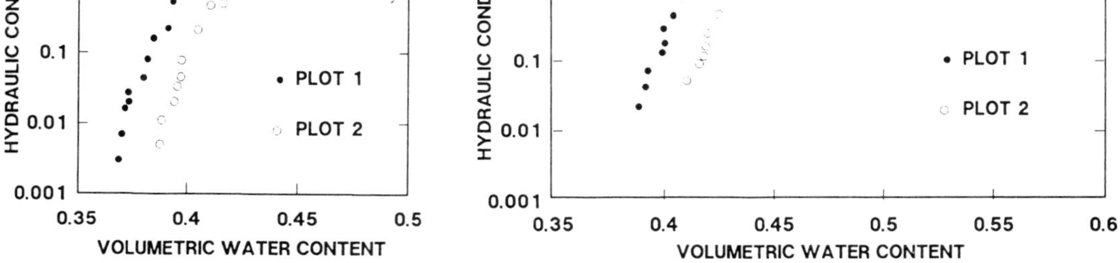

Figure 5 Conductivity curves (K - θ) from plots 1 and 2, for 0.3 m (top) and 0.75 m (bottom)

Table 2 van Genuchten (1980) parameters for the pasture site

Depth (m)	n	$\alpha(cm)^{-1}$-1)	ℓ	θ_s	θ_r	K_s	E(%)
0.30-Pl1	1.50	0.075	5.40	0.448	0.305	546	12.81
0.30-Pl2				0.468	0.325		
0.50-Pl1	1.62	0.097	1.98	0.486	0.279	676	14.24
0.50-Pl2				0.466	0.299		
0.75-Pl1	1.33	0.462	-1.77	0.573	0.304	1096	19.01
0.75-Pl2				0.593	0.324		
1.05-Pl1	1.20	0.664	-4.91	0.565	0.355	422	18.84
1.05-Pl2				0.590	0.380		
1.35-Pl1	1.37	0.020	3.65	0.526	0.420	18	8.50
1.35-Pl2				0.541	0.435		
SAND	2.68	0.145	(0.5)	0.43	0.045	297	
CLAY	1.09	0.008	(0.5)	0.38	0.068	2	

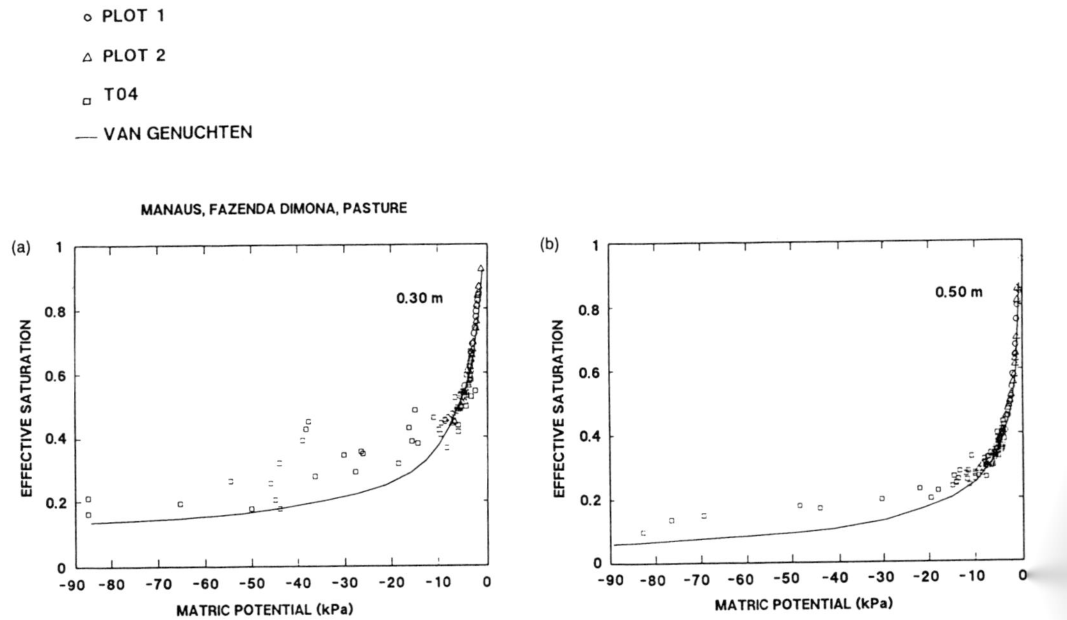

Figure 6 (a-b) Field water release data expressed in terms of effective saturation, and fitted van Genuchten curves for 0.3 m (a), 0.5 m (b).

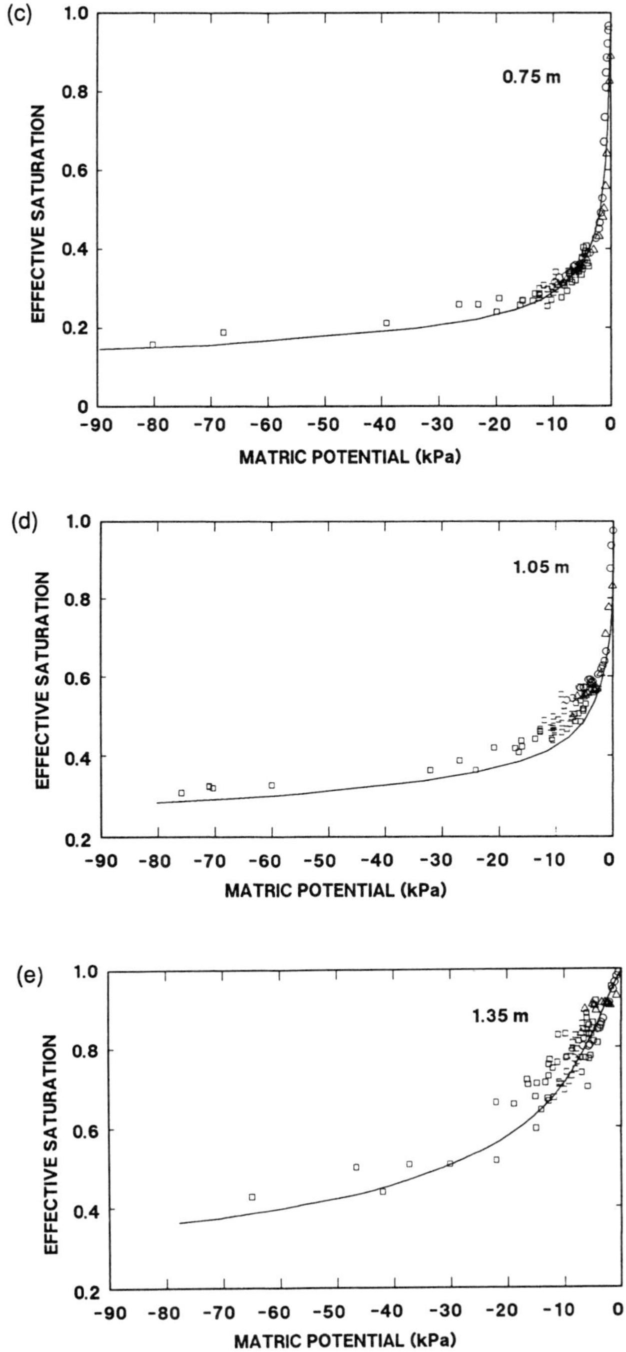

Figure 6 (c-e) Field water release data expressed in terms of effective saturation, and fitted van Genuchten curves for 0.75 m (c), 1.05 m (d) and 1.35 m (e).

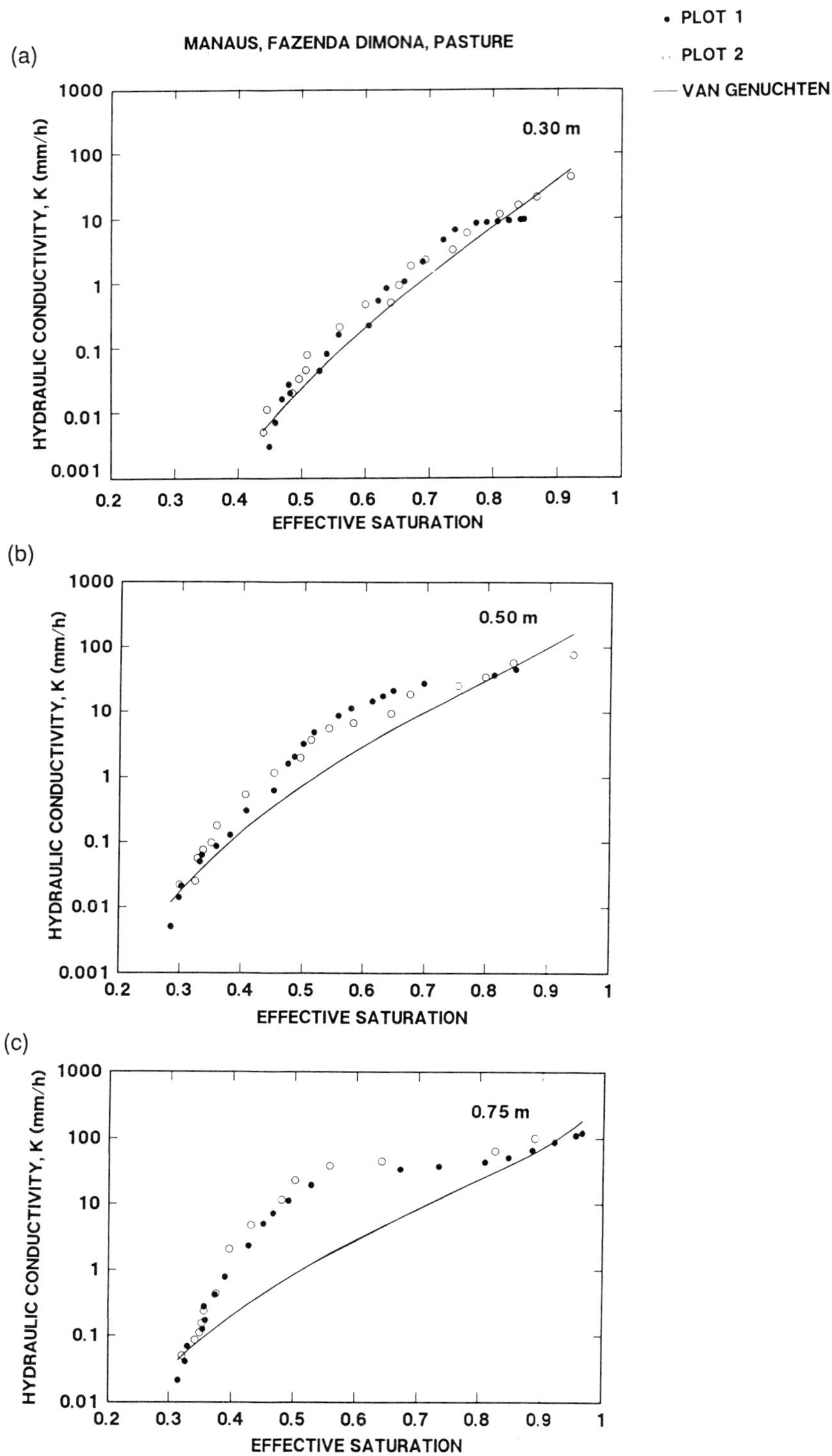

Figure 7 (a-c) Unsaturated hydraulic conductivity data expressed in terms of effective saturation, and fitted van Genuchten curves for 0.3 m (a), 0.5 m (b), 0.75 m (c).

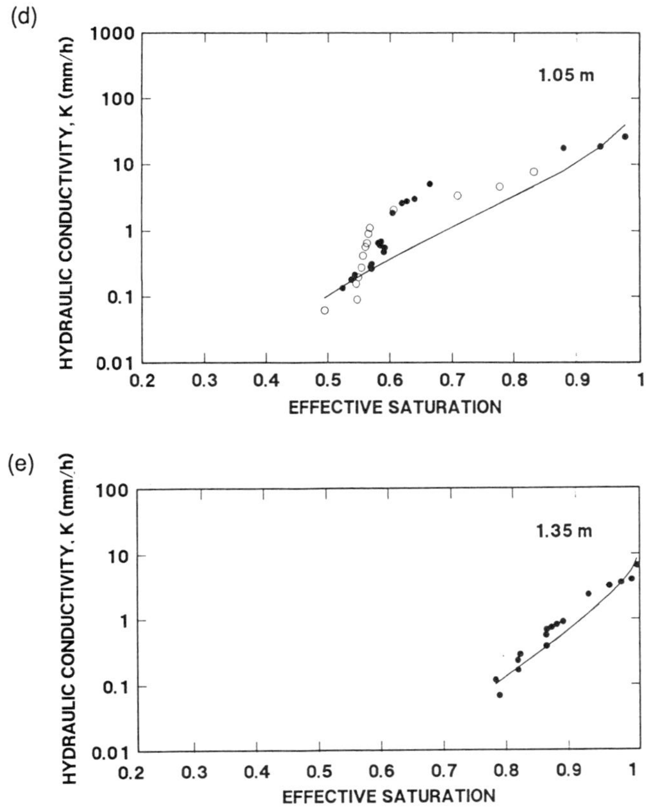

Figure 7 (d-e) Unsaturated hydraulic conductivity data expressed in terms of effective saturation, and fitted van Genuchten curves for 1.05 m (d) and 1.35 m. (e)

Figures 6 and 7 show the fitted water release and conductivity curves respectively, together with the field data, but with the water content expressed as effective saturation. For the water release curves, the results of the optimisation procedure were very good at all depths. For the conductivity curves, the fit was excellent for the 0.3 m depth and became progressively worse with depth before improving at 1.05 m and 1.35 m. The poorness of fit corresponded with the macro/meso-porosity, with the fitted curve being unable to follow the field data as these pores emptied. At 0.75 m, where the fit was poorest, 60% of the effective porosity was drained between 0 and –3 kPa. It was possible to improve the fit to the hydraulic conductivity curves at these depths by using a much higher value of θ_r, but this resulted in poor fits to the water release curve. The fits were also poor for the 0.3 and 1.05 m depths if the value of K_s measured using the permeameter was adopted directly, and not optimised.

DISCUSSION

Comparative measurements

Nortcliff and Thornes (1981), measured the saturated hydraulic conductivity of forested slope soils from Reserva Ducke using laboratory measurements on soil cores. Reserva Ducke (2° 57'S, 59° 57'W) is about 60 km to the south of the study area and the soils studied were more sandy and had a higher bulk density. For the layers between 0.15 m and 0.6 m, and between 0.9 m and 1.15 m depth, the mean saturated hydraulic conductivities were 1567 mm h^{-1} and 217 mm h^{-1}, both of which are much higher than at Fazenda Dimona. Medina and Leite (1985) measured average values of final infiltration at the soil surface in the forest, using a double ring infiltrometer, and obtained a mean rate of 223 mm h^{-1}. The great variability of the data can be explained by the difference in methods used, and the variation of soil physical properties down the slope.

Reliability of IPM data

The IPM relies on the supposition that the flow is essentially vertical and that the material is laterally isotropic. The presence of large macropores, as in these soils, may make this supposition questionable depending of the orientation of the pores (Field *et al.*, 1984). Beven and Germann (1982) suggest that theories based on Darcy's law may not be valid when significant amounts of water are conducted by macropores. Calculations based on Poiseuille's law are in error because water flow in macropores is not laminar (Chen and Wagenet, 1992) and conductivities calculated in this way are often over-estimates (Smettem, 1987). These limitations may lead to errors in the measurements of hydraulic conductivity close to saturation. However, after the macropores have emptied, the inherent supposition of the IPM method is valid and the results can be considered to be reliable.

van Genuchten parameters

The K_s values estimated by the optimisation procedure resulted in a fairly good description of the unsaturated properties of the media. However, the optimised K_s values were almost six and 25 times greater than the values measured at 0.3 m and 1.05 m respectively. The apparently severe over-estimation of K_s by the optimisation procedure may not be a serious problem as, in practice, the hydraulic conductivity near saturation is so high within the upper 0.8 m of the profile that this condition will be attained rarely. There is no evidence of surface runoff in the forest on this soil type (Nortcliff and Thornes, 1981). In 1993 the authors observed directly the effects the very high infiltration rate in the forest. During a rainfall event with an intensity of 84 mm h^{-1}, sustained for 15 minutes, no surface ponding or runoff was observed anywhere on undisturbed soils, but ponding did occur on paths, where the soil had been compacted. In the pasture, surface ponding occurred during and after rainfall events. It appears that the apparently high average surface conductivity of 66 mm h^{-1} is misleading: seven out of the ten surface K_s measurements were less than 50 mm h^{-1} and surface redistribution (small scale surface runoff) occurs from areas where the conductivity is low. However, water content measurements showed that the water lost from high points in the micro-topography infiltrated in lower areas, often only a few metres away (Hodnett *et al.*, 1995). The net effect of this redistribution process may be viewed as an enhancement of the surface hydraulic conductivity (on a larger scale than that of the field measurements). If field measurements of surface hydraulic conductivity (measured over areas of 0.03 m^2 - 0.12 m^2) are to be used in SVAT models the processes of surface redistribution must accounted for by using a surface detention parameter to allow for the temporary storage of water on the soil surface.

The van Genuchten (1980) model requires very high values of K_s to achieve a good fit because the formulation cannot deal with the extreme macropore effects seen in these soils which have 'hybrid' properties. This hybrid behaviour, combining the properties of sandy and clay soils, can be seen by comparing the parameter values from this study with the values for soils of various textures summarised by van Genuchten *et al.* (1991). In some cases, the values found were completely different from the average values for any of the soil textural groups (nb: the soils in this study have a clay content of between 75% and 85% at all depths).

For the soil in this study, the saturated water content (θ_s) varied from 0.45 to 0.59 MVF, compared with 0.38 MVF given for the clay textural category. θ_r was far higher than the published data, ranging from 0.28 MVF to 0.44 MVF compared with the maximum θ_r for any soil texture, of 0.109 MVF.

Although the values of θ_s and θ_r found in this study were much greater than those found in the literature, the values of (θ_s-θ_r) are of greater importance and interest (the absolute water content is of little importance). For the soil in this study, (θ_s-θ_r) ranged from 0.268 MVF at 0.75 m depth to 0.106 MVF at 1.35 m, even though the soil was nominally the same texture throughout the profile. However, at all depths, (θ_s-θ_r) was lower than that calculated for any of the 12 textural groups given by van Genuchten *et al.* (1991), where the lowest value was 0.28 MVF for a sandy clay.

The shape parameter, n, varied between 1.204 at 1.05 m and 1.618 at 0.50 m. These values tended to correspond to the less clayey end of the textural range. The values for the parameter α corresponded to those for sandy loam, or loam texture for the 0.3 m and 0.5 m depths and to clay for the 1.35 m depth. However, for the 0.75 m and 1.05 m depths, the α parameter values were 2–3 times greater than any of the values presented by van Genuchten *et al.* (1991).

Clothier and Smettem (1990) showed that the unsaturated properties of the Manwatu soil befitted its matrix character, but its saturated properties were texturally uncharacteristic and reflected its biogenic macroporosity. This is the case for the soils in this study, although it appears that the matrix behaviour, although clay-like, is not typical of (temperate) clay soils, with a field water content range (after macro and meso-pore drainage) of only 0.05 - 0.10 MVF. The data presented by Hodnett *et al.* (1996) strongly indicate that, at depths below 2 m, the available water may be less than 0.05 MVF. This is presumed to be a result of mineralogy (kaolinite and sesquioxides of iron and aluminium).

For soils with biogenic macroporosity, Clothier and Smettem (1990) strongly rather than the use of simpler methods based on predictions from more easily measured soil properties such as particle size distribution (which take no account of soil structure). The hybrid character of the soils also means that models such as that of van Genuchten (1980) may not be suitable because they were formulated to deal with a uni-modal pore size distribution. Recently, Othmer *et al.* (1991), Ross and Smettem (1993) and Durner (1994) have proposed formulations for use in soils with heterogeneous porous systems and these will probably produce better results in all layers.

It is clear that the application of published average parameters for the clay textural class to these Amazonian soils will lead to a large over-estimation of their water availability and will seriously mis-represent their hydraulic properties. Sanchez (1976) described oxisols as acting like sands in terms of water movement at low tensions but holding water like clays at higher tensions.

Origin of macroporosity

The macro- and meso-pores are responsible for the very high hydraulic conductivities in the range from 0 to -3 kPa. According to Chauvel *et al.* (1987b), the macro- and meso-porosity is the result of the intense biological activity under the original forest. The main agents of macropore creation are ants, termites, worms and roots, and their activity and relative importance varies with depth. Below 1.4 m depth, faunal activity is minimal and the porosity essentially consists of very fine fissures which incompletely separate micro-aggregates which are created by shrinkage and swelling and by the processes of formation of the clay minerals themselves. It is notable that, at this site, the depth at which the macropore effects became minimal (1.35 m) was very similar to the 1.4 m observed by Chauvel *et al.* (1987b).

For the pasture it appears, from the high saturated hydraulic conductivities and from the large macropore volumes indicated by the water release data, that much of the soil structure created over many thousands of years beneath forest is fairly stable and remains intact after clearance. Mechanical forest clearance using bulldozers affects the hydraulic properties to a depth of 0.3 - 0.4 m (Dias and Nortcliff, 1985) but at this pasture site clearance was carried out by cutting and burning only. It is expected that the forest and pasture soil properties differ significantly only in the top 0.1 - 0.15 m.

CONCLUSIONS

The mean saturated hydraulic conductivity (K_s) measured at the surface in the pasture using the ring permeameter was 66 mm h^{-1}, but ranged from 18 to 220 mm h^{-1}. At 0.3 m, K_s was 97 ±46 mm h^{-1} and at 1.05 m it was 17 ±7 mm h^{-1}. Although not measured directly, the saturated conductivity at 0.5 m and 0.75 m is expected to be higher in view of the greater macroporosity at these depths.

The hydraulic properties of the pasture soils varied markedly with depth, mainly due to vertical changes in the macro and meso-porosity. This is largely of biogenic origin (ants, termites, worms and roots) and remains unchanged after the forest is cleared. Although the hydraulic conductivity of the forest soil was not measured, it appears from the water release data and by inference from published data, that the hydraulic properties of the forest and pasture soils differ significantly only in the top 1 - 0.15 m. Despite the limitations caused by macropores, the IPM method proved to be suitable for measuring the hydrologically important part of the unsaturated hydraulic conductivity curve in these unusual soils. Ideally a combination of techniques should be used to determine the macro-, meso- and micro-porous sections of the curve.

The hydraulic conductivity is high at, and close to, saturation because of the influence of the macro- and large meso-pores, which are particularly important between 0.4 and 1.1 m depth. At 0.75 m, where the effects were strongest, 60% of the total effective porosity emptied between 0 and -3 kPa. Below about 1.25 m, the macro/meso-pore effects appear to be negligible and the soil physical properties are more dominated by the particle size distribution. K_s is significantly lower (between 10 and 20 mm h^{-1}), but the unsaturated conductivity at a matric potential of -6 kPa for example, is about ten times higher than in the layers above.

The values of K_s produced by the optimisation procedure to obtain van Genuchten (1980) parameters from IPM and other field data were much higher than the measured values. However, this apparent over-estimation is unlikely to cause a serious problem in modelling applications: saturated conditions are unlikely to occur often because the measured values of K_s are so high. However, extrapolations close to saturation must be treated with caution due to the inability of the van Genuchten (1980) formulation to handle the bi-modal pore size distribution of this soil. Models which have been developed to deal with multimodal pore size distributions would be able to produce better results in all layers.

The optimised parameters resulted in good fits for the water release curves at all depths. For the conductivity curves, the results were best in the layers with the least macro/meso-pore effects. The parameter values were different from those found in the literature for clay soils and it is suggested that the use of parameters derived for soils from temperate areas are likely to result in erroneous predictions of soil water behaviour in these tropical clay soils. It is notable that the values of (θ_s - θ_r) for all depths in this study were less than for any of the 12 soil textural classes presented by van Genuchten *et al.* (1991).

Oxisols similar to those examined in this study are widespread in Amazonia. For GCM modelling purposes, the use of the parameters presented here will give a better representation of the properties of oxisols than the use of parameters from the soils of temperate regions. It is clear that more research is required to improve the representation of oxisols and, more importantly, to investigate the properties of the other soil types which are common in Amazonia. It is also important to obtain the soil hydraulic parameters to depths of 4 m or more in materials not generally regarded as soil (Hodnett *et al.*,1996).

REFERENCES

Beven, K., and Germann, P. 1982. Macropores and water flow in soils. *Wat. Resour. Res.*, **18**, 1311-1325.

Brewer, R. 1964. *Fabric and mineral analysis of soils.* John Wiley, New York.

Brooks, R. H. and Corey, A. T. 1964. Hydraulic properties of porous media. Hydrol. Paper 3, 27 pp., Colorado State University, Fort Collins.

Carsel, R.F. and Parrish, R.S. 1988. Developing joint probability distributions of soil water retention characteristics. *Wat. Resour. Res.*, **24**, 755-769.

Chauvel, A. 1982. Os latossolos amarelos, álicos, argilosos dentro dos ecossistemas das bacias experimentais do INPA e da região vizinha. *Acta Amazonica* **12**, 47-60.

Chauvel, A., Guillaumet J-L. and Schubart, H.O.R. 1987b. Importance et distribution des racines et des êtres vivants dans un latossol argileux sous forêt amazonienne. *Revue d'Ecologie et de Biologie du Sol,* **24**, 19-48.

Chauvel, A., Grimaldi, M. and Tessier, D. 1991. Changes in soil pore space distribution following deforestation and revegetation: an example from the Central Amazon Basin, Brazil. *For. Ecol. & Manage.*, **38**, 259-271.

Chen C., and Wagenet, R.J. 1992. Simulation of water and chemicals in macropore soils. Part 1. Representation of the equivalent macropore influence and its effect on soil-water flow. *J. Hydrol.* **130**, 105-126.

Clothier, B.E. and Smettem, K.R.J. 1990. Combining laboratory and field measurements to define the hydraulic properties of soil. *Soil Sci. Soc. Am. J.*, **54**,299-304.

Correa, J.C. 1984. Características físico hídricas dos solos latossolo amarelo, podzólico vermelho-amarelo e podzol hidromórfico do estado do Amazonas. *Pesquisa Agropecuaria Brasileira* **19**, 347-360.

Correa, J.C. 1985. Efeito de métodos de cultivo em algumas propriedades físicas de um latossolo amarelo muito argiloso do estado do Amazonas. *Pesquisa Agropecuaria Brasileira*, **20**, 1317-1322.

Dias, A.C.C.P. and Nortcliff, S. 1985. Effects of two land clearing methods on the physical properties of an Oxisol in the Brazilian Amazon. *Trop. Agric. (Trinidad)*, **62**, 202-212.

Durner, W. 1994. Hydraulic conductivity estimation for soils with heterogeneous porous structure, *Wat. Resour. Res.*, **30**, 211-223.

Field, J. A., Parker, J.C. and Powell, N.L. 1984. Comparisions of field and laboratory measured and predictive hydraulic properties of a soil with macropores. *Soil Sci.*, **138**, 385-396.

Hodnett, M.G., da Silva, L.P. da Rocha H.R. and Cruz Senna, R. 1995. Seasonal soil water storage changes beneath central Amazonian rainforest and pasture. *J. Hydrol.*, **170**, 233-254.

Hodnett, M.G., Oyama M.D. and Tomasella, J. 1996. Comparisons of long-term soil water storage under pasture and forest in three areas of Amazonia. This publication.

Kool, J.B., Parker, J.C. and van Genuchten, M. Th. 1985. Determining soil hydraulic propertiesfrom the one-step outflow experiments by parameter estimation: I - Theory and numerical studies. *Soil Sci. Soc. Am. J.*, **49**, 1348-1354.

Luxmoore, R.J. 1981. Micro-, meso-, and macroporosity of soil. *Soil Sci. Soc. Am. J.*, **45**, 671-672.

Medina, B.F. and Leite, J.A. 1985. Influência de trés sistemas de manejo e duas coberturas vegetais na infiltração de água em um latossolo amarelo em Manaus - AM. *Pesquisas Agropecuarias Brasileiras, Brasilia,* **20**, 1323- 1331.

Mualem, Y. 1976. A new model for predicting the hydraulic conductivity of unsaturated porous media. *Wat. Resour. Res.*, **12**, 513-522.

Nortcliff, S. and Thornes, J.B. 1981. Seasonal variation in the hydrology of a small forested catchment near Manaus, Amazonas, and the implications for its management. In: *Tropical Agricultural Hydrology*, (Eds: R. Lal and E. W. Russell). John Wiley. 37-57.

Othmer, H., Diekkrüger, B. and Kutilek, M. 1991. Bimodal porosity and unsaturated hydraulic conductivity. *Soil Sci.*, **152**, 139-150.

Perroux, K.M. and White, I. 1988. Designs for disc permeameters. *Soil Sci. Am. J.*, **52**, 1205-1215.

Philip, J.R. 1969. Theory of infiltration. *Adv. Hydrosci.*, **5**, 215-296.

Ranzani, G. 1980. Identificação e caracterização de alguns solos da Estação Experimental de Silvicultura Tropical do INPA. *Acta Amazonica*, **10**, 7-41.

Rosenbrock, H.H. 1960. An automatic method for finding the greatest or least value of a function. *Computer J.*, **3**, 175-184.

Ross, P.J. and Smettem, K.R.J. 1993. Describing soil hydraulic properties with sums of simple functions. *Soil Sci. Am. J.*, **57**,26-2

Russo, D. 1988. Determining soil hydraulic properties by parameter estimation: On the selection of a model for hydraulic properties. *Wat. Resour. Res.*, **24**, 453-459.

Sanchez, P.A. 1976. *Properties and management of soils in the tropics*. John Wiley, New York.

Sellers, P.J., Mintz, Y., Sud, Y. and Dalcher, A. 1986. A simple biosphere model (SiB) for use within general circulation model. *J. Atmos. Sci.*, **43**, 506-531.

Smettem, K.R. 1987. Characterization of water entry into a soil with a contrasting textural class: Spatial variability of infiltration parameters and influence of macroporosity. *Soil Sci.*, **144**, 167-174.

Stephens, D.B. and Rehfeldt, K.R. 1985. Evaluation of closed-form analytical models to calculate conductivity in a fine sand. *Soil Sci. Soc. Am. J.*, **49**, 12-19.

van Genuchten, M. Th. 1980. A closed form equation for predicting hydraulic conductivity in unsaturated soils. *Soil Sci. Soc. Am. J.*, **44**, 892-898.

van Genuchten, M. Th., Leij, F.J. and Yates, S.R. 1991. The RETC code for quantifying the hydraulic functions of unsaturated soils. Report EPA/600/2-91/065. U.S. Environmental Protection Agency, Ada, Oklahoma, USA.

Verseghy, D.L. 1991. CLASS - A Canadian land surface scheme for GCMS. I. Soil Model. *Int. J. Climatol.*, **11**, 111-133.

Watson, K.K. 1966. An instantaneous profile method for determining the hydraulic conductivity of unsaturated porous materials. *Wat. Resour. Res.*, **2**, 709-715.

White, I. and Sully, M.J. 1987. Macroscopic and microscopic capillary length and time scale for field infiltration. *Wat. Resour. Res.*, **23**, 1514-1522.

Wooding, R.A. 1968. Steady infiltration from a shallow circular pond. *Wat. Resour. Res.*, **4**, 1259-1273.

Wösten J.H.M. and van Genuchten, M. Th. 1988. Using texture and other soil properties to predict the unsaturated soil hydraulic functions. *Soil Sci. Soc. Am. J.*, **52**, 1762-1770.

Wright, I.R., Gash, J.H.C., da Rocha, H.R., Shuttleworth, W.J., Nobre, C.A., Maitelli, G.T., Zamparoni, C.A.G.P. and Carvalho, P.R.A. 1992. Dry season micrometerology of central Amazonian ranchland, *Q. J. Roy. Met. Soc.*, **118**, 1083-1099.

RESUMO

As curvas de retenção e de condutividade hidráulica de um oxissol (75% - 85% de teor de argila) em uma fazenda agropecuária na Amazônia Central foram medidas *in situ*. A condutividade hidráulica saturada foi medida em 3 profundidades utilizando um permeâmetro anular; a condutividade não-saturada e a curva de retenção foram medidas em 5 profundidades, utilizando o método do perfil instantâneo (MPI). Este método produziu bons resultados, porém para um intervalo limitado de potencial matricial, pois a drenagem virtualmente cessou no potencial de -10 kPa (após 10 dias). A curva de retenção até o potencial matricial de -80 kPa foi determinada a partir das medidas de longo prazo de teor de umidade e potencial em um ponto não modificado na pastagem. O limite de retirada de umidade foi obtido a partir do mesmo conjunto de dados de teor de umidade.

A condutividade hidráulica saturada (K_s) variou fortemente com a profundidade. K_s foi maior na profundidade de 0,30 m (97 mm h^{-1}) que na superfície (66 mm h^{-1}), onde ocorreu compactação pelo gado, e o mínimo ocorreu na profundidade de 1,05 m (17 mm h^{-1}). Os altos valores de K_s resultam da intensa macroporosidade, particularmente entre 0,40 e 1,10 m. Uma importante característica das curvas foi a acentuada diminuição da condutividade à medida que os poros se esvaziaram, entre 0 e -3 kPa. Os efeitos dos macroporos diminuíram com a profundidade, tornando-se praticamente desprezíveis abaixo de 1,30 m.

As curvas de retenção e condutividade não-saturada medidas utilizando o MPI foram ajustadas ao modelo de Van Genuchten (1980). Afim de garantir a melhor descrição possível das curvas para todo intervalo de umidade observado no campo (além dos medidos pelo MPI), os dados da curva de retenção obtida no campo e os valores do limite de retirada foram incluídos na otimização. Os valores dos parâmetros de Van Genuchten para 5 profundidades são apresentados. As curvas de retenção foram bem descritas em todas as profundidades. Os ajustes das curvas de condutividade foram bons onde os efeitos de macroporosidade foram menores, mas foram menos satisfatórios nas camadas com forte efeito de macroporosidade. Os parâmetros de Van Genuchten obtidos foram bastante diferentes dos valores publicados para solos argilosos. O uso de tais valores de parâmetros (geralmente obtidos em solos de regiões temperadas) poderia superestimar a disponibilidade de água e subestimar K_s, e não deveria ser aplicado para esses solos tropicais caoliníticos micro-agregados.

6 Temperature and moisture profiles in soil beneath forest and pasture areas in eastern Amazonia

J.R.S. de SOUZA[1], F.M.A. PINHEIRO[1], R.L.C. de ARAUJO[2], H.S. PINHEIRO JR.[1] and M. G. HODNETT[3]

[1]*Federal University of Para, Belém, Brazil*

[2]*Federal University of Amazonas, Manaus, Brazil*

[3]*Institute of Hydrology, Wallingford, UK*

INTRODUCTION

This work is part of a study on the environmental impacts of deforestation on the microclimate in the Amazon region. It presents the results of monitoring soil moisture and temperature at depths of up to 4 metres. At this depth, the rooting densities of most native plants are expected to be small. This so-called 'weathered zone', plays an important role in solar heat storage and the energy balance at the surface. On the other hand, its moisture content is constantly changing through precipitation and evaporation processes. This soil moisture variation lead to changes in the thermal properties of this layer since the specific heat capacity of the water is much greater than that of the solid fraction of the soil.

The removal of the Amazonian rainforest and its replacement by pasture, substantially alters the microclimate at the surface, and it is necessary to determine the impact of such practice on the regional soils. This has motivated this study of the 0–4 m depth zone of soils beneath areas of contasting surface coverage.

Most soil temperature studies have been carried out in agricultural situations and are limited down to a depth of 1 m (de Vries, 1975; Ribeiro & Santos, 1975; Ghuman & Lal, 1982; Santos *et al.*, 1988). Conventional thermal probes, on the other hand, seek to determine deep geothermal flow and avoid layers influenced by solar heating; they usually log at depths greater than 30 metres. The work reported here extends the soil temperature depth zone of study, adapting the thermal sounding to this layer of significant temperature and moisture variability.

In this paper we will compare results from the two ABRACOS sites in the municipality of Marabá (Pa). The first is an area of primary forest which has been subjected to little alteration by human activity. It is part of the Reserva Companhia Vale do Rio Doce (CVRD) and is situated some 65 km south of Marabá. The second is Fazenda Boa Sorte, a deforested area that was converted into pasture about 14

Amazonian deforestation and climate. Edited by J.H.C. Gash, C.A. Nobre, J.M. Roberts and R.L. Victoria.

years ago. It is currently used for cattle-raising and is located 35 km north of Marabá, along the PA-222 highway.

In general, the uppermost layers of Amazonian topsoil are porous and thus permit rapid infiltration of rainfall. In deforested areas, however, soil compaction may occur, leading to greater run-off and also elevation of surface temperature due to increased exposure to solar radiation (Souza *et al.*, 1993). This process of heating of the soil, lowering of groundwater level and loss of porosity of Amazonian soils constitutes an imbalance that can lead to the formation of blocks of lateritic rock, thus making the soils useless for agricultural and other economic purposes. This study of shallow temperature and moisture profiles also seeks to contribute to wider environmental studies on the surface energy balance (Shuttleworth *et al.* 1984) for the two types of groundcover vegetation in Amazonia.

MATERIALS AND METHODS

The pasture and forest soils were classified respectively as a medium to clayey textured red/yellow podzolic soil and a medium textured yellowish cambisol. In the pasture the texture at the surface was sandy loam, becoming a sandy clay at depth. In the forest, the soil was more silty throughout and had a high stone content in the upper 0.6 m. Below this depth, the texture was silt loam to silt,but the material was saprolite (heavily weathered rock). More detailed descriptions are given by Hodnett *et al.* (1996).

To evaluate temperature and moisture variability in the soil at the two selected sites, measurements were made weekly during the period from August 1991 to September 1993. In all, a total of 131 temperature and 81 moisture records were analysed.

Heat transmission in soil depends both on its temperature gradient and its thermal conductivity, the latter being intrinsically linked to its moisture content (Decico, 1974; Conaway & Beck, 1977; Lee, 1977; Marangoni & Hamza, 1983). For this reason the present paper analyses the bi-monthly averages and highlights the quarters of extreme precipitation, both rainy and dry.

In general the groundwater level in Amazonia can vary from the surface to a depth of tens of metres depending on the time of year, i.e. on its precipitation gains and evaporation losses (Souza *et al.*, 1984). Soil moisture also depends on soil physical-chemical characteristics and vegetation cover, even in regions otherwise lithologically similar.

Besides specifically monitoring soil moisture and temperature, the ABRACOS experiment provided simultaneous precipitation and solar radiation data collected from micro-meteorological towers installed next to the forest and pasture experimental soil plots.

Temperature profiles were obtained by means of a thermistor probe along PVC-lined boreholes filled with water. This method of utilizing a thermistor probe is quite well known (Beck, 1965; Raspet *et al.*, 1966; Robertson *et al.*, 1966) and it is possible to obtain a high degree of precision, of approximately ±0.05 °C (Krcmar & Masin,

1970). When using thermistor probes for shallow temperature profiles, special care was taken with the time interval necessary for thermal balance between the water in the borehole and the subsoil, as well as between this same fluid and the probe thermistor, at each measurement depth (Araújo *et al.*, 1985).

The holes for temperature profile measurement were drilled with with an auger operating through a 45 mm diameter steel cutting tube. Samples were collected over 0.5 m intervals and transported from the field to Belém in airtight plastic bags to prevent loss of humidity. Alumimium tubing of 45 mm diameter was used to line the holes at both sites. The tubes were sealed at their lower ends and cut off almost level with the soil surface. They were closed at the top, except when they were being used during the logging. To facilitate lateral heat transmission from the surrounding medium, the tube was maintained constantly full of water. Temperature measurements were made at depth intervals of 0.2 m, between 0.2 and 3.8 m. The measurements at the two locations were not carried out on the same day because of the large distance between them. Nevertheles, efforts were made to maintain a routine of measurements on successive days, carrying out periodical comparative monitoring on a weekly basis at both sites.

The samples obtained during installation were weighed in their field state, and after oven drying, to obtain their original moisture content. The samples were also submitted to a thermal conductivity test in their humid state by means of a needle system, similar to that described by Von Herzen & Maxwell (1959), Langseth (1965), Smith (1973), Carvalho (1986) and Souza *et al.* (1989). These samples were also described lithologically at the laboratories of the Center for Geosciences at UFPa. Thermal conductivities of the collected samples were measured in the laboratory and ranged from 1.16 to 2.91W m^{-1} $^{o}C^{-1}$.

For the soil water content measurements, six access tubes were installed using the same method as for the thermal logging tubes. Each hole was lined with a 45 mm diameter aluminium tube which allowed measurements of water content to be made to a depth of 3.6 m. The tubes were within an area about 50 m square in the forest and 10 m square in the pasture. Weekly moisture measurements were taken at the following levels: 0.1, 0.2, 0.4, 0.6, 0.8, 1.0, 1.2, 1.5, 1.8, 2.1, 2.4, 2.7, 3.0, 3.3 and 3.6 m using a Didcot Instruments IH2 neutron probe. The methods of calibration and data processing are described by Hodnett *et al.* (1996a). Water contents were averaged at each measurement depth in the six tubes and then expressed as means for successive 0.6 m layers.

Analysis of the lithological samples collected during the drilling, was carried out as described by Souza *et al.* (1989). An automatic data collection system for precipitation (Didcot Instruments tipping bucket raingauge) and solar radiation flux (Kipp & Zonen pyranometer) was used. This system integrates accumulated precipitation and radiation for each hour. These hourly values were used to calculate the daily averages used in the analysis of the soil temperature and moisture profiles.

DISCUSSION OF THE RESULTS

SOIL TEMPERATURE

The results of soil temperature monitoring in both forest and pasture are presented in Figures 1 to 3, which show the bi-monthly averages of the temperature profiles for

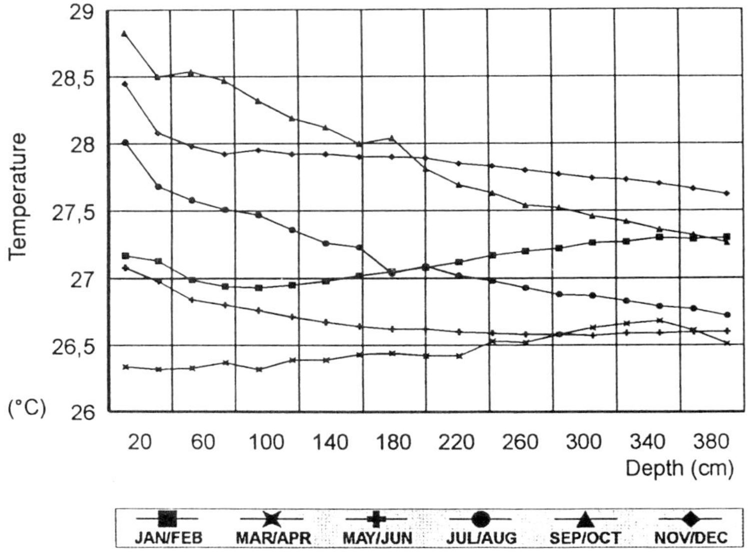

Figure 1 Bi-monthly mean soil temperature profiles beneath the pasture

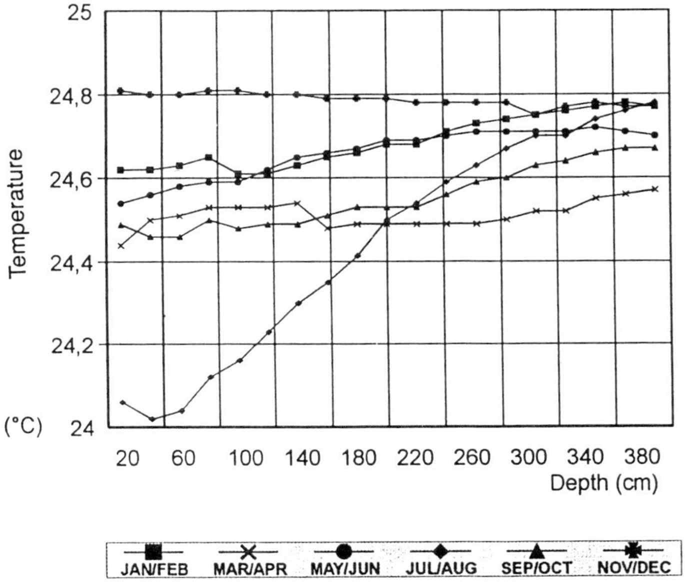

Figure 2 Bi-monthly mean soil temperature profiles beneath the forest

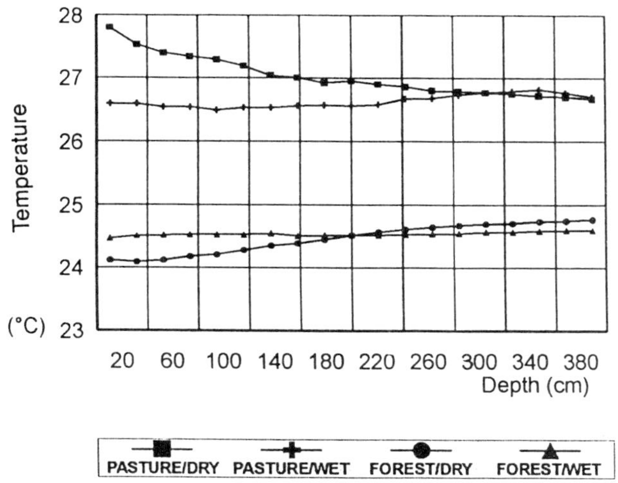

Figure 3 Mean temperature profiles for the dry and wet quarters for both sites

both pasture and forest. Figure 1 shows that for the pasture, in the layers closest to the surface (to depths of approximately 2.0 m), the profile with the highest temperatures occurred during the two-month period September-October and the lowest temperatures occurred during March-April. The temperature extremes of the profiles were 28.8 °C and 26.3 °C at 0.2 m and between 27.6 °C and 26.5 °C at a depth of 3.8 m. Figure 4 shows that the variations of the incoming solar radiation flux, measured during the experiment, and the local sunshine brightness normals at the surface, have coherent behaviour. These variations determine the thermal regimes of

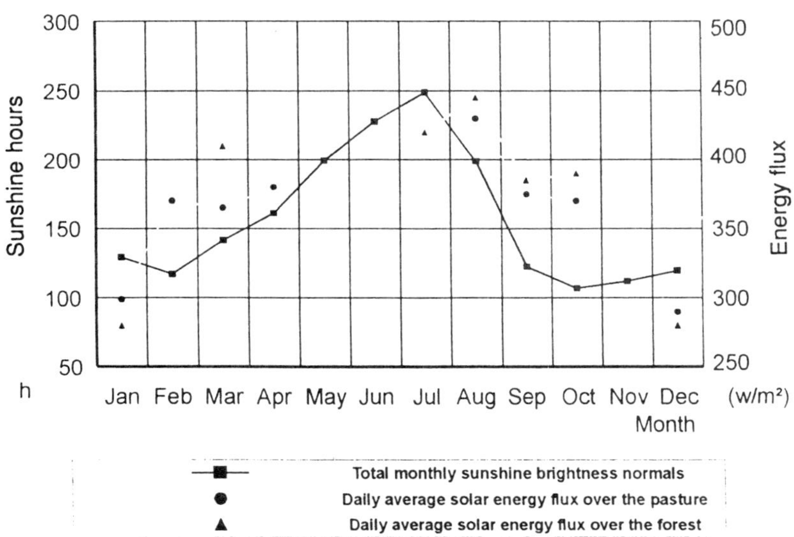

Figure 4 Solar radiation at Marabá.

the soils as demonstrated in the dislocation sequence of the temperature profiles shown in Figure 1. It can be seen that the reaction of the thermal response of the pasture soil presents a delay of approximately two months in relation to the radiation maxima and minima.

Figure 2 shows the results for the forest where, as expected, the ampltude of temperature variation was much lower than in the pasture. The bi-monthly average profiles varied between 24.1 °C and 24.8 °C at 0.2 m, and between 24.6 °C and 24.8 °C at a depth of 3.8 m. Contrary to what occurs in the pasture, the annual temperature variation underneath the forest is greatly attenuated by depth.

At the forest site the profile with the highest temperatures occurred in November-December. The profile with the lowest temperatures, for the top two metres of soil, occurred in July-August. If one considers the thermal regime observed in the forest soil in relation to the variation of solar radiation at the surface, its thermal response shows a delay of approximately five months with respect to the maxima and minima of the heat source.

The delays in the thermal responses of the soils may be explained in both cases in terms of the water stored during earlier rainy months. As long as there is abundant moisture to be evaporated at the surface, the soils remain cool. During the months of July and August there is a combination of high solar energy flux and low moisture content in the soils, which leads to their temperature rise.The longer delay of the thermal response of the forest soil must be accounted for by its lower albedo, larger energy storage in the biomass and other physiological differences with respect to the pasture.

Interestingly, there is an important drop in forest soil temperature in July-August, despite the high rate of radiation incident on the surface vegetation.The radiation which penetrates the forest canopy to the surface is strongly attenuated by the vegetation (Shuttleworth *et al.*, 1984) and less than 10% of this total radiation reaches the surface. In addition, precipitation during this period was insignificant (Figure 5). Our conclusion is, therefore, that increase in temperature under the forest surface is principally due to rainfall which transports heat from the canopy to the soil surface. Moreover, the existence of an adequate supply of moisture in the soil propitiates its increased thermal conductivity (Araújo*et al.*, 1995). This explains the almost uniform temperatures in these profiles from the surface down to 3.8 m during each two-monthly period.

Figure 3 summarizes the seasonal behavior of the thermal regimes of the pasture and forest soils. Here, the profile averages during the extreme quarters are presented in relation to rainfall. It is interesting to note the significant difference between the average temperatures in the pasture and forest: roughly 3.7 °C during the dry and 2.1 °C during the rainy period for both at a depth of 0.2 m. At 3.8 m however, the difference remains stable at 2.0 °C, regardless of season. It should be noted that the temperature gradient during the dry season is negative at the pasture and positive at the forest site. This indicates that the average heat flow is downward under the pasture and upward towards the atmosphere under the forest. Considering the very small temperature gradients during the rainy season at both sites, their heat flows during

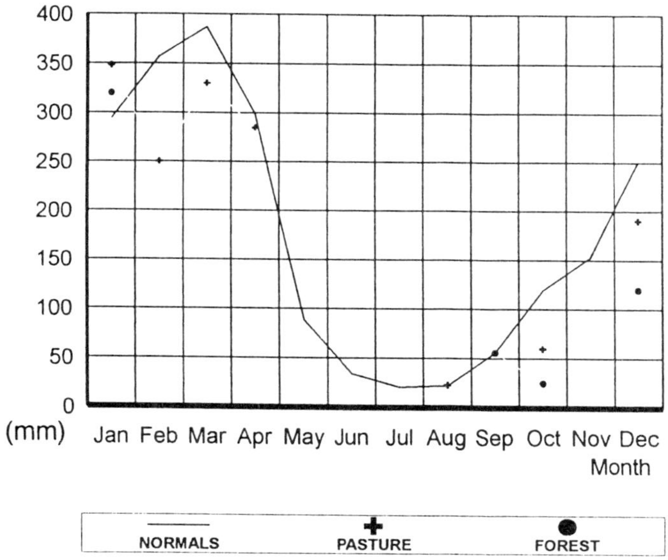

Figure 5 Monthly rainfall totals at Marabá

this period are minimal. At depths above 2.0 m, however, temperature under the pasture drops while under the forest, as has already been reported, it rises.

SOIL MOISTURE

Figures 6 and 7 present the bi-monthly averages (over two years) of the soil moisture profiles under pasture and forest respectively. These show the mean soil water content

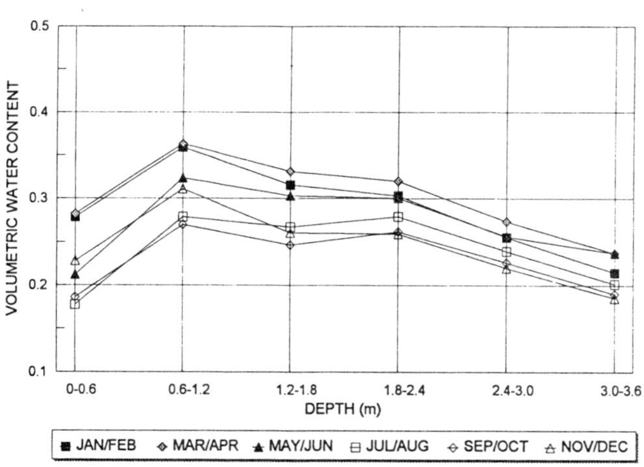

Figure 6 Bi-monthly mean soil water content profiles beneath the pasture (September 1991 - August 1993)

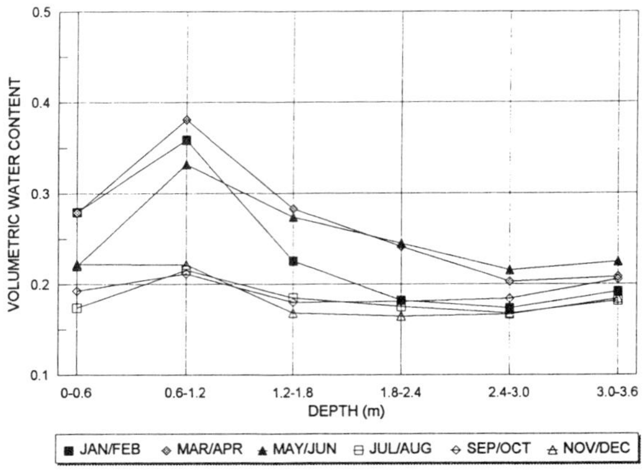

Figure 7 Bi-monthly mean soil water content profiles beneath the forest (September 1991 - August 1993)

conditions in the profile corresponding to the periods for the temperature data shown earlier. These two figures clearly show the tendency of soil moisture in each profile to reach its maximum in the layer between 0.6 and 1.2 m.

At the pasture site (Figure 6), the maximum bi-monthly average water content was during March-April. The minimum bi-monthly average water content was during September-October for the profile above 1.8 m and in November-December for the profile below. At the forest site (Figure 7), the maximum bi-monthly soil water content was also during March-April. In the top layer, the minimum was in July-August, but in the layer from 1.2 - 1.8 m and below, it was also in November-December. These data suggest that moisture gains at both sites react quickly to increases in precipitation during the wettest months (January to April). Moisture minima occur with a delay of roughly four months at the surface. Below 2.4 m, the delay is about eight months, giving a four- month recharge period and and an eight-month depletion period. The data also indicated that the rainfall infiltrated into all layers, implying that these soils have not lost their porosity through compaction. The pasture area therefore remains permeable 15 years after the forest removal.
Comparison of the pasture and forest data, expressed as bi-monthly means over the two years studied, indicates that there were larger changes of water content (drying from the March-April wettest conditions) in the forest than the pasture, except below 2.4 m. Here, there were slightly greater water content changes beneath the pasture than beneath the forest. It is not known if this condition is typical. The pattern of soil water content changes in the pasture was very similar in the two years studied but there was a large difference between the years in the forest. The bi-monthly averages for the forest for 1991/2 and 1992/3 are shown in Figure 8 to illustrate this. It can be seen that the changes of water content were much larger in the first year because the maximum water contents (in March/April) were much higher throughout the profile. The profile was wetted to 3.6 m and below in the 1991/2 wet season,but in the second

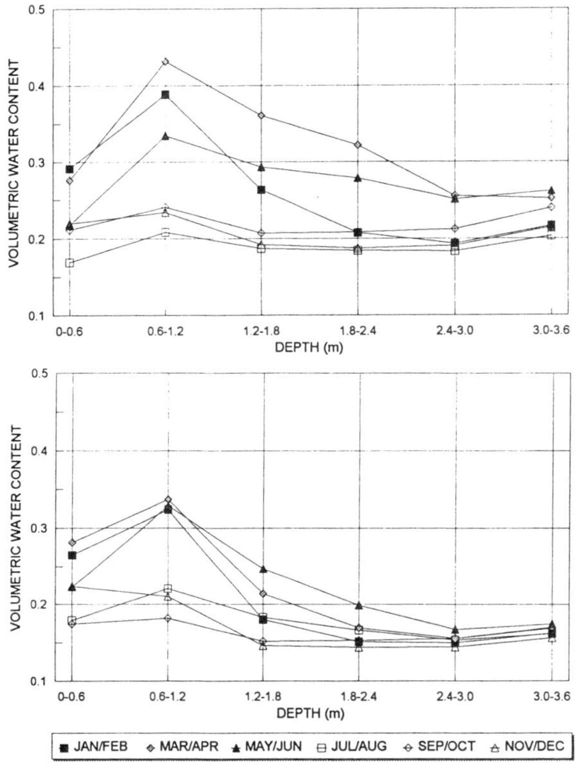

Figure 8 Bi-monthly mean soil water content profiles beneath the forest, September 1991 - August 1992 (top) and September 1992 - August 1993 (bottom)

year, the profile did not wet up much below a depth of 2 m and the water content changes in the lower profile were small because there was much less water available for abstraction.

The forest data for 1991/2 (Figure 8 top) indicate that the water content changes will be much larger in the forest than the pasture throughout the profile, if the entire profile is rewetted in the wet season. Further studies of the type described by Hodnett *et al.* (1996b) are required to identify how frequently the forest profile does not re-wet completely, leaving a soil moisture deficit to be carried over from one year to the next. There are strong indications that the forest is capable of abstracting water from much greater depths and it probably does so under these conditions. This point is discussed in greater detail by Hodnett *et al.* (1996a). If the lack of full wetting of the profile is a relatively uncommon occurrence, then long-term bi-monthly mean data would indicate larger changes beneath forest, compared to pasture. It is also of note that the use of bi-monthly means to express the soil water content data leads to an under-indication of the changes which actually occurred.

As shown in Figure 5, the precipitation regime during the period studied is typical of normal climatological behaviour for the area of Marabá (DNMET, 1992). Average precipitation at both sites was slightly below normal, except for January. In general, rainfall over the pasture was greater than over the forest but there are parts of the rainfall records missing at key times. It is not possible at this stage to determine whether the difference in soil water response between the sites was due to the differences in rainfall or evapotranspiration. During the period of this experiment the dry quarters had an average rainfall rate of 1.26 mm day^{-1} and during the rainy quarters that rate was 7.56 mm day^{-1}. Considering that these precipitation values are measured above the vegetation, the relative moisture deficiency at the forest site is partly due to the interception of rainfall and partly due to the smaller albedo of the forest and consequently its increased evapotranspiration (Silva*et al.*, 1992; Fisch*et al.*, 1992).

During the rainy season the average quarterly temperature at depths above 3 m fell under the pasture and rose under the forest. In the pasture, this behaviour may be explained in terms of an increase in the latent heat flux from the pasture to the atmosphere. In the forest, the heat gain is possibly due to the transportation of thermal energy in the infiltrating water, although the very small temperature changes suggest that rainfall does not play an important role. However, it may be that the temperature of the rainwater is close to the temperature of the forest soil so that little heat transfer occurs. If the rainwater temperature is typically about 25°C, for example, it would heat the forest soil and cool the pasture. It is of note that in the pasture, the temperatures throughout the profile reach a minimum when the profile is at its wettest, in March-April. The temperature changes at depth may indeed reflect the water flux through the profile.

CONCLUSION

Although restricted to a period of only 26 months, the observations permit an evaluation of the seasonal variability of thermal and hydrological regimes operating under pasture and forest in the area of Marabá, since the observed rainfall and solar radiation were similar to normal climatological behaviour for that area.

Temperature gradients under the pasture presented significant differences in relation to those under the forest. Average heat flow under the pasture is directed downward to the deeper layers during the dry quarters, reversing in direction during the rainy quarters. Average heat flow under the forest is quite small and always directed towards the atmosphere regardless of the season.

It was observed that average temperatures under the forest during the rainy season can drop near the surface and show very small alterations at depths greater than 3 m. The profiles show greater changes under the pasture at all depths studied. During the rainy season the average quarterly temperature at depths above 3 m fell under the pasture and rose under the forest. This behaviour can be explained in terms of a increase in the latent heat flux from the pasture to the atmosphere while the heat gain in the forest soil is possibly due to the transportation of thermal energy in the infiltrating water, by the heavy rains during this period.

It may be noticed that the dislocations of the average bi-monthly temperature profiles accompany the incidence of solar radiation increases on the surface with a delay of roughly two months at the pasture and five months at the forest site. To understand this thermal inertia, one must consider the variations in thermal conductivity of soil, resulting from their greater or lesser seasonal moisture content as well as the attenuation and storage of heat in the biomass. This may explain the difference in delay of thermal reaction of the forest soil in relation to that of the pasture.

Moisture minima occur roughly four months after the maxima at the surface. Below 2.4 m, the minimum occurs about eight months later, giving a four-month recharge period and an eight-month depletion period in the lower profile. The data also indicated that the rainfall infiltrated into all layers, implying that these soils have not lost their porosity through compaction. Therefore the pasture area remains permeable 15 years after the forest removal.

The pattern of soil water content changes in the pasture was very similar in the two years studied but there was a large difference between the years in the forest. There were much smaller water content changes during the second dry season because the profile had not wetted to the same extent as in in the previous wet season and less water was available.

The temperature profiles indicated that there is some transfer of heat into the lower profile in the forest through the deep drainage flux but it appears to be small. The lowest temperatures in the profile in the pasture occurred when the profile was wettest and reached a maximum when the profile was in its driest state.

REFERENCES

Araújo, R.L.C., Makino M. and Souza J.R.S. 1985. Estudo da Variação Temporal de Perfís Geotérmicos na Camada Intempérica na Amazônia. Parte I. Anais do III Simpósio de Geologia da Amazônia. Vol. 3, 195-206.

Araújo, R.L.C., Carvalho, J.S. and de Souza, J.R.S. 1995. A influência da Umidade Sobre os Valores de Condutividade Térmica da Subsuperfície. Submited to the Revista Brasileira de Geofísica.

Beck, A.A. 1965. Technigues for Measuring Heat Flow on Land. Ed. Lee in Terrestrial Heat Flow Geophysical Monograph Series. no. 8. 24-57.

Carvalho, H.S., Lobo, P.F.S., Campos, C.N.P. and Zembruski, S.G. 1986. Heat flow and Hydrodynamics Study in the Medium Amazon Basins. Proc. Int. Meeting on Geothermics and Geothermal Energy, 107, Guarujá-S.P-Brazil.

Conaway, J.G. and Beck, A.A. 1977. Fine-Scale Correlation Between Temperature Gradient Logs and Lithology. *Geophysics*, **42**, 1401-1410.

Decico, A. 1974. Equacões para o Cálculo da Condutividade Térmica de Alguns Solos em Função da Densidade e Umidade. *Ciência e Cultura*, **26**, 1178-1182.

de Vries D.A. 1975. Heat Transfer in Soils. In: (Eds.) de Vries, D.A. and Afgan, N.H.*Biosphere. Part 1 - Transfer Processes in the Plant Environment.* John Wiley & Sons. 1-27.

DNMET, 1992. Normais Climatológicas (1961-1990), Ministério de Agricultura e Reforma Agrária-Secretaria Nacional de Irrigação-Departamento Nacional de Meteorologia-Brasília (DF)

Fisch, G., Bastable, H.G., Shuttleworth, W.J., Dallarosa, R.L.G. and Nobre, C.A., 1992. Observações do Balanço de Radiação e Albedo em Florestas e Pastagens na Amazônia Tropical. *Anais do VII,*

1, 349-353.
Ghuman, B.S. and Lal, R. 1982. Temperature regime of a tropical soil in relation to surface condition and air temperature and its Fourier analysis. *Soil Science*, **134**, 134-140.
Hodnett, M.G., Oyama, M.D. and Tomasella, J. 1996a. Comparisons of long-term soil water storage under pasture and forest in three areas of Amazonia. this volume
Hodnett, M.G., Tomasella, J. and Oyama, M.D. 1996b. Deep soil water uptake by forest in Central Amazonia: Predictions from long-term daily rainfall data using a simple water balance model. This volume.
Krcmar, B.K. and Masin, J. 1970. Prospecting by geothermics method. *Geophysical Prospecting*, **XVIII**, 255-260.
Langseth, M.G. 1965. Techniques of measuring heat flow through the ocean floor. In: *Terrestrial Heat Flow* (Ed: Lee, W.H.), Geophysical Monograph Series, nº 8 AGU, 58-77.
Lee, T.C. 1977. Shallow-Hole Temperature Measurements - A Test Study in the Salton Sea Geothermal Field. *Geophysics*, **42**, 572-573.
Marangoni, Y.R. and Hamza, V.M. 1983. Condutividade Térmica de Sedimentos da Plataforma Continental Sudeste do Brasil. *Revista Brasileira de Geofísica*, **2**, 11-18.
Raspet, R., Swartz,J.H., Lilliard, M.E. and Robertson, E.C. 1966. Preparation of thermistor cables used in geothermal investigations. *Geological Survey Bulletin*, 1203-C, Cl-C11.
Ribeiro, M.N.G. and dos Santos, A., 1975: Observações Microclimáticas no Ecosistema Campina Amazônica. *Acta Amazônica*, **5**, 183-189.
Robertson, E.C., Raspet, R., Swartz, J.H. and Lilliard, M.E. 1966. Properties of Thermistors Used in Geothermal Investigations. *Geological Survey Bulletin*, 12038-B1-B33.
Santos, R.C., Viswanadham, Y. and dos Santos, J.M. 1988. Propriedades Térmicas do Solo. Um Estudo de Casos. *Ciência e Cultura*, **40**, 1227-1231.
Silva, L.P., Hodnett, M.G., Rocha, H.R. and Senna, R.C. 1992. A comparison of dry season soil water depletion beneath central Amazonian pasture and rainforest. Anais do VII Congresso Brasileiro de Meteorologia, **1**, 308-312.
Smith, P.J. 1973. *Topics in Geophysics*. The Open University Press, N.Y., 258 pp.
Souza, J.R.S., Rocha, B.R.P., Cohen, J.P.C. and Santos, M.A.R. 1984. Variabilidade Diária da Evapotranspiração na Orla de Belém. Anais do III Congresso Brasileiro de Meteorologia, Belo Horizonte-MG, **1**, 87-101.
Souza, J.R.S., Araújo, R.L.C. and Makino, M., 1989. Heat transfer and thermal properties of the subsoil in Belém. *Revista Brasileira de Geofísica*, **7**, 19-28.
Souza, J.R.S., Makino, M., Araújo, R.L.C. and Costa, P.N.F. 1993. Temperatura do Subsolo em Áreas de Floresta e Pastagem na Amazônia. Anais do 3º Congresso Internacional da Sociedade Brasileira de Geofísica, Rio de Janeiro-RJ; Vol. 2, 1174-1179.
Shuttleworth,W.J., Gash, J.H.C., Lloyd, C.R., Moore, C.J., Roberts, J., Marques, A. de O., Fisch, G., Silva, V. de P., Ribeiro, M.N.G., Molion, L.C.B., de Abreu Sa, L.D., Nobre, C., Cabral, O.M.R., Patel, S.R. and de Moraes, J.C., 1984. Observations of radiation exchange above and below Amazonian forest. *Quart.J.R. Meteorol. Soc.*, **110**, 1163-1169.
von Herzen, R. and Maxwell, A.E. 1959. The measurements of thermal conductivity of deep-sea sediments by a needle-probe method. *J. Geophys. Res.*, **64**, 1557-1563.

RESUMO

Durante um período de 26 meses iniciando-se em agosto 1991, perfis de temperatura do subsolo até 3,8 m de profundidade foram monitorados, semanalmente, em dois sítios próximos à Marabá. Os sítios distam entre si cerca de 100 km. Um dos sítios é coberto por floresta Amazônica primária, e o outro, por pastagem, e ambos são representativos de suas áreas adjacentes. Amostras do subsolo foram coletados, em intervalos de 0,50 m de profundidade, de furos perfurados para medição de temperatura. A análise litológica revelou uma textura de franco-arenosa a argilo-arenosa no sítio de pastagem, e de franca a franco-siltosa no de floresta, abaixo da qual o perfil foi predominantemente saprolítico.

Durante os períodos secos do ano, a temperatura sob a pastagem foi 3,7°C acima daquela sob a floresta, às 10h (hora local) e à profundidade de 0,20 m. Durante os períodos chuvosos, para todas as profundidades estudadas, as temperaturas sob a pastagem foram cerca de 2,0°C maiores que as sob a floresta. Sob a pastagem, a variação sazonal da temperatura foi maior que sob a floresta em todas as profundidades do perfil de 4 m. Os gradientes de temperatura sob a pastagem foram negativos durante as estações secas e positivos durante as estações úmidas do ano. Isto indica que, para este sítio, existe um fluxo de calor orientado para baixo nas estações secas, e fluindo na direção oposta durante as estações chuvosas. Os gradientes de temperatura sob a floresta foram pequenos, mas positivos para todas as estações, o que corresponde a um pequeno mas consistente fluxo de energia orientado para cima. Este estudo mostra grandes diferenças entre as respostas térmicas, devido às variações sazonais de aquecimento solar, do solo sob a floresta e pastagem. Estas diferenças são explicadas em termos de cobertura vegetal e umidade do solo em cada sítio.

Em ambos os sítios, análises do conteúdo de umidade foram feitas para seis camadas de 0,6 m de profundidade. Na superfície, os mínimos de umidade ocorreram, aproximadamente, 4 meses após os máximos. Abaixo de 2,4 m, o mínimo ocorreu cerca de 8 meses mais tarde, dando um período de recarga de 4 meses e um período de depleção de 8 meses no perfil mais baixo. Os dados também indicaram que a precipitação infiltrou-se em todas as camadas, implicando que estes solos não perderam sua porosidade devido à compactação. Portanto, a área de pastagem premanece permeável 15 anos após a retirada da floresta.

O padrão de variações do conteúdo de água no solo na pastagem foi muito semelhante nos dois anos de estudo, mas na floresta, houve uma grande diferença interanual. As variações durante a segunda estação seca foram muito menores, visto que a recarga do perfil não ocorreu na mesma extensão que a da estação úmida anterior, e menos água ficou disponível.

Na floresta, os perfis de temperatura indicaram que existe alguma transferência de calor no perfil inferior através de drenagem profunda, mas isto parece ser pequeno. As menores temperaturas no perfil da pastagem ocorreram no perfil mais úmido, e as maiores, no mais seco.

7 Thermal diffusivity of Amazonian soils

R.C. S. ALVALÁ[1], R. GIELOW[1], I.R. WRIGHT[2] and M.G. HODNETT[2]

[1] *Instituto Nacional de Pesquisas Espaciais, São José dos Campos, Brazil*
[2] *Institute of Hydrology, Wallingford, UK.*

INTRODUCTION

Soil temperature influences most of the processes of the soil-plant continuum. For example, it affects plant growth rate and the rate of decomposition of organic matter. It is required for the characterisation of the microclimate at the soil surface and is a component of all atmospheric models. Temperature profiles in the soil may be obtained through the integration of the heat conduction equation, which depends on the volumetric heat capacity of the soil, C, and its apparent thermal diffusivity, K, where

$$C = \rho C_b \qquad \text{J m}^{-3}\,\text{K}^{-1} \qquad (1)$$

$$K = k/C \qquad \text{m}^2\,\text{s}^{-1} \qquad (2)$$

and where k is the apparent heat conductivity in W m^{-1} K^{-1}, ρ is the bulk density of the soil in kg m^{-3} and C_b is the bulk specific heat in J kg^{-1} K^{-1}. However, these parameters vary with the composition, texture and moisture content of the soil. For a soil with a negligible organic fraction, the volumetric heat capacity, as shown by Campbell (1985) and following de Vries (1963), may be obtained from

$$C = C_m\, \rho/\rho_m + \theta\, C_w \qquad (3)$$

where C_m and C_w are the specific heats of the mineral constituents of the soil and of water respectively, ρ_m is the density of the mineral fraction and θ (kg kg^{-1}) is the volumetric moisture content of the soil.

The apparent thermal diffusivity, which is considered to be constant for a uniform soil, can be determined by several methods using measurements of transient soil temperature. These methods are based on either analytical or numerical solutions of the equation of heat conduction. Horton *et al.* (1983) evaluated six of them: amplitude, phase, arctangent, logarithmic, numerical and harmonic. However, as actual soils are vertically non-uniform, with varying composition, texture and moisture content, more elaborate methods may be used (Lettau, 1954; van Wijk and

Amazonian deforestation and climate. Edited by J.H.C. Gash, C.A. Nobre, J.M. Roberts and R.L. Victoria.

Derksen, 1963). Alternatively, the non-uniform soil may be divided into layers, each considered to be of uniform composition and structure, in which K is assumed constant.

The latter approach is used here to obtain the apparent thermal diffusivity for two deforested areas in Amazonia, each covered with pasture but with contrasting soils: clay and sand.

METHODOLOGY

The one dimensional conduction of heat through the soil is given by

$$C\left(\frac{\partial T}{\partial t}\right) = \left(\frac{\partial}{\partial z}\right)\left(CK\left(\frac{\partial T}{\partial z}\right)\right) \tag{4}$$

where T is soil temperature, t is time and z the depth (see for example Gilman, 1980). Partitioning a non-uniform soil into layers of uniform composition and texture, Equation 4 becomes

$$\frac{\partial T}{\partial t} = K_j\left(\frac{\partial^2 T}{\partial z^2}\right) \tag{5}$$

where $K_j = (k/C)_j$, the apparent thermal diffusivity of layer j, delineated by z_{j-1} and z_{j+1}, j = 1, 2, ... m, the number of layers, not necessarily of equal thickness.

The apparent thermal diffusivity can be obtained by the use of measured temperatures substituted into expressions obtained from analytical solutions of Equation 5. This is the case for the amplitude, phase, arctangent, logarithmic and harmonic methods shown by Horton *et al.* (1983) and referred to here as the 'periodic' methods. Numerical solutions of Equation 5 can also be obtained with a succession of assumed values of K and comparison made with the measured temperature profiles to find the value which minimises the differences between measured and computed profiles. Finally, K can be obtained by making it explicit in the numerical expressions for Equation 4. Five of these methods are compared in this paper, (omitting the harmonic and explicit methods).

Numerical method

Equation 5, put in the form of finite differences by means of a second order weighted scheme (Ralston and Wilf, 1964; Campbell, 1985), gives the solution for layer j, delineated by z_{j-1} and z_{j+1}, and at times t_k and t_{k+1}:

$$\frac{\left(C_{j+\frac{1}{2}}K_{j+\frac{1}{2}}\left[\frac{T_{j+1}-T_j}{z_{j+1}-z_j}\right] - C_{j-\frac{1}{2}}K_{j-\frac{1}{2}}\left[\frac{T_j-T_{j-1}}{z_j-z_{j-1}}\right]\right)}{\left[\frac{z_{j+1}-z_{j-1}}{2}\right]} = C_j\left(\frac{T_j^{k+1}-T_j^k}{\Delta t}\right) \tag{6}$$

where

$$T = pT^{k+1} + (1-p)T^k \tag{7}$$

and

$$C_{j+\frac{1}{2}} = \frac{1}{2}(C_j + C_{j+1}) \text{ , etc} \tag{8}$$

and p is a weighting factor which may vary from 0 to 1. A factor of p = 0.5 corresponds to the classic scheme of Crank-Nicholson. According to Campbell (1985), p equal to 0.6 minimises the errors in the energy balance and this value is therefore used in this study. The stability, convergence and consistency of the numerical integration (Horton *et al.*, 1983) is assured if

$$\Delta T \leq \frac{(\Delta z)^2}{2K} \tag{9}$$

To determine K corresponding to the layer j, a sequence of solutions of Equation 6 was derived with a measured initial temperature profile and with measured temperature boundary conditions. Successive constant trial values of K are used and the sum of the squared differences between the computed temperatures and their corresponding measured values is determined. The value of K which minimises this difference is considered to be the one corresponding to the layer considered, under the conditions imposed.

Amplitude

Following the procedures given by Horton *et al.* (1983) the temperatures measured at the two depths, z_1 and z_2, delimit a layer assumed to be uniform, so that

$$K = \frac{\pi}{P}\left[\frac{z_2 - z_1}{\ln(A_1/A_2)}\right]^2 \tag{10}$$

where A_1 and A_2 are the amplitudes at depths z_1 and z_2, respectively, of the corresponding transient temperature records, and P is the cyclic time period, in this case the 24 hours from 0 to 24 h local time. It should be noted that this and subsequent periodic methods result from an idealised analytical solution of the soil heat conduction equation.

Phase

$$K = \frac{P(z_2 - z_1)^2}{4\pi L^2} \tag{11}$$

where L is the time interval between measured maxima, or minima, of the soil temperature at z_1 and z_2, during the period P.

Arctangent

$$K = \frac{\pi}{P}(z_2-z_1)^2 \left\{ \arctan \left[\frac{(T_1-T_3)(T'_2-T'_4)-(T_2-T_4)(T'_1-T'_3)}{(T_1-T_3)(T'_1-T'_3)+(T_2-T_4)(T'_2-T'_4)} \right] \right\}^{-2} \quad (12)$$

where temperatures T_i and T'_i, are at depths z_1 and z_2 respectively, and are taken from four successive times, i = 1 to 4, each six hours apart during the 24 hour period P.

Logarithmic

$$K = \left[\frac{0.0121\,(z_2-z_1)}{\ln \left[\frac{(T_1-T_3)^2+(T_2-T_4)^2}{(T'_1-T'_3)^2+(T'_2-T'_4)^2} \right]} \right]^2 \quad (13)$$

where the notation is the same as for Equation 12.

To obtain reliable daily values of K with these four simple idealised periodic methods, it is essential that the daily temperature records are regular and have only one maximum and minimum during the 24-hour period considered, and that their derivatives vary slowly. In addition there should be no rain, and consequently the soil heat flux should have a slow and regular variation, especially during the daylight hours. This occurs particularly on clear days or at least when there are no strong gradients of solar radiation caused by rapidly moving convective clouds. Although some of the short period 'noise' can be accommodated by Fourier analysis and filtering (Gilman, 1977), the occurrence of rain eliminates the possibility of a constant K value throughout the day because of the change in heat capacity and conductivity with changing soil moisture (Wierenga *et al.*, 1969; Campbell, 1977).

The numerical integration used to obtain values for K with the numerical method, and also to calculate temperature profiles in any situation, are very robust. The predicted temperature profiles within a layer are always smooth, whatever the temperatures at its boundaries. However, as the objective is to obtain constant values of K for a layer during the whole 24-hour period, and the dependence on soil moisture, data from rainy days are also unacceptable for this method.

SITES, DESCRIPTION AND DATA

Data were recorded in the ABRACOS project at Fazenda Dimona, near Manaus, Amazonas and Fazenda Nossa Senhora, near Ji-Paraná, Rondônia (Gash *et al.*, 1996). Soil structure at each site (in terms of percentage clay: silt sand) is 75%: 10%: 15% at Manaus and 15%: 10%: 75% at Ji-Paraná (Wright *et al.*, 1996; Hodnett *et al.*, 1995). The data used were soil temperatures, energy fluxes and moisture profiles, measured during two dry season data missions: Mission 1 at Fazenda Dimona, 25 September to 25 October 1990 and Mission 4/5 at Fazenda Nossa Senhora, 1 April to 6 August 1993. Soil temperatures were monitored using thermistors (Campbell

Scientific, Shepshed, UK) and recorded as ten-minute averages, together with soil heat flux, using a programmable logger (Campbell Scientific, Shepshed, UK); see Wright *et al.* (1992). Soil temperatures at 10 cm are the mean of three sensors; 20 and 40 cm temperatures were from a single sensor at each depth. Soil moisture content was measured at approximately seven day intervals using a Neutron Probe (Didcot Instruments, Abingdon, UK) with each datum being the mean soil moisture estimated from eight permanently installed access tubes (Hodnett *et al.*, 1995). Radiation balance data are from the automatic weather station at each site (Culf *et al.*, 1996).

RESULTS AND DISCUSSION

Examples of typical data from wet and dry periods during 1993 at Fazenda Nossa Senhora are shown in Figures 1 and 2 which illustrate the influence of radiation

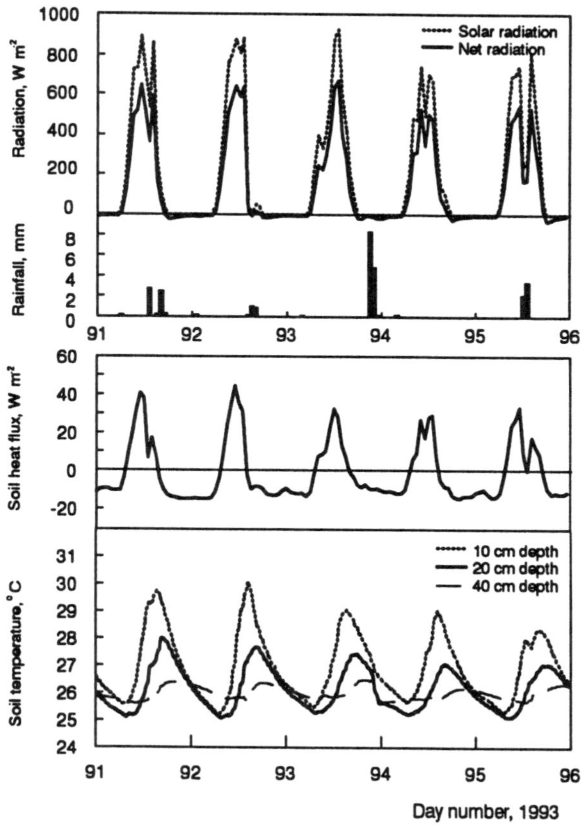

Figure 1 Wet season energy fluxes, rainfall and soil temperature measured at Fazenda Nossa Senhora (Days 91-95, 1993)

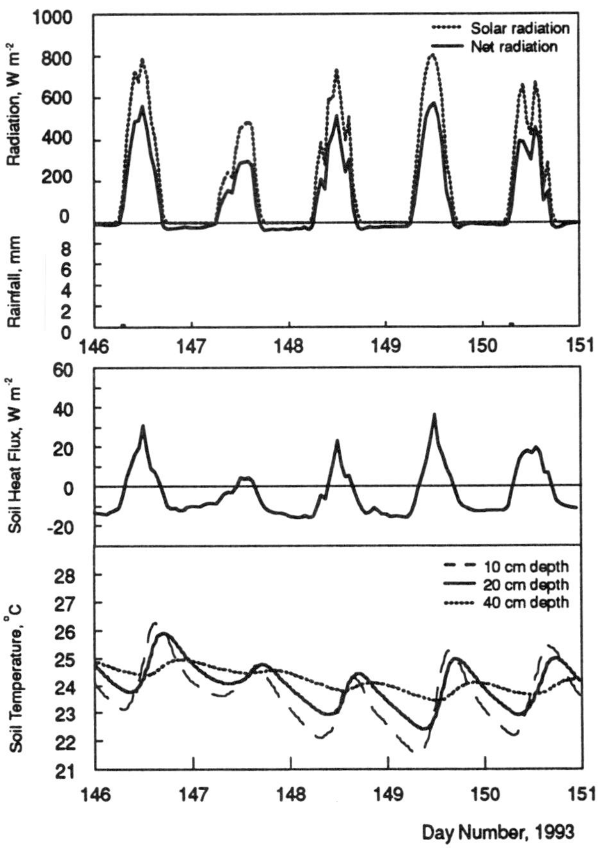

Figure 2 Dry season energy fluxes, rainfall and soil temperature measured at Fazenda Nossa Senhora (Days 146-150, 1993)

and rainfall on the soil heat flux and soil temperatures. The cloudy day (Day 150, 1993) has some insignificant rainfall (0.13 mm) at 08.00h and moderate fluctuations in cloudiness, yet presents a relatively smooth soil heat flux, and is typical of those selected to compute the apparent soil thermal diffusivity. The interaction between rainfall and soil moisture content at 10, 20 and 40 cm depths, for both sites, is shown in Figure 3. The soil water drainage after rainfall, in these periods with unsaturated soils, is rapid for the sandy soil of Fazenda Nossa Senhora, while the decrease of moisture is moderately slow during the rain-free periods. The high values of soil moisture for 10 and 20 cm depth on Day 115, 1993 are a consequence of moisture measurement very soon after the 15 mm storm. At Fazenda Dimona there was very little rain during the period shown, and although the moisture content is high, the decrease with time is slow.

Equations 6 and 10-13 were applied to estimate thermal diffusivity on nine selected days from Fazenda Nossa Senhora for which the above conditions were satisfied and

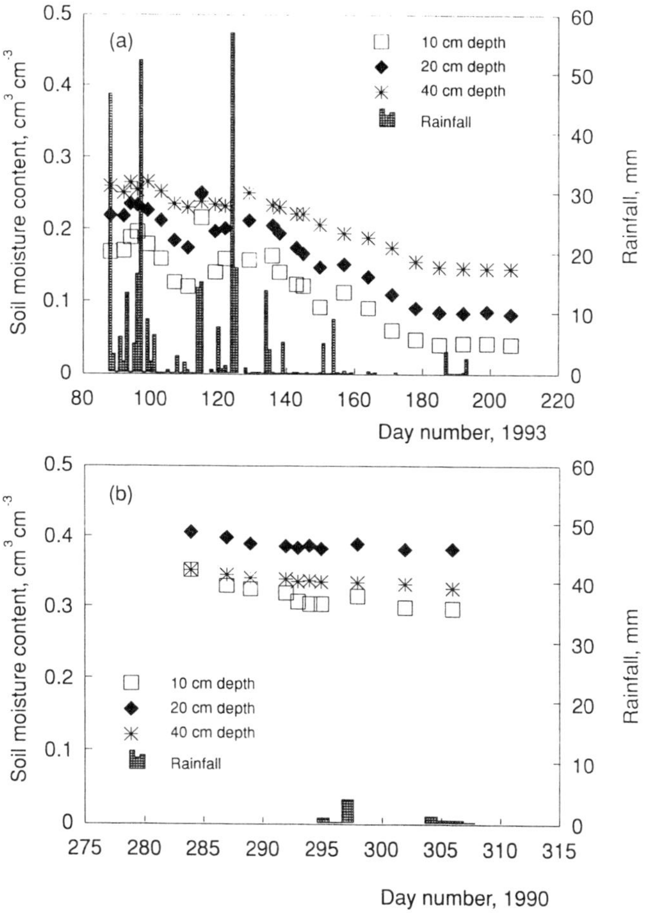

Figure 3 Soil moisture content and rainfall at (a) Fazenda Nossa Senhora and (b) Fazenda Dimona

measurements of soil moisture were available (see Table 1). Within the Fazenda Dimona data only two such days occurred and the soil moisture profiles were interpolated for the relevant period. This was possible because of the small changes in soil moisture between measurements and the absence of rain. Volumetric heat capacity was determined using Equation 3 with $C_m = 2.4$ MJ m^{-3} K^{-1} and $\rho_m = 2.65$ Mg m^{-3} for both clay and quartz, and $C_w = 4.18$ MJ m^{-3} K^{-1}. Soil bulk density ρ was taken as 1.1 for the clay and 1.3 Mg m^{-3} for the sandy soils (Wright *et al.* 1996).

The apparent thermal diffusivity computed by the amplitude method consistently showed the least scatter when compared with the other three periodic methods, and only the results from the amplitude method are presented here for comparison with the numerical method. The 'arctangent' and 'logarithmic' methods performed particularly badly. The scatter in all of the periodic methods is due to their strong sensitivity to small errors in the data. Diffusivities from the 'amplitude' method for

Table 1 Apparent thermal diffusivities and measured data for selected days

Day	Rain	Measured soil moisture				Thermal diffusivity ($\times 10^6$)			SDELQ*	rms**
		Volume fraction			Storage	Amplitude		Numerical		
	mm	m^3 m^{-3}			mm	m^2 s^{-1}			°C	°C
		10 cm	*20 cm*	*40 cm*	*0-40*	*10-20*	*20-40*	*10-40 (cm)*		
Fazenda Nossa Senhora, 1993										
11	0.51	0.120	0.174	0.230	64.4	3.092	0.862	1.410	22.160	0.395
129	0.0	0.157	0.212	0.250	77.6	2.049	1.205	1.330	1.094	0.089
145	0.0	0.121	0.166	0.22	162.9	2.212	0.862	1.530	4.165	0.170
150	0.13	0.091	0.147	0.206	53.5	1.880	0.927	1.450	2.198	0.122
171	0.0	0.059	0.109	0.174	40.1	2.084	0.983	1.670	6.649	0.217
178	0.0	0.046	0.090	0.155	33.7	1.700	0.898	1.410	1.304	0.095
192	0.13	0.041	0.083	0.146	31.1	1.722	0.958	1.410	2.539	0.134
199	0.0	0.041	0.085	0.145	31.2	2.074	1.393	1.250	3.873	0.164
206	0.0	0.039	0.081	0.145	30.4	1.828	1.062	1.373	5.129	0.190
Fazenda Dimona, 1990										
282	0.0	0.350	0.408	0.357	146.4	0.169	0.453	0.063	10.960	0.277
288	0.0	0.320	0.393	0.342	139.0	0.146	0.453	0.067	13.586	0.315

* $\Sigma(T_{measured} - T_{numerical})^2$ for depths equal to 20 cm (142 values) ** rms = $(SDELQ/142)^{0.5}$

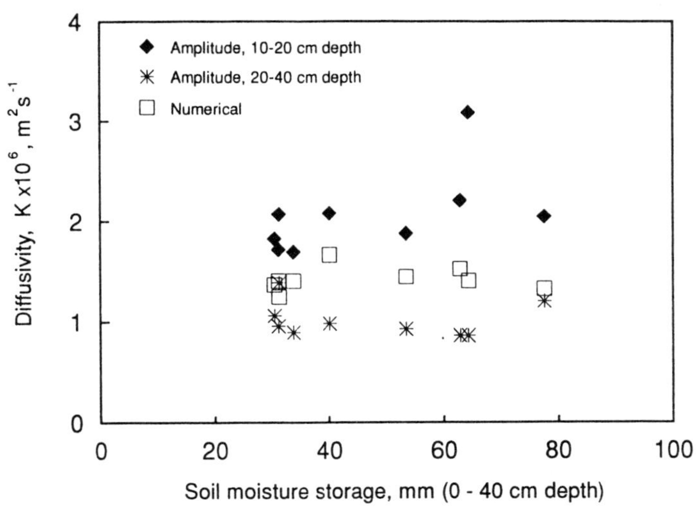

Figure 4 Calculated apparent soil thermal diffusivity for Fazenda Nossa Senhora.

Table 2 Soil water storage (0-40 cm) versus soil moisture at specific levels — linear regression parameters

Depth (cm)	Intercept	Slope	r^2	Soil moisture range
Fazenda Nossa Senhora				
15*	8.55	371.86	0.998	0.060 - 0.233
30**	-18.58	424.50	0.988	0.113 - 0.250
Fazenda Dimona				
15*	0.46	383.95	0.998	0.340 - 0.424
30**	-0.77	586.69	0.986	0.354 - 0.409

* Average between 10 and 20 cm measured values of θ
** Average between 20 and 40 cm measured values of θ

the 10 to 20 cm and 20 to 40 cm soil layers are shown in Table 1 and Figure 4. The numerical method — in principle the most appropriate if the selection conditions are satisfied — provided the most consistent results for the 10 to 40 cm soil layer (also given in Table 1 and Figure 4).

The overall water storage in the top 40 cm of soil was found to correlate linearly with the values of the soil moisture at individual depths, with correlation coefficients (r^2) between 0.986 and 0.998. Notwithstanding the underestimation of soil moisture near the surface using the neutron probe, this is a useful result for interpolating soil moisture during rain-free periods to correspond to specific K values. The parameters of these regressions are shown in Table 2.

To satisfy the stability criterion given by Equation 9, the maximum integration time interval Δt was 600 s at Fazenda Dimona and between 150 s and 200 s at Fazenda Nossa Senhora. The measurements were recorded at 600 s intervals and when the time integration increments were shorter than 600 s, the soil temperatures at the boundaries were interpolated using cubic splines between the measured values. The same method was used to interpolate vertically for the initial temperature profile. Eight depths, unequally spaced, were used for the integrations between 10 and 40 cm: 12.5, 15.0, 17.5, 20.0, 22.5, 25.0, 30.0, and 35.0 cm. At the Fazenda Nossa Senhora site (Table 1) the sums of the squared differences between 142 measured and computed soil temperatures at the 20 cm depth for each of the nine selected days vary between 1.09 and 22.16, with seven of them under 5.2. Individual root mean square (rms) differences range from 0.089 to 0.395°C. Figure 5 shows the measured and calculated soil temperatures for the selected day (Day 150, 1993).

For the sandy soil of Fazenda Nossa Senhora (Figure 4) the apparent thermal diffusivities obtained using the numerical method vary from 1.25×10^{-6} to 1.67×10^{-6} $m^2\ s^{-1}$, while the range of soil water storages (0-40 cm) is 30.4 to 77.7 mm, corresponding to soil moisture contents at 15 cm of 0.06 to 0.18 $m^3\ m^{-3}$, for a porosity

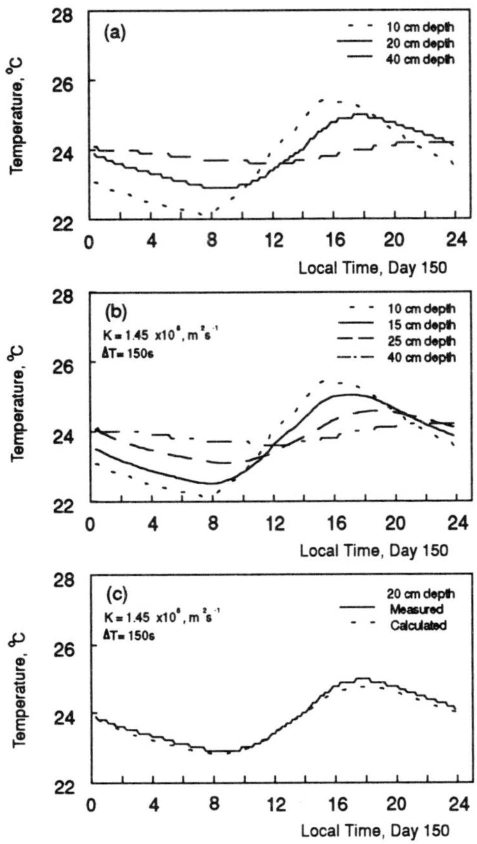

Figure 5 Soil temperatures at Fazenda Nossa Senhora for Day 150, 1993; (a) measured, (b) calculated and (c) a comparison between measured and calculated at 20 cm depth

of 51.1%. The overall mean diffusivity from this method is 1.45×10^{-6} m^2 s^{-1}. These diffusivities are consistent with the values published by Viswanadham and Mohana Rao(1972) and somewhat higher than the values of 1.0 and 0.7 x10^{-6} m^2 s^{-1} given for wet sand by Rosenberg *et al.* (1983) and Arya (1988) respectively. However, these published values are for wetter sand of lower porosity. There is a slight tendency for the diffusivities from this method to be lower when the soil moisture storage is at its lowest and highest (see Figure 4). Although this trend is not strongly evident in these results, it is consistent with a maximum diffusivity occurring at intermediate soil moisture contents (Wierenga *et al.*, 1969; Campbell, 1977).

The two estimates for diffusivity for the clay soil at Fazenda Dimona using the numerical method (Table 1) are 0.063×10^{-6} and 0.067×10^{-6} m^2 s^{-1}, corresponding to soil moisture contents of 0.36 and 0.38 m^3 m^{-3} for the 15 and 30 cm depths respectively, with a porosity of 59%. The individual r.m.s. differences for these K values are 0.277 and 0.315°C. These diffusivities seem particularly low and are not consistent with the estimates from the 'amplitude' method. This suggests heterogeneities in the 10 to 40 cm layer allowing the thinner layers of the 'amplitude' method to give more plausible values. Wright *et al.* (1996) show that porosity changes from 1.06 to 1.15 in the top 30 cm of this soil. Values for clay soil of 0.2 and 0.18 - 0.51×10^{-6} m^2 s^{-1} are given by Rosenberg *et al.* (1983) and Arya (1988) respectively, although diffusivities for the clay soil of Manaus might be expected not to conform to published values because of its unusually light texture and high porosity (Tomasella and Hodnett, 1996).

These results indicate the need for further work, especially on the moisture content dependence of apparent thermal diffusivity for these and other Amazonian soil types. For successful prediction of the soil surface heat flux boundary condition, more attention must be given to the performance of theory in the shallowest layers of the soil: not only over 24-periods when total soil heat flux is small, but also for individual hours to give accurate estimation of the daytime surface energy budget.

ACKNOWLEDGEMENT

The authors would like to thank Miss Ângela Y. Harada for help with the calculations and graphical presentation.

REFERENCES

Arya, S.P. 1988. *Introduction to micrometeorology*. Academic Press (London).

Campbell, G.S. 1977. *An Introduction to environmental biophysics*. Springer, New York.

Campbell, G.S. 1985. *Soil physics with BASIC*. Elsevier, Amsterdam.

Culf, A.D., Esteves, J.L., Marques Filho, A. de O. and da Rocha, H.R. 1996. Observations of radiation, temperature and humidity over forest and pasture in Amazonia. This volume.

Gash, J.H.C., Nobre, C.A., Roberts, J.M. and Victoria, R. L. 1996. An overview of the ABRACOS project. This volume.

Gilman, K. 1977. Movement of heat in soils. *IH Report No. 44*, Institute of Hydrology, Wallingford, UK.

Gilman, K. 1980. Estimation of soil heat flux in an upland drainage basin. *Hydrol. Sci. Bull.* **25**, 435-451.

Hodnett, M.G., Pimentel da Silva, L., da Rocha, H.R. and Cruz Senna, R. 1995. Seasonal soil water storage changes beneath central Amazonian rainforest and pasture. *J. Hydrol.* **170**, 233-254.

Horton, R., Wierenga, P.J. and Nielsen, D.R. 1983. Evaluation of methods for determining apparent thermal diffusivity of soils near the surface. *Soil. Sci. Soc. Am. J.* **47**, 25-32.

Lettau, H., 1954. Improved models of thermal diffusion in the soil. *Trans. Am. Geophys. Union* **35**, 121-132.

Ralston, A. and Wilf, H.S. 1964. *Mathematical methods for digital computers*. John Wiley & Sons, New York.

Rosenberg, N.J., Blad, B.L. and Verma, S.B. 1983. *Microclimate: the biological environment.* John Wiley & Sons, New York.

Tomasella, J. and Hodnett, M.G. 1996. Measurements and modelling of soil hydraulic properties of oxisols in central Amazonia. This volume.

Viswanadham, Y. and Mohana Rao N.J., 1972. The thermal diffusivity of soil at tropical stations in the southern hemisphere. *Pure & Appl. Geophys.* **101**, 247-265.

de Vries, D.A. 1963. Thermal properties of soils. In: *Physics and plant environment.* Ed: W.R. van Wijk, North Holland, Amsterdam, 102-142.

Wierenga, P.J., Nielsen, D.R. and Hagan, R.M. 1969. Thermal properties of a soil based upon a field and laboratory measurements. *Soil Sci. Soc. Am. J.* **33**, 354-360.

van Wijk, W.R. and Derksen, W.J. 1963. Sinusoidal temperature variation in a layered soil. In: *Physics and plant environment.* Ed: W.R. van Wijk, North Holland, Amsterdam, 179-209.

Wright, I.R., Gash, J.H.C., da Rocha, H.R., Shuttleworth, W.J., Nobre, C.A., Maitelli, G.T., Zamparoni, C.A.G.P. and Carvalho, P.R.A. 1992. Dry season micrometeorology of central Amazonian ranchland. *Q. J. R. Meteorol. Soc.* **118**, 1083-1099.

Wright, I.R., Nobre, C.A., Tomasella, J, da Rocha, H.R., Roberts, J.M., Vertamatti, E., Culf, A.D., Alvalá, R.C.S., Hodnett, M.G. and Ubarana, V.N. 1996. Towards a GCM surface parameterisation for Amazonia. This volume.

RESUMO

A difusividade térmica do solo (K) – igual à razão entre a condutividade térmica (k) e a capacidade térmica "bulk" (C) – é um importante parâmetro para a determinação da distribuição da temperatura e do fluxo de calor no solo. Este parâmetro, que contribui para a caracterização do microclima do solo, depende do conteúdo de umidade, bem como da composição e da textura do solo. Medidas da temperatura do solo às profundidades de 10, 20 e 40 cm, sob diferentes condições de umidade, foram utilizadas para determinar K em duas áreas desmatadas, com solos argiloso e arenoso, da Fazenda Dimona (Manaus) e da Fazenda Nossa Senhora (Ji-Paraná), respectivamente. Compararam-se vários métodos para a obtenção de K, incluindo uma solução numérica da Equação de Condução de Calor. Utilizaram-se as temperaturas do solo nas profundidades de 10 a 40 cm como condições de contorno, e um perfil de temperatura inicial medido, para obter valores otimizados de K que ajustem a temperatura em 20 cm. Os valores de K são apresentados como função do armazenamento de água no solo até a profundidade de 40 cm. Encontraram-se também relações lineares entre este armazenamento e a umidade em diferentes profundidades. Obteve-se um valor médio de K, para nove dias selecionados e utilizando o método numérico, igual a $1,45\times10^{6}$ m^2 s^{-1} para o solo arenoso, valor este consistente com aqueles obtidos pelos métodos analíticos. Com uma amostra de dois dias para o solo argiloso, obtiveram-se valores menos consistentes de K, variando de 0,07 a $0,45\times10^{6}$ m^2 s^{-1}.

8 Observations and modelling of rainfall interception at two experimental sites in Amazonia

V.N. UBARANA
Fundação Cearense de Meteorologia e Recursos Hídricos, Fortaleza, Brazil

INTRODUCTION

The interception of rainfall by tall vegetation is an important process in the hydrological cycle and has been the subject of research for several decades. The relationship of the process to the physical and morphological aspects of the vegetation cover was considered initially by Horton (1919). Several studies on the subject were published in the 1970s and 1980s (see Valente, 1990) which confirmed Rutter's (1967) hypothesis that the evaporation rate for water intercepted by vegetation cover can be higher than the transpiration rate in the same climatic conditions. The evaporation of intercepted water can therefore form a large component of the water balance and cannot be omitted from the water budget, especially in periods immediately after rain when the vegetation is wet.

Rain falling onto a vegetation canopy can be divided into two components. The first component remains on the canopy and is subsequently evaporated back to the atmosphere, while the second drains off the vegetation cover to the soil. Since the first component does not reach the ground and is lost from soil water recharge, it is known as the interception loss. Interception loss can vary significantly from one forest to another. A recent review of the subject (Clarke 1987) is discussed by Lloyd and Marques (1988). They argue that although reported interception losses can vary widely — from 3% of the gross rainfall in Nigeria to more than 63% in forests in Thailand — this variation is largely the result of inadequate sampling of throughfall. Values higher than 30% have been reported for temperate forests by Gash and Stewart (1977) and Gash *et al.*, (1980). The large variation in these values is due to both the different morphology of forest canopies in each case and the different types of dominant precipitation regimes in each area. At Reserva Ducke, near Manaus, Lloyd *et al.* (1988) measured the total interception loss as 8.9 ± 3.6% (of 4804 mm gross rainfall) over a two-year period. Using an analytical model of rainfall interception (Gash, 1979), Lloyd *et al.* also found that modelled interception loss varied from 10.3% to 12.6%, based on 13 years of daily rainfall records.

The present paper describes the results of a field experiment to determine the interception loss at two *terra-firme* forest sites in Brazil. In addition to producing

Amazonian deforestation and climate. Edited by J.H.C. Gash, C.A. Nobre, J.M. Roberts and R.L. Victoria.

estimates of the interception loss of different Amazonian forest sites to supplement the previous work carried out in Central Amazonia by Lloyd and Marques (1988), the data have also been used to evaluate the performance of the numerical Rutter model which simulates the interception process.

EXPERIMENT

Water which reaches the soil by dripping from the forest canopy, together with that which reaches the soil directly without interacting with the canopy (free throughfall), is known as throughfall. The component of the rainfall which runs down the trunks and stems of the trees is called the stemflow. The sum of the stemflow and the throughfall is the amount of rainfall which reaches the ground and is known as the net precipitation. Therefore, interception loss can be calculated as the difference between field measurements of gross rainfall and the net precipitation. For this study measurements were made at two of the ABRACOS forest sites: Reserva Jaru close to Ji-Paraná in Rondônia, and Reserva Vale do Rio Doce near Marabá in Pará: for site details see Gash *et al.* (1996). Net precipitation was measured as throughfall and stemflow separately. At Reserva Jaru the mean height of the forest canopy was 33 m. At Reserva Vale do Rio Doce the continuous forest canopy was lower, with a height of around 20 to 25 m, but there were occasional mature Brazil nut trees (*Bertholletia excelsa* HBK) which reached a height of 50 m. The forest vegetation at Reservas Jaru and Vale do Rio Doce is described in detail by McWilliam *et al.* (1996) and Sá *et al.* (1996), respectively.

The instrumentation was based on the methodology of Lloyd and Marques (1988) and was installed during ABRACOS Mission 3 in August and September 1992. At each site the throughfall and stemflow were sampled in a rectangular plot measuring 100 m by 4 m which contained a representative number of dominant trees compared to the immediately surrounding area. The throughfall was sampled by 30 gauges which consisted of a funnel with a diameter of 12.7 cm connected to a two litre bottle. Stemflow was sampled by ten gauges fixed to live trunks within the same plot, and selected to represent the distribution of trunks in the vicinity of the plot. Ideally, in an experiment such as this, data would be collected from the gauges after every storm, but, in remote tropical forest sites such as these, practicalities such as difficult site access and transport limitations usually prevent such frequent data collection (Jackson, 1971). In this experiment data from all gauges were collected routinely every week. Occasionally the collection period was longer, the longest period being 31 days. However, during some intensive campaign periods during the wet season the data were collected immediately after large rainfall events. These extra data allowed the calculation of model parameters such as the canopy storage capacity as discussed below. After each data collection the throughfall gauges were randomly relocated amongst the 505 possible sampling locations within the plot.

Gross rainfall at each site was measured using a tipping bucket raingauge (Didcot Instruments, Abingdon, UK) located on the ground but connected by a tube to a funnel mounted above the canopy on top of the 52 m high towers. The raingauge was constructed in this way to avoid premature and erroneous tips which can occur in a tower-mounted gauge, caused by movement of the tower. Hourly totals of rainfall, together with meteorological data necessary for the calculation of evaporation, were recorded at each site by the ABRACOS automatic weather station (Culf *et al.*, 1996). There were no measurements of litter interception as part of this work. A more detailed description of the field instrumentation, and the experiment in general, is given by Ubarana (1994).

The results obtained from approximately two years of data collection (30 August 1992 to 19 August 1994) are presented for Reserva Jaru, Ji-Paraná. During this period there were 78 collection periods, totalling 610 days and varying in length from 2 to 31 days: most were of 7 days duration. For Reserva Vale do Rio Doce, Marabá, data are presented for the period 24 September 1992 to 10 January 1994, with a total of 39 collection periods representing 430 days. Again, most periods were of 7 days duration but varied in length from 3 to 28 days. Therefore, with missing data and periods of unreliable measurement, a data recovery of over 85% was achieved at both sites.

THE RUTTER MODEL

Details of the model are given by Rutter *et al.* (1971, 1975) and Ubarana (1994). The model requires parameters which describe the morphology of the specified vegetation. These are the canopy capacity (S_c), the trunk storage capacity (S_t), the free throughfall coefficient (p), which defines the proportion of the rain which falls directly on the soil without striking the canopy, and the coefficient which defines the quantity of rain that is diverted to the trunks (p_t).

For calculation of the relatively rapid drainage dynamics, the Rutter model is operated with a five-minute time-step and driving variables are uniformly distributed over each hour.

Drainage (D) only occurs when the canopy is saturated. Formally, this relationship is given by Equations (1) and (2).

$$D = D_3 \exp [b (M_c - S_c)], \text{ if } M_c \geq S_c \quad (1)$$

$$D = 0, \qquad \text{if } M_c < S_c \quad (2)$$

where M_c is the depth of water on the canopy and D_s is the amount of drainage when $M_c = S_c$. The parameter b represents the rate of increase of drainage with depth of water on the canopy. The model calculates the total interception loss by integrating the throughfall, stemflow and the evaporation of intercepted water. The evaporation of intercepted water from the wet canopy is calculated from a reduced version of the Penman-Monteith equation (Equation 3) in which the bulk surface resistance has been set to zero,

$$E_w = \frac{\Delta R_n + \rho c_p \delta e^* / r_a}{\lambda(\Delta + \gamma)} \tag{3}$$

and where Δ is the rate of change of the saturation vapour pressure with temperature, R_n is the net radiation, ρ and c_p are the density and the specific heat capacity of the air respectively, δe^* is the water vapour pressure deficit and r_a is the aerodynamic resistance to the transport of water vapour. λ is the latent heat of vaporisation of water, γ is the psychrometric constant and E_w is the evaporation. The requirement for detailed hourly meteorological data is the principal limitation of the Rutter model. However, in the context of climate modelling, in which hourly climate is derived, this may not be such a problem.

The model calculates the aerodynamic resistance at each time-step from the hourly forcing data by using the relationship $r_a = f/u$, where f is a constant (Rutter *et al.*, 1971) and u is the windspeed. In the absence of experimental data, f is frequently computed from the vegetation height and the aerodynamic roughness. The value of $f=34.2$ estimated directly from the momentum flux by Shuttleworth *et al.* (1987) for Reserva Ducke, Manaus, was used by Ubarana (1994) for that site, however, new data from ABRACOS suggests that a better value for Amazonian forest is $f=33$ for a standard reference height of 25 m above the zero plane displacement (Wright *et al.*, 1996). This new value has been used in the present work.

With respect to the five-minute time-step, the procedure of assuming that the rainfall occurs uniformly throughout each hour can lead to a systematic increase in the duration of storms in the model (Lloyd *et al.*, 1988). For tropical areas, where storms are typically of short duration, this systematic error could result in a significant overestimation of interception. The problem has been addressed by using the procedure of Lloyd *et al.* (1988): each hour of rainfall is divided into two halves and it is assumed that the rainfall occurs completely in the first half hour if the following hour is dry, or completely in the second half hour if the preceding hour is dry.

Table 1 Values of vegetation parameters used in the Rutter model

Parameter	Unit	Reserva Jaru	Reserva Vale
S_c – Canopy capacity	mm	1.03	1.25
S_t – Trunk capacity	mm	0.09	0.10
D_s – Drainage when $M_c = S_c$	mm min^{-1}	0.0014	0.0014
b – Drainage increase with M_c variation	dimensionless	5.25	5.25
p – Free throughfall coefficient	dimensionless	0.031	0.044
p_t – Trunk deviated precipitation coefficient	dimensionless	0.010	0.023

MODEL PARAMETERS

Canopy capacity and drainage

The data collected after individual rain events during the wet season, rather than the routine weekly data collection which took place throughout the experiment, allowed a value for the canopy capacity, S_c, to be calculated. The methodology adopted was similar to that used by Lloyd *et al.* (1988) and Gash and Morton (1978). Storm events were chosen with at least six dry hours separating them from adjacent events and with gross rainfall greater than 1.5 mm and less than 15 mm. There were 37 such events in the dataset from Reserva Jaru, Ji-Paraná, and 18 from Reserva Vale do Rio Doce, Marabá. Linear regressions of throughfall against gross rainfall were computed for each site. For Jaru the regression gave a slope of 0.96±0.17 and an intercept of –0.99 ± 0.15 giving a value of 1.03 mm of gross rainfall when the throughfall is zero. For Vale do Rio Doce, the regression had a slope of 0.87± 0.20 and an intercept of -1.09 ± 0.22 giving a value of 1.25 mm of gross rainfall when the throughfall is zero. The small correction which was applied by Lloyd *et al.* (1988) to take account of evaporation during rainfall was considered to be negligible and was ignored. Consequently, the values of S_c used in the model were 1.03 mm for Jaru and 1.25 mm for Vale do Rio Doce.

The values of the drainage terms D_s and b in Equation (1) were those derived for Reserva Ducke by Lloyd *et al.* (1988) using a formulation first used by Rutter *et al.* (1975) for a Corsican pine forest.

Free throughfall coefficient

Values for the free throughfall coefficient (p), which expresses the amount of gross rainfall that falls to the ground directly, were obtained from anascope readings (Ford, 1976). In Reserva Jaru 3.1% of the canopy was considered to be clear sky, giving a value for p of 0.031. In Reserva Vale do Rio Doce a value of p = 0.044 was obtained. Using the same method at Reserva Ducke, Manaus, Lloyd *et al.* (1988) obtained a lower value of p=0.08.

Trunk storage capacity

The trunk storage capacity, S_t, was calculated in a similar way to the canopy capacity with regressions of stemflow against gross rainfall for each trunk in the plot. The mean intercept of these regressions gave a value for S_t, and the mean slope gave a value for p_t, the portion of the gross rainfall which is diverted onto the trunks. Thirty two events were used in the calculation of S_t and p_t for Jaru and ten events for Vale do Rio Doce. For Jaru the mean relationship between stemflow, S_f, and gross rainfall, P_G, was $S_f = 0.010(\pm0.004)P_G - 0.09(\pm0.09)$, and for Vale do Rio Doce the equation was $S_f = 0.023(\pm0.01)P_G - 0.10(\pm0.07)$.

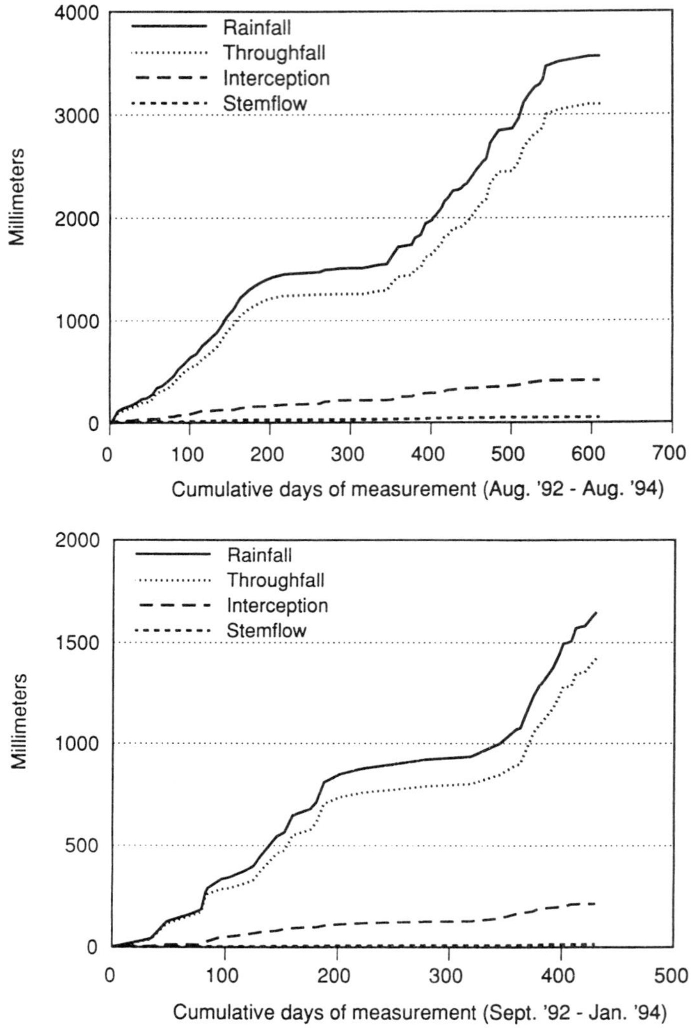

Figure 1 Cumulative gross rainfall, throughfall, interception loss and stemflow plotted against the cumulative number of days of observation at (a) Reserva Jaru, (b) Reserva Vale do Rio Doce

RESULTS

The cumulative values of gross rainfall, throughfall, interception loss and stemflow measured during the experiment are plotted in Figures 1a for Reserva Jaru and in 1b for Reserva Vale do Rio Doce. During the 78 collection periods at Jaru there was a total of 3563.8 mm of gross rainfall. The total measured throughfall was 3098.8±187.3 mm or 87.0% (±5.3%) of the gross rainfall whilst the trunk gauges measured 51.1 ± 13.7 mm or 1.4% (±0.4%) of the gross rainfall. The total measured interception loss was 414.2 ± 211.1 mm which is equivalent to 11.6 % (±5.9%) of the gross rainfall.

Table 2 Measured and modelled results expressed as percentages of the gross rainfall

Component of gross rainfall	Reserva Jaru		Reserva Vale do Rio Doce	
	Rutter	Observed	Rutter	Observed
Total interception loss	15.6	11.6	14.1	12.9
Total throughfall	83.4	87.0	84.7	86.2
Total stemflow	1.0	1.4	1.2	0.8

At Reserva Vale do Rio Doce, Marabá, the total gross rainfall over the 39 data collection periods was 1649.9 mm. The total throughfall was 1423.0 ± 102.0 mm or 86.2% (±6.2%) of the gross rainfall and the total measured stemflow was 13.6 ± 5.4 mm , 0.8% (±0.3%) of the gross rainfall. The measured interception loss was 213.3 ± 78.5 mm or 12.9% (±4.8%) of the gross rainfall.

Table 2 summarises the values of interception loss, throughfall and stemflow calculated by the Rutter model and observed in the field for each site. The Rutter model estimated a total interception loss of 557.7 mm for Jaru, equal to 15.6% of the gross rainfall. The throughfall was calculated as 2971.9 mm (83.4%) and the stemflow as 34.2 mm (1.0%). In Vale do Rio Doce the model gave a value of 232.9 mm (14.1%) for the total interception loss from a calculated throughfall and stem flow of 1397.3 mm (84.7%) and 19.7 mm (1.2%), respectively. The cumulative measured and modelled interception loss are shown in Figures 2a and 2b for Jaru and Vale do Rio Doce respectively. The cumulative measured and modelled throughfall at each site are shown in Figures 3a and 3b, and for stemflow in Figures 4a and 4b.

In their study of interception loss in Reserva Ducke, Manaus, Lloyd and Marques (1988) analysed throughfall variability by expressing each individual throughfall catch as a percentage of the gross rainfall and studying the frequency distribution. A

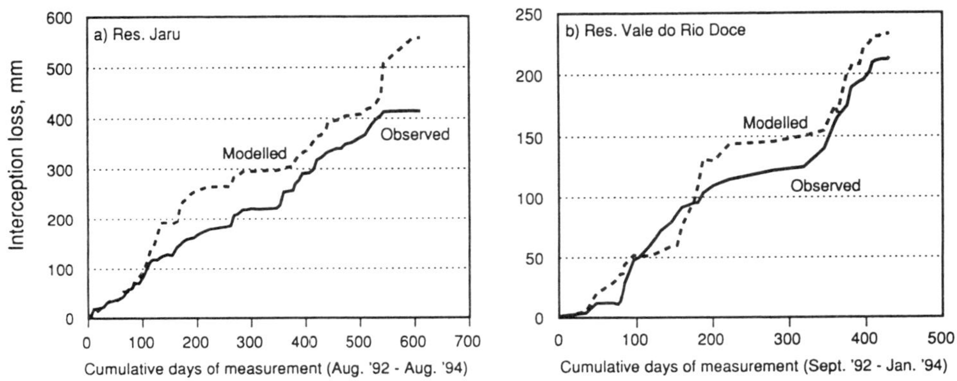

Figure 2 Cumulative observed and modelled interception loss at (a) Reserva Jaru, (b) Reserva Vale do Rio Doce

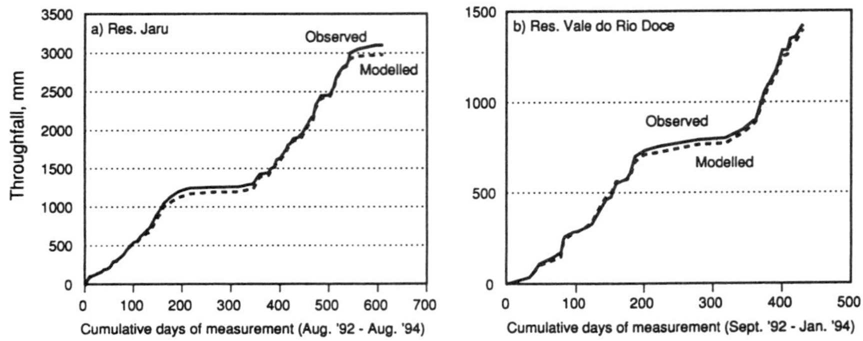

Figure 3 Cumulative observed and modelled throughfall at (a) Reserva Jaru, (b) Reserva Vale do Rio Doce

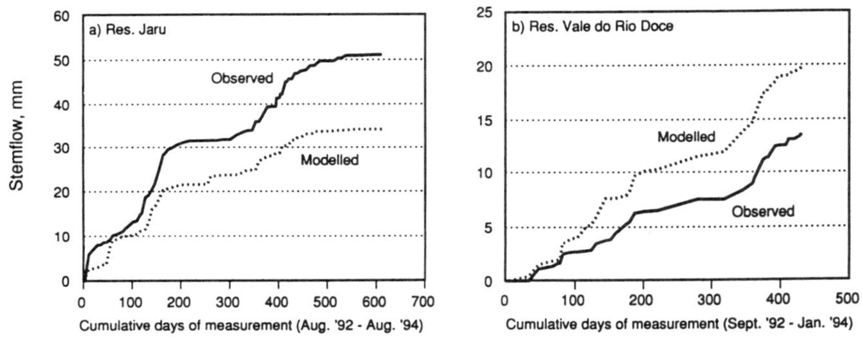

Figure 4 Cumulative observed and modelled stemflow at (a) Reserva Jaru, (b) Reserva Vale do Rio Doce

similar analysis of the data from Jaru has been carried out in the present work. The random sampling procedure utilised 501 of the possible 505 sampling points. The frequency distribution is illustrated in Figure 5, and shows that 59% of the sampling positions collected less than the gross rainfall measured during the same period and accounted for only 43% of the total measured throughfall. Conversely, 41% of the sampling points were sited under points of concentration, or 'drip points', collecting more water than the gross rainfall and accounting for 57% of the total throughfall.

The variation of interception loss with different gross rainfall rates was also studied. Figure 6 shows that for high gross rainfall rates of over 40 mm h^{-1}, the interception loss is insignificant in comparison with the gross rainfall. In contrast, when the rainfall rate is less than 2 mm h^{-1} the interception loss can be as much as 60 to 70% of the gross rainfall.

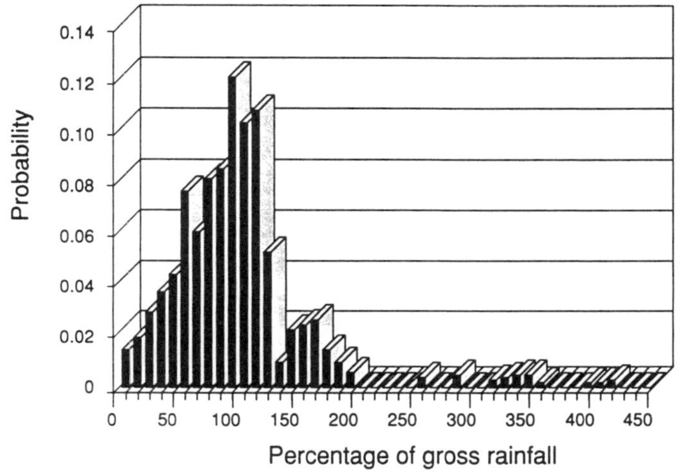

Figure 5 The frequency distribution of gauge catch as a proportion of gross rainfall for the Reserva Jaru site

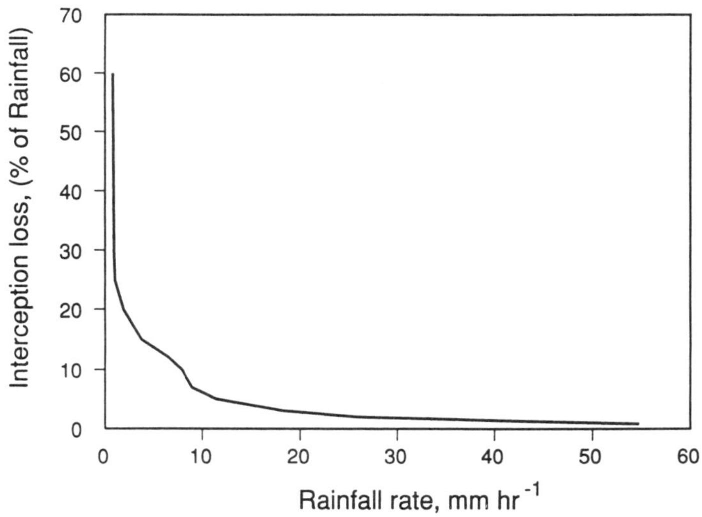

Figure 6 The values of interception loss calculated by the Rutter model for different gross rainfall rates

DISCUSSION

Figures 1a and 1b show the interception losses measured at Reserva Jaru and Reserva Vale do Rio Doce, are similar to those obtained by Lloyd and Marques (1988) for

Reserva Ducke, Manaus. They studied 225 sample days from a period lasting from 1983 until 1985 and obtained a value for the interception loss of 8.9% of the gross rainfall, a value which is close to the values obtained in the present work: 11.6% at Jaru and 12.9% at Vale do Rio Doce.

Notwithstanding the relatively high errors associated with the measured parameters, the difference between the present study and previous work may be interpreted in terms of the parameter differences between sites. The lower values of free throughfall parameter, p, found in the two sites studied, compared to that for Reserva Ducke, Manaus are consistent with the higher rates of interception loss. A lower value of p means that the effective precipitation is greater and therefore the interception loss is higher. Also the higher interception losses at Jaru and Vale do Rio Doce are consistent with the higher estimates of canopy capacity than at Manaus. Prediction of interception loss using the Rutter model is more sensitive to S_c than p, and the value of the S_c in the model can be adjusted so that the model gives the observed values for interception loss (11.6% for Jaru and 13% for Vale do Rio Doce). It was found that the values for S_c required to give these observed interception losses are 0.69 mm for Jaru and 0.60 mm for Vale do Rio Doce. Although these values are lower than the 0.74 obtained by Lloyd *et al.* (1988), the Rutter model also overestimated the interception loss at Reserva Ducke.

The throughfall frequency distribution results indicated a similar conclusion to that made by Lloyd and Marques (1988). It is difficult to conduct this type of experiment successfully in tropical forest because the irregular distribution of throughfall will lead to significant errors in the average throughfall if sampling is inadequate. At the Jaru site, for instance, 41% of the throughfall sampling points were considered to be beneath "drip points".

The present work and the work of Lloyd and Marques (1988) shows that although the actual quantity of interception loss in the Amazonian forest is high, it is a smaller proportion of the gross rainfall compared with other types of tall vegetation such as temperate forests. However, the value of the interception loss measured during this experiment varies quite strongly between collection periods. There is a strong reduction in the proportion of rainfall intercepted during heavy storms in comparison to lighter rainfall events. Figure 6 shows that in tropical forests the canopy reaches saturation quickly. This is due to both the rainfall intensity and, according to Valente (1990), the foliar structure of the vegetation which consists of big leaves with a waxy layer which tends to decrease the water retention capacity. The orientation and shape of leaves in tropical forests also facilitates fast drainage of water from the canopy.

Within the limitation of requiring detailed hourly data, the Rutter model simulated the interception process satisfactorily for the sites of this study. The performance of the model was similar to that obtained by Lloyd *et al.* (1988) for Reserva Ducke, Manaus. Considering that the errors associated with the experimental derivation of the model parameters are relatively large, the difference of approximately 3% of gross rainfall between the modelled and observed values of interception loss at both sites is acceptable. The good agreement between the modelled and observed throughfall components, as shown in Figure 3b for Vale do Rio Doce for example, suggests the values used for the parameters controlling this aspect of the model were accurately calibrated. The calibration of the stemflow was not as successful and this component will require better parameterisation in future work. Other improvements which need to be made to the model are changes to the drainage parameterisation for tropical forests and accurate calculation of the parameter f which controls the value of the aerodynamic resistance. The model is physically realistic and was able to simulate the observed data relatively well.

ACKNOWLEDGMENTS

This study was supported financially by the UK Overseas Development Administration, CAPES and FAPESP, Brazil. The work was an integral part of the author's MSc. thesis carried out at the Instituto Nacional de Pesquisas Espaciais (INPE), Brazil. Thanks are due to all members of the ABRACOS team and especially to Dr J.H.C. Gash and Dr C.A. Nobre.

REFERENCES

Clarke, R.T. 1987. The interception process in tropical rain forests: A literature review and critique. *Acta Amazonica*, **17**, 225-237.

Culf, A.D., Esteves, J.L., Marques Filho, A.O. and da Rocha, H.R. 1996. Radiation, temperature and humidity over forest and pasture in Amazonia. This volume.

Ford, E.D. 1976. The canopy of a Scots pine forest: description of a surface of complex roughness. *Agric. Meteorol.*, **17**, 9-32.

Gash, J.H.C. 1979. An analytical model of rainfall interception in forests. *Q.J.Roy. Meteorol. Soc.*, **105**, 43-55.

Gash, J.H.C. and Morton, A.J. 1978. An application of the Rutter model to the estimation of the interception loss from Thetford Forest. *J. Hydrol.*, **38**, 49-58.

Gash, J.H.C. and Stewart, J.B. 1977. The evaporation from Thetford Forest during 1975. *J. Hydrol.*, **35**, 385-396.

Gash, J.H.C., Wright, I.R. and Lloyd, C.R. 1980. Comparative estimates of interception loss from three coniferous forests in Great Britain. *J. Hydrol.*, **48**, 89-105.

Gash, J.H.C., Nobre, C.A., Roberts, J.M. and Victoria, R. L. 1996. An overview of the ABRACOS project. This volume

Horton, R.E. 1919. Rainfall interception. *Monthly Weather Review*, **49**, 603-623.

Jackson, I.J. 1971. Problems of throughfall and interception assessment under tropical forest. *J. Hydrol.*, **12**, 234-254.

Lloyd, C.R. and Marques, A.O. 1988. Spatial variability of throughfall measurements in Amazonian rainforest. *Agric. Forest. Meteorol.*, **43**, 277-294.

Lloyd, C.R., Gash, J.H.C., Shuttleworth, W.J. and Marques, A. O. 1988. The measurements and modelling of rainfall interception by Amazonian rainforest. *Agric. Forest Meteorol.*, **43**, 277-294.

McWilliam, A-L.C., Cabral, O.M.R., Gomes, B.M., Esteves, J.L. and Roberts, J.M. 1995. Forest and pasture leaf-gas exchange in south-west Amazonia. This volume.

Rutter, A.J. 1967. An analysis of evaporation from a stand of Scots pine. In: W.E. Soper and H.W. Lull (Eds), International Symposium on Forest Hydrology. Pergamon press, Oxford. 403-417.

Rutter, A.J., Kershaw, K.A., Robins, P.C., and Morton, A.J. 1971. A predictive model of rainfall interception in forest. I Derivation of the model from observations in a plantation of Corsican Pine. *Agric. For. Meteorol.*, **9**, 367-389.

Rutter, A.J., Morton, A.J. and Robins, P.C. 1975. A predictive model of rainfall interception in forest. II Generalization of the model and comparison with observations in some coniferous and hardwood stands. *J. Appl. Ecol.*, **14**, 567-588.

Sá, T.D.A., de Costa, J. de P. and Roberts, J.M. 1996. Forest and pasture conductances in southern Pará, Amazonia. This volume.

Shuttleworth, W.J., Gash, J.H.C., Lloyd, C.R., Moore, C.J.,Roberts, J.M., Molion, L.C.B.,Nobre C.A.,de Abreu Sa, L.D., Marques, A. de O., Fisch, G., Januario, M., Fattori, A.P., Ribeiro, M.N.G., Cabral ,O.M.R. 1987. Amazonian Evaporation. *Revista Brasileira de Meteorologia*, **2**, 179-191.

Ubarana, V.N. 1994. Experimentos observacionais e modelagem das perdas por interceptação da precipitação na Floresta Amazônica. MSc Thesis. Instituto Nacional de Pesquisas Espaciais. INPE 5583TDI/545. São José dos Campos, SP, Brazil.

Valente, F.M.R.T. 1990. Interceptação da precipitação por cobertas arbóreos: as condicionantes fisicas e a modelação do processo. Universidade Técnica de Lisboa., Portugal.

Wright, I.R., Nobre, C.A., Tomasella, J, da Rocha, H.R., Roberts, J.M., Vertamatti, E., Culf, A.D., Alvalá, R.C.S., Hodnett, M.G. and Ubarana, V. 1996. Towards a GCM surface parameterisation for Amazonia. This volume.

RESUMO

Um experimento observacional da interceptação de chuva pelo dossel da vegetação foi conduzido em dois sítios de floresta na Amazônia: na Reserva Jaru em Ji-Paraná e na Reserva Vale do Rio Doce em Marabá. A metodologia utilizada é aquela descrita por Lloyd e Marques (1986). Na Reserva Jaru a componente da interceptação foi de 11,6% da precipitação total durante o período de observação (agosto de 1992 a agosto de 1994). Na Reserva Vale do Rio Doce, 12,9% da precipitação total foi interceptada pelo dossel durante o período de observação (setembro de 1992 a janeiro de 1994). O modelo de interceptação de Rutter tambem foi aplicado aos dados observacionais e superestima a interceptação da vegetação nos dois sítios experimentais: 15,6% em Ji-Paraná e 14,1% em Marabá. Tendo em vista os erros envolvidos na determinação dos parametros, o modelo possuiu uma performance satisfatoria, mas que pode ser melhorada modificando-se a parametrização de drenagem e melhorando-se a calibração da componente do escoamento.

9 The isotopic composition of atmospheric water vapour inside a canopy in Amazonian forest: vertical and diurnal variation

A. RIBEIRO[1], R.L. VICTORIA[1,2], L.A. MARTINELLI[1], M.Z. MOREIRA[1] and J.M. ROBERTS[3]
[1] *Centro de Energia Nuclear na Agricultura, USP, Piracicaba, SP, Brasil*
[2] *Departamento de Fisica e Meterologia ESALQ/USP, Piracicaba, SP, Brasil*
[3] *Institute of Hydrology, Wallingford, UK*

INTRODUCTION

The Amazon basin has an area of approximately 6.4 million km^2 (7 million km^2 including the Tocantins-Araguaia basin), and comprises almost one-third of South America. Its average annual precipitation is about 2200 mm, resulting in a total volume of about 14.1×10^{12} m^3 of water per year for the entire basin. Approximately 6×10^{12} m^3 returns to the Atlantic ocean via the Amazon river, and the balance returns to the atmosphere as forest evapotranspiration (Richey *et al.* 1989, Salati *et al.* 1989). Therefore, at any place within the basin, the atmospheric water vapour is a mixture of the zonal flux derived from the Atlantic ocean and the forest evapotranspiration (Marques, 1976). The importance of evapotranspiration for the recycling of water in the Amazon basin has been shown by several authors. The estimates for the evapotranspiration contribution vary from 48 to 80% (Villa Nova et *al.*, 1976; Molion, 1979; Salati *et al.*, 1979; Jordan and Heuveldop, 1981; Marques *et al.*, 1980; Ribeiro *et* al., 1982; Leopoldo *et al.*, 1982a, 1982b; Shuttleworth *et al.*, 1984).

Evapotranspiration has two components: plant transpiration and direct water evaporation. Very little is known about the relative contribution of each process to the total evapotranspiration flux. Recent studies (Gat and Matsui, 1991; Victoria *et al.*, 1992), based on isotope data, have suggested that during the dry season in the central part of Amazonia, direct evaporation may contribute up to 40% of the total evapotranspiration flux. Varzea lakes, rivers, and canopy interception are possible sources for direct evaporation. Indeed, recent work has shown that a significant amount of water is intercepted by the canopy. Lloyd *et al.* (1988) found an average interception of 9% of the total rainfall for forest near Manaus. Ubarana (1995; 1996b) found higher values (12.5%) in Ji-Paraná (western Amazonia), and 13% in Marabá (eastern Amazonia). This water may be entirely or in part re-evaporated to

Amazonian deforestation and climate. Edited by J.H.C. Gash, C.A. Nobre, J.M. Roberts and R.L. Victoria.

the atmosphere and although the above estimates are fairly accurate, it is unknown if the process is of significant magnitude to alter the isotopic composition of the water vapour within and above the forest. In other words, knowing how the isotopic composition of water vapour changes with height and time and what mechanisms control the changes at a local scale might be of help in understanding the changes in the isotopic signals at the regional scale shown by Victoria *et al.* (1992).

We propose to test the hypothesis that the water vapour isotopic composition within and above the canopy have distinct controlling mechanisms. While the water vapour within the canopy reflects mostly condensation and evaporation processes, above the canopy the composition reflects the processes of transpiration and exchange with the upper atmosphere. We intend to do this using the water stable isotopes as tracers, through analysis of the variations in the isotopic composition of the atmospheric water vapour in the forest canopy. Both variations with time and height will be addressed. The rationale in using the isotope technique is based on the fact that direct evaporation generates vapour that is isotopically lighter than transpired water. It should also be remembered here that the condensation process leaves behind isotopically lighter water vapour if the process does not continue to completion. Further, because it is well established that plant transpiration is non-fractionating after a steady-state equilibrium is reached, transpired water has an isotopic composition similar to the soil water, and ultimately to the local rainfall (Salati *et al.*, 1979; Bariac *et al.* 1989; Bariac *et al.* 1991; Flanagan *et al.*, 1991; Yakir, 1992). Nevertheless, Zundel *et al.* (1978) studying the isotopic behaviour of leaf water in tropical trees, observed that a steady state is rarely achieved during a daily cycle, which might indicate that the isotopic composition of transpired water could differ from soil water at certain times of the day.

MATERIAL AND METHODS

The samples were collected at the Reserva Jaru in Rondonia. They were taken at different times of the day (05.30, 08.30, 13.30, 17.00, and 23.00 h), and at different heights (0.5, 10, 30 and 50 m) within and above the forest, using previously evacuated 500 ml glass containers. Samples were taken for three consecutive days, from 21 to 23 May, 1993. The samples at different heights were collected through inert plastic hoses attached to a 50 m high aluminium tower, as used by the ABRACOS project for micrometeorological measurements (Gash *et al.*, 1996). The 13.30 and 05.30 sampling coincided with the maximum and minimum peaks of available energy. The samples were prepared for isotopic analysis using a sample preparation technique for isotope analysis in micro quantities of water, based on the techniques proposed by Epstein and Mayeda (1953), Bariac *et al.* (1982) and Ferhi *et al.* (1983). The analysis for ^{18}O were performed in a Finnigan Delta E Mass Spectrometer, and for D in a Varian Mat GD150, both fitted with a double inlet and double collector.

The results are given in δ‰ deviation from the international standard SMOW (Standard Mean Ocean Water), defined as:

$$\delta\,{}^{0}/_{00} = \frac{(R_{sample} - R_{smow})}{R_{smow}} \times 1000$$

Where R is the ratio D/H, or $^{18}O/^{16}O$.

The atmospheric water vapour concentration was determined by psychrometry, using the wet and dry bulb temperatures in a fully ventilated ventilated shield.

RESULTS AND DISCUSSION

The results are plotted in a diagram of $\delta D \times \delta^{18}O$ in Figure 1. The Meteoric Water Line (MWL), which represents the regression of $\delta D \times \delta^{8}O$ for precipitation waters of the world and is defined by the equation $\delta D = 8 \times \delta^{18}O + 10$, is also shown. Major deviations from this line indicate an evaporation process intervening. During the process of evaporation from a free water body, the remaining liquid will become enriched with the heavier isotopes (^{18}O and deuterium) while the resulting vapour will be depleted in those isotopes (Craig, 1961; Gat, 1971). As a consequence, the remaining fraction of a water body that suffered strong evaporation will not fall on the MWL but below it. To maintain a mass balance, the generated water vapour would be isotopically depleted but fall above the MWL. This is not the case for the data shown in Figure 1, where most fall below the MWL and seem to be following a

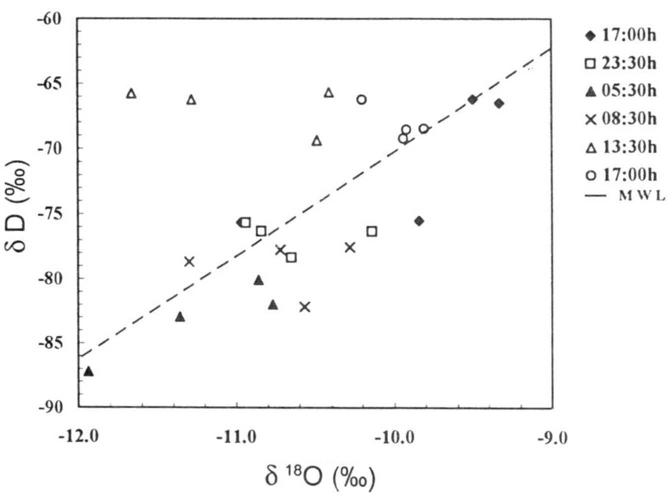

Figure 1 $\delta^{18}O$ vs δD showing the results for all samples. The dotted line is the Meteoric Water Line (MWL) defined as $\delta D = 8*\delta^{18}O + 10$.

cycle of depletion in isotopes during the night and enrichment during the day. The data suggest that there is not a major evaporation source and that changes in isotopic composition should be attributed to another process. Since they are somehow following the MWL, we will analyse just the δD composition to try to understand the phenomenon. The analysis would be similar if $\delta^{18}O$ were to be used. Also, the results for the same sampling time but for different days showed similar behaviour and will be pooled together.

The δD values vary with the sampling height, as well as during a daily cycle. At night there was a clear decrease in the isotopic composition of the water vapour for all heights (Figure 2). This trend was reversed during the day when the isotopic values become heavier. There are two hypotheses to explain the observed decrease in the isotopic values: (i) during the night there is no transpiration but only evaporation so the night-time decrease in the δD values is due to the injection of evaporated water into the atmosphere, and (ii) during the night, with the decrease in temperature, there is an increase in the condensation process and the remaining water vapour is isotopically lighter. One way to test our hypotheses is to look to the night-time variation of the water vapour concentration. If it is increasing, then the evaporation process is prevailing, and consequently the first hypothesis would be correct. On the other hand, if the vapour concentration is decreasing, condensation would be the dominant process and then the second hypothesis would be correct. It is clear from Figure 3 that the condensation is responsible for the observed decrease in the isotopic value, since there is a sharp decrease in the vapour concentration in all forest strata. During the day both evaporation and transpiration are active, and this can be seen from the increase in the vapour concentration during the day in all strata (Figure 3). In this case, the increase in the isotopic values clearly indicates that the re-evaporation and/or transpiration process is higher than the evaporation. From this point the vapour concentration would start to decrease as the condensation process closes the cycle.

Another important trend observed was the isotopic variation with sampling height. By the end of the night, the boundary layer is fully formed, gases produced during

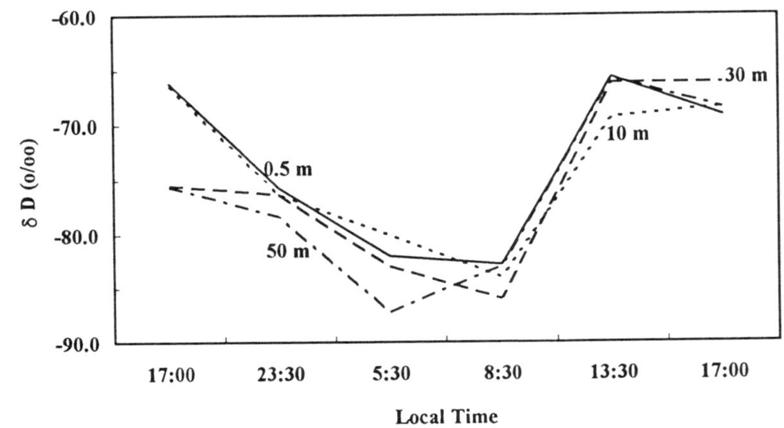

Figure 2 Variation of δD with sampling time during a daily cycle

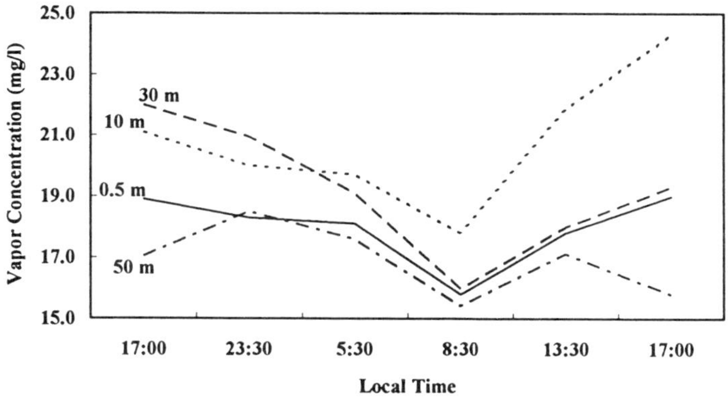

Figure 3 Variation of the vapour concentration with sampling time during a daily cycle

the night are trapped inside the forest, and the exchange with the atmosphere is less intense. This has already been shown in several studies, of which that for CO_2 is a good example (Quay *et al.*, 1989). In our case, a gentle trend can be seen at 05.30 in the morning; the isotopic values become progressively lighter up through the forest canopy (Figure 4). This may indicate that the condensation process mentioned above is occurring preferentially at the canopy level where the lighter isotopic values, together with a decrease in the vapour concentration, were observed (Figure 5). At 08.30, three hours later, the same trend persisted for the vapour concentration but it was not clear for the isotopes. At 13.30 the isotopic values became heavier due to transpiration and the isotope gradient practically disappeared (Figure 4). At this time, although the available energy was the highest possible, the water vapour concentration for 10, 30 and 50 m is smaller than at other times, with the exception of the 08.30 sampling (Figure 5). The most likely explanation is that convective mixing in the boundary layer is coupling the canopy to the deeper atmosphere (Martin *et al.* 1988).

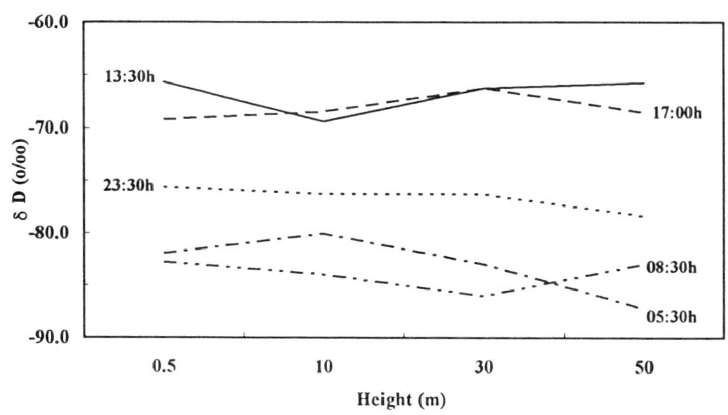

Figure 4 Variation of δD with sampling height during a daily cycle

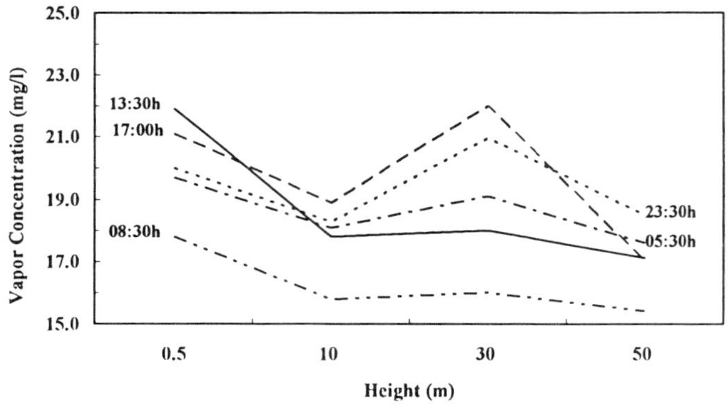

Figure 5 Variation of the vapour concentration with sampling height during a daily cycle

The 0.5 m layer seems to be unaffected. At 17.00 the lack of any clear trend for the isotopic concentration still persisted, i.e. there was no clear gradient between all forest strata. Finally, at 23.00 the isotopic value became lighter again, indicating the beginning of the condensation process.

To test our results further and to explain the variations in the isotopic composition of the vapour, a simple Rayleigh distillation model, considering only the condensation and re-evaporation of condensed water, was used. It was considered that the entire condensed or re-evaporated fraction does not interfere with the continuity of the process, i.e. it is an open process. The condensation process was modelled using the Rayleigh equation:

$$\delta = \delta_0 + \varepsilon \ln f$$

where δ_0 is the initial isotopic composition of the vapour (^{18}O or D), δ is the isotopic composition at a time when f is the residual vapour fraction, and ε is the fractionation factor, defined as $\varepsilon=\alpha-1$. α is an inverse function of the temperature, as shown by several authors (Dansgaard, 1964; Bottinga and Craig, 1968; Majoube, 1971) and defined by the following equations:

$$\alpha_{18(v,l)} = \exp\left(- \frac{1.1370 \times 10^3}{T^2} + \frac{0.4156}{T} + 2.0667\times10^{-3}\right)$$

$$\alpha_{D(v,l)} = \exp\left(- \frac{24.8448 \times 10^3}{T^2} + \frac{76.2480}{T} - 52.6120\times10^{-3}\right)$$

which were used to calculate the values of α. The temperature was taken as the average of the observed ones at the two sampling times. The residual vapour fraction was calculated as the ratio of the concentrations at the sampling times.

To model the re-evaporation process the following mass balance equation was used:

$$\delta_1 = \frac{\delta_0 Q_0 + \delta_e Q_e}{Q}$$

where δ_1, δ_0, δ_e are the modelled, previous sampling and re-evaporated isotopic composition of water vapour, respectively. The Q_1, Q_o and Q_e were the respective concentrations of water vapour. To calculate the δ_e values it was necessary to know the average isotopic composition of the condensed water (δ_c) and the remaining fraction of water (δ_r) during the re-evaporation process. The isotopic composition of re-evaporated water was then defined as:

$$\delta_e = \frac{\delta_c Q_c + \delta_r Q_r}{Q_e}$$

Starting with the value ford and vapour concentration for the first sampling time, the model was run in sequence, and the result was compared with the observed data. Figure 6 shows the differences between modelled and observed values of $\delta^{18}O$ (a) and D(b) for different times and heights, for three consecutive days, for all individual samples. The condensation model was used for the points 1, 2, 3 and 6 and the re-evaporation model for points 4, 5 and 7.

The fit between observed and modelled data is good for the night samples, suggesting that condensation is responsible for the drop in the isotopic composition of the vapour. At 08.30 h (point 3), the fit is still good, with the exception of the sample at canopy level (30 m) and above it (50 m). At those heights the observed values are heavier than the calculated, indicating a possible source of isotopically heavier vapour. The re-evaporation of the previously condensed water vapour and the transpiration are likely candidates in this case because the highest foliar index and the transpiration flux are observed at the canopy level (Roberts *et al.*, 1993).

At 13.30 h the data show a good fit for the re-evaporation model, with the exception of the canopy level, where the observed isotopic composition is higher than the modeled one. This enrichment, which is the maximum observed, is suggesting the existence of an isotopically heavier source, probably the transpired water. Zundel *et al.* (1978) studying the isotopic composition of leaf water of tropical trees showed heavier values during the morning, reaching a maximum between 12 and 14 h and decreasing thereafter. Also, according to Roberts *et al.* (1993), the maximum transpiration rate for Manaus is at around 13.00 h.

This same tendency was observed at 17.00 h (point 5), the actual isotopic composition is higher than the modeled, suggesting the existence of an isotopically heavier source, likely the transpiration. It should also be pointed out here that the best fit was observed for the lower levels (0.5, 10 m), which indicates a small exchange with the atmosphere above.

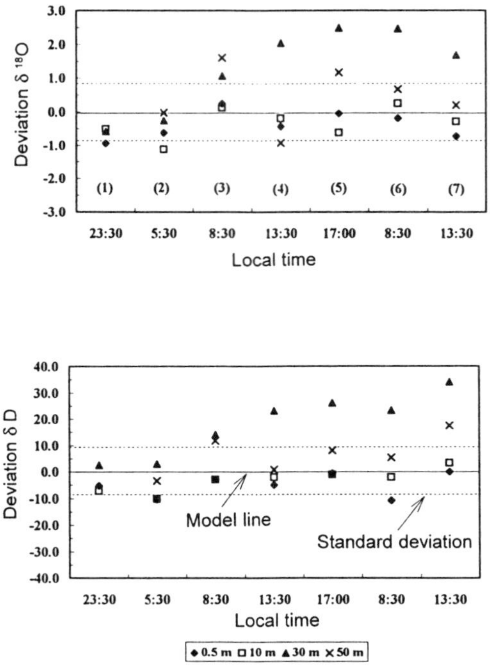

Figure 6 Difference between observed and modelled values for $\delta^{18}O$ (a) and δD (b). The dotted line represents the standard deviation.

CONCLUSIONS

The data suggest that transpiration does play a significant role in controlling the isotopic composition of the water vapour at the canopy level, during the day. Furthermore, it appears that the processes that are changing the isotopic composition under and above the canopy are decoupled. Under the canopy, a simple condensation-re-evaporation process is able to explain the changes. The vapour at the lower levels under the canopy appears to be locally generated and does not have strong exchanges with the atmosphere above. At the higher levels (canopy level and above) the changes in isotopic composition indicate a more complex pattern of mixture, where transpiration and exchanges with the upper atmosphere are important.

However, care should be taken not to extrapolate these results for the entire Amazon basin and for different seasons, because only three days of sampling were used. It will be necessary to repeat the experiment at other sites and at other times to infer that similar mechanisms operate at the regional scale.

ACKNOWLEDGEMENTS

This study was supported by FAPESP (Brazil) and by the ODA ABRACOS (UK) project.

REFERENCES

Bariac, T., Jusserand, C. and Letolle, R. 1982. A routine method for ^{18}O analysis of water vapour. *Intl. J. Appl. Radiat. Isot.* **33**, 1365-68.

Bariac, T., Jusserand, C. and Mariotti A. 1989. Evolution spatio-temporelle de la composition isotopique de l'eau dans le continuum sol-plante-atmosphere. *Geochimica et Cosmochimica Acta* **54**, 413-24.

Bariac, T., Deleens, E., Gerbaud, A., Andre, M. and Mariotti A. 1991. La composition isotopique (^{18}O, ^{2}H) de la vapeur d'eau transpiree: etude en conditions asservies. *Geochimica et Cosmochimica Acta* **76**, 3391-3402.

Bottinga, Y. and Craig, H. 1969. Oxygen isotope fractionation between CO_2 and water and the isotopic composition of marine atmosphere CO_2. *Earth Planet. Sci. Letters* **5**: 285-95.

Craig, H., 1961. Isotopic Variations in Meteoric Waters. *Science*, **113**: 1702-1703.

Dansgaard, W. 1964. Stable isotopes in precipitation. *Tellus*, **16**: 436-68.

Epstein S. and Mayeda T. 1953. Variation of ^{18}O content of waters from natural sources. *Geochim. Cosmochim. Acta* **4**, 213-24.

Flanagan, L., Comstock, J. and Ehleringer, J. 1991. Comparison of modeled and observed environmental influences on the stable oxygen and hydrogen isotope composition of leaf water in *Phaseolus vulgaris* L.P. *Physiol.* **96**: 588-96.

Ferhi, A., Bariac, T. and Lètolle, R. 1983. $^{18}O/^{16}O$ ratio in leaf water and cellulose of aquatic and terrestrial plant. In: *Paleoclimates and Paleowaters. A Collection of Environmental Isotopic Studies.* Proceed. Symp., IAEA, Vienna, 85-99.

Gash, J.H.C., Nobre C., Roberts J., Victoria R. L. 1996. An overview of ABRACOS project. This volume.

Gat, J. 1971. Comments on Stable Isotope Method in Regional Groundwater Investigations. *Water Resour. Res.*, **7**: 980-993.

Gat, J.R. and Matsui, E. 1991. Atmospheric water balance in the Amazon Basin: an isotopic evapotranspiration model. *J. Geophys. Res.*, **96**: 13,179-13,188.

Jordan, C.F. and Heuveldop, J. 1981. The water budget of an Amazonian rain forest. *Acta Amazônica,* **11**: 87-92.

Leopoldo, P.R., Franken, W. and Salati, E. 1982a. Balanço hídrico de pequena bacia hidrográfica em floresta Amazônica de terra-firme. *Acta Amazônica*, **12**: 333 - 338.

Leopoldo, P.R., Franken, W., Matsui, E. and Salati, E. 1982b. Estimativa da evapotranspiração de floresta Amazônica de Terra-Firme. *Acta Amazônica*, **12**: 23-28.

Lloyd, C.R., Gash, J.H.C., Shuttleworth, W.J. and Marques, A.O. 1988. The measurements and modeling of rainfall interception by Amazonian rainforest. *Agric. and Forest Meteorol.*, **43**: 277-294.

Majoube, M. 1971. Fractionment en Oxygen-18 et en Deutérium entre l'eau et sa vapeur. *J. Chem. Phys.*, **68**: 1423-26.

Marques, J. 1976. Contribuição ao estudo hidrológico da Bacia Amazônica. Piracicaba, Universidade de São Paulo, 118p. (Master Thesis).

Marques, J., Salati, E. and Santos, J.M., 1980. Cálculo da evapotranspiração real na bacia Amazônica através do método aerológico. *Acta Amazônica*, **10**: 357-361.

Martin, C.L., Fitzjarrald, D., Garstang, M., Oliveira, A.P., Greco S. and Browell E. 1988. Structure and Growth of mixing layer over the Amazonian rain forest. *J. Geophys. Res.*,**93**,1361-1375.

Molion, L.C. 1979. A climatonomic study of the energy and moisture fluxes of the Amazon basin

with considerations of deforestation effects. Madison Univ. of Wisconsin ,156p. (PhD Thesis).
Quay, P., King, S. and Wilbur, D. 1989. $^{13}C/^{12}C$ of atmospheric CO_2 in the Amazon Basin: forest and river sources. *Journal of Geophysical Research*, **94**, 18,327-18,336.
Ribeiro, M.N.G., Salati, E., Villa Nova, N.A. and Demetrio, C.G.B. 1982. Radiação solar disponível em Manaus (AM) e sua duração com o brilho solar. *Acta Amazônica*, **12**, 339-346.
Richey, J. E., Mertes, L.A., Victoria, R.L., Forsberg, B.R., Dunne, T., Oliveira, E. and Tancredi, A. 1989. Sources and routing of Amazon River floodwave. *Global Biogeochemical Cycles* **3**: 191-204.
Roberts, J., Cabral, M.R.O., Fisch, G., Molion, L.C., Moore, C.J. and Shuttleworth, W.J. 1993. Transpiration from an Amazonian rainforest calculated from stomatal conductance measurements. *Agric. and Forest Meteorol.*, **65**, 175-196.
Salati, E., Dall'Olio, A. Gat, J. and Matsui, E. 1979. Recycling of water in Amazon basin: an isotope study. *Wat. Resourc. Res.*, **15**, 1250-1258.
Salati, E., Victoria, R. L.,Martinelli, L.A., and Richey, J.E. 1989. Deforestation and its possible role in changes in the Brazilian Amazon. In: R. DeFries (Ed.)*Global Change and Our Common Future: Papers from a Forum*. National Academy Press, Washington , 1989.
Shuttleworth, W.J., Gash, J.H.C., Lloyd, C.R., Moore, C.J., Roberts, J., Marques, A.O., Fisch, G.F., de Paula Silva, V., Ribeiro, M.N.G., Molion, L.C., Abreu Sa, L.D., Nobre, C., Cabral, M.R.O., Patel, S.R. and Moraes, J.C. 1984. Eddy correlation measurements of energy partition for Amazonian forest. *Q. J. R. Meteorol. Soc.*, **110**, 1143-1162.
Ubarana, V.N. 1995. Experimentos observationais e modelagem das perdas por interceptação da precipitação na Floresta Amazônica, Instituto Nacional de Pesquisas Espacias, Brazil, 94p. (Master Thesis).
Victoria, R.L., Martinelli, L.A., Mortatti, J. and Richey, J. 1992. Mechanism of water recycling in the Amazon Basin: isotopic insights. *AMBIO*, **20**, 384-387.
Villa Nova, N.A., Salati, E. and Matsui, E. 1976. Estimativa da evapotranspiração na bacia Amazônica. *Acta Amazônica*, **6**, 215-228.
Yakir, D. 1992. Water compartmentalization in plant tissue: isotopic evidence. **In**: *Water and Life* (Eds. SGN Somero, CB Osmond and CL Bolis), Springer-Verlag, Berlin, pp 205-222.
Zundel, G., Miekeley, W., Breno, M.G. and Förstel, H. 1978. The $H_2^{18}O$ Enrichment in the Leaf Water of Tropic Trees: Comparison of Species from the Tropical Rain Forest and the Semi-Arid Region in Brazil. *Rad. and Environm. Biophys.* **15**, 203-212.

RESUMO

A composição isotópica da água pode ser usada para inferir os processos aos quais uma parcela de água foi submetida. Por exemplo, o processo de evaporação parcial altera a composição isotópica da água, gerando vapor mais leve que a fonte. Portanto, processos de evaporação-condensação podem alterar a composição isotópica do vapor d'água atmosférico, dentro e sobre o dossel de uma floresta, e ser útil para distinguir os mecanismos que a controlam. Neste trabalho, uma análise da composição isotópica e da concentração de vapor na floresta, para três dias consecutivos, é apresentado. O experimento foi conduzido na Reserva Biológica do Jaru, em Ji-Paraná, Rondônia. Os resultados mostraram que a composição isotópica do vapor de água, tanto para o ^{18}O como para o D, seguem um ciclo. Durante a noite, o processo de condensação prevalece tornando a composição isotópica do vapor de água cada vez mais leve. Os resultados apresentam bom ajuste ao modelo de condensação de Rayleigh. Durante o dia é observado um processo inverso, os valores isotópicos se tornando pesados novamente. Provavelmente isso ocorre devido a evaporação da água previamente condensada e transpiração das plantas, fontes de vapor de água com uma composição isotópica mais pesada. Os dados também sugerem que, durante o dia, a composição isotópica do vapor ao nível da copa das árvores é controlada principalmente pelo processo de transpiração. Finalmente os dados também sugerem que os processos que controlam a composição isotópica do vapor de água dentro e acima da copa das árvores são independentes.

10 Radiation, temperature and humidity over forest and pasture in Amazonia

A.D. CULF[1], J.L. ESTEVES[2], A. de O. MARQUES FILHO[3] and H.R. da ROCHA[4]
[1]*Institute of Hydrology, Wallingford, UK*
[2]*INCRA, Ji-Paraná, Brazil*
[3]*INPA, Manaus, Brazil*
[4]*Universidade do São Paulo, Brazil*

INTRODUCTION

One of the major objectives of ABRACOS is to provide reliable long term measurements of near surface climate at the centre of large Amazonian clearings and to compare these with similar measurements above adjacent uncleared forest (Shuttleworth *et al.*, 1991). Some preliminary observations for one such paired site were discussed by Bastable *et al.* (1993), and Maitelli and Wright (1996) who also included data from a weather station in an urban situation. Culf *et al.* (1996) discuss long term measurements of albedo from each of the sites. Here longer term observations of radiation components and near surface climate variables for all three of the ABRACOS paired sites are presented.

SITES AND INSTRUMENTATION

The three paired forest and pasture sites at Manaus, Marabá and Ji-Paraná with which this paper is concerned are fully described by Gash *et al.* (1996). Automatic weather stations were installed at both the forest and pasture sites in all three locations. The first installation was at the Manaus sites (Reserva Ducke and Fazenda Dimona) where the data record begins in October 1990. The record from the Marabá sites (Reserva Vale do Rio Doce and Fazenda Boa Sorte) begins in July 1991 and from the Ji-Paraná sites (Reserva Jaru and Fazenda Nossa Senhora) in November 1991. In this paper data up to the end of 1993 are considered. The weather stations comprise upward and downward facing solarimeters (Kipp and Zonen, Delft, The Netherlands), a net radiometer (REBS, Seattle, USA), wet and dry bulb platinum resistance thermometers (Didcot Instrument Company, Abingdon, UK) in an aspirated psychrometer screen (Institute of Hydrology, UK), an anemometer, wind vane and 0.2 mm resolution tipping bucket raingauge (all Didcot Instrument Company, Abingdon, UK) and two soil heat flux plates (Thornthwaite, Elmer, New

Amazonian deforestation and climate. Edited by J.H.C. Gash, C.A. Nobre, J.M. Roberts and R.L. Victoria.

Jersey, USA). Hourly average values from these instruments were recorded by a CR10 datalogger (Campbell Scientific, Shepshed, UK) which sampled the instruments at 0.1 Hz. The weather stations were visited weekly to download the data, and to carry out any necessary servicing.

Although the sites are fully described elsewhere (Culf *et al.*, 1996; Gash *et al.*, 1996), and the aerodynamic roughness lengths of each site are discussed by Wright *et al.* (1996), a short summary of the site characteristics relevant to the present study is given here for convenience.

a. Reserva Ducke, Manaus

Reserva Ducke (2° 57'S, 59° 57'W) is an area of protected primary forest, 25 km north-east of Manaus at 80 m above mean sea level. The mean forest canopy is 35 m high, but some trees reach up to 40 m. The site is surrounded by undisturbed forest for at least 5 km. The automatic weather station is mounted at the top of a 45 m aluminium tower.

b. Fazenda Dimona, Manaus

Fazenda Dimona (2°19'S, 60°19'W) is a cattle ranch located about 100 km north of Manaus. The ranch is a 10 km^2 clearing at an altitude of 120 m above mean sea level. The automatic weather station is mounted at the surface with the windspeed being measured at a height of 2 m and the reflected solar radiation at a height of 1.2 m.

c. Reserva Vale do Rio Doce, Marabá

This site (5°45'S, 49°10'W) is in the forest reserve of the Companhia Vale do Rio Doce 50 km to the south of Marabá. The forest reserve is 17,000 hectares in area and is to a large degree uncut, but is almost completely surrounded by cleared areas. The automatic weather station is mounted on top of a 52 m tower positioned to the east of a mature Brazil nut tree (*Bertholletia excelsa* HBK) with a height of 49 m, but the continuous forest canopy in the area surrounding the tower is much lower, around 20 to 25 m.

d. Fazenda Boa Sorte, Marabá

Fazenda Boa Sorte (5°10'S, 48°45'W) is a cattle ranch 50 km to the north-east of Marabá at 170 m above mean sea level. There is little forest remaining in this area and the fetch is pasture or abandoned pasture for several km in all directions. The automatic weather station is mounted at the top of a 5 m aluminium tower with the windspeed being measured at a height of 5.4 m and the reflected solar radiation at a height of 3.9 m.

e. Reserva Jaru, Ji-Paraná

Reserva Jaru (10°05'S, 61°55'W) is a forest reserve owned by the Brazilian Environmental Protection Agency (IBAMA) and is located about 80 km north of Ji-Paraná at 120 m above sea level. In the sector from the north west clockwise to south-south east, the predominant wind direction, the fetch is undisturbed forest for several tens of km. In the remaining sector the undisturbed fetch is shorter, of the order of 1 km. The Ji-Paraná river is about 300 m to the west of the tower and forms the western boundary of Reserva Jaru. The mean height of the forest canopy is 33 m. The automatic weather station was mounted on the top of a 52 m tall tower.

f. Fazenda Nossa Senhora, Ji-Paraná

This site (10°45'S, 62°22'W) is a cattle ranch 220 m above sea level about 50 km east-north-east of Ji-Paraná. The ranch is situated in a strip of cleared land about 4 km wide and several tens of kilometres long, in the centre of an area of about 50 km in radius which has undergone large scale clearance. The automatic weather station instruments are mounted at the same heights as at Fazenda Boa Sorte.

THE RADIATION BUDGET

Figure 1 shows monthly mean values of the measured components of the radiation budgets at each site. These values, along with values of mean temperature and humidity discussed below, are also given in Tables 1 to 6. The net all-wave radiation at the forest sites varies between 9.7 and 13.3 MJ $m^{-2}day^{-1}$ and is, in general, strongly correlated with the incoming solar radiation. The net all-wave radiation is generally less over the pasture sites than over the forest, ranging from 7.4 to 11.3 MJ $m^{-2}day^{-1}$. The low values in August and September at Fazenda Nossa Senhora may be due to the large amount of smoke which is common in the area at that time of the year when a great deal of burning of overgrown pastures takes place. In the Marabá region, on the other hand, there has been little burning over the duration of ABRACOS with the exception of the 1991 dry season. Table 7 shows that the net all-wave radiation over the pasture sites is, on average, 1.27 MJ $m^{-2}day^{-1}$, or 11.1%, less than over the forest sites.

Figure 1 also shows the monthly mean values of incoming solar radiation. The average values and differences between the forest and pasture sites are given in Table 8. The irregular nature of the plot for Reserva Vale do Rio Doce in the first half of the year is because for March, April and May only data from one year of observations were available. Table 8 shows that average values range from 15.71 MJ m^{-2} day^{-1} at Reserva Ducke to 17.42 MJ m^{-2} day^{-1} at Reserva Jaru. On average, the difference between the incoming solar radiation at the pasture and forest sites is insignificant, being at most only about 3.3% of the measured values. This difference accounts for only 7% of the average difference in net radiation between forest and pasture sites. However, the cumulative difference graph plotted in Figure 2 shows that whilst for Manaus and Marabá there is little difference in incoming solar radiation between the forest and pasture sites, during the dry season at Ji-Paraná there is a systematic difference.

A decrease in cloud cover close to large rivers in Amazonia has been observed in satellite images (Molion and Dallarosa, 1990, for example). However, the Ji-Paraná river close to the Reserva Jaru tower is only some 200 m in width compared with

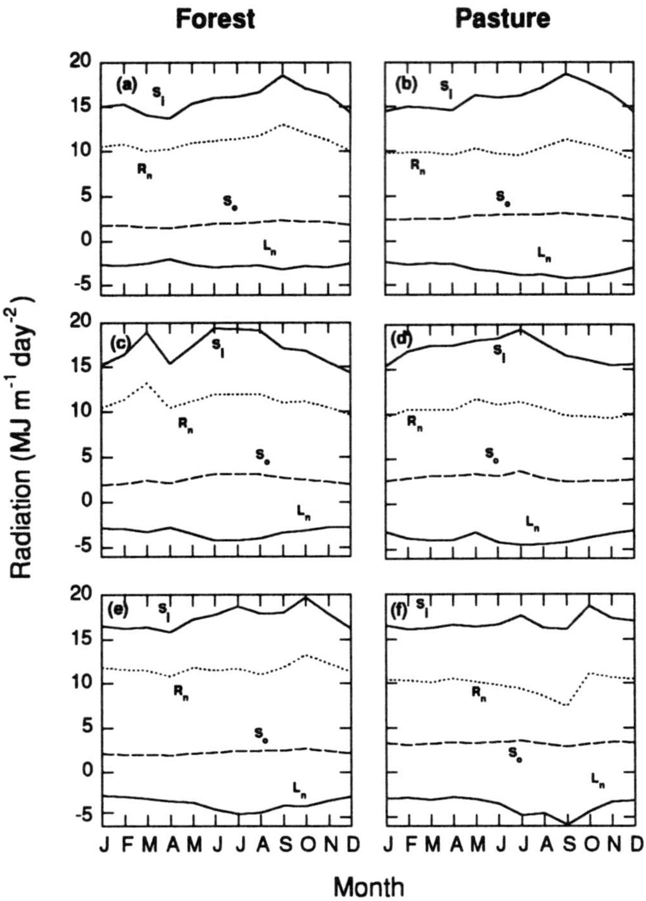

Figure 1 Monthly average values of incoming solar radiation, S_i (upper solid line), net all-wave radiation, R_n (dotted line), reflected solar radiation, S_o (dashed line) and net longwave radiation, L_n (lower solid line). (a) Reserva Ducke, Manaus, (b) Fazenda Dimona, Manaus, (c) Reserva Vale do Rio Doce, Marabá, (d) Fazenda Boa Sorte, Marabá, (e) Reserva Jaru, Ji-Paraná and (f) Fazenda Nossa Senhora, Ji-Paraná.

the several kilometres required for a river to affect atmospheric boundary layer structure and hence cloud growth. In the present case the observation of less incoming solar radiation at Fazenda Nossa Senhora in comparison to Reserva Jaru can be interpreted as the effect of increased cloud cover over the pasture area during the dry season. This increase may occur because the reduced evaporation from the pasture under conditions of water stress leads to a higher sensible heat flux. The convective boundary layer grows more quickly and so the mixed layer over the pasture may reach the lifting condensation level before the mixed layer over the forest, resulting in longer periods of cloud cover. This interpretation is supported by

Table 1 Monthly average values of net all-wave radiation (R_n), incoming solar radiation (S_i), reflected solar radiation (S_o), net longwave radiation (L_n), temperature (T) and specific humidity (q) for Reserva Ducke, Manaus.

Month	Radiation Components (MJ $m^{-2}day^{-1}$)				T (°C)	q (g kg^{-1})
	R_n	S_i	S_o	L_n		
1	10.51	14.98	1.80	-2.70	25.7	16.9
2	10.81	15.27	1.77	-2.73	25.4	16.9
3	10.01	14.08	1.56	-2.53	24.8	17.0
4	10.25	13.75	1.48	-2.02	24.6	17.0
5	10.95	15.38	1.75	-2.68	25.6	17.2
6	11.18	16.04	1.94	-2.93	26	17.0
7	11.39	16.15	1.98	-2.78	25.7	16.3
8	11.76	16.66	2.14	-2.76	25.6	16.0
9	13.03	18.52	2.35	-3.14	26.6	16.3
10	12.06	17.05	2.20	-2.79	26.2	16.5
11	11.28	16.35	2.13	-2.95	26.3	16.6
12	10.05	14.37	1.79	-2.52	25.5	17.0

Table 2 As for Table 1 but for Fazenda Dimona, Manaus

Month	Radiation Components (MJ $m^{-2}day^{-1}$)				T (°C)	q (g kg^{-1})
	R_n	S_i	S_o	L_n		
1	9.76	14.54	2.39	-2.41	25.6	17.1
2	9.83	14.99	2.44	-2.69	25.2	17.1
3	9.82	14.90	2.49	-2.55	24.8	17.3
4	9.55	14.58	2.50	-2.57	24.7	17.4
5	10.28	16.32	2.83	-3.24	255	17.5
6	9.65	16.02	2.90	-3.47	25.1	17.1
7	9.49	16.26	2.91	-3.90	24.6	16.3
8	10.37	17.15	2.95	-3.84	24.5	16.3
9	11.30	18.64	3.06	-4.24	26.6	17.0
10	10.72	17.69	2.85	-4.09	25.4	16.9
11	10.03	16.41	2.68	-3.65	25.6	16.9
12	9.01	14.47	2.36	-3.07	25.5	17.5

Table 3 As Table 1 but for Reserva Vale do Rio Doce, Marabá

Month	Radiation Components (MJ $m^{-2}day^{-1}$)				T (°C)	q (g kg^{-1})
	R_n	S_i	S_o	L_n		
1	10.45	15.25	1.95	-2.90	25.1	16.4
2	11.38	16.38	2.04	-2.92	24.9	16.2
3	13.20	18.90	2.40	-3.30	25.2	16.0
4	10.40	15.40	2.10	-2.80	25.3	16.9
5	11.20	17.30	2.70	-3.50	26.3	16.7
6	11.97	19.33	3.17	-4.20	26.4	15.2
7	11.90	19.26	3.12	-4.27	26.1	14.0
8	11.96	19.10	3.10	-4.05	25.7	14.2
9	11.02	17.06	2.67	-3.37	26.4	15.3
10	11.13	16.85	2.49	-3.17	26.1	15.6
11	10.47	15.50	2.23	-2.80	25.7	16.1
12	9.65	14.41	1.97	-2.82	25.3	16.6

Table 4 As Table 1 but for Fazenda Boa Sorte, Marabá

Month	Radiation Components (MJ $m^{-2}day^{-1}$)				T (°C)	q (g kg^{-1})
	R_n	S_i	S_o	L_n		
1	9.65	15.35	2.60	-3.10	25.0	16.5
2	10.44	17.01	2.85	-3.76	25.1	16.5
3	10.47	17.60	3.12	-4.00	25.1	16.3
4	10.42	17.64	3.18	-4.03	25.6	17.1
5	11.65	18.20	3.25	-3.15	25.8	16.7
6	11.00	18.40	3.10	-4.20	25.4	14.7
7	11.33	19.42	3.68	-4.50	25.4	13.0
8	10.63	17.90	2.87	-4.40	26.0	14.2
9	9.79	16.44	2.49	-4.20	26.2	15.6
10	9.68	16.01	2.58	-3.73	25.9	15.7
11	9.47	15.40	2.60	-3.29	25.8	16.3
12	9.98	15.58	2.70	-2.92	25.5	16.7

Table 5 As Table 1 but for Reserva Jaru, Ji-Paraná

Month	Radiation Components (MJ $m^{-2}day^{-1}$)				T (°C)	q (g kg^{-1})
	R_n	S_i	S_o	L_n		
1	11.78	16.55	2.11	-2.61	24.6	16.8
2	11.57	16.33	2.00	-2.73	24.3	16.8
3	11.44	16.44	2.00	-2.95	24.7	16.9
4	10.81	15.86	1.90	-3.21	24.9	17.2
5	11.80	17.30	2.16	-3.41	25.1	16.6
6	11.45	17.85	2.25	-4.15	25.4	15.4
7	11.70	18.75	2.45	-4.65	23.2	13.4
8	10.99	17.94	2.45	-4.50	24.3	14.2
9	11.80	18.05	2.50	-3.75	25.3	15.6
10	13.30	19.78	2.67	-3.81	25.7	16.3
11	12.30	17.87	2.40	-3.17	25.0	16.6
12	11.30	16.26	2.20	-2.76	24.8	16.9

Table 6 As Table 1 but for Fazenda Nossa Senhora, Ji-Paraná

Month	Radiation Components (MJ $m^{-2}day^{-1}$)				T (°C)	q (g kg^{-1})
	R_n	S_i	S_o	L_n		
1	10.36	16.61	2.60	-2.92	24.5	16.3
2	10.28	16.18	3.05	-2.90	24.5	16.5
3	10.10	16.33	3.21	-3.08	24.4	16.3
4	10.51	16.66	3.36	-2.85	24.7	15.6
5	10.12	16.45	3.27	-3.02	24.2	15.6
6	9.74	16.64	3.38	-3.51	23.3	14.1
7	9.35	17.65	3.55	-4.80	25.4	11.9
8	8.50	16.30	3.20	-4.60	23.7	12.8
9	7.35	16.16	2.87	-5.89	24.5	14.7
10	11.13	18.72	3.20	-4.38	25.0	15.7
11	10.65	17.40	3.40	-3.35	24.8	16.1
12	10.45	17.10	3.35	-3.20	24.7	16.5

Table 7 Average values of net radiation at the forest and pasture sites

Location	Forest $MJ\ m^{-2}\ day^{-1}$	Pasture $MJ\ m^{-2}\ day^{-1}$	Change over pasture $MJ\ m^{-2}\ day^{-1}$	(%)
Manaus	11.11	9.98	-1.13	-10.2
Marabá	11.23	10.37	-0.96	-7.7
Ji-Paraná	11.69	9.88	-1.81	-15.5
Average	11.34	10.08	-1.27	-11.1

Table 8 Average values of incoming solar radiation at the forest and pasture sites

Location	Forest $MJ\ m^{-2}\ day^{-1}$	Pasture $MJ\ m^{-2}\ day^{-1}$	Change over pasture $MJ\ m^{-2}\ day^{-1}$	(%)
Manaus	15.71	16.00	+0.29	+1.8
Marabá	17.06	17.08	+0.02	+0.1
Ji-Paraná	17.42	16.85	-0.57	-3.3
Average	16.73	10.08	-0.09	-0.5

the observation that deeper convective boundary layers sometimes occur over the pasture than over the forest (Nobre *et al.*, 1996) and by satellite observations showing enhancement of cumulus clouds over pasture in Rondônia (Cutrim *et al.*, 1995). However, smoke from the pasture burning may also have had the effect of reducing the input of solar radiation over the pasture areas and the observations should be interpreted with caution.

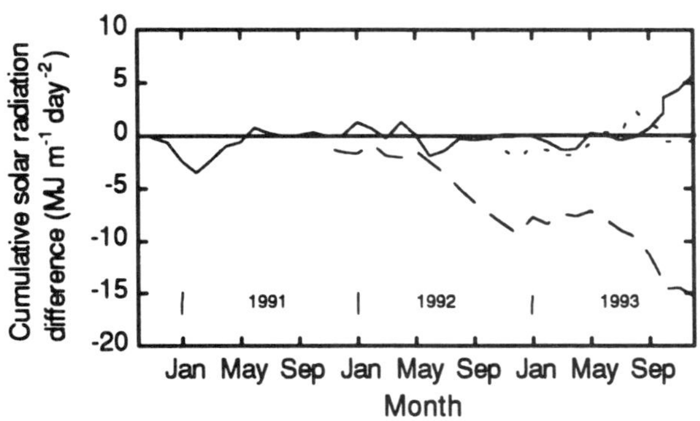

Figure 2 The difference between forest and pasture incoming solar radiation plotted cumulatively for each of the paired sites. Solid line: Manaus. Dotted line: Marabá. Dashed line: Ji-Paraná.

It is interesting to note from Figure 2 that no similar systematic difference in incoming solar radiation between the forest and pasture sites is observed for either the Manaus or the Marabá sites. Probably the difference is only observed in Ji-Paraná because there the forest and pasture areas are both large, and so separate boundary layers may be expected to develop above the two individual vegetation types. In contrast, as discussed by Gash *et al.* (1996), the site at Manaus is a small pasture surrounded by a large area of forest, whilst at Marabá the forest area is relatively small and is surrounded by a large area of pasture. At both of these sites the boundary layer growth will be dominated by the sensible heat flux from the larger component of the surface cover and so no separate boundary layers, which may lead to differences in cloud cover, may develop.

Figure 1 also shows monthly mean values of reflected solar radiation. However, for most purposes, reflected solar radiation is best discussed in terms of albedo, the ratio of reflected solar radiation to incoming solar radiation. Seasonal variations of albedo and the differences in albedo between the forest and the pasture sites have been discussed in detail by Culf *et al.* (1995)and may be summarised as follows. Figure 3 shows monthly average values of albedo for the three forest sites and the three pasture sites. The values for each individual site are given in Culf *et al* (1995).

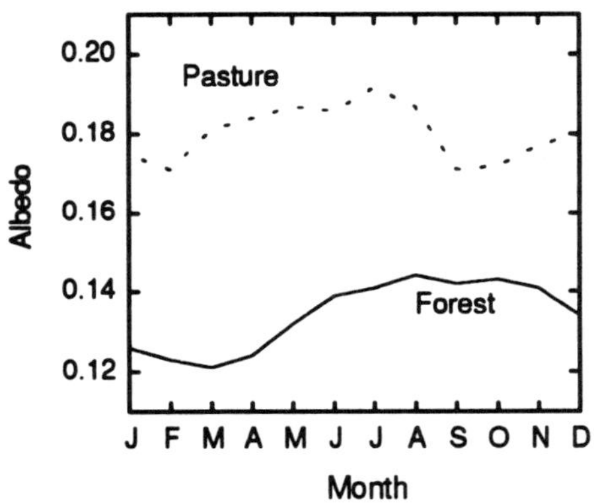

Figure 3 Monthly mean values of forest (solid line) and pasture (dotted line) albedo. The value plotted is the mean of values from the three sites in each case.

Table 9 Average values of reflected solar radiation over forest and pasture sites

Location	Forest MJ m^{-2} day^{-1}	Pasture MJ m^{-2} day^{-1}	Change over pasture MJ m^{-2} day^{-1}	(%)
Manaus	1.91	2.70	+0.79	+41.4
Marabá	2.50	2.93	+0.43	+17.2
Ji-Paraná	2.36	3.26	+1.00	+42.0
Average	2.22	2.96	-0.09	+33.3

Table 10 Average values of net upward longwave radiation

Location	Forest MJ m^{-2} day^{-1}	Pasture MJ m^{-2} day^{-1}	Change over pasture MJ m^{-2} day^{-1}	(%)
Manaus	2.71	3.31	+0.60	+22.1
Marabá	3.34	3.77	+0.43	+12.9
Ji-Paraná	3.48	3.71	+0.23	+6.6
Average	3.18	3.60	+0.42	+13.2

As would be expected, the albedo of the pasture sites is higher than that of the forests. However, Culf *et al.* (1996) showed that there is some variation between the sites with the forest albedo measurements from Reserva Vale do Rio Doce being significantly higher than those from the other two forest sites. The albedo from all the forest sites showed a seasonal variation with the highest albedos occuring at the same time as the driest soil moisture conditions. At the pasture sites no such strong seasonal variation was observed but Wright *et al.* (1996) show that the albedo there is related to the leaf area index. The average values of reflected solar radiation at each site and the average increases in the amount of solar radiation reflected from the surface over the pasture areas are given in Table 9. The table shows that on average there is an increase in reflected solar radiation over the pasture sites of 0.74 MJ m^{-2} day^{-1} or 33.3%, accounting for 58% of the decrease in net all-wave radiation observed over the pasture sites when compared with the forests.

Neither the upward or downward components of longwave radiation were measured during ABRACOS but it is possible to determine the net longwave radiation at each site by considering the difference between the net all wave radiation and the net solar radiation. Figure 1 also shows the net longwave radiation for each of the paired sites. As is usually found, the net longwave radiation is negative, acting to cool the surface, and is consistently more negative over the pasture sites than over the forests. The average values are given in Table 10. The table shows that, on average, the net upward longwave flux is increased over the pasture sites, when compared with the forest sites, by 0.42 MJ m^{-2} day^{-1} or 13.2%. This increase in the net upward longwave flux accounts for 33% of the observed decrease in net all-wave radiation over the pasture sites.

TEMPERATURE

The mean monthly temperatures and the mean daily maximum and minimum temperatures for each of the six sites are shown in Figure 4. The mean temperature is relatively constant throughout the year at Manaus and at Marabá ranging from 24.5 to 26.6°C. At Ji-Paraná the average temperature ranges from 22.1 to 25.7°C with a pronounced drop during June, July and August. This drop in temperature

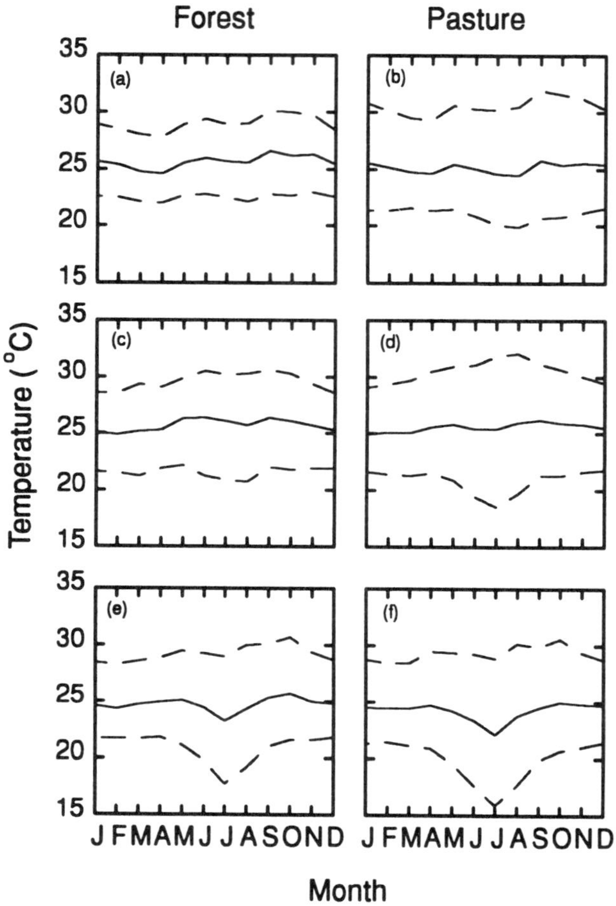

Figure 4 Mean monthly temperatures (solid line) and mean daily maximum and minimum temperatures (dashed lines). (a) Reserva Ducke, Manaus, (b) Fazenda Dimona, Manaus, (c) Reserva Vale do Rio Doce, Marabá, (d) Fazenda Boa Sorte, Marabá, (e) Reserva Jaru, Ji-Paraná and (f) Fazenda Nossa Senhora, Ji-Paraná.

during the dry season may be surprising but it is due to the influence of cold fronts, or 'friagems', which can penetrate this far north during these dry season months. Following the passage of such a front, the maximum temperature may be 10 °C lower than the maximum temperature on the day preceeding the event. The temperature may take two to four days to return to typical values before the passage of the front. There were six of these events in 1992 and nine in 1993.

Figure 5 shows the mean diurnal temperature ranges of the six sites. The range was larger for the pasture sites than for the forests. Such a result is expected since the lower aerodynamic roughness of the pasture sites (Wright *et al.*, 1996) leads to less efficient turbulent transfer and allows the development of strong temperature inversions during the night and super-adiabatic surface layers during the day. There

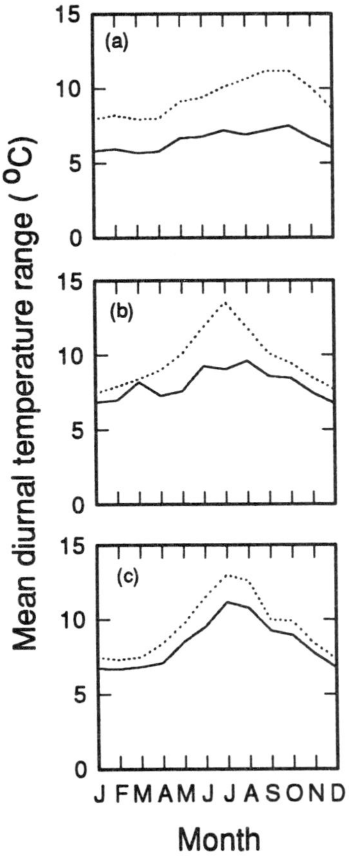

Figure 5 Mean diurnal temperature ranges. Forest sites: solid lines. Pasture sites: dashed lines. (a) Manaus, (b) Marabá, (c) Ji-Paraná.

is an increase in the difference in diurnal temperature range between forest and pasture during the June to August dry season at both Marabá and Ji-Paraná. This increase is presumably due to the reduction in evaporation, and hence increase in sensible heating of the air, at the pasture sites during these dry periods as reported by Wright *et al.* (1996). At the Manaus sites the difference in mean diurnal temperature range between the forest and pasture sites is larger throughout the year than at the other two sites but is at a minimum during the wettest part of the year (February and March) and at a maximum towards the end of the driest period (September). The large difference between forest and pasture values throughout the year is probably because the Fazenda Dimona weather station is located closer to the surface than the other two fazenda weather stations. For typical conditions during the driest part of the year (Bastable *et al.*, 1993; Wright *et al.*, 1992) the diurnal range would be reduced by around 3°C if the temperature measurement was made at the same height as at Fazenda Boa Sorte and Fazenda Nossa Senhora. An additional effect is that the Fazenda Dimona site has the shortest aerodynamic roughness length of any of the pasture sites (Wright *et al.*, 1996) implying that the turbulent mixing there is less efficient than at Fazenda Boa Sorte or Fazenda Nossa Senhora. However, this low roughness is only likely to increase the diurnal temperature range by about 1 °C over the value that would be observed if the aerodynamic roughness was of a similar value to the other pasture sites. The diurnal temperature range of the forest sites is larger at Ji-Paraná and Marabá than at Manaus throughout the year, with the largest range occuring at Ji-Paraná.

SPECIFIC HUMIDITY AND SPECIFIC HUMIDITY DEFICIT

Figure 6 shows the mean monthly specific humidity and mean daily maximum and minimum specific humidity. The specific humidity at Manaus is relatively constant throughout the year with a slight drop in June, July and August corresponding to the dry season. Values range from 16.0 to 17.2 g kg^{-1}. At Marabá and Ji-Paraná there is

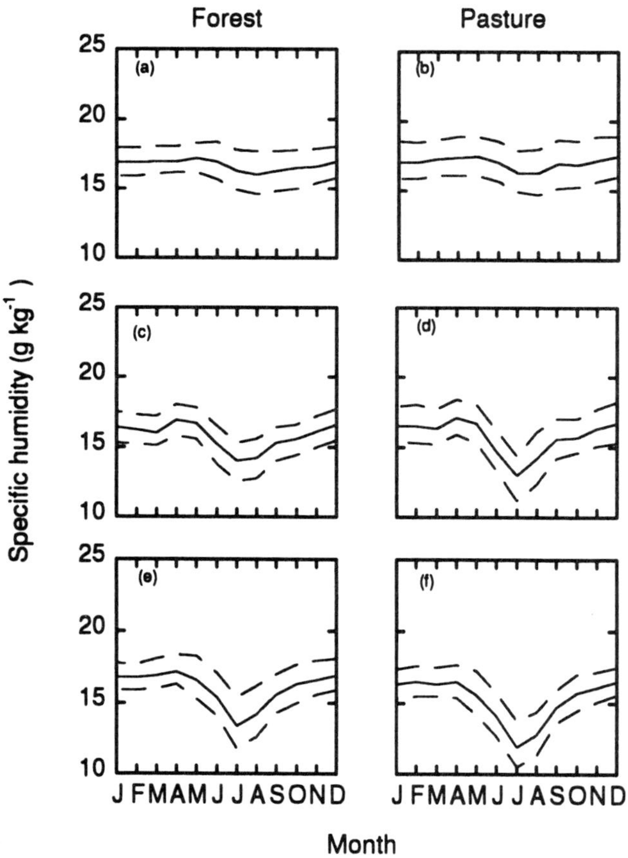

Figure 6 Mean monthly specific humidities (solid lines) and mean daily maximum and minimum specific humidities (dashed lines). (a) Reserva Ducke, Manaus, (b) Fazenda Dimona, Manaus, (c) Reserva Vale do Rio Doce, Marabá, (d) Fazenda Boa Sorte, Marabá, (e) Reserva Jaru, Ji-Paraná and (f) Fazenda Nossa Senhora, Ji-Paraná.

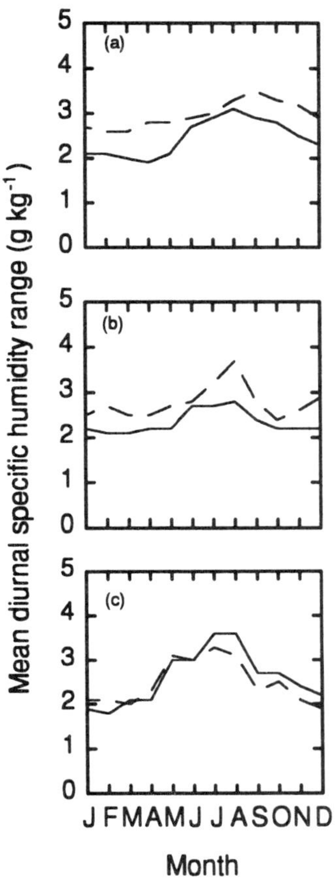

Figure 7 Mean diurnal specific humidity ranges. Forest sites: solid lines. Pasture sites: dashed lines. (a) Manaus, (b) Marabá, (c) Ji-Paraná.

a much more marked seasonal cycle corresponding to the increased severity of the dry season at these sites and, at Ji-Paraná, the influence of the dry air of the friagem mentioned in Section 4. The values at Ji-Paraná range from 11.9 to 17.2 g kg^{-1} and at Marabá from 13 to 17.1 g kg^{-1}. The average diurnal range of specific humidity for each site is shown in Figure 7. On average the range is about 2.5 g kg^{-1} for all the sites. The greatest seasonal variation occurs at Ji-Paraná where the range exceeds 3.5 g kg^{-1} during the dry season.

In addition to the specific humidity, the specific humidity deficit is of interest. Figure 8 shows monthly mean values of specific humidity deficit and the mean daily maxima and minima. The largest values of mean deficits occur in Marabá which, on average, has the most pronounced dry season. A mean deficit of 7.3 g kg^{-1} occurs at Reserva Vale do Rio Doce, whilst the highest mean value at the other forest sites is 5.3 g kg^{-1} at Reserva Ducke. At Fazenda Boa Sorte the deficit has a mean value of 7.8 g kg^{-1} in July, with an average maximum value of 19.4 g kg^{-1}.

During a short period of observations at Fazenda Dimona at the end of 1990,

Bastable *et al.* (1993) observed that the air often became saturated during the night. Figure 8 shows that this phenomenon occurs at Fazenda Dimona throughout the year. At the other two pasture sites, saturation is common during the first half of the year but less common during the dry season. Figure 9 shows the proportion of days of each month when the air becomes saturated for at least one hour. Saturation occurs much more frequently at Fazenda Dimona (85-100% of days) than at any of the other sites, although, as for the temperature observations, it should be noted that the weather station at Fazenda Dimona is mounted closer to the ground than at the other two pasture sites.

The air becomes saturated at Reserva Jaru during the night on about 80% of days during the first half of the year, and about 40 to 60% of days during the second half

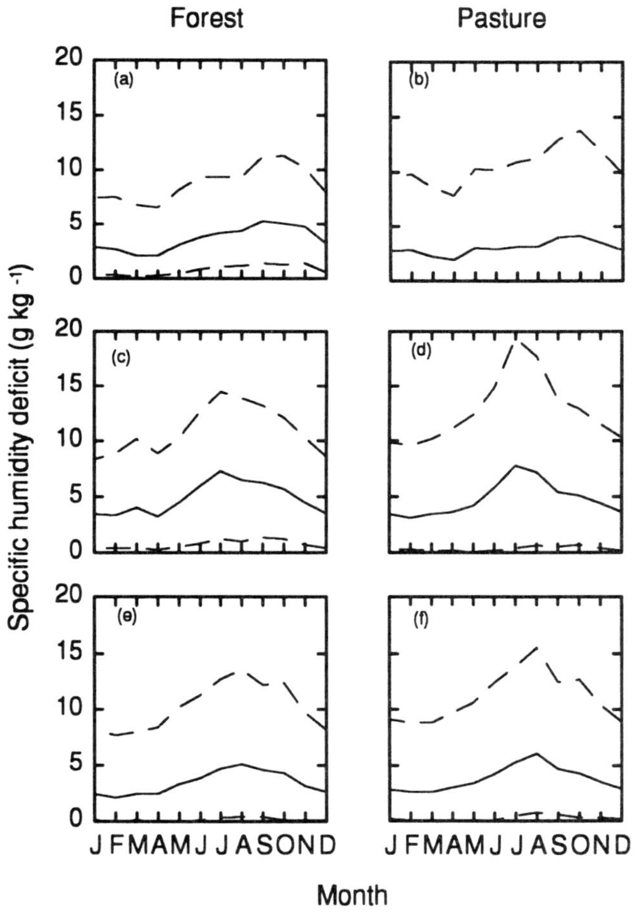

Figure 8 Mean monthly specific humidity deficits (solid lines) and mean daily maximum and minimum specific humidity deficits (dashed lines). (a) Reserva Ducke, Manaus, (b) Fazenda Dimona, Manaus, (c) Reserva Vale do Rio Doce, Marabá, (d) Fazenda Boa Sorte, Marabá, (e) Reserva Jaru, Ji-Paraná and (f) Fazenda Nossa Senhora, Ji-Paraná.

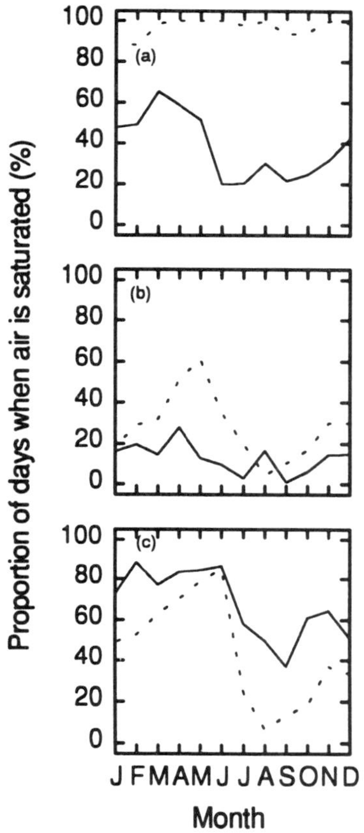

Figure 9 The mean proportion of days per month when the specific humidity deficit was zero for at least one hour. Solid lines: forest sites. Dotted lines: pasture sites. (a) Manaus, (b) Marabá, (c) Ji-Paraná.

of the year. These high proportions may be due to the proximity of the Reserva Jaru site to the Ji-Paraná river which is only about 300 m from the tower. Although at the other forest sites saturation is not so common, it still occurs on up to 60% of days at Reserva Ducke during the wettest part of the year but on less than 25% of nights at Reserva Vale do Rio Doce.

CONCLUDING REMARKS

This paper presents some of the results of a long-term programme of climatological observations over Amazonia. The long term measurements of radiation components at the surface have shown that on average the effect of replacing forest by pasture in Amazonia is to reduce the net radiation at the surface by about 11% as a result of both changes in albedo and in the longwave radiation balance. Wright *et al.* (1992) emphasise the importance of such reductions in net radiation because they found, for the Manaus sites, that the ratio of evaporation to net radiation was similar at the forest and pasture sites during the wet season. Evaporation differences between the sites were therefore attributable to differences in net radiation. The data also illustrate the fact that large scale deforestation can produce feedbacks in the atmosphere, with results from Ji-Paraná showing a reduction in incoming solar radiation over an extensive pasture area during the dry season which is possibly due to increased cloud cover. Spatial variability of climate in Amazonia is also illustrated, with strong seasonality in humidity and temperature observed at Marabá and Ji-Paraná which is not observed in Manaus at the centre of Amazonia. This unique dataset should prove to be extremely valuable for validating the predictions made by climate and weather forecasting models and, subsequently, for improving their performance.

ACKNOWLEDGEMENTS

Many people were involved in the collection of the data presented in this paper. We would like to express our thanks to all the past and present participants in ABRACOS, and in particular to the field observers without whom this work would have been impossible.

REFERENCES

Bastable, H.G., Shuttleworth, W.J., Dallarosa, R.L.G., Fisch, G. and Nobre, C.A. 1993. Observations of climate, albedo and surface radiation over cleared and undisturbed Amazonian forest. *Int. J. Climatol.* **13**, 783-796.

Culf, A.D., Fisch, G. and Hodnett, M.G. 1995. The albedo of Amazonian forest and Ranchland. *J. Climate* **8**, 1544-1554.

Cutrim, E., Martin, D.W. and Rabin, R. 1995. Enhancement of cumulus clouds over deforested lands in Amazonia. *Bull. Am. Met. Soc.* **76**, 1801-1805.

Gash, J.H.G., Nobre, C.A., Roberts, J.M. and Victoria, R.L. 1996. An overview of ABRACOS. This volume.

Maitelli, G.T. and Wright, I.R. 1996. The climate of a riverside city in the Amazon Basin: urban-rural differences in temperature and humidity. This volume.

Molion, L.C.B. and Dallarosa, R.L.G. 1990. Pluviometria da Amazônia: são os dados confiaveis? *Climanálise* **5**, 40-42.

Nobre, C.A., Fisch, G., da Rocha, H.R., Lyra, R.F. da F., da Rocha, E.P., da Costa, A.C.L. and Ubarana, V.N. 1996. Observations of the atmospheric boundary layer in Rondônia. This volume.

Shuttleworth, W.J., Gash, J.H.C., Roberts, J.M., Nobre C.A., Molion, L.C.B. and Ribeiro M.N.G. 1991. Post deforestation Amazonian climate: Anglo-Brazilian research to improve prediction. *J. Hydrol.* **129**, 71-85.

Wright I.R., Nobre, C.A., Tomasella, J., da Rocha, H.R., Roberts, J.M., Vertamatti, E., Culf, A.D., Alvalá, R.C., Hodnett, M.G. and Ubarana, V. 1996. Towards a surface parameterisation for Amazonia. This volume.

Wright, I.R., Gash, J.H.C., da Rocha, H.R., Shuttleworth, W.J., Nobre, C.A., Maitelli, G.T., Zamparoni, C.A.G.P. and Carvalho, P.R.A. 1992. Dry season micrometeorology of central Amazonian ranchland. *Q. J. Roy. Met. Soc.* **118**, 1083-1099.

RESUMO

Estações meteorológicas automáticas vêm coletando dados em áreas experimentais (floresta e pastagem) do Projeto ABRACOS desde 1990, em Manaus, e desde 1991, em Ji-Paraná e Marabá. Este trabalho apresenta alguns dos resultados dessas observações para períodos entre 26 e 39 meses. Médias mensais do saldo de radiação, radiação solar incidente e refletida, e saldo de radiação de onda longa são apresentadas para cada área. Em média, o saldo de radiação é 11% mais baixo nas áreas de pastagem em relação às áreas de florestas. Mostrou-se que essa diferença pode ser atribuído principalmente à diferença de albedo e saldo de radiação de ondas longas. Em Ji-Paraná, entretanto, existem diferenças sistemáticas na radiação solar incidente, que podem ser devido ao aumento de nebulosidade sobre a pastagem durante a estação seca. Médias mensais de temperatura, umidade e déficit de umidade específica também são apresentadas. Existe pouca variação sazonal de temperatura em Manaus ou Marabá, porém existe um resfriamento acentuado durante a estação seca emm Ji-Paraná. Tanto Marabá quanto Ji-Paraná possuem um ciclo sazonal de umidade marcante que não se observa em Manaus.

11 The climate of a riverside city in the Amazon Basin: urban-rural differences in temperature and humidity

G.T. MAITELLI[1] and I.R. WRIGHT[2]
[1] *Universidade Federal de Mato Grosso, Cuiabá, Brasil*
[2] *Institute of Hydrology, Wallingford, UK*

INTRODUCTION

The climate of cities continues to be a focus of activity as urbanisation accelerates throughout the world and it is well known that, in general, an urban area is warmer than its rural surroundings. This difference in air temperature is called the urban temperature excess or urban heat island intensity (UHII) (Yamashita, 1990).

The urban heat island effect is the result of direct warming by human activities such as manufacturing, transportation, heating and air-conditioning of buildings, and by man-made changes in the surface which modify the overall energy balance. The strong absorption, storage and subsequent release of solar energy by buildings and road surfaces, the increased wind resistance of buildings, the reduced evapotranspiration due to the removal of vegetation and the impermeability of the surface are important factors that keep urban air warmer and less humid than the air over the surrounding rural areas (Oke, 1987; Arya, 1988). The horizontal and vertical structure of the urban heat island effect of mid-latitude cities is well documented by Bornstein (1968), Oke (1971), Kopec (1970), Unwin (1980), Johnson (1985), Mattsson and Barring (1985), Ross and Oke (1988), Wang *et al.* (1990), Carnahan and Larson (1990) and Moreno-Garcia (1994). There is increasing interest in urban climates in tropical areas see for example the work of Nieuwolt (1966, 1990), Nakamura (1967), Jauregui (1973), Jauregui *et al.* (1992), Hannel (1976), Sham (1973, 1979), Oguntoyinbo (1986), Duchon (1986), Padmanabhamurthy (1986), Oke (1986), Adebayo (1987a,b) and Yamashita *et al.* (1988).

Over the last ten years studies of climate and micrometeorology have been reported for rural sites in the equatorial region of Amazonia (Shuttleworth *et al.*, 1984; Wright *et al.*, 1992; Bastable *et al.*, 1993; Wright *et al.*, 1996). Also Fisch (1990) and Fisch *et al.* (1995) have reported general urban-rural micrometeorological comparisons, however there have been no studies about the effect on climate of urban development

Amazonian deforestation and climate. Edited by J.H.C. Gash, C.A. Nobre, J.M. Roberts and R.L. Victoria.

within Amazonia. The Anglo BRazilian Amazonian Climate Observation Study (ABRACOS) aims, as just one of its objectives, to provide evidence for a systematic difference between the climate of riverside urban developments in Amazonia, such as Manaus, and adjacent forested areas, Shuttleworth *et al.* (1991).

This paper presents results from an investigation into the effects of urbanisation on the temperature and humidity within the equatorial city of Manaus, based on data collected by ABRACOS automatic weather stations (AWS) at urban and rural locations. The urban heat island effect is investigated by comparing the variations in temperature and humidity of the urban atmosphere to that of two examples of the surrounding rural climate.

STUDY AREA

Manaus city (3° 06' 07" S, 60° 10' 30" W) is in the Amazonian rainforest zone within the humid equatorial environment of Brazil. The size and population of the city is increasing; in 1970 the population was about 300,000 whereas the estimated figure for 1991 was one million.

Manaus is situated on a fluvial terrace to the north of the Rio Negro, almost at its confluence with the Rio Solimoes where they combine to become the River Amazon. The city ranges in elevation from about 20 m above mean sea level (AMSL) in the valley area of the central part of the city to 40 m AMSL. The AWS was located in a flat area of sparse scrub vegetation, just outside of the densest area of commercial and residential development and two kilometres from the city centre. Figure 1 shows the distribution of land-use in Manaus in 1990 together with the location of the urban AWS.

The forest site is located in the Reserva Florestal Ducke, a protected area of rainforest, 25 km north-east of Manaus. The pasture site is located at the Fazenda Dimona about 80 km north of Manaus, see Gash *et al.* (1996). The clearing is approximately 3 km by 4 km, and was converted from forest to pasture about 12 years before the study. A description of the characteristics of the rural forest site is given by Shuttleworth *et al.* (1984) and (1985), and of the rural pasture site by Wright *et al.* (1992).

There is very little difference between in the topography of the urban and rural sites. The major difference between the city and its environs which could have implications on the atmospheric temperature and humidity conditions is that of type of surface cover and the proximity of the river. While soil and vegetation is the dominant ground cover in both rural areas, buildings and paved surfaces dominate the city surface.

INSTRUMENTATION

The urban automatic weather station (AWS) was installed 1.5 m above ground, at a site representative of, rather than central to, the urban area of Manaus, see Figure 1.

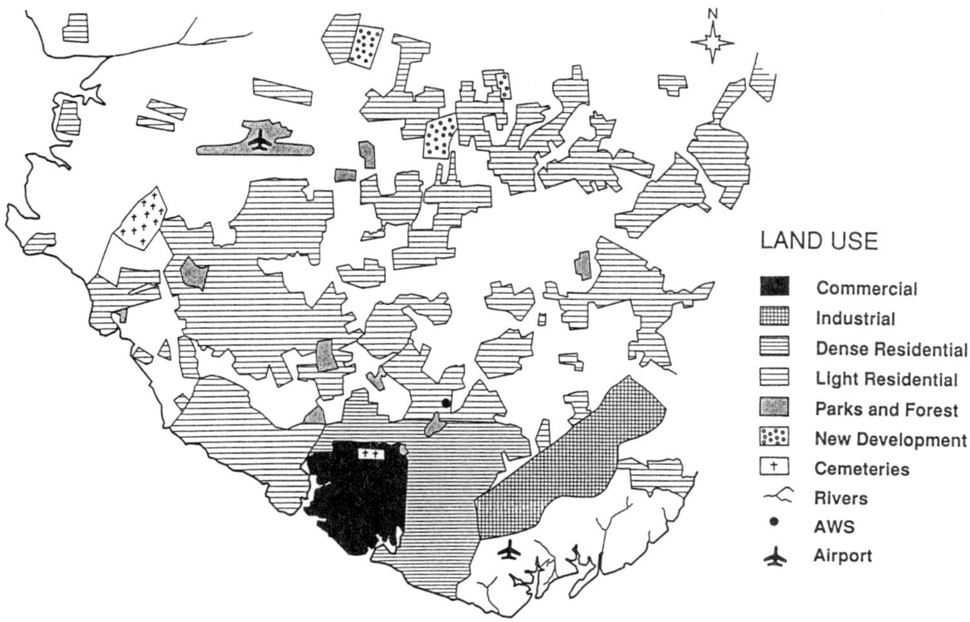

Figure 1 The location of the ABRACOS AWS within Manaus and the surrounding land uses

Of the rural stations, one was installed in the Reserve Ducke forest, at the top of a 45 m scaffolding tower, the other at 1.5 m over typical ranchland vegetation at Fazenda Dimona. Each AWS (Didcot Instruments, Abingdon, UK) recorded hourly data on a programmable logger (Campbell Scientific, Shepshed, UK) including the temperature, humidity and windspeed used in this study. Temperature and humidity were measured using calibrated platinum resistance thermometers mounted in a shielded and aspirated psychrometer screen (Institute of Hydrology, UK). A detailed description of the AWS instrumentation is given by Bastable *et al.* (1993).

RESULTS AND DISCUSSION

The temperature and humidity records analysed here for the city of Manaus cover the period from November 1990 to September 1992. Monthly mean values of daily maximum, minimum and mean temperature and humidity were computed and are shown in Figure 2, together with the rainfall for the same period. For a 14 month period from February 1991 to March 1992, mean monthly and hourly temperatures representing urban Manaus were compared with those for the two rural sites to investigate the overall urban heat island intensity.

In published literature UHII for a single pair of weather stations has been defined in various ways to describe the overall temperature difference between urban and rural sites. In this study we use two definitions:

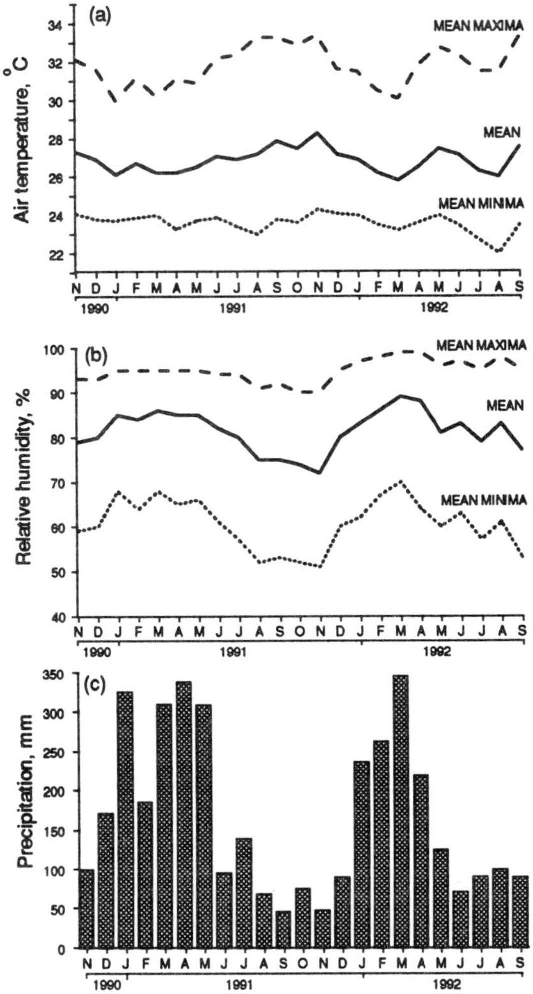

Figure 2 Manaus climate: November 1990 to September 1992. Monthly means, maxima and minima of (a) temperature and (b) relative humidity, together with (c) rainfall totals

(1) The mean difference between urban (u) and rural (r) monthly temperature statistics, which gives three different UHIIs with respect to the mean, maximum or minimum monthly temperatures:

$$UHII_{1,x} = \left[\frac{1}{D} \sum_{d=1}^{d=D} (T_{u,d,x} - T_{r,d,x}) \right]_m$$

where d is the day number, D is the number of days in the month m, and x denotes the monthly mean, maximum or minimum temperature.

(2) The monthly mean difference between urban and rural hourly temperatures, calculated via values for separate hours, thus giving 24 values for each month with which to observe the distribution of urban-rural differences:

$$UHII_2 = \left[\frac{1}{D}\sum_{d=1}^{d=D}(T_{u,h}-T_{r,h})\right]_{h,m}$$

where h denotes hourly mean temperatures. Whereas $UHII_1$ gives more information on the overall range and seasonality of urban-rural differences, $UHII_2$ illustrates the diurnal differences. $UHII_2$ is also much less sensitive to a small amount of missing data. In this study there are only 14 complete months of data for estimates of $UHII_1$, yet it is possible to calculate $UHII_2$ representing a 20 month period. Without the missing data $UHII_{1,mean}$ equals $UHII_2$ over the same period.

When a spatial network of information is available, it is common to quote UHII as the difference between the mean rural temperature and the spatial maximum temperature over the city, wherever that may occur (Oke, 1987).

Although these definitions of urban heat island intensity have been used frequently in other studies (Nkemdirim and Truch, 1978; Lee, 1979; Unwin, 1980; Adebayo, 1987b; Yamashita, 1990 and Yague *et al.*, 1991); it is necessary to take care when interpreting the results. Not only is the rural vegetation type and heterogeneity important, it has been shown that different definitions of UHII can give different results and both urban and rural weather stations can have widely differing positional characteristics. For example, an urban station can be either located at the centre of a city for maximum effect, or placed to represent the general city climate. Also the type of rural environment will have a profound effect on the comparison with urban climates. In this study the urban AWS was not installed close to the city centre but located to represent the general heat island effect.

Mean monthly temperature

The annual climate of the Manaus area has distinct wet and dry periods: this is clearly illustrated by the monthly rainfall figures recorded at the Manaus site (Figure 2c). The wet period begins around December and extends into May with the dry period covering the remaining months. The monthly mean temperatures ranged from 25.8°C to 28.2°C (Figure 2b) and the monthly mean of the daily maxima range from 30.0°C to 33.4°C, and the minima from 22.0°C to 24.3°C. The overall highest hourly mean temperature, 39.0°C, was recorded in May 1992 and the lowest, 20.4°C, in September 1992.

Figure 3 shows the differences in monthly mean, mean maximum and mean minimum temperature between the urban and rural sites. The predominantly positive difference between these monthly mean statistics show the higher overall temperatures in the city compared to those observed at both rural sites. Figure 3 also shows a noticeable contrast in the seasonality of the mean monthly differences in temperature statistics between city and the two rural sites. During the 1991 dry season, the differences in urban-forest maxima and minima initially diverge as the

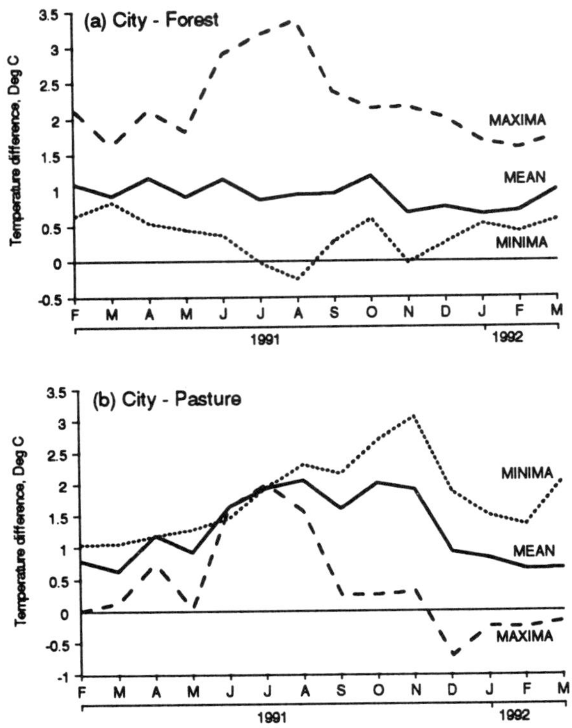

Figure 3 The difference in monthly mean, maximum and minimum temperature between the Manaus site and (a) the forest and (b) the pasture sites

city becomes warmer by day and colder at night (May–July). In contrast, the pasture differences show the opposite trend: dry season evaporation is less, and produces an energy balance which is more similar to that of the city. Wright *et al.* (1992) describe the rapid change in the energy balance of the pasture site during the 1990 dry season, compared to the more constant annual energy balance of the forest. Table 1 gives the overall means of the data shown in Figure 3, weighted to represent a 12 month period. These overall means give the UHII corresponding to Method 1 above: the UHIIs for the second method are also given in Table 1. For the pasture,

Table 1 Urban heat island intensities for the city of Manaus

Mean temperature difference, K		City–Pasture	City–Forest
Mean monthly mean	$UHII_{1,mean}$	1.37	0.93
Mean monthly maxima	$UHII_{1,max}$	0.48	2.25
Mean monthly mimima	$UHII_{1,min}$	1.85	0.33
Mean Diurnal hourly	$UHII_2$	1.37	0.96

the UHII based on minima ($UHII_{1,min}$) is larger than that based on maxima, and is typically found for mid-latitude cities: the forest comparison does not show the same relationship.

Mean diurnal temperature

Figure 4 shows the frequency distribution of monthly mean temperature differences for each hour in the day, and therefore shows for each site, the distribution of the data used to calculate $UHII_2$. The city–forest distribution ($n = 439$) has a mean of 0.96 K and a median of 0.7 K, and the city-pasture values ($n = 464$) have a mean of 1.37 K and a median of 1.4 K. $UHII_{1,mean}$ for forest (0.93 K) is 0.03 different from the forest $UHII_2$ (0.96 K) because of the different periods represented: 14 and 20 months respectively. Comparing the city and forest, 3% of the data were missing (from December 1991). For the pasture, the method of calculation has made no difference to the value of UHII ($UHII_{1,mean} = UHII_2 = 1.37$ K) even though 8% of the data were missing (December 1991 and May/June, 1992).

Figure 5 shows the mean diurnal temperature and windspeed of Manaus city compared with those for the two rural sites. The city has a greater capacity to store heat in its structure during the day compared to the pasture and, to a lesser extent, the

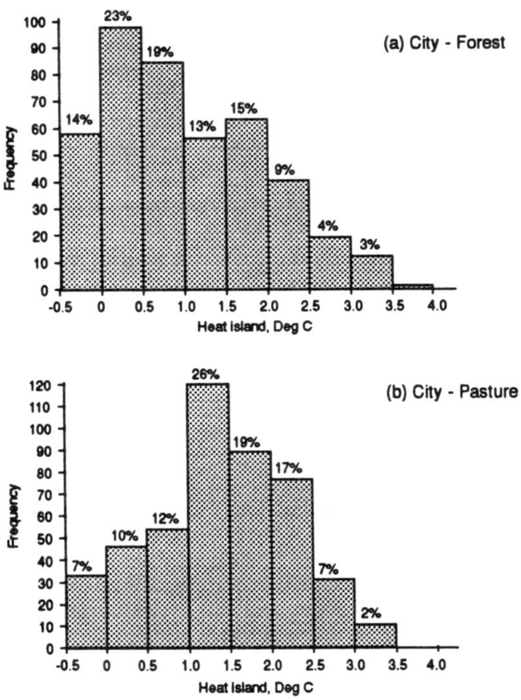

Figure 4 The frequency distribution of mean monthly difference (24 hourly values for each month), between Manaus and (a) the forest and (b) the pasture site.

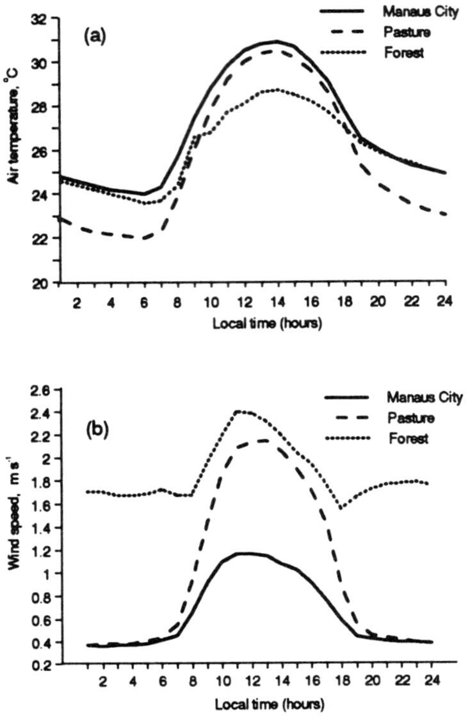

Figure 5 The overall mean diurnal (a) temperature and (b) windspeed of Manaus city and the two rural sites

forest. At the pasture site, the diurnally stored heat is much smaller compared to the city because of the small biomass and the decoupling effect of the aerodynamically smoother surface. At night the radiative cooling, the resultant stable temperature inversion and the small reserve of stored heat energy allows the pasture temperature to fall 2 K below that of the city. Then, during the first three hours of daylight, the city and pasture temperatures converge as solar energy goes into heating their surfaces. In contrast to this, in the forest much energy goes to evaporation and, to a smaller extent, into biomass heat storage (Moore and Fisch, 1986) and temperatures diverge during the first hours of daylight. The higher overall windspeeds over the forest (Figure 5b) will tend to keep the air cooler over the forest compared to the city.

Figure 6 shows the difference in the overall mean diurnal temperature together with similar graphs for the separate months of April and September, representing the wet and dry seasons respectively. Whereas the city–forest diurnal differences are similar in both seasons, particularly during the day, the city–pasture differences are more affected by seasonality. In the dry season less solar energy is used for pasture evaporation (Wright *et al.*, 1992; Wright *et al.*, 1996) and consequently more energy is available for heating and cooling the air and soil. This accounts for the relative similarity between the city and the pasture around midday in the dry season. In the wet season, evaporation and limited heat storage suppresses

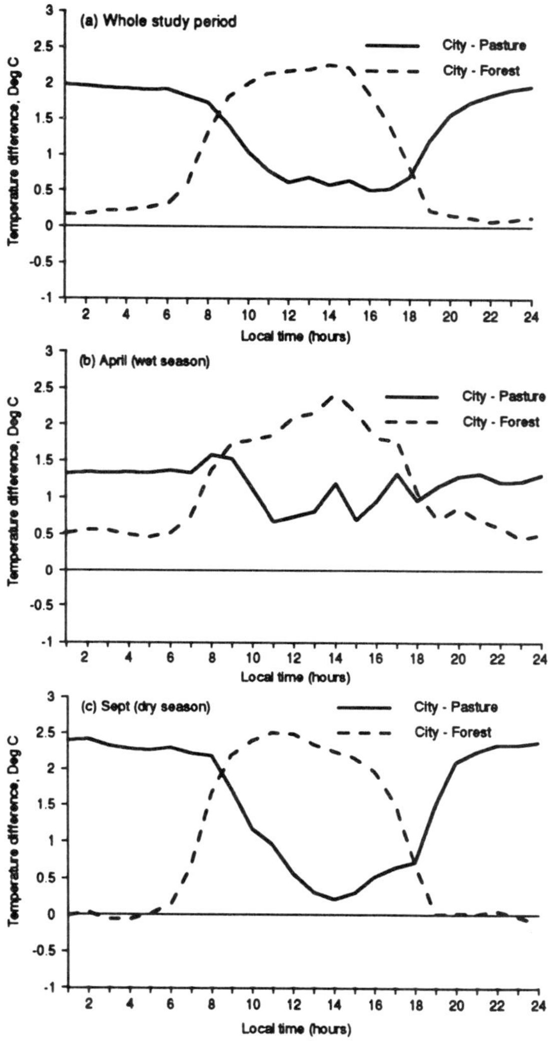

Figure 6 Mean diurnal temperature difference between Manaus city and the two rural sites for (a) all data, (b) the wet season month of April and (c) the dry season month of September.

temperatures in the pasture compared with those in the city.

Effective albedo for cities can be very variable and is currently not known for Manaus city. Flat concrete can have a high albedo (up to 0.27, Arya, 1988), but for a larger city area albedo can be considerably reduced by vegetation and the density of radiation trapping 'canyons' created by streets of tall buildings. In view of the mixed structure of Manaus, it is likely that the effective albedo of the city is similar to the value of 0.18 found for the pasture (Culf *et al.*, 1995) rather than the value of 0.13 found for the forest (Culf *et al.*, 1995). When compared to the forest, the higher sensible heat fluxes in the city create higher temperatures even though the total available energy in the city is smaller due to the higher albedo and surface temperatures.

Greco *et al.* (1992) compared the pressure gradients between forest sites in the Manaus region and interpreted them in terms of temperature differences between these sites. They also discussed the mesoscale circulation between the forest and river/city (Manaus) complex in terms of temperature differences derived from airborne remote sensing and radiometric surface temperatures. They found the river to be 2-3K cooler than the forest during the day and 6-8K warmer at night. Therefore, the city–forest differences reported here, positive during the day and near zero at night, suggest complex circulations as the three large and mainly independent microclimates interact, with circulation being generally towards the city by day and towards the river at night.

Humidity

Mean monthly relative humidities observed in Manaus during the dry season (Figure 2b), were between 72%rh and 88%rh, increasing to 80%rh to 90%rh during the wet season. Overall minimum hourly values were close to 50%rh.

Figure 7 shows the mean diurnal humidity for Manaus city and the two rural sites

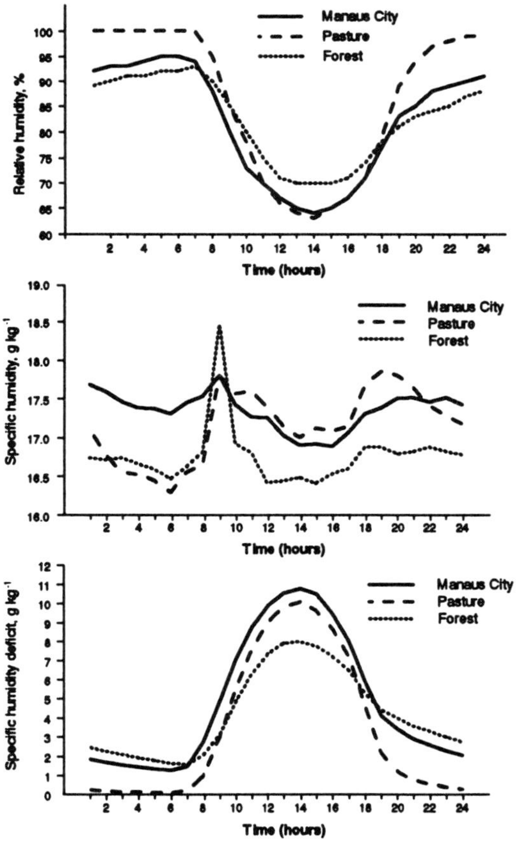

Figure 7 Mean diurnal (a) relative humidity, (b) specific humidity and (c) specific humidity deficit for Manaus city and the two rural sites.

in three ways: relative humidity, specific humidity and specific humidity deficit. Figure 7 clearly shows the regular occurrence of saturated air at the pasture site during the night: early morning mist and heavy dew were commonly observed. With the exception of this high relative humidity in the pasture, induced by night-time radiation losses, Figure 7a and 7c show a tendency for the forest to be more humid than the other two sites. Although the city is generally drier in terms of relative humidity (%rh) and specific humidity deficit, the higher city temperatures allow the air to contain more water vapour as shown by the specific humidity in Figure 7b. Notwithstanding the early morning surge in forest specific humidity, the diurnal pattern is similar to that of the city, indicating similar interactions with the convective and nocturnal boundary layers. Diurnal specific humidities at the pasture site are less similar, probably due to the aerodynamically smoother surface and the stronger temperature inversions which result in less efficient mixing with higher atmospheric layers.

Figure 8 shows the differences in relative humidity and specific humidity deficit with respect to the two rural sites. In terms of relative humidity, at night the city was 5%rh - 9%rh less humid than the pasture yet about 3%rh more humid than the forest. During the day the city relative humidity was similar to that of the pasture but 3%rh to 6%rh drier than the forested area. The mean difference in specific humidity deficit

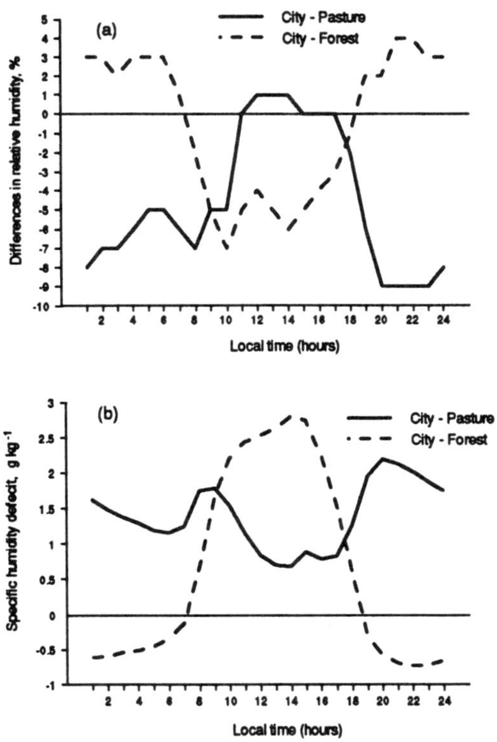

Figure 8 Mean diurnal difference in (a) relative humidity and (b) specific humidity deficit, between Manaus city and the two rural sites.

between the city and the forest was 0.65 g kg^{-1}, and 1.41 g kg^{-1} when compared to the pasture. City atmospheres are usually drier than those of rural areas during the afternoon and early evening, but are often more moist than rural areas at night. This pattern, shown by the forest differences in Figure 8, has generally been attributed to differences in dewfall, evapotranspiration, eddy diffusion, and the urban heat island effect, (Henry and Dicks, 1985). Contrary to this, the pasture site loses heat more rapidly at night and stays more humid, in terms of specific humidity deficit, compared to the city during all hours.

CONCLUSIONS

The analyses carried out on aspects of temperature and relative humidity in the city of Manaus show that they are affected by urbanisation and produce a noticeable heat island effect. The urban heat island intensity based on 14 months of data and given by the monthly difference in temperature statistics, was 1.37K, 0.48K and 1.85K for the mean, maxima, minima respectively when compared to the pasture, and 0.93K, 2.25 K and 0.33 K respectively when compared to the forest (Table 1). Using 20 months of data and a method of calculation that is more tolerant of missing data, the UHII was found to be 1.37 and 0.96 for the pasture and forest respectively.

In terms of relative humidity, the city was 5–9%rh less humid than the pasture at night, yet about 3%rh more humid than the forest. During the day the city was similar to the pasture, but 3%rh to 6%rh less humid than the forested area.

These results are clear evidence that the magnitude of the urban heat island intensity greatly depends on the land-use of the surrounding rural areas and, to a lesser extent, the method of calculation. Future studies of urban–rural differences in temperature and humidity must concentrate on a more detailed analysis of differences in land-use in the rural areas in addition to the urban-rural contrasts. More detailed observations of surface characteristics, mesoscale meteorological conditions and energy balance components are needed from tropical cities to understand the processes controlling the urban heat island effect and the effect on the internal and surrounding microclimate.

REFERENCES

Adebayo, Y.R. 1987a. The effect of urbanization on some characteristics of relative humidity in Ibadan. *J. Climatol.* **7**, 599-607.

Adebayo, Y.R. 1987b. A note on the effect of urbanization on temperature in Ibadan. *J. Climatol.* **7**, 185-192.

Arya, S.P. 1988. *Introduction to Micrometeorology*. Academic Press Inc., USA.

Bastable, H.G., Shuttleworth, W.J., Dallarosa, R.L.G., Fisch G. and Nobre, C.A. 1993. Observations of climate, albedo and surface radiation over cleared and undisturbed Amazonian Forest. *Int. J. Climatol.* **13**, 783-796.

Bornstein, R. D. 1968. Observations of the urban heat island effect in New York city. *J. Appl. Meteorol.* **7**, 575-582.

Carnahan, W.H. and Larson, R.C. 1990. An analysis of a urban heat sink. *Remote Sens. Environ.* **33**,

65-71.
Culf, A.D., Fisch, G. and Hodnett, M.G. 1995. The albedo of Amazonian Forest and ranchland. *J. Climatol.* **8**, 1544-1554.
Duchon, C. E. 1986. Temperature trends at San Juan, Puerto Rico. *Bull. Amer. Meteorol. Soc.* **67**, 1370-1377.
Fisch, G. 1990. Climatic aspects of the Amazonian Tropical Forest. *Acta Amazonica* **20**, 39-48.
Fisch, G., Ferreira, S.H.S. and Bastable, H.G. 1995. A micrometeorological comparison amongst forest, grass and urban sites in the Amazon Region. Proc. 4th Int. Conf. Southern Hemisphere Meteorology and Oceanography. Hobart, Australia, 252-253.
Gash, J.H.C., Nobre, C.A., Roberts, J.M. and Victoria, R.L. 1996. An overview of the ABRACOS project. This volume.
Greco, S., Ulanski, S., Garstang, M. and Houston, S. 1992. Low-level nocturnal wind maximum over the central Amazon Basin. *Bound.-Layer Meteorol.* **58**, 91-115.
Hannel, F.G. 1976. Some features of the Heat Island in an equatorial city. *Geografiska Annaler*, 1-**2**, 58 A, 95-109.
Henry, J.A. and Dicks, S.E. 1985. Urban and rural humidity distributions, relationships to surface materials and land use. *J. Climatol.* **5**, 53-62.
Jauregui, E. 1973. The Urban Climate of Mexico City. *Erdkunde* **XXVII**, Heft 4, 298-307.
Jauregui, E., Godinez, L. and Cruz, F. 1992. Aspects of Heat Island Development in Guadalajara, Mexico. *Atmos. Environ.* **268**, 391-396.
Johnson, D.B. 1985. Urban modification of diurnal temperature cycles in Birmingham, UK. *J. Climatol.* **5**, 221-225.
Kopec, R.J. 1970. Further observations of the urban heat island in a small city. *Bull. Amer. Meteorol. Soc.* **51**, 602-606.
Lee, D.O. 1979. Contrasts in warming and cooling rates at an urban and a rural site. *Weather* **34**, 60-66.
Mattsson, J.O. and Barring, L. 1985. Canyon geometry street temperatures and urban heat island in Malmö, Sweden. *J. Climatol.* **5**, 433-444.
Moore, C.J. and Fisch, G. 1986. Estimating heat storage in Amazonian tropical forest. *Agric. J. For. Meteorol.*, **38**, 147-169.
Moreno-Garcia, M.C. 1994. Intensity and form of the urban heat island in Barcelona. *Int. J. Climatol.* **14**, 705-710.
Nakamura, K. 1967. City temperature of Nairobi. *Japanese Progress in Climatology*. 61-65.
Nieuwolt, S. 1966. The urban microclimate of Singapore. *J. Trop. Geogr.* 30-37.
Nieuwolt, S. 1990. Urban planning and building design for urban centres in the humid tropics. *African Urban Quarterly* **5**, 96-103.
Nkemdirim, L.C. and Truch, P. 1978. Variability of temperature fields in Calgary, Alberta. *Atmos. Environ.* **12**, 809-822.
Oguntoyinbo, J.S. 1986. Some aspects of the urban climates of tropical Africa. Urban climatology and its applications special regard to tropical areas, WMO- N° 652, Geneva, 110-135.
Oke, T.R. 1971. The urban boundary layer in Montreal. *Boundary-Layer Meteorol.* **1**, 411-437.
Oke, T.R. 1986. Urban climatology and the tropical city: an introduction. Urban climatology and its applications with special regard to tropical areas, WMO- N° 652, Geneva, 1-45.
Oke, T.R. 1987. *Boundary Layer Climates*. Second Edition. Routledge, London.
Padmanabhamurty, B. 1986 .Some aspects of the urban climates of India. Urban climatology and its applications with special regard to tropical areas, WMO - N°652, Geneva, 136-163.
Ross, S. L. and Oke, T.R. 1988. Tests of three urban energy balance models. *Boundary-Layer Meteorol.* **44**, 73-96.
Sham, S. 1973. The urban heat Island: its concept and application to Kuala Lumpur. *Sains Malaysiana* **2**, 53-64.
Sham, S. 1979. Mixing depth wind speed and air pollution potential in Kuala Lumpur. *Malaysian Nature Journal* **32,** 233-245.

Shuttleworth, J.W., Gash, J.H.C., Lloyd, C.R., Moore, C.J., Roberts, J.M., Marques, A.O.F.de, Fisch, G., Silva,V.P.R.de, Ribeiro, M.N.G., Molion, L.C.B., Sa, L.D.A., Nobre, C.A., Cabral, O.M.R., Patel, S.R. and Moraes, J.C. 1984. Eddy correlation for Amazoniam forest. *Quart. J. Roy. Meteorol. Soc.* **110**, 1143-1162.

Shuttleworth, J.W., Gash, J.H.C., Lloyd, C.R., Moore, C.J., Roberts, J.M., Marques Filho, A.O. F.de, Fisch, G., Silva, V.P.R.de, Molion, L.C.B., Cabral, O.M.R., Patel, S.R. and de Moraes, J.C. 1985. Daily variations of temperature and humidity within and above Amazonian Forest. *Weather* **40**, 102-108.

Shuttleworth, J.W., Gash, J.H.C., Roberts, J.M., Nobre, C.A., Molion, L.C.B. and Ribeiro, M.N.G. 1991. Post deforestation Amazonian climate Anglo Brazilian research to improve prediction. *J. Hydrol.* **129**, 71-91.

Unwin, D.J. 1980. The synoptic climatology of Birmingham's urban heat island. *Weather* 1965-1974.

Wang, W.C., Zeng, Z. and Karl, T.R. 1990. Urban heat islands in China. *Geophys. Res. Letters* **17**, 2377-2380.

Wright, I.R., Gash, J.H.C., da Rocha, H.R., Shuttleworth, J.W., Nobre, C.A., Maitelli, G.T., Zamparoni, C.A.G.P., and Carvalho, P.R.A. 1992. Dry season micrometeorology of central Amazonian Ranchland. *Q. J. Roy. Met. Soc.* **118**, 1083-1099.

Wright, I.R., Gash, J.H.C., da Rocha, H.R. and Roberts J.M. 1996. Modelling surface conductance for Amazonian pasture and forest. This volume.

Yague, C., Zurita, E. and Martinez, A. 1991. Statistical analysis of the Madrid urban heat island. *Atmos. Environ.* **25B**, 327-332.

Yamashita, S. , Imamura, I.R. and de Monteiro, C.A.F. 1988. Observation of heat island in Patos and Campina Grande, northeast Brazil. *Latin American Studies* **10**, 145-156.

Yamashita, S. 1990. The urban climate of Tokyo. *Geogr. Rev. Japan* **63**, 98-107.

RESUMO

Foram investigadas as variações de temperatura e umidade do ar de um centro urbano na região equatorial úmida da Amazônia brasileira. Medidas horárias de temperatura e umidade, coletadas durante dois anos por uma estação meteorológica automática instalada na área urbana de Manaus foram comparadas com dados obtidos de duas estações semelhantes, localizadas em áreas rurais diferenciadas, pastagem e floresta primária, ambas localizadas a menos de 80 km de Manaus.

As médias diária e mensal de temperatura indicam, com clareza, o efeito da ilha de calor, com a temperatura na área urbana persistentemente maior que a das duas áreas rurais. As temperaturas mensais máxima, mínima e média da cidade, baseadas em dados de 14 meses completos, foram maiores que as da pastagem em 0,48 K, 1,85 K e 1,37 K respectivamente; e maiores que as da floresta em 2,25 K, 0,33 K e 0,93 K respectivamente. A intensidade média da ilha de calor, baseada em diferenças horárias de temperatura de 20 meses, teve uma média diária no valor de 1,37 K quando comparado com o sítio de pastagem, e 0,96 K quando comparado com a floresta. O ciclo diurno das diferenças de temperatura entre cidade-pastagem foram mais afetadas pela estação seca que as diferenças entre cidade-floresta.

À noite, em termos de umidade relativa (%ur), a cidade foi 5%ur a 9%ur menos úmida que a pastagem, e cerca de 3%ur mais úmida que a floresta. Durante o dia, a umidade da cidade foi similar à da pastagem, mas 3%ur a 6%ur mais seca que a área de floresta.

12 In-canopy microclimate of Amazonian forest and estimates of transpiration

O.M.R. CABRAL[1], A.-L.C. McWILLIAM[2] and J.M. ROBERTS[2]
[1]Embrapa/CNPMA, Jaguariúna, São Paulo, Brazil
[2]Institute of Hydrology, Wallingford, UK

INTRODUCTION

Roberts *et al.* (1993), showed that calculation of tropical forest transpiration is viable using a combination formula which requires only physiological and in-canopy weather data. The Canopy Layer and Total Transpiration Routine (CLATTER) requires stomatal and aerodynamic conductances for the leaves in each layer (Roberts *et al.*, 1990), estimates of leaf area index, and net radiation, which can be obtained using a Beer law hypothesis. The whole canopy transpiration is derived by summing the output from each level of the canopy. The model makes no assumption about in-canopy transfer and the gradient-diffusion hypothesis is not assumed. However, to avoid these assumptions, the model requires profiles of air temperature and humidity deficit through the canopy. Roberts *et al.* (1993) used observed profiles, so the objectives of this study are to evaluate whether the profiles can be accurately obtained from above-canopy data, and to verify how adequate is their use in a multi-layer model. Comparisons with a single-layer model (Shuttleworth, 1988) are also performed to verify the necessity of a multilayer approach.

MATERIAL AND METHODS

The air temperature and specific humidity deficit data measured in and above the canopy, as well as global and net radiation, were collected in the *terra firme* forest in the Reserva Florestal Ducke, 25 km from Manaus, during the Amazon Region Micro-meteorological Experiment (ARME) and the Anglo-Brazilian Amazonian Climate Observational Study (ABRACOS), as reported by Shuttleworth *et al.* (1985), Roberts *et al.* (1993) and Bastable *et al.* (1993).

During the 1991 mission of ABRACOS (August-September), the Reserva Ducke tower (Roberts *et al.*, 1990) was instrumented with six levels of aspirated

Amazonian deforestation and climate. Edited by J.H.C. Gash, C.A. Nobre, J.M. Roberts and R.L. Victoria.

psychrometers and anemometers at 35, 30, 25, 20, 15 and 5 m height, recording ten-minute averages of air temperature, specific humidity and wind speed; an automatic weather station was also in operation above the canopy.

The measurements made over the Ducke forest (50 days during 1991 and 37 during ARME) were split to form a set of at least 48 hours time interval between days, based on the autocorrelation coefficients (Brett and Tuller, 1991), which were calculated after the removal of the diurnal cycle, and showed significant serial correlations up to time lags of 30 hours, for both air temperature and humidity deficit.

The resulting set of days containing daylight data (06.00 to 18.00 hours) was used to fit the relationships between the air temperature (T_z) and specific humidity deficit (D_z) measured at different heights (z) in the profiles and 45 m level, using the least squares method:

$$T_z = a_z + b_z\, T_{45} \tag{1}$$

$$SD_z = a_z' + b_z'\, D_{45} \tag{2}$$

where T_{45} and D_{45} are the temperatures and deficits observed at 45 m.

The remaining days, although not entirely independent of the set used to fit the relationships, were also used to verify the goodness of fit of the adjusted equations and to calculate the hourly transpiration rates with CLATTER (Roberts *et al.*, 1993. These were then compared with the rates based on measured profiles.

RESULTS

PROFILES OF AIR TEMPERATURE

The overall average hourly values of air temperature are shown in Figure 1: these do not contain the standard errors and the apparent gradients must therefore be regarded as a qualitative description, since the differences between levels were not significant. The warmest layers recorded were in the range 25-35 m, where the major leaf area concentration occurs (McWilliam *et al.*, 1993; Roberts *et al.*, 1993) and where most of the solar radiation is absorbed.

The weak gradients in those layers during the afternoon indicate the efficient mixing reported by Shuttleworth *et al.* (1985). The decoupling at 5 m is due to the low amount of radiation reaching the forest floor — only 1% of the above-canopy value (Shuttleworth *et al.*, 1984). Similar values of temperature were recorded in the levels 45 m and 15 m, and 25 m and 35 m, which agree with the gradients reported before (Shuttleworth *et al.*, 1985; Fitzjarrald *et al.*, 1990), corresponding to an outward flux of sensible heat, also found in other tropical forests (Allen *et al.*, 1972; Pinker, 1982; Thompson and Pinker, 1975).

Figure 2 shows the average profiles of air temperature for selected hours, which exhibit a broad maximum through the canopy, although the temperature in the 30 m

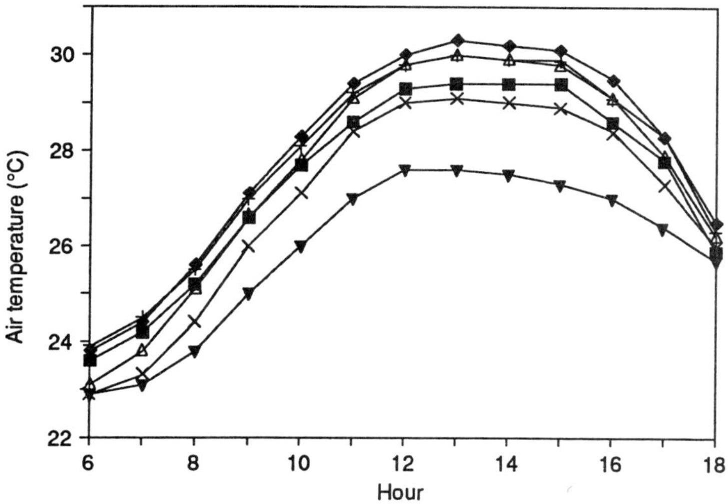

Figure 1 Hourly average of air temperatures observed above and below the tropical terra-firme forest at heights: (■) 45 m; (+) 35 m; (◆) 30 m; (Δ) 25 m; (×) 15 m and (▼) 5 m.

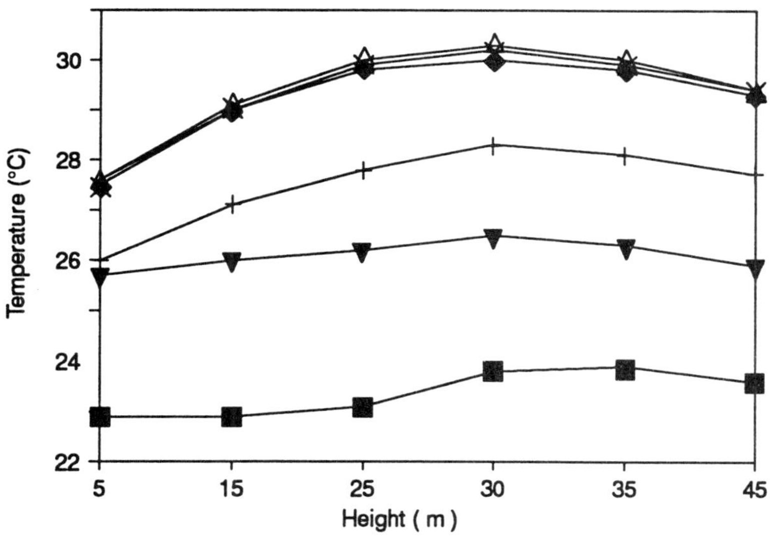

Figure 2 Hourly average profiles of air temperatures at times: (■) 6 h; (+) 10 h; (◆) 12 h; (Δ) 13 h; (×)14 h and (▼) 18 h.

layer was slightly higher during the afternoon. The forest radiative cooling (18.00 h), can be seen through the decrease in temperature of the warmest layers.

The fitted coefficients of Equation 1 for each level are shown in Table 1. The slopes around 1 indicate that the intercepts can be regarded as offsets, with the exception of the 5 m layer, whose distinctive behaviour was referred to previously.

The estimated profiles of temperature are compared with those measured in Figure 3 for each height and for both periods. The agreement is better for the top

Table 1 Coefficients (a = intercept; b = slope) of Equation 1 for temperatures (°C) at each level as a function of the air temperature observed at 45 m.

Level height (m)	a	b	r^2
35	-0.13	1.02	0.99
30	0.39	1.01	0.92
25	-1.24	1.05	0.93
15	0.71	0.96	0.89
5	7.04	0.69	0.79

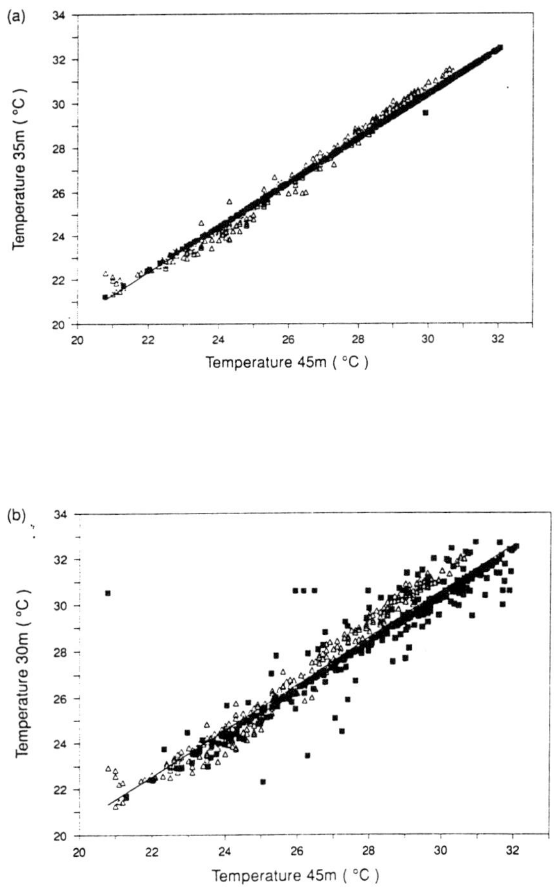

Figure 3 Air temperatures estimated and measured during ARME (1) and ABRACOS (2) at heights: (a) 35 m; (b) 30 m, compared with the air temperatures observed at 45 m over the terra-firme forest.

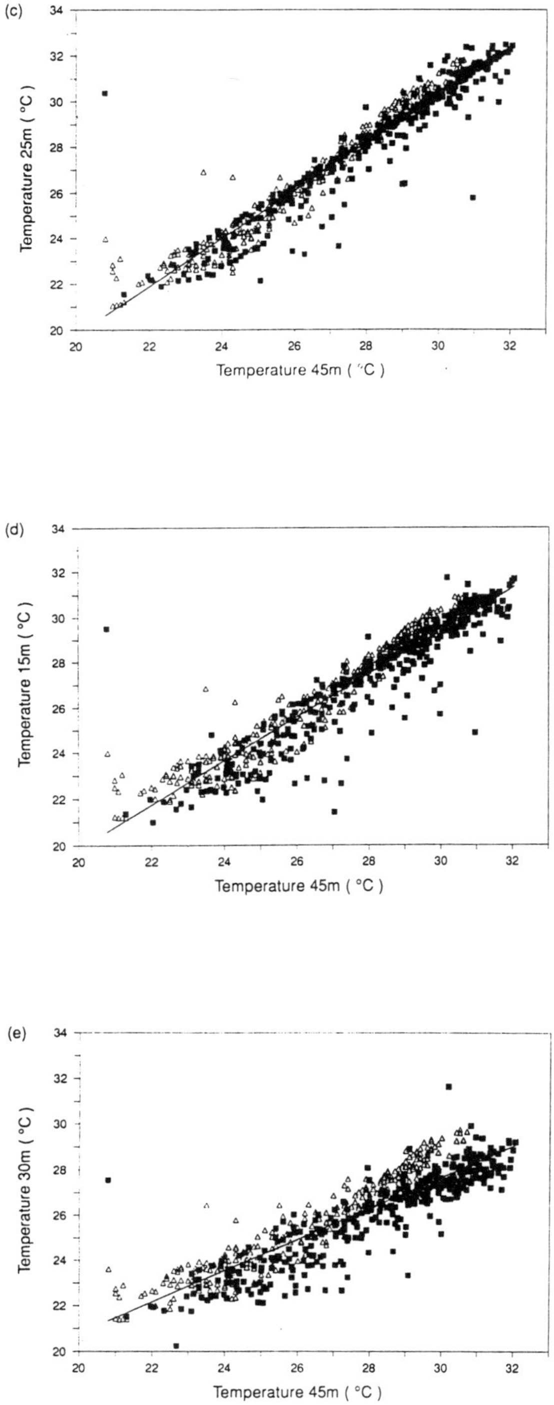

Figure 3 (continued) Air temperatures at heights: (c) 25 m; (d) 15 m and (e) 5 m.

levels, which show minor scattering, although the residuals are not randomly distributed around the curves as a consequence of the autocorrelation in the series. In this case the least squares estimates are still unbiased, but the standard errors of the regressions may be too small (Montgomery and Peck, 1982).

The 5 m height showed the greatest discrepancies: the ABRACOS maximum temperatures were consistently underestimated because the temperatures recorded during the ARME were lower, being measured at 1.3 m instead of at 5 m.

PROFILES OF SPECIFIC HUMIDITY DEFICITS

Figure 4 shows the overall average hourly values of humidity deficits. The 45 m and 25 m layers exhibited the same deficits as the 30 m and 35 m layers, and the higher deficits were also recorded at 30 m, similar to the temperature values, although the 15 m and 5 m layers were decoupled.

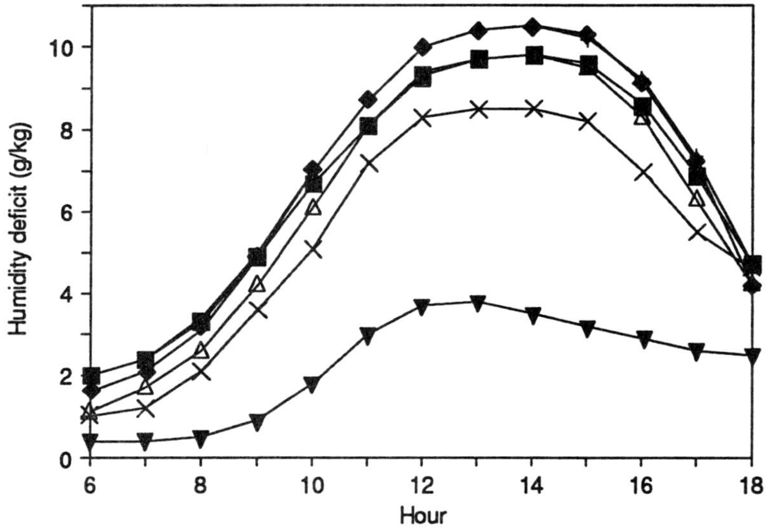

Figure 4 Hourly average of specific humidity deficits observed above and below the tropical terra-firme forest at heights: (■) 45 m; (◆) 35 m; (+) 30 m; (Δ) 25 m; (×) 15 m and (▼) 5 m.

The profiles for selected hours in Figure 5 imply that the patterns observed for temperatures are followed by the humidity deficits, but with the exception occurring at 18.00 h, which still shows the5 m layer decoupling and a minimum between 25 m and 35 m in response to the canopy cooling.

The fitted parameters of Equation 2 are shown in Table 2. The adjusted intercepts, except the 5 m level, can also be interpreted as the average offsets in relation to 45 m. Figure 6 shows the comparison between the estimated and observed deficits with differences more pronounced for deficits recorded during ARME, which were lower than ABRACOS because of the different height of the measurements, as noted above for temperature.

Table 2 Coefficients of Equation 2 (a = intercept; b = slope) for specific humidity deficits (g kg^{-1}) at each layer as a function of the deficits observed at 45 m.

Level height (m)	a	b	r^2
35	-0.17	1.08	0.99
30	-0.59	1.12	0.98
25	-0.96	1.09	0.93
15	-0.67	0.91	0.87
5	-0.49	0.40	0.68

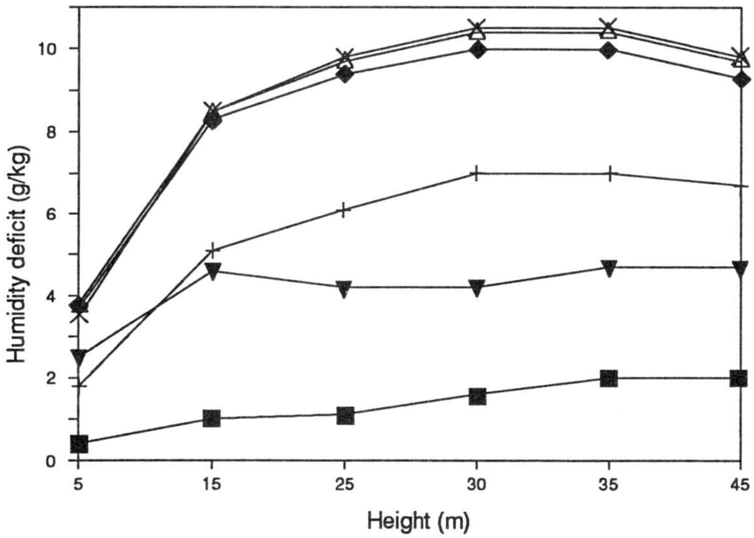

Figure 5 Hourly average profiles of specific humidity deficits at times: (■) 6 h; (+) 10 h; (◆) 12 h; (Δ) 13 h; (×) 14 h and (▼) 18 h.

The fitted equations can reproduce sufficiently accurate estimates of humidity deficits above the 25 m level but not for the lower layers, where large departures are observed, depending on the period investigated.

Attempts to fit similar relationships between specific humidity above and below canopy showed poor resolution, since the specific humidity remains more constant through the day (Shuttleworth *et al.*, 1985), and consequently the deficits are controlled by temperature changes. In this case, the estimate of humidity deficits based on temperature and specific humidity would induce greater errors.

THE FOREST TRANSPIRATION

The estimated hourly profiles of air temperature and specific humidity deficit obtained from ARME and ABRACOS above canopy data have been used to

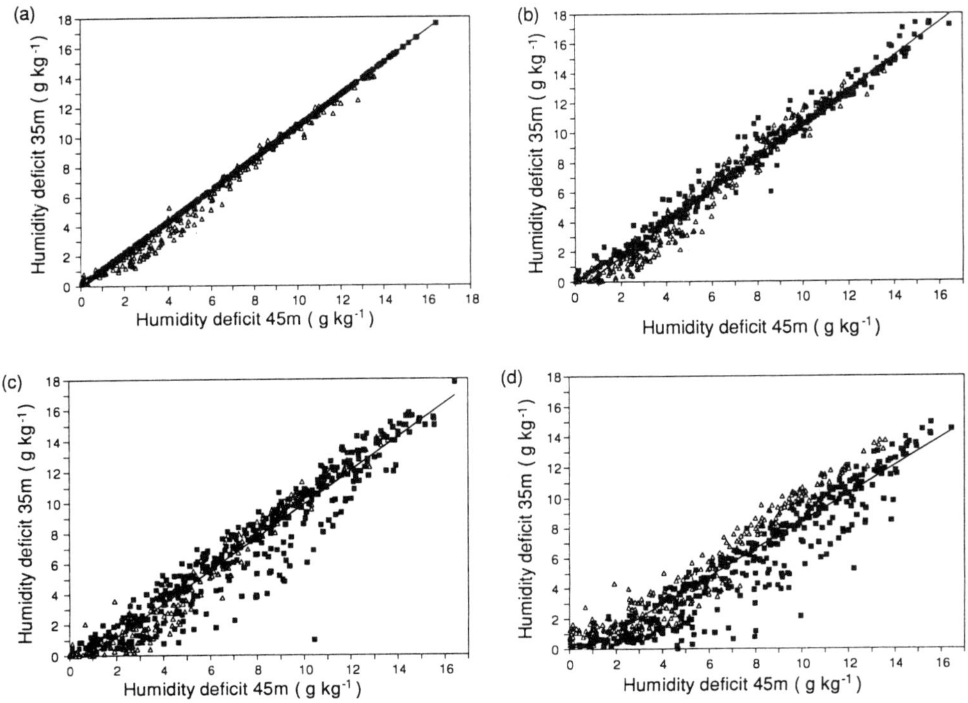

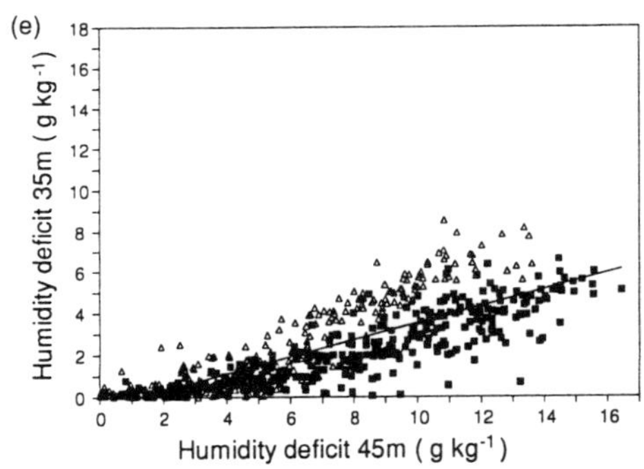

Figure 6 Specific humidity deficits estimated and measured during ARME (1) and ABRACOS (2) at heights: (a) 35 m; (b) 30 m; (c) 25 m; (d) 15 m and (e) 5 m, compared with the specific humidity deficits observed at 45 m over the terra-firme forest.

calculate the hourly rates of transpiration with CLATTER (Roberts *et al.*, 1993), and compared with those based on the measured profiles, which are assumed to represent the actual forest transpiration.

Figure 7 shows the hourly rates of transpiration and also shows the single-layer (big-leaf) estimates according to Shuttleworth (1988), where the surface conductance Model 2 given by Dolman *et al.* (1991) has been used. The agreement is better for the multilayer approach, whose scatter is lower than the single-layer results, as indicated by the residual sum of squares in Tables 3 and 4, for both periods. The transpiration rates were underestimated by the single-layer model, which is also observed in the multilayer results but only for rates above 0.5 mm h^{-1}, particularly in the ABRACOS data.

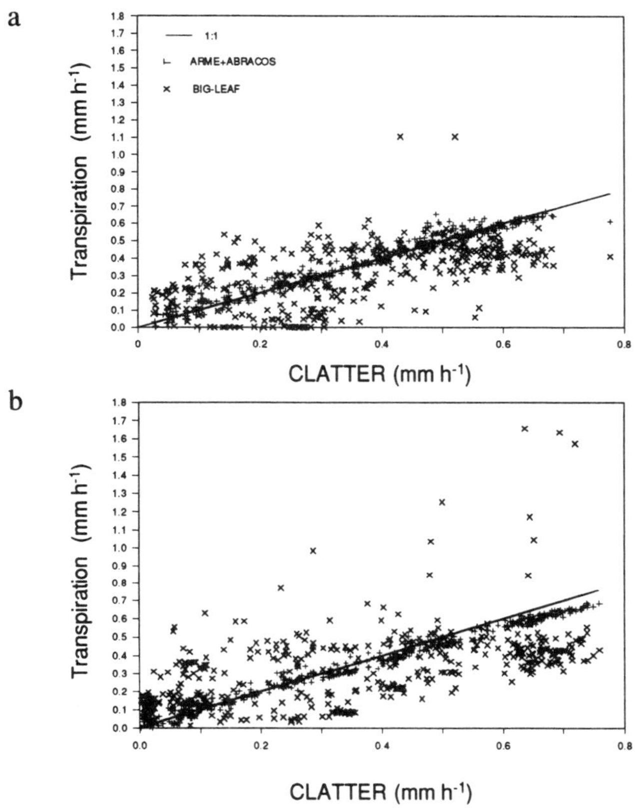

Figure 7 Hourly rates of transpiration calculated from estimated air temperatures and specific humidity deficits data (+), and from a single-layer model (×), compared with CLATTER applied to observed profiles of ARME (a) and ABRACOS (b).

Figure 8 shows the daily totals of forest transpiration and confirms the findings previously presented, i.e. both approaches tend to under-estimate the CLATTER results obtained using the measured data, whereas the discrepancy is greater for the single-layer outputs, as shown in Tables 3 and 4.

DISCUSSION

The correlation between the Amazonian forest above-canopy and canopy air temperatures and humidities reflects the more uniform structure in the leaf-area distribution leading to more uniform profiles with no marked points of inflection. These are the result of a progressive capture of radiation through the canopy (Shuttleworth, 1988). This feature was first noticed by Moore and Fisch (1986), who successfully used the hourly changes in these variables measured above the canopy, to estimate air sensible and latent heat storage in the canopy.

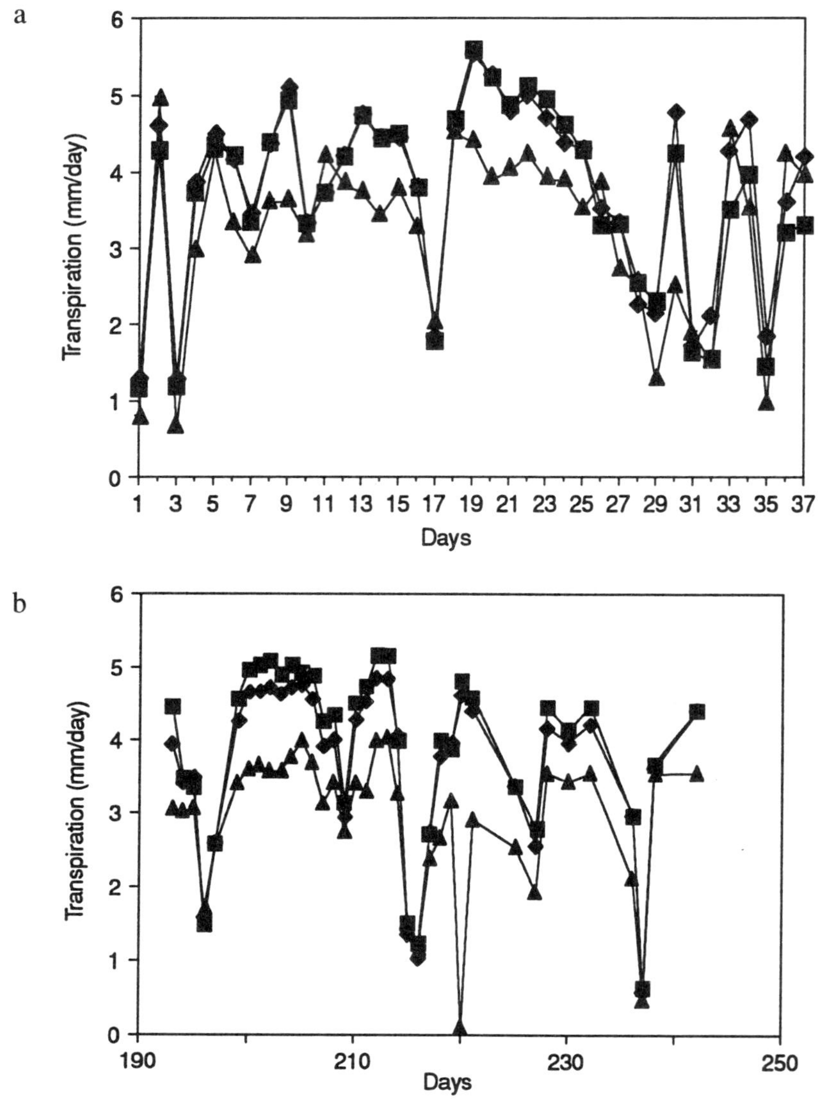

Figure 8 Daily totals of transpiration derived from the CLATTER model, based on estimated (◆) and measured (■) air temperatures and specific humidity deficits, and the big-leaf model results (▲), for ARME (a) and ABRACOS (b) periods.

The temperature profiles were estimated within 5% and the humidity deficits within 25%, albeit with disparities particularly below 25 m height. The reason for this lies in the least square method which was applied to data exhibiting serial correlation, promoting the under-estimate of maximal and the over-estimate of minimal observed values.

Notwithstanding the fact that the estimated profiles of humidity deficit showed significant deviations from the observed data, the resulting transpiration rates were

Table 3 Statistics for transpiration multilayer model based on observed and estimated profiles of temperature and specific humidity deficit, and a single-layer (Big-Leaf) model during the ARME.

	Observed profiles	Estimated profiles	Big-Leaf model
Totals (mm)	135.8	140.4	120.1
Averages (mm/day)	3.7	3.8	3.2
Standard deviations	1.2	1.2	1.2
Sum of residuals		4.6	-15.7
Sum of residuals squared		0.6	11.6

Table 4 Statistics for transpiration multilayer model based on observed and estimated profiles of temperature and specific humidity deficit, and a single-layer (Big-Leaf) model during ABRACOS.

	Observed profiles	Estimated profiles	Big-Leaf model
Totals (mm)	143.8	136.9	108.4
Averages (mm/day)	3.9	3.7	2.9
Standard deviations	1.2	1.1	0.9
Sum of residuals		-6.9	-1.0
Sum of residuals squared		0.7	20.8

within 5%, contrasting with the single-layer model whose results, when compared to the CLATTER value, under-estimated by 25% and 12% the ABRACOS and ARME data respectively.

The accuracy of the transpiration rates can be partially explained through the sensitivity analysis of CLATTER (Roberts *et al.*, 1993): to produce a 5% deviation in transpiration rates, more than a 25% deviation in humidity deficits or air temperatures would be necessary.

Based on the estimated deficits for 15 m and 5 m layers the transpiration rates at these levels were under-estimated for ABRACOS and over-estimated for ARME, which is confirmed by the totals in Tables 3 and 4, whereas these layers contribute on average only 20% and 5% respectively of the total forest transpiration (Roberts *et al.*, 1993).

For transpiration rates below 0.1 mm h^{-1} the scatter is increased and were more than 50% of the estimates based on measured profiles; however these periods have a lower weight in the daily totals and are prone to higher errors (Roberts *et al.*, 1993). For the most important part of the day, i.e. when the hourly transpiration rates were above 0.4 mm h^{-1}, the estimates are inside the ±10% range.

The results of this study clearly show that it is feasible to infer the forest transpiration from the average daytime in-canopy microclimate. A worthwhile research area now would be to test the possibility of estimating rainforest transpiration using large networks of surface sites (Fitzjarrald *et al.*, 1990), combined with standard meteorological data observed in nearby clearings, to predict the weather variables over the forest (Hutjes *et al.*, 1991; Dolman *et al.*, 1988).

CONCLUSIONS

The high degree of correlation found between the Amazon forest above-canopy and below-canopy air temperature and specific humidity deficit, expressed through empirical functions, although site specific, indicates the possibility of adequately describing the in-canopy microclimate to estimate the transpiration rates for complex forest types, such as tropical rainforest. This greatly reduces the number of parameters to be measured without generating large errors, and may be particularly valuable for extending the limited measurements available for Amazonian rainforest.

REFERENCES

Allen, L.H., Lemon Jr, E. and Müller, L. 1972. Environment of a Costa Rican forest. *Ecology* **53**, 102-111.

Bastable, H.G., Shuttleworth, W.J., Dallarosa, R.L.G., Fisch, G.F., Nobre, C.A. 1993. Observations of climate, albedo and surface radiation over cleared and undisturbed Amazonian forest. *Int. J. Climatol.* **13**, 783-796.

Brett, A.C. and Tuller, S. 1991. The autocorrelation of hourly wind speed observations. *J. Appl. Meteorol.* **30**, 823-833.

Dolman, A.J., Stewart, J.B. and Cooper, J.D. 1988. Predicting forest transpiration from climatological data. *Agric. For. Met.* **42**, 339-353.

Dolman, A.J., Gash, J.H.C., Roberts, J.M. and Shuttleworth, W.J. 1991, Stomatal and surface conductances of tropical rainforest. *Agric. For. Met.* **54**, 303-318.

Fitzjarrald, D.R., Moore, K.E., Cabral, O.M.R., Scolar, J., Manzi, A.O. and Sá, L.D. de A. 1990. Daytime turbulent exchange between the Amazon forest and the atmosphere. *J. Geophys. Res.* **95**, 16825-16838.

Hutjes, R.W.A., Klaassen, W., Kruijt, B. and Veen, A.W.L. 1991. Predicting near-surface variations over different vegetation types. *Hydrological interactions between atmosphere, soil and vegetation.*, IAHS Publ. No. 204, 75-84.

McWilliam, A.-L.C., Roberts, J.M., Cabral, O.M.R., Leitao, M.V.B.R., da Costa, A.C.L., Maitelli, G.T. and Zamparoni, C.A.G.P. 1993. Leaf area index and above-ground biomass of *terra firme* rainforest and adjacent clearings in Amazonia. *Functional Ecol.* **7**, 310-317.

Montgomery, D.C. and Peck, E.A. 1982. *Introduction to linear regression analysis.* John Wiley & Sons, New York, 504 pp.

Moore, C.J. & Fisch, G.F. 1986. Estimating heat storage in Amazonian tropical forest. *Agric. For. Met.* **38**, 147-169.

Pinker, R.T. 1982. The diurnal asymmetry in the albedo of tropical forest vegetation. *For. Sci.* **28**, 297-304.

Roberts, J.M., Cabral, O.M.R. and Aguiar L.F. 1990. Stomatal and boundary-layer conductances in an Amazonian terra firme rainforest. *J. Appl. Ecol.* **27**, 336-353.

Roberts, J.M., Cabral, O.M.R., Fisch, G., Molion, L.C.B., Moore, C.J. and Shuttleworth, W.J. 1993. Transpiration from an Amazonian rainforest calculated from stomatal conductances measurements. *Agric. For. Met.* **65**, 175-196.

Shuttleworth, W.J., Gash, J.H.C., Lloyd, C.R., Moore, C.J., Roberts, J.M., Marques Filho, A. de O., Fisch, G., Silva Filho, V. de P., Ribeiro, M. de N.G., Molion, L.C.B., de Abreu Sa, L.D., Nobre, C.A., Cabral, O.M.R., Patel, S.R. and de Moraes, J.C. 1984. Observations of radiation exchange above and bellow Amazonian forest. *Q. J. R. Meteorol. Soc.* **110**, 1163-1169.

Shuttleworth, W.J., Gash, J.H.C., Lloyd, C.R., Moore, C.J., Roberts, J.M., Marques Filho, A. de O., Fisch, G., Silva Filho, V. de P., Ribeiro, M. de N.G., Molion, L.C.B., de Abreu Sa, L.D., Nobre, C.A., Cabral, O.M.R., Patel, S.R. and de Moraes, J.C. 1985. Daily variations of temperature and humidity within and above Amazonian forest. *Weather* **40**, 102-108.

Shuttleworth, W.J. 1988. Evaporation from Amazonian rainforest. *Proc. R. Soc. London* **B233**, 321-346.

Thompson, O.E and Pinker, R.T. 1975. Wind and temperature profile characteristics in a tropical evergreen forest in Thailand. *Tellus* **27**, 562-573.

RESUMO

A estimativa da transpiração de florestas através de modelos de várias camadas tem como fator limitante a necessidade da especificação do microclima no interior da vegetação, principalmente os perfis de temperatura do ar e dos déficits de saturação. Com este objetivo, os perfis foram obtidos a partir de funções lineares empíricas dos mesmos parâmetros medidos sobre a floresta.

Os cálculos da transpiração horária e diária utilizando-se uma rotina de Penman-Monteith foram similares aos estimados através dos perfis medidos, reduzindo a quantidade de informações sem a introdução de erros apreciáveis.

13 Biomass production and accumulation in nutrient-limited rain forests: implications for responses to global change

E. MEDINA and E. CUEVAS
Instituto Venezolano de Investigaciones Cientificas, Caracas, Venezuala

The forests of northern South America are an expanse of continuous vegetation comprising the southern Orinoco basin in the north and the huge Amazon basin in the south. This area, of around 10^7 km^2, is too large for simple ecological generalizations. However, this basin has been reasonable well studied geologically, hydrologically and ecologically so it is feasible to provide some general hypothesis about the ecological properties of some widely distributed forests (Jordan, 1989; Medina and Cueva, 1989; Junk, 1989).

Since the pioneering work of H. Sioli (1954) a large number of studies have shown that central and eastern Amazon regions are characterized by comparatively poor soils, with exceptionally fertile soils along the "varzea" lakes, characteristic of the so called "white water rivers" (Fittkau *et al.*, 1975). Even this general statement needs to be qualified because since the early studies of the Amazon basin the occurrence of contrasting soils types associated with the geology of the region has been recognised (Sioli, 1975). It is not the purpose of this paper to provide a review of the ecological work performed in the Amazon basin. We refer to Jordan (1989) where the literature has been reviewed. The specific nutrient relationships of the forest of the upper Río Negro have been discussed in detail by Medina and Cuevas (1989). Most results and hypotheses discussed in the present paper derive from our experience in the upper Río Negro basin, the main northern tributary of the Amazon river. We will not consider the extensive forest areas found in the alluvial islands along the Amazonas river because their nutritional conditions are determined by the supply of sediments from geologically younger areas (Furch and Klinge, 1989; Junk, 1989).

In the Amazon basin intensive deforestation and fragmentation is taking place (Skole and Tucker, 1993) and, given the magnitude of the areas involved, it makes a significant contribution to the current increase of atmospheric CO_2 and associated climate change (Shukla, *et al.*, 1990; Houghton, 1991; Dixon *et al.*, 1994). Therefore, there is much interest in understanding the functioning of the forest and its recovery capacity, to predict the potential effects of global change.

We will show here that several widespread forest types in the Amazon basin are

Amazonian deforestation and climate. Edited by J.H.C. Gash, C.A. Nobre, J.M. Roberts and R.L. Victoria.

nutrient-stressed, their productivity and persistence are associated with the activity of root symbiosis, and perhaps the whole rhizosphere. Then we discuss the potential long-term effects of increasing atmospheric CO2 concentration on nutrient-limited tropical rainforests. Our hypothesis is that elevated atmospheric CO_2 will increase the activity of root symbiosis and improve nutrient availability in the long term. Whether this phenomenon will result in a higher sequestration of carbon or in higher turnover rates is still a matter of speculation.

ORGANIC MATTER PRODUCTION OF AMAZONIAN FORESTS: IS IT NUTRIENT-LIMITED?

The history of erosion and redeposition in the Amazon basin has led to an extensive leaching of nutrients, particularly the common cations K, Ca, and Mg, and the inorganic phosphate reserves. (Fittkau *et al.*, 1975; Furch and Klinge, 1989). This impoverishment is particularly intensive in the central and northern areas of the basin characterised by rains around or above 2000 mm a^{-1} (Salati *et al., 1978;* Matsui *et al.*, 1983). Moreover, nutrient cycling and turnover seem to be tied to organic matter dynamics (Tiessen*et al., 1994b).* This seems to indicate that biological processes are the main regulators of nutrient availability.

In some specific cases in the Amazon basin, the nutritional limitations have been clearly demonstrated, based on soil nutrient inventories, biomass accumulation and functional properties of the forests.

SOIL NUTRIENTS AND FOREST BIOMASS

Forests of the upper Río Negro have lower above ground-biomass than that reported for forest growing on less leached soils (Jordan, 1989). Only forests in extremely nutrient poor Tropaquods, also flood-prone, develop considerably higher root biomass compared to other forests (Table 1). In addition, differences between average specific leaf areas among the forests considered here indicate selection of tree species with lower nutritional requirements (Medina*et al.*, 1990). These differences have been associated with the lower soil fertility of acid tropical soils. Indeed, forests on relatively young soils of central America and Africa have much higher contents of total soil N and P. The Dipterocarp forest in Malayia has N contents much higher than both Amazonian forest considered here and has very low P content, similar to that of forests on Tropaquods.

In Amazonian *Terra Firme* forests, productivity does not appear to be limited primarily by nitrogen, but by phosphorus availability. To analyse this phosphorus limited productivity it is not enough to know the total P content of the soils. Cuevas and Medina (1988) reported root growth limited by P in forests having higher total soil P content than surrounding forests with lower total soil P. Phosphorus availability has to be evaluated by a detailed measurement of the different P fractions in the soils.

Tiessen *et al.* (1994a) conducted a detailed P fractionation in forest soils located

Table 1 Comparison of biomass accumulation and soil nutrient content in Amazonian and non-Amazonian rainforest (simplified after Jordan 1989)

	Tall caatinga San Carlos	Oxisol forest Venezuela	Evergreen forest Ivory Coast	Dipterocaarp forest Malaysia	Lowland rainforest Costa Rica	Moist forest Panama
Root biomass (mg ha^{-1})	132	56	49	20.5	14.4	11.2
Above-ground biomass (mg ha^{-1})	268	264	513	475	382	326
Specific leaf area (cm^2 g^{-1})	47	65	-	88	139	131-187
Total calcium (kg ha^{-1})	195	7	-	115	6530	22166
Total calcium calcium (kg ha^{-1})	785	1697	6500	6752	20000	-
Total phosphorus (kg ha^{-1})	36	243	600	44	7000	22

along a toposequence in San Carlos de Río Negro. The forest located at the top of the slope is highly diverse, characterized by a dense root-mat on top of a sandy mineral soil. The forest at mid-slope is dominated by taller trees (*Monopterix uacu, Micrandra sprucei or Eperua purpurea*), and the forest at the bottom of the slope has vegetation consisting of patches dominated by *Eperua leucantha, Manilkara sp., Micrandra sprucei* or *Glycoxylon inophylium* (Medina and Cuevas, 1989). Total carbon and nitrogen are higher in the mid-slope forest followed by the top-slope and the bottom-slope forests (Table 2). Total P, however, decreases from the top slope towards the bottom slope. Most P in the top slope is not available but sequestered in iron concretions. The available P is represented by the resin P (the amount of P extracted by an anion exchange resin) which increases in the opposite direction to total P.

Table 2 Total amounts of carbon, nitrogen and phosphorus fractions (g m^{-2}) in a toposequence of rainforests in San Carlos de Río Negro (from Tiessen *et al.*, 1994). Slope top corresponds to *terra firme* mixed forest, mid-slope corresponds to the Yévaro-Guaco complex and the slope bottom to tall Amazonian caatinga (Medina and Cuevas, 1989).

Site	Total C	Total N	Resin P_{in}	Alkali P_{in}	Resist P_{in}	P_{org}	Total P
Top slope (Plinthic hapludox)	21290	935	3.4	25	290	27	311
Σ concretions	3900	66	0.8	13	249	6	213
Σ fine soil	17390	849	2.6	12	41	21	98
Mid slope (Plinthic or Pteroferric hapludox)	30140	1333	4.0	13	92	15	127
Bottom slope (Tropaquod or Psammaquent)	10270	414	5.5	1.5	6.6	7	19

NUTRIENT CONCENTRATIONS IN PLANT TISSUES AND ECOSYSTEM NUTRIENT INVENTORIES

Nutrient concentration in plant tissues varies according to availability in the soil and specific requirements. The amount of nutrients contained in the biomass of a given ecosystem synthesizes both the long-term fertility of the site and the selection of species that maximize utilization of the available nutrients. In San Carlos de Río Negro, Cuevas and Medina (1988) and Medina and Cuevas (1989) showed a clear-cut example of the nutritional differentiation of Amazonian forests along the toposequence described above. Results indicate that nitrogen and biomass accumulations vary proportionately along the toposequence. The peaks in biomass and nitrogen content values were found in the Guaco-forest (*Monopterix uacu*) and both decrease steadily along the Caatinga complex. In contrast, peak P accumulation in biomass was found in the tall Caatinga, decreasing both in the forests on oxisols, and in the tropaquods with higher seasonality in soil saturation and drought. Finally the peak accumulation of K, Ca and Mg is displaced towards the tall bana.

The relatively lower organic matter productivity of the Amazonian forests in the upper Río Negro Basin is evident when rates of litter fall and cycling of N and P are compared with those of other well-documented lowland rain forests (Figure 1). Clearly, both concentration and flux of P in litterfall are at the lower end of the forests reported by Proctor (1983). The concentration and flux of N in litterfall are near the average for this data set. In addition, it should be noticed that while P flux is strongly dependent on P concentration, values for N are more scattered, showing similar rates of litter production at different N concentrations (differences in Nitrogen use

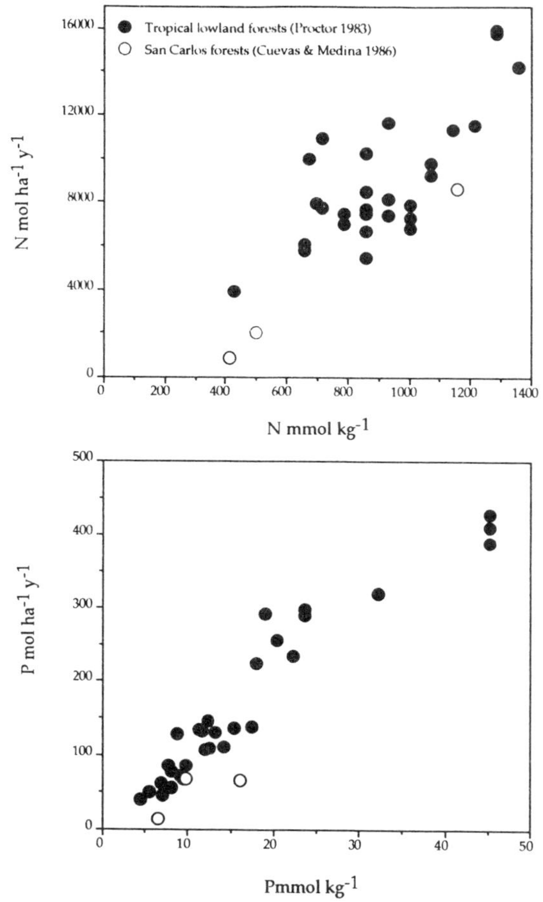

Figure 1 Nitrogen and phosphorus fluxes in tropical lowland forests (Proctor, 1983; Cuevas and Medina, 1986).

efficiency). These data support the notion that, in general, organic matter production in tropical forests is more limited by P than by N (Vitousek and Sanford, 1986). Availability of nutrients varies among Amazonian forests, as indicated by concentration and flux figures in litterfall (Table 3). Litterfall in *Terra Firme* forests has the highest N concentration, while that of the tall caatinga forest is more concentrated in C and M and that of the Bana has significantly higher concentration of K.

PHYSIOLOGICAL ADAPTATIONS TO LOW NUTRIENT AVAILABILITY

General low nutrient availability is associated with species that may be characterized by a number of correlated traits, such as low photosynthetic rate, slow growth, high carbon/nutrient ratios, and relatively high root/ shoot ratios (Chapin *et al.* 1993).

Table 3 Differences in mass, nutrient content and nutrient fluxes in fine litterfall in Amazonian forests (upper Río Negro basin; Cuevas and Medina, 1986)

Forest type	Leaf litter ($g\ m^{-2}\ a^{-1}$) (s.e.)		N	P	K	Ca	Mg
Terra firme	757 (24)	Concentration ($mg\ g^{-1}$)	**16.2**	0.3	2.4	1.7	0.7
		Flux ($g\ m^{-2}\ a^{-1}$)	12.1	0.21	1.5	1.3	0.5
Tall caatinga	399 (19)	Concentration ($mg\ g^{-1}$)	7.0	**0.5**	2.1	**7.7**	**3.1**
		Flux ($g\ m^{-2}\ a^{-1}$)	2.8	0.20	0.8	3.1	1.3
		Concentration ($mg\ g^{-1}$)	5.8	0.2	**4.7**	7.4	2.5
		Flux ($g\ m^{-2}\ a^{-1}$)	1.2	0.04	1.0	1.5	0.5

the leaf content of Mg ($r^2 = 0.90$), and Ca ($r^2 = 0.67$), while it was weakly and negatively correlated with the content of N ($r^2 = 0.48$). Correlations with P or K concentrations were not significant.

Decomposition of litter

Decomposition of litter is slower in Amazonian rainforests compared to other tropical forests on richer soils. This is a result of the low nitrogen/ lignin ratios of the leaf litter in these forests (Cuevas and Medina 1988). Both the rate of decomposition and the rate of nutrient release from litterfall appear to be dependent on litter nutrient content/ and activity of soil microorganisms C). Experiments on litter decomposition conducted in mixed forests on concretionary oxisols show that the disappearance rates of Ca and Mg resemble the disappearance of K, the most leachable element in forest ecosystems everywhere (Swift *et al.*, 1979) (Figure 2). This is not the case with low caatinga forests, where the root mat grows within a thick matrix of decomposing organic matter (Cuevas and Medina, 1988). Mycorrhizal roots attached to the litter in the *Terra firme* forest are thought to cause accelerated release of Ca and Mg.

OCCURRENCE AND ACTIVITY OF SYMBIOSIS

Occurrence of mycorrhizae, particularly, vesicular-arbuscular mycorrhizae (VAM) is widespread in Amazon basin forests (Singer and Morello, 1960; Stark and Spratt, 1977; St. John, 1980; St. John and Uhl, 1983; Moyerson, 1992). Ectomycorrhizal symbioses are less abundant in forests of the upper Río Negro, but are found commonly in legumes (Caesalpiniaceae) and Nyctagynaceae, frequently together with endomycorrhizal infections (Moyerson, 1992).

The role of mycorrhizal symbiosis in providing the host plant with nutrients and water has been thoroughly documented in the literature (Harley and Smith, 1983;

Table 4 Fine root accumulation during 3-4 months in in-growth cylinders filled with vermiculite imbibed in nutrient solutions as indicated. Values in g m^2 day^{-1}. * = differences significant at 0.01 (Kruskal-Wallis non-parametric test). (From Cuevas and Medina, 1988.)

Forest type	n	Control	$+NH_4Cl$	$+KH_2PO_4$	$+CaCl_2$
Terra firme	9	0.852	2.581	3.148*	3.874*
Tall caatinga	11	0.068	0.408*	0.118	0.173*
Low bana	9	0.058	0.255*	0.551*	0.167

Many of these properties have been documented in Amazonian forests of the upper Río Negro.

Lower nutrient concentration in biomass and nutrient flux in litterfall are not enough to prove the existence of nutrient limitation. The species occupying the low fertility soils in these areas might simply have lower nutritional requirements. Increased soil fertility in the short-term could generate marked responses in growth and nutrient uptake, while, in the long-term higher soil fertility could lead to changes in forest species composition.

Photosynthesis–nitrogen relationships

Many Amazonian tree species on low fertility soils behave as sclerophylls, having low photosynthetic rates on a per unit dry weight basis, in association with low concentrations of N and P (Medina *et al.*, 1990, Reich *et al.*, 1994). These properties are strongly correlated with the low specific leaf areas generally reported for these forests (Sobrado and Medina, 1980; Reich and Walters, 1994).

Root growth responses to fertilization and organic matter

Experiments conducted in the root mat of Amazonian forests on oxisols showed significant increases in root growth induced by the presence of organic matter (Jordan and Escalante, 1980; St. John, 1983). In-growth cylinders providing additions of N, P or Ca resulted in different amounts of fine roots produced per treatment in three contrasting forest types in San Carlos de Río Negro (Table 4). In terra firme forest, root growth in the root mat increased significantly due to addition of P and Ca. Increased root growth was observed in the tall caatinga only with added N, while in the low bana both N and P additions had significant effects. A parallel experiment measuring fine root growth in the top soil did not give significant differences among treatments (Cuevas and Medina, 1988). When leaves of tree species with different nutrient concentration were set in decomposition bags on top of the *Terra firme* root mat, it was observed that amount of root penetration was positively correlated with

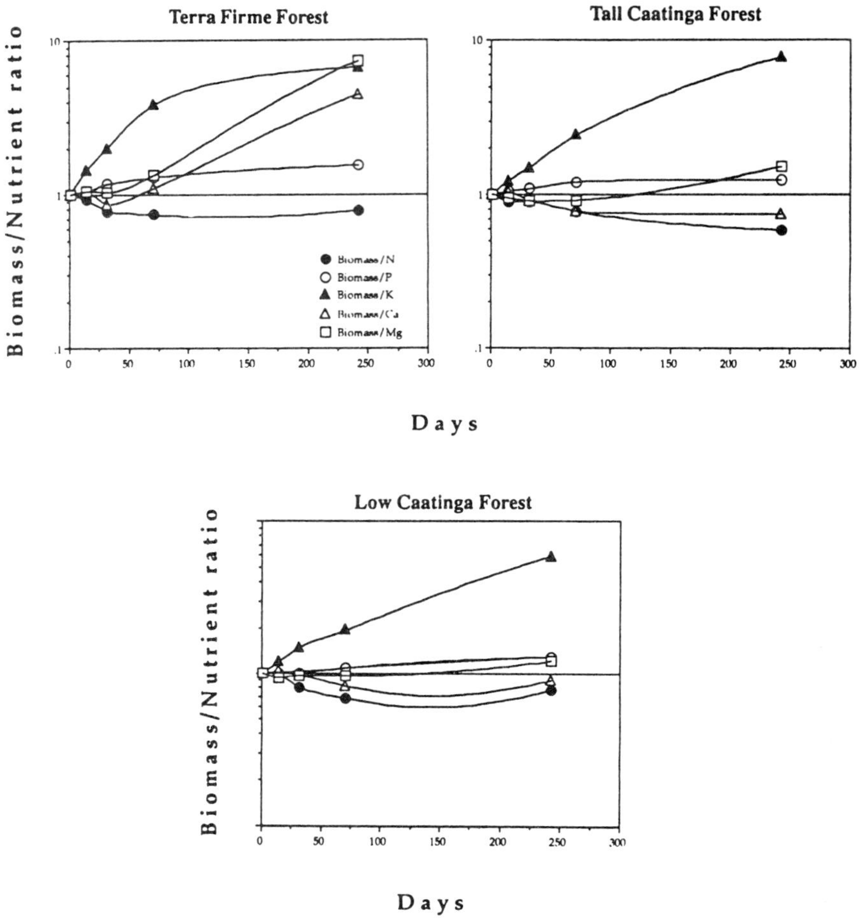

Figure 2 Biomass/nutrient ratios indicating rate of nutrient release (from Ceuvas and Medina, 1988)

Allen, 1991). Since the influential paper of Went and Stark (1968) numerous research projects have been conducted in tropical forests demonstrating the ecological importance of these mutual symbioses for productivity and persistence of forests in nutrient poor environments. It is generally accepted that mycorrhizae confer an advantage to the host plant in nutrient-stressed environments by increasing the size of the nutrient gathering surface well beyond the "nutrient depletion zone". In addition, some mycorrhizal symbioses, mostly ectomycorrhizae, are able to extract P from poorly soluble sources.

Presence of mycorrhizal symbioses contributes to a sink for photosynthetic assimilates. In general, the magnitude of the infection and the effectiveness of the mutual relationship are a compromise between photosynthetic capacity of the host plant and nutrient uptake and transference of the fungi. It has been calculated that nearly 25% of plant primary production is dedicated to maintenance and growth of mycorrhizas (Paul and Kucey, 1981; Allen, 1991). Phosphorus-rich soils reduce mycorrhizal infection in many tree species, while the contrary occurs in nutrient poor soils (Harley and Smith, 1983). We hypothesize that current productive capacity and nutrient accumulation in many Amazonian forests depends on the activity of mycorrhizal symbioses. Therefore, environmental factors that increase their activity, and/or their efficiency of nutrient capture will result in higher nutrient supply for the host plants.

Nitrogen fixation by symbiotic and free-living microorganisms constitutes a significant source of this element in many Amazonian ecosystems (Herrera and Jordan, 1981; Jordan *et al.*, 1982). The rates calculated have a large range of variation and it is not yet possible to give an overall account of how much nitrogen used in forest function derives from this biological activity. However, it is known that the activity of N_2-fixer depends on the availability of energy, in the form of respiratory substrates, and of P. Energy substrates are essentially supplied by the primary producers, while, as indicated before, P availability is associated with the activity of mycorrhizal symbiosis.

HYPOTHESES ON THE IMPACTS OF ELEVATED CO_2

Given the above depicted ecological constraints for Amazonian forests, what could we expect from global climate change, in particular the increase of atmospheric CO_2 concentration?

Global climatic change is driven mainly by increasing atmospheric CO_2 concentrations. This phenomenon is possibly leading to changes in patterns of rainfall distribution and increase in average temperature. Atmospheric CO_2 enrichment is the result of increasing fossil fuel burning and the accelerating rates of deforestation in the tropics (Houghton et al., 1991; Dixon *et al.*, 1994). Is it possible to predict the potential impact of elevated atmospheric CO_2 on biomass accumulation and productivity in nutrient-stressed forests?

Elevated CO_2 and plant growth

There is already a rich literature dealing with experiments on the impact of elevated CO_2 on plant growth and photosynthesis and general patterns of response at the level of individual plants are already known (Bazzaz, 1990). A thorough analysis of potential impact of environmental change on ecosystems taking a multiple factor approach is given by Field *et al.* (1992).

A summary of the present understanding of the effect of CO_2 on tree growth and photosynthesis is the following: all the experiments (short term < three months, or long term >three years) indicate that growth under elevated CO_2 leads to an increase in biomass accumulation and photosynthesis. Much of the biomass increase is allocated to the roots (Mauney *et al.*, 1992; Idso and Kimball, 1993), but the magnitude of the effect depends very much on nutrient availability and water stress (Bazzaz, 1990; El Kohen and Mousseau, 1994; Conroy *et al.*, 1992; Mousseau and

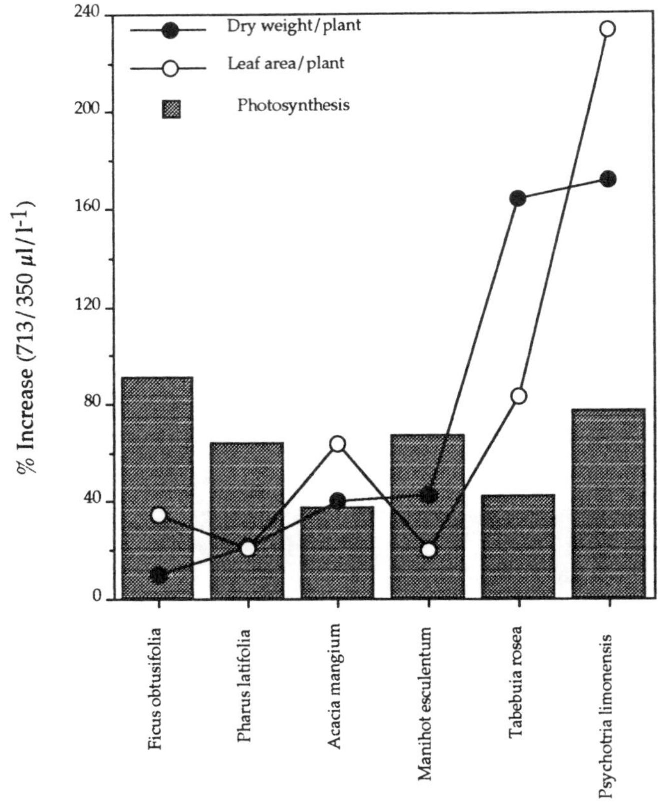

Figure 3 Relative increases in photosynthesis, leaf area and dry weight induced by elevated CO_2 in tropical species (Ziska *et al.*, 1991)

Saugier, 1992; Sage, 1994). Nutrient availability during the elevated CO_2 treatment is critical for the magnitude of the growth response and to delay the reduction in photosynthesis observed after prolonged periods. In all experiments a reduction in leaf nitrogen content has been observed, which might be explained by the reduction in specific leaf area. A reduction in leaf chlorophyll content has been also observed, but difficult to understand unless the plants under elevated CO_2 experience true N deficiency. In many experiments photosynthesis and leaf nitrogen are linearly correlated, but photosynthetic rate per unit leaf N is higher under elevated CO_2 conditions (Larigauderie *et al.*, 1988). Interestingly, elevated CO_2 increases the efficiency of light utilization by shade plants (Idso and Kimball, 1993). If the effect of increased C/N ratios in plant tissues observed experimentally is long-lasting, increased atmospheric CO_2 concentration since last century (290 to 350 ppm) should be reflected in higher C / N ratios and a decrease in $\delta^{13}C$ values. As far as we know these effects have not been measured yet (however, see Street-Perrot, 1994).

The few papers about CO_2 effects on growth of tropical plants show short-term increases in photosynthesis and biomass accumulation (Ziska *et al.*, 1991) (Figure

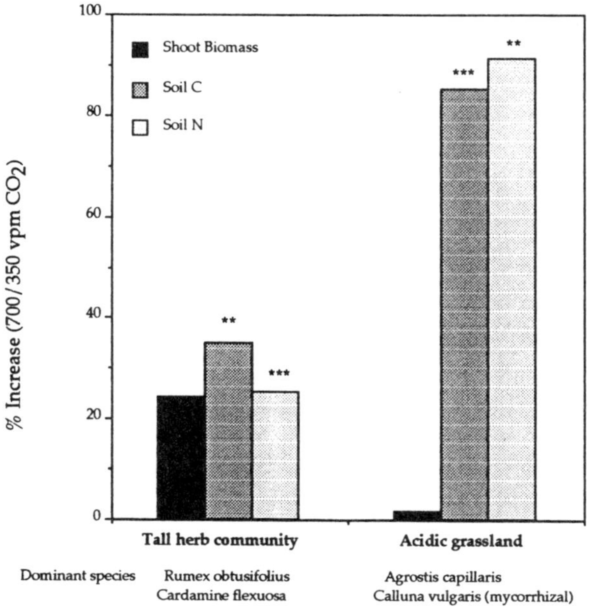

Figure 4 Relative increase in shoot biomass and soil carbon and nitrogen in grassland communties differing in nutrient availability (Diaz *et al.*, 1993)

3). However, an experiment intending to simulate complex multi-species conditions did not show differences in photosynthesis or total biomass accumulation after a period of six months (Körner and Arnone, 1992). This experiment shows that increased photosynthate due to elevated CO_2 has been transported to the roots and possibly increased the amount of root exudates. Therefore, soil respiration was nearly three times higher in the elevated CO2 chamber compared to the control. Increased biomass accumulation could not have been determined by low nutrient supply, because nutrient leaching from the high CO_2 chamber was much higher than in the control. These results indicate that under elevated CO_2 (1) soil microflora are more active in decomposing organic matter, (2) that not enough roots have been developed to pick up the nutrients released during decomposition. Unfortunately, the authors did not indicate the degree of mycorrhizal formation in their experimental plants.

Elevated CO_2 and plant-microbes interactions

A number of recent papers document the role of mycorrhiza symbiosis as carbon sinks and potential sources of soil carbon. Diaz *et al.*, (1993) showed that in non-mycorrhizal plants elevated CO_2 resulted in a higher exudate release from roots into the soils. The expanded microflora sequestered mineral nutrients creating a nutrient limitation and growth reduction in the high CO_2 treatments. However, beneficial effects were observed in the mycorrhizal species *Calluna vulgaris*. Mycorrhizae in this species provide a sink preventing nutrient sequestration by free living microorganisms. The authors interpret their results as indicative of a negative feedback of increased CO_2 in plants growing in natural communities. The authors did not measure changes in underground biomass allocation, but showed a marked increase in carbon and nitrogen associated with the microbial biomass (Figure 4). These results could represent only short-term effects. In the long-term, larger microbial biomass will increase both carbon and nitrogen resources in the soil, improving soil fertility and plant productivity. These effects may be more pronounced in this time scale in low nutrient environments. Adjustments to long-term CO_2 enrichment may not be directly controlled by CO_2, but instead may be a secondary response to greater carbon gain which CO_2 enrichment initially promotes (Sage, 1994).

Rygiewicz and Andersen (1994) showed that mycorrhizae increase carbon allocation below ground and stimulate tumover rate through respiration. The conclusion is that mycorrhizae lead to a more rapid release of carbon. However, the authors do not indicate that higher mycorrhizal activity may lead to a more active nutrient uptake that might increase productivity and carbon sequestration both in soil and plants.

Similar results were reported by Rouhier *et al.* (1994) in *Castanea sativa* cultivated in a CO_2-enriched atmosphere. Elevated CO_2 improved carbon supply to the soil

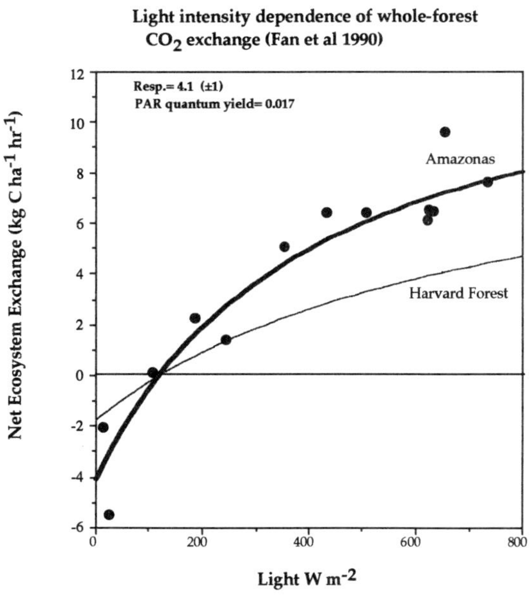

Figure 5 Net ecosystem carbon exchange measured in central Amazonian forest (Fan *et al.*, 1990) and in a temperate deciduous forest (Wofsy *et al.*, 1993)

microflora causing a higher respiration rate. During the experiment the authors did not observe increased carbon accumulation in the plant. This might be a transient phenomenon because, in the long run, nitrogen enrichment in the soil will result in an improvement of tree growth.

Sink-source relationships

It seems that many of the apparently contradictory results reported in the literature regarding the effect of elevated CO2 on plant and ecosystem production derives from varying source-sink strengths in the different experiments. The strong regulation of photosynthesis exerted by sink strength through feedback inhibition and changes in nitrogen investment in Rubisco were thoroughly reviewed by Stitt (1991). Intensified research in this area, particularly under natural conditions, looking to the patterns of biomass allocations and plant-microbe interactions will help to predict ecosystem responses to elevated CO_2.

Recycling of respiratory CO_2 and forest photosynthesis

Measurements of CO_2 concentration in the atmosphere and within forests' canopies were conducted in Central Amazonia during two contrasting seasons (Wofsy *et al.*, 1988; Fan *et al.*, 1990). Rates of CO_2 production through soil respiration were also measured, therefore carbon balances for the forest canopy could be constructed. Results showed a strong dependence of canopy carbon balance on light intensity. This emphasizes the potential sensitivity of the forest's photosynthesis to changes in cloud frequency. The rates of carbon uptake measured in Amazonian closed forest canopies are higher than the corresponding values measured in a temperate deciduous forest (Wofsy *et al.*, 1993) (Figure 5). In spite of differences in methodology, it is not obvious which is the cause of the marked differences in CO_2 uptake. Possible the higher leaf indices in the Amazonian forests are the main factor. In any case, the lower photosynthetic rates measured in individual plants (Reich *et al.*, 1994) are not reflected in the whole canopy.

Within the canopy, relatively high CO_2 concentrations, ranging from 400-700 vpm, have been documented both in Central Amazonia (Wofsy *et al.*, 1988) and in forests of the upper Río Negro Basin (Medina *et al.*, 1986). These high concentrations in the understorey are characteristic of dense forests, both in high and low latitudes (Medina and Minchin, 1980; Ethleringer (*pers. comm.*) and result from the production of CO_2 in soil respiration (root respiration + microbial organic matter decomposition). Plants in the understorey and the forest floor photosynthesize during several hours at CO_2 concentrations considerably higher than those of the open atmosphere. As a consequence, they are recycling a fraction of the CO_2 originated in the decomposition of organic matter. This was shown by measuring the $\delta^{13}C$ values of healthy leaves of tree species in different strata of the forest. The results showed that leaves, particularly those near the forest floor, have $\delta^{13}C$ values significantly lower than those of the upper canopy (Medina and Minchin, 1980). These differences have been attributed to simultaneous effects of lower light intensity (Farquhar *et al.*, 1982) and assimilation of CO_2 depleted in ^{13}C (Medina and Minchin, 1980; van der

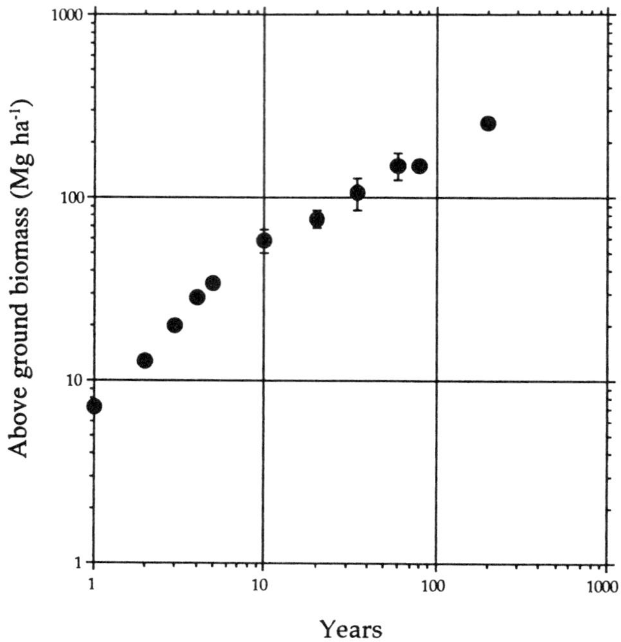

Figure 6 Above-ground biomass accumulation during secondary succession in Amazonian forest of the upper Río Negro (Saldarriaga *et al.*, 1988)

Merwe and Medina, 1989).

Recently Phillips and Gentry (1994) reported that there has been a significant increase in tree turnover (mean of % mortality and % recruitment) in tropical forests. This increase may be the result of atmospheric CO_2 enrichment. The authors suggest that a fraction of this increase results from the higher frequency of lianas, plants highly sensitive to increased CO_2. If, in addition, the most responsive trees are fast growers, there may be a displacement in species composition favouring gap specialists against slow growing shade tolerants. This situation may turn tropical forests into a net source of CO_2 because the slow growers have denser wood (Phillips and Gentry, 1994).

We do not know which is the actual situation in Amazonian forests. However, it is reasonable to predict that rates of regeneration and recruitment within natural forest are going to be markedly affected by increased atmospheric CO_2 concentrations.

Recovery capacity and biomass accumulation in Amazonian forests

The operation of compensating mechanisms for counteracting atmospheric CO_2 enrichment depends on the existence of biospheric sinks. It is the size and efficiency of those sinks that is being reduced by widespread and rapid deforestation in the tropics.

The size and the intensity of the deforestation process affect the recovery capacity of the forests (Uhl, 1982). Forests can recover from small scale disturbance such as

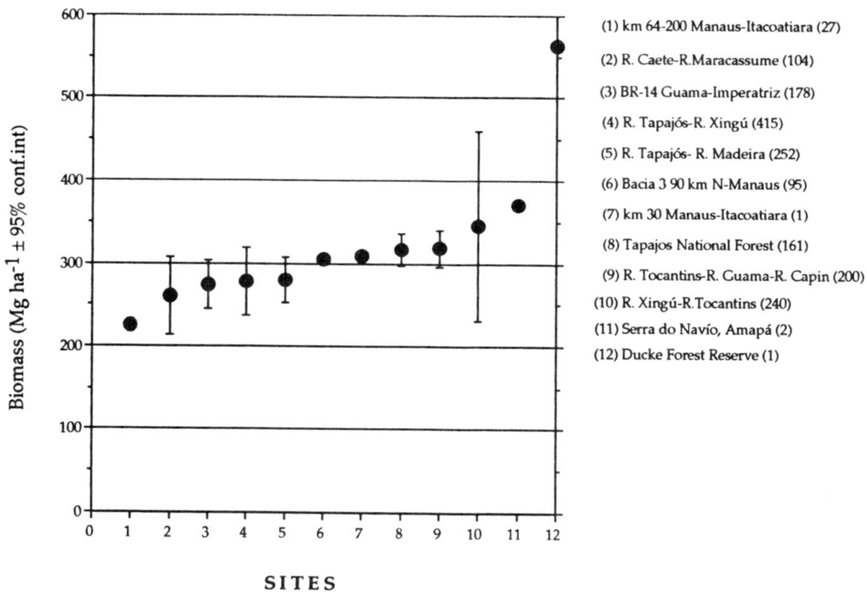

Figure 7 Above-ground biomass of Amazonian forests calculated from large inventories (Brown and Lugo, 1992)

slash-and-burn agriculture, in periods as short as 200 years, if secondary succession is allowed to run undisturbed (Saldarriaga *et al.*, 1988) (Figure 6). However, the efficiency of photosynthetically active radiation (PAR) conversion calculated for above ground biomass is lower than values reported for temperate forests (Saldarriaga and Luxmore, 1991). This lower efficiency might be related to higher maintenance costs and nutrient stress. Elevated atmospheric CO_2 could increase efficiency of PAR conversion and thereby increase the rate of recovery of forest after small to medium-scale disturbances.

Present levels of above ground forest biomass in different Amazonian regions are characterized by differences in the order of 100 Mg ha^{-1} (Brown and Lugo, 1992) (Figure 7). The forests with lower biomass might be located on less favourable landscapes, or may find themselves in a process of secondary succession after partial exploitation in the past. If the latter is the case, these differences indicate the amount of biomass that can be accumulated in large forest extensions if left undisturbed. In the case that forest productivity is limited by nutrient availability it is possible that natural forests will increase in density. In the long term, however, it is not possible to predict whether increased productivity caused by higher CO_2 will translate in higher biomass density or higher turnover rates as a result of changes in species composition.

CONCLUSIONS

In our opinion, predicted increases in atmospheric CO_2 concentration, with similar conditions of water supply and temperature, will markedly affect productivity, turnover rates and biomass allocation patterns in nutrient-limited Amazonian forests. These modifications are associated with their strong dependence of nutrient cycling and nutrient-use efficiency on associations with symbiotic and free-living micro-organisms. The extension and efficiency of nutrient supply of these associations depend on the provision of carbon compounds by the primary producers. Higher CO_2 concentrations allow, under natural conditions, enhanced photosynthetic rates and higher water- and nutrient-use efficiencies.

REFERENCES

Allen, M.F. 1991. *The ecology of mycorrhizae*. Cambridge Studies in Ecology. Cambridge University Press, UK.

Bazzaz, F.A. 1990. The response of natural ecosystems to the rising global CO_2 levels. *Ann. Rev. Ecol. Syst.*, **21**, 167-196.

Brown, S. and Lugo, A.E. 1992. Above-ground biomass estimates for tropical moist forests of the *Brazilian Amazon. Interciencia*, **17**, 8-16.

Chapin, F.S., Autumn, K. and Pugnaire, F. 1993. Evolution of suites of traits in response to environmental stress. *Am. Nat.*, **142** (Suppl.), 578-592.

Conroy, J.P., Milham, P.J. and Barlow, E.W.R. 1992. Effect of nitrogen and phosphorus availability on the growth response of *Eucalyptus grandis* to high CO_2. *Plant, Cell and Env.*, **15**, 843-847.

Cuevas, E. and Medina, E. 1986. Nutrient dynamics in Amazonian forest ecosystems 1. Nutrient flux in fine litter fall and efficiency of nutrient utilization. *Ocologia*, **68**, 466-472.

Cuevas, E. and Medina E. 1988. Nutrient dynamics in Amazonian forest ecosystems. II. Fine root growth, nutrient availability and leaf litter decomposition. *Oecologia*, **76**, 222-235.

Diaz, S., Grime J.P., Harris, J. and McPherson, E. 1993. Evidence of a feedback mechanism noting plant response to elevated carbon dioxide. *Nature*, **364**, 616-617.

Dixon, R.K., Brown, S., Houghton, R.A., Solomon, A.M., Trexler, M.C. and Wisniewski, J. 1994. Carbon pools and flux of global forests ecosystems. *Science*, **263**, 185-190.

El Kohen, A. and Mousseau, M. 1994. Interactive effects of elevated C02 and mineral nutrition on growth and CO_2 exchange of sweet chesnut seedlings *(Castanea sativa)*. *Tree Physiol*, **14**, 679-690.

Fan, S.-M., Wofsy, S.C., Bawkin, P.S., Jacob, D.J. and Fitzjarrald, D.R. 1990. Atmosphere-bio sphere exchange of CO_2 and O_3 in the Central Amazon foresfl. *Geophys. Res*, **95** (DIO), 16851-16864.

Field, C.B., Chapin, S., Matson, P.A., Mooney, H.A. 1992. Responses of terrestrial ecosystems to the changing atmosphere: a resource-based approach. *Ann. Rev. Ecol. Syst*, **23**, 201-235.

Fittkau, E.J., Irmler, U., Junk, W.J., Reiss, F. and Schmidt, G.W. 1975. Productivity, biomass/ and population dynamics in Amazonian water bodies. In: Golley, F.B. and Medina, E. (Eds.) *Tropical Ecological Systems. Ecological Studies*, 11, pp. 289-311. Springer Verlag, Berlin.

Furch, K. and Klinge, H. 1989. Chemical relationships between vegetation, soil and water in contrasting inundation areas of Amazonia. In: Proctor, J. (Ed.) *Mineral nutrients in tropical forest and savannas ecosystems*. Special Publication Series. Brit. Ecol. Soc. No. 9. Blackwell Scientific Publ. Oxford. 189-204.

Harley, J.L. and Smith, S.E. 1983. *Mycorrhizal symbosis*. Academic Press. London

Herrera, R. and Jordan, C.F. 1981. Nitrogen cycle in a tropical Amazonian rain forest: the Caatinga of low mineral nutrient status. In: Clark, F.E.and Rosswall, (Eds.) *Terrestrial nitrogen cycles.*

Ecological Bulletin (Stockholm), **33**, 493-505.

Houghton, R A. 1991. Tropical deforestation and atmospheric carbon dioxide. *Climatic Change*, **19**, 99-118.

Idso, S.B. and Kimball, B. 1993. Tree growth in carbon dioxide enriched air and its implication for global carbon cycling and maximum levels of atmospheric CO_2. *Global Biogeochem. Cycles*, **7**, 537-55.

Jordan, C.F. (Ed.) 1989. *An Amazon rainforest: the structure and function of a nutrient stressed ecosystem and the impact of slash-and-bum agriculture.* Man in the Biosphere series Vol. 2. UNESCO, Paris, and the Parthenon Publ. Group. Ltd, UK.

Jordan, C.F. and Escalante, G. 1980. Root productivity in an Amazonian rain forest. *Ecology*, **6**, 14-18.

Jordan, C.F., Caskey, W., Escalante, G., Heffera, R., Montagnini, F., Todd, R. and Uhl, C. 1982. The nitrogen cycle in a "Tierra Firme" rainforest in the Amazon territory of Venezuela. *Plant and Soil*, **97**, 325-332.

Junk, W.J. 1989. The use of Amazonian floodplains under ecological perspective. *Interciencia*, **14**, 317-322.

Körner, C. and Arnone III, J. Responses to elevated carbon dioxide in artificial tropical ecosystems. *Science*, **257**, 1672-1675

Larigauderie, A., Hilbert, D.S.W. and Oechel, W.C. 1988. Effect of CO_2 enrichment and nitrogen availability on resource acquisition and resource allocation in a grass, *Bromus mollis.Oecologia*, **77**, 544-549.

Matsui, E., Salati, E., Ribeiro, M.N.G., Reis, C.M. dos, Tancredi, A.C.S.N. and Gat. J.R. 1982. Precipitation in the Central Amazon basin: the isotopic composition of rain and atmospheric moisture at Belém and Manaus. *Acta Amazônica*, **13**, 307- 369.

Mauney, J.R., Lewin, K.F., Hendrey, G.R. and Kimball, B.A. 1992. Growth and yield of cotton exposed to free-air CO_2 enrichment (FACE). *Crit. Rev. Plant Sci.*, **11**, 213-22.

Medina, E. and Minchin, P. 1980. Stratification of d^{13} C values of leaves in Amazonian rain forests. *Oecologia*, **45**, 377-378.

Medina, E. and Cuevas, E. 1989. Patterns of nutrient accumulation and release in Amazonian forests of the upper Rio Negro basin. In: Proctor, J. (Ed.)*Mineral nutrients in tropical forest and savannas ecosystems.* Special Publication Series. Britisch Ecological Society No. 9. Blackwell Scientific Publ. Oxford. 217-240.

Medina, E., Garcia, V., Cuevas, E. 1990. Sclerophylly and oligotrophic environments: relationships between leaf structure, mineral nutrient content and drought resistance in tropical rain forests of the upper Río Negro Region. *Biotropica*, **22**, 51-64.

Moyerson, B. 1992. Importancia relativa de las ectomicorrizas respecto a las endomicorrizas en un bosque tropical Huvioso sobre suelo arenoso podsolizado (Caatinga Amazonica) en San Carlos de Río Negro, T.F. Amazonas. *Tesis Magister Scientiarum*. IVIC. Caracas.

Mousseau, M. and Saugier, B. 1992. The direct effect of increased CO_2 on gas exchange and growth of forest tree species. *J. Exp. Bot.*, **43**, 1121-1130.

Paul, E.A. and Kucey, R.M.N. 1981. Carbon flow in plant nutrient associations. *Science*, **213**, 473-474.

Phillips, O.L. and Gentry, A.H. 1994. Increasing turnover through time in tropical forests. *Science*, 263, 954-958.

Proctor, J. 1983. Tropical forest litterfall; (1) problems of data comparison. In: Tropical rain forest ecology and management. (eds.) S.C. Sutton, T.C. Whitmore and I.C. Chadwick. Spec. Publ. Ser. B.E.S., Blackwell. 267-273.

Reich, P.B. and Walters, M.B. 1994. Photosynthesis-nitrogen relationships in Amazonian tree species. II. Variation in nitrogen vis-a-vis specific leaf area influences mass- and mass-area based experessions. *Oecologia*, **97**, 73-81.

Reich, P.B., Walters, M.B., Ellsworth, D.S and Uhl, C. 1994. Photosynthesis-nitrogen relationships in Amazgroan., tree species. 1. Patterns among species and communities. *Oecologia*, **97**, 62-72.

Rouhier, H., Billes G., El Kohen A., Mousseau M. and Bottner P. 1994. Effect of elevated CO_2 on carbon and nitrogen distribution within a tree *(Castanea sativa* Mill.)-soil system. *Plant and Soil,* **162**, 281-292. 1994.

Rygiewicz, P.T. and Andersen, C.P. 1994. Mycorrhizae alter quality and quantity of carbon allocated below ground. *Nature*, **369**, 58-60.

Sage, R.F. 1994. Acchmation **of** photosynthesis to increasing atmospheric CO_2: the gas exchange perspective. *Photos. Res.,* **39**, 351-368.

Salati, E., Marques, J. and Molion, L.C.B. 1978. Origem e distribúção das chuvas na Amazonia. *Interciencia*, **3**,200-206.

Saldarriaga, J.G., West, D. C., Tharp, M.L. and Uhl, C. 1988. Long-term chronosequence of forest succession in the upper Rfo Negro of Colombia and Venezuela. *J. Ecol.* **76**, 938-958.

Saldarriaga, J.G. and Luxmore, R.J. 1991. Solar energy conversion efficiencies during succession of a tropical rain forest in Amazonia. *J. Trop. Ecol.* **7**, 233-242.

Singer, R. and Morello, J.H. 1960. Ectotrophic forest tree mycorrhizae and forest communities. *Ecology*, **41**, 54-551.

Shukla, L., Nobre, C. and Sellers, P. 1990. Amazon deforestation and climate change. *Science,* **247**, 1322-1325.

Sioli, H. 1954. Gewisserchemie und Vorginge in den Boden im Amazonasgebiet. *Naturwissenschaften*, **41**, 456-457.

Sioli, H. 1975. Tropical rivers as expressions of their terrestrial environments In: Golley, F.B. and Medina, E. (Eds.) *Tropical Ecological Systems*. Ecological Studies 11. Springer Verlag Berlin. 275-288.

Skole, D. and Tucker C. 1993. Tropical deforestation and habitat fragmentation in the Amazon: satellite data from 1978 to 1988. *Science*, **260**, 1905-1910.

Sobrado, M. and Medina, E. 1980. General morphology, anatomical structure and nutrient content of sclerophyflous leaves of the "Bana" vegetation of Amazonas. *Oecologia* (Berlin), **45**, 341-345.

Stark, N. and Spratt, M. 1977. Root biomass and nutrient storage in rainforest oxisols near San Carlos de Río Negro. *Trop. Ecol.*, **18**, 1-9.

St. John, T.V. 1980 A survey of mycorrhizal infection in an Amazonian rain forest. *Acta Amazonica,* **10**, 5287-533.

St. John, T.V. and Uhl, C. 1983. Mycorrhizae in the rain forest at San Carlos de Río Negro. *Acta Cient. Ven.,* **34**, 233-237.

St. John, T.V. 1983. Response of tree roots in decomposing organic matter in two lowland Amazonian rain forests. *Can. J. Forest Res.* **13**, 346-349.

Stitt M. 1991. Rising CO_2 levels and their potential significance for carbon flow in photosynthetic cells. *Plant, Cell and Env.,* **14**, 741-762.

Street-Perrott, F.A. 1994. Palaeo-perspectives: changes in terrestrial ecosystems. *Ambio.* **23**, 37-43.

Swift, M.J., Heal, O.W. and Anderson, J.M. 1979. *Decomposition in terrestrial systems.* Blackwell, Oxford.

Tiessen, H., Chac6n, P, and Cuevas, E. 1994a. Phosphorus and nitrogen status in soils and vegetation along a toposequence of dystrophic rainforests on the upper Río Negro. *Oecologia* (Berl.), **99**, 145-150.

Tiessen, H., Cuevas E., and Chac6n P. 1994b. The role of soil organic matter in sustaining soil fertility. *Nature,* **371**, 783-785.

Uhl, C. 1982. Recovery following disturbances of different intensities in the Amazon rain forest of Venezuela. *Interciencia,* **7**, 19-24

van der Merwe, N.J. and Medina, E. 1989. Photosynthesis and $^{13}C/^{12}C$ ratios in Amazonian rain forests. *Geochim. Cosmochim. Acta,* **53**, 1091-1094.

Vitousek, P.M. and Sanford, Jr. R.L. 1986. Nutrient cycling in moist tropical forest. *Ann. Rev. Ecol. Syst.,* **17**, 137-167.

Went, F. and Stark, N. 1968. The biological and mechanical role of soil fungi. *Proc. Nat. Acad. Sci. USA,* **60**, 497-504.

Wofsy, S.C., Harriss, R.VC. and Kaplan, W.A. 1988. Carbon dioxide in the atmosphere over the Amazon basin. *J. Geophys. Res.*, **93** (D2), 1377-1387.
Wofsy S.C. *et al.* 1993. Net exchange of CO_2 in a mid-latitude forest. *Science*, **260**, 1314-1317.
Ziska, L.H., Hogan, K.P., Smith, A.P. and Drake, B.G. 1991. Growth and photosynthetic response of nine tropical species with long-term exposure to elevated carbon dioxide. *Oecologia*, **86**, 383-389.

RESUMO

Geralmente, acredita-se que os ecossistemas de floresta na Bacia Amazônica têm um baixo teor de nutrientes. Como exceções, temos as várzeas ao longo do Río Solimões e florestas nas encostas da Amazônia Ocidental. O teor de nutrientes das florestas na Amazônica varia em função da química do solo e das inundações. Na maior parte das florestas de *Terra Firme* em oxissóis, a disponibilidade de fósforo é muito baixa, enquanto nas florestas inundáveis sobre areia branca (Igapós do Río Negro), a mineralização do nitrogênio parece ser o fator limitante. Além disso, a disponibilidade de cátions como K, Ca e Mg é particularmente baixo nas florestas de *Terra Firme* se comparados às de Caatinga.

Processos nos ecossistemas como produção, acumulação e decomposição de biomassa, desenvolvimento da área foliar, e reciclagem de nutrientes parecem ser fortemente influenciados pela disponibilidade de nutrientes. As relações fotossíntese-nitrogênio das principais espécies da floresta são comparáveis às da vegetação esclerófila de climas mais secos. Contudo, conforme esperado para climas úmidos, o tamanho da folha é maior. A adição de nutrientes provoca respostas distintas no crescimento de raízes finas, conforme o tipo de floresta. As florestas de *Terra Firme* respondem mais a P, Ca e Mg, enquanto as de Caatinga respondem mais a N. O fornecimento de nutrientes é limitado também pelas taxas de mineralização, e no caso de P, pela incorporação físico-química. Sob estas condições, as associações e simbioses entre o sistema radicular e microorganismos parecem ser cruciais para a persistência e resistência do ecossistema (por exemplo, recuperação após corte e queimada).

Mudanças globais na bacia Amazônica consistirão em aumento geral da concentração do CO_2 atmosférico, em pequeno aumento da temperatura média do ar, e em profundas mudanças no uso do solo. Neste artigo discutimos as características eco-fisiológicos relevantes dos ecossistemas de floresta da Amazônia, a fim de estimar sua produtividade potencial e capacidade de recuperação dentro dos cenários atual e futuro de perturbações antropogênicas.

Sustentamos que o aumento da concentração de CO_2 na atmosfera e o aumento da temperatura, poderão intensificar a extração de nutrientes pelas raízes e pelas simbioses de micorrizas (devido a um melhor suprimento de carboidratos), e a liberação de nutrientes pela matéria orgânica (devido a maiores taxas de decomposição). Como os processos no ecossistema são limitados pelo fornecimento de nutrientes, esperamos uma maior produtividade, uma maior dinâmica da biomassa, e, possivelmente, uma maior acumulação de biomassa em áreas de floresta natural.

14 Forest and pasture conductances in southern Pará, Amazonia

T.D. de A. SÁ[1], J.de P.R. da COSTA[2] and J.M. ROBERTS[3]
[1] *EMBRAPA/CPATU, Belém, Pará, Brazil*
[2] *Departamento de Meteorologia, UFPA, Belém, Pará, Brazil*
[3] *Institute of Hydrology, Wallingford, UK*

INTRODUCTION

Predictions of the effect of land-use change in Amazonia on global cycles requires an understanding of the water, carbon and nutrient cycles of the native forest and of replacement vegetation. Considering the role of vegetation in transferring water vapour to the atmosphere through transpiration, it is important to improve the representation of vegetation in atmospheric general circulation models (GCMs), and to assess the stomatal conductance, which controls water vapour loss through the stomata of the leaves, and the surface conductance, which represents the restriction on transpiration from the surface as a whole (Dolman *et al.*, 1991; Wright *et al.*, 1996).

During the last decade, progress has been made in specifying the leaf gas exchange behaviour of Amazonian rainforest species through the height profile of the forest canopy and their relationships with weather variables and soil water content (Roberts *et al.*, 1990; 1993). However, these earlier plant physiological studies and the associated micrometeorological research (Shuttleworth *et al.*, 1984) were carried out at a site in central Amazonia (Reserva Ducke, Manaus), where the dry season is limited to two or three months (Gash *et al.*, 1996). The need was recognised to extend these studies into forested and adjacent cleared areas in other regions, under different continental conditions and differing in the extent of their dry seasons, so providing a wider range of information to be used in GCMs (Shuttleworth *et al.*, 1991; Roberts and Cabral, 1993).

This paper reports some of the results of the evaluation of seasonal variation of stomatal conductance and its relationship to meteorological variables and soil moisture for species of rainforest and adjacent pasture in southern Pará State, Brazil, where there is a well-defined dry season and cattle ranching is becoming a widespread land-use (Uhl *et al.*, 1988). Information on the leaf area index for these forest and pasture sites is presented by Roberts *et al.* (1996).

Amazonian deforestation and climate. Edited by J.H.C. Gash, C.A. Nobre, J.M. Roberts and R.L. Victoria.

MATERIAL AND METHODS

FOREST SITES AND SPECIES

The forest study was conducted 50 km south of Marabá at a reserve of the Companhia Vale do Rio Doce (CVRD), an area of 17 000 ha, mostly unexploited and surrounded by cleared areas (Gash *et al*., 1996). The vegetation is "terra firme" forest with a continuous canopy around 20 m high, where *Sagotia brachysepala* is the most common tree species, but there is also a higher, discontinuous canopy consisting almost entirely of mature Brazil nut trees (*Bertholletia excelsa*) up to 40-50 m high (Silva, 1988; Salomão, 1991). The soil in the experimental area is a shallow clay-textured and silt-rich Cambisol (Inceptsol-Dystrochrepts), with a high concentration of gravel down to 70 cm depth (Rodrigues *et al.*, 1994).

Physiological measurements were made on 11 tree species and a liana species accessible from a 52 m high sectional aluminium tower and on seven understorey species (Figure 1, Tables 1 and 2).

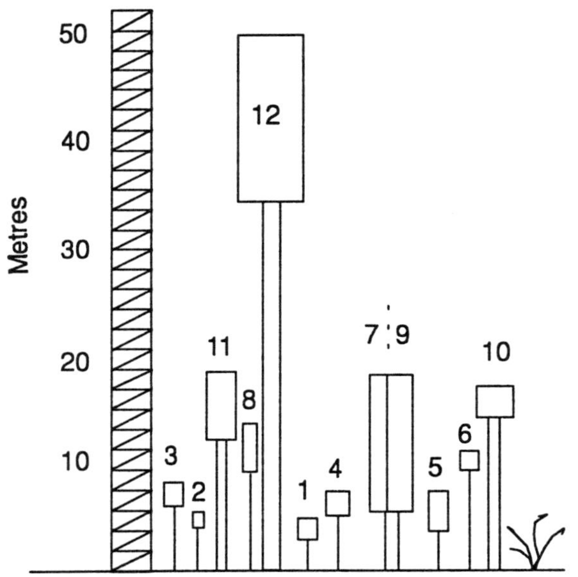

Figure 1 Profile of trees, liana and ground species sampled in Reserva Vale do Rio Doce, Marabá, Pará

PASTURE SITES AND SPECIES

The pasture study was conducted at a cattle ranch (Fazenda Boa Sorte) at Bom Jesús do Tocantins (São Raimondo), 50 km northeast of Marabá. The pasture area is covered mainly with the grass *Panicum maximum*, but shows bare soil patches at most times of the year. Babaçu palms (*Orbygnia martiniana*) are found scattered in this area. The soil in the experimental area is a deep well-drained medium/clay-

Table 1 Trees and liana surrounding the tower at Reserva Vale do Rio Doce, Marabá, Pará, and levels sampled

Tree/Level	Species (Family)	Height *m*	d.b.h.[1] *cm*	CL[2] *m*
1/ 2,3	*Guarea guidonia* (L.) Sleumer (Meliaceae)	4.0	8	1.8
2/3	*Iryanthera juruensis* Warb. (Myristicaceae)	5.5	18	0.5
3/4	*Protium apiculatum* Swartz (Burseraceae)	7.0	17	1.5
4/4	*Siparuna guianensis* Aubl. (Monimiaceae)	7.3	21	1.7
5/ 4,5	*Brosimum lactescens* (S.More) C.C. Berg (Moraceae)	7.4	32	3.7
6/4-5	*Iryanthera juruensis* Warb. (Myristicaceae)	9.2	31	0.6
7/4-7	*Sagotia brachysepala* (Muell Arg.) R. Secco (Euphorbiaceae)	17.0	47	11.1
8/6-8	??? (Polygonaceae) (Liana species)			
9/8	*Guarea guidonia* (L.) Sleumer (Meliaceae)	17.0	10	11.1
10/9	*Pourouma guianensis* Aubl. (Moraceae)	15.6	81	2.2
11/8-11	*Inga alba* (SW) Willd (Leg. Mimosaceae)	18.2	74	6.5
12/20-24	*Bertholletia excelsa* H.B.K. (Lecythidaceae)	49.4	683	15.9

[1] Diameter at breast height (1.3m) [2] Canopy Length

Table 2 Species sampled on the forest floor at Reserva Vale do Rio Doce, Marabá, Pará

Number	Species (Family)
1	*Rinorea passoura* (DC.) Kuntze (Violaceae)*
2	*Astrocaryum mumbaca* Mart. (Palmae)*
3	*Duguetia flagellaris* Huber (Anonaceae)*
4	*Paragonia piramidata* (L. Rich.) Bur. (Bignoniaceae)*
5	*Geonoma maxima* (Point) Kunth. (Palmae)
6	*Adiantum latifolium* Lam. (Polypodiaceae)
7	*Pouroma guianensis* Aubl. (Moraceae)

* Most frequently sampled

textured red-yellow podzol (Ultisol-Haplustults; Rodrigues *et al.*, 1994). Stomatal conductance measurements were made on six species: *Panicum maximum*, *Orbygnia martiniana* and four shrubs (Table 3).

Table 3 Species sampled on the pasture at Fazenda Boa Sorte, Bom Jesús do Tocantins, Pará.

Number	Species (Family)
1	*Panicum maximum* Jacq. (Gramineae)*
2	*Orbygnia martiana* Barb.Rodr. (Palmae)*
3	*Sida rhombifolia* L. var. canarensis (Willd.) K.Schum (Malvaceae)*
4	*Tabebuia* sp. (Bignoniaceae)*
5	*Amphilophium* sp. (Bignoniaceae)
6	*Calopogonium mucunoides* Desv. (Fabaceae)

* Most frequently sampled

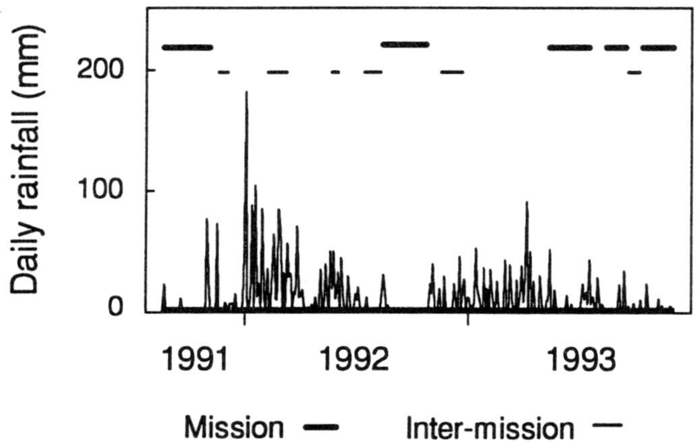

Figure 2 Daily rainfall during the study period: upper horizontal bars represent ABRACOS Missions and lower horizontal bars Inter-missions

MEASUREMENT PERIODS

Physiological measurements were made during 11 periods, including five during ABRACOS missions (M2 to M6) and in six additional inter-mission periods (I-mA to I-mF; Table 4). These studies made datasets available representing different seasons during a 27-month period (Figure 2): M2, I-mD and I-mF took place in the dry season, I-mB, I-mC, I-mE and M4 in the wet season and the remaining during transition periods.

METEOROLOGICAL VARIABLES AND SOIL MOISTURE

Meteorological variables and soil moisture content were routinely monitored during the ABRACOS project and were used to interpret physiological responses. Meteorological variables were derived from the automatic weather stations (AWS) permanently installed at the top of sectional aluminium towers, 52 m high at the forest site and 5 m high at the pasture site, as described by Culf *et al.* (1996). The data included hourly average values of global solar radiation (R_s), air temperature (T) and specific humidity deficit (D). Soil moisture storage was measured at approximately weekly intervals using a neutron probe (Didcot Instrument Co., Abingdon, UK). The readings were made in six neutron probe access tubes at each site, at 10 cm intervals down to 3.6 m depth (Hodnett *et al.*, 1996). Soil moisture storage totals in the upper 1 m and 2 m layers of soil for each site were used to interpret physiological responses.

STOMATAL CONDUCTANCE

Stomatal conductance (g_l) was measured at both sites with a dynamic diffusion porometer (Model AP4, Delta-T Devices, Burwell, Cambs., UK), described by Monteith *et al.* (1988). At the beginning of the study g_l was measured on both abaxial and adaxial leaf surfaces to determine the species behaviour. It was observed that it was not necessary to monitor the adaxial surface of all forest canopy species assessed, but some of the forest ground species (*Astrocaryum mumbaca* and *Paragonia piramidata)* and some of the pasture species (*Panicum maximum*, *Sida rhombifolia* L. var. *canarensis* and *Tabebuia* sp.) exhibit relatively high values on both surfaces, with a markedly seasonal pattern: these species must be monitored on both leaf surfaces. At both sites daily courses of g_l were measured, at 1 to 2 hour intervals, from 08.30 until 16.00 h. The number of measurement sets was often restricted, particularly during the wet season, because leaves were still wet late into the morning, and sometimes because of rain showers during the day. Calibration followed the manufacturer's instructions.

At Reserva Vale do Rio Doce measurements were made throughout the forest canopy on three leaves per individual per level. On most of the days measurements were made on all trees and levels (Table 1), but on the days when measurements of leaf water potential were made, they were limited to trees 7, 11 and 12, and mainly to Levels 6, 7, 10, 20, 23 and 24. Measurements were made on five to ten individuals of the selected species (Table 2).

At Fazenda Boa Sorte, measurements were made on five to ten randomly-selected individuals of the studied species (Table 3), although measurements were mainly restricted to Species 1 to 4. Large changes in vegetation cover occurred in the pasture during the observation period, ranging from periods when the grass had green, expanded, vigorous leaves (January 1992) to periods when L^* was only around 0.5 (June 1992).

In all situations, measurements were made about halfway along the leaves. Before each set of measurements the porometer was switched on and allowed to equilibrate for about ten minutes, and it was calibrated after every run, according to the manufacturer's instructions. Photosynthetic photon flux density (Q_p) was monitored simultaneously with a quantum sensor adjacent to the measurement.

LEAF WATER POTENTIAL

Leaf water potential (ψ) measurements were made with a Schölander-type pressure chamber system (Schölander *et al.*, 1965) at Level 10 on the tower (19 m). Measurements were limited to tree species 7, 11 and 12 (Table 1) and Levels 6, 7, 10, 20, 23 and 24) and were made over 17 days, distributed during M2, I-mB, I-mD, M3, I4, I5 and I6 (Table 4). Three leaves near to those measured for g_l were excised, immediately placed in a plastic box containing wet paper towels, and transported quickly to the pressure chamber. In cases when the petioles were too short to pass through the lid of the chamber, it was necessary to cut away part of each side of the lamina to extend the petiole to fit the chamber.

Table 4 Days of the year when plant physiological measurements were made

Year	1991		1992					1993			
Mission	2					3		4	5		6
I-mission		A	B	C	D		E			F	
Forest:											
Tower	222*	316	20*	86	168*	54*	342	98	170	241	280*
	224*		24*		170*	62*	346	102*	171*	244	285
	226*					75*	349	114*	172		291
	228*						351	118	174*		294*
	241							119			295
Ground	227	317	24	86	173	58	348	104	167	242	279
	228					60		105			288
	243					74		111			296
								116			
Pasture:	234	315	18	87	167	255	342	110	169	240	284
	236		21		169	269	352	112	173	243	293
	244		25			270		113			
						273					

* Days when ψ was monitored

RESULTS

RAINFALL AND SOIL MOISTURE

Figures 2 and 3 show that the distribution of missions and inter-missions has allowed data sets corresponding to a wide range of rainfall and soil moisture availability to be obtained. Low rainfall periods were mostly represented by M2, I-mD, M3 and M6, while I-mB, I-mC, I-m-E and M4 occurred during high rainfall periods. The remaining observational campaigns took place in transitional periods (Figure 2). No rainfall was recorded during M2 at either forest or pasture sites, while during I-mB and I-mC plant physiology measurements were curtailed because of persistent heavy rain.

Figure 3 shows that the lowest soil moisture storages were found at both sites during M2, M3, I-mF and M6, while the highest values occurred during I-mB, I-mC and

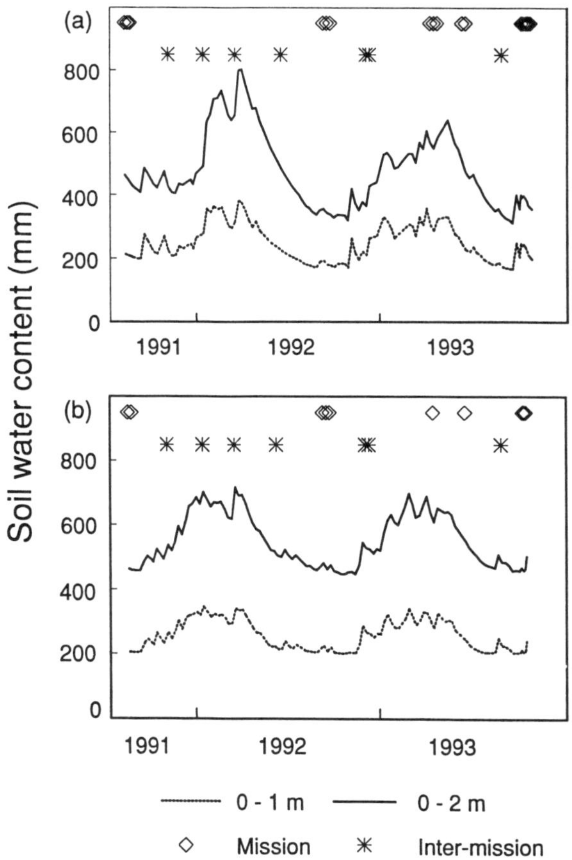

Figure 3 Soil moisture storage in the first 1 m and 2 m depth in the soil in Marabá sites: (a) forest and (b) pasture. Upper filled circles represent ABRACOS Missions, and diamonds Inter-missions.

M4. Although low rainfall was recorded during I-mD, soil moisture storage was still not very low, particularly at the forest site, because of rainfall in the preceding months.

SOLAR RADIATION, HUMIDITY DEFICIT AND TEMPERATURE ABOVE THE CANOPY

Table 5 gives the average midday value for above-canopy solar radiation (R_s), specific humidity deficit (D) and air temperature (T), for days when physiological measurements were made during seven experimental missions. Trees around the

Table 5 Average midday values of solar radiation R_s (W m^{-2}), specific humidity deficit D (g kg^{-1}) and temperature T (° C) for physiology measurement days at the forest tower, forest plots (Reserva Vale do Rio Doce) and pasture (Fazenda Boa Sorte). Data are from seven experimental missions.

Mission	R_s	D	T
		Forest tower	
M2	731	13.5	30.2
ImB	227	2.5	25.2
ImD	816	14.7	30.2
M3	697	8.8	26.6
M4	794	8.4	29.3
M5	824	11.9	30.0
M6	697	10.5	30.0
		Forest plots	
M2	714	15.5	30.6
ImB	285	2.5	24.8
ImD	706	13.4	29.9
M3	878	11.5	30.0
M4	727	8.7	29.9
M5	841	12.2	29.2
M6	547	11.2	30.0
		Pasture	
M2	796	17.5	32.6
ImB	679	5.9	28.0
ImD	844	13.8	30.7
M3	853	12.8	30.6
M4	657	9.0	29.9
M5	805	13.4	31.6
M6	654	13.0	31.5

forest tower, forest understorey in forest plots and plants in the pasture were measured. Differences in midday values reflect the maximum differences observed between experimental missions and at different sites during missions.

R_s differences within periods were more evident during I-mB: because of thick cloud, lower values were recorded at the forest site, compared to the values in the pasture site. This occurred when physiology measurements were made through the canopy as well as when they were carried out at ground level.

D exhibited similar trends in all three sets of data, with some differences in magnitude emerging during some measuring periods. Particularly during M2 the values at the pasture site were considerably higher than those from the forest, while during I-mB, although the values found at the pasture were the lowest among all the study periods, those found at the forest site were consistently lower.

Trends in *T* were similar to those found in *D*, with only small differences between periods.

PLANT PHYSIOLOGY MEASUREMENTS

The maximum stomatal conductance values are presented for all eleven measurement periods, for a wide range of species, and for both forest and pasture sites, but the other results are limited to parts of the periods (M2, I-mB, I-mD, M3, M4, M5 and M6) and to some of the studied species. The analysis of the values measured in trees surrounding the tower was limited to the days, species and heights when leaf water potential measurements were also available.

FOREST AROUND THE TOWER

Seasonal and daily variation of stomatal conductance and water potential

Figure 4 shows the comparison between average hourly values of g_l and ψ, measured between 10.00 h and 14.00 h, for three of the species surrounding the tower at Reserva Vale do Rio Doce in seven measurement periods: it shows the pattern of variation associated with relative position in the forest canopy. The lower canopy species *Sagotia brachysepala* had g_l values lower than the other species in all study periods, few of the values exceeding 100 mmol m^{-2} s^{-1} (Figure 4a). The values of g_l in *Inga alba* and *Bertholletia excelsa* (Figure 4a) were similar during I-mB and I-mD but differed in the other periods. There was a tendency in *Bertholletia excelsa* for values to exceed those of *Inga alba* during the drier periods. Apart from low values (around 100 mmol m^{-2} s^{-1}) for both species in I-mD and for Inga in M6, g_l was above 200 mmol m^{-2} s^{-1}, and in I-mB and M4 in *Bertholletia excelsa* it exceeded 300 mmol m^{-2} s^{-1}. Figure 4b shows that *Sagotia brachysepala* and *Inga alba* followed similar seasonal patterns of ψ with less negative values than were found in *Bertholletia excelsa*. As with g_l, smaller differences in ψ were observed during I-mD than in other periods.

Figure 5 shows diurnal patterns of variation of g_l and ψ during periods with relatively high (M4, Figure 5a) and low (M6, Figure 5b) soil moisture storage. The values of g_l in *Inga alba* in M4 were consistently higher than those found in this tree

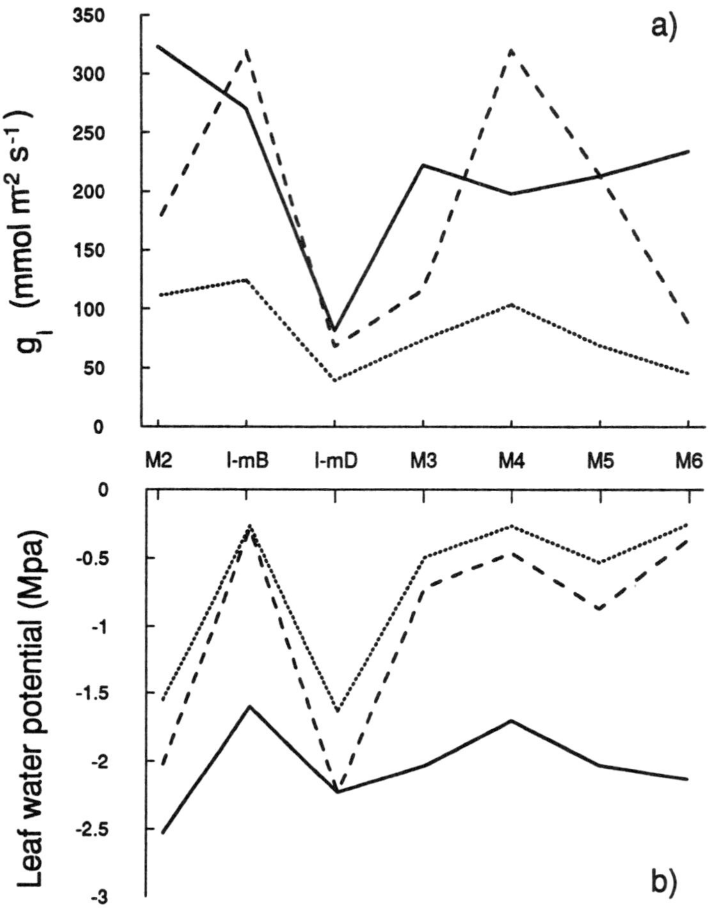

Figure 4 Average values of g_l (a) and ψ (b) from 10.00 - 14.00 h for *Sagotia brachysepala* (·····), *Inga alba* (– – –) and *Bertholettia excelsa* (——) during M2, I-mB, I-mD, M3, M4, M5 and M6.

during M6. The range of g_l values in *Sagotia brachysepala* and *Bertholletia excelsa* was similar in M4 and M6, but there were differences in the times when peaks occurred. During M4 *Inga alba* and *Bertholletia excelsa* showed a sharp decline in g_l at noon, as has been reported for several tropical species (Roy and Salager, 1992), followed by recovery during early afternoon; in *Sagotia brachysepala* the fall and recovery of g_l occurs later. During M6 the daily pattern of variation in g_l differed in all species, with no match in fall and recovery times. The diurnal course of ψ showed similar trends in M4 and M6, but differences in magnitude were evident: the unexpectedly lower values in *Inga alba* during M4 and considerably lower values in *Bertholletia excelsa* during M6 were associated with lower moisture conditions in the air and the soil.

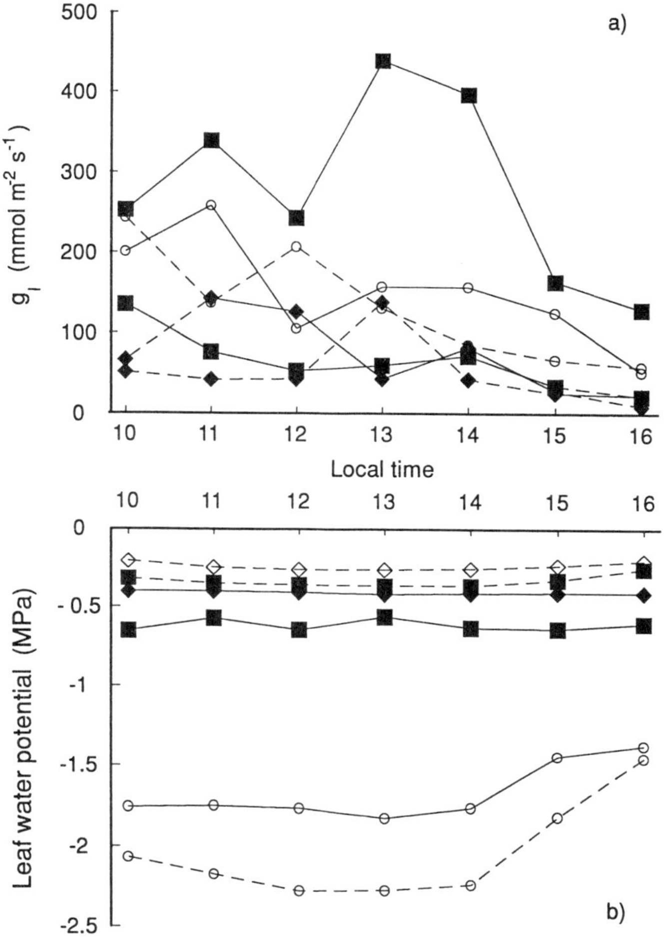

Figure 5 Hourly values of g_l (a) and ψ (b) from 10.00 h to 16.00 h for *Sagotia brachysepala* (◆), *Inga alba* (■) and *Bertholettia excelsa* (o), during M4 (——) and M6 (- - -).

Humidity deficit effects on g_l

The interpretation of the influence of specific humidity deficit (D) on g_l was made by separating the data into four 200 W m^{-2} classes of above-canopy solar radiation (R_s). The relationship between D and g_l in each class was then investigated by linear regression analysis. Table 6 presents the coefficients of linear regression for *Sagotia brachysepala*, *Inga alba* and *Bertholletia excelsa* using data from M2, M3, M4, M5 and M6. For *Inga alba* a statistically significant negative relationship exists between D and g_l for all radiation classes, which was also the case for *Bertholletia excelsa*, except in the highest radiation class. In *Sagotia brachysepala* however, the only significant relationship between D and g_l was when R_s was above 500 W m^{-2}. Because there are no significant differences between radiation classes in any of the

Table 6 Linear regression for g_l (mmol m^{-2} s^{-1}) measured in three trees surrounding the forest tower, with above-canopy D (g kg^{-1}) in four R_s classes.

Above-canopy R_s (W m^{-2})		Species		
		Sagotia brachysepala	*Inga alba*	*Bertholletia excelsa*
>700	r	0.19 N.S.	0.58 ***	0.14 N.S.
	a	—	413 ± 86	—
	b	—	-21 ± 3.8	—
501-700	r	0.17 N.S.	0.49 **	0.37 *
	a	—	397 ± 130	324 ± 118
	b	—	-22 ± 8.2	-13 ± 5.0
301-500	r	0.46 *	0.64 *	0.80 ***
	a	70 ± 19	277 ± 74	332 ± 66
	b	-2.5 ± 1.1	-15 ± 4.0	-20 ± 3.4
≤ 300	r	0.83 *	0.61 **	0.80 ***
	a	147 ± 30	406 ± 106	348 ± 89
	b	-9.6 ± 2.9	-28 ± 10	-25 ± 4.1

r = correlation coefficient; a = intercept of regression line (±S.E.);
b = slope of regression line (±S.E.). * P<0.05; ** P<0.01; *** P<0.001; N.S. = not significant.

species in the slope and intercepts of the regressions, data for each species has been pooled over all radiation classes: the relationships are shown in Figure 6.

Maximum stomatal conductance

Table 7 shows the values of g_{lmax} for five trees (all levels) obtained in all eleven measuring periods. The two lower canopy trees (*Guarea guidonia* at Levels 2 and 3 and *Sagotia brachysephala* at Levels 4 to 7) showed g_{lmax} values ≤ 300 mmol m^{-2} s^{-1} in all periods, although it is clear that a seasonal trend exists which is associated with soil moisture storage. The highest recorded values in seven out of eleven periods were from *Bertholletia excelsa*, which showed values above 600 mmol m^{-2} s^{-1} in M2, I-mA, I-mB, I-mE and M4, between 400 and 600 mmol m^{-2} s^{-1} in I-mC and M6, between 200 and 400 mmol m^{-2} s^{-1} in I-mD, M3 and M5, and values below 200 mmol m^{-2} s^{-1} only in I-mF. During M3, M5 and I-mF the higher values were observed in *Pouroma guianensis*, a pioneer species, which during I-mF was the only species to present a g_{lmax} value greater than 100 mmol m^{-2} s^{-1}. During most of the periods g_{lmax} values of *Inga alba* followed trends similar to those observed in *Bertholletia excelsa*.

Table 7 Values of g_{lmax} (mmol m^{-2} s^{-1}) for all missions and inter-missions in species of the forest (canopy and understorey) and pasture.

Site/species	M2	Im-A	Im-B	Im-C	Im-D	M3	Im-E	M4	M5	Im-F	M6
Forest canopy											
G. guidonia	143	91	182	183	86	29	140	155	114	24	175
S. brachysepala	203	140	215	266	134	138	154	232	118	56	201
I. alba	460	293	510	332	321	229	395	600	346	88	181
P. guianensis	712	630	677	523	324	283	710	597	243	81	442
B. excelsa	567	294	337	292	149	420	420	271	520	589	364
Forest understorey											
R. passoura	95	113	132	152	110	34	153	232	29	33	330
A. mumbaca	113	128	614	188	206	66	149	500	64	94	466
D. flagellaris	64	66	184	53	65	41	150	146	30	32	155
P. piramidata	20	44	216	136	73	28	68	271	16	17	178
Pasture											
P. maximum	387	119	255	290	134	306	162	351	106	540	230
O. martiana	473	-	310	810	585	370	489	850	479	397	360
S. rhombifolia	-	-	833	850	318	165	-	1070	536	644	477
Tabebuia spp.	-	-	449	-	670	-	-	930	663	825	409

FOREST UNDERSTOREY SPECIES

Seasonal and daily variation of stomatal conductance

Particular attention was paid to the most common species of the forest understorey at Reserva Vale do Rio Doce, *Duguetia flagellaris*. The majority of values fell below 100 mmol m^2 s^{-1} and a seasonal relationship emerged which was related to soil moisture storage. Figure 7 shows a weak but significant positive relationship between hourly average values of g_l and soil moisture storage between 0 and 1 m depth ($r = 0.399$, $a = -13.4 \pm 35$, $b = 0.238 \pm 0.11$). No statistically significant relationship was found between g_l and Q_p or D in *Duguetia flagellaris*.

Maximum stomatal conductance

Table 7 gives the maximum values of stomatal conductance g_{lmax} measured in four forest understorey species (*Rinorea passoura*, *Astrocaryum mumbaca*, *Duguetia flagellaris* and *Paragonia piramidata*) during all eleven periods. Clear trends associated with soil moisture availability are found in all species. The consistently higher values found in *Astrocaryum mumbaca* in periods when relatively high soil moisture levels occur is, to some extent, associated with the tendency of this species to have high g_l values in both upper and lower leaf surfaces when soil moisture is high.

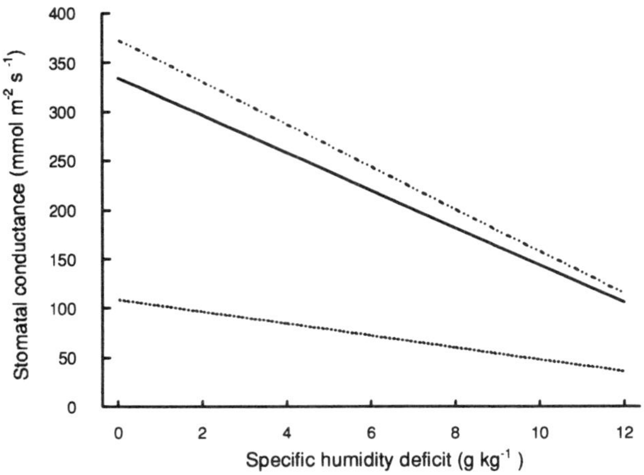

Figure 6 Relationship between g_l and D in *Sagotia brachysepala* (······), *Inga alba* (—··–) and *Bertholettia excelsa* (——): data for each species are pooled over all radiation classes.

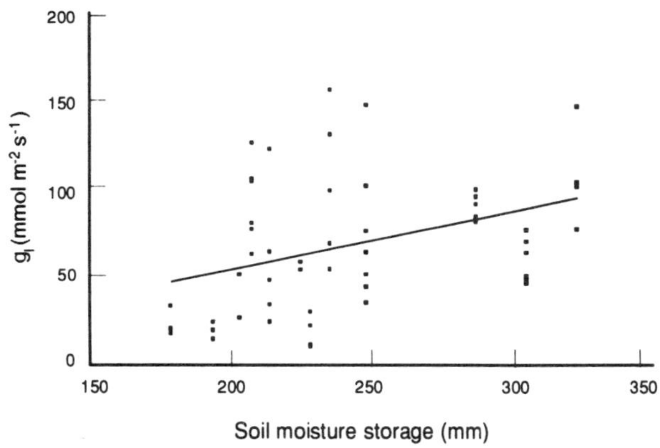

Figure 7 Relationship between hourly average values of g_l and soil moisture (0-1m depth) of the forest understorey species *Duguettia flagelaris*.

PASTURE

Seasonal and daily variation of stomatal conductance

Panicum maximum is by far the major component of the vegetation found at the pasture site at Fazenda Boa Sorte, and the greatest effort was directed at analysing and interpreting data from this species. Figure 8 shows average values of g_l and D for times between 10.00 to 14.00 h during M1, I-mB, I-mD, M3, M4, M5 and M6: there is a clear indication of a negative relationship between g_l and D. The lowest

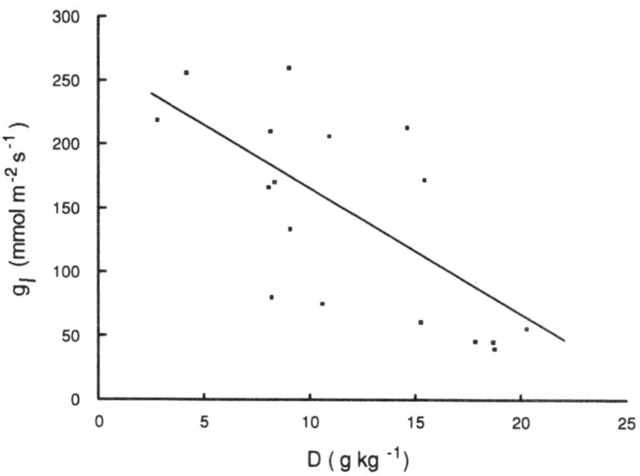

Figure 8 Relationship between average values from 10.00 to 14.00h for g_l of *Panicum maximum* and *D*, during seven measurement periods at the pasture site, Fazenda Boa Sorte.

average values observed in I-mB, I-mD, M3 and M6, all below 100 mmol m^{-2} s^{-1}, coincided with periods when *D* was higher than 15 g kg^{-1}. On the other hand the highest g_l values found (greater than 200 mmol m^{-2} s^{-1}), in I-mB and M3, are associated with *D* lower than 10 g kg^{-1}. Analysis of the relationship between g_l and *D* using data from three radiation classes (≤ 500, 501-700 and >700 W m^{-2}) revealed statistically significant negative relationships in the three radiation classes, but differences in slope and intercept were not significant (Table 8).

Table 8 Linear regression for g_l (mmol m^{-2} s^{-1}) in *Panicum maximum*, with above-canopy *D* (g kg^{-1}), in three R_s classes.

Above-canopy R_s (W m^{-2})		
> 700	r	0.552 ***
	a	271±75
	b	-10±2.8
501-700	r	0.462*
	a	211±69
	b	-6.8±2.6
≤ 500	r	0.665**
	a	260±58
	b	-12±3.4

r = correlation coefficient; a = intercept of regression line (±S.E.);
b = slope of regression line (±S.E.). * P<0.05; ** P<0.01; *** P<0.001; N.S. = not significant.

Maximum stomatal conductance

Table 7 shows the maximum stomatal conductance values (g_{lmax}) for four species: *Panicum maximum*, *Orbygnia martiana*, *Sida rhombifolia* and *Tabebuia sp.* The considerable variation in g_{lmax} values amongst the studied species and periods is, to some extent, associated with the availability of soil water and atmospheric deficit, although differences in g_{lmax} are also associated with the state of the vegetation. By the end of M2 an accidental fire burnt most of the vegetation, followed by a rapid recovery during the next rainy season (Fisch *et al.*, 1994). *Panicum maximum* fluctuated between periods in which it was particularly dry (mainly in I-mD and M5) and others (I-mB, I-mC and M4) in which it was particularly green, with well-expanded leaves. This species also exhibited a seasonal change in stomatal behaviour, with stomata closing more frequently in the upper leaf surface during particularly dry periods. Except in I-mF, g_{lmax} was below 400 mmol m^{-2} s^{-1} in all observation periods.

Orbygnia martiana (babaçu), a widely distributed palm found in this region (May *et al.*, 1985), showed a marked seasonal pattern of variation, with g_{lmax} data ranging from around 300 to over 800 mmol m^{-2} s^{-1}. Figure 15 shows that the two shrubs (*Sida rhombifolia* and *Tabebuia sp.*), also have relatively high g_{lmax} values in most of the periods. Particularly during wet periods, *Sida rhombifolia*, shows high g_l values on both leaf surfaces.

DISCUSSION

The seasonal variation of meteorological variables and soil moisture availability observed between the eleven measuring periods proved to have a large effect on stomatal conductances in all species studied, as well as on the water status of forest canopy species.

Considering the stomatal behaviour of the species most commonly found in the study areas, stomatal conductance was not high in all situations. In the forest area, the most common canopy species *Sagotia brachysephala* (Salomão, 1991), with its crown concentrated within lower canopy levels (mainly Levels 4 to 7), showed g_l values below 150 mmol m^{-2} s^{-1} during most of the studied situations (Figure 4), with g_{lmax} exceeding 200 mmol m^{-2} s^{-1} only during M2, I-mC and M4 (Table 7). This agrees with the values found for forest species of similar canopy position in other sites in Amazonia (Roberts *et al.*, 1990, McWilliam *et al.*, 1996) and other tropical forests (Oberbauer *et al.*, 1987). Among the forest understorey species, the most common (*Duguetia flagellaris*) had g_l values on most occasions lower than 100 mmol m^{-2} s^{-1} (Figure 7), which are comparable with the values found for many understorey plants in rainforests (Pearcy and Calkin, 1983; Langenheim *et al.*, 1984; Oberbauer *et al.*, 1987; Roberts *et al.*, 1990; McWilliam *et al.*, 1996), and g_{lmax} reaching an unexpected higher value (330 mmol m^{-2} s^{-1}) during M6, under relatively low soil moisture storage conditions (Table 7). In the pasture area, *Panicum maximum* showed an expected consistent seasonal pattern of g_l (Table 7), but with values relatively lower than found in grass species at the other ABRACOS sites (McWilliam *et al.*, 1996), based on a consideration of either average g_l or g_{lmax} values (maximum 540 mmol m^{-2} s^{-1}, observed during I-mF).

Seasonal patterns were evident for g_l values measured both in the forest (canopy and understorey) and in the pasture species, with higher values predominant during wet periods, mainly during I-mB, I-mC, and M4. Some exceptional stomata behaviour was nevertheless observed in some cases, for example with the species *Pouroma guianensis* (I-mE) and *Bertholletia excelsa* (M2) in the forest canopy, *Duguetia flagellaris* (M6) in the forest understorey, and *Panicum maximum* (I-mF and M6) in the pasture. These presented relatively high values of either average g_l or of g_{lmax} under the low soil moisture storage conditions prevailing during these measuring periods. This may be partially explained by the phenological stages of these species during these periods, as well as by differences in atmospheric conditions.

To some extent leaf water potential values showed a less pronounced seasonal difference in the three species analysed in this paper. As happened with g_l, the smaller difference among species was observed during I-mD (Table 7), when D and R_s were relatively high (Figure 4) and soil moisture storage was falling (Table 5). A clearer pattern, however, was found with respect to tree position within the forest canopy, with lower ψ values found in the upper canopy species *Bertholletia excelsa* (ranging from -1.0 to -3.3 MPa), and higher values (-0.2 to -2.5 MPa) in *Sagotia brachysephala*, which has its crown at lower levels (Figure 1; Figures 7-9). This agrees with the findings of McWilliam *et al.* (1996) in a southwest Amazonian forest and results from other tropical forests (Oberbauer *et al.*, 1987).

FOREST STOMATAL CONDUCTANCE

During most of the measurement periods the vertical pattern of distribution of g_l (Figures 4 and 5) followed the profiles found in other tropical forests characterised by decreasing ranges from the upper canopy to the forest understorey (Pearcy and Calkin, 1983; Langenheim *et al.*, 1984; Fanjul and Barradas 1985; Oberbauer *et al.*, 1987; Roberts *et al.*, 1990; McWilliam *et al.*, 1996). Such patterns have been shown to be related to the decline in light levels from the top of the canopy to the ground level (Yoda, 1974; Chazdon and Fetcher, 1984; Oberbauer and Strain, 1986; Yates *et al.*, 1989; Whitmore, 1990; Smith *et al.*, 1992).

Some variations were also found in g_l values between leaves from different levels within the crown of the upper canopy species *Bertholletia excelsa*, which is reported to exhibit height-related anatomical differences (Medri and Lleras, 1979). However there may also be a seasonal variation in leaf age.

Daily courses of g_l showed seasonal differences in the time of decrease and recovery of stomatal opening, around the time when evaporation would be high, and also in the relative position of different species with respect to the average g_l values. During the relatively wet M4 (Figure 5), *Inga alba* (a medium canopy species) showed consistently higher values, followed by *Bertholletia excelsa* (an upper canopy species) and *Sagotia brachysephala* (a lower canopy species). During the relatively dry M6 (Figure 5) the higher values were found in *Bertholletia excelsa*, followed in most hours by *Inga alba*.

A statistically significant negative relationship was found between g_l and D for R_s values below 500 W m^{-2} for all three canopy species; for R_s values up to 700 W m^{-2} in *Bertholletia excelsa*, and for all R_s classes in *Inga alba* (Table 6, Figure 6). This denotes the close relationship between atmospheric moisture deficit and g_l in canopy species, which has been reported for Amazonian forest species (Roberts *et al.*, 1990; McWilliam *et al.*, 1996) and in other forest biomes elsewhere (Whitehead *et al.*, 1981; Chiariello, 1984; Landsberg, 1986; Hinckley and Braatne, 1994).

Seasonal fluctuations in g_{lmax} values (reflecting soil moisture availability) for five species with different canopy positions (Table 7) show a more consistent distribution for the lower canopy species (*Guarea guidonia* and *Sagotia brachysephala*) and medium canopy (*Inga alba*) species and a less consistent pattern for the upper canopy species (*Bertholletia excelsa*) and the pioneer species (*Pouroma guianensis*). The lower level tree *Guarea guidonia* has been reported as having a competitive advantage by an ability to develop suppressed seedlings under the very low sunlight found close to rainforest floors, expressed by its ability to develop leaf structural features adapted to the prevailing light conditions (Fisher, 1986).

A comparison between the g_{lmax} values observed in these canopy trees with those found in the literature for tropical forests shows that the values are within the limits found for tropical forest species elsewhere (reviewed by Roberts *et al.*, 1990; McWilliam *et al.*, 1996).

With respect to the forest understorey, the most common species found in the Reserva Vale do Rio Doce forest, *Duguetia flagellaris*, shows seasonal variation in g_l, weakly but significantly related to soil moisture availability in the first metre depth (Figure 7), with g_l values falling within the ranges mostly found in tropical forest understorey species (Roberts *et al.*, 1990; McWilliam *et al.*, 1996). Diurnal fluctuation of g_l values, on the other hand, seems also to be related to atmospheric variables, as it has different patterns before and after midday, if values from the wet M4 and from the relatively dry M6 are compared.

The pattern of g_{lmax} values measured in four understorey species over eleven contrasting periods (Table 7) showed that although all follow the same trend, with higher values coinciding mostly with wet periods, the values found during these periods with high soil moisture storage in the palm *Astrocaryum mumbaca* were consistently higher — twice those of the species with the next highest g_{lmax} during M4 and three times the second species during I-mB. This suggests that in places where this species is more frequent in the understorey, the relative contribution of this lower vegetation layer may be substantially higher during the wet season.

FOREST LEAF WATER POTENTIAL

The magnitude and range of ψ in three forest tree species at different vertical positions in the canopy indicate a strong relationship between height in the canopy and ψ range (Figures 4 and 5), with medium level *Inga alba* showing a wider range (from about -1.5 to -3.0 MPa), the upper canopy species *Bertholletia excelsa* with a range shifted towards lower values (-1.0 to about -3.3 MPa), and the lower canopy species *Sagotia brachysephala* having values shifted towards higher values (about -0.2 to -1.5 MPa). These values are within the limits mostly found in tropical forest species corresponding to those canopy positions (McWilliam *et al.*, 1996). Even lower ψ values have been reported (Carvalho *et al.*, 1993) for groups of tree species in primary and secondary forests in eastern Amazonia.

PASTURE STOMATAL CONDUCTANCE

As in some forest canopy species, g_l in *Panicum maximum*, the commonest species in the pasture, was negatively correlated to D, for all R_s classes studied (Table 8). This explains the seasonal distribution of these values: higher g_l values coincided with periods when air humidity was higher (low D values, Figure 8). This relationship was clear for all studied seasons, and for the different leaf conditions, which changed from very dry, in periods such as I-mD, to very green and well developed during periods such as I-mB.

In contrast to the dominant grasses at the other ABRACOS pasture sites, for example *Brachiaria brizantha* in Ji-Paraná, where g_l values above 400 mmol m^{-2} s^{-1} were common during M4, corresponding to a wet season (McWilliam *et al.*, 1996), average g_l values in Fazenda Boa Sorte were mostly lower than 250 mmol m^{-2} s^{-1} for *Panicum maximum*, even during wet periods (Figure 8). This may be related to the low technology management adopted on this farm, associated with the soil conditions. Leaf area index in this pasture is only moderate, even during the wet season, and in the dry season falls to the lowest value observed at all the ABRACOS pasture sites (Roberts *et al.*, 1996). During most of the study periods a relatively high concentration of weedy shrubs grew in the pasture, suggesting competition with the grass for water, nutrients and in some extreme cases even for light. There is evidence that *Panicum maximum* is sensitive to light reduction (Fleischer *et al.*, 1984; Toledo and Torres, 1990).

The seasonal pattern of g_{lmax} distribution for *Panicum maximum*, *Orbygnia martiana* (a widespread palm) and two shrub species (*Sida rhombifolia* and *Tabebuia sp.*) during eleven periods (Table 7) gives an idea of the seasonal stomatal behaviour of these species, which is related to environmental processes as well as to their phenology and physiological status, since by the middle of M2 all the vegetation at Fazenda Boa Sorte was burnt (Fisch *et al.*, 1994) and allowed to recover without any management from then on, which means that the root systems of all grass and non-grass components of the pasture were allowed to developed continuously during the study period. From Table 7 it is evident that, particularly during the wet periods, g_{lmax} in the non-grass species is consistently higher, suggesting that if these components become more predominant as the pasture degrades, the gas exchange pattern of these pasture areas would be modified.

CONCLUSIONS

1 The highest average values of stomatal conductance (g_l) in the forest were observed both in the emergent *Bertholletia excelsa* and in the main canopy species, *Inga alba* and *Pouroma guianensis*. g_l was substantially less in lower canopy and understorey trees.

2 g_l fell during the day in association with increasing atmospheric humidity deficit; the biggest change was in the emergent and upper forest canopy species, with initially higher g_l earlier in the day.

3 Apart from the weak positive association of g_l and soil water content in the understorey species, *Duguetia flagelaris*, no relationship was found between g_l of the forest species and soil water content.

4 Maximum g_l of *Panicum maximum*, the dominant grass species in the pasture, was around 400 mmol m^{-2} s^{-1}. Much higher values were observed in the shrub species growing in the pasture: this has implications for changes in pasture transpiration if shrub invasion becomes a predominant feature.

5 No clear relationships were observed between g_l and soil water conditions in the pasture, but g_l of *Panicum maximum* was negatively correlated with increasing atmospheric humidity deficit.

ACKNOWLEDGEMENTS

The authors give thanks for the scientific support received from the Universidade Federal do Pará (UFPA), Centro de Pesquisa Agroflorestal da Amazônia Oriental (EMBRAPA/CPATU), and Companhia Vale do Rio Doce (CVRD).

Acknowledgement is also given here to all the people involved in collecting the data presented in this paper, as well as the local field observers and coordinators. Special thanks also go to Sr. Manuel dos Reis Cordeiro for his valuable contribution in identifying the studied plant species.

REFERENCES

Carvalho, C. J. R. de, Sá, T. D. de A., Nepstad, D. C., Jipp, P. and Vieira, S. 1993. Relações hídricas de componentes da vegetação de floresta primária, floresta secundária e pastagem degradada da Amazônia oriental. In: Reunião de Ecofisiologia de Florestas Tropicais. Kourou, SILVOLAB-Guyane.

Chazdon. R.L. and Fetcher, N. 1984. Light environments of tropical forests. In: Medina, E., Mooney, H.A. & Vázquez-Yánez, C. (Eds.), *Physiological ecology of plants of the wet tropics*. Junk, The Hague. 27-36.

Chiariello, N. 1984. Leaf energy balance in the wet lowland tropics. In: Medina, E., Mooney, H.A. & Vázquez-Yánez, C. (Eds.), *Physiological ecology of plants of the wet tropics*. Junk, The Hague, 85-99.

Culf, A.D., Esteves, J.L., Marques Filho, A.O. and Rocha, H.R. da. 1996. Observations of radiation, temperature and humidity over forest and pasture in Amazonia. This volume.

Dolman, A.J., Gash, J.H.C., Roberts, J. and Shuttleworth, W.J. 1991. Stomatal and surface conductance of tropical rainforest. *Agric. For. Met.* **54**, 303-318.

Fanjul, L. and Barradas, V.L. 1985. Stomatal behaviour of two heliophile understorey species of a tropical deciduous forest in Mexico. *J. Appl. Ecol.*, **22**, 943-954.

Fisch, G., Wright, I.R. , and Bastable, H.G. 1994. Albedo of tropical grass: a case study of pre- and post burning. *Int. J. Climatol.* **14**, 103-107.

Fisher, J.B. 1986. Sun and shade effects on the leaf of Guarea (Meliaceae): plasticity of a branch analogue. *Bot. Gaz.* **147**, 84-89.

Fleischer, J.E., Masuda, Y. and Goto, I. 1984. The effect of light intensity on the productivity and nutritive value of Green Panic (*Panicum maximum* var. *trichoglume* cv. Petrie). *J. Jap. Soc. Grassl. Sci.* **30**, 191-194.

Hinckley, T.M. and Braatne, H. 1994. Stomata. In: *Plant-environment interactions*. Wilkinson, R. E. (Ed.), Marcel Dekker, New York, 323-355.

Hodnett, M.G., Tomasella, J. and Oyama, M. 1996. Comparisons of long-term soil water storage under pasture and forest in three areas of Amazonia. This volume.

Gash, J.H.C., Nobre, C.A., Roberts, J.M. and Victoria, R.L. 1996. An overview of the ABRACOS project. This volume.

Landsberg, J.J. 1986. *Physiological ecology of forest production*. Academic Press, London.

Langenheim, J.H., Osmond, C B., Brooks, A. and Ferrer, P.J. 1984. Photosynthetic responses to light in seedlings of selected Amazonian and Australian rainforest species. *Oecologia* **63**, 215-224.

McWilliam, A.-L. C., Cabral, O. M. R., Gomes, B. M., Esteves, J. and Roberts, J. M. 1996. Forest and pasture leaf-gas exchange in Southwest Amazonia. This volume.

May, P. H., Anderson, A. B., Frazão, J. M. F., and Balick, M. J. 1985. Babassu palm in the agroforestry systems in Brazil's mid-north region. *Agrofor. Syst.* **3**, 275-295.

Medri, M.E. and Lleras, E. 1979. Ecofisiologia de plantas da Amazônia- 2. Anatomia foliar e ecofisiologia de *Bertholletia excelsa* Humb. & Benpl. (Castanha-do-pará)- Lecythidaceae. *Acta Amaz.* **9**, 15-23.

Monteith, J.L., Campbell, G.S. and Potter, E.A. 1988. Theory and performance of a dynamic porometer. *Agric. For. Met.* **44**, 27-38.

Oberbauer, S.F. and Strain, B.R. 1986. Effects of canopy position and irradiance on the leaf physiology and morphology of *Pentaclethera macroloba* (Mimosaceae). *Amer. J. Bot.* **73**, 409-416.

Oberbauer, S.F., Strain, B.R. and Riechers, G.H. 1987. Field water relations of a wet-tropical forest tree species, *Pentaclethra macroloba* (Mimosaceae). *Oecologia* **71**, 369-374.

Pearcy, R. W. and Calkin, H. W. 1983. Carbon dioxide exchange of C3 and C4 species in the understorey of a Hawaiian forest. *Oecologia* **58**, 26-32.

Roberts, J. M. and Cabral, O. M. R. 1993. ABRACOS: a comparison of climate, soil moisture and physiological properties of forests and pastures in the Amazon Basin. *Commonw. For. Rev.* **72**, 310-315.

Roberts, J. M., Cabral, O. M. R. & Aguiar, L. F. de 1990. Stomatal and boundary-layer conductances in an Amazonian terra firme rain forest. *J. Appl. Ecol.* **27**, 336-353.

Roberts, J., Cabral, O. M. R., Fisch, G., Molion, L. C. B., Moore, C. J. and Shuttleworth, W. J. 1993. Transpiration from Amazonian rainforest calculated from stomatal conductance measurements. *Agric. For. Met.* **65**, 175-196.

Roberts, J.M., Cabral, O.M.R., da Costa, J.P., McWilliam, A-L. C. and Sa, T.D. de A. 1996. An overview of the leaf area index and physiological measurements during ABRACOS. This volume.

Rodrigues, T. E., Martorano, L. H. and Costa, J. de P. R. da 1994. Caracterização dos solos nas áreas experimentais do ABRACOS na região de Marabá, PA. Belém, EMBRAPA/CPATU.

Roy, J. and Salager, J-L. 1992. Midday depression of net CO_2 exchange of leaves of an emergent rain forest tree in French Guiana. *J. Trop. Ecol.* **8**, 499-504.

Salomão, R. de P. 1991. Uso de parcelas permanentes para estudos da vegetação da floresta tropical úmida. I. Município de Marabá, Pará. *Bol. Mus. Para. Emilio Goeldi, sér. Bot.* **7**, 543-604.

Schölander, P. F., Hammel, H. T., Bradstreet, E. D. and Hemmingsen, E. A. 1965. Sap pressure in vascular plants: negative hydrostatic pressure can be measured in plants. *Science* **148**, 339-346.

Shuttleworth, W. J., Gash, J.H.C., Lloyd, C.R., Moore, C.J., Roberts, J.M., Marques, A.O., Fisch, G.F., Silva, V de P., Ribeiro, M.N.G., Molion, L.C.B., Sa, L de A., Nobre, C.A., Cabral, O.M.R., Patel, S.R. and de Moraes, J.C. 1984. Eddy correlation measurements of energy partition for Amazonian forest. *Quart. J. Roy. Met. Soc.* **110**,1143-1162.

Shuttleworth, W. J., Gash, J. H. C., Roberts, J. M., Nobre, C. A., Molion, L. C. B. and Ribeiro, M. de N. G. 1991. Post-deforestation Amazonian climate: Anglo-Brazilian research to improve prediction. *J. Hydrol.* **129**,71-85.

Silva, M. F. F. 1988. Relatório final do Projeto Carajás. Sub-projeto Inventário Botâncico. Convênio MPEG/CVRD, Contrato 16/83.

Smith, A. P., Hogan, K. P. and Idol, J. R. 1992. Spatial and temporal patterns of light and canopy structure in a lowland tropical moist forest. *Biotropica* **24**, 503-511.

Toledo, J. M. and Torres, F. 1990. Potential of silvopastoral systems in the rain forest. In: Moore, E. Agroforestry land-use systems. Waimanolo, Nitrogen Fixing Tree Assoc., 35-52.

Uhl, C., Buschbacher, R. and Serrão, E. A. S. 1988. Abandoned pastures in Eastern Amazonia. 1. Patterns of plant succession. *J. Ecol.* **76**, 663-681.

Whitehead, D., Okali, D. U. U., and Fasehun, F. E. 1981. Stomatal response to environmental variables in two tropical forest species during dry season in Nigeria. *J. Appl. Ecol.* **18**, 571-587.

Whitmore, T. C. 1990. *An Introduction to Tropical Rain Forest.* Oxford, Clarendon Press.

Wright, I. R., Nobre, C. A., Tomasella, J., Rocha, H. R. da, Roberts, J. M., Vertamatti, E., Culf, A.D., Alvalá, R. C., Hodnett, M.G and Uberana, V. 1996. Towards a GCM surface parameterisation for Amazonia. This volume.

Yates, D. J., Unwin, G. L., and Doley, D. 1989. Rainforest environment and physiology. *Proc. Ecol. Soc. Aust.* **15**, 31-37.

Yoda, K. 1974. Three-dimensional distribution of light intensity in a tropical rain forest of west Malaysia. *Jap. J. Ecol.* **24**, 247-254.

RESUMO

Foram realizados monitoramentos de condutância estomática (g_l) e de potencial da água em folhas (ψ) através de um dossel de floresta tropical pluvial, dominada por castanheiras-do-brazil (*Bertholletia excelsa*) e monitoramentos em uma área de pastagem adjacente de *Panicum maximum*, com palmeiras babaçu (*Orbygnia martiana*) e miscelânea de arbustos, ambos os locais situados próximo a Marabá, Pará.

Padrões verticais de distribuição de g_l em espécies da floresta concordam com perfís encontrados em outras florestas tropicais, com valores e amplitudes decrescendo do topo do dossel em direção ao sub-bosque. Relação negativa significativa foi encontrada entre g_l e déficit de umidade específica (D) para espécies do dossel, em diferentes faixas de radiação solar (R_s), e uma relação significativa positiva foi encontrada entre teor de água armazenada no solo na camada superior (1m) em *Duguetia flagellaris*, a espécie mais comum no sub-bosque. Valores de g_{lmax}

encontrados na palmeira do sub-bosque *Astrocaryum mumbaca* foram até três vezes superiores aos encontrados em outras espécies desse nível. A distribuição de ψ em árvores do dossel sugere uma estreita relação entre faixas de variação de ψ e altura no dossel.

Em contraste com outras áreas de pastagem do ABRACOS, os valores de g_l na gramínea *Panicum maximum* não excederam 250 mmol m^{-2} s^{-1} durante a maioria do tempo, mesmo durante as estações chuvosas. Relação significativa negativa entre valores de e g_l de D foi encontrada nessa espécie. Valores de g_{lmax} medidos em componentes não gramínea dessa pastagem excederam em muito os observados em *Panicum maximum*, notadamente durante períodos úmidos.

15 Forest and pasture leaf-gas exchange in south-west Amazonia

A.-L.C. McWILLIAM[1], O.M.R. CABRAL[2], B.M. GOMES[3], J.L. ESTEVES[4] and J.M. ROBERTS[1]

[1] *Institute of Hydrology, Wallingford, UK*
[2] *EMBRAPA-CPAA, Manaus, Brazil*
[3] *Federal University of Rondônia, Ji-Paraná, Brazil*
[4] *INCRA, Ji-Paraná, Brazil*

INTRODUCTION

Pioneering studies in a previous Anglo-Brazilian initiative (Roberts *et al.*, 1990) produced detailed measurements of stomatal conductance made throughout the rainforest canopy at the Reserva Florestal Ducke, Manaus. These showed strong vertical gradients of stomatal conductance and consistent daily patterns although seasonal differences were small. Functional relationships between stomatal conductance and weather variables were derived for different seasons having differing soil water conditions. The value of these data has been extended beyond a qualitative description of the physiological behaviour within a rainforest canopy.

Some of the understanding provided by this work has entered into the Simple Biosphere Model (SiB), a functional vegetation parameterization for incorporation in GCMs (Sellers *et al.*, 1989). Additionally, the physiological data set from the Reserva Ducke has been used in the development of CLATTER, a multi-layer canopy transpiration model which gave an excellent agreement with eddy correlation measurements in a comparison over a range of seasons at the Reserva Ducke (Roberts *et al.*, 1993). The agreement of a model derived from stomatal conductance data taken from a small number of trees with data from a device sampling over the order of a few hectares in such an apparently heterogeneous forest is, at first, surprising. An interpretation of this result is that in the horizontal plane one assemblage of species behaves much as another and as such can be regarded as a functional group. Such an interpretation enhances considerably the case for using vertical profiles of plant physiological measurements from a single tower at a rainforest site to represent the surrounding vegetation.

It is, of course, unreasonable to assume that this horizontal homogeneity extends to vegetation in the same region but experiencing quite different seasonal soil water balances because of different forest positions within the local topography. Neither should natural forest vegetation in different Amazonian regions be expected to behave similarly, given that there is considerable regional variation in seasonal soil water

Amazonian deforestation and climate. Edited by J.H.C. Gash, C.A. Nobre, J.M. Roberts and R.L. Victoria.

supply and substantial variation in physiognomy, which gives equally large scope for physiological differences.

A central objective of the ABRACOS project is to extend the understanding of physiological responses of primary rainforest to other regions of Amazonia and to compare those with the responses of the tropical pasture which is replacing the forest. There is as yet a dearth of information on the physiological behaviour of species which comprise such pastures. In particular, ABRACOS has allowed the possibility of examining the physiological behaviour of vegetation growing in regions experiencing well developed dry seasons. Longer and more severe dry seasons have been predicted as a consequence of large scale Amazonian deforestation (Nobre *et al.*, 1991).

In this paper we present the seasonal patterns of leaf gas exchange and plant water relations of species comprising the vertical profile at the ABRACOS primary rainforest and adjacent pasture sites in south-west Amazonia, a region which experiences a significant dry season. We relate the responses of stomatal conductance and photosynthesis of forest and pasture species to associated weather and soil variables and functional relationships are derived.

MATERIALS AND METHODS

SITES AND SPECIES

Forest

This experimental site was in *terra firme* forest in Reserva Jaru (10°05'S 61°55'W). The mean canopy height was 33 m, but the tallest emergent trees reached 44 m. Plant physiological measurements were made on six tree species and two Liana species accessible from the 52 m sectional aluminium tower in the forest, and on two ground species, tree and palm (Figure 1, Table 1).

Pasture

Fazenda Nossa Senhora da Aparecida (10°45'S 62°22'W) is approximately 80 km east of Reserva Jaru; it is a cattle ranch created by clearing primary forest about 12 years ago. The dominant grass species was *Brachiaria brizantha* (A. Rich.) Stapf., a perennial grass whose common names are "palisade grass" (Samoa), "signal grass" (East Africa) and "estrella de Africa" (Latin America). This is an African species that is fairly drought-tolerant (Skerman and Riveros, 1990) and has been widely used in the New World tropics as forage grass.

Measurement periods

Plant physiological measurements were made at both sites during three experimental missions spanning three different seasons within a one-year period. ABRACOS Mission 3 (M3) was in the dry season (August-September 1992), Mission 4 (M4) in the wet season (March-April 1993) and Mission 5 (M5) in the transition period between the wet and dry seasons (June 1993).

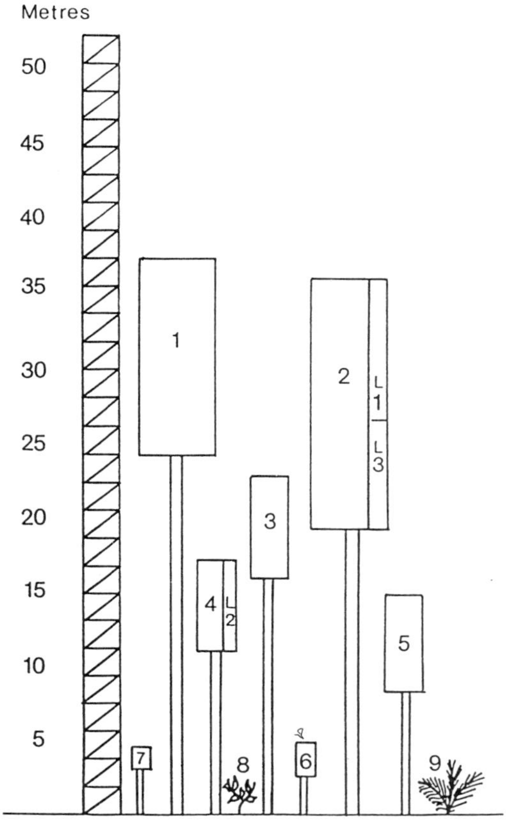

Figure 1 Schematic representation of trees, lianas and palms sampled in the Reserva Florestal Jaru, Rondonia. Table 1 gives the names of the species sampled.

Weather and soil moisture

Data from automatic weather stations at both the forest and pasture sites at Ji-Paraná and soil moisture data collected using a neutron probe have been obtained routinely throughout the ABRACOS Project. Full details of these measurements are given elsewhere in this volume. Weather and soil moisture data for the periods when physiological measurements were made were used to interpret trends in physiological behaviour. Automatic weather station data that were used from each site comprised hourly average values of solar radiation (R_s), air temperature (T), specific humidity deficit (D) and rainfall. Soil water content (mm) of the upper 1 metre of soil was derived from neutron probe measurements.

Gas exchange measurements

Stomatal conductance (g_l) and net photosynthesis (A) were measured using an infrared gas analyzer and leaf chamber (LCA-3 carbon dioxide analysis system and PLC-B leaf chamber, The Analytical Development Company Ltd, Hoddesdon, UK).

At Reserva Jaru gas exchange measurements were made on 18 days throughout the forest canopy, in general every two hours from 08.00 h until 16.00 h. Three to five leaves were sampled for each species.

Table 1 Description of trees surrounding tower at Reserva Jaru, Rondônia, and level at which they were sampled

Tree/Level	*Species* (Family)	Height m	dbh*[1] cm	Canopy length m
1/18,16	*Cedrella odorata* (Meliaceae)	36.0	100	12.6
2/16,13,12	*Inga* sp. (Leguminosae-Mimoseaceae)	35.1	39	16.2
L1/16	*Dioclea cf bicolor Bth.* (Leguminosae-Papilloniaceae)	-	-	-
L3/16,14	*Strychnos amazonicus* Krukoff (Loganiaceae)	-	-	-
3/10	*Protium polybotrium* (Turcz) Engl (Burseraceae)	22.1	26.5	5.8
4/8,5	*Leonia glycicarpa* Ruiz (Violaceae)	16.6	17	5.4
L2/8	*Derris pterocarpa* (DC) Killip (Leguminosae-Papilloniaceae)	-	-	-
5/7	*Guarea* cf *da visii* Sandwit (Meliaceae)	14.2	-	5.6
6/2	*Hirtella hispidula* Steud. (Crysobalanaceae)	4.5	2	2
7/1	*Erythroxyllum cf. macrophyllum* Cav. (Erythroxyllaceae)	4.5	2.5	1
8/0	*Theobroma microcarpum* Mart. (Sterculaceae)	1	≤ 1	-
9/0	*Maximiliana maripa* (Corre Serra) Drude (Palmae)	1	≤ 1	-

Note: Tree number refers to the species in Figure 1; L relates to liana species.
*[1] dbh = diameter at breast height (*c.* 1.3 m)

At Fazenda Nossa Senhora, five plots within the pasture area were randomly selected, and within each plot three to five leaves were sampled at hourly intervals throughout the daylight period. Fully illuminated leaves were measured in most cases (upper canopy leaves). On some days measurements of both sun and shade leaves were measured. In the case of the grass and palm species, because a single blade did not fully cover the aperture in the leaf chamber and depending on the species, two or three leaf blades were used to fully occupy the leaf chamber aperture without overlapping.

Leaf water potential (ψ)

Leaf water potential (ψ) was measured using a Schölander type pressure chamber (Schölander *et al.*, 1964) at approximately the same times as the g_l and A measurements. Some pre-dawn measurements of ψ were made, but not routinely. Three to five leaves were sampled in the same vicinity as the measurement of g_l and A, wrapped in cling-film and transported quickly (under 60 seconds) to the pressure chamber. For measurements at the forest site leaf petioles were sealed into the lid of the pressure chamber with a compressible gland. In some cases where the petioles were short it was necessary to extend them downwards into the lamina by cutting away some of the lamina each side of the main midrib. Leaves at the pasture site were accommodated in a slit rubber bung, which sealed the leaf into the lid of the pressure chamber. The values presented here have not been adjusted for the height of the leaf sampling. The low latitude of the Ji-Paraná forest and pastures sites (10°S) means that fluctuations in daylight duration from dawn to the times given for sampling over the seasons when measurements were made are all less than 30 minutes.

RESULTS

WEATHER AND SOIL MOISTURE

M3 commenced at the end of the dry period (Figure 2) after almost two months without rain. In the pasture, apart from a storm of 17 mm in early July (Day 187), there had been no significant rainfall for 35 days prior to the first measurements. The first measurements were made at the forest two days after a storm (28 mm), before which it had not rained in nearly three months.

M4 was towards the end of the wet season. In the pasture 134 mm of rain fell in the 30 days before the first measurements during this mission, and the soil in the 0-1 m layer was substantially wetter than in Mission 3 (Figure 2). In the forest, 322 mm rain fell in the 30 days leading up to the first measurements, and soil moisture content increased from 200 to 300 mm.

M5 represented the beginning of the dry season, and in both the pasture and the forest there was a marked decrease in both rainfall (16 mm and 54 mm rain in the previous 30 days for pasture and forest respectively) and soil moisture content (both reduced to 200 mm soil moisture content in the top metre of soil).

Figure 3 shows the hourly average values during the daylight periods for solar radiation (R_s), specific humidity deficit (D) and air temperature (T) at the forest and pasture sites for the periods in which the plant physiological measurements were made. Maximum R_s was similar between seasons and sites, except at the forest in M5, where there were lower values than at the pasture.

D was consistently lower at the forest than at the pasture site (maximum values were 12.5 g kg^{-1} at the forest, 19 g kg^{-1} at the pasture); at both sites D was lowest during the wet season. The highest temperatures occurred during the driest periods.

During M5 plant physiological measurements were also carried out on two days when a cold spell occurred. Figure 3a shows how R_s, T and D were reduced on these days, with only R_s showing a diurnal fluctuation although this was very much reduced in comparison with the average conditions experienced in each season.

FOREST

Stomatal conductance

Figure 4a shows the diurnal patterns of g_l for the main species surrounding the tower

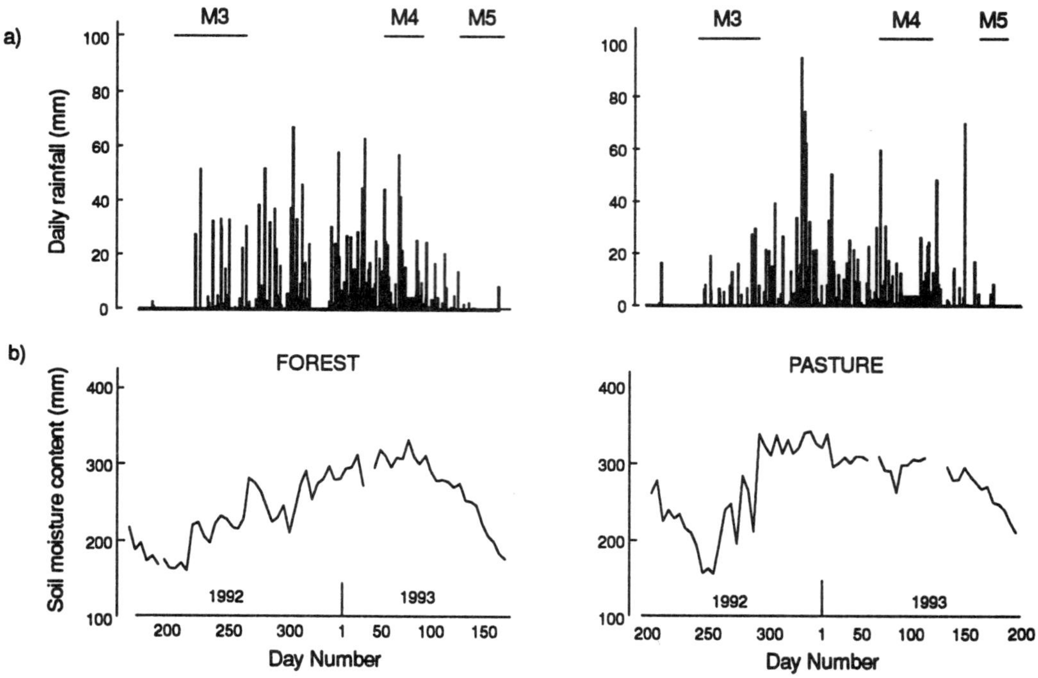

Figure 2 Seasonal fluctuations in (a) daily rainfall and (b) soil moisture content in the top 1 m of soil, at a primary rainforest site and recently created pasture in Rondonia. Bars represent the mission periods when physiological measurements were made.

at Reserva Jaru in the three mission periods. There is a suggestion of seasonal variation in g_l in the forest, but the trend is not statistically significant. In M3, apart from *Leonia glycicarpa*, g_l of all the tree species (mean of four similar days), are lower than the other study periods in M4 and M5, even though the soil moisture content in M5 is similar to that found in M3. It was not possible, generally, to use the IRGA before 08.00 h each day because the leaves were wet with dew. Maximum g_l for all species during the drier season of M3 occurred at 09.00 h, and it declined throughout the daylight period (Figure 4a). For *Inga*, level 14, this pattern occurred for all missions. The maximum g_l at this time varied between the missions: the lowest was in M3 (346 mmol m^{-2} s^{-1}), the highest in M5 (445 mmol m^{-2} s^{-1}). The same higher g_l was found for the *Inga* at level 12 in M5. During M4, the leaves on the *Inga* at level 16 were not accessible, therefore it was only possible to sample at level 12. Instead, the tallest emergent tree next to the tower, *Cedrella odorata*, had leaves during this mission, whereas they were not present during M3 and M5. *Cedrella odorata* showed similar diurnal patterns of g_l as *Inga*. For the within canopy species, there was no difference between season for *L. glycicarpa* (level 8). Conductances for the ground species *Theobroma microcarpum* were higher during M4 and M5 and consistently higher than the other two within-canopy species

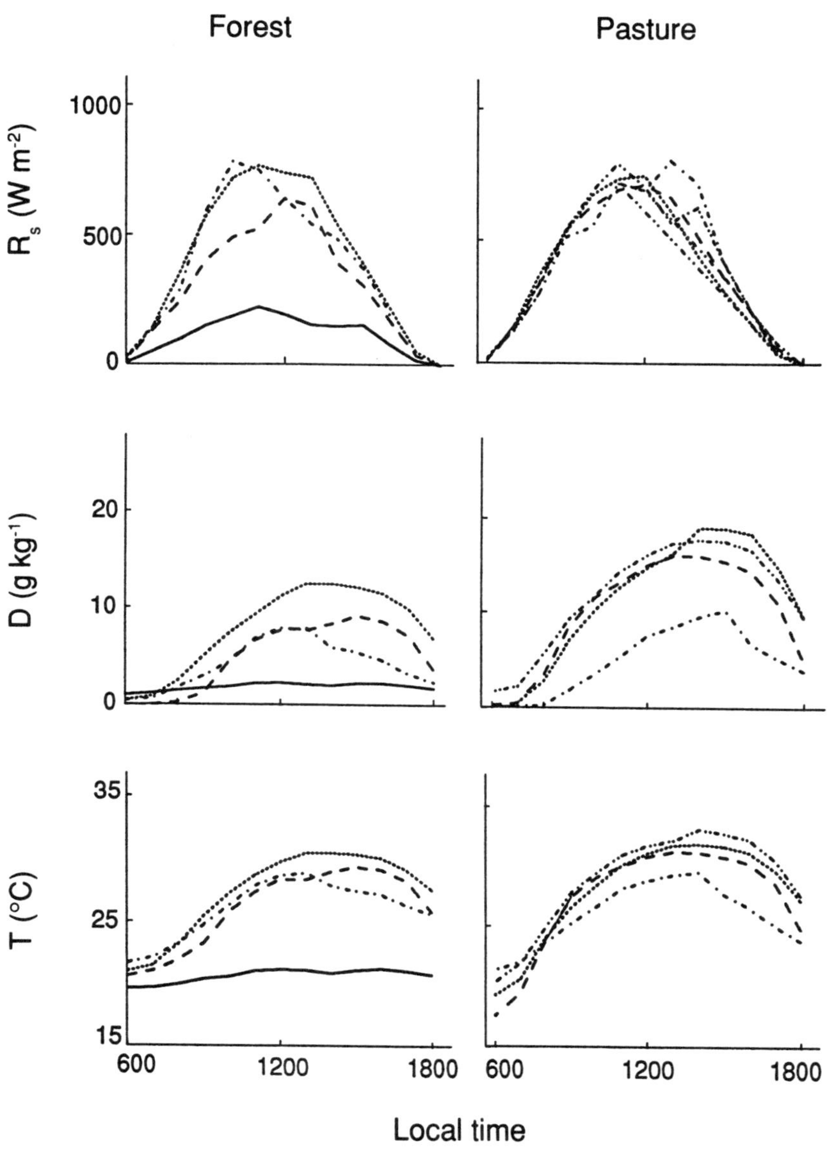

Figure 3 Hourly averages of solar radiation (R_s), specific humidity deficit (D) and temperature (T) measured on the days when physiological data were being collected.
Forest (left): M3 (·····), M4 (·-·-·), M5 (— — —) and M5-cold days (—); pasture (right): M3 (·····) 14 August 1992 and (·-·-·) 25 August 1992, M4 (·-·-·) and M5 (— — —).

(Figure 4a). This species also showed little decline throughout the diurnal period during M4 and M5, remaining around 200 mmol m^{-2} s^{-1}. During M3, the g_l of this species was at a maximum at the begining of the day (198 mmol m^{-2} s^{-1}) and declined steadily throughout the day to 71 mmol m^{-2} s^{-1}.

Net photosynthesis

A was markedly higher in *Inga* (level 16) in M5 (maximum of 8.4 μmol m^{-2} s^{-1}) than in M3, but by 13.00 h had reached the same rates as found in M3 and for *C. odorata* in M4 (7 μmol m^{-2} s^{-1}) and declined to 1.3 μmol m^{-2} s^{-1} by 17.00 h (Figure 4b). The same species at level 12, however, showed the highest *A* rates in M3 (7.6 μmol^{-2} s^{-1} 12-13 h) and the lowest in M5 (reaching only 3.2μmol m^{-2} s^{-1}). The same pattern was

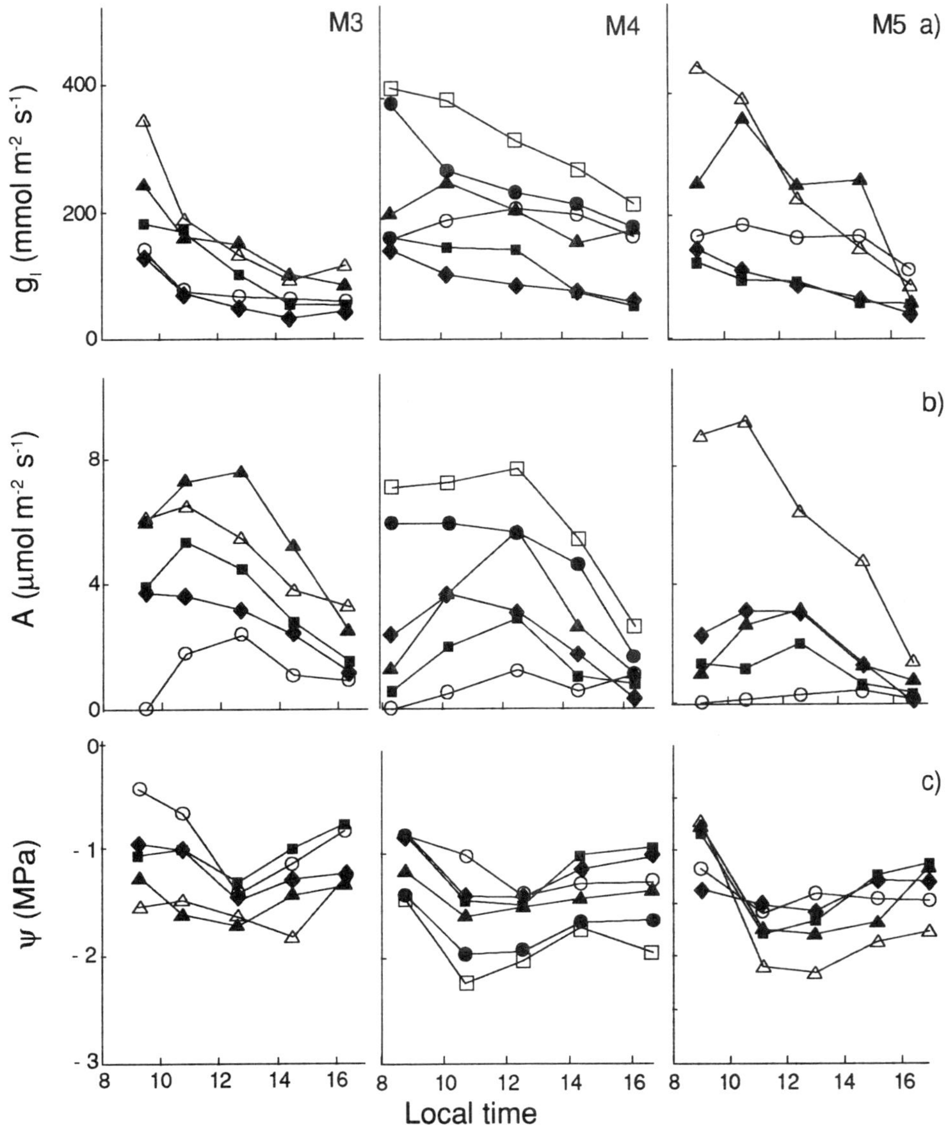

Figure 4 Diurnal variation in (a) g_l, (b) *A* and (c) *ψ*, in the species around the sampling tower at Reserva Florestal Jaru, Rondônia, during M3, M4 and M5. Data points are the mean of four climatically similar days, each representing the mean of 3 to 5 leaves. *Cedrella odorata*, 33 m M4 only (□); *C. odorata* 29 m M4 only (●); *Inga*, measured at 29 m (Δ); *Inga* 22 m (▲); *Protium polybotrium*, 18 m (◆); *Leonia glycicarpa*, 14 m (■); *Theobroma microcarpum* 0-2 m (○).

found in *L. glycicarpa* and *T. microphyllum*. Comparing M3 and M4 (i.e. the wettest and driest measurement periods) no statistical difference in *A* could be found between the two seasons for any of the species. *A* for *Protium polybotrium* did not vary between any of the seasons and had a mean maximum of 3.6μmol m^{-2} s^{-1} around 10.00 h.

Leaf water potential (ψ)

Lowest ψ was measured in the emergent trees *Inga* and *Cedrella odorata* (Figure 4c), and the values increased lower down in the forest canopy. The most negative values of ψ were observed in *Inga* (M5, level 14, -2.16 ± 0.16) with the least negatve in *T. microcarpum* (M4, ground level, -1.41 ± 0.21). ψ did not vary between seasons (Figure 4c). Minimum ψ reached were -1.87 ± 0.37 (*Inga*), -1.94 ± 0.59 (*C. odorata*) and -2.16 ± 0.16 (*Inga*) for M3, M4 and M5, respectively. In M3 the leaves tended to reach their minimum ψ later in the day (12.00-13.00 h) than for the leaves in M4 and M5 (10.00-11.00 h). For *Inga* and *C. odorata* this may be related to the higher g_l attained by these species between 09.00 and 10.00 h in the morning. Pre-dawn ψ in *Inga* during M3 showed that leaf turgor was nearly fully regained by the end of the dark period (0.53 ± 0.13 at 05.30 h, 6 September 1992: data not shown).

g_l and water potentials in forest on cold days

g_l was not significantly different in any of the species during the cold period at the forest (Figure 5) but *A* was markedly reduced in all species, presumably because of low solar radiation conditions which prevailed during the cold days. Maximum hourly solar radiation at midday reached only around 200 W m^{-2} which would not be light-saturating for *A*. Instead of a maximum *A* of 9.6 μmol m^{-2} s^{-1}, the rate was reduced to 6.9 μmol m^{-2} s^{-1} for *Inga* (level 16). In all other species (except *T. microphyllum*, which had extremely low *A* during all of the missions), *A* was reduced by approximately 30% in the cold conditions (Figure 5). ψ remained low throughout the daylight period, and becoming only slightly less negative by the end of the day.

PASTURE

Stomatal conductance

The g_l measurements for three missions for *Brachiaria brizantha* are shown in Figure 6. Figure 6a shows two days during Mission 3; 14 August was prior to the first rains and 25 August was two days after the first major storm in at least 30 days (42 mm of rain fell in the week before these measurements). The maximum g_l (mean ± STD of five days) was increased significantly from a maximum g_l of 249 $mmol^{-2}$ s^{-1} at midday (14 August 1992) to 459 mmol m^{-2} s^{-2} (25 August 1993). Mission 4 took place during the rainy season of this area (see Figure 3) and g_l was at its maximum in this species during this period. In the 30 days before these measurements 134 mm of rain fell. During the wet season, the maximum g_l reached — given as an average of five days — was 621 mmol m^{-2} s^{-1} near midday (Figure 6b), while after two months

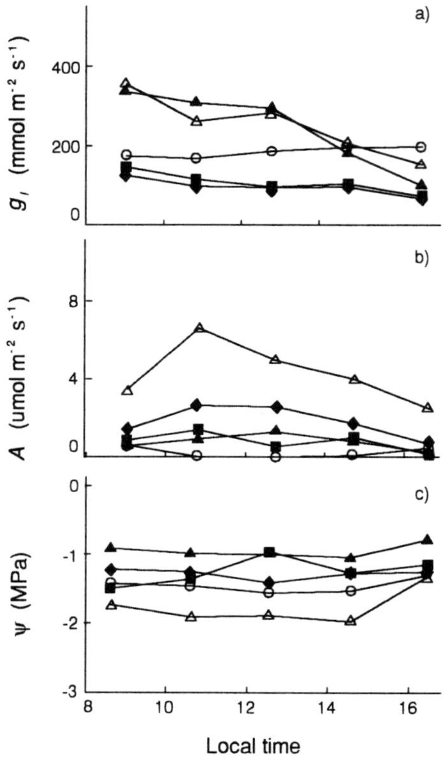

Figure 5 Diurnal variation in (a) g_l (b) A and (c) ψ, in species around the sampling tower, during a cold spell (10-11 June 1993, M5) in Reserva Florestal Jaru, Rondonia. *Inga* at 29 m (Δ); *Inga* at 22 m (▲); *Protium polybotrium* at 18 m (◆); *Leonia glycicarpa* at 14 m (■); *Theobroma microcarpum*, 0-2 m (○).

of more limited rainfall (the onset of the dry season with only 16 mm rain in the previous 30 days) g_l reached a maximum of only 174 mmol m^{-2} s^{-1} at around 10.00 h in the morning. This was even lower than the g_l at the end of the dry season in 1992.

Net photosynthesis

In *B. brizantha, A* followed the same diurnal pattern as g_l but was not so strongly affected by the seasons. During both the dry periods of August 1992 and June 1993 peak values were around 15 µmol m^{-2} s^{-1} (Figure 6c and 6d). After periods of rain A increased significantly to a maximum of 19.6 and 21.6 µmol m^{-2} s^{-1} at 11.00-12.00 h in August 1992 and March 1993, respectively.

Leaf water potential (ψ)

The diurnal pattern of ψ mirrored the diurnal patterns of g_l in all the seasons, except that the most negative values occurred between 13.00 and 14.00 h, instead of 11.00-12.00 h for the peak g_l values. There was a fall in ψ between the 14 and 25 August 1992 (Figure 6e) but the change is not significant. The least negative ψ was observed in the wet season of 1993 (-1.3 MPa at 1400 h; Figure 6f) but became more negative (-1.8 MPa) again with the drying out of the season. In both the wetter and drier seasons ψ in *B. brizantha* recovered turgor by the end of the night period (Figure 6f).

POSSIBLE FACTORS CONTROLLING g_l

Forest

g_l in the *Inga* (levels 12 and 16, M3 and M5) and *Cedrella odorata* (levels 16 and 18, M4) was negatively correlated with *D*, although for *C. odorata* the spread of the points was greater, suggesting less control by *D* (Figure 7, Table 2). No correlation

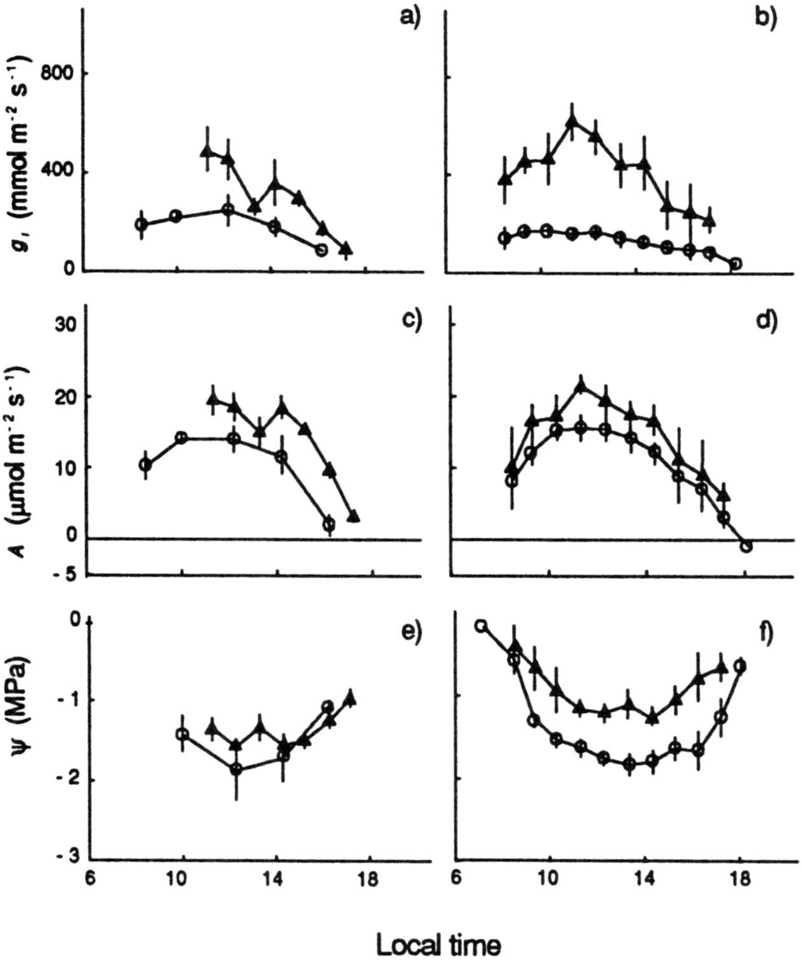

Figure 6 Diurnal variation in (a) g_l (c) *A* and (e) ψ, in *Brachiaria brizantha* during M3 (○ 14 August 1992; ▲ 25 August 1992) and (b) g_l (d) *A* and (f) ψ during M4 (▲) and M5 (○). For M3 bars represent the standard deviation of the mean of five leaves sampled; for M4 and M5 they represent the standard deviation of the mean of five similar days sampled, in which five leaves were sampled for each hour on each day.

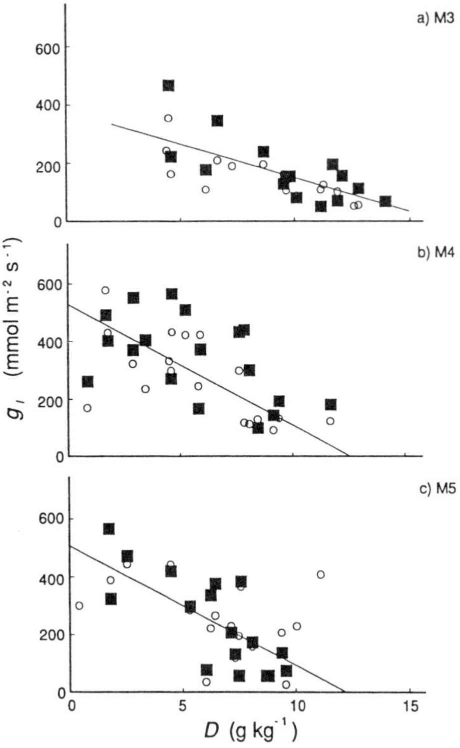

Figure 7 The relationship between g_l and D for canopy species *Inga* (a and c), (■ level 16, ○ level 12) and *Cedrella odorata* (b), (■ level 18, ○ level 16)

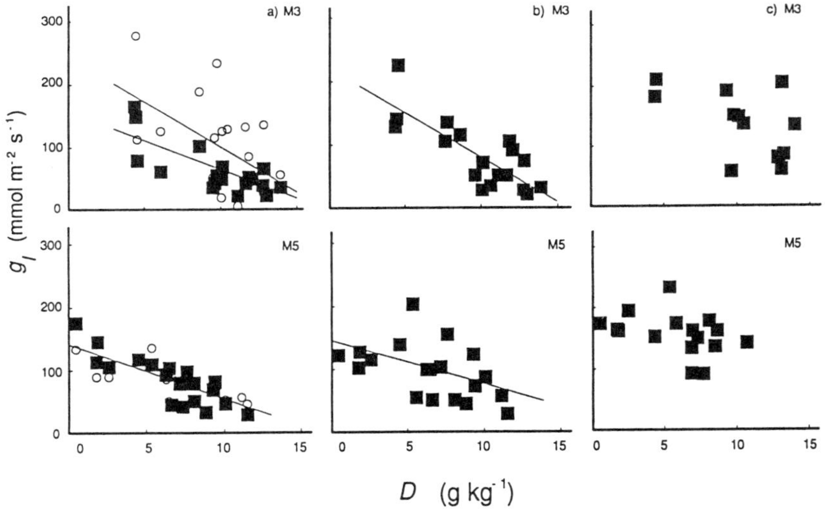

Figure 8 The relationship between g_l and D for (a) within-canopy species *Protium polybotrium* (○) and *Leonia glycicarpa* (■), (b) Liana *Derris pterocarpa* and (c) ground species *Theobroma microcarpum*, during mission periods M3 and M5

Table 2 Linear regression statistics for g_l (mmol m^{-2} s^{-1}) with D (g kg^{-1}) for tropical tree species and liana surrounding tower in Reserva Jaru, Rondônia.

Species		M3	M4	M5
Cedrella odorata	*r*	-	0.80^{***}	-
(L16)	*a*	-	530±94	-
	b	-	42±8.3	
Inga sp. L16	*r*	0.75^{**}	-	0.72^{***}
	a	449±77	-	530±107
	b	-29±7	-	52±11
L12	*r*	0.70^{***}	N.S.	0.67^{**}
	a	311±54	-	488±103
	b	-17±4.2	-	-38±11
Protium polybotrium	*r*	0.76^{***}	N.S.	0.81^{***}
	a	160±26	-	151±23
	b	-9.8±2	-	-9.8±1.7
Leonia glycicarpa	*r*	0.56^{*}	N.S.	0.72^{***}
	a	244±66	-	132±24
	b	-14±5.3	-	-7.5±1
Theobroma	*r*	0.56^{*}	N.S.	-
microcarpum	*a*	236±48	-	-
	b	-9.6±4.3	-	-
Derris pterocarpa	*r*	0.79^{***}	N.S.	0.52^{*}
(Liana species)	*a*	221±34	-	146±39
	b	-14±2.7	-	-6.9±2.8

r correlation coefficient; *a* intercept of regression line (±S.E.); *b* slope of regression line (±S.E.).
*** $P < 0.001$, ** $P < 0.01$, * $P < 0.05$; N.S.: not significant.

was found with A, ψ or T, but this may be as a result of poor sample size and more complicated sampling conditions than were found for the grass species, where a much wider range of light environments could be measured. During the wettest season (M4) no correlations were found between the rest of the species and D (Figure 8). During M3 and M5, g_l was negatively correlated with D for the within-canopy species, but in the ground species *Theobroma microcarpum* g_l was not sensitive to D at all in any season. These correlations were based upon D values from the automatic weather station, and not from within-canopy values. During the cold days encountered in M5, no correlations were found between g_l and D.

Pasture

Figure 9 shows the relationships between g_l and various environmental and leaf parameters in *Brachiaria brizantha*. Table 3 gives the equations of these regressions and shows that g_l is positively correlated with Q_p and negatively with ψ in both M4 and M5. These missions were compared because they represent the most extreme g_l and leaf water potential responses, and the largest sample sizes. The coefficients

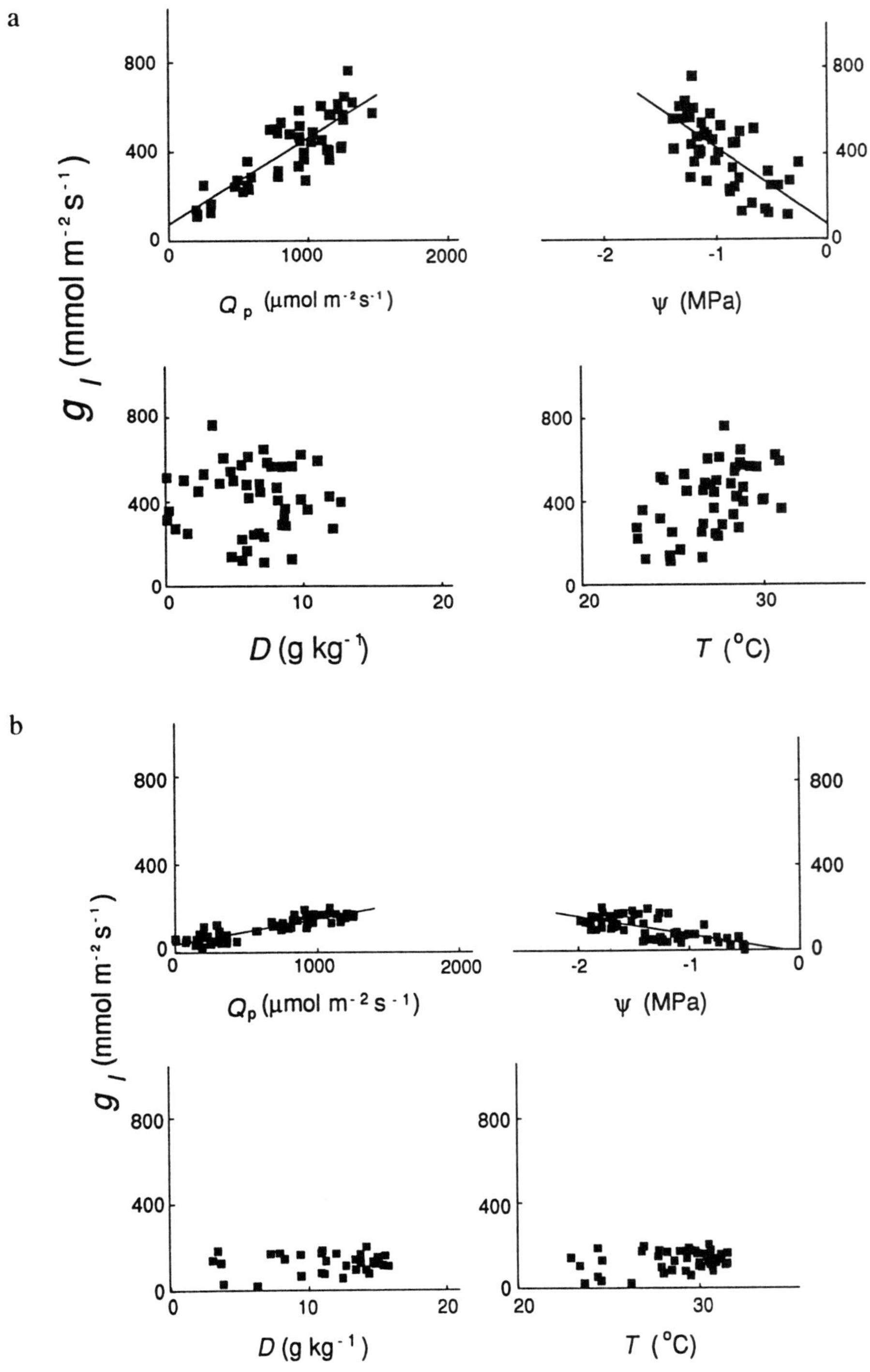

Figure 9 The relationship between g_l and Q_p, ψ, D and T in *Brachiaria brizantha* during (a) M4 and (b) M5

Table 3 Linear regression statistics for (a) g_l (mmol m^{-2} s^{-1}) and (b) A, with environmental and physiological variables for *Brachiaria brizantha*.

Independent variable		M4	M5
(a)			
Q_p	r	0.84***	0.90***
(µmol m^{-2} s^{-1})	a	76.9±88.5	36.9±23.0
	b	0.383±0.038	0.113±0.007
Leaf water potential	r	0.69***	0.65***
(MPa)	a	69.0±117	-14.4±39.7
	b	-362±57	-87±13.2
D (g kg^{-1})		N.S.	N.S.
T (°C)		N.S.	N.S.
(b)			
Q_p	r	0.90***	0.99***
(µmol m^{-2} s^{-1})	a	2.49±2.4	1.04±1.0
	b	0.014±0.001	0.015±0.001

r correlation coefficient; a intercept of regression line (±S.E.); b slope of regression line (±S.E.).
*** $P < 000.1$; N.S.: not significant.

for the positive linear regression of A with Q_p did not differ significantly between the M4 and M5 (Table 3). g_l was not found to be correlated with D or T.

DISCUSSION

During the wet season, the pasture had higher maximum values of g_l and A compared with the forest. While on an single day g_l and A reached as high as 766 ±99 mmol m^2 s^{-1} (11.00 h) and 24 ±0.77 µmol m^2 s^1 in the pasture (mean ±STD five plots of leaves on a single day), maximum g_l and A in the forest were 566 ±76 mmol m^2 s^{-1} (*Cedrella odorata* in M4, mean ±STD of five leaves) and 12.6 ±2.3 µmol m^2 s^{-1}, (measurements of g_l and A were not from the same time of day). During Mission 3, after several months with little rainfall and therefore with reduced soil water content, g_{lmax} and A in the forest were 468 ±123 mmol m^2 s^{-1} and 8.1 ±1.4 µmol m^2 s^{-1}, while at the same time the equivalent values in the pasture were substantially lower: 224 ±22 mmol m^2 s^{-1} (g_l) and 14 ±0.47 µmol m^2 s^{-1} (A).

Minimum ψ in the canopy species in the forest was -2.41 MPa (Inga, M5) and -2.64 MPa (*Cedrella odorata*, M4), with no seasonal differences. In the pasture, ψ did not reach such low values even in the drier periods: -1.38 MPa was the minimum during the wet season, and -1.97 MPa was observed during the dry periods. The pasture was extremely sensitive to the drying out of the soil, whereas in the forest it was difficult to find a significant difference between the seasons in g_l, ψ and A. Presumably, much of this is associated with the deeper roots of the tree species, and different physiological responses between the two plant groups.

Forest

The values of g_l, A and ψ compared well with the limited range of values already found for tropical forest species (see Table 3 in Roberts *et al.*, 1996). From that table it can be seen that g_{lmax} for canopy species covers a wide range of values, presumably mostly as a result of differences in species, leaf age, tree age and soil conditions. For the canopy, values range from tree species in central Amazon of 185 mmol $m^{-2}s^{-1}$ (*Bocoa viridiflora*; Roberts *et al.*, 1990) to as high as 1200 mmol m^{-2} s^{-1} for both *Tectona grandis* and *Gmelina arborea* during the wet season in Nigeria (Whitehead *et al.*, 1981; Grace *et al.*, 1982). However the species studied by Whitehead *et al.* and Grace *et al.* are characterized by having huge entire leaves which would have a relatively small leaf boundary-layer conductance (Grace, 1983). This would mean that high conductances would be necessary to sustain high leaf transpiration and adequate leaf cooling. Even though there was considerable leaf to leaf variation throughout the diurnal period, the decline in g_l (from about 09.00 h) throughout the day is exhibited by nearly all species in all seasons. However, as found by Roberts *et al.* (1990), g_l for the upper canopy species changed substantially more than for the within-canopy species.

Until recently there have been few measurements of canopy A from tropical forest, but those few found A_{max} similar to that measured in this study: 9.4 μmol $m^{-2}s^{-1}$ (Pearcy, 1987); 8 - 14 μmol $m^{-2}s^{-1}$ (Koch *et al.*, 1994), and 12.6 - 18.9 μmol $m^{-2}s^{-2}$ for diverse tropical trees (Koyama, 1981). The values of A_{max} for both *C. odorata* and *Inga* were similar. Considering values obtained from many different canopy species it appears that the maximum value obtained may not be related to the particular species but rather as a result of the position it is in. Species that occupy a particular niche may behave in a similar fashion, like a functional group (Körner, 1993).

The within-canopy and ground species had the lowest g_l and A reflecting low photosynthetic capacities. This decline in A from the top of the canopy to ground level is probably a consequence of the decline in light levels and reflects the variation in the photosynthetic light response of leaves growing in various canopy levels (Mooney *et al.*, 1984). Light levels at the ground level of many rainforests can be less than 1% of that reaching the top of the canopy.

Statistically there was no evidence that the forest gas exchange was affected by the seasonal changes in rainfall and therefore soil moisture in this part of Amazonia. During the peak of the wet season, the tallest emergent tree (*C. odorata*) had a new set of leaves, and leaves were absent from this species during all other measurement periods. The g_{lmax}, A and ψ of this species was similar to that measured in *Inga* after the soil had commenced drying out. The highest g_{lmax} was measured in *Inga* during M5; but the higher g_l compared with the same species in M3 may be more a consequence of leaf age rather than a seasonal difference. *Inga* leaves measured in M5 were fully expanded but fresher and newer than those present in M3. During M4 the leaves of *Inga* at the highest level were senescing and were too few and inaccessible to measure. Lower g_l in older leaves have been reported by Reich and Borchert (1988). They found that in five tropical tree species studied, the g_l was influenced by leaf age. But the relationship between g_l and leaf age appears to be species-specific since although three of the species studied showed a lowering of g_l with leaf age, in one species (*Tabebuia rosea*) there was an increase in g_l with increasing leaf age along with the lowering of leaf water potential. The decline in g_l with leaf age in certain species, while not apparent in some, has been attributed to the different water use efficiencies of the different species (Reich and Borchert, 1988). In *T. rosea*, *A* declines while transpirational water loss increases with leaf age, thereby decreasing the water use efficiency of this species. In the present study there was no apparent change in the relationship between *A* and g_l between the seasons (Figure 10) but in the pasture, *A* is only slightly decreased compared with the large decrease in g_l with increasing soil moisture deficit, which increased water use efficiency of the grass species. Decreased *A* and ψ also implies the onset of stress in plants but there is no evidence from values observed in this study that these parameters changed seasonally for any species at the forest site.

Pasture

In contrast to the gas-exchange measurements from the forest, the gas-exchange in the pasture, and in particular g_l, was significantly affected by a reduction in soil moisture content in the pasture. In an area of Senegal where the average annual rainfall does not exceed 750 mm, Boyer (1986) recorded ψ of -1.9 MPa for *B. brizantha*, stomata closed between -1.4 and -1.7 MPa, corresponding to g_l of between 80 and 140 mmol m^{-2} s^{-1}. These values compare well with the measurements made on *B. brizantha* in this study. Here, at their most stressed, the g_l of this species reached a maximium of 200 mmol m^{-2} s^{-1} (Figure 6a and b), and ψ reached -1.8 MPa.

Even though *B. brizantha* is a C_4 grass (Boyer, 1986; Hattersley, 1992) maximum *A* obtained (22 μmol $m^{-2}s^{-1}$) was more comparable to that of a C_3 grass. For some C_4 type grasses, *A* of up to 60 μmol m^{-2} s^{-1} has been reported (Nobel, 1991) but the high *A* recorded for many C_4 species are from irrigated and fertilized crops rather that those exposed to more harsh environments. Even though the water may not have been limiting for *B. brizantha* during the rainy season, the soils are known to be poor in this area and therefore certain nutrients may well have been limiting. Two C_4 tropical grasses, *Themeda triandra* and *Pennisetum mezianum,* have been extensively studied in Nairobi National Park by Kinyamario and Imbamba (1992). Highest *A* rates observed at midday were approximately 30 μmol m^{-2} s^{-2} in *P. mezianum.* As with the *B. brizantha* in this study, the lowest rates were observed during the dry season. However, this may suggest that the species was stressed and therefore was not able to economise the fixing of C while having fairly high g_l values. The range of diurnal g_l values between dry and wet seasons was also comparable to that shown by *B. brizantha*. During the rainy period maximum g_l was 542 and 446 mmol m^{-2} s^{-1} for *T. triandra* and *P. mezianum*, respectively, and was reduced to g_{lmax} of 200 mmol m^{-2} s^{-2} for both species during the dry season. Much lower ψ was experienced by these savanna species compared with *B. brizantha* (as low as -4 MPa, but generally around -2.5 MPa during the wet season), suggesting that *B. briantha* was not as water stressed as these species, or had better control on its transpirational water loss. Since ψ and g_l are good indicators of a plant's reaction to drought stress, the data presented here suggest good water use efficiency of *B. brizantha*. During the drier periods, g_l was reduced by approximately 70%, while at the same time A was reduced by only 32% (i.e. the plants were still fixing carbon with less water loss), and maintained not too negative ψ.

INFLUENCES ON g_l IN FOREST AND PASTURE

The g_l of tropical rainforest species has been found to respond to D and Q_p in a similar way to that of temperate species (Chiariello, 1984). Fetcher (1979) found g_l and ψ in several rainforest species were sensitive to changes in D but they did not respond to Q_p. Whitehead *et al.* (1981) found that in *Gmelina arborea* and *Tectona grandis* 80% of the g_l was controlled by either D or Q_p. In the present study, g_l was negatively correlated with D in the canopy species in all seasons, whereas it only correlated with the within-canopy species in M3 and M5 (i.e. the drier seasons). However, during M5 there were days when a cold front came up from the south of Brazil causing marked decreases in T, D and R_s. On these days, g_l was completely decoupled from D and yet still showed the same g_l rates and diurnal pattern of closure. The apparent insensitivity of the stomata to deficit on these cold days suggests the possibilty of an internal circadian rhythm for g_l. g_l and D did not correlate at all for *Theobroma microcarpum*, which was the ground species. Fanjul and Barradus (1985) also found little response between g_l and D in the wet season in trees growing in tropical forest, but the relationship became apparent as the soils dried out. In the present study no correlations were found between g_l and Q_p, T or ψ. In the case of Q_p, the lack of relationship may be more a consequence of the poor sample size and the difficulty of exposing the species to extreme enough light environments within the time available.

In contrast to the forest, g_l in the pasture was not correlated with D, but was strongly correlated with Q_p and ψ (Figure 9). It has been suggested that D has a larger

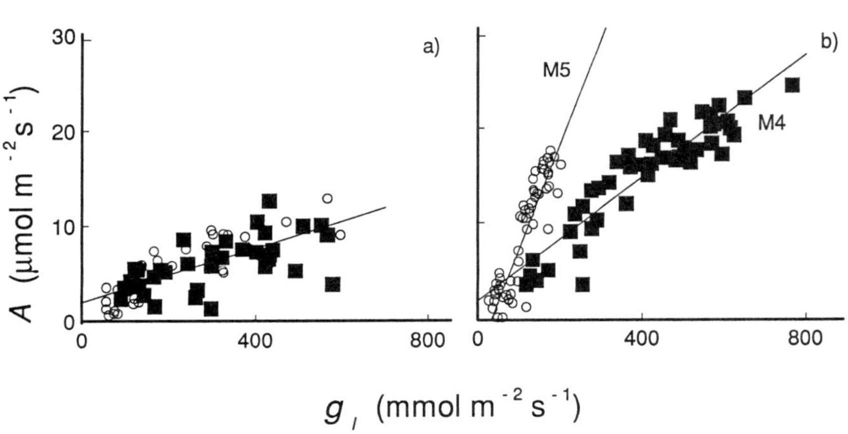

Figure 10 The relationship between A and g_l in: (a) canopy species *Cedrella odorata*, M4 (■) and *Inga*, M5 (○) from Reserva Florestal Jaru, (b) pasture species *Brachiaria brizantha*, (■) M4 and (○) M5.

influence on transpiration (and therefore g_l) in aerodynamically "rough" vegetation such as tropical rainforest (*cf* pasture, which is "smooth") than does Q_p (Mansfield and Davies 1985). There is a paucity of data on the responses of g_l to D in tropical pastures. However, relationships between g_l and ψ have been recorded in *B. Brizantha* growing in Senegal (Boyer, 1986), although the linear relationship between the two parameters trailed off at ψ below -1.8 MPa. However, the g_l in *B. decumbens* was found to be independent of changes in ψ (Jones *et al.*, 1980; Hesla *et al.*, 1985).

A and g_l were strongly coupled in both wet and dry seasons in *B. brizantha* while the coupling was not so strong in the canopy species in the forest (Figure 10). In the pasture the slope of the response changed markedly from the wet to the dry season suggesting increased water use efficiency of *B. brizantha* with increased drying out, i.e. in the dry season more CO_2 was fixed for less water lost.

CONCLUSIONS

Maximum stomatal conductance observed at the forest site, Reserva Jaru (400 mol m^{-2} s^{-1}) was measured for upper canopy species and declined to the lowest values observed which were for forest floor species. The highest values occur around mid-morning and decline steadily throughout the rest of the day. This decline was particularly marked in the upper canopy species. This diurnal decline in g_s is associated with increasing humidity deficit of the atmosphere. Seasonal changes in soil moisture content had little influence on g_s although one upper canopy species became leafless in the dry season. In contrast to species in the forest maximum g_s of the predominant species in the pasture, *Brachiaria brizantha,* was, under non-limiting conditions, higher than in the forest (600 mmol m^{-2} s^{-1}) and declined substantially in association with declining soil moisture levels and showed no association with atmospheric humidity deficit.

Maximum leaf photosynthesis, *A* (20 µmol m^{-2} s^{-1}) of the C4 pasture species was, as expected, higher than the maximum values observed in the forest species (12 µmol m^{-2} s^{-1}). In contrast to the g_s of the pasture, *A* was insensitive to changes in soil water content. For this reason water use efficiency increased with lowered soil water content. For the forest *A* was, like g_s, unaffected by changes in soil water content so that water use efficiency remained constant over a wide range of conditions.

ACKNOWLEDGEMENTS

The authors would like to thank IBAMA for the permission to use the Reserva Florestal Jaru for the forest measurements. We would also like to thank those from INCRA, the University of Rondônia (especially Leice, Marcia and Paulo), Ivanildo Alves Trinidade (EMBRAPA Belem), and Markus Liedgens, all of whom helped collect the data on gas exchange and water potentials over the three mission periods, and Manoel dos Reis Cordeiro (EMBRAPA-CPATU, Belém) for the identification of forest species.

REFERENCES

Bastable, H.G., Shuttleworth, W.J., Dallarosa, R.L.G., Fisch, G. and Nobre, C.A. 1993. Observations of climate, albedo, and surface radiation over cleared and undisturbed amazonian forest. *International J. Climatol.* **13**, 783-796.

Boyer, J. 1986. Comportement hydrique de deux plantes fourragères cultivées au Senegal (*Andropogon gayanus* et *Brachiaria brizantha*) en fonction des contraintes exercées par le milieu. *Revue de Elevage et de Medecine Veterinarie des Pays Tropicales* **39**, 443-451.

Chiarello, N.R. 1984. Leaf energy balances in the wet lowland tropics. In: E. Medina, H.A. Mooney and C. Vazquez-Yanez (Eds.) *Physiological Ecology of plants of the wet tropics*, Dr Junk Publishers, The Hague, 85-98.

Fanjul, L. and Barradus, V.L. 1985. Stomatal behaviour of two heliotropic understorey species of a tropical deciduous forest in Mexico. *J. Appl. Ecol.* **22**, 943-954.

Fetcher, N. 1979. Water Relations of Five Tropical Tree Species on Barro Colorado Island, Panama. *Oecologia* **40**, 229-233.

Grace, J., Okali, D.U.U. and Fasehun, F.E. 1982. Stomatal conductances of two tropical trees during the wet season in Nigeria. *J. Appl. Ecol.* **19**, 659-670.

Grace, J. 1983. *Plant-Atmosphere Relationships*, Chapman and Hall, London.

Hattersley, P.W. 1992. C4 photosynthetic pathway variation in grasses (Poaceae): its significance for aris and semi-arid lands. In: Chapman, G.P. (Ed.), *Desertified Grasslands: Their Biology and Management*. Academic Press, London, 181-212.

Hesla, B.I., Tieszen, H.L. and Boutton, T.W. 1985. Seasonal water relations of savanna shrubs and grasses in Kenya, East Africa. *J. Arid Environ.* **8**, 15-31.

Jones, C.A., Pena, D. and Carabaly, A. 1980. Effects of plant water potential, leaf diffusive resistence, rooting density and water use on the dry matter production of several tropical grasses during short periods of drought stress. *Trop. Agric.* **57**, 211-219.

Kinyamario, J.I. and Imbamba, S.K. 1992. Savanna at Nairobi Park, Nairobi. In: S.P Long, M.B. Jones and M.J. Roberts (Eds.) *Primary Productivity of Grass Ecosystems of the Tropics and Sub-tropics*, Chapman and Hall, London, 25-69.

Koch, G.W., Amthor, J.S. and Goulden, M.L. 1994. Diurnal patterns of leaf photosynthesis, conductance and water potential at the top of a lowland rain forest canopy in Cameroun: measurements from the *Radeau des Cimes*. *Tree Physiology* **14**, 347-360.

Körner, Ch. 1993. Scaling from Species to Vegetation: The Usefulness of Functional Groups. In: E.D. Schulze and H.A. Mooney (Eds.) *Biodiversity and Ecosystem Function*, Springer Verlag, Berlin, 117-140.

Koyama, H. 1981. Photosynthetic rates in Lowland rain forest trees of Peninsular Malaysia. *Japanese J. Ecol.* **31**, 361-369.

Mansfield, T.A. and Davies, W.J. 1985. Mechanisms for leaf control of gas exchange. *Bioscience* **35**, 158-164.

Mooney, H.A., Field, C. and Vazquez-Yanes, C. 1984. Photosynthetic characteristics of wet tropical forest plants. In: E. Medina, H.A. Mooney and C. Vazquez-Yanes (Eds.) *Physiological Ecology of plants of the wet tropics*, Dr Junk Publishers, The Hague, 139-154.

Nobel, P.S. 1991. *Physicochemical and Environmental Plant Physiology*. Academic Press. New York. 635 pp.

Nobre, C.A., Sellers, P.J. and Shukla, J. 1991. Amazonian deforestation and regional climate change.

J. Climatol. **4**, 957-988.
Pearcy, R.W. 1987. Photosynthetic gas exchange resposes of Australian tropical forest trees in canopy, gap and understorey micro-environments. *Functional Ecol.* **1**, 169-178.
Reich P.B. and Borchert R. 1988. Changes with Leaf Age in Stomatal Function and Water Status of Several Tropical Tree Species. *Biotropica* **20**, 60-69.
Roberts, J., Cabral, O.M.R. and de Aguiar, L.F. 1990. Stomatal and boundary-layer conductances in an Amazonian terra firme rain forest. *J. Appl. Ecol.* **27**, 336-353.
Roberts, J., Cabral, O.M.R., Fisch, G., Molion, L.C.B., Moore, C.J. and Shuttleworth, W.J. 1993. Transpiration from an Amazonian rainforest calculated from stomatal conductance measurements. *Agric. For. Met.* **65**, 175-196.
Rowntree, P.R. 1991. Atmospheric parameterization schemes for evaporation over land: Basic concepts and climate modelling aspects. In: Schmugge and André, J.C. (Eds.) *Land Surface Evaporation*, Springer Verlag, New York, 5-29.
Schölander, P.F., Hammel, H.T., Hemmingsen, E.D. and Bradstreet, E.D. 1964. Hydrostatic pressure and osmotic potential in leaves of mangroves and some other plants. *Proceedings of the National Academy of Sciences of the USA*, **52**, 119-125.
Sellers, P.J., Shuttleworth, W.J., Dorman, J.L., Dalcher, A. and Roberts, J.M. 1989. Calibrating the simple biosphere model for Amazonian tropical forest using field and remote sensing data. Part 1: Average calibration with field data. *J. Appl. Met.* **28**, 727-759.
Skerman, P.J. and Riveros, F. 1990. *Tropical grasses*. FAO Plant Production and Protection Series, no. 23, Food and Agriculture Organization, Rome.
Stephens, G.R. and Waggoner P.E. 1970. Carbon Dioxide Exchange of a Tropical Rain Forest. Part I. *Bioscience* **20**, 1050-1053.
Whitehead, D., Okali, D.U.U. and Fasehun, F.E. 1981. Stomatal response to environmental variables in two tropical forest species during the dry season in Nigeria. *J.Appl. Ecol.* **18**, 571-587.
Wright, I.R., Gash, J.H.C., da Rocha, H.R., Shuttleworth, W.J., Nobre, C.A., Maitelli, G.T., Zamporini, C.A.G.P. and Carvalho, P.R.A. 1992. Dry season micrometeorology of central Amazonian ranchland. *Q. J. Roy. Met. Soc.* **118**, 1083-1099.

RESUMO

Medições de condutância estomática, fotossíntese líquida e potencial de água nas folhas foram realizadas em todo dossel de uma floresta de *terra firme* e na pastagem adjacente (*Bracharia brizantha*) durante três períodos, abarcando várias condições de umidade do solo. Na floresta, os maiores valores de condutância estomática foram observados próximo do topo do dossel de 33 metros de altura, e diminuindo na direção da base. Durante o dia, o máximo de g_l ocorreu no meio da manhã, diminuindo no decorrer do dia. Durante a estação úmida, a pastagem apresentou um maior g_{lmax}, A e ψ que as espécies do dossel na floresta; porém, com o aumento do estresse hídrico no solo, g_{lmax}, A e ψ foram todos significativamente reduzidos em *B. brizantha*, mas não na floresta. Nas espécies que compõem o dossel da floresta, em todas as estações, g_l teve correlação negativa com o déficit de umidade específica (D); nas espécies no interior do dossel, somente nas estações secas. Na floresta, não foram encontradas relações de g_l com Q_p, ou com ψ. Em *B. brizantha*, g_l teve boa correlação com Q_p e ψ em todas as estações, mas não com D. Em *B. brizantha*, g_l foi mais sensível a baixos valores de umidade no solo que A e, consequentemente, a eficiência do uso da água aumentou em solos secos.

16 An overview of the leaf area index and physiological measurements during ABRACOS

J.M. ROBERTS[1], O.M.R. CABRAL[2], J.P. da COSTA[3], A.-L.C. McWILLIAM[1] and T.D. de A. SÁ[4]
[1] *Institute of Hydrology, Wallingford, UK*
[2] *EMBRAPA-CPAA, Manaus, Brazil*
[3] *Universidade Federal do Pará, Belém, Brazil*
[4] *EMBRAPA-CPATU, Belém, Brazil*

INTRODUCTION

Evaporation is to an extent dependent on land surface features and albedo, and surface roughness and surface conductance are the principal vegetation characteristics influencing any interactions between the land surface and climate (Rowntree, 1991). The structure of vegetation and the amount and spatial distribution of its foliage are the essence of albedo and surface roughness, although changes in leaf physiology might contribute to leaf absorbance or posture changes which in turn might influence surface albedo. However, because surface conductance integrates both the extent of the canopy and the biological behaviour of the different foliage elements in that canopy, it constitutes the main physiological interaction between vegetation and climate.

Surface conductance is easily measured, using techniques which average over large spatial areas as well as over the whole height profile of the vegetation. To create realistic models of bulk surface conductance it is important that this type of measurement is complemented by more detailed studies which can allow the relative contributions of changes in leaf area to be separated from the effects of leaf gas exchange.

For this reason a comprehensive programme of physiological studies has been undertaken within the ABRACOS at project all three pairs of sites. Location details of the ABRACOS sites are given by Gash *et al.* (1996) and of the instrumentation for the measurements of surface climatology by Culf *et al.* (1996), micrometeorology by Wright *et al.* (1992, 1996) and soil moisture by Hodnett *et al.* (1996). This paper presents information on the leaf area index for the ABRACOS sites and overviews the information now available about leaf stomatal conductance.

Amazonian deforestation and climate. Edited by J.H.C. Gash, C.A. Nobre, J.M. Roberts and R.L. Victoria.

MATERIALS AND METHODS

VEGETATION

Full details of the trees surrounding the forest tower at the Reserva Ducke are given by Moore and Fisch (1986) and Roberts *et al.* (1990). Further details from another forest site in the Manaus area selected to represent that at the Reserva Ducke and used for destructive estimation of L^* are given by McWilliam *et al.* (1993). This last paper also gives information about species composition at Fazenda Dimona, the Manaus pasture site. Full details of the floristics of both the Marabá forest and pasture sites are given by Sá *et al.* (1996) while McWilliam *et al.* (1996) present details of the vegetation composition at the Ji-Paraná forest and pasture sites.

LEAF AREA INDEX – FOREST (L^*)

Manaus

For the forest near Manaus there are three separate estimates of L^*:

(1) L^* and its vertical distribution was measured using destructive sampling in a 400 m^2 plot in September and October 1990 (McWilliam *et al.*, 1993).

(2) Roberts *et al.* (1993) made litter leaf area measurements at the Reserva Ducke over a two-year period from 1983 to 1985 under the assumption that the average leaf life span was 12 months and that the annual cumulative litter leaf area equates to the canopy L^*. Over this period litter was collected at approximately two-weekly intervals in 20 trays with bases perforated for drainage (0.11 m^2). Area was measured with a leaf area machine (Li-3100, LICOR Inc, Lincoln, Nebraska, USA).

(3) Another approach adopted by Roberts *et al.* (1993) was to calculate the vertical distribution of leaf area from published information for the Manaus area (Klinge, 1973; Klinge *et al.*, 1975). Conversion of this published information of vertical distribution of leaf fresh weight to leaf area required assumptions about leaf fresh weight to dry weight ratios and the specific leaf area appropriate to canopy layers (Medina and Klinge, 1983).

Marabá

Litter collections have also been made at the Reserva Vale do Rio Doce, Marabá for three years beginning in August 1991. Eighteen litter collection trays (0.34 m^2) were used and litter was collected at two-weekly intervals. Total area of the foliage portions of the litter collected in each tray was measured with the LICOR Li-3100 leaf area machine. Total dry weight of leaf litter in each tray was also determined. A close relationship was found between litter leaf area and dry weight at Marabá in data pooled from all litter trays for all the bi-weekly samples over the period when these data are available (August 1991 to January 1993). The regression equation

$$\text{area (m}^2) = -1.58 + 8.49\ \text{weight (kg)} \qquad (r^2 = 0.977)$$

can serve as a good estimator of litter leaf area from leaf litter weight only. In addition to total leaf area and dry weight in each tray since December 1992 individual leaves have had their area and dry weight determined to derive specific leaf area, σ_f (m^2 kg^{-1}). In June and December 1992 samples of healthy fully expanded foliage for each species accessible from the towers at Marabá were selected. The area of each leaf was measured with a leaf area machine and the oven dry weight obtained and σ_f was calculated for each leaf.

Ji-Paraná

Litter collections were made for 12 months beginning in July 1992 using similar trays to those at Marabá and also collecting at two weekly intervals. Area was measured using a leaf area measurement system (Delta-T Devices, Cambridge, UK).

LEAF AREA INDEX – PASTURE

Sampling was done at Fazenda Dimona, Manaus in 1990, and at Fazenda Nossa Senhora, Ji-Paraná and Fazenda Boa Sorte, Marabá in both 1992 and 1993. On each sampling occasion twelve plots of grass and shrubs (above-ground parts only) were sampled randomly in the clearing areas using a 0.5 m^2 or 1 m^2 quadrat. The total fresh weight of the samples was measured, and a measured 10% sample taken. This subsample was divided into living and dead material and the living material further subdivided into stem and leaf material. Leaf area and biomass were measured as for the forest samples.

LEAF PHYSIOLOGY

Manaus – forest and pasture

Substantial information exists for leaf stomatal conductance (g_s) for the Reserva Ducke from physiological studies made during ARME by Roberts *et al.* (1990). Pasture leaf stomatal conductance (g_s) and net photosynthesis (A) were measured in July and August 1991 (ABRACOS Mission 2) at Fazenda Dimona using a portable infra-red gas analyser (Model LCA3, Analytical Development Company, Hoddesdon, Hertfordshire, UK). Measurements were made on both grass species (*Brachiaria humidicola* and *B. decumbens*) and also on three resprouting tree species growing in the pasture. Measurements were made every two hours beginning at 08.00 until 16.00 h on 25 randomly selected leaves of each pasture species. Measurements were made on the resprouts on separate days to the measurements on the *Brachiaria* species but in a period when weather conditions and patterns were similar from day to day. Twenty-five measurements (five leaves from five individuals) were made on fully expanded, fully exposed leaves every two hours from 08.00 until 16.00 h.

Marabá – forest and pasture

Full details of the measurements of stomatal conductance (g_s) made in the Reserva Vale do Rio Doce and Fazenda Boa Sorte are given by Sá *et al.* (1996).

Ji-Paraná – forest and pasture

McWilliam *et al.* (1996) have fully described the measurements of stomatal conductance (g_s) and leaf net photosynthesis (A) made at the Reserva Jaru and Fazenda Nossa Senhora sites.

RESULTS

LEAF AREA INDEX – FOREST

Table 1 summarises the various estimates of leaf area index for the three ABRACOS sites. Although various methods have been used to measure and estimate leaf area index and a number of assumptions have been made, there is overall consistency in the estimates for each site. Figure 1 shows the cumulative litter leaf area index for three years at Marabá and indicates the consistency in amounts falling each year with slight increases of fall in the dry season around August and September. The estimate for forest at Manaus derived by Roberts *et al.*, (1993) from data published by Klinge (1973) and Klinge *et al.*, (1975) is higher than the other two estimates, particularly that of McWilliam *et al.*, (1993). It may be relevant that these last authors sampled in September and October 1990, the dry season at Manaus, when the canopy quantity should be least. The use of leaf fresh weight to derive estimates of leaf area should be used with caution because of the potential differences in fresh:dry weight ratio depending on the water content at the time of sampling. The fresh weights of foliage given by McWilliam *et al.* (1993) for Manaus in central Amazonia were equivalent to a L^* of 5.7, which was 9% greater than the 12.9 t ha^{-1} measured by Higuchi *et al.* (1994) in Southern Pará, also close to Marabá. Using these values it can be estimated

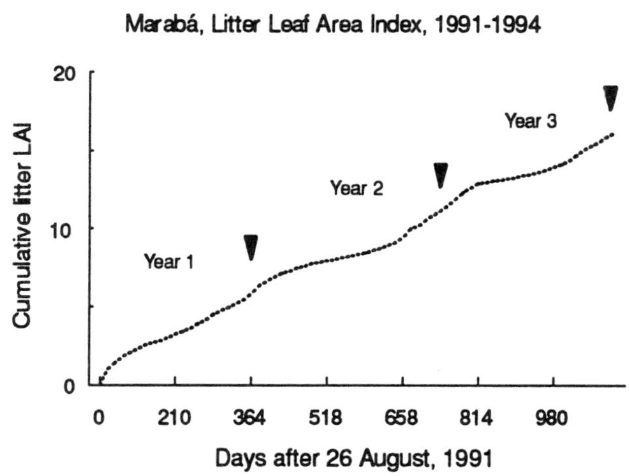

Figure 1 Cumulative leaf litter area index for three years from August 1991 at the Reserva Vale do Rio Doce, Marabá.

Table 1 Leaf area index (L^*) of ABRACOS forest sites

Site	Ji-Paraná	Manaus	Marabá
Method			
Litter area[1]	4.63	6.1	5.38±0.43
Destructive sampling	-	5.7±0.56 [2]	-
Literature	-	6.61[3]	5.3[4]
Optical method	4.4 [5]		

[1] This study
[2] McWilliam *et al.*, 1993.
[3] Roberts *et al.*, 1993, Klinge, 1973. Klinge *et al.*, 1975.
[4] Higuchi *et al.*, 1994.
[5] J. Grace (pers. comm., 1994).

that the L^* for the plot of Higuchi *et al.* was 5.3 — identical to the average of the values estimated from litter leaf areas for the ABRACOS Marabá site. At Ji-Paraná there was excellent agreement between the estimate from litter fall and the gap fraction method used by Grace (Table 1). Overall, these estimates indicate a difference between sites with the Reserva Ducke at Manaus having around 35% more foliage than the forest at Ji-Paraná and about 15 per cent more than at the Marabá forest site.

There is much less information about the vertical distribution of L^* for the ABRACOS sites with the only direct measurements being those described by McWilliam *et al.* (1993). Figure 2 shows good correspondence between the downward cumulative values determined by McWilliam *et al.* compared with estimates made by Roberts *et al.* (1993) from published values of Klinge (1973) and Klinge *et al.*, (1975). The vertical distribution of the foliage is important in multi-layer canopy transpiration modelling and we return to it later in discussions of effective leaf area index.

Figure 3 shows specific leaf area σ_f of living foliage taken from around the tower at Marabá and values obtained from the destructive sampling at Manaus by McWilliam *et al.*, 1993. There is a decline in σ_f from values of between 4 and 6 at the tops of the canopies to values exceeding 12 at the canopy bases. Figure 4 shows that the bulk of litter is in the same range of categories. The range of specific leaf areas for living leaves and the majority of litter leaves correspond to the range (8-27 $m^2 kg^{-1}$) which has been cited by Medina (1986) for studies in Malaysia and Central America. This also shows that the frequency distribution of litter foliage over the range of σ_f is very different and possibly offers a means of allocating the estimated total L^* to arbitrary canopy layers. Clearly, analysis of the σ_f of litter offers a way of identifying the source of leaf fall in terms of canopy position and thus the temporal variation in

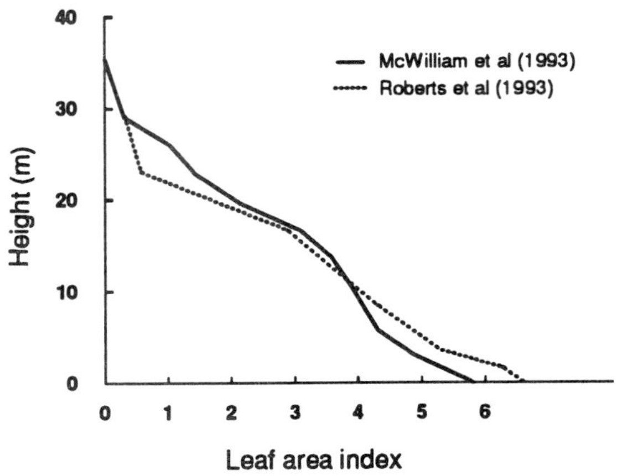

Figure 2 Cumulative leaf area index derived from literature values by Roberts *et al.*, 1993 (- - -) and determined destructively by McWilliam *et al.*, 1993 (——).

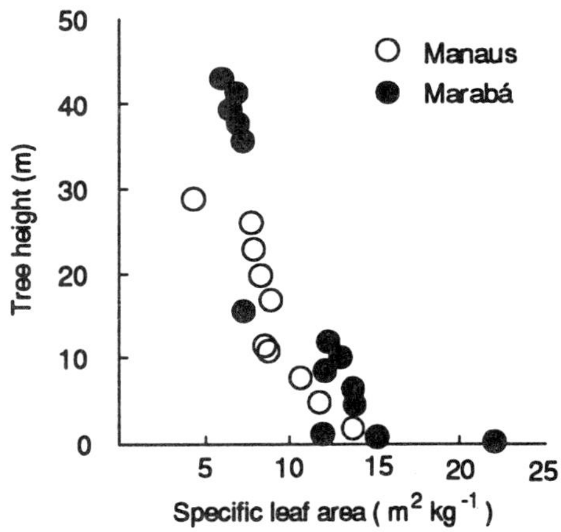

Figure 3 The relationship between specific leaf area (living leaves) with height in the canopy for the forests at Reserva Ducke, Manaus and Reserva Vale do Rio Doce, Marabá.

L^* at different heights. However there was no significant difference between months in the mean value of σ_f of the litter foliage, implying that leaf fall is more or less equally distributed over the canopy each month.

LEAF AREA INDEX – PASTURE

Table 2 gives the mean L^* with standard deviations for all the pasture samples taken during the ABRACOS experiment, only limited sampling was done at Fazenda

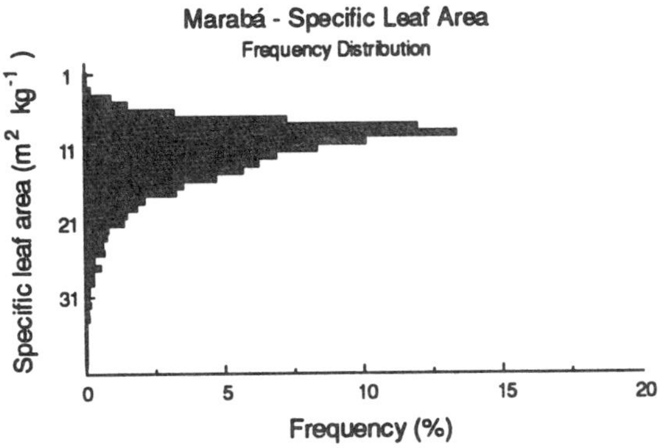

Figure 4 Frequency distribution of specific leaf area in the leaf litter fraction measured over 22 months at the Reserva Vale do Rio Doce, Marabá.

Dimona, Manaus. The samples relating to the mean values for Manaus were taken in the dry season of September and October 1990 and the transition period between the wet and dry season in 1991. It is expected that substantially higher values would have been observed in the wet season. During the wet season at Ji-Paraná L^* was close to 4 but fell substantially in the dry season to below 2. In the dry season at Marabá, however, there is a much more dramatic fall-off in leaf cover and the values for September and December 1992 show that this situation can persist for many weeks. Clearly there is a very wide scope for surface conductance and transpiration to be strongly influenced by changes in pasture leaf area index over and above any changes caused by alteration in leaf stomatal conductance.

Table 2 Leaf area index (L^*) of ABRACOS pasture sites[1]

Date	Ji-Paraná	Manaus	Marabá
September 1990	-	1.22 ± 0.61	-
July 1991	-	1.86 ± 0.99	-
January 1992	-	-	1.42 ± 0.62
June 1992	-	-	0.89 ± 0.47
August 1992	1.55 ± 0.72	-	-
September 1992	1.66 ± 0.57	-	0.49 ± 0.43
December 1992	-	-	0.57 ± 0.23
April 1993	3.90 ± 1.02	-	1.64 ± 0.6
June 1993	2.55 ± 0.78	-	0.71 ± 0.35
October 1993	-	-	0.88 ± 0.5

[1] including green stem area

STOMATAL CONDUCTANCE – FOREST

Figure 5 shows the diurnal trends of mean stomatal conductance (g_s) of species around the forest tower at the Reserva Ducke, Manaus. There are a number of features which are similar to the responses shown for species accessible from the forest towers at Ji-Paraná and Marabá by McWilliam *et al.* (1996) and Sá *et al.* (1996) respectively. The species with the highest g_s are those in the uppermost canopy layers and g_s falls systematically to the lowest values which are from ground layer species. The highest values of the upper and middle canopy species occur at mid-morning and fall off steadily throughout the day. This diurnal pattern was also reflected in the pattern of surface conductance reported for the forest at the Reserva Ducke by Shuttleworth (1988) as well as being observed in the studies of g_s at Ji-Paraná and Marabá (McWilliam *et al.*, 1996; Sá *et al.*, 1996). The steady decline throughout the day from peak values at mid-morning is associated with increases in atmospheric specific humidity deficit, *D*. A significant negative relationship was found by Roberts *et al.* (1990) for four species from the upper and middle canopy and similar relationships have been shown for the Reserva Jaru forest, Ji-Paraná (McWilliam *et al.*, 1996) and the Reserva Vale do Rio Doce, Marabá (Sá *et al.*, 1996). Figure 6 shows a comparison of the relationship between g_s and *D* for an upper and lower canopy species representative of each of the ABRACOS forest sites. There is similarity in the two categories and a larger change in g_s for a change in *D* is observed in the upper canopy species.

There is very little evidence that declining soil moisture content influences g_s substantially at any of the ABRACOS forest sites. At the Reserva Ducke, Roberts

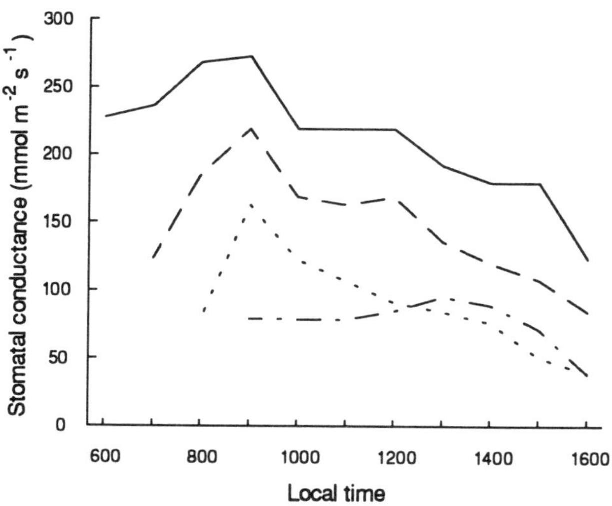

Figure 5 The diurnal variation of stomatal conductance of four species from around the sampling tower, Reserva Ducke, Manaus. *Piptadenia suaveolens*, 25.6 m (——); *Licania micrantha*, 25.6m (— —); *Naucleopsis glabra*, 17m (- -) and a palm, *Scheelea sp*, 0-3m (. - . - . -).

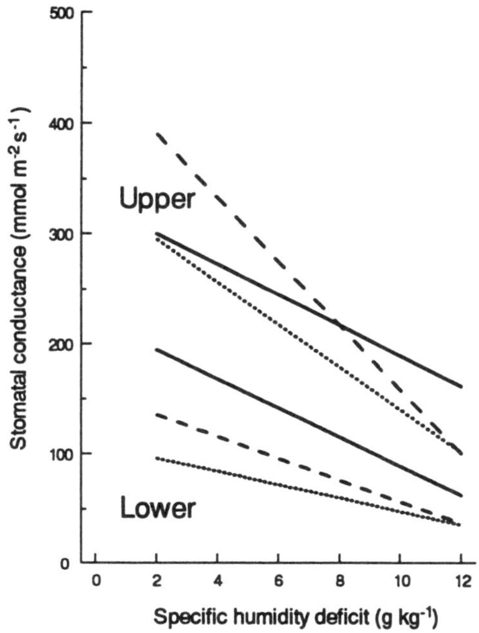

Figure 6 The relationship between stomatal conductance and air specific humidity deficit for an upper and a lower canopy species from each of the forests at Reserva Ducke, Manaus (——); Reserva Jaru, Ji-Paraná (.....) and Reserva Vale do Rio Doce, Marabá (- - -).

et al., (1990) showed that seasonal differences in the relationship between g_s and D changed such that higher intercepts and slopes were apparent in the wet season and that this response was most marked when solar radiation was highest. However observations made in the dry season also coincided with the period, particularly in the upper canopy, when leaves were senescing and later replaced. This might have been a contributing factor to the behaviour of the leaf conductance.

STOMATAL CONDUCTANCE – PASTURE

McWilliam *et al.* (1996) and Sá *et al.* (1996) have presented diurnal and seasonal patterns of g_s for pasture species at Fazenda Nossa Senhora, Ji-Paraná and Fazenda Boa Sorte, Marabá respectively. At Fazenda Dimona, Manaus, limited studies were made on the two predominant pasture species *Brachiaria humidicola* and *B. decumbens* as well as shrubs in July and August 1991. Over this period substantial soil water abstraction had occurred which, coupled with limited rainfall, led to a substantial soil water deficit (Hodnett *et al., 1996).* The midday values of g_s for *Brachiaria decumbens* were particularly high in the July measurement period, reaching over 1 mol m^{-2} s^{-1}. The diurnal pattern of g_s was more symmetrical around

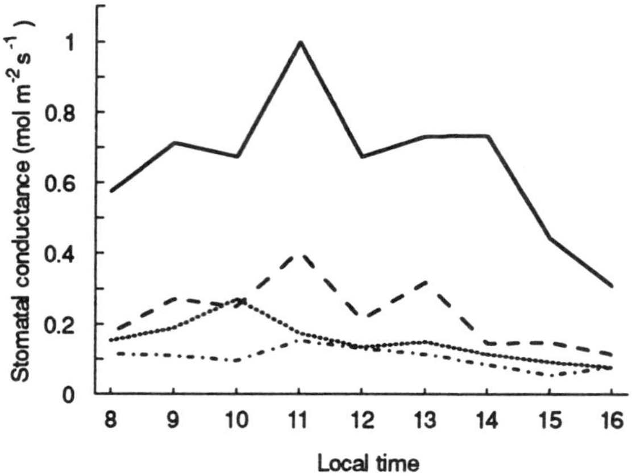

Figure 7 Diurnal averages of stomatal conductance for *Brachiaria decumbens* (——) and *B. humidicola* (— —) in July (- - -) and August (. _ . _ . _) 1991 in pasture at Fazenda Dimona, Manaus.

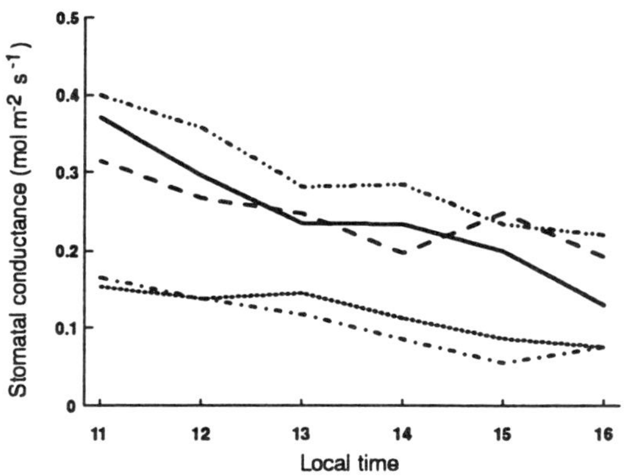

Figure 8 Diurnal changes in stomatal conductance of three shrub species, *Melastoma species 1*, (— —); *Melastoma species 2* (.. _ .. _ .. _); *Lacre sp.*, (————); and two grass species *Brachiaria decumbens* (. . . . . .) and *B. humidicola* (. _ . _ . _) in the pasture at Fazenda Dimona, Manaus.

noon in comparison with forest conductances, implying that radiation has a greater role to play than for forest g_s. By August there had been a substantial fall in g_s in both *Brachiaria decumbens* and *B. humidicola*. While there is consistency in the data for g_s from the three forest sites, with respect to magnitude and responses for the pasture sites we see different responses, particularly with respect to seasonal changes in soil water availability. At two of the sites — Fazenda Dimona, and Fazenda Nossa

Senhora—g_s falls in response to declining soil water but this is not so clearly the case at Fazenda Boa Sorte. Figure 7 shows the response of the two pasture species *Brachiaria humidicola* and *B. decumbens* to lowering of soil water storage between July and August 1991 at Fazenda Dimona. At the same time it is valuable to see the comparable values for woody shrubs which remain less affected by changes in soil water content, presumably because of their deeper roots. Figure 8 shows that the g_s of the three shrub species are almost double that of the *Brachiaria* species in August 1991.

Although maximum values are broadly similar, the behaviour of g_s for *Panicum maximum* at the Marabá pasture (Fazenda Boa Sorte) in response to changes in soil moisture content is different from the *Brachiaria* species at the other two clearing sites at Manaus and Ji-Paraná. Maximum g_s values for *P. maximum* occur when soil moisture is least and this relationship is interpreted by regarding soil moisture as having very little influence on conductance, whereas the high irradiance associated with dry periods does influence the level of g_s.

DISCUSSION

Estimates of leaf area index (L^*) for the ABRACOS forest sites ranges from 4.4 at Ji-Paraná to above 6 at Manaus with Marabá intermediate between these (5.4). Medina and Klinge (1983) have reported L^* of 5.2 and 5.1 for two forests in Venezuela. However information collated by Cannell (1982) shows that the values from Venezuela and those reported here for three sites in Brazil are substantially lower than those reported for example for rainforest sites in Panama, Thailand and Malaysia. Presumably the nutrient deficient soils at the Venezuela and Brazil sites are responsible for these differences. The use of annual fall of litter for estimating L^* assumes that the average leaf life span is 12 months. Medina and Klinge (1983) believe this to be reasonable but expect that the average leaf life span decreases with increasing dryness of the site. The two sites with the lower values of L^* are both drier than Manaus, but the L^* values determined from annual leaf litter area are supported by other independent estimates. Reich *et al.* (1991) also suggest that the average leaf life span in tropical forests is around 12 months but clearly the questions of the longevity and phenology of leaves in tropical forests should be priorities for future studies.

Although we have observed differences in L^* between the ABRACOS forest sites this does not necessarily mean that there will be site differences in forest transpiration on this basis alone. Roberts *et al.* (1993) have developed a multi-layer transpiration model, CLATTER, which provides a useful framework in which to examine the contribution of different canopy layers to total forest transpiration. Figure 9 shows, as a percentage of total for the forest at Manaus, L^* in each canopy layer and the transpiration from that layer. It is clear that the lower canopy layers contribute far less effectively to total transpiration in comparison with layers higher in the canopy. There are a number of reasons for this. In Figure 5 we have already shown a large

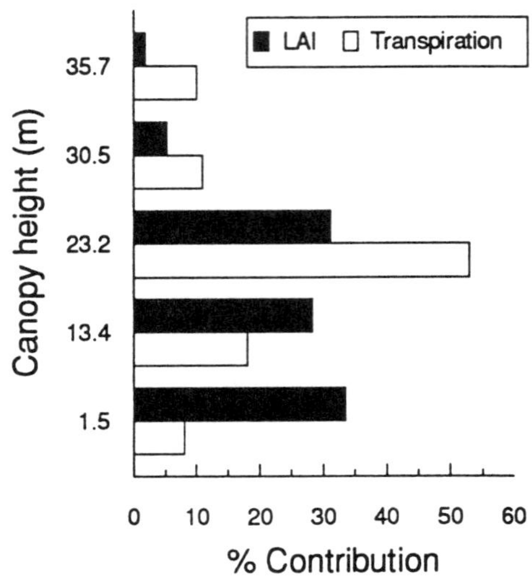

Figure 9 Percentage of total leaf area index and total transpiration calculated from the CLATTER model for the five layers of the forest canopy, Reserva Ducke, Manaus, Amazonas.

reduction in g_s towards the base of the forest canopy. Roberts *et al.* (1990) have shown a parallel fall in leaf boundary-layer conductance, while atmospheric humidity deficit and temperature are also lower at the canopy base. Available energy would also decline to very small levels at the bottom of the canopy. In less dense canopies, such as at Reserva Jaru, Ji-Paraná, depending on the particular canopy arrangement and the consequent modification of conductances and microclimate, the net effect on transpiration of a canopy with a substantially lower L^* may be far less as a percentage than the difference in L^* might suggest. Kelliher *et al.* (1995) arrive at a similar conclusion and also point out that where a substantially lower L^* is the case the contribution of soil evaporation might become important as a compensation for reduced transpiration. However in the case of the forests we have studied in Amazonia soil evaporation does not make a substantial contribution to vapour fluxes from the forests.

A systematic increase in specific leaf area (σ_f) has been shown in the living foliage from the top to the bottom of the rainforest canopies at Manaus and Marabá. Currently we do not know by how much σ_f changes as a leaf senesces and falls as litter and if changes in σ_f are greater in some parts of the canopy than in others. Nevertheless we have shown a wide span of σ_f in the litter fraction at Marabá which parallels that in the living foliage. The systematic variation of σ_f with canopy position has been demonstrated in a range of forest types, e.g. chestnut coppice, UK (Ford and Newbould, 1971); Sitka spruce plantation, UK (Norman and Jarvis, 1974); mixed hardwood forest, USA (Hutchison *et al.*, 1986) and temperate *Nothofagus* rainforest, New Zealand (Hollinger, 1989). There is less comprehensive information available from rainforests but the trends are the same. Data for rainforests have been presented by Oberbauer (1987), Medina (1986), Medina and Klinge (1983), Myers *et al.* (1987) and Yoda (1974). Because it has been shown to be closely correlated with several aspects of leaf physiology the systematic variation in σ_f hrough the canopy space has considerable potential as a simple means of categorizing the likely activity of a fraction of canopy. This might even be done from leaf litter. Development of methods to more readily and accurately scale physiological activity from leaves to canopies will remain a key research area (Jarvis, 1995). We believe that the vertical variation of σ_f and its persistence as a signature in forest litter has an important role to play in such scaling exercises. This study has not provided information about the seasonal fluctuations in L^* at the ABRACOS forest sites but we assume the seasonal reduction from the maximum value is around 10 to 15 per cent. Measurements of photon flux density, solar or net radiation at the forest floor and above the canopy has been made at all three forest sites. No situation has occurred where substantial changes in the receipt of radiation at the forest floor has increased significantly, implying that L^* fluctuations are rather small.

The situation is different at the three ABRACOS pasture sites. L^* has not been measured at all three pasture sites in the wet season but we consider the value of 3.79 at Fazenda Nossa Senhora, Ji-Paraná, at the end of the wet season likely to be typical. However, the fall in L^* during the dry season, which differs at all three pasture sites, constitutes a substantial fraction of the maximum L^* and has a considerable influence on the seasonal change in surface conductance of the pasture sites.

The maximum g_s values observed at the three ABRACOS forest sites are quite similar and all sites show similar relationships with g_s falling in association with increasing atmospheric humidity deficit. There is also a very consistent pattern in behaviour through the canopy space, with the highest values at the top and a systematic decline towards the ground. There are now a great number of published values of g_s for different species in tropical forests around the world and many of these have been collated in Table 3. Mostly the values fall into a broad range which covers the values measured at the ABRACOS forest sites. Only the values published by Whitehead *et al.* (1981) and Grace *et al.* (1982) for *Tectona* and *Gmelina* are especially high. Most leaf photosynthesis values reported for tropical forest trees are quite moderate. Only a limited number of studies of leaf water potential have been made and it is notable that values as low as –4.0 MPa can be sustained at the top of some of the rainforest in Brazil. Although some types of tropical rain forest have been relatively well studied when Bruijnzeel and Proctor (1995) extensively reviewed hydrological and biogeochemical data from Tropical Cloud Montane Forests and they drew attention to the very limited amount of physiological data for this vegetation type.

Many published studies in Table 3 can be used to illustrate a negative correlation between g_s and atmospheric humidity deficit. This link is very relevant to modelling stomatal behaviour and is strongly involved in maintaining rainforest transpiration at moderate levels. A feature of our studies in Brazil (Roberts *et al.*, 1990; McWilliam *et al.*, 1996 and Sá *et al.*, 1996) has been that species with initially the highest values

Table 3 Published values of stomatal conductance (g_s), net photosynthesis (A) and leaf water potential (ϕ) with authors

Country/Species	g_s (mol m^{-2} s^{-1})	A (μmol m^{-2} s^{-1})	ϕ (MPa)	*Reference*
AUSTRALIA				
Agyrodendron paraltum	0.170	9.4	-	Pearcy (1987)
BRAZIL				
Piptadenia suaveolens	0.290	-	-3.80	Roberts *et al.* (1990)
Licania micrantha	0.220	-	-	"
Bocoa viridiflora	0.185	-	-2.60	"
Naucleopsis glabra	0.160	-	-	"
Cedrella odorata	0.552	12.6	-2.64	McWilliam *et al.* (1996)
Inga spp.	0.556	12.9	-2.41	"
Bertholettia excelsa	0.400			Sá *et al.* (1996)
CAMEROON				
Sacoglottis gabonensis	0.320	11.0	-	Koch *et al.* (1994)
Dialium pachyphyllum	0.32	10.0	-	"
COSTA RICA				
Goethalsia meiantha		6.5		Stephens and Waggoner (1985)
Pentaclethra macroloba	0.160			Oberbauer *et al.*(1987)
JAMAICA				
Clethra occidentalis	0.280			Aylett (1985)
Cyrilla racemosa	0.300			"
Hedyosmum arborescens	0.104			"
MALAYSIA				
various species		12.6-18.9		Koyama (1978)
MEXICO				
Jacquinia pungens	0.188			Fanjul and Barradas (1985)
Coccoloba liebmanni	0.236			"
Piper auritum	0.700			Chiarello (1984)
Piper hispidum	1.000			Mooney *et al.* (1984)
NIGERIA				
Wet season				
Tectona grandis	1.2		-1.0	Grace *et al.* (1982)
Gmelina arborea	1.0		-1.0	"
Dry season				
Tectona grandis	1.2		-1.2	Whitehead *et al.* (1981)
Gmelina arborea	0.6		-2.0	"

Table 3 continued

PANAMA		
Anacardium excelsum	0.300	Meinzer *et al.* (1993)
Trichilia cipo	0.390	Fetcher (1979)
Cordia alliodora	0.260	"
THAILAND		
Hopea ferrea	0.420	Pitman (1996)
VENEZUELA		
Podocarpus rospigliossi	0.135	Meinzer *et al.* (1984)
" *oleifolius*	0.155	"

of g_s have shown the greatest degree of reduction in g_s in association with changes in D. This type of response has been referred to earlier by Morison (1987) and will also contribute to reducing differences between species at levels of D common during the main daylight period. Unfortunately many of the studies listed in Table 3 deal only with individual species and the extent of the existence of responses referred to above cannot be judged. Also, many of the studies have been made on seedlings or saplings which may account for effects of soil moisture on g_s which have been observed in some cases. In our studies, however, the influence of soil moisture on g_s is only observed in a few individual species. We regard the influence of changes in soil water content in controlling leaf conductance as secondary to the controls associated with irradiance and atmospheric humidity deficit. Nepstad *et al.* (1994) have reported the considerable rooting depths of rainforests in the Amazon region. These deep roots would ensure an adequate soil water supply even in years with protracted dry seasons. There is a need to extend the type of studies we have reported to a wider range of primary rainforest types and involving measurements through the whole canopy profile.

Studies made at the three ABRACOS rainforest sites show substantial similarity in behaviour of different species at comparable positions in the canopies. In contrast the behaviour of species in the pasture do differ. *Brachiaria brizantha* at Fazenda Nossa Senhora, Ji-Paraná and *B. humidicola* and *B. decumbens* at Fazenda Dimona, Manaus all have very high g_s when soil water is not limiting but this falls sharply when soil moisture supply becomes limited. None of these species seems to have a response in which g_s is related to atmospheric humidity deficit. The behaviour of *Panicum maximum* at the pasture site at Marabá (Fazenda Boa Sorte) is quite different. The maximum observed g_s value (Sá *et al.*, 1996) is substantially lower than the *Brachiaria* species at the other two sites. However in *Panicum maximum* changes in g_s are more associated with changes in D than with fluctuations in soil water content. At the Marabá pasture the change in L^* is the greatest of the three pasture sites and there reductions in surface conductance and transpiration would be particularly influenced by the reduction in L^*. There is clear evidence from both the Manaus and Marabá pastures that shrubs which invade the pastures have g_s which are maintained during periods when soil water deficits develop demonstrating that the physiology and micrometeorology of vegetation during secondary forest development deserves much more detailed study. In parts of eastern Amazonia particularly, secondary forest functions as a fallow vegetation in cropping systems which are in many cases sustainable. Understanding the nutrient and biomass accumulation and the associated hydrological processes has considerable applied value. Additionally, understanding the particular stresses endured by individual shrubs invading abandoned pastures is relevant to practical problems of secondary forest establishment and other enterprises, e.g. agroforestry, which require tree species to establish on abandoned cleared areas in Amazonia. It is paradoxical that although tropical rainforest is arguably the most difficult vegetation to investigate we are probably in a better position to predict the behaviour of many of the forests in Amazonia than we are of the types of pastures that replace them.

It is important to note that all the ABRACOS pasture and forest sites are located on relatively flat areas representative of plateau sites. Clearly there will be occasions when the responses of vegetation on different slope sequences will be different to those that we have observed in ABRACOS. The most likely difference would be expected in pasture growing in valley bottoms where the substantial responses to soil moisture deficits we have shown in both g_s and L^* are unlikely to occur. It is also important to be aware that the forest physiology measurements are made on species in long-established situations. The behaviour patterns for species invading gaps in the forest canopy may be different.

CONCLUSIONS

(1) Differences in L^* were observed for all three ABRACOS forest sites which were confirmed by a number of methods. The Reserva Ducke, Manaus had the highest value 6.1, Reserva Vale do Rio Doce was 5.4 and the Reserva Jaru was lowest with a value around 4.4. Seasonal fluctuations on these values are thought to be small.

(2) Very large seasonal fluctuations were observed at the ABRACOS pasture sites with the lowest value of around 0.5 at Fazenda Boa Sorte, Marabá, being the lowest dry season value. The maximvalues of L^* in the pasture sites are thought to be between 3.5 to 4.

(3) Maximum values of g_s in the upper levels of the forest canopies at the three ABRACOS forest sites were around 350-400 mmol $m^{-2} s^{-1}$. Values declined down through the forest canopies and generally were at maximum values at mid-morning and falling for the rest of the day. In all three forests and for a majority of the species studied there were significant negative association of g_s with atmospheric humidity deficit (D).

(4) g_s in the three pasture sites, under non-limiting conditions, was substantially higher than the species at the top of the forest canopies, sometimes exceeding 1 mol m^{-2} s^{-1}. However g_s of the pasture grasses of *Brachiaria* in the Ji-Paraná and Manaus pastures responded sensitively to lowering of soil water availability. *Panicum maximum* growing in the pasture at Marabá had an initially lower g_s than the *Brachiaria* species but responded less to drying soil.

ACKNOWLEDGEMENTS

We would like to acknowledge the scientific support received from the Instituto Nacional de Pesquisas Espaciais, the Instituto Nacional de Pesquisas da Amazônia, INCRA (Ji-Paraná), IBAMA (Ji-Paraná), EMBRAPA-CPAA (Manaus), EMBRAPA-CPATU (Belém), the Universidade Federal de Rondônia, the Universidade Federal de Pará and the Companhia Vale do Rio Doce.

REFERENCES

Aylett, G.P. 1985. Irradiance interception, leaf conductance and photosynthesis in Jamaican upper montane rain forest trees. *Photosynthetica*, **19**, 323-327.

Bruijnzeel, J. and Proctor, J. 1995. Hydrology and biogeochemistry of Tropical Montane cloud Forests: What do we really know? In: *Tropical Montane Cloud Forests*. (Eds.: L.S. Hamilton, J.O. Juvik, F.S. Scatena, Springer, Berlin. 38-78.

Cannell, M.G.R. 1982. *World forest biomass and primary production data*. Academic Press, London. 391 pp.

Chiarello, N. 1984. Leaf energy balance in the wet lowland tropics. In: *Physiological ecology of plants in the wet tropics*, (Eds.) E. Medina, H.A. Mooney and C. Vasquez-Yanes, Junk, The Hague, 27-36.

Culf, A.D., Esteves, J.L., Marques, A. de O. and da Rocha, H.R. 1996. Radiation, temperature and humidity over forest and pasture in Amazonia. This volume.

Fanjul, L. and Barradas, V.L. 1985. Stomatal behaviour of two heliophile understorey species of a tropical deciduous forest in Mexico. *J. Appl. Ecol.*, **22**, 943-954.

Fetcher, N. 1979. Water relations of five tropical tree species on Barro Colorado Island, Panama. *Oecologia*, **40**, 229-233.

Ford, E.D. and Newbould, P.J. 1971. The leaf canopy of a coppiced deciduous woodland. 1. Development and structure. *J. Ecology*, **59**, 843-862.

Gash, J.H.C., Nobre, C.A., Roberts, J.M. and Victoria, R.L., 1995. An overview of ABRACOS. This volume.

Grace, J., Okali, D.U. and Fasehun, F.E. 1982. Stomatal conductance of two tropical trees during the wet season in Nigeria. *J. Appl. Ecol.*, **19**, 659-670.

Higuchi, N., Marden dos Santos, J., Imanaga, M. and Yoshida, S. 1994. Above-ground biomass estimate for Amazonian dense tropical moist forests. *Memoirs of the Faculty of Agriculture, Kagoshima University*, **30**, 43-54.

Hodnett, M.G., Oyama, M.D. and Tomasella, J. 1996. Comparisons of long-term soil water storage behaviour under pasture and forest in three areas of Amazonia. This volume.

Hollinger, D.Y. 1989. Canopy organization and foliage photosynthetic capacity in a broad-leaved evergreen montane forest. *Functional Ecology*, **3**, 53-62.

Hutchison, B.A., Matt D.R., McMillen, R.T., Gross, L.J., Tajchman, S.J. and Norman, J.M. 1986. The architecture of a deciduous forest in Eastern Tennessee, USA. *J. Ecol.*, **74**, 635-646.

Jarvis, P.G. 1995. Scaling processes and problems. *Plant, Cell and Environment*, **18**, 1079-1089.

Kelliher, F.M., Leuning, R., Raupach, M.R. and Schulze, E-D. 1995. Maximum conductances for evaporation from global vegetation types. *Agric. For. Meteorol.*, **73**, 1-16.

Klinge, H.1973. Struktur und artenreichtum des zentralamazonischen regenwaldes. *Amazoniana*, **4**, 283-292.

Klinge, H., Rodriques, W.A., Brunig, E.F. and Fittkau, E.J. 1975. Biomass and structure in a central Amazonian Rain Forest. In: *Tropical Ecological Systems. Trends in Terrestrial and Aquatic Research* (Eds.) F.B. Golley and E. Medina, Springer, Berlin. 115-122.

Koch, G.W., Amthor, J.S. and Goulden, M.L. 1994. Diurnal patterns of leaf photosynthesis, conductance and water potential at the top of a lowland rain forest canopy in Cameroon: measuremnt from the *Radeau des cimes. Tree Physiology*, **14**, 347-360.

Koyama, H. 1978. Photosynthesis studies in Pasoh forest. *Malay. Nat. J.*, **30**, 253-258.

McWilliam, A-L.C., Roberts, J.M., Cabral, O.M.R., Leitão, M.V.B.R., da Costa, A.C.L., Maitelli, G.T. and Zamparoni, C.A.G.P. 1993. Leaf area index and above-ground biomass of *terra firme* rainforest. *Functional Ecology*, **7**, 310-317.

McWilliam, A-L.C., Cabral, O.M.R., Gomes, B.M., Esteves, J.L. and Roberts, J.M. 1996. Forest and pasture leaf gas exchange in south-west Amazonia. This volume.

Medina, E. 1986. Forests, savannas and montane tropical environments. In: *Photosynthesis in contrasting environments* (Eds.) N.R. Baker and S.P. Long. Elsevier, Amsterdam, The Netherlands. 139-171.

Medina, E. and Klinge, H. 1983. Productivity of tropical forests and tropical woodlands. In: *Encyclopaedia of Plant Physiology Volume 12D. Physiological Plant Ecology IV. Ecosystem Processes: Mineral Cycling, Productivity, and Man's influence.* (Eds.) O.L. Lange, P.S. Nobel, C.B. Osmond and H. Ziegler, Springer, Berlin. 281-303.

Meinzer, F.C., Goldstein, G. and Jaimes, M. 1984. The effect of atmospheric humidity on stomatal control of gas exchange in two tropical coniferous species. *Can. J. Botany*, **62**, 591-595.

Meinzer, F.C., Goldstein, G., Holbrook, N.M., Jackson, P., and Cavalier, J. 1993. Stomatal and environmental control of transpiration in a lowland tropical forest tree.*Plant, Cell and Environment*, **16**, 429-436.

Mooney, H.A., Field, C., Vasques-Yanes, C. and Chu, C. 1983. Environmental controls on stomatal conductance in a shrub of the humid tropics. *Proc. Nat. Acad. Sciences*, **80**, 1295-1297.

Moore, C.J. and Fisch, G. 1986. Estimating heat storage in Amazonian tropical forest. *Agric. For. Meteorol.*, **38**, 147-169.

Morison, J.I.L. 1987. Intercellular CO_2 concentration and stomatal responses to CO_2. In: *Stomatal Function.* (Eds.) E. Zeiger, G.D. Farquhar and I.R. Cowan , Stanford University Press, Stanford, USA, 229-251.

Myers, B.J., Robichaux, R.H., Unwin, G.L. and Craig, I.E. 1987. Leaf water relations and anatomy of a tropical rainforest tree species vary with crown position. *Oecologia*, **74**, 81-85.

Nepstad, D.C., Carvalho, C.R. de, Davidson, E.A., Jipp, P.H., Lefebvre, P.A., Negreiros G.H., da Silva, E.D., Stone, T.A., Trumbore, S.E. and Vieira, S. 1994. The role of deep roots in the hydrological and carbon cycles of Amazonian forests and pastures. *Nature*, **372**, 666-669.

Norman, J.M. and Jarvis, P.G. 1974. Photosynthesis in Sitka Spruce (*Picea sitchensis* (Bong.) Carr.). III. Measurements of canopy structure and interception of radiation. *J. Appl. Ecol.*, **11**, 375-398.

Oberbauer, S.F., Strain, B.R. and Reichers, G.H. 1987. Field water relations of a wet tropical forest tree species, *Pentaclethra macroloba* (Mimosaceae). *Oecologia*, **71**, 369-374.

Pearcy, R.W. 1987. Photosynthetic gas exchange responses of Australia rain tropical forest trees in canopy gap and understory microenvironments. *Functional Ecology*, **1**, 169-178.

Pitman, J.I. 1996. Stomatal behaviour of the dipterocarp *Hopea ferrea* Lanessan in tropical dry evergreen forest, Thailand. *Journal of Applied Ecology*, (in press).

Reich, P.B., Uhl, C., Walters, M.B. and Ellsworth, D.S. 1991. Leaf life span as a determinant of leaf structure and function among 23 tree species in Amazonian forest communities. *Oecologia*, **86**, 16-24.

Roberts, J., Cabral, O.M.R. and de Aguiar, L.F. 1990. Stomatal and boundary-layer conductances in an Amazonian *terra firme* rainforest. *J. Appl. Ecol.*, **27**, 336-353.

Roberts, J.M., Cabral, O.M.R., Fisch, G., Molion, L.C.B., Moore, C.J. and Shuttleworth, W.J. 1993. Transpiration from an Amazonian rainforest calculated from stomatal conductance measurements. *Agric. For. Meteorol.*, **65**, 175-196.

Rowntree, P. R. 1991. Atmospheric parameterization schemes for evaporation over land: basic concepts and climate modelling aspects. In: Schmugge, T.J. and Andre, J.C. (Eds.), *Landsurface evaporation: measurement and parameterization*. Springer-Verlag, Berlin, 5-29.

Sá, T.D. de A., da Costa, J.P and Roberts, J.M. 1996. Forest and pasture conductances in southern Pará, Amazonia. This volume.

Shuttleworth, W.J. 1988. Evaporation from Amazonian rainforest. *Proc. R. Soc. Lond.* ***B*** **233**, 321-346.

Stephens, G.R. and Waggoner, P.E. 1970. Carbon dioxide exchange of a tropical rain forest. Part 1. *Bioscience*, , 1050-1053.

Whitehead, D., Okali, D.U.U. and Fasehun, F.E. 1981. Stomatal response to environmental variables in two tropical forest species during the dry season in Nigeria. *J. Appl. Ecol.*, **18**, 571-587.

Wright, I.R., Gash, J.H.C., da Rocha, H.R., Shuttleworth, W.J., Nobre, C.A., Maitelli, G.T., Tamponini, C.A.G.P. and Carrelho, P.R.A. 1992. Dry season micrometeorology of central Amazonian ranchland. *Q. J. R. Met. Soc.* **118**, 1083-1099.

Wright, I.R., Gash, J.H.C., da Rocha, H.R. and Roberts, J.M. 1996. Modelling surface conductance for Amazonian pasture and forest. This volume.

Yoda, K. 1974. Three-dimensional distribution of light intensity in a tropical rainforest of West Malaysia. *Japanese J. Ecol.* **24**, 247-254.

RESUMO

A magnitude e flutuação na condutância de superfície (g_c) reflete duas importantes propriedades da vegetação, a saber, índice de área foliar (L^*) e condutância estomatal (g_s). Este artigo apresenta informações a respeito de L^* para os sítios de floresta e pastagem do projeto ABRACOS, e fornece uma visão geral dos dados de g_s disponíveis, provenientes dos estudos do ARME e ABRACOS. Para as florestas, o valor de L^* foi maior em Manaus (5,7), intermediário em Marabá (5,3) e menor em Ji-Paraná (4,4). Para os sítios de pastagem, o valor de L^* foi menor que para os sítios de floresta, e também apresentou diferenças entre os sítios. Nos sítios de pastagem do ABRACOS, foram observadas grandes variações sazonais de L^*, as quais contribuíram substancialmente para a redução da condutância de superfície (g_c) na estação seca.

Enquanto há pouca evidência, em todos os sítios de floresta do ABRACOS, de que os valores de g_s da floresta são influenciados pela redução do conteúdo de água no solo, há evidência de que eles respondem à radiação e ao déficit de umidade do ar, e de que as respostas são semelhantes para diferentes florestas. A resposta de g_s à redução da umidade do solo ocorre, de fato, nos sítios de pastagem em Manaus e Ji-Paraná; mas com menor expressão na pastagem em Marabá, onde o maior decréscimo em L^* foi, provavelmente, o principal determinante na redução da condutância do dossel. Nos sítios de pastagem de Manaus e Ji-Paraná, tanto g_s quanto L^* diminuíram com a redução co conteúdo de água no solo. No sítio de pastagem de Marabá, g_s das gramíneas apresenta menor resposta à baixa disponibilidade de água no solo, e L^* decresce para os níveis mais baixos encontrados em todos os sítos do ABRACOS.

17 Carbon dioxide flux over Amazonian rain forest in Rondônia

J. GRACE[1], J. LLOYD[2], J. McINTYRE[1], A.C. MIRANDA[3], P. MEIR[1] and H.S. MIRANDA[3]

[1] Institute of Ecology and Resource Management, University of Edinburgh, UK;
[2] Institute of Advanced Studies, Australian National University, Canberra, Australia;
[3] Laboratorio de Ecologia, Universidade Nacional de Brasilia, Brasilia, Brazil.

INTRODUCTION

The role of tropical forests in the carbon and water balance of the planet has been the subject of much speculation. About 0.17 x 10^{12} m^2 of tropical forest is converted annually to less carbon-rich forms of land use such as agriculture, corresponding to a transfer to the atmosphere of 1.1-3.6 Pg of carbon per year (Houghton *et al.*, 1987; Houghton, 1991; FAO, 1993). Nevertheless most of the world's tropical forest is mature and undisturbed, and very little is known about its carbon balance. Undisturbed rain forests occupy 10×10^{12} m^2, or 7% of the land cover (FAO, 1993), of which by far the largest is in the Amazon basin.

It is widely believed that undisturbed ecosystems are in a 'steady state' with gains in plant carbon exactly balanced by losses due to death and decomposition. However, the environment to which vegetation is exposed has never been constant and currently the terrestrial biosphere may be undergoing fertilisation as a result of increasing concentrations of CO_2 coupled with higher deposition rates of nitrogen (Körner, 1993; Gifford, 1994). If this is the case, undisturbed tropical forest may be a large sink of CO_2 because of its large area. Here we review recent measurements of carbon dioxide and water vapour flux over tropical rain forest in Brazilian Amazonia, to test the hypothesis that such forest sequesters carbon from the atmosphere, and to examine the influence of climatological variables on these fluxes.

METHODS

Two field campaigns were conducted during the dry and wet seasons (September 1992, April-June 1993) at Reserva Jaru, Rondônia, Brazil. Details of this site are given in Gash *et al.*, 1996.

Hourly means of solar and net radiation, aspirated wet and dry bulb temperature, heat flux to the soil, wind speed and direction were measured using an automatic weather station (Didcot Instruments, Abingdon, UK) mounted 15 m above the 30 m-

Amazonian Deforestation and Climate. Edited by J.H.C. Gash, C.A. Nobre, J.M. Roberts and R.L. Victoria.

tall forest (Culf *et al.*, 1996).

Fluxes of carbon dioxide, water vapour and sensible heat were measured by eddy covariance systems mounted 15 m above the 30 m-tall forest (Grace *et al.*, 1995). The hardware of the eddy covariance system comprised an infra-red gas analyser (IRGA), a sonic anemometer, and a system for pumping air (Moncrieff *et al.*, 1995). The fast-responding closed-path IRGA was used to measure CO_2 and water vapour (Li 6262, Li-Cor, Lincoln, Nebraska, USA). Air from above the canopy (53 m above the ground, 20 m above the canopy) was drawn through 6 mm diameter tubing (Dekabon 1300, Deane & Co., Glasgow, UK) at a flow rate of 4 dm^3 min^{-1} into the IRGA. The calibration of the IRGA was checked every few days using a calibration cylinder prepared in our laboratory and checked against air streams made using precision gas-mixing pumps (Wosthoff OHG, Bochum, Germany). The vertical component of wind was measured at the same location using a 3-dimensional sonic anemometer (A1002R, Solent, Lymington, UK).

The software, EDDYSOL, first carries out co-ordinate rotation to obtain the true values of the vertical wind component, w. Then, corrections are made for air density effects caused by simultaneous transfer of heat and water vapour (Webb *et al*, 1980), although for the closed system it is assumed that temperature fluctuations have been removed by the time the air reaches the sample cell inside the IRGA. Finally, it calculates half-hour averages of fluxes in real time, storing the raw data (if required) and processed results on a laptop computer. Post-processing corrections were subsequently made for the non-ideal frequency response of the system, using transfer functions. The corrections are system-specific, as they depend on such things as the response time of the instruments and the path length of the sonic anemometer (Moncrieff *et al.*, 1995). They are largest at low wind speeds and under stable meteorological conditions. Corrections are usually less than 20%.

Stomatal and aerodynamic conductances were found as described in detail by Grace *et al.* (1995).

CO_2 profiles were sampled for isotopic analysis by pumping air at 1 dm^3 min^{-1} from 1, 16, 32 or 50 m above the ground using a vacuum pump, passing it through three ethanol/dry ice traps to remove water, then through three liquid nitrogen traps to remove CO_2. The resulting solid CO_2 was then redistilled into a liquid N_2-cooled glass tube which was sealed using a flame torch. About 100 μmol was obtained in each tube. Tubes were transferred to the laboratory and the isotopic composition was determined using a SIRA VG-24 mass spectrometer.

RESULTS

1993 DATA SET

The 1993 data were more or less continuous for 44 days, with only short interruptions for calibration and system maintenance. In this time, there were considerable variations in the weather, with four *friagem* periods (when cold air from the south

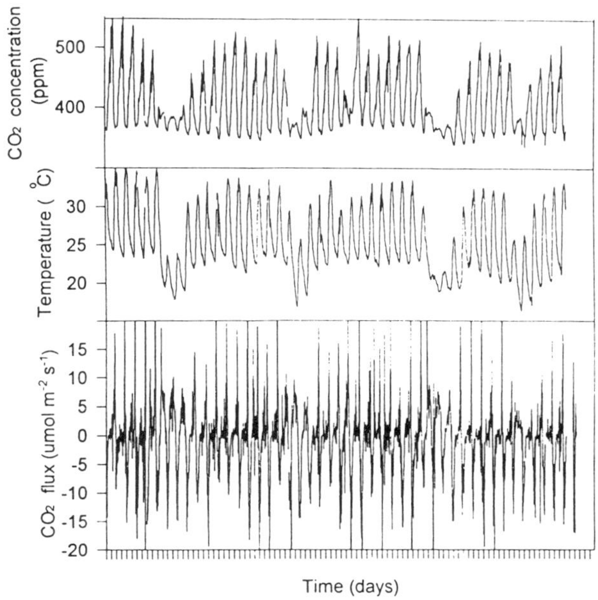

Figure 1 Raw data collected over a 44-day period beginning 9th May 1993. Upper panel: concentrations of CO_2; centre panel: temperature of the air as indicated by the sonic anemometer; lower panel: fluxes of CO_2, the sign convention is that a positive sign denotes flux from the vegetation to the atmosphere.

penetrates to this latitude) during which the temperature fell and the solar radiation was low (Figure 1). The CO_2 concentrations at 45 m were higher than expected, with a mean and standard deviation of 405 ± 48 ppmv (μmol CO_2 per mol air), and showed a well-defined diurnal cycle which was suppressed during the *friagem*. The minimum CO_2 concentrations were less than 350 ppmv. The CO_2 flux data are more variable than we have found in other vegetations using the same measuring system (see Miranda *et al.*, 1996). Frequently, there was little CO_2 leaving the canopy at night, and a large efflux in the early morning. This pattern led us to explore the storage of CO_2 inside the canopy-space.

CONCENTRATIONS IN THE CANOPY

The concentration of CO_2 built up from the ground upwards, starting in the late afternoon. In the early morning it frequently decreased very rapidly, as the wind began to ventilate the canopy. This ventilation of the canopy rarely occurred at night, because of atmospheric stability (Figure 2). Here, we show one especially dramatic occasion, when a storm began in the early part of the night. We saw a distinct accumulation of CO_2 in the lower stratum until 20.00 hours, then a thorough mixing brought about by violent winds. Measurements had to be abandoned for several hours but were resumed at 01.00 hours, when the process of accumulation was prevented

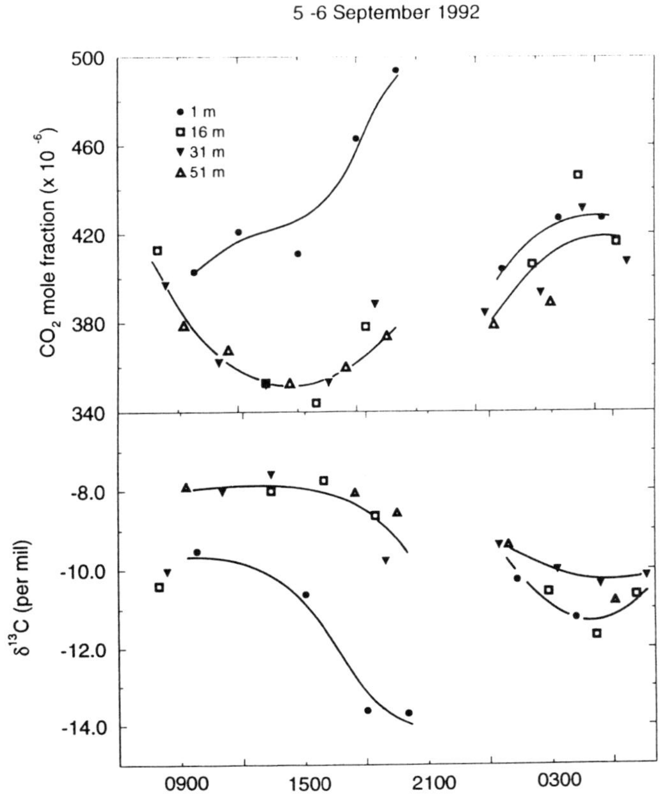

Figure 2 Concentrations of CO_2 and the δ³C signal over a 24 hour period during which a violent storm occurred (20.00 h to midnight). Symbols denote height above the canopy as follows: (●) 1 m, (□) 16 m, (▲) 32 m and (▽) 50 m.

by high winds occurring in the wake of the storm. A similar pattern is seen in the δC signal, values below -8 indicating a contribution of respiratory CO_2.

Another, more typical, example of in-canopy CO_2 concentrations is given by Kruijt *et al.*, 1996. It shows evidence of nocturnal mixing inside the canopy, even when meteorological conditions are stable.

STORAGE AND PHYSIOLOGICAL FLUXES

The fluxes measured above the canopy with the eddy covariance system are the result of two processes, viz. (a) physiological activity of the ecosystem (photosynthesis and respiration of leaves, and respiration of leaf, stems, wood and components of the soil) and (b) changes in storage of CO_2 inside the canopy-space, brought about by variations in the ventilation of the canopy, itself under meteorological control. The hour by hour changes in storage ΔC (μmol m^{-2} s^{-1}) can be estimated from the concentration profiles by first fitting curves to describe the shape of the profile, and then subtracting curves over a suitably short time interval. The relationships between these components are expressed as:

$$F_n = F_{eco} + \Delta C$$

where F_n is the flux measured by the eddy covariance system and F_{eco} is the flux caused by the physiology of the ecosystem. Examples of this calculation, over a three day period, are shown in Figure 3. On windy days and nights (Figure 4) F_{eco} does not

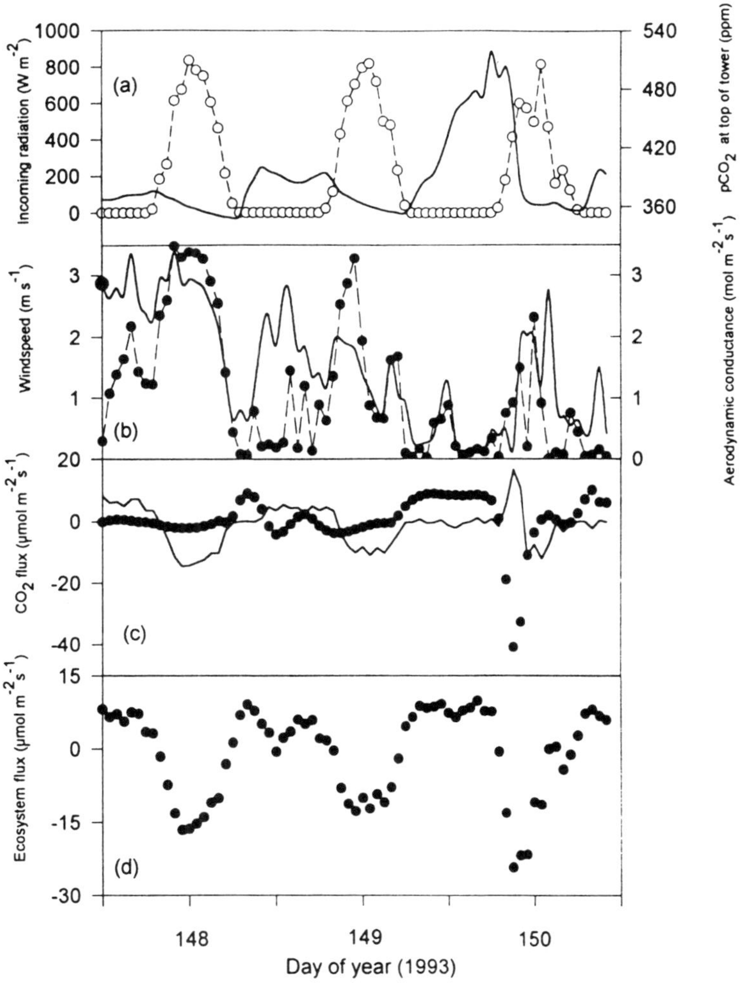

Figure 3 An example of the CO_2 fluxes, and their dependency on the environment. Panel (a) shows the incoming solar radiation (o) and the partial pressure of CO_2 at the top of the tower. Panel (b) shows the windspeed and aerodynamic conductance (•). Panel (c) shows the flux above the canopy as measured by the eddy covariance sensor (solid line) and the change in storage (•). Panel (d) shows the ecosystem flux obtained from the data shown in panel (c). Reproduced with permission of Blackwell Science Ltd. from Grace *et al.* (1995).

differ from F_n, but when wind speeds are low (commonly at night) the ΔX values were appreciable and large differences were seen between F_{eco} and F_n. On most days the calculated ecosystem flux was strongly correlated with the incoming solar radiation, as shown especially well on day 150 (Figure 3).

CONDUCTANCES

It is useful to examine contrasting days in detail (Figures 4a and b), noting the behaviour of the ecosystem stomatal conductance (found by inversion of the Penman-Monteith equation). Day 148 (1993) was just after a *friagem* period, characterised by high winds, low temperatures and relatively low leaf-air vapour pressure deficits

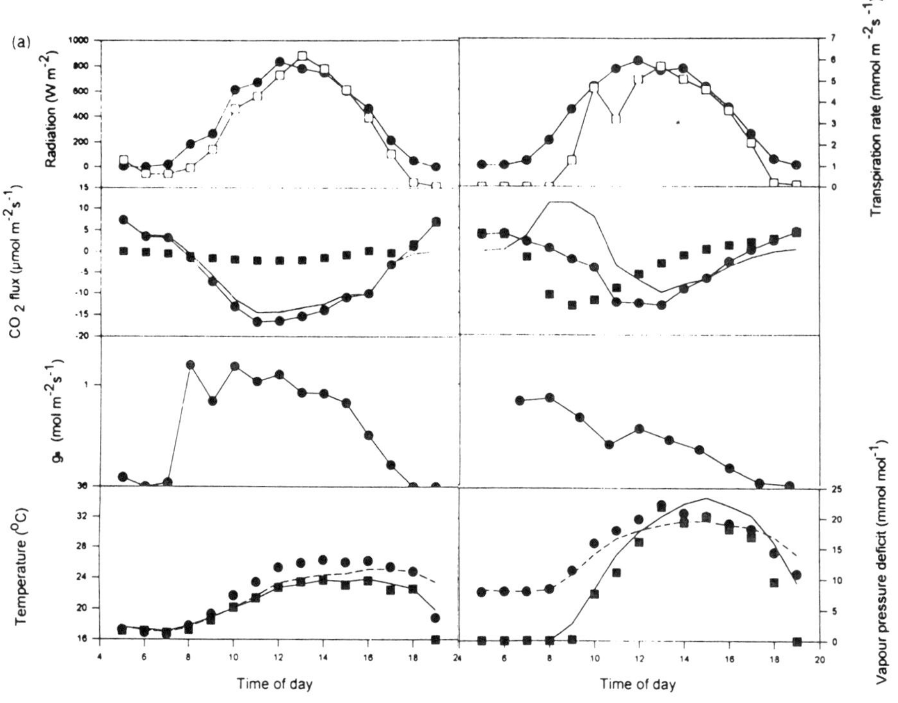

Figure 4 Fluxes on two contrasting days: (a) day 148, cool windy and bright, following a night of near-neutral stability and (b) day 154, warm with stable conditions at night. Symbols for the top panel: radiation (□), transpiration (•); symbols for the second panel are: flux above the canopy as measured with the eddy covariance sensor (line), storage flux (■), ecosystem flux (•); lower panel: leaf temperature (•), leaf-to-air vpd (■), air temperature (broken line), vpd (solid line). Reproduced with permission of Blackwell Science Ltd. from Grace *et al.* (1995).

(vpd). As a result of the high winds, the storage term was small. The stomatal conductances were high, as much as 1 mol m^{-2} s^{-1} , and remained so all morning, reducing appreciably at 15.00 hours. Ecosystem fluxes were strongly negative (uptake into the canopy), in this case reaching 17 µmol m^{-2} s^{-1} at 11.00 hours.

Day 153 was a more typical day, occurring after CO_2 had built up inside the canopy the night before, and the storage flux was large. Maximum rates of CO_2 influx into the canopy were lower (14 µmol m^{-2} s^{-1}), perhaps because of lower stomatal conductances. Stomatal conductances began at 0.8 mol m^{-2} s^{-1}, peaked at a very early stage in the morning, and declined as the day continued, possibly in response to the much higher VPD. This pattern of stomatal behaviour is also seen in porometer data from Reserva Ducke by Roberts *et al.* (1990). Absolute canopy conductances obtained in this study are somewhat higher than those reported at Ducke by Shuttleworth (1988) as are the leaf scale measurements (McWilliam *et al.*, 1996).

LIGHT RESPONSE CURVES

For the purposes of modelling carbon balance over a whole year, an elite data set was assembled, choosing only those hours when the wind direction was from the north (good fetch, and avoiding the river), conditions were perfectly dry (avoiding water on the leaves) and radiation fluxes were steady (avoiding the problem of a non-linear and dynamic response to fluctuating irradiance as clouds pass rapidly across the sun).

These data, when plotted against photon irradiance (PAR, µmol photons m^{-2} s^{-1}) provide a light response curve, from which a number of inferences regarding the physiological functioning of the forest may be drawn (Figure 5). The ecosystem 'dark respiration', obtained at PAR=0, is 6-7 µmol CO_2 m^{-2} s^{-1}. The rate at which carbon

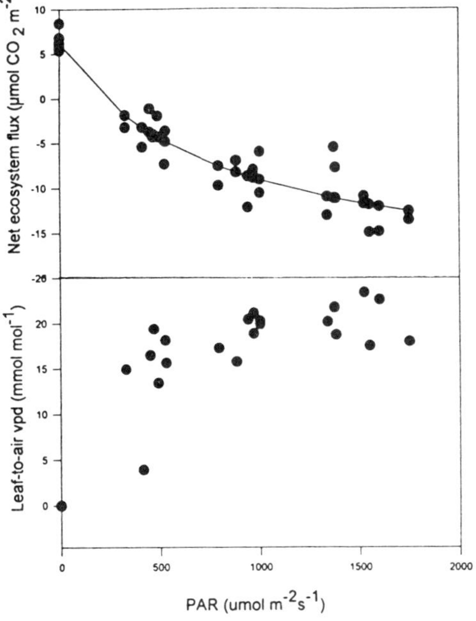

Figure 5 Ecosystem flux in relation to photosynthetically-active radiation (PAR), using only the elite data set.

assimilation increases with photon irradiance is an estimate of the efficiency of utilisation of incident radiation: the figure is 0.024 carbon atoms per photon, or, expressed as a quantum requirement, 40 photons are needed to capture one atom of carbon. The light compensation point (PAR at which the ecosystem is in a state of carbon balance), is 260 μmol photons m^{-2} s^{-1}. The maximum assimilation rate, estimated from this graph is 15 μmol CO_2 m^{-2} s^{-1}. This light response curve suggests an essential similarity between the gas exchange at this site, and that of a denser forest at Reserva Ducke (Fan *et al.*, 1990).

ACCUMULATION OF CARBON, OBSERVATION AND MODEL

The fluxes shown previously (Figure 1) may be integrated to explore the hypothesis that the forest accumulates carbon. In fact, both the dry- and wet-season data show accumulation of carbon except for the periods when the solar radiation was extremely low, and the average rate was 0.05 mol C m^{-2} d^{-1} (Figure 6).

The carbon assimilatory mechanism of the ecosystem was modelled, by combining existing models as follows to make a 'big-leaf' model with a minimal number of parameters. First the aerodynamic and bulk-stomatal conductances were calculated from above-canopy water and heat fluxes, as already mentioned, but allowing for the effects of atmospheric stability and incorporating the 'excess resistance' (Monteith & Unsworth, 1990). A combined stomatal conductance/biochemical model was then fitted to the corrected flux data (Lloyd *et al.* 1995) . This model used an empirical stomatal submodel in which stomatal conductance is a function of PAR and VPD (Lloyd, 1991) combined with well known chloroplast biochemical equations (Farquhar *et al.*, 1991), but incorporating physiological temperature sensitivities (Harley *et al.*, 1986). Internal conductances for diffusion within the leaf are included (Lloyd *et al.*, 1992). Respiration from soil, trunks and branches was specified in the model from

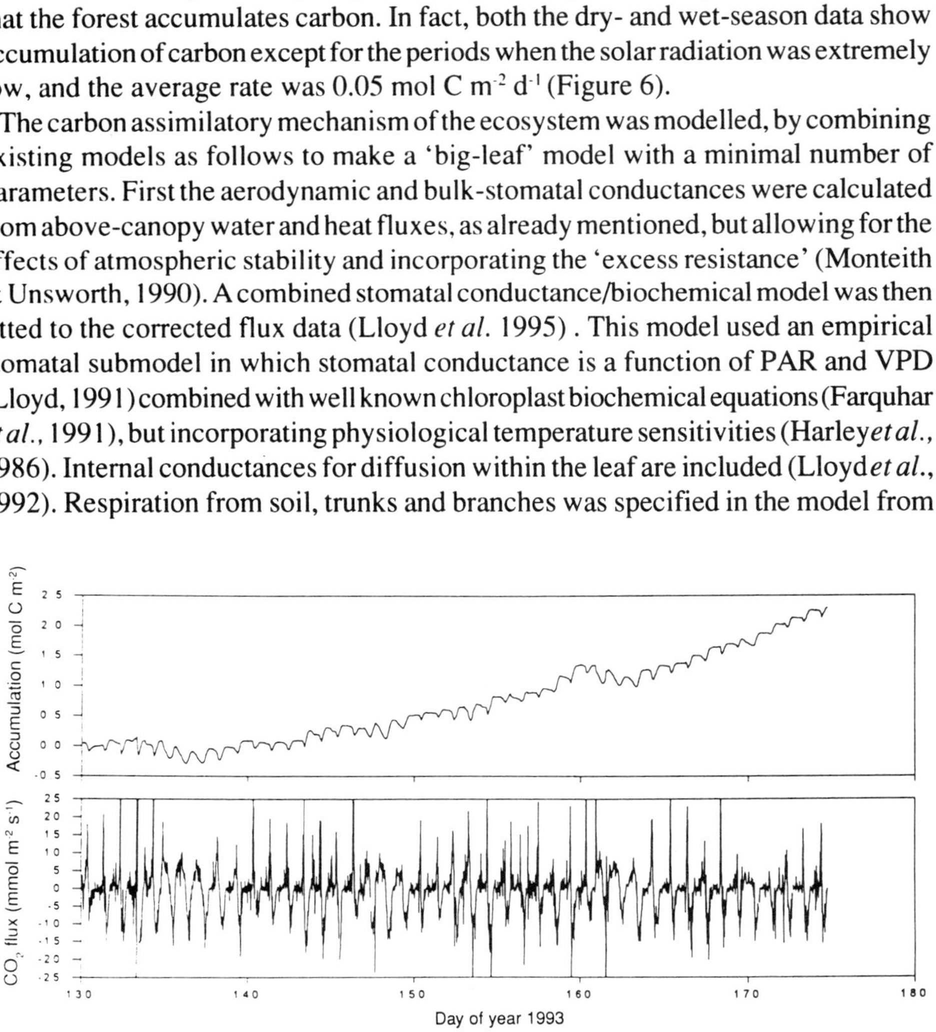

Figure 6 Carbon dioxide fluxes (as in Figure 1), and accumulated fluxes

direct measurements made in the field (Meir *et al.*, 1996).

The model was run using one year's climatological data from Reserva Jaru (1 July 1992 - 30 June 1993) we ran the model. Over this period it estimated an accumulation of 7.6 mol C m^{-2} y^{-1} for the area of rainforest examined. If the assumption is made that this site is representative of average forest in the Amazon basin ($5 \times 10^{12} m^2$), this accumulation would be 40×10^{12} mol C y^{-1} or 0.50 Pg C y^{-1}. For the world's intact tropical rainforest the corresponding figure would be 0.9 Pg C y^{-1}. Similar conclusions about the sink strength of the rain forest are drawn from using the SiB model (da Rocha *et al.*, 1996).

The estimates for the region as a whole must be regarded as *provisional*, as they do not take account of the spatial variation in forest types and climate. Improved estimates will become available after data are collected at several sites. It is however noteworthy that a recent analysis from the GASLAB global flask monitoring network shows the tropical biota to be a carbon sink (Enting *et al.*, 1995).

The model provides information about temperature and radiation sensitivity, and it was run under a range of conditions. The year of the measurements (1992/93) was cooler than average, and may have had a reduced irradiance as a result of aerosols from the eruption of Mount Pinatubo. The model is especially sensitive to temperature, and when half-a-degree is added to bring the data nearer to 'average' conditions, the forest accumulates less carbon (Figure 7).

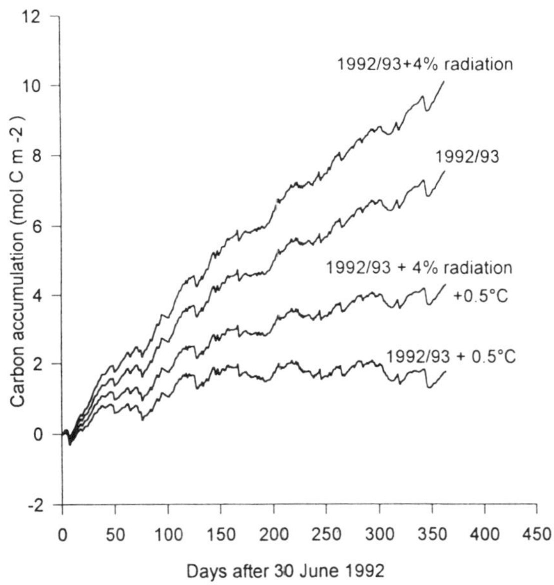

Figure 7 Sensitivity testing of the fitted model (Lloyd *et al.*, 1995). Runs were: standard run using the existing climatic data (1992/93), the effect of increasing radiation alone by 4%, the effect of increasing temperature alone by 0.5°C, and the effect of temperature and radiation increases together.

The picture that emerges, is that the carbon balance of the forest is very temperature sensitive, and as we show elsewhere in this volume, the important component of this temperature sensitivity is the soil respiration (Meir *et al.*, 1995). It seems likely that the inter-annual variation of temperature in Brazil is sufficient to bring about major shifts in the carbon balance of the region.

ACKNOWLEDGEMENTS

We thank W. Coupland, I. Enting, P. Groeneveld, P.G. Jarvis, P. McCracken, A. McWilliam, J. Massheder, J. Moncrieff and C.S. Wong for discussion and practical help. We thank those organisations in Brazil that facilitated the field work: INCRA, IBAMA, INPE & University of Brasilia, Federal University of Rondônia. The work was supported by the TIGER programme of the Natural Environment Research Council. Jon Lloyd was assisted by the Australia/New Zealand/UK Tripartate agreement on climate change.

REFERENCES

Culf, A.D., Esteves, J.L., de O.M. Filho, A. and da Rocha, H.R. 1996. Radiation, temperature and humidity over forest and pasture in Amazonia. This volume.

Enting, I.G., Trudinger, C.M. and Trivett, N.B.A. 1995. A synthetic inversion of the concentration and $d^{13}C$ of atmospheric CO_2. *Tellus.*

Fan, S., Wofsy, S.C., Bakwin, P.S. and Jacob, D.L. 1990. Atmosphere-Biosphere exchange of CO_2 and O_3 in the Central Amazon forest. *J. Geophys. Res.* **95**, 16851-16864

FAO. 1993. *Third Interim Report on the Sate of Tropical Forests.* FAO, Rome.

Farquhar, G.D., von Caemmerer, S. and Berry, J.A. 1991. A biochemical model of photosynthetic CO_2 assimilation in leaves of C3 species. *Planta* **149**, 78-90.

Gash, J.H.C., Nobre, C.A., Roberts, J.M. and Victoria, R.L. 1996. An overview of ABRACOS. This volume

Gifford, R.M. 1994. The global carbon cycle: a viewpoint on the missing sink. *Australian J. Plant Physiol.* **21**, 1-15.

Grace, J, Lloyd, J., McIntyre, J., Miranda, A.C., Meir, P., Miranda, H., Moncrieff, J.M., Massheder, J., Wright, I.R. and Gash, J. 1995. Fluxes of carbon dioxide and water vapour over an undisturbed tropical rainforest in south-west Amazonia. *Global Change Biology* **1**, 1-12.

Harley, P.C., Tenhunen, J.D. and Lange, O.L. 1986. Use of an analytical model to study limitations on net photosynthesis in *Arbutus unedo* under field conditions. *Oecologia* **70**, 393-401.

Houghton, R.A., Boone, R.D., Fruci, J.R., Hobbie, J.E., Melillo, J.M., Palm, C.A., Peterson, B.J., Shaver, G.R., Woodwell, G.M., Moore, B., Skole, D.L. and Myers, N. 1987. The flux of carbon from terrestrial ecosystems to the atmosphere in 1980 due to changes in land use: Geographic distribution of the global flux. *Tellus* **39b**, 122-139.

Houghton, R.A. 1991. Tropical deforestation and atmospheric carbon dioxide. *Climatic Change* **19**, 99-118.

Körner, C. 1993. CO_2 fertilization: the great uncertainty in future vegetation development. 53-70. In: A.M. Solomon, H.H. Shugart (Eds.) *Vegetation Dynamics and Global Change.* Chapman & Hall, London.

Kruijt, B., Lloyd, J., Grace, J., McIntyre, J.A., Faraquar, G.D., Miranda, A.C. and McCracken, P.

1996. Sources and sinks of CO_2 in Rondonia tropical rainforest. This volume.
Lloyd, J. 1991. Modelling stomatal responses to environment in *Macadamia integrifolia*. *Aust. J. Plant Physiology* **18**, 649-660.
Lloyd, J., Syvertsen, J.P., Kreidemann, P.E. and Farquhar, G.D. 1992. Low conductances for CO_2 diffusion from stomata to the sites of carboxylation in leaves of woody species. *Plant, Cell & Environment* **15**, 873-900.
Lloyd, J., Grace, J., Miranda, A.C., Meir, P., Wong, S.C., Miranda, H., Wright, I., Gash, J.H.C. and McIntyre, J. 1995. A simple calibrated model of Amazon rainforest productivity based on leaf biochemical properties *Plant, Cell & Environment* **18**, 1129 - 1145.
McWilliam, A-L.C., Cabral, O.M.R., Gomes, B.M., Esteves, J.L. and Roberts, J.M. 1996. Forest and pasture leaf-gas exchange in south-west Amazônia. This volume.
Meir P, Grace, J. and Miranda, A. 1996. Soil respiration measurements in the Brazil forest and cerrado vegetation during the wet season. This volume.
Miranda, A.C., Miranda, H.S., Grace, J., Lloyd, J., McIntyre, J.A.,Meir, P Riggan, P., Lockwood, R. & Brass, J. 1996. Fluxes of CO_2 over a cerrado *sensu stricto* in Central Brazil. This volume.
Moncrieff, J.B., Massheder, J.M., Verhoef, A., Elbers, J., Heusinkveld, B., Scott, S., de Bruin, H., Monteith, J.L. and Unsworth, M.H. 1990. *Principles of Environmental Physics*. Arnold, London.
Roberts, J., Cabral, O.V.R. and de Aguiar, L.F. 1990. Stomatal and boundary layer conductances in an amazonian terra firme rain forest. *J. Appl. Ecol.* **27**, 336-353.
da Rocha, H.R., Sellers, P.J., Collatz, G.J., Wright, I.R. and Grace, J. 1996. Estimate of water vapour and carbon exchanges in an Amazonian rain forest by the SiB2 model. This volume.
Shuttleworth, W.J. 1988. Evaporation from Amazonian rain forest. *Proc. Roy. Soc., London* **233**, 321-346.
Webb, E.K., Pearman, G.I. and Leuning, R. 1980. Correction of flux measurements for density effects due to heat and water vapour transfer. *Q. J. R. Met. Soc.* **106**, 85-100.

RESUMO

Fluxos de dióxido de carbono sobre uma floresta nativa, localizada na Reserva Biológica do Jaru, em Rondônia, foram medidas nas estações úmida e seca. Ao mesmo tempo, perfis de CO_2 no interior do dossel foram registrados. Tipicamente, a ventilação natural do dossel foi baixa à noite, e o CO_2 oriundo da respiração acumulou-se próximo à superfície do solo, atingindo concentrações superiores a 500 ppmv. No início da manhã, com o aumento do transporte turbulento, detetou-se um grande efluxo de CO_2. A partir da variação horária dos perfis de concentração de CO_2, foi possível calcular a variação do CO_2 armazenado por m^2 da superfície. Esta taxa pode ser subtraída do fluxo sobre a floresta para revelar o comportamento diurno da atividade fisológica do dossel, isto é, *o fluxo de carbono do ecossistema*. A condutância estomática do ecossistema foi obtida a partir de taxas medidas de evapotranspiração. Ela variou de 0,4 a 1,0 mol m^{-2} s^{-1} no início da manhã, diminuindo os valores com o transcorrer do dia. O fluxo de carbono do ecossistema mostrou uma grande e curvilínea relação com a radiação incidente. Essa função pode ser interpretada fisiologicamente a fim de fornecer uma respiração noturna de 6-7 μmol m^{-2} s^{-1}, ponto de compensação de luz de 260 μmol quanta m^{-2} s^{-1}, exigência quântica de cerca de 40 fótons por átomo de carbono, e taxa máxima de assimilação de carbono de 14-17 μmol m^{-2} s^{-1}. O ecossistema retirou carbono da atmosfera na maior parte dos dias, sendo a média de 0,05 mol m^{-2} d^{-1} A fim de estimar o armazenamento anual de carbono, um modelo matemático que incorpora processos meteorológicos, fisiológicos e bioquímicos foi empregado, utilizando dados micrometeorológicos. Foi sugerido um acúmulo anual de carbono na ordem de $7{,}6 \pm 1{,}8$ mol C m^{-2} a^{-1} (ou 0,9 ton C ha^{-1} a^{-1}). A taxa de acumulação no modelo é muito sensível à temperatura, devido ao efeito da temperatura sobre a respiração.

18 Soil respiration in a rainforest in Amazonia and in cerrado in central Brazil

P. MEIR [1], J. GRACE[1], A. MIRANDA[2] and J. LLOYD [3]

[1] *University of Edinburgh, UK*

[2] *Universidade de Brasilia, Brazil*

[3] *Institute of Advanced Studies, Australian National University, Canberra, Australia*

INTRODUCTION

The respiratory component of the terrestrial carbon cycle is dominated by the evolution of carbon dioxide from the soil (Bolin, 1983; Houghton and Woodwell, 1989). Despite its importance, this flux remains poorly quantified, especially in tropical regions (Singh and Gupta, 1977; Fung *et al.*, 1987; Raich and Schlesinger, 1992). There is a recognised need to improve our understanding of the main sources of carbon dioxide in tropical forests, and how the underlying emission processes, such as soil carbon efflux, may change in concert with climatic variables (Townsend *et al.*, 1992; Gifford, 1994).

Carbon dioxide is produced in soils by the respiration of roots, microbes and soil fauna, and by chemical oxidation of carbon compounds (Lundegårdh, 1927). The rate of transfer of carbon dioxide to the atmosphere is controlled by its rate of production, soil-atmosphere gradients in temperature and CO_2 concentration, soil physical properties, and ambient pressure fluctuations. Although it comprises several processes — both biological and physical — the overall carbon release is generally referred to as soil respiration.

Here we report the measurement of soil efflux of CO_2 from two Brazilian vegetation types: tropical rainforest and tropical dry forest or cerrado (*sensu stricto*). The effects of the different processes that together result in soil respiration (e.g. Edwards and Harris, 1977; Ewel *et al.*, 1987) are aggregated in this study in order to estimate the process as a whole. For the rainforest site a yearly carbon budget has been developed and the role soil respiration plays in determining this is discussed.

METHODS

The main field site was a rainforest at the Reserva Jaru, Rondônia, Brazil (10° 4.84' S, 61° 56.6' W, 120 m above sea level). It is a palm-rich forest classed as *Floresta*

Amazonian deforestation and climate. Edited by J.H.C. Gash, C.A. Nobre, J.M. Roberts and R.L. Victoria.

Ombrofila Aberta, with a leaf area index of about four, an estimated above-ground biomass of 140-180 t ha^{-1} and a height of 30 m. The study site is described elsewhere (e.g. Grace *et al.*, 1996). The surface soil is a loamy sand, giving way to a sandy-clay-loam at 0.6 m, over granitic bed-rock at a depth of 1.2 to 4 m (Hodnett *et al.*, *1996*). The depth of the water table below the surface varies from 1 m in the wet season to 4 m in the dry season. Fieldwork was conducted during September 1992 (dry season) and May to June, 1993 (wet season).

Shorter field campaigns were also undertaken during April and July 1993 (early to late dry season) in cerrado vegetation, 1000 m above-sea-level at the Reserva Ecológica de Aguas Emendades, DF, Brazil. It is an example of cerrado *sensu stricto* protected from fire for at least eight years. The canopy height is 10 m, with a leaf area index of about one. The soil is a dystrophic, deep and well-drained red-yellow latisol with pH < 5 (EMBRAPA, 1987; Miranda *et al.*, *1996*).

Efflux of CO_2 from the soil surface was measured by recording the accumulation of CO_2 in a chamber sealed to the soil surface and connected in closed circuit to a Licor LI 6200 infra-red gas analyser (Licor Inc., Nebraska, USA) with a drying column containing magnesium perchlorate for stabilising humidity. Soil respiration is calculated as a function of atmospheric pressure, the increase in CO_2 concentration in the enclosed air during measurement, the area of ground under the chamber, and the system volume (e.g., Nakayama, 1990). The chamber was a perspex box, with drilled perspex sheet baffles behind the inlet and in front of the outlet. A small aperture at the side allowed venting to the atmosphere (by attaching narrow-bore tubing), or measurement of pressure differentials (by attaching a digital manometer). The chamber was fitted into a water trough attached to the perimeter of a steel collar inserted 1-2 cm into the soil. If the collar could not be inserted without some disturbance to the soil, damp sand was used to seal the outside edges to the soil surface. The perspex chamber was sealed to the collar by adding water to the trough, carefully avoiding any spillage. The basal area of the whole chamber was 0.042 m^2; the volume (depending on the depth of collar insertion) was approximately 0.0073 m^3.

The closed infra-red gas analyser system did not suffer from the anomalous suction induced fluxes of CO_2 from the soil associated with some open-chamber techniques (e.g. Kanemasu *et al.*, 1974). The short deployment time during measurement appeared to avoid problems associated with static pressure fluctuations resulting from weak turbulence at the forest floor (e.g. Hutchinson and Livingston, 1993; Baldocchi and Myers, 1991). Very low wind speeds at ground level are characteristic of mature broadleaf forest (Odum *et al.*, 1970; Hanson *et al.*, 1993) and major turbulence events reaching the bottom of the canopy are probably rare.

After checking that chamber CO_2 concentration was close to ambient levels (400-600 ppm, depending on time of day), measurements were made by logging data for 60 or 100 second periods, using a five-second time step, and then repeating the procedure. Data were subsequently checked for linearity of the CO_2 concentration rise with time. Measurements not fulfilling this requirement (approx. 10%) were discarded. To start with, a second reading was taken at each microsite 10-20 minutes after the first to ensure that deployment of the collar had not caused an initial flush of CO_2 by disturbing diffusion resistances at the soil surface (Mosier, 1989); almost no sites behaved in this way. Pressure differences between inside and outside the chamber were not detectable.

Twenty-four-hour measurements were made for individual microsites, the collar being left in the soil for the whole of these periods. Sampling of the forest floor was necessarily restricted to relatively flat patches of ground. Notwithstanding this, a wide range of microsites was chosen in an attempt to represent soil diversity with respect to distance from tree trunks, litter depth, and topography.

Soil temperature at 1cm depth was measured at the same time using an amplified Cu-Cn thermocouple (sensitivity ± 0.2°C). Soil temperature at 1cm depth was also logged elsewhere, using a Campbell Scientific 21X micrologger. The lag between changes in air and soil temperature, and the shaded environment of the forest floor, meant that chamber emplacement had no measurable effect on soil surface temperature during the period.

RESULTS

Mean respiration rates were higher, and the range of rates was greater for the rainforest (RF) : mean = 5.5 ± 1.6 μmol $m^{-2}s^{-1}$ than for the cerrado (C): mean = 3.0 ±0.8. Twenty-four-hour measurement sequences for individual microsites indicated a strong temperature-respiration relationship for both RF (data from 2 nights in May 1993) and C (data from one night in April). The diurnal temperature cycles are

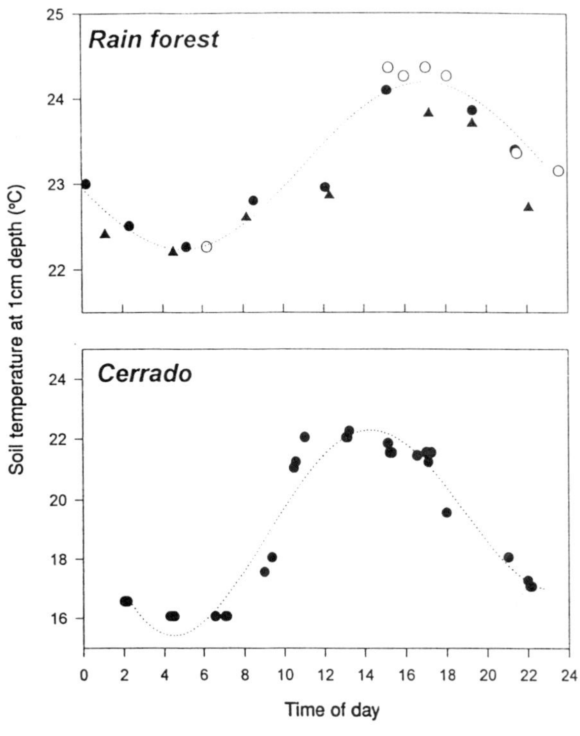

Figure 1 Diurnal temperature cycles for soil temperature during 24-hour measurements of soil respiration in C (April 1993) and RF (May-June 1993). Dotted lines do not represent the modelled temperature cycle but are plotted to emphasise the cyclic temperature pattern for the diurnal measurements. Symbols represent data from different microsites.

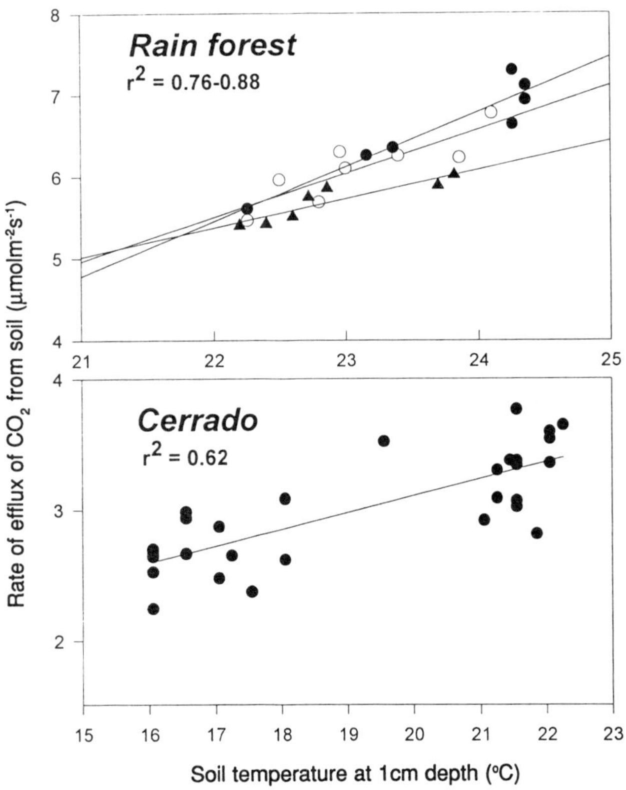

Figure 2 Soil respiration as linear functions of temperature (for 24-hour measurements only); C in April, 1993, RF in May-June 1993. Significance is not attached to the best fit line as not all measurements are independent. Symbols represent data from different microsites.

shown in Figure 1. The relationship between soil respiration rate and temperature for these measurements was adequately described by a straight line with r^2 values of 0.88 - 0.76 for RF and 0.62 for C (Figure 2: regression statistics are not given as data for 24-hour sequences are time series and therefore do not satisfy the criterion of independence). There was no evidence for a diurnal cycle in soil respiration except from the effect of temperature. A larger set of measurements for RF (n=42) and C (n=10), obtained throughout the 1993 field campaigns, are plotted describing respiration as an exponential function of temperature. This gave highly significant regressions for each forest (Figure 3). Also marked in Figure 3 are mean data points for a smaller set of soil respiration rates obtained using the same method at a different time of year: for RF in September 1992 (n=9) and for C in July 1993 (n=8). The absolute value of the July mean for cerrado suggests that respiration rate is positively related to soil moisture content (Edwards, 1975; Wiant, 1967). For RF, the 95% confidence limits for the regression and the mean for September 1992 overlap (Figure 3a). The regression equation for RF including the September 1992 data is not statistically different from the regression for the main 1993 data set. A comparison of the two exponential functions (Figure 4) shows the RF to have a higher overall respiration rate, but also a higher Q_{10} value (2.3) than the cerrado (1.6): Q_{10} is a commonly-used term describing the relative increase in the rate of a biological process for a temperature increase of 10°C.

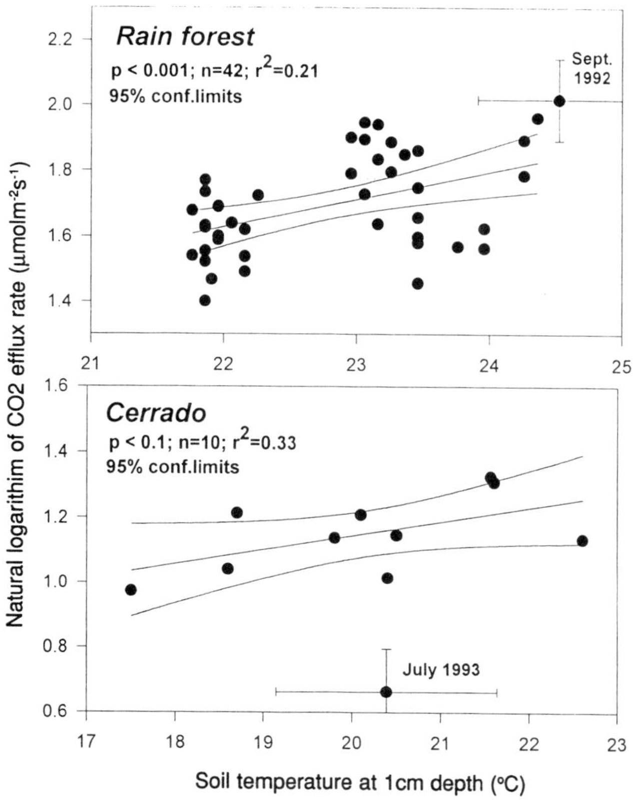

Figure 3 Soil respiration measurements for all microsites plotted as a regression (with 95% confidence limits) of the natural logarithm of CO_2 efflux against temperature for C (April 1993) and RF (May-June 1993). Data points for different times of the year are marked as means with 95% confidence limits (September 1992 for RF, July 1993 for C).

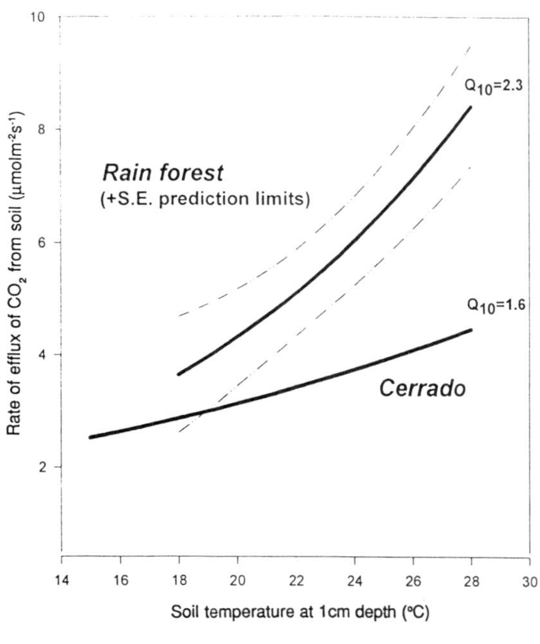

Figure 4 Soil respiration as an exponential funtion of temperature for RF and for C. S.E. limits of prediction are plotted for RF; Q_{10} values are given for both.

SCALING UP

Scaling up the data was restricted to the RF site. Soil respiration values for the night were initially summed with preliminary estimates for leaf and wood respiration to obtain total forest respiration values. Good agreement was found with data obtained independently using the eddy covariance technique (Grace*et al.*, 1996) for night-time respiration of the whole canopy (Table 1).

Table 1 Comparison of night values for whole canopy respiration measured using eddy covariance and chamber techniques. Dataare for the same measurement period (May-June 1993).

Component of canopy	Measurement method	
	Chamber (process level) Respiration (μmol m^{-2} s^{-1})	Eddy covariance (canopy level) Respiration (μmol m^{-2} s^{-1})
Leaf	1.0 ± 0.1	---
Wood	0.7 ± 0.3	---
Soil	5.5 ± 0.7 (S.E.)	---
Canopy (total)	7.2 ± 1.0	6.6 ± 0.2 (S.E.)

Soil temperature values (T_s) at 1cm depth were estimated from above-canopy dry bulb temperatures (T_c) measured from the top of a micrometeorological tower at the site. T_s fitted the measured soil temperature values with an r^2 value of 0.84. A year's T_c data (Culf*et al.*, 1996) were then used to estimate soil temperature and thence soil respiration in hourly time steps from 1 July 1992. This yielded a soil carbon efflux value of 19.5 (± S.E. 14%) t C $ha^{-1}y^{-1}$, and was combined with the results from running a canopy scale model for the same site, using the same (Culf*et al.*,1996) meteorological data set (Grace *et al.*, 1996), to build a tentative forest carbon budget (Figure 5). In this description soil respiration represents a very large proportion of the total respiratory component of the carbon budget for this forest (84%). The balance of gross primary productivity and total respiration shows the forest to be a net sink at 1.1 t C $ha^{-1}y^{-1}$ for the year 1992-93, representing only 6% of the total efflux of carbon dioxide from the soil.

It is possible, however, that year-to-year climatic anomalies may result in fluctuations in the carbon balance (Sarmiento, 1993). In line with this view, it is desirable to make some estimate of the sensitivity of the carbon balance to changes in temperature. Re-running the prediction of soil respiration for the same period with T_c + 0.5°C up to T_c + 2.0°C (a reasonable estimate for temperature increase in a 2 x CO_2 world (Houghton *et al.*, 1990; Houghton *et al.*, 1992)) the forest appears to change from a net sink of 1.1 t C $ha^{-1}y^{-1}$ at T_c(1992-93) to a weak net source of less than 0.1 t C $ha^{-1}y^{-1}$ at T_c+ 2.0°C. Even the smallest temperature increase examined here yields a 24% reduction (from 1.1 to 0.86 ton C $ha^{-1}y^{-1}$) in the sink strength of the forest that is due to soil respiration alone. Inclusion of estimates of biomass respiration would further increase the magnitude of this reduction. It should be noted that although these

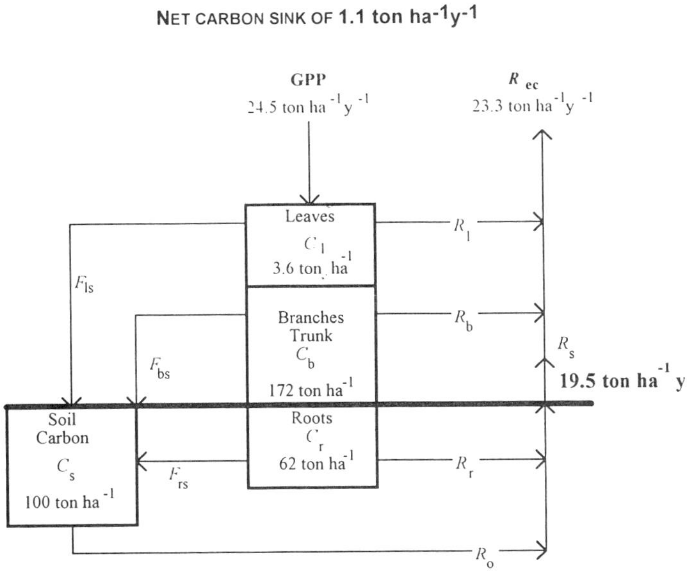

Figure 5 Carbon flows in a tropical rainforest at Reserva Jarú, Rondônia, Brazil, based on measurements in 1992/93. Vegetation-atmosphere fluxes measured using eddy covariance; leaf, stem and soil respiration measured by 'closed' infra-red gas analyser. C_b = carbon in branches and trunks; C_l, C_r, C_s = carbon content in leaves, roots and soil; R_b, R_{ec}, R_l, R_o, R_R and R_S = respiratory fluxes from trunk, ecosystem as a whole, leaves, soil carbon, roots and soil as a whole; F = fluxes describing shedding of dead material. Carbon stores for biomass are based on Brown *et al.*, 1992 (for roots, Deans, D., pers. comm.) and for soil carbon Post *et al.* (1992). All compartments are in ton ha^{-1} y^{-1}. Ecosystem carbon sequestation (sink strength, shown here as GPP - R_{ec}) was measured directly by eddy covariance and subsequently scaled up using a model as 1.1 ton C ha^{-1} y^{-1}. Soil respiration contributes approximately 84% of the total forest efflux of carbon dioxide.

these results are in line with the canopy scale model for this forest (Grace *et al.* 1995), no other component of the forest carbon cycle (such as biomass respiration, changes in phenology, etc.) is changed in the simulation. Predictions of longer-term effects of soil respiration on the carbon cycle must include feedback effects on net primary production and carbon storage (Trumbore *et al.*, 1990; Jenkinson *et al.*, 1991; Gifford, 1994).

Table 2 Effect on soil respiration of increasing canopy temperature by 0-2°C and its consequent effect on the forest carbon budget. Negative signs signify a carbon sink; no other component of the carbon cycle is changed in these calculations.

Temperature change (°C)	Soil respiration ($t\ ha^{-1}\ y^{-1}$) ± S.E.%	New net C balance ($Pg\ C\ ha^{-1}\ y^{-1} \times 10^{-10}$)
0	19.52 ± 14.1%	−1.1
+0.5	19.78 ± 13.8%	−0.84
−1.0	20.04 ± 13.6%	−0.58
−2.0	20.67 ± 31.2%	+0.05

DISCUSSION

The soil respiration rates obtained for the cerrado are consistent with previous studies for physiognomically similar vegetation types (Schulze, 1967; Raich and Schlesinger, 1992). The rates reported here for RF are higher than some other estimates for Amazonian forest soils (e.g. Medina, 1980; Keller *et al.*, 1986). But at a soil temperature of 20°C the estimate for this forest is similar to that of Wofsy *et al.* (1988) at Reserva Ducke, Manaus (4.5 *versus* 4.2 mol $m^{-2}s^{-1}$), although they do not state soil temperature. Laying aside the differences in methodology between different studies, it is clear that the diversity of soil types across Amazonia (Jordan, 1985) will lead to a corresponding range of soil respiration rates. The site-to-site variation found in both RF and C is a common feature of soil respiration studies (Schlesinger, 1977; Singh and Gupta, 1977; Nakayama 1990): it is a descriptor of the biological variation in the soils related to nitrogen, carbon and moisture levels. The greater variation in RF values reflects the difference in temperature sensitivity of respiration in the two soils, possibly coupled with a more heterogeneous soil type.

Scaling up chamber measurements in time and space should be done with caution (Raupach and Denmead, 1993; Huchinson and Livingstone, 1993). An adequate sample of natural variation is required, as are 24-hour measurements. Year-round monitoring is also highly desirable. This is particularly important in environments where marked seasonal temperature changes occur, since the CO_2 efflux rate is affected by temperature effects on the diffusion coefficient of CO_2 (Reiners, 1973, cited by Schlesinger, 1977) as well as by the physiological source processes. At the Jaru site soil temperatures averaged 22.9 °C (±95% c.f. 0.02°C), suggesting that this issue is less critical here. Furthermore, the close adherence of respiration to soil temperature fluctuations facilitates the construction of a relatively simple predictive model (Edwards and Sollins, 1973). The estimates made here are strengthened to a certain extent by the inclusion of data for the previous September (1992), as these values are within the 95% confidence limits for the temperature-respiration regression (Figure 3a) and inclusion of the September data makes no significant change to the regression. However it is clear that a larger measurement programme would settle concerns over the representation of spatial variability and temperature response, and the possible effects of an annual (or longer-term) forest cycle on the soil respiratory account sheet.

Estimation of the temperature sensitivity of soil respiration is central to exercises involving scaling up and prediction. The errors incurred here are likely to be greater for the brief cool periods during 1992-93 (Culf *et al.*, *1996*), as the range experienced (16-25°C) departed from the regression in Figure 3a to a larger extent at lower temperatures. There is a need to improve this with measurements of soil respiration, canopy and soil temperature at other times of the year. Lloyd and Taylor (1994) argue against the use of an exponential model to describe the temperature-respiration relationship, though over the relatively narrow temperature range experienced for the Jaru site, an error of this type is within the margin of measurement error. The Q_{10} value reported here (2.3) is higher than Fung *et al.* (1987) give for their very broad classification "tropical/subtropical woody vegetation" (1.4-1.6), but is less than the global values (2.5) given by Raich and Schlesinger (1992). The possible future effects of soil respiration in the tropics on the terrestrial carbon budget have been emphasised recently (Townsend *et al.*, 1992; Gifford, 1994). The warm and fluctuating temperatures at tropical latitudes make soil respiration a highly labile and responsive component of the terrestrial carbon cycle. The results presented here lend support to this thesis.

CONCLUSION

The short-term measurements described are comparable both with previous work and with parallel measurements at the same site using micrometeorological techniques. Soil respiration is shown to be a large, temperature-sensitive contributor to the carbon budget of rainforest. Tentative scaling up and prediction exercises reveal the potential for this component of the carbon cycle to govern the behaviour of the forest as a net sink or source for carbon dioxide.

ACKNOWLEDGEMENTS

We thank J. McIntyre, M.P. Rocha and S. Wilson for practical help at various stages, D. Deans for root biomass estimates, and organisations in Brazil that facilitated the field work: INCRA, IBAMA, INPE and the University of Brasilia. Jon Lloyd was assisted by the Australia/New Zealand/UK Tripartite Agreement on Climate Change. The work formed part of the TIGER programme of the Natural Environment Research Council.

REFERENCES

Baldocchi, D.D. and Meyers,T.P. 1991. Trace gas exchange above the floor of a deciduous forest 1. Evaporation and CO_2 efflux. *J. Geophys. Res.* **96B**, 7271-7285.

Bolin, B. 1983. Changing global biogeochemistry. In: *Oceanography. The present and the future.* Brewer, P. (Ed.). Springer Verlag, New York, 305-26.

Brown, S.; Gillespie, A.J.R. and Lugo, A.E. 1992. Biomass estimation methods for tropical forests with applications to forest inventory data. *For. Sci.* **35**, 881-902.

Culf, A.D. , Esteves, J.L., Marques Filho, A. de O. and da Rocha, H.R. 1996. Radiation, temperature and humidity over forest and pasture in Amazonia. This volume.

Edwards, N.T. and Harris,W.F. 1977. Carbon cycling in a mixed deciduous forest floor. *Ecology* **58**, 431-437.

Edwards, N.T. 1975. Effects of temperature and moisture on carbon dioxide evolution in a mixed deciduous forest floor. *Soil Sci. Soc. Amer. J.* **39**, 361-365.

Edwards, N.T. and Sollins, P. 1973. Continous measurement of carbon dioxide evolution from partitioned forest floor components. *Ecology* **54**, 406-412.

EMBRAPA. 1987. Levantamento y reconhecimento dos solos do Distrito Federal, Boletim

Técnico **5**, EMBRAPA-SNLCS.

Ewel, K.C., Cropper, W.P. and Gholz, H.L. 1987. Soil CO_2 evolution in Florida slash pine plantations II. Importance of root respiration. *Can. J. For. Res.* **17**, 330-333.

Fung, I.Y., Tucker, C.J. and Prentice, K.C. 1987. Application of advanced very high resolution radiometer vegetation index to study atmosphere-biosphetre exchange of CO_2. J. *Geophys. Res.* **93(D3)**, 2999-3015.

Grace, J., Lloyd, J., McIntyre, J., Miranda, A., Meir, P., Miranda, H., Nobre, C., Moncief, J., Massheder, J., Mahli, Y., Wright, I. and Gash, J. 1995. Carbon dioxide uptake by and undisturbed tropical rainforest in south-west Amazonia. 1992-1993. Science **270**, 778-780.

Gifford, R.M. 1994. The global carbon cycle: a viewpoint on the missing sink. *Aust. J. Plant Physiol.* **21**, 1-15.

Hanson, P.J., Wullschleger, S.D., Bohlman, S.A. and Todd, D.E. 1993. Seasonal and topographic patterns of forest floor CO_2 efflux from an upland oak forest. *Tree Phys.* **13**, 1-15.

Hodnett, M.G., Oyama, M. and Tomasella, J. 1996. Comparisons of long-term soil water storage behaviour under pasture and forest in three areas of Amazonia. This volume.

Houghton, R.A. and Woodwell, G.M. 1989. Global climatic change. *Sci. Am.* **260**, 36-44.

Houghton, J.T., Jenkins, G.J. and Ephraums, J.J. 1990. *Climate Change: The IPCC Scientific Assessment.* Report of the Intergovernmental Panel on Climate Change, Cambridge University Press.

Houghton, J.T., Callander, B.A. and Varney, S.K. 1992. *Climate Change 1992: The supplementary report to the IPCC Scientific Assessment.* Report of the Intergovernmental Panel on Climate Change, Cambridge University Press.

Huchinson, G.L. and Livingston, G.P. 1993. Use of chamber systems to measure trace gas fluxes. In: *Agricultural ecosystem effects on trace gases and global change*, L.A. Harper. ASA Spec. Publ. **55**, ASA / CSSA / SSSA, Madison, USA.

Jenkinson, D.S., Adams, D.E. and Wild, A. 1991. Model estimates of CO_2 emissions from soil in response to global warming. *Nature* **351**, 304-306.

Jordan, C.F. 1985. Soils of the Amazon rainforest. In: *Key Environments: Amazonia*, G.T. Prance and T.E. Lovejoy (Eds), Pergamon Press, Oxford, 83-94.

Kanemasu, E.T., Powers, W.L. and Sij, J.W. 1974. Field chamber measurements of CO_2 flux from soil surface. *Soil Science* **118**, 223-237.

Keller, M., Kaplan, W.A. and Wofsy, S.C. 1986. Emissions of N_2O, CH_4 and CO_2 from tropical forest soils. *J. Geophys. Res.* **91(D2)**, 11791-11802.

Lloyd, J. and Taylor, J.A. 1994. On the temperature dependence of soil respiration. *Functional Ecology* **8**, 315-23.

Lundegårdh, H. 1927. Carbon dioxide evolution and crop growth. *Soil Sci.* **23**, 417-453.

Medina, E., Klinge, H., Jordan, C.F., and Herrera, R. 1980. Soil respiration in Amazonian rainforests in the Rio Negro basin. *Flora* **170**, 240-250.

Miranda, A.C., Miranda, H.S., Lloyd, J., Grace, J., McIntyre, J.A., Meir, P., Riggan, P., Lockwood, R. and Brass, J. 1996. Carbon dioxide fluxes over a cerrado *sensu stricto* in Central Brazil. This volume.

Mosier, A.R. 1989. Chamber and isotope techniques. In: *Exchange of trace gases between terrestrial ecosystems and the atmosphere*, M.O. Andreae and D.S. Schimel (Eds.), John Wiley and Sons, Chichester, UK, 175-187.

Nakayama, F.S. 1990. Soil respiration. In: *Instrumentation for studying vegetation canopies for remote sensing in optical and thermal infrared regions*, N.S. Goel and J.M. Norman, (Eds.), Remote Sensing Reviews, Vol. 5, Issue 1, Harwood Acad. Publ., New York, 175-187.

Odum, H.T., Lugo, A., Cintronn, G. and Jordan, C.F. 1970. Metabolism and evapotranspiratoin of some rainforest plants and soil. In: *A tropical rainforest*, H.T.Odum and R.F. Pigeon (Eds.), Division of Technical Information, US Atomic Energy Commission, Washington, DC, 1103-1164.

Post, W.M., Emanuel, W.R., Zinke, P.J. and Stangenberger, A.G. 1982. Soil carbon pools and world life zones. *Nature* **298**, 156-159.

Raich, J.W. and Schlesinger, W.H. 1992. The global carbon dioxide flux in soil respiration and its relationship to vegetation and climate. *Tellus* **44B**, 81-99.

Raupach, M.R. and Denmead, O.T. 1993. Methods for measuring atmospheric gas transport in agricultural and forest systems. In: *Agricultural ecosystem effects on trace gases and global change*, L.A. Harper. ASA Spec. Publ. **55**, ASA / CSSA / SSSA, Madison, USA.

Reiners, W.A. 1973. Terrestrial detritus and the carbon cycle. *Brookhaven Symp. Biol.* **24**, 303-327.

Sarmiento, J.L. 1993. Atmospheric CO_2 stalled. *Nature* **365**, 607-698.

Schlesinger, W.H. 1977. Carbon balance in terrestrial detritus. *Ann. Rev. Ecol. Syst.* **8**, 51-81.

Schulze, E.D. 1967. Soil respiration of tropical vegetation types. *Ecology* **48**, 652-653.

Singh, K.P. and Gupta, S.R. 1977. Plant decomposition and soil respiration in terrestrial ecosystems. *Bot. Rev.* **43**, 449-528.

Townsend, A.R., Vitousek, P.M. and Holland, E.A. 1992. Tropical soils could dominate the short-term carbon cycle feedbacks to increased global temperatures. *Climatic Change* **22**, 293-303.

Trumbore, S.E., Bonani, G. and Wolfi, W. 1990. The rates of carbon cycling in several soils from AMS ^{14}C measurements of fractionated soil organic matter. In: *Soils and the greenhouse effect*. A.F. Bouwman (Ed.), John Wiley and Sons, New York, 405-414.

Wiant, H.V. 1967. Influence of temperature on the rate of soil respiration. *J. Forestry* **65**, 489-490.

Wofsy, S.C., Harriss, R.C. and Kaplan, W.A. 1988. Carbon dioxide in the atmosphere over the Amazon basin. *J. Geophys. Res.* **93(D2)**, 1377-1387.

RESUMO

A componente de respiração do ciclo de carbono terrestre é dominado pela emissão do dióxido de carbono vindo do solo (Bolin, 1983; Houghton e Woodwell, 1989). Apesar de sua importância, a magnitude desse fluxo permanece pobremente quantificado, especialmente em regiões tropicais (Singh e Gupta, 1977; Fung *et al.*, 1987; Raich e Schlesinger, 1992). Existe uma reconhecida necessidade em melhorar a compreensão das fontes principais de dióxido de carbono em florestas tropicais, e de que modo os processos de emissão, como o efluxo de carbono do solo, podem alterar em função das variáveis climáticas (Townsend *et al.*, 1992; Gifford, 1994).

Dióxido de carbono é produzido nos solos pela respiração de raízes, microrganismos, fauna do solo, e pela oxidação química dos compostos de carbono (Lundegårdh, 1927). A taxa de transferência do dióxido de carbono para a atmosfera é controlada pela sua taxa de produção, pelos gradientes de temperatura e concentração de dióxido de carbono entre solo-atmosfera, pelas propriedades físicas do solo, e pelas flutuações de pressão do ambiente. Embora abrangendo, necessariamente, vários processos, tanto biológicos quanto físicos, o processo global de liberação de carbono é geralmente referido como respiração do solo.

Nesta comunicação, apresentamos as medidas de efluxo de CO_2 para dois tipos de vegetação do Brasil: floresta tropical e cerrado (*sensu stricto*). Os efeitos dos diferentes processos que conjuntamente resultam na respiração do solo (e.g., Edwards e Harris, 1977; Ewel *et al.*, 1987) estão agregados neste estudo a fim de estimar o processo como um todo. Para o sítio de floresta tropical um balanço anual de carbono é desenvolvido e o papel da respiração do solo nesse balanço é discutido.

19 Sources and sinks of CO_2 in Rondônia tropical rainforest

B. KRUIJT[1], J. LLOYD[2], J. GRACE[1], J.A. McINTYRE[1], G.D. FARQUHAR[2], A.C. MIRANDA[3] and P. McCRACKEN[1]

[1] *Institute of Ecology and Resource Management, University of Edinburgh, UK*

[2] *Research School of Biological Sciences, Australian National University, Canberra*

[3] *Departamento de Ecologia, Universidade Federal de Brasilia*

INTRODUCTION

Assessment of the carbon balance of the Amazon basin involves extrapolating and integrating local observations of carbon exchange to the whole region. Because the number of measurement sites is limited, only a rough distinction of a few forest types can be used to weight the areal integrals (Wright *et al.*, 1996), which potentially biases our estimates. It is therefore useful in extrapolating to use models with parameters that depend to some degree on the structural, taxonomic and edaphic characteristics of the ecosystem. Such parameterization allows more widely available information about the land cover to be used when scaling up, but it requires that the magnitude of the components of the forest carbon budget be known. This paper contributes to this by estimating the vertical distribution of carbon sources and sinks in the rainforest at Reserva Jaru, Rondônia, using methods other than traditional extensive leaf-scale gas analysis.

Components of the budget in the Reserva Jaru have been investigated during the ABRACOS/TIGER field campaigns in 1992 and 1993. These included the net CO_2 flux above the canopy (Grace *et al.*, 1995a & b, 1996a), soil and wood respiration measurements (Meir *et al.*, 1996) and leaf-scale photosynthesis (McWilliam *et al.*, 1996). Changes in canopy air CO_2 storage were taken into account in calculating net fluxes, and were calculated from changes in concentration measured along a vertical profile within and above the canopy.

The CO_2 concentration profiles, however, also contain sub-canopy-scale information. Vertical concentration differences arise from differences in the balance between supply or demand of carbon by the ecosystem components (plants, soil, dead material), and from turbulent transport by air to or from the vegetation layers. If the profile and the turbulent transport are known in sufficient detail, it should be possible to infer the vertical distribution of carbon sources and sinks.

Turbulence is more complex inside the forest canopy than well above it. Above twice the canopy height, relatively simple 'K-theory' satisfactorily predicts fluxes of constituents like water vapour and CO_2 from wind shear and the local gradient of the

Amazonian deforestation and climate. Edited by J.H.C. Gash, C.A. Nobre, J.M. Roberts and R.L. Victoria.

constituent. At lower heights, however, K-theory is increasingly less valid, since the gradients change with height over distances similar to the scale of the turbulent motions, typically of the same order as the canopy height (Legg *et al.* 1986). One consequence of this is that K-theory cannot explain frequently observed counter-gradient fluxes (Wilson, 1989). This implies that in estimating the turbulent transport, either higher-order transport terms should be taken into account, or at least the rapid transfer of constituents over large distances should be estimated explicitly.

The latter approach was adopted here to model the turbulent transport and infer the source/sink distribution of carbon using CO_2 concentration profiles, in the form described by Raupach (1987, 1989a, 1989b) as the relatively simple Lagrangian Localised Near Field (LNF) theory. The model, outlined in more detail in Appendix 1, was parameterized by multi-layer turbulence measurements inside the forest canopy at the Reserva Jaru site.

Measurements of within-canopy concentration profiles and turbulence in tropical rainforest are relatively scarce (Shuttleworth *et al.*, 1985; Cabral *et al., 1996;* Pinker and Holland, 1988). Typically, such data show strong persistent gradients from the top of the canopy down to the forest floor. The existence of a temperature inversion below the upper canopy is often evident, suggesting a decoupling of the lower canopy air from the conditions above. On the other hand, the available data suggest that the microclimate inside a tropical rainforest canopy is not essentially different from that in other vegetation types.

A different, and perhaps complementary, approach to source/sink identification would be to use the difference in isotopic composition between CO_2 released during respiration and that in the atmosphere. This difference exists because photosynthesis discriminates against the heavier ^{13}C stable isotope in CO_2 in air and in favour of ^{12}C (Farquhar *et al.*, 1989). There is little or no discrimination during respiration. So, whilst respiration releases ^{13}C-depleted CO_2, photosynthesis tends to *enrich* the air with ^{13}C, while through mixing with tropospheric air the canopy air composition will tend towards the relatively undepleted background level. Analysis of the stable isotope composition can thus provide some insight into the strength and location of sources and sinks, but clearly several other unknowns (mixing, discrimination ratio) need to be taken into account for a qualitative assessment.

The degree of ^{13}C depletion in a sample, denoted $\delta^{13}C$, is computed as the normalised deviation of the $^{13}C/^{12}C$ ratio from a standard ratio found in a fossil formation as 0.01124 (the PDB standard). The discrimination, usually denoted as Δ, is expressed as the difference between the isotope ratios in the air and in plant material, normalised by the plant's ratio (Lloyd & Farquhar, 1994). In a simpler analysis Farquhar *et al.* (1982) showed that Δ depends linearly on the ratio of chloroplastic CO_2 concentration C_c to the CO_2 in the air C_a. This ratio in turn is a function of photosynthesis A and the total air-to-chloroplast conductance for CO_2, g:

$$\Delta = a + (b - a)C_c/C_a = b - (b - a)(A/gC_a) \quad (1)$$

where *a* is the fractionation during diffusion from the leaf surface to the sites of carboxylation within the chloroplast (*ca* 4 per mil) and *b* is the fractionation during carboxylation of Rubisco (*ca* 29 per mil).

For tropical rainforests in Amazonia, vertical gradients of both CO_2 and its carbon isotopes may be appreciable (Van der Merwe & Medina, 1989). Similarly, large vertical gradients in $\delta^{13}C$ of leaf dry matter are observed (Medina & Minchin, 1980). Under debate still is the degree to which the greater depletion of ^{13}C observed in leaf dry matter from lower forest strata is a consequence of more depleted source CO_2 (resulting from a greater contribution from the soil respiratory source) or whether gradients in photosynthetic physiology within the plant canopy also contribute to this phenomenon (Francey & Farquhar, 1982). Within the Amazon basin, Martinelli *et al.* (1991) observed a marked decline of understorey leaf $\delta^{13}C$ ratios when sampling in a transect from the Atlantic ocean towards the interior of the basin. They attributed this to differences in the degree of refixation of nocturnally respired CO_2 present in the nocturnal boundary layer that breaks up later in the morning in the inland basin. Alternatively, some of the variation may be due to effects of soil nutrient or hydrological characteristics on discrimination. For example, Medina and Minchin (1980) found that, on average, $\delta^{13}C$ for upper canopy leaves in a southern Venezuelan podsol forest was 1.8‰ depleted compared to an adjacent forest on lateritic soil.

In this study, measurements have been taken of isotopic ratios in air and plant tissue along a vertical gradient in Rondônia rainforest. These will provide a qualitative indication of source/sink strength, and also address the questions, outlined above, about the degree of recycling of carbon in the rainforest.

Summarising, the aim of this study is to identify the source or sink strengths of carbon at various levels inside the Reserva Jaru tropical rainforest, using two different approaches. Firstly, the usefulness of Lagrangian (LNF) theory (Raupach, 1987, 1989a) to infer them from vertical profiles of CO_2 concentration and turbulence statistics, is investigated. Secondly, vertical profiles of stable carbon isotope ratios are assessed to provide information about the distribution of plant physiological activity at various heights in the canopy.

METHODS

Site

The site and tower in the Reserva Jaru have been described in detail in other papers in this volume. The forest is closed, on average 30 m tall but with emergents up to 45 m. Because of the importance of emergents for turbulence, an effective forest height was chosen arbitrarily as 40 m. The vertical distribution of leaf area was estimated visually (Figure 1) to serve as a qualitative reference.

CO_2 profiling system

Air was sampled from five heights (1, 7.5, 15, 30 and 50 m) to analyse the CO_2 partial pressure (Figure 1) as described by Grace *et al.* (1995b). Since only one gas-analyser was available, lines were sampled asynchronously within 15-minute periods, but apart from that sampling was continuous. Off-line, values were interpolated to give average half-hourly profiles.

Stable isotopes

Using a similar, but separate sampling system at 1, 16, 32 and 50 m, CO_2 was sampled and isolated using a cryogenic trapping system similar to the laboratory system described by Hubick *et al.*, 1988. Tubes sampling from these heights were flushed for two minutes before trapping the CO_2 in liquid nitrogen. The air was passed through an ethanol/dry ice slurry before the liquid nitrogen trap to remove any water vapour present. The flow rate of air through the traps was about 10^{-3} m^3 min^{-1}. After the field work, samples were transferred to Canberra to analyse concentrations of ^{12}C and ^{13}C using a SIRA VG-24 mass spectrometer. Just as with the overall CO_2 sampling, the measurements were not synchronous at all heights, and interpolated profiles have been computed.

Turbulence measurements

At 50 m (z/h=1.25) a fixed eddy covariance system was continuously measuring fluxes of water vapour, CO_2 and momentum, using a three-dimensional sonic

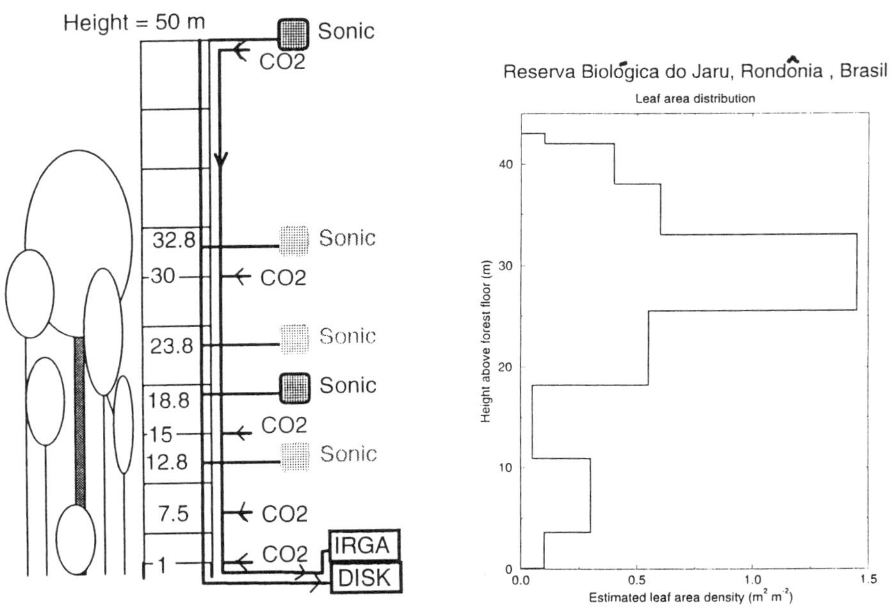

Figure 1 Schematic representation of the forest and the instrument tower used in this study (left). The positions of the sonic anemometer are shown by rounded boxes. The heights of instruments are given inside the tower. The arrows show the positions of CO_2 and isotope sampling points. A *visual* assessment of the leaf area density profile is given on the right.

anemometer (Solent), an infrared gas analyser (LiCor 6262) and a portable computer running EddySol software (Moncrieff *et al.*, 1996, Grace *et al.*, 1995b). Apart from storing half-hourly fluxes, this system is also capable of storing the 20 Hz raw measurements of the three components: windspeed, water vapour and CO_2 partial pressure. During a three-week period in June 1993, a second and similar sonic anemometer with its own computer and EddySol program, but without a gas analyser, was deployed at four heights within the canopy: 12.8 m, 18.8 m, 23.8 m and 32.8 m (Figure 1). Sampling was nearly continuous for two to three days at each level. In this way, paired data sets of simultaneous 20 Hz measurements were obtained, with one time series within the canopy and the other above.

For this study, only the raw series of the vertical (w) component of windspeed has been analysed. Values of the standard deviation of vertical windspeed σ_w and horizontal windspeed U were computed, as well as the skewness, and autocorrelation functions of w, using programs based on work by McCracken (1993). Values of the Lagrangian time scale T_l were derived by integrating the autocorrelation functions between zero lag and the lag of the function's first zero-crossing, and subsequent multiplication by U/σ_w (Equation a4).

The half-hourly in-canopy values of σ_w and T_l were normalised to the corresponding value measured at 50 m, and the normalised values were averaged for each level. Finally, these values have been fitted to simple power functions of relative height in the canopy, using a non-linear least squares procedure.

Inverse Lagrangian modelling

The background of the model is described in Appendix 1. The resulting operational model, as used here, requires input of concentrations at a sufficient number of heights within the canopy, and values of σ_w and T_l, also at several heights. The turbulence profile should preferably be parameterized to a simple, continuous model, and Raupach (1989a) suggests using simple power laws to represent the variation of σ_w and T_l with height.

It is useful to reiterate the underlying assumptions of the model. In principle, LNF theory requires locally homogeneous, stationary turbulence, with Gaussian velocity distributions. Especially within plant canopies, however, the last two conditions are often not met. Windspeed distributions, both vertical and horizontal, are mostly skewed, but Raupach (1989a) argues that in general it is reasonable to use the model within canopies. A comparison with a stochastic random-flight dispersion model showed that model results are not very sensitive to velocity skewness. Raupach (1989a) also showed that using Equation (a4) for T_l instead of indirectly derived values significantly improved model results compared to measurements, which provides some confidence in assuming simple similarity of T_l with the Eulerian time scale T_E. Stationarity can be expected to be a real problem when profiles are rapidly changing, for example early in the morning. Whether the model still holds inside a tropical rainforest may become clear in this paper.

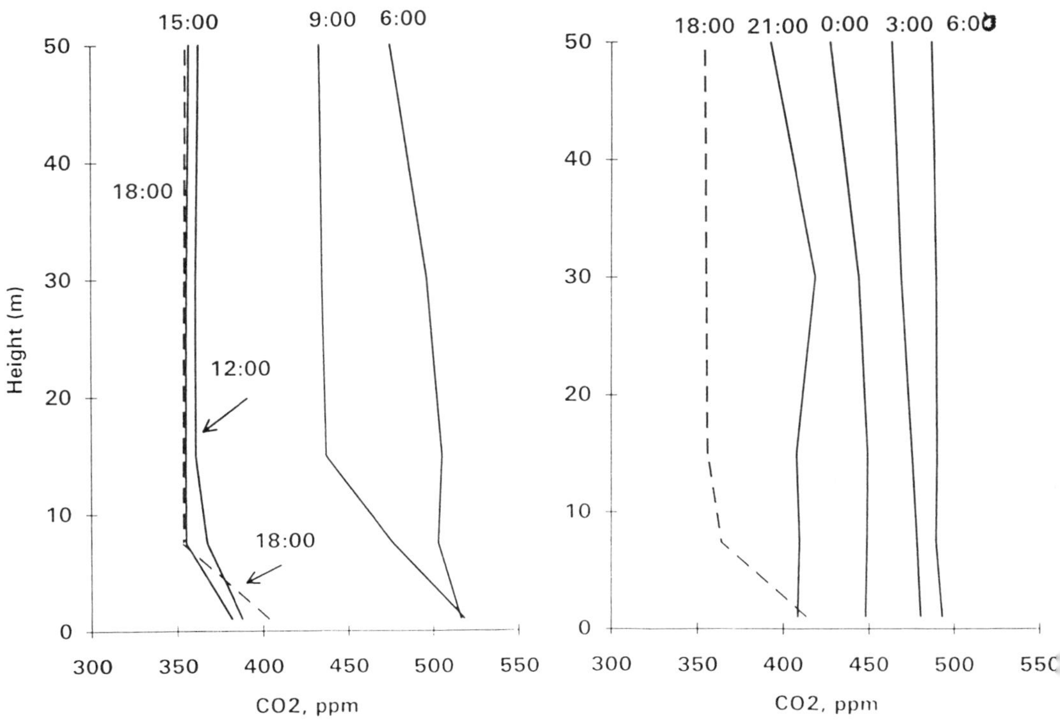

Figure 2 *Example of the daily course of CO_2 concentration profiles, for daytime (left) and night-time (right). Times are indicated right of the top of a few profiles. The dashed line represents the profile at 18:00 h, and is the same at the left-hand side and the right-hand side.*

RESULTS

Profiles

Figure 2 shows the development of the CO_2 concentration profile during a typical day. Just before dawn concentrations are highest, and almost constant with height throughout the canopy and above, up to the top of the tower. After dawn, the profile is rapidly depleted in all but the lowest measurement point (1 m above the ground). From about noon till sunset, the profile is fairly constant, with low CO_2 concentrations (about 350 ppm) from 7.5 m upward. Close to the ground, high concentrations persist, up to 400 ppm. During the night, the profile first rapidly 'straightens' by CO_2 enrichment of the canopy air space above 7.5 m, followed by a steady enrichment, at similar rates throughout the canopy, sometimes to very high values just before sunrise.

Stable isotope profiles

Measurements of stable isotope concentrations were made during the dry season in September 1993 and during the wet season in May 1994. No concurrent isotope profiles are available for the period when turbulence measurements were made. Figure 3 shows four instances of the vertical distribution of isotope ratios, with the associated overall CO_2 concentrations, for morning and afternoon in both the dry and the wet season. Clearly, the high concentrations near the ground shown before are present again, and they are associated with a strong depletion of ^{13}C in the air, with the strongest effect in the morning and in the wet season. Figure 4 summarises the analysis of isotopic ratios on plant material, sampled at various heights in the canopy. The low values for understorey plants are striking, indicating very strong ^{13}C depletion of the tissues. Again, the wet season values are lower than the dry season values. Litter and roots give somewhat less extreme values, since they represent plant material from all heights.

In general, the values for stable isotope ratios are similar to, or more negative than, previous measurements (Medina *et al.*, 1986; van der Merwe and Medina, 1988). The stronger ^{13}C depletion of plant tissue in the wet season suggests a rapid turnover of biomass, since with a low turnover rate such rapid changes in tissue composition could not occur.

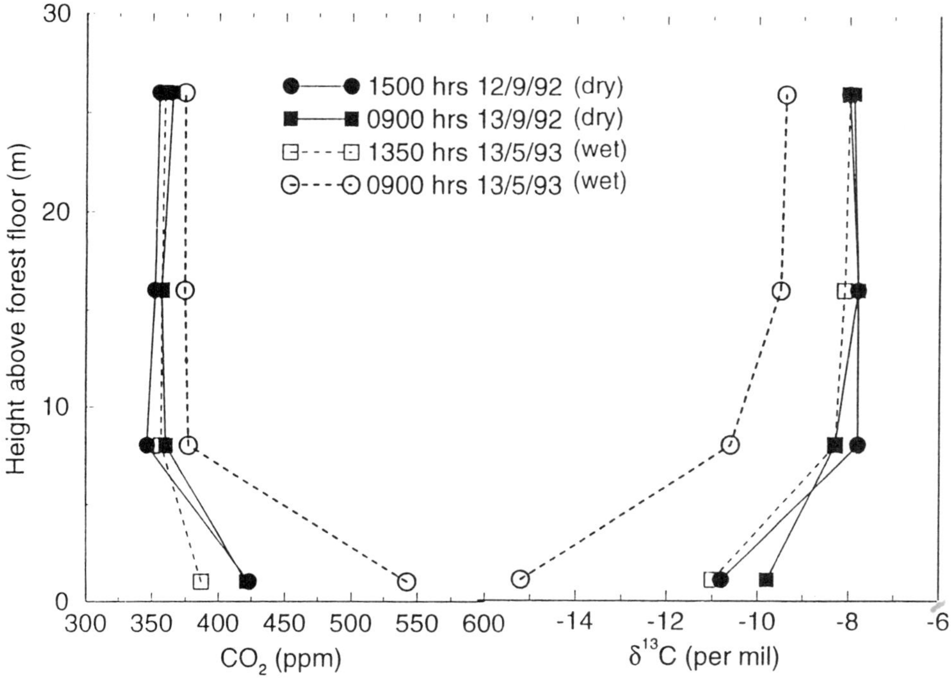

Figure 3 Simultaneously measured profiles of CO_2 concentration (left) and carbon isotope ratio (right), for two days in different seasons and for two different times of the day.

Discrimination and C_c/C_a ratios.

If photosynthetic discrimination by a canopy were invariant with foliage height and invariant with time, then, providing that fractionation did not occur during respiration, a straight line relationship between isotopic composition of CO_2 and its partial pressure would occur (Keeling, 1958). Nevertheless, variations in isotopic discrimination might be expected to change with photon irradiance, and hence within canopies (Francey and Farquhar, 1982).

Further, diurnal variations in isotopic discrimination are also likely as the ratio C_c/C_a changes with light and humidity (Lloyd *et al.*, 1996). Even so, despite these complications a very good correlation (r^2>0.95) could be shown between $\delta^{13}C$ and the CO_2 concentration. This correlation was used to predict profiles of isotope ratios in the air for the whole wet season. These profiles for air and plant tissue have been averaged, weighting with modelled canopy photosynthesis rate (Lloyd *et al.*, 1996), and the results are plotted in Figure 5 and compared with the average profile of plant tissue isotope ratio. The $\delta^{13}C$ profile of canopy air shows little variation with height, except close to the forest floor. This can be explained by a similar profile shape of C_a. However, the $\delta^{13}C$ profile of foliage does show an appreciably progressive ^{13}C depletion with depth into the canopy, and the combined $\delta^{13}C_{air}$ and $\delta^{13}C_{plant}$ profiles indicate an increasing discrimination (Δ) with depth, except for the very lowest level. Keeping in mind that daytime C_a generally varies little with depth, with only an increase near the forest floor, we can see from Equation (1) that the variation in Δ indicates that the *A*/*g* ratio decreases with depth. Since radiation decreases with depth as well, both *A* and *g* can be expected to decrease, but apparently *A* does so faster than *g*.

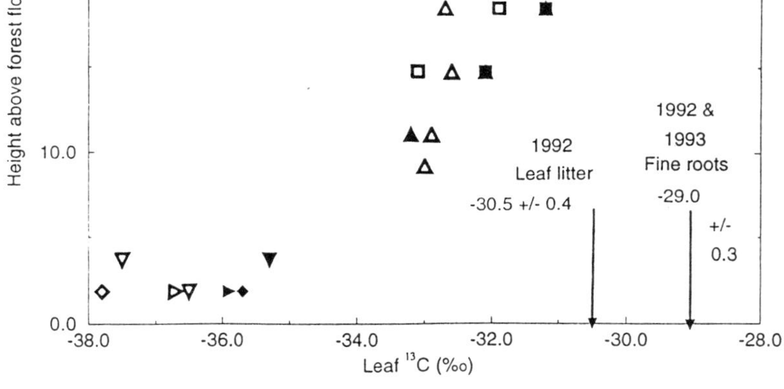

Figure 4 Isotope ratios measured in foliage taken from different heights and different seasons in the forest, as well as from leaf litter and fine roots in the soil (indicated by arrows to the abscissa). The different symbols refer to different species.

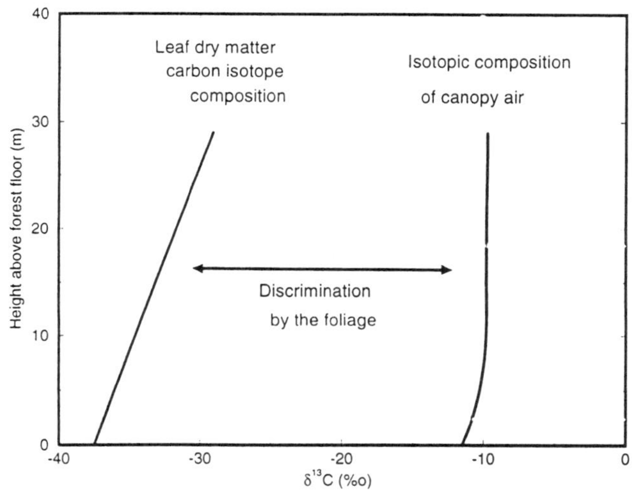

Figure 5 Comparison of averaged and smoothed (see text) $\delta^{13}C$ profiles for foliage (left) and canopy air (right), indicating the amount of discrimination by stomata and physiology

Turbulence

A small part of the time series for vertical windspeed at two levels is shown in Figure 6. Typically two types of motion seem to be present: one component fluctuating rapidly and one very slowly. The time window used to calculate T_l is 102 seconds, less than the typical period of the slow component. A longer time window did not seem to change the values of the integral time scales. Still, the slow component is interesting, since it may represent processes that are not taken into account by the model. Intermittent turbulence (Finnigan and Brunet, 1995) is one such process, but the regular pattern in Figure 6 rather suggests a phenomenon like gravity waves, that may be associated with stable conditions near the ground. Gravity waves, unlike intermittent turbulence, do not contribute to mixing. Unfortunately, it is hard to decide which process is active here without more detailed analysis of the patterns in the raw turbulence data.

Overall, individual half-hourly in-canopy values of σ_w and T_l showed little correlation with the corresponding 50 m values. After separating daytime and night-time values, however, relationships were clearer and mean normalised values could be computed for each level with acceptable standard errors, for daytime. At night, the small standard deviations and large time scales inside the canopy seemed effectively decoupled from the above canopy flow, and the vertical distribution functions have no relationship with the values at 50 m. It is possible that this decoupling is the result of stable stratification above the canopy whilst the cooling canopy causes instabilities inside the canopy (Grace *et al.*, 1996b). For night-time we used:

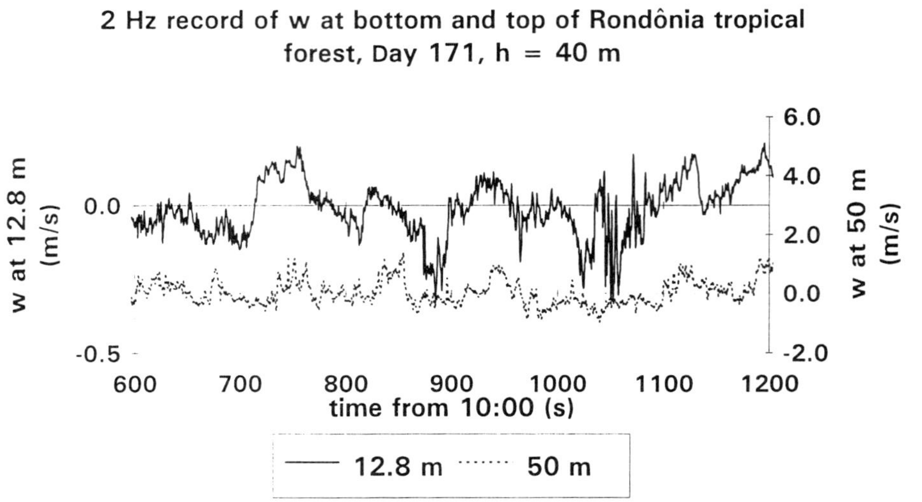

Figure 6 Arbitrary daytime 2 Hz (block averaged) record of raw measurements of vertical wind velocity *w*, at the bottom and top of the tower

$$\sigma_w(z) = 0.05 - 0.012\,(^z/_h) \qquad \text{m s}^{-1}$$
$$T_l(z) = 14 + 9\,(^z/_h)^2 \qquad \text{s} \qquad (2)$$

In daytime, Lagrangian time scales were on average constant with height and could be described by

$$T_l(z) = T_l(50)\ 0.65 \qquad \text{s} \qquad (3)$$

Unfortunately, there were no turbulence data available for close to the ground, and this caused considerable uncertainty in the fitting of functions for height dependence of s_w. Three different, but plausible functions for daytime conditions are listed in Table 1 and plotted in Figure 7. The main uncertainty is whether s_w approaches zero near the ground, or stays finite. As is shown in the next section, this makes considerable difference for the predictions of the LNF model.

It is useful to compare the results of the present analysis with the simple scaling laws suggested by Legg *et al.* (1986) and Raupach (1989b). These laws predict that, in a near-neutral atmosphere above a canopy, σ_w is greater by a factor of about 1.25 than the friction velocity u_*, and declines exponentially with depth into the canopy, whilst T_l is nearly constant with height up to $2h$ and adopts values such that the ratio $(T_l u_* / h)$ is of the order of 0.3. Figure 8a shows the above-canopy relationship of σ_w and u_*. It is clear that the ratio of both variables is generally larger than 1.25. This is not an uncommon observation and is probably due to the frequently unstable

Table 1 Average predicted partitioning of CO_2 fluxes over the canopy layers and soil for daytime, using two different configurations of the inverted Lagrangian model and three different parameterizations of the s_w profiles. Shown are the fluxes F_c for each level i, normalised by either the total canopy flux ($F_{c,total}$) or the lower canopy flux ($F_{c,1}$). The standard deviations of the average relative fluxes are shown in brackets.

a $\sigma_w(z) = \sigma_w(50)\ 0.91\ (z/h)^{1.4}$ σ_w zero at ground

	2 layers	2 layers plus soil	
i: Layer	$F_{c,i}/F_{c,total}$	$F_{c,i}/F_{c,1}$	$F_{c,i}/F_{c,total}$
0: soil	-	-0.12 (0.95)	0.09 (0.55)
1: 0-20 m	-0.096 (0.12)	1	-0.22 (1.0)
2: 20-40 m	1.1 (0.12)	-11.2 (63.7)	1.12 (0.33)

b $\sigma_w(z) = \sigma_w(50)\ \left(0.15 + 0.85\ (z/h)^{1.4}\right)$ σ_w small but constant near ground

	2 layers	2 layers plus soil	
i: Layer	$F_{c,i}/F_{c,total}$	$F_{c,i}/F_{c,1}$	$F_{c,i}/F_{c,total}$
0: soil	-	-0.98 (0.10)	-1.3 (7.2)
1: 0-20 m	-0.23 (0.024)	1	1.3 (7.2)
2: 20-40 m	1.23 (0.024)	-0.20 (0.56)	0.96 (0.6)

c $\sigma_w(z) = \sigma_w(50)\ \left(0.23 + 0.97\ (z/h)^{3.5}\right)$ σ_w larger and constant near ground

	2 layers	2 layers plus soil	
i: Layer	$F_{c,i}/F_{c,total}$	$F_{c,i}/F_{c,1}$	$F_{c,i}/F_{c,total}$
0: soil	-	-0.94 (0.051)	-5.9 (27.3)
1: 0-20 m	-0.30 (0.030)	1	6.5 (28.9)
2: 20-40 m	1.30 (0.030)	-0.017 (0.27)	0.43 (1.9)

atmosphere over the forest. Figure 8b shows the relationship of $(T_l\, u_* / h)$ within the canopy with that above. The fitted equations show that the Lagrangian time scales are essentially constant with height within the canopy, but are substantially larger at

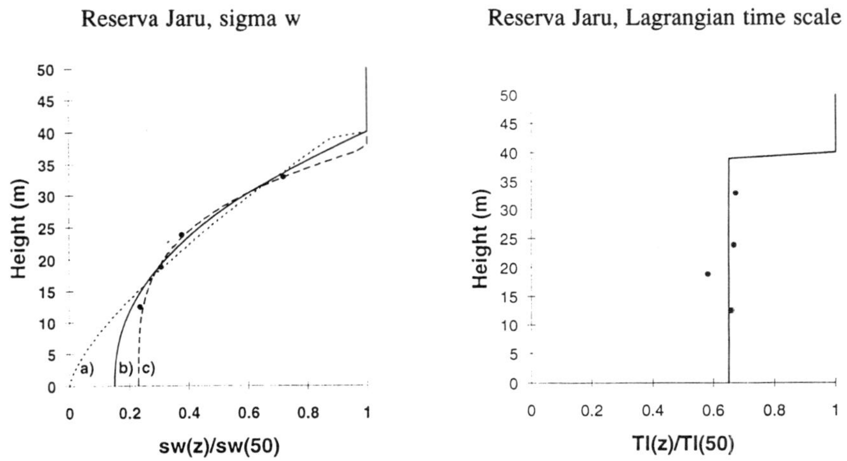

Figure 7 (left) Three fitted normalised daytime profiles of vertical velocity standard deviation. The labels at the bottom of the curve refer to the fitted equations **a**, **b** and **c** in Table 1. (right) The fitted Lagrangian time scales profile. In each graph the solid dots represent the averaged normalised values derived from measurements. The associated standard errors (not shown) are typically an order of magnitude smaller than the means.

the top of the tower, which is still well below $2h$. Furthermore, the ratios inside the canopy are about five times smaller than the value of 0.3 predicted by the scaling laws. The possible reasons for this will be discussed below.

Derived source/sink distributions

Two days in June 1993, for which good profile data were available for both turbulence and CO_2 concentrations, were used for simulation runs with the inverse Lagrangian model. The model was initially set up to predict CO_2 source strengths for only two levels: the lower canopy half and the upper canopy half. No more levels were selected to minimise instabilities in the calculations. Several different power functions for the height dependence of σ_w were applied (Table 1), and the results used for a sensitivity analysis. First, we concentrate on the set of results that predicted total canopy fluxes most closely, using the power function that forces σ_w through zero at the ground (Table 1c).

In Figure 9a we compare the sum of the two sources, i.e. the total predicted canopy flux, with the flux measured by eddy covariance. Neither of these is corrected for change in storage (Grace *et al.*, 1995b). The data agree reasonably well. The modelled flux is more variable: it seems to overestimate early morning CO_2 efflux, for example. Figure 9b shows the predicted fluxes for the two compartments of the canopy. This simulation predicts a small efflux from the lower canopy half, compensated in daytime by an uptake in the upper half that exceeds the total canopy uptake. Also in Figure 9b is the CO_2 efflux from the soil, derived from directly measured soil efflux (chamber measurements by Meir *et al.*, 1996): in daytime this

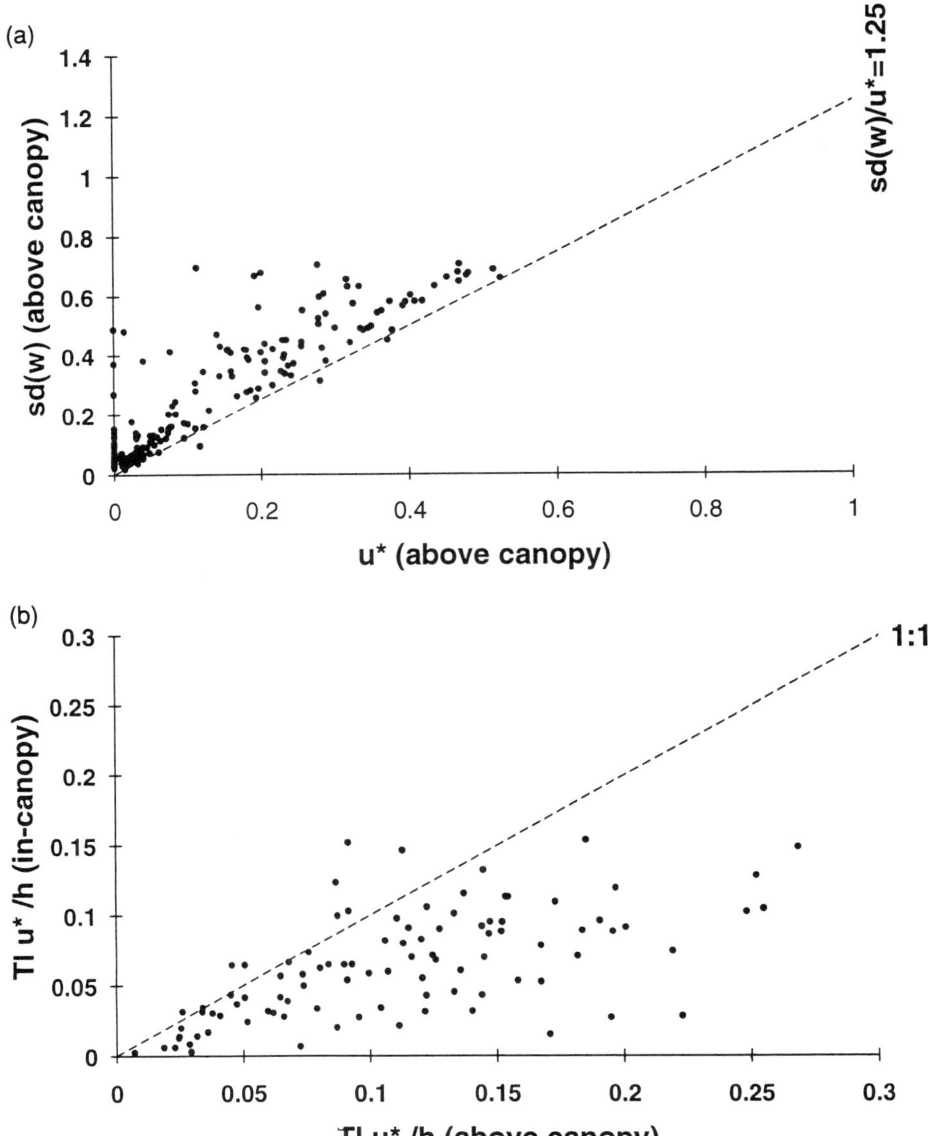

Figure 8 (a) The relationship of friction velocity (u_*) and vertical velocity standard deviation — here *sd(w)* — for the 50 m level. (b) The relationship of the 50 m Lagrangian time-scale normalised according to Legg *et al.* (1986) with its value within the canopy.

is about five times as great as the efflux from the first 20 metres of the canopy. If we accept both the model results and the soil efflux data, the conclusion must be that there is substantial uptake of CO_2 (of the order of 5 μmol m^{-2} s^{-1}) in the lower canopy. The generally observed low radiation fluxes, and indeed photosynthesis rates themselves in the lower canopy (McWilliam *et al.*, 1996), do not support this conclusion.

Since there was considerable uncertainty in the turbulence functions, it seems useful to explore the sensitivity of the model to them. Also, the ability of the model to correctly resolve efflux from the soil was explored, by adding a soil layer to the model. Three different functions describing different behaviour of σ_w, shown in Table 1 and plotted in Figure 7 were applied to the two versions of the inverse model, while keeping T_l and the observed CO_2 profiles the same.

Table 1 shows the average predicted partitioning of CO_2 fluxes over the canopy layers and soil (if applicable) for daytime, by showing the fluxes for each level, normalised by either the total canopy flux or the lower canopy flux. It also shows the standard deviations of those relative fluxes, and if a standard deviation is large the flux is less dependent of the normalising flux. For a two-layer canopy, increasing σ_w close to the ground and at the same time assuming less change in σ_w with height increases the efflux from the lower canopy. This would suggest less compensating uptake of CO_2 released from the soil if not the total predicted canopy flux in those cases would be very large and greatly exceed the flux measured by eddy covariance (predicted total fluxes are not shown in Table 1). In a 'three-layer' setup, the soil flux predicted with the same turbulence regime as used in Figure 9 is very small and negative: it is an uptake (Table 1a). Clearly adding a soil layer does not solve our problems here. Increasing σ_w close to the ground increases the soil efflux, but at the same time an equally large compensating uptake in the lower canopy is simulated. Whilst the soil efflux and uptake in the lower canopy can grow to large (opposite) values, the predicted overall canopy fluxes are of the same order as the measured canopy fluxes, and are quite independent of the soil and lower canopy fluxes. Summarising, of the σ_w functions that fit the data reasonably well, some of them lead to unrealistic source/sink distributions whilst others yield overall fluxes that are too high.

DISCUSSION

The concentration profiles, isotope ratios and modelling effort all contribute information about the location and strength of sources and sinks of CO_2 in the forest. The change in time of concentration profiles provides complementary information to the modelling using profile shape and turbulence. The time changes indicate an early morning flush of CO_2 from the upper canopy. This is also reflected in the eddy covariance measurements and the model results. At night, weak mixing, probably a result of instability, may be responsible for the CO_2 concentrations increasing at similar rates at all heights while the main source of CO_2 is respiration from the soil (Meir *et al.*, 1996).

Contrary to the observations of night-time soil efflux and mixing, the night-time sources and sinks predicted by the model are weak and hardly significant. An explanation for this discrepancy might be the highly intermittent (inhomogeneous) nature of night-time turbulence, in which circumstances the Lagrangian scheme is not valid.

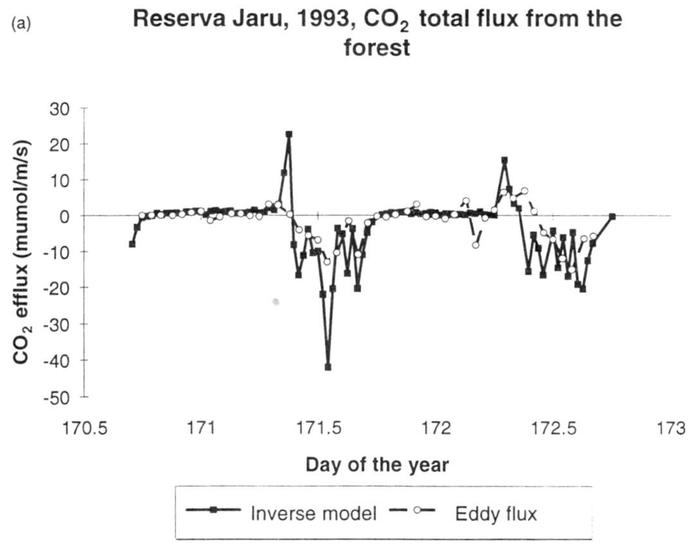

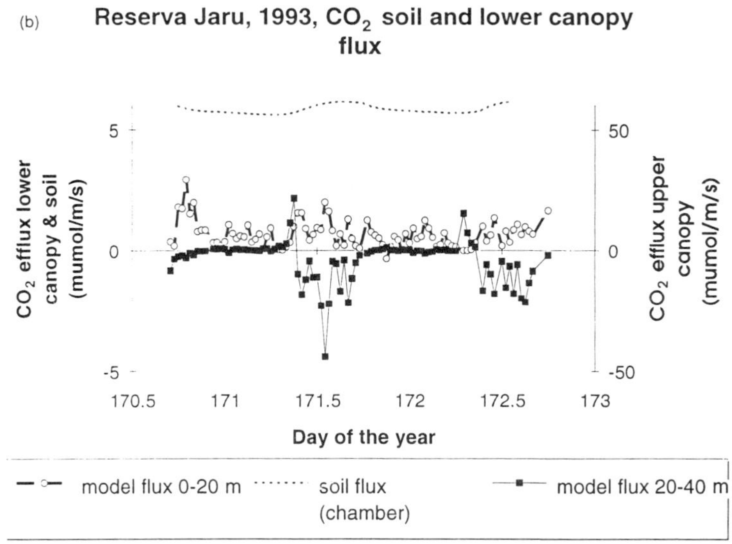

Figure 9 (a) Comparison of two-day time series of overall canopy CO_2 flux measured by the inverse Lagrangian model and Eddy covariance. (b) Comparison of modelled lower and upper canopy CO_2 efflux, also compared with the soil efflux extrapolated from average soil efflux chamber measurements.

The daytime source/sink distributions predicted by the model, using various alternative assumptions for the distribution of σ_w, are either not realistic in shape, or the magnitude of the fluxes is too large. A realistic source/sink profile would show substantial soil efflux, a relatively small sink in the lower canopy and high uptake in the upper canopy half. None of the model configurations investigated showed such a distribution, and in fact the sensitivity analysis suggests that it could only be realised in a situation where σ_w *decreases* with height in the lower canopy, which is hard to imagine in a real daytime situation.

Strongly compensating fluxes in the lower canopy are, as argued before, not realistic, but two observations in this study might challenge this argument. Firstly, we observed large negative $\delta^{13}C$ values in plant material near the ground. This suggests that some of the carbon being taken up by those plants originated from respiration, probably from the soil. However, we also saw a large difference between isotopic ratios in air and plants, which shows stronger discrimination than higher up in the canopy. The assimilation may also involve a relatively large amount of previously respired CO_2 molecules, but that does not necessarily mean assimilation itself is large (Lloyd *et al.*, 1996). In other words, low $\delta^{13}C$ values can result from little assimilation of (respired) CO_2 that is strongly fractionated, and is not an indicator of strongly compensating fluxes between soil and lower canopy.

Secondly, the model predictions are not surprising in view of the typical daytime shape of the CO_2 profiles, if simple flux-gradient theory were to apply. The persistence of high CO_2 concentrations close to the ground suggests very weak turbulence and absence of coupling with the air above, and a relatively strong source of CO_2 near the ground. The equally persistent kink in the profile, just above the lowest level, implies that the CO_2 is somehow removed from that level efficiently, either by high uptake or by turbulence. As mentioned before, it has been shown repeatedly that flux-gradient theory does not apply inside a canopy, and this argument rather illustrates similarities between the LNF scheme and flux-gradient theory, which both seem to fail in this dense vegetation. The essential difference between flux-gradient theory and LNF theory is in accounting for near-field fluxes. Two studies (Dolman and Wallace, 1991; van den Hurk and McNaughton, 1995) have shown that the inclusion of the near-field effect makes little difference in predicted overall canopy fluxes.

Since the Lagrangian time scales were small, the model was rerun with much larger time scales, to see if this might bring the results closer to expectation. Surprisingly, the reverse was true. The extreme values in the lower canopy only increased. The observed low values for $(T_l u_* / h)$ may be explained by the denseness of the forest, so that turbulent eddies do not penetrate to the very bottom. Indeed, if in the ratios h is replaced by $(h\text{-}d)$, where d is the zero-plane displacement, assumed at $0.75h$, then they are very close to the 0.3 observed by Legg *et al.* (1986). The difference between the ratio at 50 m and those below may be explained by the poor definition of h. In fact, the forest may be seen as a two-stage canopy, where crowns emerging above the 30 m high main canopy reach up to 45 m, and may pose a different set of conditions to the turbulence at 50 m as opposed to the lower levels.

The strong CO_2 gradient near the ground was derived from the difference between the measurement points at 1 m and 7.5 m height. It may be questioned whether the persistently high concentrations at 1 m were real. As far as we know, possible systematic errors can be ruled out. All concentrations in the profile were measured using the same gas analyser. Theoretically, there might be some accumulation of CO_2 due to near-soil drainage flows converging below the tower. The occurrence of the right topography for this is highly unlikely, though, and so are drainage flows in daytime at the bottom of a dense vegetation. Another possible artefact may arise from the rigid windowing of the canopy in the model (Raupach, *pers. comm.*). Improvements to this have been suggested by van den Hurk & McNaughton (1995).

If we accept the observed concentration and turbulence profiles, the source/sink distribution that we might find acceptable (high soil efflux, low uptake in the lower canopy and high uptake in the upper canopy) would probably only be simulated by a model that takes account of events of rapid displacement over distances larger than half the canopy height. In principle, the near-field transport in LNF theory allows for such transport, but for it to predict the 'acceptable' distribution the near-field should be much larger relative to the far-field. Since the near-field and the far-field fluxes both scale with T_l, this is hard to achieve in LNF theory by adjusting the turbulence statistics. Such non-local transfer is associated with dominance of canopy-scale gusts and large velocity skewness (Finnigan and Brunet, 1995). Turbulence in these cases is locally non-homogeneous and violates the assumptions of LNF theory.

Baldocchi (1992) also reached the conclusion that a simple Lagrangian model should somehow represent inhomogeneous turbulence, when he unsuccessfully tried to model observed concentration profiles inside a soya bean crop using LNF theory. The effect of large skewness was, however, tested by Raupach (1989a) using a random-walk alternative to LNF theory: contrary to conclusions here the effects were small.

Whether in the present case the canopy turbulence was dominated by canopy-scale gusts as well, or whether other processes were active, remains to be investigated, (concluded from Figure 6). If gravity waves were present, they will have contributed to measurements of σ_w, leading to an overestimate of the mixing. However, the sensitivity tests showed that lowering σ_w does not improve our results. The occurrence of intermittent turbulence, which was not taken into account in the model, seems to provide the best explanation for the present results, but we stress that the occurrence of such features in our data have not been demonstrated yet.

CONCLUSIONS

Modelling the location of sources and sinks by combining profiles of CO_2 concentration and turbulence statistics predicted either too little CO_2 efflux from the soil or too much uptake in the lower canopy. Isotope analysis does suggest a higher degree of recycling of CO_2 in the lower canopy, but this is not necessarily related to a high fixation rate.

The discrepancies between model results and expectation may be explained by the neglect of intermittent or slow, but intense, turbulent events sweeping though the canopy at regular intervals and displacing CO_2 to the upper canopy. As a first step to improve this, a random-walk, stochastic Lagrangian model should probably be applied to the present situation. The study stresses the need for more within-canopy turbulence and physiological work in tropical rainforests, including CO_2 flux measurements at various heights and analysis of the spatial structure of turbulence.

ACKNOWLEDGEMENTS

We are very grateful to Michael R. Raupach (CSIRO, Canberra) for sending us the code of his model and guiding us with helpful advice throughout the analysis. We thank the personnel of INCRA and IBAMA at Ji-Paraná for their help and for providing transport and accommodation during the field work, and several anonymous referees for their constructive comments. The work was supported by NERC-TIGER grantsGST/02/597 and GST/02/984.

REFERENCES

Baldocchi, D.D. 1992. A Lagrangian random-walk model for simulating water vapour, CO_2 and sensible heat flux densities and scalar profiles over and within a soya bean canopy. *Boundary-Layer Meteorol.* **61**, 113-144.

Cabral, O.M.R., McWilliam, A-L.C. and Roberts, J.M. 1996. The in-canopy microclimate of Amazonian forest and estimates of transpiration.This volume.

Dolman, A.J. and Wallace, J.S., 1991. Lagrangian and *K*-theory approaches in modelling evaporation from sparse canopies. *Q. J. Roy. Met. Soc.* **117**, 1325-1340.

Farquhar, G.D., Ehlringer, J.R. and Hubick, K.T., 1989. Carbon isotope discrimination and photosynthesis. *Ann. Rev. Plant Physiology and Plant Molecular Biol.* **40**, 503-537.

Farquhar, G.D., O'Leary, M.H. and Berry, J.H., 1982. On the relationship between carbon isotope discrimination and intercellular carbon dioxide concentration in leaves. *Australian J. Plant Physiol.* **7**, 315-327.

Finnigan, J.J. and Brunet, Y. 1995. Turbulent airflow in forests on flat and hilly terrain. In: *Wind and Trees,* (Eds.) M. Coutts and J. Grace. Cambridge University Press, UK. 3-39.

Francey, R.J. and Farquhar, G.D. 1982. An explanation of $^{13}C/^{12}C$ ratios in tree rings. *Nature* **297**, 28-31.

Grace, J., Lloyd, J., McIntyre, J., Miranda, A.C., Meir, P., Miranda, H., Nobre, C., Moncreiff, J., Mahli, Y. Wright, I.R., and Gash, J., 1995a. Carbon dioxide uptake by an undisturbed tropical rainforest in south-west Amazonia, 1992to 1993. *Science,* **270**, 778-780.

Grace, J., Lloyd, J., McIntyre, J., Miranda, A.C., Meir, P., Miranda, H., Moncrieff, J.M., Massheder, J., Wright, I.R., and Gash, J., 1995b. Fluxes of carbon dioxide and water vapour over an undisturbed tropical forest in south-west Amazonia. *Global Change Biology,* **1**, 1-12

Grace, J., Lloyd, J., McIntyre, J., Miranda, A., Meir, P. and Miranda, H. 1996a. Carbon flux over Amazon rainforest in Rondônia. This volume.

Grace, J., Mahli, Y., Lloyd, J. McIntyre, J. Miranda, A.C., Meir, P. Miranda, H. 1996b. The use of eddy covariance to infer the net carbon dioxide uptake of Brazilian fainforest. *Global Change Biology,* **2**, in press.

Hubick, KT., Shorter, R. and Farquhar, G.D., 1988. Heritability and genotype x environment interactions of carbon isotope discrimination and transpiration efficiency in peanut (*Arachis hypogaea* L.). *Autralian J. Plant Physiology* **15**, 799-813.

Keeling, C.D., 1958. The concentration and isotopic abundances of carbon dioxide in rural areas. *Geochim. Cosmochim. Acta* **13**: 322-334

Legg, B.J., Raupach, M.R. and Coppin, P.A., 1986. Experiments on scalar dispersion within a plant canopy, part III: An elevated line source. *Boundary-Layer Meteorol.* **35**, 227-302.

Lloyd, J. and Farquhar, G.D. 1994. ^{13}C discrimination during CO_2 assimilation by the terrestrial biosphere. *Oecologia* **99**, 201-215

Lloyd, J., Grace, J., Miranda, A.C., Meir, P., Wong, S.C., Miranda, H.S., Wright, I.R., Gash, J.H. and McIntyre, J., 1995. A simple calibrated model of Amazon rainforest productivity based on leaf biochemical properties. *Plant, Cell & Environment* **18**, 1129-1145.

Lloyd, J., Kruijt, B., Holliger, D.Y., Grace, J., Wong, S.-C., Francey, R.J, Kelliher, F., Farquhar, G.D., Shulze, E.-D., Miranda, A.C., Miranda, H.S., Wright, I.R. and Gash, J.H.C., 1996. Vegetation effects on the isotopic composition of atmospheric CO_2 at local and regional scales: theoretical aspects and a comparison between rainforest in Amazonia and a boreal forest in Siberia. In press.

Martinelli, L.A., Devol, A.H., Victoria, R.L. and Richey, J.E., 1991. Stable carbon isotope variation in C_3 and C_4 plants along the Amazon river. *Nature* **353** 57-59.

McCracken, P.J., 1993. *Turbulent Exchange of Momentum and Carbon Dioxide of a Sitka spruce Plantation* PhD thesis, University of Edinburgh.

McWilliam, A-L.C. Cabral, O.M.R., Gomes, B.M., Esteves, J.L. and Roberts, J.M., 1996. Forest and pasture leaf-gas exchange in south-west Amazônia. This volume.

Medina, E., Montes, G., Cuelvas, E. and Rokzandic, S. 1986. Profiles of CO_2 concentration and $\delta^{13}C$ values in tropical rainforests of the upper Rio Negro Basin, Venezuela. *J. Trop. Ecol.*, **2**, 207-217.

Medina, E. and Minchin, P., 1980. Stratification of $\delta^{13}C$ values of leaves in Amazonian rainforests. *Oecologia* **45**, 377-378.

Meir, P., Grace, J., Lloyd, J. and Miranda, A.C., 1996. Soil respiration in a rainforest in Amazonia, and in Cerrado in Central Brazil. This volume.

Moncrieff, J.B., Massheder, J.M., De Bruin, H., Elbers, J., Friborg, T., Heusinkveld, B., Kabat, P., Scott, S., Soegaard, H. and Verhoef, A., 1996. A system to measure surface fluxes of momentum, sensible heat, water vapour and carbon dioxide. *J. Hydrol.* in press.

Pinker, R.T. and Holland, J.Z., 1988. Turbulence structure of a tropical forest. *Boundary-Layer Meteorol.* **43**, 43-63.

Raupach, M.R., 1987. A Lagrangian analysis of scalar transfer in vegetation canopies. *Q. J. Roy. Met. Soc.* **113**, 107-120.

Raupach, M.R., 1989a. A practical Lagrangian method for relating scalar concentrations to source distributions in vegetation canopies. *Q. J. Roy. Met. Soc.* **115**, 609-632.

Raupach, M.R., 1989b. Applying Lagrangian fluid mechanics to infer scalar source distributions from concentration profiles in plant canopies. *Ag. and For. Meteorol.* **47** 85-108.

Shuttleworth, W.J., Gash, J.H.C., Lloyd, C.R., Moore, C.J., Roberts, J., De O. Marques Filho, A., Fish, G., De Paula Silva Filho, V., De Nazaré Gões Ribeiro, M., Molion, L.C.B., De Abreu Sá, L.D., Nobre, J.C.A., Cabral, O.M.R., Patel, S.R. and De Moraes, J.C., 1985. Daily variations of temperature and humidity within and above Amazonian forest. *Weather*, **40**, 102-108.

van den Hurk, B.J.J.M. and McNaughton, K.G., 1995. Implementation of near-field dispersion in a simple two-layer surface resistance model. *J. Hydrol.*, **116**, 293-311.

Van der Merwe, N.J. and Medina, E., 1989. Photosynthesis and $^{13}C/^{12}C$ ratios in Amazonian rainforests. *Geochim. Cosmochim. Acta* **53**, 1091-1094.

Wilson, J.D., 1989. Turbulent transport within the plant canopy. In: *Estimation of Areal Evapotranspiration*, (Eds.) T.A. Black, D.L. Spittlehouse, M. Novak & D.T. Price. IAHS press, Wallingford, UK.

Wright, I.R., Gash, J.H.C., da Rocha, H.R. and Roberts, J.M., 1996. Modelling surface conductance for Amazonian pasture and forest. This volume.

APPENDIX 1

The background of the Lagrangian model and the inversion technique to infer sources from profiles is described in detail by Raupach (1989a, 1989b). The treatment will not be repeated here, but a brief outline is useful.

A Lagrangian model attempts to explicitly track parcels of air in time. Hence, if the source/sink strengths everywhere in the canopy were known, the concentration profile could be derived from all probabilities for parcels of air to move between pairs of points inside the canopy. The spread of a 'cloud' along height z from a source at any height in the canopy, at time t after release can be expressed in terms of the autocovariance of vertical windspeed:

$$d\sigma^2_z/dt = 2\sigma^2_w \int_0^t r_L(s)ds \qquad \text{(a1)}$$

where σ_w is the standard deviation of vertical windspeed w. The Lagrangian autocorrelation function $r_L(s)$ in homogeneous, stationary turbulence usually declines exponentially from unity at zero lag (s), to zero at a finite lag. The integral in Equation (a1) has units of time and is interpreted as the *Lagrangian time scale* T_l. It is a measure of the average persistence time of the turbulent eddies. Only over periods of time much greater than T_l the integral does not increase with time, and as Raupach (1989a) shows, the dispersion is then *diffusive*. This is labelled the *far field* and fluxes can be described in terms of a diffusivity K and the local concentration gradient as

$$F_c = K \cdot \partial c/\partial z = \sigma_w{}^2 \cdot T_l \cdot \partial c/\partial z \qquad \text{(a2)}$$

Since fluxes represent the integrated sources and sinks in the canopy, source/sink strengths will be reflected in the curvature of the gradient if the far field is dominant. Conversely, at shorter time scales, in the *near field*, the dispersal is non-diffusive. The spread of the cloud is nearly linear in time, more rapid than in the far field. Now, the dispersal must be modelled the Lagrangian way, i.e. from the transition probabilities of air from all canopy levels to all other canopy levels. Combining the two dispersion fields, the concentration profile $C(z)$ can be described as:

$$C(z) = C_0 + \int_{z_2=0}^{h} J(z, z_2)S(z_2)dz_2 \qquad \text{(a3)}$$

where C_0 is the background concentration, and J the time integrated probability of particles to originate from a source S at height z_2 and pass through height z. Raupach (1989a) works out an analytical approximation for this probability as a function of distance (z-z_2) and σ_w and T_l in z_2.

Raupach (1989b) uses this analysis to show that the inverse of Equation (a3), inferring source/sink strengths, is possible using the same model. To do this, (a3) is stated in a discrete form, defining a set of matrix equations, which can be solved for either concentrations C_i or source strengths S_i. He further shows, that the solution is reasonably stable if the number of levels with measured concentrations exceeds the number of levels for which a source strength is required.

In practice T_l cannot be measured by a fixed turbulence instrument. The integral of the vertical windspeed autocorrelation yields the *Eulerian* time scale T_E, which can only be related to T_l empirically, and only if the anisotropy of the turbulence is not too great. Raupach (1989b) suggests that:

$$T_l = T_E \, U/\sigma_w \qquad \text{(a4)}$$

provides a reasonable estimate, and this was used here.

RESUMO

A distribuição vertical das fontes e sorvedouros de CO_2 de uma floresta em Rondônia foi estudado a fim de entender melhor a troca total de CO_2 no dossel e as taxas de reciclagem de carbono dentro do ecossistema. Foram medidos a distribuição vertical de concentração de CO_2, a estrutura da turbulência (variância da velocidade vertical e escalas de tempo Lagrangeanos), e os perfis da razão de isótopos estáveis de carbono do ar e planta. Os dois primeiros perfis foram combinados como entrada de um modelo de dispersão Lagrangeano, o qual infere possíveis localizações de fontes e sorvedouros. As razões de isótopos fornecem a informação da fração de carbono que é reciclada dentro do dossel, bem como da atividade fisiológica das plantas. Os perfis de concentração mostraram claramente padrões diários, com uma alta e persistente concentração de CO_2 próximo à superfície. As escalas de tempo Lagrangeanos derivados de medidas de turbulência foram surpreendentemente baixos e pouco correlacionados com a turbulência acima do dossel. As razões de isótopos próximos à superfície foram das mais baixas medidas na Amazônia, sugerindo alta reciclagem. Entretanto, esses valores podem ser explicados, ao menos parcialmente, por alta discriminação resultante de baixa atividade fotossintética. O modelo Lagrangeano inverso calculou, de forma realística, os fluxos de CO_2 em todo dossel e provavelmente o efluxo do solo. Porém, a extração calculada na metade mais baixa do dossel foi excessivamente alta. Uma possível explicação para este comportamento, aparentemente contraditório, consiste em se haver desprezado os vórtices turbulentos coerentes com a escala do dossel, os quais misturam CO_2 intermitentemente em todo dossel. O estudo salienta a necessidade de uma compreensão mais detalhada da turbulência em dosséis na floresta tropical.

20 Carbon dioxide fluxes over a cerrado *sensu stricto* in central Brazil

A.C. MIRANDA[1], H.S. MIRANDA[1], J. LLOYD[2], J. GRACE[3], J.A. MCINTYRE[3], P. MEIR[3], P. RIGGAN[4], R. LOCKWOOD[4] and J. BRASS[5]

[1]*Universidade de Brasilia, Brazil*
[2]*Australian National University, Canberra, Australia*
[3]*University of Edinburgh, Edinburgh, UK*
[4]*USDA Forest Service, California, USA*
[5]*NASA Ames Research Centre, California, USA*

INTRODUCTION

The future of the Amazonian rainforest has been a matter of worldwide concern because drastic changes like replacing the forest by pasture may impose significant changes in the global climate and affect the precipitation regime in nearby ecosystems (Salati and Vose, 1984; Nobre *et al.*, 1991). In the central part of Brazil, bordering the Amazonian forest, there is a distinctive savanna-like vegetation with over a thousand endemic species, the *cerrado*. The possible coupling between these two large biomes is still a subject under study, but it seems likely that changes in Amazonia may affect the functioning and composition of cerrado ecosystems (through reducing the precipitation, for example); and *vice versa*, changes in the energy and mass transfer from the cerrado to the atmosphere may affect the weather regime in Amazonia.

The objective of the present work is to examine the rate of net CO_2 assimilation at intervals over the course of the year to obtain a first estimate of the carbon sink strength of a very common physiogonomic form of cerrado vegetation: the cerrado *sensu stricto*.

THE CERRADO

Cerrado vegetation occupies an area of about 2.0 x 10^6 km^2 in central Brazil (Figure 1). Although the Amazonian rainforest occupies a larger area (3.5 x 10^6 km^2) the cerrado is an important biome not just because of its vast area but also because it is species rich.. Two hundred and seventy species of grasses (Filgueiras, 1991) and about 800 woody species are known to exist (Heringer *et al.*, 1977) and the number of vascular plant species is likely to be over 2500 (Coutinho, 1990). Except for certain areas of tropical rain forest, the cerrado is the richest vegetation in the world with

Amazonian deforestation and climate. Edited by J.H.C. Gash, C.A. Nobre, J.M. Roberts and R.L. Victoria.

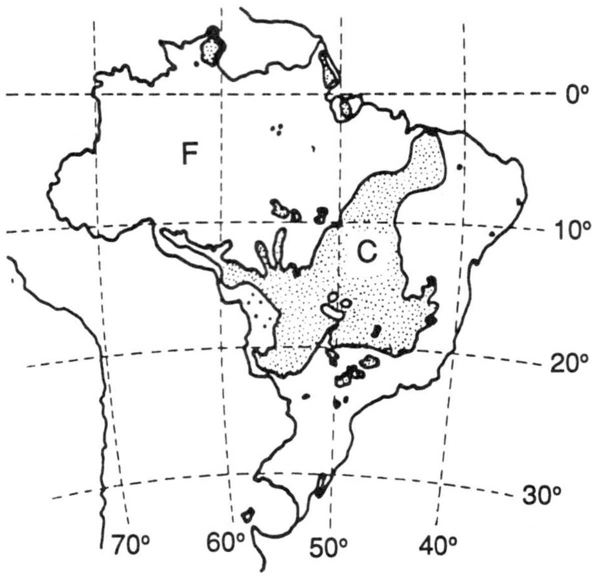

Figure 1 Distribution of cerrado (C) and tropical forest (F) in Brazil

over 450 vascular species per hectare (Eiten, 1994). An extensive description of cerrado vegetation has been given by Eiten (1972, 1992) and Ratter (1992).

Cerrado is characterised by a ground layer of grasses, small palms, shrubs and twisted or leaning trees. Although botanists may describe as many as 26 floristic physiogonomic forms (Eiten, 1983) the cerrado is usually divided into five forms: campo limpo, campo sujo, campo cerrado, cerrado *sensu stricto* (cerrado *s.s.* or simply cerrado) and cerradão. All campo forms together occupy 11.6% of cerrado, cerrado *sensu stricto* occupies 53% and cerradão is 8.3%. Several minor forms of vegetation account for the remainder (Dias, 1992). Campo limpo and campo sujo are both dominated mainly by C_4 grasses but they are distinct in appearance since the flora of the campo limpo has no woody plants at all, while there are at least some scattered shrubs in areas of campo sujo. Where the wood layer becomes denser and the view more closed, the vegetation is called campo cerrado if it is an open scrub, or cerrado *sensu stricto* if closed scrub: both may have scattered trees up to 8 to 10 m high. Cerradão can be either an open-canopy arboreal woodland with trees up to 25 m high and a dense grass underlayer (few shrubs) or a closed-canopy forest with a dense woody underlayer and sparse grasses (Figure 2). Fire is a common event in all forms of cerrado, and more frequent where the grassy layer is more abundant.

The major part of the cerrado vegetation has a tropical climate with monthly average temperatures above 18 °C. The annual rainfall ranges from 600 to2200 mm. Usually more than 85% of the precipitation occurs from October to April (wet season). Table 1 gives some climatic variables in the core area of the cerrado and Table 2 describes the vegetation in the wet and dry seasons.

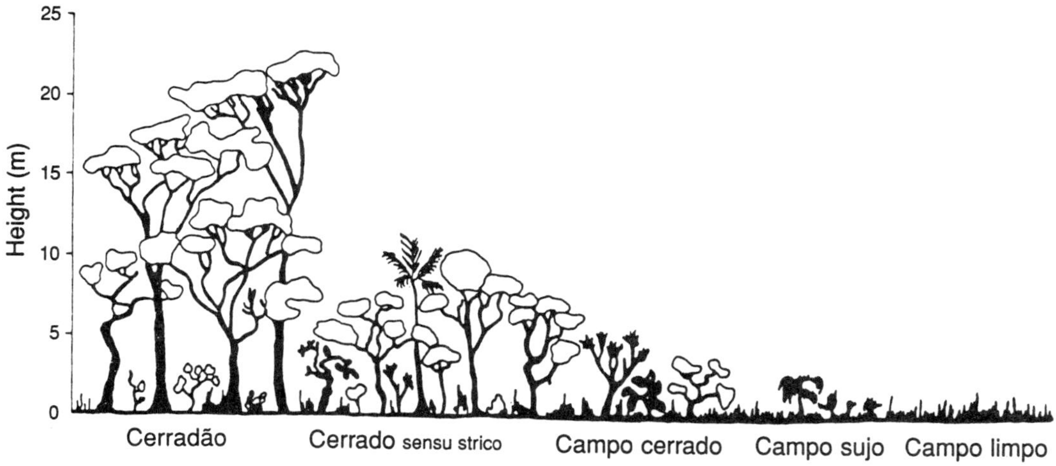

Figure 2 A gradient from campo limpo to cerradão, physiognomic forms of the cerrado vegetation

Table 1 Mean climatological data from the weather station at Brasilia (DF) for the period 1963 to 1990 (DNMET, 1992; CEPA/DF, 1984)

Month	Pa avg (hPa)	T avg (°C)	T max (°C)	T min (°C)	P (mm)	Rh (%)	S (hours)	Cloud (0 - 10)	u (m/s)
Jan	885.9	21.6	26.9	17.4	241.4	76	157.4	7.0	1.30
Feb	885.4	21.8	26.7	17.4	214.7	77	157.5	7.0	1.28
Mar	885.6	22.0	27.1	17.5	188.9	76	180.9	7.0	1.44
Apr	886.4	21.4	26.6	16.8	123.8	75	201.1	6.0	1.72
May	887.6	20.2	25.7	15.0	39.3	68	234.3	5.0	1.30
Jun	889.0	19.1	25.2	13.3	8.8	61	253.4	3.0	1.78
Jul	889.2	19.1	25.1	12.9	11.8	56	265.3	3.0	0.88
Aug	888.2	21.2	27.3	14.6	12.8	49	262.9	3.0	1.01
Sep	887.2	22.5	28.3	16.0	51.9	53	203.2	4.0	0.88
Oct	885.8	22.1	27.5	17.4	172.1	66	168.2	7.0	1.11
Nov	884.8	21.7	26.6	17.5	238.0	75	142.5	8.0	1.01
Dec	884.8	21.5	26.2	17.5	248.6	79	138.1	8.0	1.34
Total or mean	886.6	21.2	26.6	16.1	1552.1	67	2364.8	5.7	1.16

Pa avg = average atmospheric pressure; Ta avg = average air temperature; Tmax = maximum air temperature; Tmin = minimum air temperature; P = precipitation; Rh = relative humidity; S = sunshine hours; u = mean wind speed at 2.0 m (mean of 6 years - 1973/79). Cloud is given in tenths of the sky obscured by clouds.

Table 2 Summary of biophysical and physiological parameters for Amazonian rainforest and cerrado. Sources: Culf *et al.* (1995), Grace *et al.* (1995), Miranda *et al.* (1996), Meir *et al.* (1996).

	Amazonian rainforest (Reserva Jaru)	Cerrado (wet season)	Cerrado (dry season)
L*/	4.0	1.0	0.4
Height (m)	35	8 - 10	8 - 10
Above biomass (t ha^{-1})	300	40	30
g_s (mol m^{-2} s^{-1})	0.4 - 0.8	0.2 - 0.4	0.1 - 0.2
g_a (mol m^{-2} s^{-1})	0 - 3.0	0 - 3.0	0 - 3.0
z_o	4	1.2	1.2
$\lambda E/R_n$	0.7	0.6	0.4
F_{net} (μmol m^{-2} s^{-1})	15	12	4
F_{res} (μmol m^{-2} s^{-1})	5.5	4.1	2.1
F_{soil} (μmol m^{-2} s^{-1})	5.5	2.8	-
α (photon $atom^{-1}$)	20 - 40	68	77
albedo	0.13	0.13	0.18

L*/, leaf area index; g_s and g_a, surface and aerodynamic conductance; z_o, zero plane displacement; F_{net} and F_{res}, maximum net CO_2 uptake and maximum CO_2 loss at night; F_{soil}, soil respiration by the chamber method; a, apparent quantum efficiency at low irradiance.

MATERIAL AND METHODS

THE EXPERIMENTAL SITE

The measurements of radiation and fluxes of sensible heat, evapotranspiration and carbon dioxide over cerrado *sensu stricto* were carried out in the core area of the cerrado province, at the Reserva Ecológica de Águas Emendadas, DF, Brazil (15° 33' S, 47° 36' W, 1000 m above-sea-level) about 50 km northeast from Brasília. The Reserve of 15000 ha contains many forms of cerrado vegetation. The site used has more than 40 woody species with *Sclerolobium paniculatun* Vog. the dominant species, four palm species and an understorey of grass, of which five are especially common. The leaf area index (L*/) in the wet season is made up of trees (L*/ = 1.0), shrubs (L*/ = 0.1) and grasses (L*/ = 0.4). In the dry season the corresponding figures are 0.4, 0.1 and 0.2. All the common grassses except *Echinolaena inflexa* (Poir.) Chase. have been found to have the C_4 photosynthetic pathway; the trees and shrubs are C_3. A 12 m scaffolding instrument tower was placed in an area of gentle slope (<4°) where the vegetation is a tropical xeromorphic semi-deciduous braodleaf tree-and-scrub woodland form of cerrado *s.s.* (Eiten, 1984). The canopy surface is modulated by shrubs and trees 8 to 10 m tall, with occasional emergent trees reaching 12 m. The tower is surrounded by cerrado *s.s.* but *vereda* (a permanent grassy marsh with buriti palms - *Mauritia vinifera* Mart.) limits the wind fetch in the southwest direction to about 2 km. In the direction of the prevailing winds — north and north-east— the fetch is from 3 to 6 km. Grasses, thin-stemmed shrubs and semi-shrubs, small acaulescent palms and litter cover the soil surface, a dystrophic, deep and well-drained red-yellow latosol with a pH around 4. An accidental fire reached the site in 1987, but only grasses and thin plants from the ground layer were killed.

INSTRUMENTATION AND THEORY

Radiation sensors were installed over the cerrado at a height of 13.5 m on an arm extending 5 m from the tower. Measurements of net radiation (R_n), were made with a Funk Type radiometer (Swissteco, Melbourne, Australia). Shortwave, or solar radiation (S) was measured with a Moll-Gorezynski solarimeter (Kipp and Zonen CM3, Delft, The Netherlands) and two pyranometers (Licor LI200B, Nebraska, USA). One pyranometer was mounted facing upwards, and the other facing downwards, to measure incoming and reflected solar radiation respectively.

On the ground, about 10 m from the tower, soil heat flux (G) was measured by using four soil heat flux discs (Solar Radiation Instruments, Melbourne, Australia) placed about 0.03 m below the soil surface. Data from these sensors were collected every minute by two data loggers (21X, Campbell Scientific Inc., Logan, USA) and half-hourly interval averages were calculated and stored in solid state memory pods (SM 192 Storage Module, Campbell Scientific Inc., Logan USA). Both data loggers were synchronised in time with the eddy covariance system so that all averaged fluxes referred to the same half-hour interval.

Fluxes of carbon dioxide, latent and sensible heat were measured by using the eddy covariance method (proposed by Swinbank, 1951) which is much more feasible following the development of fast sensors and data acquisition systems (Moncrieff *et al.*, 1996). Derived quantities include the aerodynamic and stomatal ecosystem conductances obtained from the observed friction velocity and water vapour fluxes (discussed in detail by Grace *et al.*, 1995).

A three-dimensional ultrasonic anemometer (A1002R, Solent, Lymington, UK) was installed at a height of 15 m to measure the three spatial components of the wind velocity (u, v and w) and the air temperature. Water vapour and CO_2 were sampled using a fast-responding closed-path infra-red gas analyser (Li 6262, Li-Cor, Lincoln, Nebraska). Air was drawn at 4 l min^{-1} from 50 mm below the sonic path, down through a 6 m tube, 5.5 mm in diameter (Dekabon 1300, Deane & Co., Glasgow, UK). Readings of the windspeeds, air temperature and gas concentrations were made at 21 Hz using a 386 laptop computer loaded with the software Eddisol to calculate the fluxes in real time as half-hour averages (Moncrieff *et al.*, 1996). Co-ordinate rotations of the sonic anemometer data were performed by the software to correct for any streamline distortion at the site and to obtain w as the perpendicular component of windspeed.

The fluxes are calculated as

$$F_c = \overline{w' \rho'_c} + F_{correction}$$

where w' and ρc' are the fluctuating quantities of *w* and the density of carbon dioxide (or water vapour) in the air, measured at the same instant as *w*, and $F_{correction}$ is a term made up several corrections. The equations and corrections (transfer functions) used to calculate the fluxes over the cerrado followed exactly the same procedure described by Grace *et al.* (1995).

Net ecosystem surface conductance (g_s) was estimated by inversion of the Penman-Monteith equation using measurements of transpiration rate, latent and sensible heat fluxes and vapour pressure deficit as measured by the tower instruments. The vapour pressure within the canopy and its temperature were calculated from aerodynamic resistances and latent and sensible heat fluxes (Grace *et al.*, 1995). Canopy-to-air vapour pressure deficit was then estimated as the difference between the saturation vapour pressure at the estimated canopy temperature and the estimated vapour pressure at the canopy surface.

The accuracy of the measuring system was assessed from the water vapour fluxes. When the net radiation was plotted against the algebraic sum of evaporation, sensible heat flux and soil heat flux, the correlation coefficient was $r = 0.81$ and the slope of the relation was not significantly different from unity (Miranda *et al.*, 1996). It is concluded that the energy budget is essentially closed and that measurement errors are usually less than 10% for water vapour fluxes.

RESULTS AND DISCUSSION

The results refer to datasets of three to five days of continous measurements obtained during eight campaigns covering the wet and dry seasons of 1993. Half-hourly data are presented elsewhere (Miranda *et al.*, 1996). In this paper we are concerned with 72-hour integrals for each measurement period, to give the net CO_2 flux (NCF).

NCF above the cerrado community was markedly seasonal (Figure 3). During March and April, in the wet season, the vegetation has many green leaves (L*/= 1.0) and lost water vapour at a maximum rate of about 10 mmol m^{-2} s^{-1} (440 Wm^{-2}). At this time of the year the cerrado assimilated carbon dioxide from the atmosphere during the day at a maximum rate of about 10 μmol m^{-2} s^{-1}, and released to the atmosphere 5 μmol m^{-2} s^{-1} during the night. In the dry season in September (L*/ = 0.4), evapotranspiration rates and net CO_2 fluxes were reduced to about half of the wet season values. The maximum rates of CO_2 fluxes determined in this work are comparable to 13 μmol m^{-2} s^{-1} for the *Nothofagus* cool temperate rainforest in New Zealand (Hollinger *et al.*, 1994), 17 μmol m^{-2} s^{-1} for the humid Amazonian forest at Rondônia (Grace *et al.*, 1995) but are smaller than the rates determined by Wofsy *et al.* (1993) for deciduous temperate forest (25 μmol m^{-2} s^{-1}). Net ecosystem dark respiration (which includes soil, stems and leaves) fluctuated seasonally between 1.6 and 2.3 μmol m^{-2} s^{-1}. These rates, considering the corresponding environmental temperatures, were similar to the ones at Harvard Forest (2 - 3 μmol m^{-2} s^{-1}) and rather smaller than the ones for tropical rainforests of Brazil and New Zealand *Nothofagus* (5 - 7 μmol m^{-2} s^{-1}).

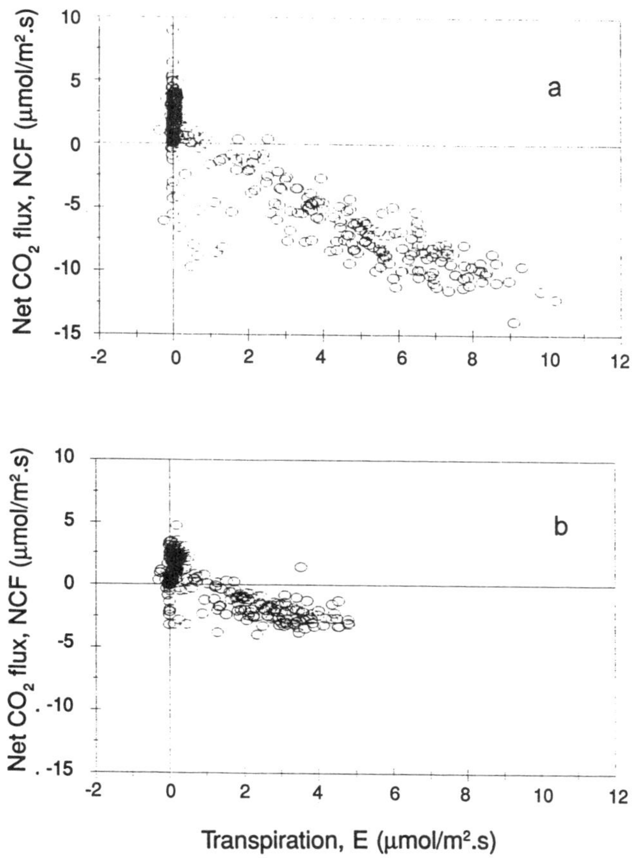

Figure 3 Contrasting behaviour in wet and dry season. Relation between net CO_2 flux and evapotranspiration for cerrado vegetation during the wet (a) and dry (b) season of 1993 at Reserva Ecológica de Águas Emendadas, Brasília, DF.

The measured values of evapotranspiration were used in the Penman-Monteith equation to calculate the surface conductance of the cerrado vegetation. The relationship between net ecosystem CO_2 flux and surface conductance for both seasons is shown in Figure 4. Although the surface conductance can be about 0.4 mol m^{-2} s^{-1} in the wet season it is usually less than 0.2 mol m^{-2} s^{-1} in the dry season. The physiological functioning of the cerrado vegetation depends upon season. There is a reduction of the leaf area index during the dry season for grasses and shrubs. The reduction of 50% in the values of the surface conductance may well be a consequence of the death of grasses and the abscission of leaves from some shrubs and trees species in the dry season. Associated with this change in the vegetation there is also a change in the albedo from 0.13 in the wet season to 0.18 in the dry season.

During the wet season the cerrado was a carbon sink fixing about 55 kg of CO_2 ha^{-1} day^{-1}, but became a source of CO_2 in September in the dry season. With the rains

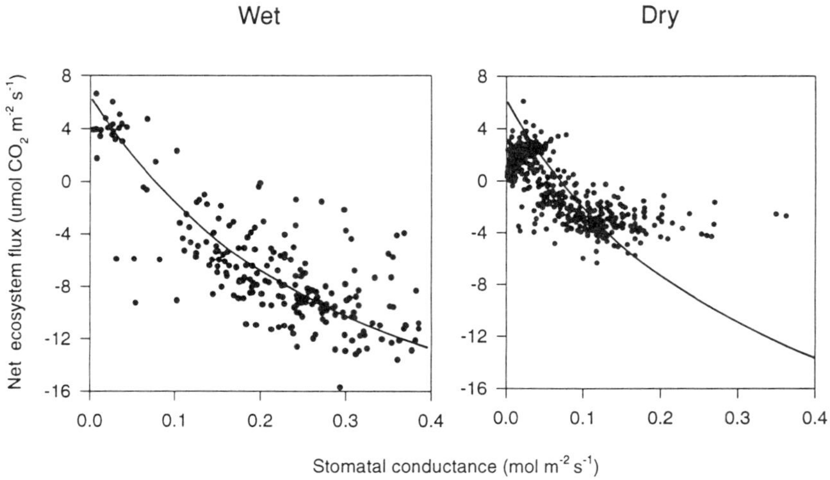

Figure 4 Relation between net ecosystem flux and stomatal conductance for cerrado vegetation during the wet (a) and dry (b) season of 1993 at Reserva Ecológica de Águas Emendadas, Brasília, DF. The curve drawn through the points in (a) is reproduced in (b) to highlight the difference between (a) and (b).

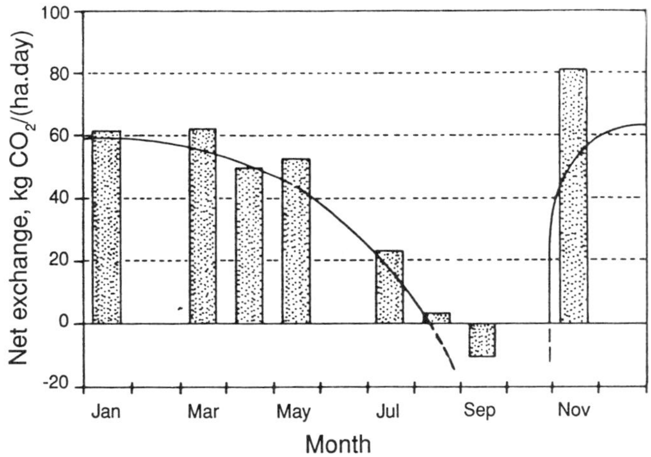

Figure 5 Summary of eight measurement campaigns in 1993, each campaign consisting of several days' measurement. Net ecosystem CO_2 exchange (positive values = assimilation) measured above a cerrado *sensu stricto* at Reserva Ecológica de Águas Emendadas, Brasília, DF.

in November, the cerrado was turned into a sink of CO_2 from the atmosphere (Figure 5). A rough estimate of the net carbon assimilation over the whole year may be obtained by numerical integration of the solid line in Figure 5: the result is a net gain of 2.5 t C ha^{-1} y^{-1}. This is equivalent to a biomass accumulation rate of 5.0 t dry weight ha^{-1} y^{-1}. The experimental site had been burned seven years previously by a small fire and may be functioning as a carbon sink on an annual basis while the vegetation is still recovering from the fire. However, there are no data to estimate how long this kind of vegetation would take to recover from a fire: the recovery time may depend on the intensity and the damage caused. It is also possible that the cerrado is accumulating carbon in the soil and, in this case, the sink strength may well depend on the length of the dry season. The length of the dry season in central Brazil, on the other hand, may be affected by deforestation or drastic changes in land use in Amazonia. Therefore, research at several other sites is needed to understand the functioning of the cerrado ecosystems, and their role in atmospheric carbon balance.

A comparison between rainforest and cerrado is made in Table 2, using data from previous work. The rainforest is taller, rougher, and has appreciably higher stomatal and aerodynamic conductances than the cerrado. It absorbs more radiation than the cerrado, and a larger fraction of the radiation is used in transpiration. Such comparative data for Brazilian vegetation are required to model the interdependence of cerrado and rainforest.

CONCLUSIONS

(1) The cerrado *sensu stricto* at Águas Emendadas is a net absorber of CO_2 in the wet season, but becomes less active aas the dry season continues. In 1993 the site was briefly a source of CO_2 in the dry season. Overall, the vegetation absorbed about 2.5 t C ha^{-1} y^{-1} in 1993.

(2) The site had an albedo that varied seasonally, from 0.13 in the wet season to 0.18 in the dry season. The stomatal conductance also varied seasonally, ranging from a maximum of 0.2 - 0.4 mol m^{-2} s^{-1} in the dry season, to 0.1 to 0.2 in the wet season. As a result of these changes, the amount of water transpired into the atmosphere declined markedly in the dry season. $\lambda E/R_n$ was 0.6 in the wet season and 0.4 in the dry season.

(3) Cerrado *sensu stricto* is a fire dominated ecosystem and a complete analysis of carbon, water and energy fluxes requires data from more sites, representing vegetation at different times following burning.

(4) For the carbon balance of the cerrado as a whole, attention should be paid to the other four physignomic forms of cerrado, as cerrado *sensu stricto* occupies only 53% of the entire cerrado.

ACKNOWLEDGEMENTS

We thank Dr W. B. da Cruz for useful discussions and M. P. Rocha and Pedro Olavo for practical help at various stages of this work. We are grateful to SEMATEC and the University of Brasilia for facilitating the field work. We acknowledge financial support from the Natural Environment Research Council through its TIGER (Terrestrial Initiative in Global Environmental Research) programme. J. Lloyd's travel was assisted by the Australia/New Zealand/UK Tripartite Agreement on Climate Change and the Royal Society of London/Australian Academy of Science Exchange Programme.

REFERENCES

CEPA/DF. 1984. *Características hidroclimáticas do Distrito Federal*. Secretaria de Agricultura do Distrito Federal. GDF, Brasília.

Coutinho, L. M. 1990. O Cerrado: ecologia do fogo. *Ciência Hoje*, **12**,22-30.

Culf, A.D., Fisch, G. and Hodnett, M.G. 1995. The albedo of Amazonian forest and ranchland. *J. Climatol.*, **8**, 1544-1554.

Dias, B.F.S. 1992. Cerrado: uma caracterização. In: *Alternativas de Desenvolvimento dos Cerrados: manejo e conservação dos recursos naturais renováveis*. (Ed. B.F.S. Dias). FUNATURA/ IBAMA, Brasília, pp.15 - 26.

DNMET. 1992. *Normas climatológicas (1961 -, 1990)*. Ministério da Agricultura e Reforma Agrária. Secretaria Nacional de Irrigação. Departamento Nacional de Meteorologia. SPI/EMBRAPA, Brasília.

Eiten, G. 1972. The cerrado vegetation of Brazil. *Bot. Rev*. **38**, 201-341.

Eiten, G. 1983. *Classificação da Vegetação do Brasil*. CNPq, Brasilia.

Eiten, G. 1984. Vegetation of Brasilia. *Phytocoenologia* 12, 271-292.

Eiten, G. 1992. Natural Brazilian vegetation types and their causes. *Anais da Academia Brasileira de Ciências* **64**, 35-65.

Eiten, G. 1994. Vegetação do Cerrado. In: M. N. Pinto (Ed.). *Cerrado*. SEMATEC, Brasilia. pp. 17-76.

Filgueiras, T.S. 1991. A floristic analysis of the Gramineae of Brazil's Distrito Federal and a list of the species occurring in the area. *Edinburgh J. Bot.*, **48**, 1-8.

Grace, J., Lloyd, J., McIntyre, J., Miranda, A. C., Miranda, H. S.; Wright, I. and Gash, J. 1995b. Fluxes of water vapour and carbon dioxide over an undisturbed tropical forest in south-west Amazonia. *Global Change Biol.*, **1**, 1-12.

Heringer, E. P., Barroso G. M., Rizzo, J. A. and Rizzini, C. T. 1977. A flora do Cerrado. In *4 Simpósio sobre o Cerrado*. Editora USP. pp. 211-232.

Hollinger, D.Y., Kelliher, F. M., Byers, J.N., Hunt, J.E., McSeveny, T.M. and Weir, P.L. 1994. Carbon dioxide exchange between an undisturbed old-growth forest and the atmosphere. *Ecology*, **75**, 143-150.

Meir, P., Grace, J., Miranda, A. and Lloyd, J. 1996. Soil respiration in a rainforest in Amazonia and in cerrado in central Brazil. This volume.

Miranda, A.C., Miranda, H.S., Lloyd, J., Grace, J., Francey, R.J., McIntyre, J.A., Meir, P., Riggan, P., Lockwood, R. and Brass, J. 1996. Fluxes of carbon, water and energy over Brazilian cerrado: An analysis using eddy covariation and stable isotope. *Plant Cell and Environment* (submited).

Moncrieff , J.B., Massheder, J.M., De Bruin, H., Elbers, J., Friborg, T., Heusinkveld, B., Kabat, P., Scott, S., Soegaard, H. and Verhoef, A. 1995. A system to measure surface fluxes of momentum, sensible heat, water vapour and carbon dioxide. *J. Hydrol.* (in press).

Nobre, C.A., Sellers, P.J. and Shukla, J. 1991. Amazonia deforestation and regional climate. *J. Climate*, **4**, 957-988.

Ratter, J.A. 1992. Transitions between cerrado and forest vegetation in Brazil. In *Nature and dynamics of forest-savanna boundaries*. Eds. P.A. Furley, J. Proctor and J. A. Ratter. Chapman

& Hall, London. pp. 417-429.
Salati, E. and Vose, P. B. 1984. Amazon Basin: a system in equilibrium. *Science* **225**, 129-225.
Swinbank, W. C. 1951. The measurements of vertical transfer of heat and water vapour by eddies in the lower atmosphere. *J. Meteorol.*, **8**, 135-145.
Wofsy, S. C., Goulden, M. L., Munger, J. W., Fan, S. M., Bakwin, P. S., Daube, B. C., Bassow, S. L. and Bazzaz, F. A. 1993. Net exchange of CO_2 in a mid-latitude Forest. *Science* **260**, 1314-1317.

RESUMO

A técnica de covariância de fluxo turbulento ("eddy covariance") foi utilizada para medir a troca líquida de dióxido de carbono entre a atmosfera e a vegetação de cerrado (*sensu stricto*). O experimento foi realizado em um sítio localizado na Estação Ecológica de Águas Emendadas, próximo à Brasília, DF. A vegetação foi queimada 7 anos antes das medidas por um fraco incêndio. Um conjunto de dados de 3 a 5 dias de medições contínuas foram obtidos ao longo de oito campanhas de campo realizadas nas estações úmida e seca de 1993. Na estação úmida (de outubro a abril), a vegetação de cerrado assimilou dióxido de carbono da atmosfera, durante o dia, a uma taxa máxima de 12 μmol m^{-2} s^{-1}; durante a noite, em torno de 2,8 μmol m^{-2} s^{-1} foram liberados. O cerrado foi um sorvedouro de carbono durante a estação úmida, fixando, em média, 55 kg de CO_2 ha^{-1} dia^{-1} em dias secos e sem nebulosidade dos meses de janeiro, março, abril e maio. Na estação seca (de junho a outubro), as taxas de troca líquida estiveram, em julho, em torno da metade das observadas na estação chuvosa, reduzindo-se a quase zero em agosto, e o cerrado tornou-se uma fonte de CO_2 em setembro. Com as chuvas de novembro, o cerrado começou a funcionar novamente como um sorvedouro de carbono. Parece que essa forma fisionômica do cerrado pode desempenhar um papel importante no balanço de carbono na atmosfera, e parece provável que a troca líquida anual de CO_2 seja dependente de eventos de incêndios e de duração da estação seca.

21 Estimation of the leaf area index and total biomass of tropical regenerating forests: comparison of methodologies

M. HONZÁK[1], R.M. LUCAS[2], I. do AMARAL[3], P.J. CURRAN[2], G.M. FOODY[1] and S. AMARAL[4]

[1]*Department of Geography, University of Wales, Swansea, UK*
[2]*Department of Geography, University of Southampton, UK*
[3]*Instituto Nacional de Pesquisas da Amazonia, Manaus, Brazil.*
[4]*Instituto Nacional de Pesquisas Espaciais, São José dos Campos, Brazil*

INTRODUCTION

Tropical regenerating forests are a sink for atmospheric carbon dioxide and, as such, restore some of the carbon lost through tropical deforestation (Detwiler and Hall, 1988; Brookfield, 1992). The quantity of carbon held in regenerating forest is, however, uncertain at the regional/global scales, since the extent of different forest regeneration stages and their associated carbon stocks are unknown. The regional extent of regenerating forest cover in the tropics may, however, be mapped and several stages of regeneration distinguished using remotely sensed data (Curran and Foody, 1994). Ideally, differentiation of regeneration stages using remotely sensed data requires that each stage defined differs in terms of contained carbon (which represents approximately 50% of the biomass) and that these carbon stocks can be quantified.

A study by Lucas *et al.* (1993) demonstrated that early stages of regeneration could be defined on the basis of forest age, as estimated by comparing time-series of Landsat sensor data. However, regenerating forests of similar age may not necessarily support the same quantities of carbon or accumulate carbon at the same rate. An alternative approach to defining regeneration stage is therefore required. One such approach is to establish relationships between the remotely sensed signal and measured properties of forests which relate to biomass and which also vary as forests regenerate.

A number of previous studies have related remotely sensed data to measured properties of tropical forests, although many have been inconclusive (e.g. Foody and Curran, 1994; Sader *et al.*, 1989; Wu, 1990; Box *et al.*, 1989; Cook *et al.*, 1989), particularly when using data acquired by optical satellite sensors. This is partly because only the spectral reflectance from the uppermost layers of the canopy is detected by such sensors. For this reason direct relationships between forest biomass

Amazonian deforestation and climate. Edited by J.H.C. Gash, C.A. Nobre, J.M. Roberts and R.L. Victoria.

and the remotely sensed reflectance cannot be expected (Wickland, 1991). The upper surface of forest canopies is, however, composed mainly of leaves, and the early stages of succession are characterised by an increase in the leaf area index (LAI), which occurs regardless of location or species composition (Folster *et al.*, 1976; Uhl, 1987; Saldarriaga *et al.*, 1988). The LAI is defined as the projected one-sided area of leaf per ground area.

The increase in LAI as forests regenerate leads to an increase both in the absorption of red (R, 0.40-0.70 μm) radiation by leaf pigments and in the scattering of near infrared (NIR, 0.71-1.20 μm) radiation (Curran, 1983). This relationship suggests that changes in LAI over time may be manifested within the R or NIR images obtained by satellite sensors. This has already been demonstrated for forests in temperate regions where strong correlations between the LAI, R and NIR radiance and indices such as the Normalised Difference Vegetation Index (NDVI = [NIR-R]/[NIR+R]) have been obtained (Peterson *et al.*, 1987; Curran *et al.*, 1992).

Assuming that similar relationships between the R and NIR radiance and LAI exist for tropical forests, several stages of regeneration may be distinguished using remotely sensed data. However, for assessing regional/global carbon distribution, each stage needs to be unique in terms of its carbon and so relationships between LAI and biomass are also required. The aim of this study was to define such a relationship which could subsequently be exploited to estimate the spatial extent of biomass, and hence carbon, using remotely sensed data.

METHODS

The study was undertaken for areas of tropical forest regeneration north of Manaus in Brazilian Amazonia (Figure 1). Data from radiation sensors and hemispherical cameras were used to estimate the canopy gap fraction, from which the LAI was derived. Total biomass was estimated using forest inventory data and a suite of pan-tropical allometric regressions.

STUDY SITE

The regenerating forest sites selected for study were Reserva Ducke, the Biological Dynamics of Forest Fragments Project (BDFFP) research area of the Instituto Nacional de Pesquisas da Amazônia (INPA) and the Smithsonian Institute, and nearby farms 100 km north of Manaus. The age of all forests within the BDFFP area and on surrounding farms was determined to within 1-2 years by comparing a time-series of Landsat Multispectral Scanner System (MSS) and Thematic Mapper (TM) scenes, which were available for most years between 1974 and 1991 (Lucas *et al.*, 1993). The two regenerating forests sampled at Reserva Ducke were aged 21 and 30 years, from INPA records.

DATA COLLECTION AND ANALYSIS

Within the BDFFP research area and surrounding farms, thirteen 10×100 m (0.1 ha) plots were located within large tracts of forests of known age that were relatively homogeneous, structurally and floristically. A 20×15 m plot was sampled within two-year-old forest at the BDFFP area; two 10×100 m plots were sited within the regenerating forests at Reserva Ducke.

Estimation of the LAI

The LAI was estimated from canopy gap fraction, defined as the fraction of view in some direction that is not blocked by foliage (Welles, 1990). The gap fraction was

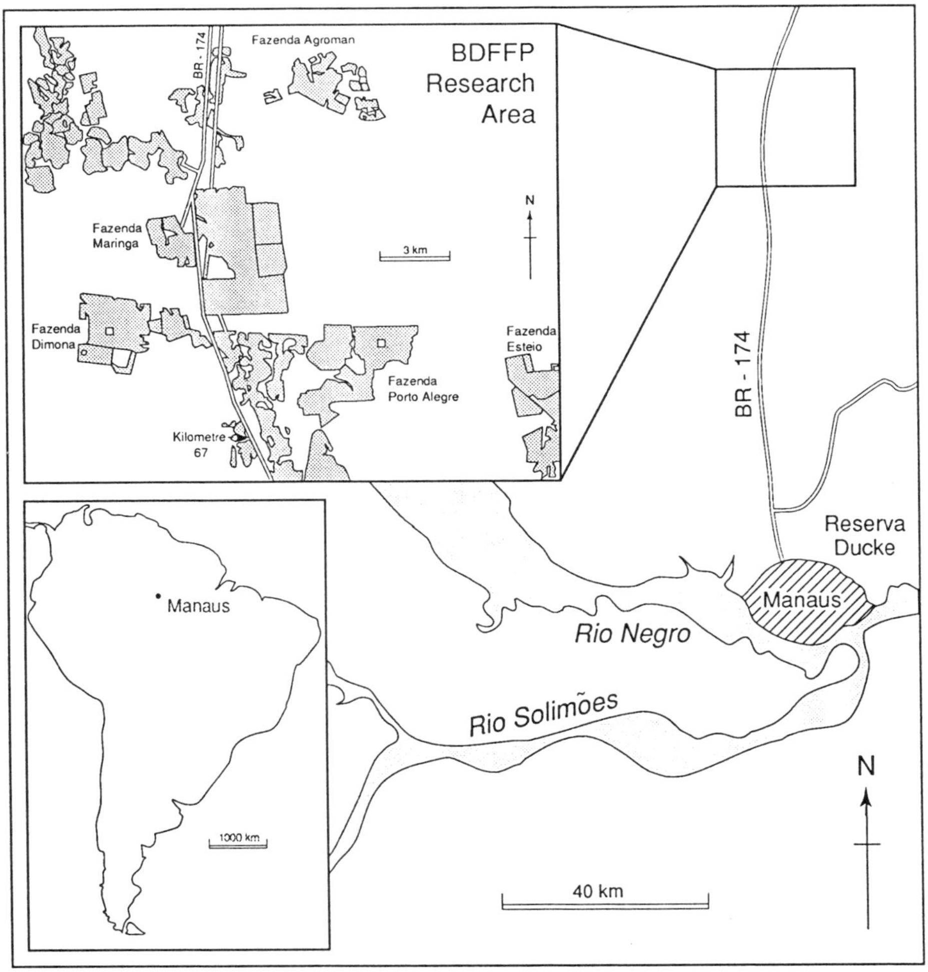

Figure 1 Location of the BDFFP research area and Reserva Ducke

derived from below-canopy light radiation measurements and hemispherical photographs.

Measurements of the below canopy radiation environment were recorded, in mainly sunlit conditions, using a DECAGON sunfleck ceptometer (model SF-80, Decagon Devices, Inc.). This instrument combined a microprocessor with a 1 m probe, equipped with 80 light sensors at 1 cm intervals, and measured sunflecks and photosynthetically active radiation (PAR).

Using the ceptometer in sunfleck mode, the number of sensors registering sunflecks was recorded and a percentage sunfleck fraction was calculated. To determine a manual threshold for a sunfleck to be registered, measurements were recorded at a site below the forest canopy where light conditions were considered typical of the forest as a whole and where at least one sensor recorded a sunfleck with direct light penetration. This threshold is calculated as half of the highest light reading from any sensor on the probe. The establishment of this manual threshold allowed gap fraction to be related directly to sunfleck fraction (Decagon Devices, 1989).

Transmitted PAR, which also relates to canopy gap fraction (Norman and Campbell, 1991), was determined as the ratio of below-canopy PAR to PAR measured simultaneously outside the forest canopy. Prior to the calculation of tranmitted PAR, the diffuse PAR, determined by taking measurements when the ceptometer was shaded (Pierce and Running, 1988), was subtracted from both the incoming above and below-canopy PAR (Nel and Wessman, 1993).

Within each plot, and at 10 m intervals along the centre transect, measurements of sunfleck fraction and PAR were taken at 15° intervals within a 360° circle and averaged to take account of the large spatial variability of below-canopy PAR and sunflecks beneath the canopy (Chason *et al.*, 1991). The centre transect of each plot was sampled up to five times within two hours of solar noon, to obtain measurements of below-canopy PAR and sunfleck fraction at different solar zenith angles. The solar zenith angle at each measurement time and location was then calculated using the procedure formulated by Michalsky (1988). Throughout the measurement period, the average incoming PAR was recorded each minute using a ceptometer in an open area and measurements of diffuse PAR were taken at regular intervals.

Hemispherical photographs of the canopy were taken at the same 10 m intervals within each plot to define canopy gap fraction on the basis of the geometric distribution of canopy openings (Rich, 1990). Photographs were taken at the same site as the ceptometer readings, either early in the morning or in the evening, where possible before sunrise or after sunset. At each point, one photograph was taken at the exposure registered by a hand-held light meter when pointed to the open sky in the direction of the zenith, and two more were taken at aperture settings bracketing the first (Rich, 1990).

Since estimates of canopy gap fraction were obtained at a number of zenith and azimuth angles (by measuring sunfleck fraction and transmitted PAR several times a day at the same place) and from different segments within each hemispherical photograph, the LAI (L) was derived for measurements of the canopy gap fraction at specific solar zenith angles (f_i) by minimising the following function:

$$F = \Sigma (\ln f_i + k_i L)^2$$

where k_i is the extinction coefficient: the fraction of the leaf area projected onto a horizontal surface from the direction of the incident beam of radiation (Norman, 1979; Campbell, 1986). This inversion of gap fraction data considers the ellipsoidal leaf angle distribution proposed by Campbell (1986).

Estimation of total biomass

Within each plot, the diameter at breast height (D, cm) of all trees (where $D > 3$ cm) was measured and the height of a random sample of trees (270 individuals) was estimated using an inclinometer (Hamilton, 1975). All trees measured were identified to genus, and where possible to species.

The total biomass was estimated for each plot using the allometric regression equations of Brown *et al.* (1989) and Deans *et al.* (1996) (Table 1). Those of Brown *et al.* (1989) may be used to estimate the total above-ground biomass (TAGB), for trees within the moist tropical life zone, as a function of D and height (H, m) with or without wood density (S, Mg m^{-3}). The regression equations were derived from measurements of 168 trees in Amazonia and southeast Asia, ranging in diameter from 5 to 130 cm. The regenerating forests at Manaus, however, contained trees of only a relatively small diameter range ($D < 35$ cm). The relationship of TAGB with D, H and S within the 0-35 cm diameter range was therefore redefined using the Brown *et al.* (1989) dataset. Since the below-ground component was not determined using these regressions, total biomass was estimated by expressing below-ground

Table 1 Allometric regression equations for the estimation of biomass using measurements of diameter (D), height (H) and wood density (S)

Publication	Source	*n*	Regression equation
Brown *et al.* (1989)	Amazonia South-east Asia	168	$TAGB = \exp(-2.409 + 0.952 \ln(D^2HS))$
			$TAGB = \exp(-3.114 + 0.972 \ln(D^2H))$
Brown *et al.* (1989) for trees of $D < 35$ cm		142	$TAGB = \exp(-2.327 + 0.937 \ln(D^2HS))$
			$TAGB = \exp(-3.068 + 0.957 \ln(D^2H))$
Deans (1996)	West Africa (Cameroon)	6	$TB = 0.025\ D^2H\ S/0.45$
		14	$TB = 0.025\ D^2H$
Deans (1996)		6	$TB = 0.764\ \pi\ (D/2)^2\ S/0.45$
		14	$TB = 0.75\ \pi\ (D/2)^2$

biomass as a percentage of the above-ground component. Estimates of this percentage for tropical forests range from 8% (Uhl, 1987) to 25% (Saldarriaga *et al*., 1988). However, for this study, an average estimate of 19%, obtained for forests in Amazonia ranging from nine years of age to maturity, was used (Saldarriaga *et al*., 1988). The regression equations of Deans *et al*. (1996) were generated by destructively harvesting and weighing the above- and below-ground components of six *Terminalia*, four *Trema*, three *Musanga* and one *Triplochiton* tree (with diameters from < 5 cm to 20 cm); these are regenerative forest species native to West Africa. These regression equations may be used to estimate total biomass directly as a function of D, H and S.

The wood density of over 1200 tropical tree species has been determined (Chudnoff, 1980; Reyes *et al*., 1992). Where wood density values were available for species at the site, the allometric regression equations of Brown *et al*. (1989) and Deans *et al*. (1996), which take account of wood density, were applied as biomass estimates from such regression equations: these are more accurate (Reyes *et al*., 1992). These regression equations also account for the varying amount of carbon stored per unit wood volume: this is typically greater for primary forest species and least for secondary ones (Lugo and Brown, 1992). Where the wood density was unknown, the regression equation was based on D, with or without H.

Two estimates of height (H) were obtained for all trees. The first (H_0) was based upon a general allometric relationship, defined using D and H measurements for all species at the study site, and a more specific regression equation which was formulated using D and H measurements for species at the study site with similar wood density (Table 2). The second estimate of height (H_B) was based upon the published transformed non-linear regression of Brown *et al*. (1989): this estimates H from D (Table 2) and is based upon data representing 3824 trees within the moist life zone.

Table 2 Allometric regression equations relating height (H) to tree diameter (D) as defined within this study and by Brown *et al*., 1989

Group	*n*	Height estimation equation
Cecropiaceae	88	$H=\exp(1.102+0.626 \ln D)$
Clusiaceae	29	$H=\exp(1.395+0.539 \ln D)$
Melastomataceae	37	$H=\exp(1.550+0.482 \ln D)$
Annonaceae, Simaroubaceae	21	$H=\exp(1.371+0.598 \ln D)$
Other	95	$H=\exp(1.292+0.594 \ln D)$
All species (This study)	270	$H=\exp(1.387+0.539 \ln D)$
All species (Brown *et al.* 1989)	3824	$H=\exp(1.071+0.567 \ln D)$

For the estimation of biomass, H_B alone was used as input to the original allometric regression of Brown *et al.* (1989). Both H_O and H_B were used as input to this regression recalculated for trees < 35 cm, and also that of Deans *et al.* (1996): thus two estimates of biomass were derived using each equation, and another estimate was made on the basis of cross-sectional area (Deans *et al.*, 1996). In total, six different methods were used to estimate the total biomass for each plot.

RESULTS

For each plot at the study site, three estimates of LAI and six estimates of total biomass were derived, but large differences were seen in both parameters.

LAI

The estimates of gap fraction, based on sunfleck fraction, transmitted PAR and hemispherical photographs, generated different values of LAI for each plot (Table 3). Average LAI estimates for all plots ranged from 2.97-6.18 (based on sunfleck fraction) to 4.21-9.60 (based on transmitted PAR). Estimates derived from hemispherical photographs showed the least variation, with average LAI ranging from 3.45 to 4.79. However, the trend in LAI estimates obtained using the three methods was consistent. By averaging the LAI estimates for all forests of similar age, a general increase in LAI with forest age was evident for forests up to 21 years old, although a decline in LAI was noted for forests aged between 4 and 8 years, 9 and 16 years and 21 and 30 years (Figure 2).

TOTAL BIOMASS

Total biomass estimates for the 16 plots varied considerably, depending upon which allometric regression equation and height measurement was used (Table 4). By applying a non-linear model to regress biomass accumulation against forest age, the difference between biomass estimates was shown to be greatest within the older forests (Figure 3). For the 30-year-old forests, the biomass derived using Method 3 was 435 Mg ha^{-1}, almost double the 257 Mg ha^{-1} obtained using Method 4. By assuming the rate of biomass accumulation to be linear over the period of 30 years and by taking the lowest and highest estimates (based on Methods 3 and 4 respectively), the rate of accumulation was 8.6-14.5 Mg ha^{-1} year.

The biomass estimated using Method 2 was consistently lower than that obtained using Method 1. This observation indicates that the recalculation of the regression of Brown *et al.* (1989) for trees with D < 35 cm slightly reduces the biomass estimate. However, by using H_O instead of H_B, the biomass estimate increased by 17-30% (Table 4). The biomass estimate based on cross-sectional area was the largest, but only for forests younger than 20 years.

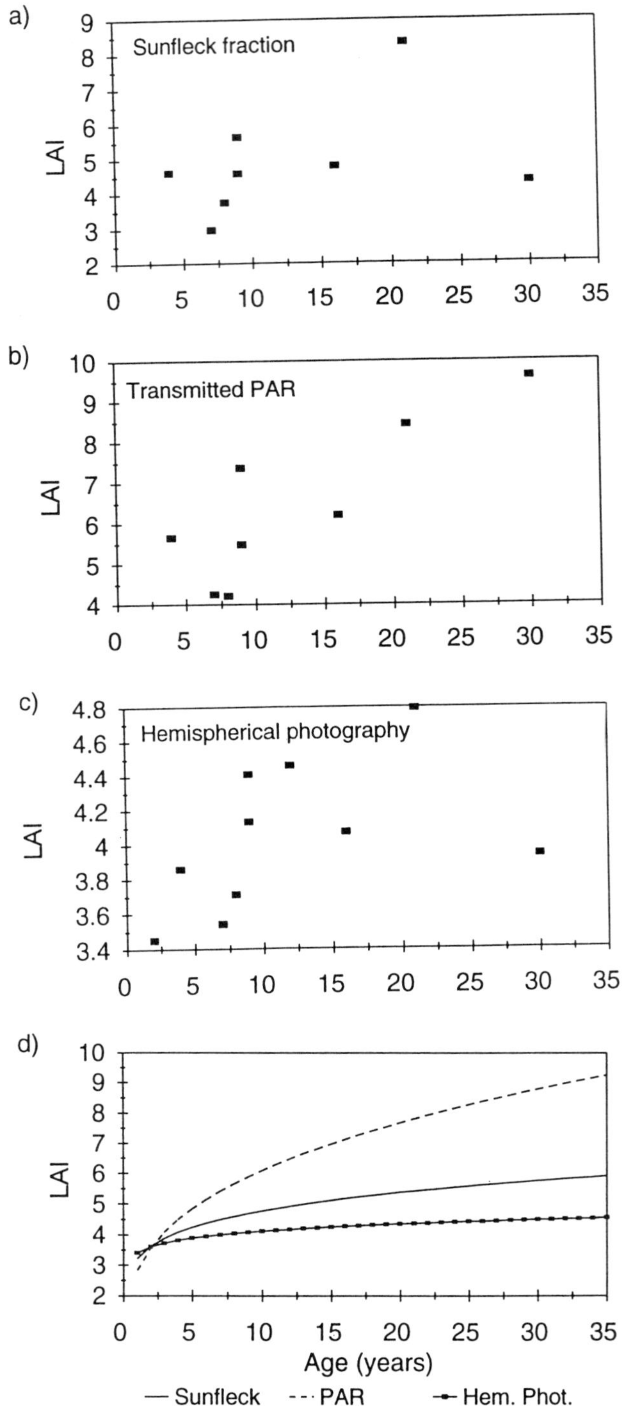

Figure 2 The relationship between forest age and the LAI as defined using (a) sunfleck fraction, (b) transmitted PAR and (c) hemispherical photographs; (d) shows the allometric regressions relating age to the LAI defined using these methods.

Table 3 Estimates of LAI, based on sunfleck fraction s, transmitted PAR and hemispherical photographs, H phot. for sixteen plots located within regenerating forests near Manaus

Age years	Location	Dominant genera	Density trees ha^{-1}	No. of zenith angles*	LAI Sunflecks	LAI P A R	LAI H. phot.
2	Fazenda Dimona	*Cecropia*	4300	0	-	-	3.45
4	Fazenda Esteio	*Vismia, Bellucia*	5280	3	4.62	5.65	3.86
7	Kilometre 67	*Cecropia, Laetia*	3830	3	2.97	4.25	3.54
8	Fazenda Maringa	*Vismia, Cecropia*	2110	1	3.76	4.21	3.71
9	Fazenda P. Alegre	*Cecropia*	1820	4	5.03	5.04	4.00
9	Fazenda P. Alegre	*Cecropia*	1290	3	4.27	5.81	4.62
9	Fazenda P. Alegre	*Cecropia*	1710	3	4.80	5.27	4.27
9	Fazenda P. Alegre	*Cecropia*	1720	4	4.17	5.64	4.23
9	Fazenda P. Alegre	*Cecropia*	1620	3	4.69	5.63	3.54
9	Fazenda Esteio	*Vismia, Cecropia*	2240	4	5.78	6.86	4.55
9	Fazenda Esteio	*Cecropia, Guatteria*	2820	2	5 .00	6.63	4.49
9	Fazenda Esteio	*Miconia, Cecropia*	2520	2	6.18	8.61	4.19
12	Kilometer 67	*Cecropia, Laetia*	2240	0	-	-	4.46
16	Fazenda Agroman	*Cecropia, Laetia*	2210	3	4.80	6.20	4.07
21	Reserve Ducke	*Miconia, Tapiria*	1770	1	8.33	8.43	4.79
30	Reserve Ducke	*Miconia, Croton*	1830	1	4.33	9.60	3.94

* Ceptometer measurements

DISCUSSION

Accurate estimates of LAI and total biomass of tropical forests are required to derive relationships between these properties and the remotely sensed radiance. However, there are considerable uncertainties associated with estimating them from field measurements. This discussion aims to raise awareness of how estimates of forest properties vary and to offer some explanation of the different estimates of LAI and total biomass.

LAI

The variation in LAI determinations for each plot can be attributed to differences in the gap fraction estimated from light sensor data and hemispherical photographs. The most accurate method for estimating canopy gap fraction, and hence LAI, cannot be determined as no direct measurements of LAI were obtained and allometric relationships between LAI and tree diameter or sapwood cross sectional area were unavailable.

The leaf biomass of regenerating forests increases to a maximum during the first

Table 4 Estimates of total above-ground biomass (TABG) for sixteen plots located within regenerating forests near Manaus.

Method Age (years)	**1**	**2**	**3**	**4**	**5**	**6**
2	76	75	88	52	62	120
4	47	47	61	29	39	87
7	89	87	106	59	73	134
8	115	110	139	76	97	146
9	164	157	196	119	150	188
9	159	153	190	120	152	176
9	133	128	160	97	123	159
9	145	139	172	106	133	169
9	141	135	166	103	129	160
9	118	114	145	83	108	153
9	143	139	176	101	131	191
9	145	140	178	104	134	181
12	135	130	165	97	125	174
16	147	141	176	104	132	182
21	291	267	334	193	245	252
30	389	351	435	257	323	290

Method **1** Brown *et al.* (1989) TAGB and height estimation equation (H_B)

Method **2** Brown *et al.* (1989) TAGB data (re-computed for D < 35 cm) and height estimation equation (H_B)

Method **3** Brown *et al.* (1989) TAGB data (re-computed for D < 35 cm) and height estimation equations derived from our data (H_O)

Method **4** Deans (1996) TB estimation equations and Brown *et al.* (1989) height estimation equation (H_B)

Method **5** Deans (1996) TB estimation equations and height estimation equations derived from our data (H_O)

Method **6** Deans (1994) TB cross sectional area estimation equations

Method **1**
$$TAGB=\exp(-2.409+0.9522\ln(D^2HS))$$
$$TAGB=\exp(-3.1141+0.9719\ln(D^2H))$$

Methods **2** & **3**
$$TAGB=\exp(-2.327+0.9366\ln(D^2HS))$$
$$TAGB=\exp(-3.068+0.9566\ln(D^2H))$$

Methods **4** & **5**
$$TB=0.02489\ D^2H\ S/0.45$$
$$TB=0.02477\ D^2H$$

Method **6**
$$TB=0.7631\ \pi\ (D/2)^2S/0.45$$
$$TB=0.7489\ \pi\ (D/2)^2$$

15 years of regeneration and then declines slightly (Brown and Lugo, 1990). Assuming a correspondence with leaf biomass, the LAI of regenerating forests at 12-15 years should be roughly equivalent to that recorded for mature forests, and certainly should not exceed it. Previous studies have estimated the LAI of Amazonian forests to range from 5.1 (Williams *et al.*, 1972) to 7.5 (Saldarriaga, 1985) with a mean of 6.2 (McWilliam *et al.*, 1993). Recent work by Roberts *et al.* (1996) reported

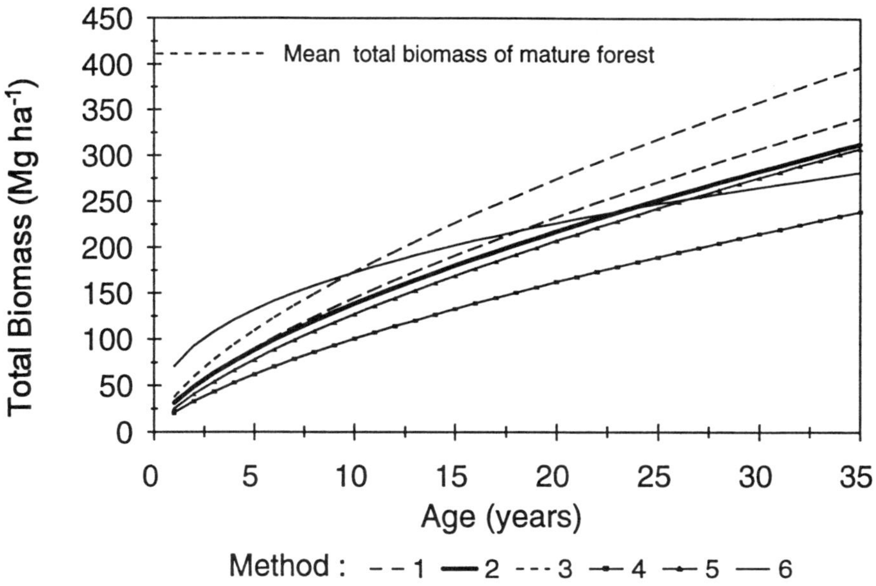

Figure 3 Allometric regressions relating forest age to total biomass, derived using methods 1-6 (Table 3).

values of 4.4 and 5.3 for the ABRACOS sites in Ji-Paraná and Marabá, while a value of 5.7 was reported for forest near Manaus (McWilliam *et al.*, 1993). However, values as high as 22.4 have been recorded for Panamanian tropical moist forests (Golley *et al.*, 1975). In this study, the LAI derived using transmitted PAR for the 21 and 30 year old forests at Reserve Ducke were greater than the maximum LAI values recorded for mature forests in Amazonia and may therefore be overestimated. The two estimates of LAI derived from sunfleck fraction and transmitted PAR may, however, be less accurate compared to those obtained for younger forests as transmitted PAR (and sunfleck fraction) were recorded at only two solar zenith angles. Measurements were also taken under partially cloudy conditions which may account for the low LAI estimates for the 30 year old forest derived from sunfleck fraction. Based on the gap fraction derived from hemispherical photographs the LAI for all forests was less than 5.0 and such estimates may be more comparable with those obtained previously for tropical forests.

The study has demonstrated some of the limitations in using light sensors for estimating the LAI of forests, especially in tropical regions. The model of Norman and Campbell (1991), for example, requires a large spread in the range of solar zenith angles, while Nel and Wessman (1993) indicated that, ideally, ten or more data points are required, which must include the extremes of the solar zenith angles and range from 30° to 50°. This frequency of measurement is difficult in tropical forest regions, because of the build up of cloud and haze during the day. Consequently, most measurements of radiation can only be taken in the morning and early afternoon. For this reason, the use of hemispherical photographs or light sensors that measure multiple angle light transmission simultaneously (e.g. the LICOR LI2000 Plant Canopy Analyzer) is recommended for tropical forests. Previous studies (e.g. Pierce and Running, 1988; Welles, 1990) have discouraged the use of sunfleck fraction to estimate the LAI of forests. In particular sunflecks appear larger and less distinct under taller forest canopies because penumbral effects diminish the influence of more distant foliage (Norman and Campbell, 1991; Welles, 1990). Even so, the LAI estimates, based on sunfleck fraction, for most forests at Manaus are below those obtained previously for mature forest and follow similar patterns with age to those derived from other methods. The accuracy of LAI estimates from sunfleck fraction is, however, likely to decline during the succession, as the replacement of pioneer dominants by a greater diversity of late secondary and primary forest species changes the canopy from a single layer to one that is taller and multi-layered.

When recording below-canopy and incoming PAR simultaneously, differences in the sensitivity of ceptometers at certain light intensities may lead to inaccuracies in the estimates of transmitted PAR. During a later study in Cameroon, the absolute PAR values recorded by two ceptometers became increasingly divergent from 700 $\mu mol\ m^{-2}\ s^{-1}$. A comparison of PAR measurements derived for a range of illumination intensities by different sensors was therefore undertaken to generate a sensor inter-calibration curve. Inter-calibration was also applied for different ceptometers during the Brazilian study but for only one light intensity. The divergence in values with increasing light intensities was therefore not discovered and the large estimates of LAI derived using transmitted PAR for all forests may have resulted partly from differences in the sensitivity of the ceptometers used to measure PAR outside and below the canopy.

Perhaps the most appropriate method for estimating LAI is to derive canopy gap fraction from hemispherical photographs. In this study, all photographs of the forest canopy were of excellent quality, especially as most were taken shortly before sunrise or after sunset and under clear sky conditions. Canopy gap fraction was also estimated from a large number of zenith and azimuth angles from within the hemisphere. Furthermore, the negatives for each sample point were scanned to generate a fine spatial resolution 1024×1024 pixel digital image, which allowed many of the smaller foliage elements to be discriminated using standard thresholding procedures.

Future efforts aimed at defining the LAI for tropical forests should therefore consider the use of several procedures for collecting information on the radiation environment or define relationships with LAI measured using direct or semi-direct techniques before undertaking a more general survey. A number of different radiative transfer models should also be inverted to estimate LAI as Nel and Wessman (1993) concluded that when a smaller number of gap fraction measurements are recorded, inversion of both the Beer-Lambert Law and a modified version of this model (Norman, 1988) generated estimates of forest LAI of greater accuracy compared to the model of Norman and Campbell (1991). Furthermore, the assumption of randomness in the distribution of foliage elements in space within most inversion models should be recognised, especially as this condition is unlikely given the increasing diversity of species and canopy form as forests age. This assumption may, however, be valid for the younger forests where several pioneer species, such as *Cecropia*, dominate the upper canopy.

TOTAL BIOMASS

The large differences in total biomass estimated using the pan-tropical regressions of Brown *et al.* (1989) and Deans *et al.* (1996) can be attributed to the different origins of the datasets used for their derivation. The regression equation calculated using the Brown *et al.* (1989) dataset, for trees with D <35 cm, was based upon diameter measurements collected for both primary and secondary forest species. As species with higher wood density were used in deriving this regression equation, overestimates of biomass can be expected. In contrast, the regressions of Deans *et al.* (1996) were derived solely from secondary forest species with lower wood densities, and they may provide a better representation of biomass for this study.

The lower estimates of biomass were obtained when height was defined using the regression equation of Brown *et al.* (1989) rather than using the locally-derived regression equation. However, the height of mature forest species with large wood densities is less compared to secondary forest species with lower wood densities and of the same diameter. The inclusion of mature forest species in the formulation of the regression equation therefore leads to an underestimate of the height of secondary forest species and hence a reduction in biomass estimates. The locally-derived estimates of height may be more appropriate as these were derived using data for secondary forest species and were based upon both species specific and general regression equations.

The most accurate estimate of biomass is difficult to ascertain. However, the suitability of the allometric regressions for estimating total biomass may be judged against data collected previously on tropical forest biomass. Studies have shown that the total biomass estimated for mature tropical forests close to Manaus ranged from 275 Mg ha^{-1} to 483 Mg ha^{-1}, with a mean of 362 Mg ha^{-1} (Lucas *et al.*, 1996). Both the original and recalculated regression equations of Brown *et al.* (1989), regardless of the height estimates used, may therefore overestimate the biomass as estimates for the 30-year-old forest at Reserve Ducke approached or exceeded the mean estimate for mature forest. The regression equations of Deans *et al.* (1996) generated estimates of total biomass for the 30-year-old forest that were less than the mean biomass estimate for mature forests. However, the biomass obtained using the locally-derived height estimate exceeded the minimum biomass estimate for mature forest. This observation suggests that the allometric regression equations of Deans *et al.* (1996), which uses height estimates derived from the regressions of Brown *et al.* (1989), generates the most representative estimate of biomass. Even so, the biomass estimates obtained using locally-derived estimates of height are not unrealistic.

RELATIONSHIP BETWEEN TOTAL BIOMASS AND LAI.

On the basis of comparisons with biomass estimates obtained for tropical forests in Amazonia, the total biomass defined using the regression equation of Deans *et al.* (1996), with height estimated from the regression relationship of Brown *et al.* (1994), was the most accurate for regenerating forests at Manaus. Likewise, the estimates of LAI obtained using gap fraction data from hemispherical photographs were considered to be representative for these forests. On these assumptions, a relationship between LAI and total biomass was formulated. Based on 15 10 x 100 m plots, the LAI was shown to increase positively with total biomass ($r = 0.61$, signficance level = 95 %). This observation suggests estimates of biomass may be estimated for regenerating forests younger than 30 years by exploiting relationships with LAI.

CONCLUSIONS

Canopy gap fraction estimates from hemispherical photographs may be more reliable and consistent than those obtained from line radiation sensors such as the ceptometer although future studies should compare gap fraction estimates derived using sensors, such as the LICOR LI2000, which measure radiation flux within a hemisphere. Assuming the biomass of regenerating forests to be less than that of mature forest, the most realistic estimates of biomass were generated using the allometric regression equation of Deans *et al.* (1996) which was derived using data on regenerating forest species only. This result suggests that allometric regression equations need to be both derived from and applied to forests of similar regeneration stage and, ideally, stand-specific allometric regression equations need to be generated. Similarly, regression equations relating tree height to diameter need to be defined for tree species with similar wood densities rather than from a large dataset containing height and diameter data for species with a large range of wood densities.

The relationship of LAI with the spectral radiance recorded by satellite sensors has yet to be defined. However, if a relationship exists, as it does for temperate forests, different stages of forest regeneration, which are relatively distinct in terms of their biomass and carbon stocks, could be distinguished. The number of regeneration stages that can be defined using the relationship between LAI and the remotely-sensed radiance should also be greatest during the first 20-30 years of regeneration when LAI increases most rapidly and the relationship between LAI and biomass is strongest. This first 20-30 years of regeneration also represents the period of greatest carbon sequestration (Brown and Lugo, 1990).

The use of remotely-sensed data to estimate the properties of vegetation requires accurate remotely sensed data and ground data on which to base the predictive relationship. Experience gained in temperate environments has indicated that the accuracy of remotely sensed data may be greater than that of ground data on forest properties (Curran and Williamson, 1985). This is a particular problem in the tropics where the need for remotely sensed estimates of forest parameters is pressing, few representative forest datasets exist and the logistics of collecting such data appear overwhelming. This study has illustrated the problems associated with estimating the LAI and biomass of tropical forests and serves to identify the collection of accurate biophysical data as a major challenge to be faced in the remote sensing of tropical forests.

ACKNOWLEDGEMENTS

The authors acknowledge the financial support provided by the Natural Environment Research Council (NERC) through its Terrestrial Initiative in Global Environmental Research (TIGER) programme, awards 90 and GST/02/604. Assistance was provided by the Instituto Nacional de Pesquisas Espaciais (INPE), Instituto Nacional de Pesquisas Amazônia (INPA) and the World Wide Fund for Nature (WWF)/ Smithsonian Institute. We especially thank Bruce Nelson (INPA), Evlyn Novo, Thelma Krug and Carlos Nobre (INPE), Claude Gascon and Rita Mesquita (WWF/ Smithsonian Institute) and Emma Underwood (University of Southampton).

REFERENCES

Box, E.O., Holben, B.N. and Kalb, V. 1989. Accuracy of the AVHRR vegetation index as a predictor of biomass, primary production and net CO_2 flux. *Vegetatio* **80**, 71-89.

Brookfield, H. 1992. 'Environmental colonialism', tropical deforestation and concerns other than global warming. *Global Environ. Change* **21**, 93-6.

Brown, S., Gillespie, A.J.R. and Lugo, A.E. 1989. Biomass estimation methods for tropical forests with applications to forest inventory data. *For. Sci.* **35**, 881-902.

Brown, S. and Lugo, A.E. 1990. Tropical secondary forests. *J. Tropical Ecol.* **6**, 1-32.

Campbell, G.S. 1986. Extinction coefficients for radiation in plant canopies calculated using an ellipsoidal inclination angle distribution. *Agric. For. Met.* **36**, 317-321.

Chason, J.W., Baldocchi, D.D. and Huston, M.A. 1991. A comparison of direct and indirect methods for estimating forest canopy leaf area. *Agric. For. Met.* **57**, 107-128.

Chudnoff, M. 1980. *Tropical Timbers of the World.* US Department of Agriculture, Forest Service, Forest Products Laboratory.

Cook, E.A., Iverson, L.R. and Graham, R.L. 1989. Estimating forest productivity with Thematic Mapper and biogeographical data. *Remote sensing of Environment* **28**, 1-17.

Curran, P.J. 1993. Multispectral remote sensing for the estimation of green leaf area index. *Phil. Trans. Roy. Soc. Lond.* ***A* 30G**, 257-270.

Curran, P.J. and Williamson, H.D. 1995. The accuracy of ground data used in remote sensing investigations. *International J. Remote Sensing* **6**, 1637-1651.

Curran, P.J., Dungan, J.L. and Gholz, H.L. 1992. Seasonal LAI in slash pine estimated with Landsat TM. *Remote Sensing of Environment* **39**, 3-13.

Curran, P.J. and Foody, G.M. 1994. The use of remote sensing to characterise the regenerative states of tropical forests. In: Foody, G.M. and Curran, P.J. (Eds.), *Environmental Remote Sensing from Regional to Global Scales*, John Wiley, Chichester, 44-83.

Deans, J.D., Moran, J. and Grace, J. 1996. Biomass relationships for tree species in regenerating semi-deciduous moist tropical forest in Cameroon. *For. Ecol. and Manage.* In press.

Decagon Devices 1989. *Sunfleck Ceptometer User's Manual*. Delta-T Devices, Cambridge.

Detwiler, R.P. and Hall, C.A.S. 1988. Tropical forests and the global carbon cycle. *Science* **239**, 42-47.

Fölster, H., de Lassalas, G. and Khana, P. 1976. A tropical evergreen forest with perched water table, Magdalena Valley, Colombia. *Oceologia Plantarum* **11**, 297-320.

Foody, G.M. and Curran, P.J. 1994. Estimation of tropical forest extent and regenerative stage using remotely sensed data. *J. Biogeog.* **21**, 223-244.

Golley, F.B., McGinnis, J.T., Clements, R.G., Child, G.I. and Duever, M.J. 1975. *Mineral Cycling in a Tropical Moist Forest Ecosystem*. Elsevier Scientific Publishing Company, Oxford.

Goward, S.N., Dye, D.G. and Tucker, C.J. 1985. North American vegetation patterns observed by NOAA-7 AVHRR. *Vegetatio* **64**, 3-14.

Hamilton, G.J. 1975. Forest Mensuration Handbook. Her Majesty's Stationery Office, London. *Forestry Commission Booklet* **39**, 275pp.

Lucas, R.M., Honzak, M., Foody, G.M., Curran, P.J. and Corves, C. 1993. Characterising tropical secondary forests using multi-temporal Landsat sensor imagery. *Int. J. Remote Sensing* **14**, 3061-3067.

Lucas, R.M., Curran, P.J., Honzak, M., Foody, G.M., do Amaral, I. and Amaral, S. 1996. Disturbance and recovery of tropical forests. Balancing the carbon account. *This volume*.

Lugo, A.E. and Brown, S. 1992. Tropical forests as sinks of atmospheric carbon. *For. Ecol. and Manage.* **54**, 239-255.

McWilliam, A.L.C., Roberts, J.M., Cabral, O.M.R., Leitão, M., de Costa, A.L., Maitelli, G.T. and Zamparoni, C. 1993. Leaf area index and above ground biomass of *terra firme* rain forest and adjacent clearings in Amazonia. *Functional Ecol.* **7**, 310-317.

Michalsky, J.J. 1988. The astronomical almanac's algorithm for approximate solar position (1950-2050). *Solar Energy* **40**, 227-235.

Nel, E.M. and Wessman, C.A. 1993. Canopy transmittance models for estimating forest leaf area index. *Canadian J For. Res.* **23**, 2579-2586.

Norman, J.M. 1979. Modelling the complete crop canopy. In: Barfield, B.J. and Gerber, J. (Eds.), *Modification of the Aerial Environment of Crops*, American Society of Agricultural Engineers, St. Joseph, MI, 249-277.

Norman, J.M. 1988. *Crop Canopy Photosynthesis and Conductance from Leaf Measurements*. Workshop prepared for LI-COR, Inc., Lincoln, NE.

Norman, J.M. and Campbell, G.S. 1991. Canopy structure. In: Pearcy, R.W., Ehleringer, J., Mooney, H.A. and Rundel, P.W. (Eds.), *Plant Physiological Ecology, Field Methods and Instrumentation*, Chapman and Hall, London, 301-325.

Peterson, D.L., Spanner, M.A., Running, S.W., and Teuber, K.B. 1987. Relationship of Thematic Mapper Simulator data to leaf area index of temperate coniferous forests. *Remote Sensing of Environment*, **22**, 323-475.

Pierce, L.L. and Running, S.W. 1988. Rapid estimation of coniferous forest leaf area index using a portable integrating radiometer. *Ecology*, **69**, 1762-1767.

Reyes, G., Brown, S., Chapman, J. and Lugo, A.E. 1992. *Wood Densities of Tropical Tree Species*. United States Department of Agriculture, General Technical Report SO-88, 15pp.

Rich, P.M. 1990. Characterising plant canopies with hemispherical photographs. *Remote Sensing Reviews*, **5**, 13-29.

Roberts, J., Cabral, O.M.R., Leitao, M., da Costa, J.P., McWilliam, A.-L.C., and Sá, T.D. de A. 1996. An overview of the leaf area index and physiological measurements during ABRACOS. *This volume*.

Sader, S.A., Waide, R.B., Lawrence, W.T. and Joyce, A.T. 1989. Tropical forest biomass and successional age class relationships to a vegetation index derived from Landsat TM data. *Remote Sensing of Environment*, **28**, 143-156.

Saldarriaga, J.G. 1985. *Forest Succession in the Upper Rio Negro of Colombia and Venezuela*. PhD dissertation, University of Tennessee, Knoxville, USA.

Saldarriaga, J.G., West, D.C., Tharp, M.L. and Uhl, C. 1988. Long-term chronosequence of forest succession in the upper Rio Negro of Colombia and Venezuela. *J. Ecol.* **76**, 938-958.

Uhl, C. 1987. Factors controlling succession following slash-and-burn agriculture in Amazonia. *J. Ecol.* **75**, 377-407.

Welles, J. 1990. Some indirect methods of estimating canopy structure. *Remote Sensing Reviews* **5**, 31-43.

Wickland, D.E. 1991. Global ecology: the role of remote sensing. In: Esser, G. and Overick, D. (Eds.), *Modern Ecology: Basic and applied aspects*, Elsevier Scientific Publishing Company, Amsterdam, 725-747.

Williams, W.A., Loomis, R.S. and Alvim, P. de T. 1972. Environments of evergreen rainforest on the lower Rio Negro, Brazil. *Tropical Ecol.* **13**, 65-78.

Wu, S. 1990. Assessment of tropical forest stand characteristics with multipolarisation SAR data acquired over a mountainous region in Costa Rica. *IEEE Transactions on Geoscience and Remote Sensing* **28**, 752-755.

RESUMO

Estimativas da extensão e da biomassa associada aos diferentes estágios de regeneração, obtidas via sensoriamento remoto, são necessárias para estimar os estoques de carbono associados a florestas tropicais em regeneração. Os estágios de regeneração podem ser definidos a partir de relações entre a radiância obtida através de sensoriamento remoto e as propriedades medidas dos dosséis de florestas as quais variam durante a sucessão e as quais estão relacionadas com biomassa. Em um sítio próximo Manaus, o IAF foi estimado para alguns estágios de regeneração utilizando-se dados de "gap fraction" (fração de falhas do dossel), derivados a partir de "sunfleck fraction" (fração de áreas iluminadas ou manchas de radiação solar no solo sob o dossel), PAR transmitido e fotografias hemisféricas, e o modelo de transferência de radiação de Norman e Campbell (1991). A biomassa total foi também estimada para cada estágio utilizando a regressão alométrica de Brown et al. (1989) e Deans *et al.* (1996) e medições de diâmetro de árvores, altura e densidade de madeira. Entretanto, grandes diferenças nas estimativas de IAF e biomassa total foram obtidas utilizando cada método. O estudo conclui que as estimativas mais realistas e consistentes de IAF foram derivados com as estimativas de "gap fraction" do dossel a partir de fotografias hemisféricas. As regressões alométricas de Deans *et al.* (1996) foram consideradas as que geraram as estimativas de biomassa mais representativas. Uma significativa correlação positiva entre essas estimativas de IAF e biomassa foi observada, indicando que quaisquer relações definidas entre IAF e radiância obtida por sensoriamento remoto podem ser exploradas para estimar a extensão, a biomassa e os estoques de carbono associadas aos diferentes estágios de regeneração.

22 Disturbance and recovery of tropical forests: balancing the carbon account

R.M. LUCAS[1], P.J. CURRAN[1], M. HONZAK[2], G.M. FOODY[2], I. do AMARAL[3] and S. AMARAL[4]
[1] *University of Southampton, UK*
[2] *University of Wales, Swansea, UK*
[3] *Instituto Nacional de Pesquisas da Amazônia, Manaus, Brazil*
[4] *Instituto Nacional de Pesquisas Espaciais, São Paulo, Brazil*

INTRODUCTION

The recent widespread conversion of mature tropical forests to agriculture and other land uses has resulted in a large reduction in biotic carbon reserves and contributed to 30% of the annual increase in atmospheric carbon dioxide (Bueno and Helene, 1991). Current estimates suggest that between 1.2 and 2.6 Pg of carbon is transferred to the atmosphere annually through worldwide conversion of tropical forests (Houghton *et al.*, 1990; Brown *et al.*, 1993; Sampson *et al.*, 1993; Dixon *et al.*, 1994). These losses of terrestrial carbon have, however, been partly offset by sequestration of atmospheric carbon by regenerating forests (Brown and Lugo, 1990).

An assessment of the carbon stocks of both mature and regenerating forests is required if changes in the carbon flux resulting from tropical land cover transformations are to be modelled (Curran *et al.*, 1996). Carbon stocks may be estimated from forest biomass by applying a conversion factor of approximately 0.5 (Atjay *et al.*, 1979; Iverson *et al.*, 1992; Brown *et al.*, 1993; Dixon *et al.*, 1994) and the biomass of forest stands may be established through, for example, destructive harvesting of trees or timber volume sampling (Brown and Iverson, 1992; Detwiler and Hall, 1988; Bueno and Helene, 1991; Houghton *et al.*, 1987). Estimates of the biomass of mature tropical forests are, however, scarce, and due to the spatial heterogeneity of these forests, are often unrepresentative of the forest as a whole (Brown and Iverson, 1992). As a result, there is considerable uncertainty as to the quantity of carbon held within mature closed forests and also in the proportion of this carbon that is released to the atmosphere through biomass burning and decomposition (Smith *et al.*, 1993; Sampson *et al.*, 1993). The fraction of this carbon restored to the terrestrial system by tropical regenerating forests is also unknown as few studies have assessed the rate and quantity of carbon uptake by these forests.

Models of tropical carbon flux require spatial estimates of the extent of both mature and regenerating forest. Several research groups have estimated the spatial extent of tropical closed forest for regions such as Amazonia (e.g. INPE, 1990; INPE, 1992;

Amazonian deforestation and climate. Edited by J.H.C. Gash, C.A. Nobre, J.M. Roberts and R.L. Victoria.

Joint Research Centre, 1992; Skole and Tucker, 1993) by using remotely sensed data. However, up-to-date estimates of the regional extent of regenerating forests are generally unavailable due mainly to the rapidly changing nature of land cover in the tropics, although Brown and Lugo (1990) suggest that regenerating forests represent over 31% of the tropical closed forest area and are forming at a rate of 90×10^3 km^2 y^{-1}.

The larger expanses of regenerating forest in Brazilian Amazonia correspond closely to the areas of deforestation. The loss of mature forests in this region accelerated after the 1960s and, as a consequence, many of the regenerating forests are less than 30 years old and at a stage when the rate of carbon assimilation into the biomass is at a maximum (Brown and Lugo, 1990; Whitmore, 1990). These younger regenerating forests represent a significant sink for atmospheric carbon. The expansion and growth of these forests can be mapped using data from remote sensing satellites which have been observing the entire tropical belt for at least the past 20 years (Foody and Curran, 1994). Archives of Landsat MSS (Multi-spectral Scanning System) and TM (Thematic Mapper) data extend back to 1972 and 1982 respectively (Sader *et al.*, 1990) giving a unique record of land cover transformations (e.g. mature forest loss, agricultural expansion and regeneration) in the tropics.

A number of studies (Mausel *et al.*, 1993; Moran *et al.*, 1994) have involved the use of a time-sequence of remotely sensed data to describe the extent and succession of tropical regenerating forests. Time sequences have also been used to provide estimates of the age of regeneration. For example, a study undertaken at a site north of Manaus (2°20'S, 60°00'W) in the Brazilian state of Amazonas used a time-sequence of Landsat MSS/TM data to age forests regenerating on abandoned cattle pastures (Lucas *et al.*, 1993). However, for carbon flux models, different stages of forest regeneration need to be defined as a function of biomass, and hence carbon content. The research described in this chapter, which is an extension of the time series analysis undertaken by Lucas *et al.* (1993), aimed first to define several different stages of tropical forest regeneration at the Manaus site and then to map the extent of each stage using remotely sensed data. The ultimate aim of the research was to assign estimates of biomass to each of the regeneration stages defined and thereby generate spatial estimates of biomass and carbon.

FOREST LOSS AND REGENERATION ON CATTLE PASTURES

In Brazilian Amazonia, the 15 million hectares of cattle pasture (Smith *et al.*, 1995) represent approximately 65% of the estimated 23 million hectares of land that has been deforested (excluding fragmented forest: Skole and Tucker, 1993). The reductions in carbon stocks resulting from this land cover transformation are perhaps the most extreme, as most farms (or fazendas) were established on land covered previously with relatively undisturbed closed mature forest (Brown and Lugo, 1990).

The majority of the cattle pastures in Brazil were formed following the introduction of fiscal incentives in the late 1960s (Serrão *et al.*, 1978; Fearnside, 1988). These incentives were also largely responsible for the establishment of cattle ranches along the BR 174 Highway which was constructed in the early 1970s to connect Manaus with the Roraima state capital, Boa Vista. As with many of the pastures in Brazilian Amazonia, the withdrawal of tax incentives in the late 1980s and the general decline in land productivity through inadequate management strategies led to the widespread degradation and eventual abandonment of many fazendas near Manaus and this, in turn, provided sites for forest regeneration (Fearnside, 1988; Corves and Bax, 1992).

STUDY SITE

The main study sites were the BDFFP (Biological Dynamics of Forest Fragments Project) research area of the Brazilian National Institute for Amazonian Research (INPA) and the Smithsonian Institution, and regenerating forest blocks within the Reserva Ducke, which are both located to the north of Manaus (see Figure 1 of Honzak *et al.*, 1996). Before the field campaign, the approximate age of the regenerating forests at the BDFFP site (within Fazendas Porto Alegre, Esteio, Maringa and Agroman) was determined initially from a time-series of Landsat TM data for the years 1985, 1988, 1989 and 1991 (Lucas *et al.*, 1993) and then increased in accuracy by using Landsat MSS data from 1976-1984 and a SPOT HRV image from 1986. The ages of forests in the Reserva Ducke were determined from INPA records.

FIELD DATA COLLECTION

Inventory plots (10 × 100 m) were located within a range of regenerating forests of known age. Thirteen of the plots were located within forests at the BDFFP site (which ranged in age from 4 to 16 years) while a further two plots were located in the Reserva Ducke in stands aged at 21 and 30 years respectively. Within each plot, data were collected on tree diameter (for all trees greater than 3 cm in diameter) and species composition. A more detailed description of the field data collection is presented by Honzak *et al.* (1996).

CLASSIFICATION BASED ON FOREST INVENTORY DATA

To express the floristic composition of each plot, an Importance Value Index (IVI), which is a combined measure of the percentage relative frequency, abundance and dominance for each species or genus identified (Krebs, 1972), was calculated. Three types of forest regeneration at the study site were then classified according to the IVI recorded for genera within each plot (Table 1). In two plots, the IVI was greatest for species of the genera *Vismia* (>30%) and the combined IVI for the genera *Miconia*

Table 1 The major secondary forest types identified within the BDFFP site near Manaus

Forest type	Main genera	IVI
F_{VMB}	*Vismia*	> 30 %
	Miconia	> 8 %
	Bellucia	> 13 %
F_{CP}	*Cecropia*	> 39 %
F_{Mix}	*Cecropia*	> 20 % but < 39 %

and *Bellucia* (Family: *Melastomataceae*) exceeded 20%. In this study, these forests are referred to as F_{VMB}. In many plots, however, the pioneer tree genus *Cecropia* predominated (IVI >20 %). Within six plots, the IVI for *Cecropia* exceeded 39 % and trees of this genus formed an umbrella over an understorey comprised mainly of species of the genera *Aparisthmium*, *Inga*, *Pourouma*, *Guatteria*, *Laetia* and *Byrsonima*. Within three plots, the IVI for *Cecropia* ranged between 20-30%, although the umbrella of *Cecropia* was either penetrated by other fast growing genera (e.g. *Goupia*) or experienced natural dieback and thereby exposed an understorey of other secondary forest species (e.g. *Laetia*, *Goupia* and *Miconia*).

Consequently, where the IVI for *Cecropia* exceeded 39%, forests were classified as F_{CP}, whilst remaining forests, which represented the changeover from early pioneer to slower growing secondary and primary forest species and therefore a later stage of forest regeneration, were classified as F_{Mix}. The forests sampled within the remaining two plots at the BDFFP site contained a large number of genera although the IVI of few exceeded 10 %. These forests were classified as F_{Mix} as their overall species composition was typical of this forest type. The two forests sampled at the Reserva Ducke were considered to be a later successional stage of F_{Mix}.

CLASSIFICATION USING REMOTELY SENSED DATA

Following the definition of regenerating forest types on the basis of the IVI, the potential for estimating the spatial extent of each type using remotely sensed data, either singly or in a time-series, was assessed.

SPECTRAL SEPARABILITY OF FOREST TYPES

As an initial step to the classification of a single-date Landsat TM image (8 August, 1991) of the BDFFP site, image statistics relating to each of the three forest types were obtained by identifying image pixel coordinates which corresponded to the field coordinates of the plot locations (determined using a Magellan Global Positioning System NAV 5000 PRO). The spectral separability of the three successional forest types within TM Channels 3 (visible), 4 (near-infrared) and 5 (mid-infrared) was then

assessed on the basis of a transformed divergence algorithm which provided a measure of the statistical separation of pairs of classes as a covariance-weighted distance of between-class means (Singh, 1984). This analysis suggested that forest types F_{VMB}, F_{CP} and F_{Mix} were spectrally indistinguishable from mature forest in the visible channels but separable from mature forest in the near and mid-infrared channels. A feature space plot of the near and mid-infrared wavebands (Figure 1) also demonstrated that as regenerating forests colonised pasture, the near and mid-infrared radiance of both F_{CP} and F_{VMB} first increased to a maximum (due perhaps to increases in leaf cover) and thereafter declined steadily (due to increased shadowing) and eventually merged with that of mature forest. The analysis also illustrated a greater similarity in the near and mid-infrared radiance of both F_{VMB} and F_{Mix} to mature forest, compared to that of F_{CP}. This may be a function of the smaller leaf size and more vertical orientation of leaves within the upper canopy of both the mature forest and F_{VMB}/F_{Mix} compared to the larger leaf size (often >1 m) and more horizontal leaf orientation characteristic of the upper canopy of F_{CP}.

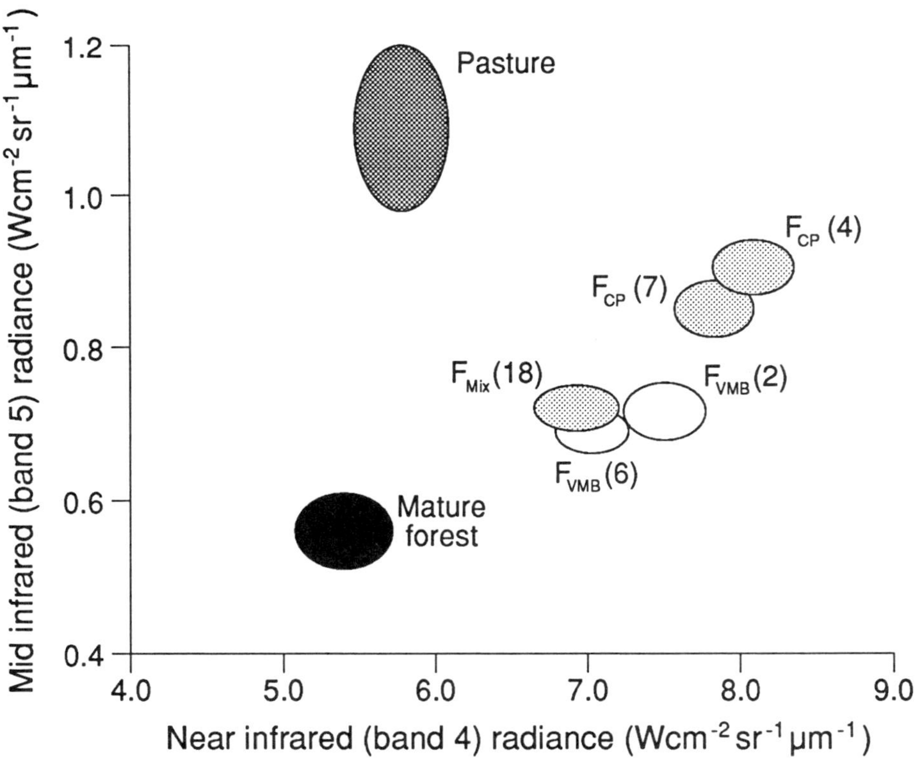

Figure 1 Near and middle infrared response of different stages of forest regeneration, mature primary forest and pasture at Manaus, as defined using the Landsat TM. Figures in brackets represent the forest age in years.

CLASSIFICATION OF F_{CP} USING SINGLE-DATE LANDSAT TM DATA

The pattern of spectral signatures for the different regenerating forest types within the feature space plot of Landsat TM Channels 4 and 5 suggested that their discrimination within the image may be achieved through classification. A texture-based maximum likelihood classification algorithm based on a causal Markov random field model (Palubinskas, 1988) was therefore applied to Channels 3, 4 and 5 of the 1991 Landsat TM image. Based on the knowledge obtained through both field inventories and observations, training statistics were derived (from raw data values) for 11 land cover classes, including the three forest regeneration types, and used in the classification (Foody *et al.*, 1996).

To test the accuracy of the classification, further sites representing each of the 11 land cover classes defined were identified within the image, and the accuracy of their classification was expressed as a percent correct allocation. The overall percent correct allocation for the 11 classes, was 86.5 %, while unweighted and weighted kappa coefficients were 0.83 and 0.93 respectively. These accuracies were consistently larger compared to conventional per-pixel maximum likelihood and minimum distance algorithms performed on the same data set (Foody *et al.*, 1996). However, the classification was unable to provide a reliable discrimination of F_{VMB} from F_{Mix} although the separation of F_{CP} from both F_{VMB} and F_{Mix} was achieved on the basis of textural and spectral measures alone. Additional sources of data which could be used to discriminate F_{VMB} and F_{Mix} within Landsat TM imagery were therefore investigated.

DISCRIMINATION OF F_{Mix} FROM F_{VMB}

Previous studies on forest regeneration in Amazonia indicated that *Cecropia* and *Vismia* species are widespread throughout the region (Richards, 1952; Fearnside, 1988; Whitmore, 1990), with the former commonly occupying agricultural sites abandoned within several years of initial forest clearance, and the latter favouring the more intensively used lands (Uhl *et al.*, 1981; Uhl and Jordan, 1984; Uhl *et al.*, 1988; Buschbacher *et al.*, 1988).

The intensity of land management is partly a function of the period of land occupation. The length of time pastures were used at the BDFFP site before abandonment was therefore determined by tracing land use within a time-sequence of Landsat MSS/TM and SPOT HRV imagery recorded between the early 1970s and 1991. This analysis indicated that the majority of pastures which were abandoned within 1-3 years of clearance corresponded to the areas classified as F_{CP} by Foody *et al.* (1996). It was therefore assumed that F_{Mix}, being a more advanced successional stage of F_{CP}, would also only occupy land abandoned within the first three years of clearance, whilst F_{VMB} would dominate areas used for longer periods. On this basis, F_{VMB} and F_{Mix} were distinguished and all forest types were subsequently aged using the time-sequence of remotely sensed data. In the final classification, 11 classes representing three forest types of different age were mapped for the BDFFP site alone (Figure 2).

ESTIMATION OF BIOMASS AND CARBON STOCK

The total biomass was estimated for each of the fifteen plots by using allometric regressions derived by Deans *et al.* (1996). These regressions were generated by destructively harvesting and weighing the above and below-ground components of six *Terminalia*, four *Trema*, three *Musanga* and one *Triplochiton* tree (with diameters ranging from <5 to 20 cm) which are regenerating forest species native to West Africa (Honzak *et al.*, 1996). The regressions estimate total biomass as a function of diameter at breast height (D, cm) and total height (H, m) with or without wood density (S, Mg m^{-3}). The values of S were taken from Reyes *et al.* (1992) and H was estimated from D by using the regression defined by Brown *et al.* (1989) for forests located within the tropical moist life zone (Honzak *et al.*, 1996).

To scale-up plot biomass to the larger areas of forest, the estimates obtained from the thirteen plots at the BDFFP site were assumed, on the basis of plot location alone, to indicate the biomass associated with seven of the eleven forest classes defined in Figure 2. The biomass of the five unsampled forest classes was estimated from regressions relating forest age to biomass as defined (for the inventoried forests) using the allometric regressions of Deans (1996). The appropriate biomass estimates for all 11 forest classes were then scaled-up to the area of each forest class to generate spatial estimates of biomass for all regenerating forests at the BDFFP site (Figure 3). The carbon associated with each regeneration stage was then approximated by assuming carbon stocks to be half of the biomass (Iverson *et al.*, 1994).

The estimates of carbon for each stage of regeneration can be aggregated to assess the magnitude of the carbon sink associated with all regenerating forests at the BDFFP site. Similarly, both carbon stocks held within the undisturbed mature forests and the losses of carbon resulting from the conversion of these forests to pasture can be approximated on the basis of published estimates of mature forest biomass (e.g. McWilliam *et al.*, 1993). By using these spatial estimates of carbon, the fluxes of carbon associated with tropical deforestation and regeneration can be modelled.

BALANCING THE CARBON ACCOUNT

In the simplest terms, the release of terrestrial carbon associated with the conversion of mature forests to agriculture can be quantified as a product of the carbon per unit area of forest and the total area of forest loss. However, the complexity and spatial variability of the structure of mature tropical forests, which results partly from the large diversity and partly from a dynamic natural succession, has prevented the assignment of a value or range of values which accurately define their biomass. For example, estimates of total above-ground biomass for moist closed forests near Manaus range from 275 Mg ha^{-1} (Rodrigues, 1967) to 563 Mg ha^{-1} (Lechthaler, 1956; Figure 4). This large variation in estimated biomass for mature forests is due mainly to differences in both the location of inventories and sampling methodologies (Brown and Iverson, 1992). The biomass estimate of Lechthaler (1956) at the

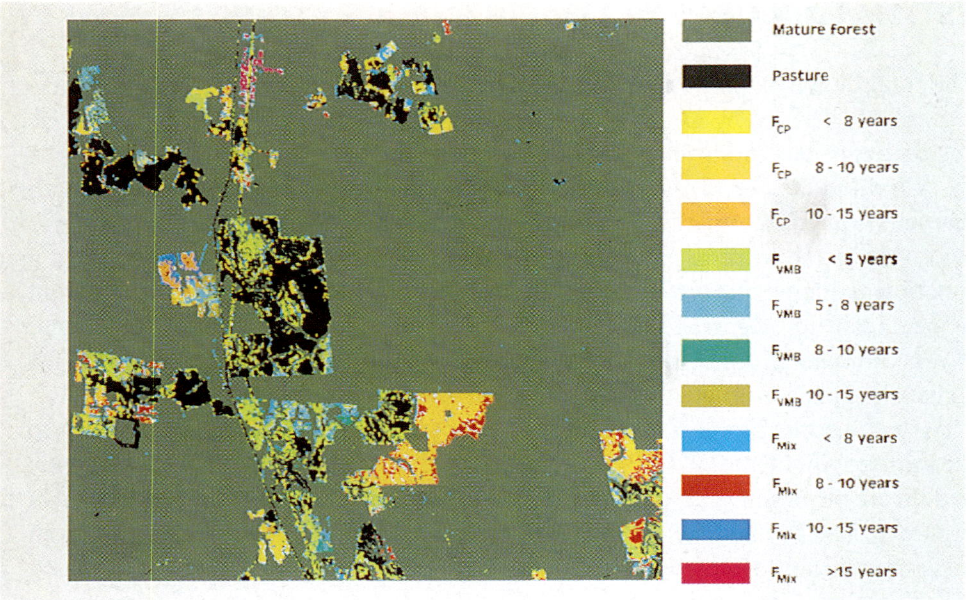

Figure 2 The classification of regenerating forest types at Manaus. F_{CP} was discriminated from other forest types on the basis of spectral and textural differences in Landsat TM Channels 3, 4 and 5. A time sequence of Landsat TM imagery was analysed to define both land management history, which was used subsequently to distinguish F_{VMB} and F_{Mix}, and the age of regrowth.

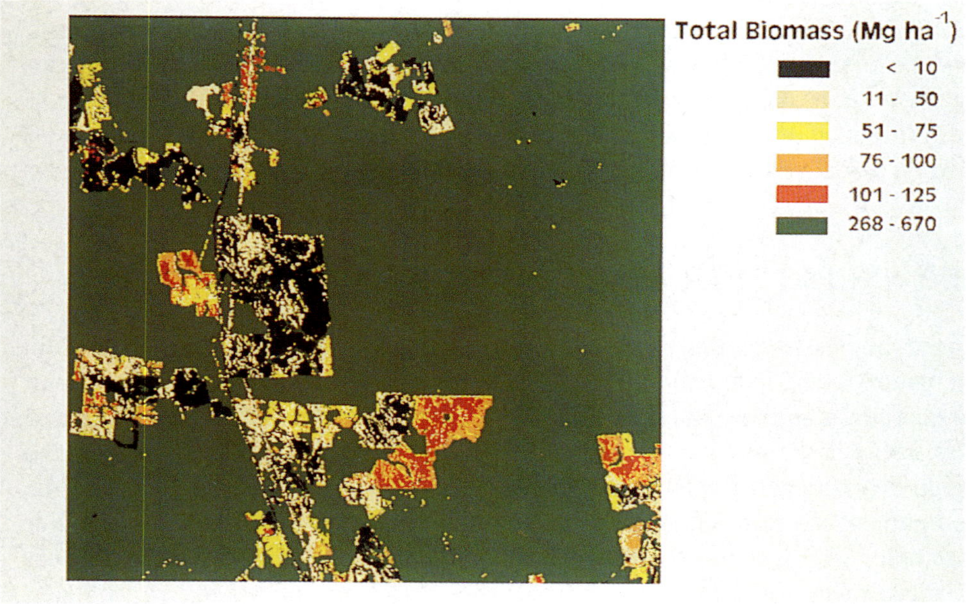

Figure 3 Estimates of total biomass stocks associated with regenerating forests at the BDFFP site, Manaus, as defined using the allometric regressions of Deans *et al.* (1996) and the height estimation equation of Brown *et al.* (1989).

Reserva Ducke, for example, is one of the largest recorded for mature Amazonian forests and may be biased due to the inclusion of big trees within the sample or the selection of an unrepresentative plot (Brown and Iverson, 1992). The wide range of biomass estimates for mature forests near Manaus therefore prevents an accurate assessment of the carbon stocks associated with mature forests and the carbon losses resulting from the conversion of these forests to cattle pasture and agricultural lands. Furthermore, there are considerable uncertainties in the quantity of below ground biomass of mature forests and the fluxes of carbon resulting from the decomposition of this component following forest removal. The rate of carbon release to the atmosphere during the conversion process is also variable with large and immediate losses resulting from biomass burning and more gradual releases occurring during long-term decomposition (Fearnside, 1985).

Because of the uncertainties in the carbon stocks of mature forests, the time-period required for regenerating forests to recover carbon to pre-disturbance levels is unknown. However, one approach used to estimate the time period required for

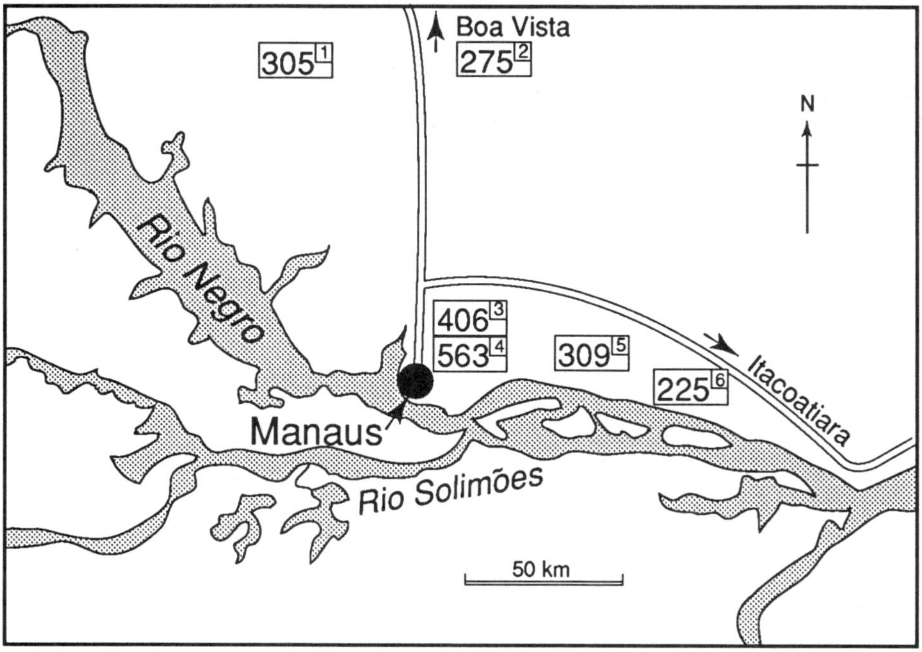

1 Forest managment basin 90 km N of Manaus (Higuchi *et al.*, 1985)
2 Km 42 Boa Vista Rd. (McWilliam *et al.*, 1993)
3 Adjacent to Walter Egler Forest Reserve (Klinge *et al.*, 1975)
4 Ducke Forest Reserve (Lechthaler, 1956)
5 Km 30, Itacoatiara Rd. (Prance *et al.*, 1976)
6 Km 64 - 200, Itacoatiara Rd. (Rodrigues, 1967)

Figure 4 Estimates of total above-ground biomass for mature primary forests at Manaus (Mg ha^{-1}).

complete carbon restoration at Manaus was to compare the range of biomass estimates obtained for mature forests with those derived for regenerating forests of known age. Within the 30-year old forest plot at the Reserva Ducke, for example, the total above ground biomass estimated using the allometric regressions of Deans *et al.* (1996) was 175 Mg ha^{-1}. By assuming that the total above ground biomass estimate of 275 Mg ha^{-1} (McWilliam *et al.*, 1993) is typical for mature forest at Manaus and that a linear rate of biomass accumulation is maintained, regenerating forests may potentially restore biomass (and hence carbon) to pre-disturbance levels within 50 years. However, the period required for forests to recover their pre-disturbance carbon stocks is extended considerably if the mature forest biomass estimate of 563 Mg ha^{-1} is considered true, particularly as the rate of biomass accumulation for regenerating forests declines after 30 years (Uhl, 1987; Saldarriaga *et al.*, 1988; Brown and Lugo, 1990).

DISCUSSION

The contribution of tropical regenerating forests to the reduction of carbon dioxide levels in the atmosphere is uncertain because knowledge of the biomass accumulated within different stages of forest regeneration is limited and there is little information on the spatial extent of each stage at both a local and regional level. This study has defined a methodology for estimating the age and extent of several types of forest regeneration by using remotely sensed data and suggested procedures for assigning biomass values to each of these forest types. Estimates of the biomass (and hence carbon stocks) associated with all regenerating forests at Manaus have been made. However, the errors associated with the classification of forest regeneration types and the estimation of biomass are likely to be large. The aim of this discussion is to provide an understanding of the main reasons for these errors.

CLASSIFICATION USING REMOTELY-SENSED DATA

Time sequences of remotely sensed data provide a unique insight into the dynamics of forest regeneration and agricultural land use in the tropics. This record of land transformation is, however, often incomplete because remotely sensed data may be unavailable for periods of up to several years. Changes in land cover may therefore remain undetected and the accuracy of estimates of forest age or periods of land use is also reduced. Even so, remotely sensed data often represent the only information source available for land cover mapping and monitoring, especially in areas undergoing rapid changes in land use.

In the analysis of time-sequences of remotely sensed data, the errors of class assignment associated with the classification of single date imagery and their subsequent comparison are compounded. This aggregation of errors is one of the main limitations of a stratified approach to the classification of land covers. In this study, additional errors were also introduced by superimposing the two-tiered classification of F_{CP}, F_{Mix} and F_{VMB} onto the age class map. In particular, the discrimination of F_{VMB} and F_{Mix} was based entirely on the assumption that the period of land use prior to abandonment was significant in determining the species composition of the regenerating forests. To a certain extent, this assumption was justified on the basis of previous studies of the species composition of forests regenerating on abandoned pastures (e.g. Uhl *et al.*, 1988) and also on the observed correspondence of the area occupied by F_{CP} (as determined from the texture-based maximum likelihood classification of Landsat TM data) and the land used for periods of less than three years. However, this assumption introduces the greatest uncertainty into the classification of regenerating forest types.

DEFINITION OF REGENERATION STAGES FROM FIELD DATA

The classification of F_{CP}, F_{VMB} and F_{Mix} using remotely sensed data also assumes that only these three distinct forest types existed at the BDFFP site. However, it is likely that a more mixed community of forests became established on many pastures. During the field campaign, only seven different forests, which colonised pastures abandoned either within 1-3 years of clearance or after five years of clearance were sampled. The classification of forest types therefore did not account for forests which occupied land used for 3-5 years and which may have been comprised of a mix of F_{CP} and F_{VMB} or were dominated by other pioneer genera. The classification of F_{Mix} from field data also assumes that these forests were originally of type F_{CP} whereas the only evidence for this was the large IVI recorded for *Cecropia* compared to other secondary forest species or observations of dieback of this genus within the forests sampled.

ESTIMATION OF THE BIOMASS OF REGENERATING FORESTS

The procedure for scaling-up biomass estimates from plots to the area of the forest assumes that forests of the same age and with similar histories of land management will support the same levels of biomass. Evidence for the homogeneity of these regenerating forests was obtained by comparing species data from five plots located within nine year old forest F_{CP} at Fazenda Porto Alegre. These forests were established on land cleared of mature forest in 1983 and abandoned in the same year. The comparison revealed that within all five plots, the IVI exceeded 39% for trees of the genus *Cecropia* and was similar for understorey genera. The similarity of the IVI for three of these plots is illustrated in Figure 5. Furthermore, the estimates of total biomass for the five plots, derived using the allometric regressions of Deans *et al.* (1996), ranged from 97 to 119 Mg ha^{-1}, which were similar to the 82 to 103 Mg ha^{-1} range obtained using the same allometric regressions for three plots within nine-year-old regenerating forest (F_{Mix}) at Fazenda Esteio.

The forests at Fazendas Porto Alegre and Esteio colonised land which was not used intensively prior to abandonment, and the entire pasture was abandoned at one time. The scaling-up procedure may not, however, apply so well to forests regener-

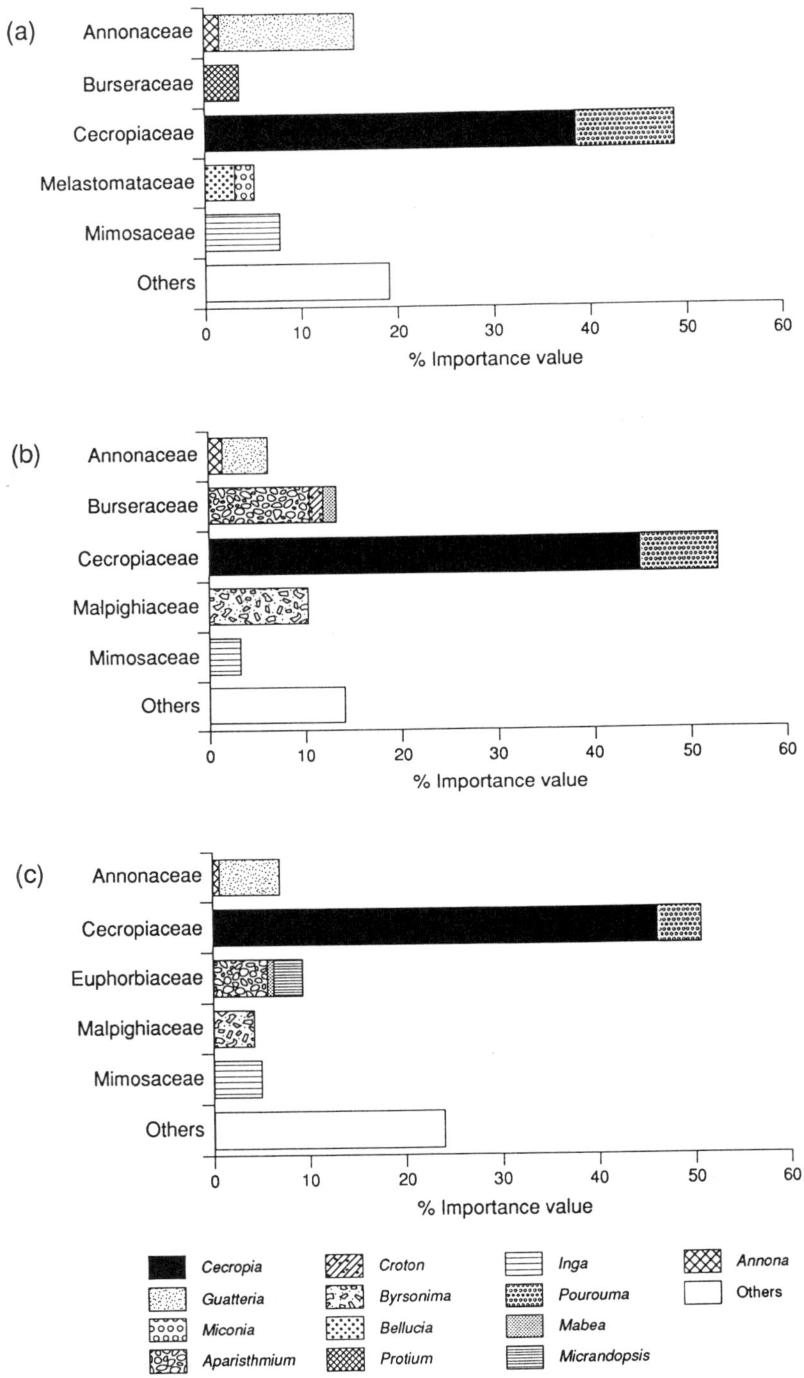

Figure 5 The similarity in percentage IVI of tree genera within (a) Plot 3, (b) Plot 4 and (c) Plot 5 in 9-year-old forest at Fazenda Porto Alegre demonstrates the homogeneity of regenerating forests establishing on many abandoned pastures. This forest colonised a pasture which was cleared of primary forest without the use of fire and was abandoned within a year of formation.

ating on sites which were used intensively for many years. For example, a comparison of the regenerating forest area within the time-sequence of remotely sensed data at Manaus indicated that the invasion of forests onto the more degraded pastures was gradual and often initiated from the boundary with the mature forests. The more random nature and timing of forest regeneration on these more intensively used pastures suggests that the structural and floristic homogeneity of these forests (e.g. F_{VMB}) is likely to be reduced compared to forests colonising less intensively used pastures (e.g. F_{CP}). Furthermore, the heterogeneity of all forest types is likely to increase with age as the dominance by a few pioneer genera is reduced through the influx of late regeneration and mature forest species. The species composition and growth of forests is also likely to vary according to factors such as soil type and topography. The procedure for scaling-up plot biomass estimates to the forest level should therefore be applied with caution.

CONCLUSION

This study has demonstrated the benefits, and also the limitations, of integrating remotely sensed data with forest inventory data to describe tropical forest regeneration on abandoned pastures and agricultural lands. In particular, time sequences of Landsat MSS/TM and SPOT HRV data can be used to map and describe in detail the expansion and utilisation of the deforested area and the regeneration of forests.

The analysis of field data suggested that three types of forest regeneration (F_{CP}, F_{Mix} and F_{VMB}) were common to the study site and that the composition of these forests was partly influenced by the history of land usage with F_{CP} colonising pastures abandoned within several years of clearance and F_{VMB} establishing on the more intensively used pastures. Furthermore, the study demonstrated that relatively homogenous tracts of forests may establish, particularly when entire pastures are abandoned shortly after clearance. The apparent homogeneity of forests of the same age and establishing themselves on land with a similar history of land usage provided the basis for scaling-up biomass estimates from plots to the area of the forest.

The estimates derived using the regressions of Deans *et al.* (1996) were considered, on the basis of previous studies (Uhl and Jordan, 1984; Uhl *et al.*, 1988), to be representative of the biomass of regenerating forests at Manaus and were therefore used to quantify carbon stocks within all regenerating forests at the BDFFP site. This analysis concluded that at least 50 years are required for regenerating forests to restore biomass and carbon stocks to pre-disturbance levels.

As carbon is released to the atmosphere through burning of fossil fuels and the removal of the world's forests, so there is compensation through the recovery of carbon by regenerating forests, enhanced levels of photosynthesis and carbon uptake and expansion of the forested area. Although the magnitude of this compensation is uncertain, remotely sensed data provide a tool for determining baseline information on the incremental growth of regenerating vegetation, the areas of forest loss and the extent of the remaining mature forest cover. Increasingly, more accurate plot biomass data and actual measurements of carbon flux are becoming available and these can be used in the future to provide a better understanding of the transfer of carbon between the terrestrial system and the atmosphere.

ACKNOWLEDGEMENTS

The authors acknowledge the financial support provided by the Natural Environment Research Council (NERC) through its Terrestrial Initiative in Global Environmental Research (TIGER) programme awards 90 and GST/02/604 and the assistance provided by the Instituto Nacional de Pesquisas Espaciais (INPE), Instituto Nacional de Pesquisas da Amazônia (INPA), World Wide Fund for Nature (WWF) and the Smithsonian Institution, but especially Bruce Nelson (INPA), Evlyn Novo (INPE), Thelma Krug (INPE), Carlos Nobre (INPE), Claude Gascon and Rita Mesquita (WWF/Smithsonian Institute) and Emma Underwood (University of Southampton).

REFERENCES

Atjay, G.L., Ketner, P. and Duvigneaud, P. 1979. Terrestrial primary production and phytomass. In: Bolin, B. (Ed.), *The Global Carbon Cycle*, John Wiley and Sons, Chichester, 129-181.

Brown, S., and Lugo, A.E. 1990. Tropical secondary forests. *J. Trop. Ecol.* **6**, 1-32.

Brown, S. and Iverson, L.R. 1992. Biomass estimates for tropical forests. *World Resource Review* **4**, 366-384.

Brown, S., Gillespie, A.J.R. and Lugo, A.E. 1989. Biomass estimation methods for tropical forests with applications to forest inventory data. *For. Sci.* **35**, 881-902.

Brown, S., Iverson, L.R., Prasad, A. and Liu, D. 1993. Geographical ditribution of carbon in biomass and soils of tropical Asian forests. *Geocarto International* **4**, 45-59.

Bueno, M.A.F. and Helene, M.E.M. 1991. Global deforestation and CO_2 emissions: Past and present. A comprehensive review. *Energy and Environ.* **2**, 235-282.

Buschbacher, R., Uhl, C. and Serrão, E.A.S. 1988. Abandoned pastures in eastern Amazonia. II. Nutrient stocks in the soil and vegetation. *J. Ecol.* **76**, 682-699.

Corves, C. and Bax, S. 1992. Amazon skywatch. *Geogr. Mag.* **64**, 33-36.

Curran, P.J., Foody, G.M., Lucas, R.M., Honzak, M. and Grace, J. 1996. The carbon balance of tropical forests: From the local to the regional scale. In: van Gardigen, P., Foody, G.M. and Curran, P.J. (Eds.). *Scaling-up.* Cambridge University Press, Cambridge. In press.

Deans, J.D., Moran, J. and Grace, J. 1996. Biomass relationships for tree species in regenerating semi-deciduous moist tropical forest in Cameroon. *For. Ecol. and Manage.* In press.

Detwiler, R.P. and Hall, C.A.S. 1988. Tropical forests and the global carbon cycle. *Science* **239**, 42-47.

Dixon, R.K., Brown, S., Houghton, R.A., Solomon, A.M., Trextrer, M.C. and Wisniewski, J. 1994. Carbon pools and flux of global tropical forest ecosystems. *Science* **263**, 185-190.

Fearnside, P.M. 1985. Brazil's Amazon forest and the global carbon problem. *Interciencia* **10**, 179-186.

Fearnside, P.M. 1988. An ecological analysis of predominant land uses in the Brazilian Amazon. *The Environmentalist* **8**, 281-300.

Foody, G.M. and Curran, P.J. 1994. Estimation of tropical forest extent and regenerative stage using remotely sensed data. *J. Biogeography* **21**, 223-244.

Foody, G.M., Palubinskas, G., Lucas, R.M., Curran, P.J. and Honzak, M., 1996. Identifying terrestrial carbon sinks: Classification of successional stages in regenerating tropical forest from Landsat TM data. *Remote Sensing of Environment*, Submitted.

Honzak, M., Foody, G., Lucas, R.M., Curran, P.J., do Amaral, I. and Amaral, S. 1996. Estimation of the leaf area index and total biomass of tropical secondary forests: A comparison of methodologies. *This volume*.

Houghton, R.A., Boone, R.D., Fruci, J.R., Hobbie, J.E., Melillo, J.M., Palm, C.A., Peterson, B.J., Shaver, G.R., Woodwell, G.M., Moore, B., Skole, D.L. and Myers, N. 1987. The flux of carbon from terrestrial ecosystems to the atmosphere in 1980 due to changes in land use: Geographic distribution of the global flux. *Tellus* **39**, 122-139.

Houghton, J.T., Jenkins, G.J. and Ephraums, J.J. Eds., 1990. *Climate Change: The IPCC Scientific Assessment*. Cambridge University Press, Cambridge.

INPE 1990. Deforestation of the Brazilian Amazon Forest evaluated by remote sensing techniques. Instituto Nacional de Pesquisas Espaciais. *INPE Space News* **1**, 3-6.

INPE 1992. *Deforestation in Brazilian Amazon*. Instituto Nacional de Pesquisas Espaciais, Av. dos Astronautas, São José dos Campos, Brazil, 2pp.

Iverson, L.R., Brown, S., Prasad, A., Mitasova, H., Gillespie, A.J.R. and Lugo, A.E. 1992. Use of GIS for estimating potential and actual biomass for continental South and Southeast Asia. In: *Effects of Land Use Change on Atmospheric Carbon Dioxide Concentrations: Southeast Asia as a Case Study*, V.H. Dale (Ed.), Springer-Verlag, New York, Ch. 3.

Joint Research Centre 1992. Tropical ecosystem environment observations by satellite (TREES). In: *Institute for Remote Sensing Applications: Annual Report*, Joint Research Centre, Commission of the European Communities, Ispra, Italy, 82-85.

Krebs, C.J. 1972. *Ecology*. Harper and Row Publishers, New York.

Lechthaler, R. 1956. Inventario das arvores de um hectare de *terra firme* da zona Reserva Florestal Ducke, Municipio de Manaus. Instituto Nacional de Pesquisas da Amazônia, Botánica 3. *Instituto Brasiliero de Bibliografia e Documentação*, Rio de Janeiro, Brazil.

Lucas, R.M., Honzak, M., Foody, G.M., Curran, P.J. and Corves, C. 1993. Characterizing tropical secondary forests using multi-temporal Landsat sensor imagery. *Int. J. Remote Sensing* **14**, 3061-3067.

Mausel, P., Wu, Y., Li, Y., Moran, E.F. and Brondizio, E.S. 1993. Spectral identification of successional stages following deforestation in the Amazon. *Geocarto International* **4**, 61-71.

McWilliam, A.L.C., Roberts, J., Cabral, O.M.R., Leitão, M., de Costa, A.L., Maitelli, G.T. and Zamparoni, C.A.G.P. 1993. Leaf area index and above ground biomass of *terra firme* rain forest and adjacent clearings in Amazonia. *Functional Ecol.* **7**, 310-317.

Moran, E.F., Brondizio, E., Mausel, P. and Wu, Y. 1994. Integrating Amazonian vegetation, land-use and satellite data. *Bioscience* **44**, 329-338.

Palubinskas, G. 1988. A comparative study of decision making algorithms in images modelled by Gaussian Markov random fields. *Int. J. Pattern Recognition and Artificial Intelligence* **2**, 621-639.

Reyes, G., Brown, S., Chapman, J. and Lugo, A.E. 1992. *Wood Densities of Tropical Tree Species*. United States Department of Agriculture, General Technical Report SO-88, 15pp.

Richards, P.W. 1952. *The Tropical Rain Forests: An Ecological Study*. Cambridge University Press, Cambridge, 377-403

Rodrigues, W.A. 1967. Inventário florestal piloto ao longo da estrada Manaus-Itacoatiara, Estado do Amazônas: Dados preliminares. In: *Atas do Simposio Sobre a Biota Amazônica: Conservação da Natureza e Recursos Naturais*, H. Lent (Ed.), Conselho Nacional de Pesquisa, Rio de Janeiro, Brazil, **7**, 257-267.

Sader, S.A., Stone, T.A. and Joyce, A.T. 1990. Remote sensing of tropical forests: An overview of research and applications using non-photographic sensors. *Photogram. Engng & Remote Sensing* **56**, 1343-1351.

Saldarriaga. J.G., West, D.C., Tharp, M.L. and Uhl, C. 1988. Long-term chronosequence of forest

succession in the upper Rio Negro of Colombia and Venezuela. *J. Ecol.* **76**, 938-958.
Sampson, R.N., Apps, M., Brown, S., Cole, V.C., Downing, J., Heath, L.S., Ojima, D.S., Smith, T.M., Solomon, A.M. and Wisniewski, J. 1993. Workshop summary statement: terrestrial biospheric carbon fluxes — quantification of sinks and sources of CO_2. *Water, Air & Soil Poll.* **70**, 3-15.
Serrão, E.A.S., Falesi, I., de Veiga, J. and Neto, J. 1978. Productivity of cultivated pastures on low fertility soils in the Amazon of Brazil. In: Sánchez, P.A. and Tergas, L.E. (Eds.), *Pasture Production in Acid Soils of the Tropics*, Centro Internacional de Agricultura Tropical, Cali, Colombia, 195-225.
Singh, A. 1984. Some clarifications about the pairwise divergence measure in remote sensing. *Int. J. Remote Sensing* **5**, 623-627.
Skole, D. and Tucker, C. 1993. Tropical deforestation and habitat fragmentation in the Amazon: Satellite data from 1978 to 1988. *Science* **260**, 1905-1910.
Smith, N.J.H., Serrão, E.A.S., Alvim, P.T. and Falesi, I.C. 1995. Amazonia: Resiliency and dynamism of the land and its people. United Nations. University Press, Tokyo.
Smith, T.M., Cramer, W.P., Dixon, R.K., Leemans, R., Neilson, R.P. and Solomon, A.M. 1993. The global terrestrial carbon cycle. *Water, Air & Soil Poll.* **70**, 19-37.
Uhl, C. 1987. Factors controlling succession following slash-and-burn agriculture in Amazonia. *J. Ecol.* **75**, 377-407.
Uhl, C. and Jordan, C.F. 1984. Succession and nutrient dynamics following forest cutting and burning in Amazonia. *Ecology* **65**, 1476-1490.
Uhl, C., Clark, K., Clark, H. and Murphy, P. 1981. Early plant succession after cutting and burning in the upper Rio Negro region of the Amazon Basin. *J. Ecol.* **69**, 631-649.
Uhl, C., Buschbacher, R. and Serrão, E.A.S. 1988. Abandoned pastures in eastern Amazonia. I. Patterns of plant succession. *J. Ecol.* **76**, 663-681.
Whitmore, T.C. 1990. *An Introduction to Tropical Rain Forests.* Clarendon Press, Oxford.

RESUMO

A perda de carbono terrestre devido à conversão de florestas tropicais maduras para outros tipos de cobertura vegetal é parcialmente recuperada pela regeneração de florestas. Os fluxos de carbono associados à transformação da cobertura tropical, entretanto, não estão bem definidos devido à grande variabilidade das estimativas de carbono para florestas maduras e não maduras e devido ao conhecimento limitado da extensão espacial dessas florestas.

As perdas de carbono terrestre são especialmente grandes onde florestas maduras são substituídas por pastagens. Na Amazônia brasileira, as pastagens representam cerca de um terço da área desmatada, embora muitas tenham sido abandonadas e colonizadas por florestas em regeneração. Neste estudo, descrevemos alguns tipos de florestas em regeneração sobre pastagens abandonadas e terras cultivadas próximas a Manaus, e estimamos, utilizando uma série temporal dos dados do Landsat MSS/TM e SPOT HRV, a extensão espacial e a idade de cada tipo de floresta. Para cada tipo de floresta, a biomassa e o estoque de carbono foram estimados e mapeados. O estudo demonstra de que modo dados obtidos através de satélites de sensoriamento remoto e inventários sobre as florestas podem ser integrados a fim de estimar os fluxos de carbono associados à perda e subsequente regeneração de florestas tropicais.

23 Advances in land surface parameterisations for GCMs

R.E. DICKINSON
University of Arizona, USA

INTRODUCTION

Global climate models have evolved out of the technology of numerical weather prediction (NWP). NWP solves the equations for atmospheric fluid motions as an initial value problem using periodic cycles of observations to initialise with data. General Circulation Models (GCMs) were developed as versions of the NWP models that could run for many years and generate atmospheric climatic statistics. With the increased practical concerns about climate change, it became necessary to emphasise modelling the complete climate system, including land and ocean processes. Oceans have the major role of storing thermal energy whereas land especially shapes the environment in which people live.

The initial GCMs in the 1970s treated land as a simple reservoir of water and thermal energy. The first parameterisations used only a few global numbers and had no dependence on any observed data. In the 1980s, modellers began to make the land models geographically specific and to distinguish between soils and vegetation. The data used were gleaned from a wide variety of heterogeneous sources; geographical data were inferred from such sources as national atlases based on observations over different time periods. At the end of the last decade the international community began systematic programmes to improve and validate the land surface parameterisations. These included major field programmes. Satellites can now provide uniform descriptions of global land cover for use in climate models. Scientists are learning to reformulate the models to recognise the importance of spatial heterogeneities and to include vegetation canopies that jointly transpire water vapour and assimilate carbon dioxide (recent reviews include Dickinson, 1995a & b).

ATMOSPHERIC INPUTS

The recognition that GCMs should generate adequately realistic surface climates has led to an optimism that better land surface sub-models would carry the models a long way towards simulations of better surface climates. However, this optimism has been premature. The introduction of better land models in many modelling programs

Amazonian deforestation and climate. Edited by J.H.C. Gash, C.A. Nobre, J.M. Roberts and R.L. Victoria.

did not banish previously noted large departures from observed climates and, in some cases, made them worse. This, at least partial, failure has been a result of the atmospheric models providing erroneous inputs to the surface calculations.

Atmospheric models have not produced adequately realistic distributions of cloud radiative properties and precipitation. These distributions depend, in part, on coupling to the land surface, but also on many processes purely internal to the atmosphere. The difficulties with cloud radiative properties in models are most well known in terms of the very wide variety of different atmospheric fluxes at the top of the atmosphere (TOA) given by different models (Cess *et al.*, 1990). Errors in the reflection of solar radiation by clouds, revealed by TOA fluxes, map closely to errors in solar radiation at the surface. Discussions of model errors of surface fluxes have been given by Shuttleworth and Dickinson (1989), Dickinson and Kennedy (1991), Garratt *et al.*, (1993), Garratt (1994), Hahmann *et al.*, (1995) and Ward (1995).

The excess surface solar radiation predicted by the model has been an error common to many GCMs. Figure 1 (from Dickinson and Kennedy, 1991) illustrates this problem in terms of a comparison of incident solar radiation over the United States

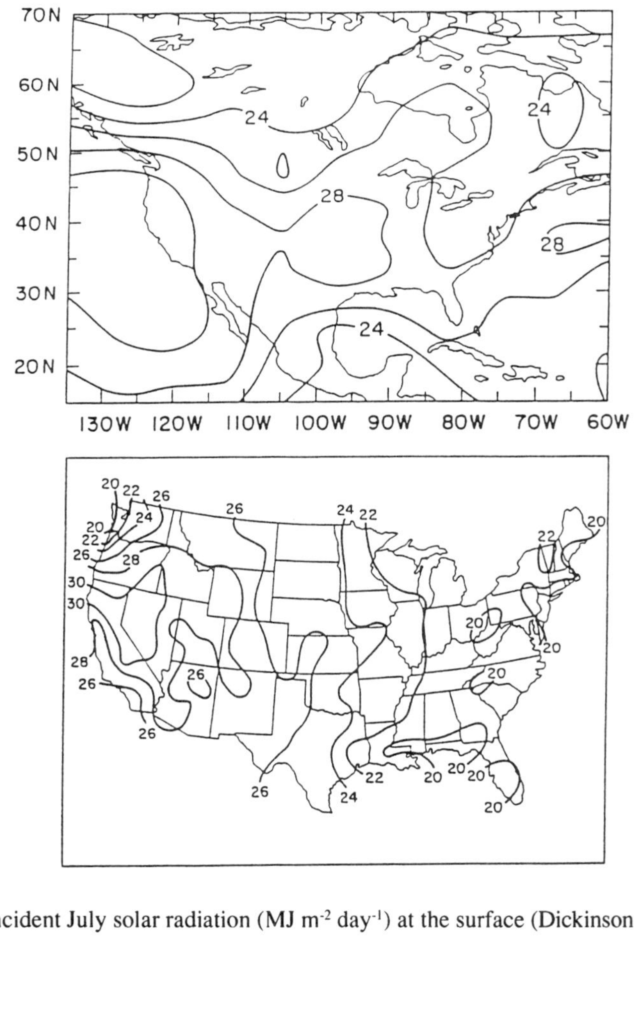

Figure 1 Incident July solar radiation (MJ m^{-2} day^{-1}) at the surface (Dickinson and Kennedy, 1991)

in July from a National Center for Atmospheric Research (NCAR) GCM, versus observations. The attenuation of solar radiation in a molecular atmosphere is now well-established in climate models and varies significantly only because of its dependence on geometry and water vapour concentrations. Outside these factors, solar radiation at the surface varies primarily according to the extent of its reflection by clouds and aerosols. The common neglect of aerosols may explain some of the excess of solar radiation. However, insufficient reflection by clouds is potentially the largest source of a high bias for solar radiation incident over land.

One reason that climate models have given unrealistic surface fluxes of radiation has been the near absence of global observational data on these fluxes. Data on TOA fluxes from National Aeronautics and Space Administration's (NASA) Earth Radiation Budget Experiment (ERBE) have helped this situation considerably. Global data on surface solar fluxes are just becoming available through the World Climate Research Programme's (WCRP's) Surface Radiation Budget (SRB) data archives (Pinker and Laszlo, 1992; Darnell *et al.*, 1992).

The physical processes that form clouds over land may differ substantially from those over the ocean, a result of different populations of cloud condensation nuclei, the strong diurnal cycle of planetary boundary layer (PBL) heating over land, and differing forcing of vertical motions and of atmospheric moisture distribution. However, as yet, little or no attention has been given to how clouds over land might differ from oceanic ones. Clouds should be substantially brighter over land than over the ocean because of smaller droplet sizes (Kiehl, 1994). Cloud radiative parameterisations of many current GCMs may have been designed largely to fit oceanic cloud regimes.

Precipitation has also been problematical in GCMs. The surface parameterisation of a GCM transports moisture from the model's surface into its lowest atmospheric layer. This moisture is then further mixed upward and distributed spatially by large-scale atmospheric winds, vertical diffusion and convection. Clouds and precipitation form at levels within an atmospheric column where the water vapour concentrations would otherwise exceed saturation. This excess water vapour results either from large-scale upward motions or through the occurrence of moist convection.

Details of precipitation are important for land process modelling. They affect the supply of water to the canopy and to the soil and determine how much water is lost to runoff. Precipitation, on a global average, balances the surface evaporation. Locally, convective precipitation depends on the stability of PBL air to moist convection and hence on the detailed modelling of surface processes. For example, studies of the climatic effects of Amazonian deforestation (Henderson-Sellers *et al.*, 1993b) indicate that, if anything, changes in surface evapotranspiration (ET) may cause even larger changes in precipitation.

GCMs can now broadly reproduce global-scale spatial and seasonal patterns of precipitation fairly well. They are largely dry when and where they are supposed to be dry, and wet when and where they are supposed to be wet. This degree of validation is usually judged by comparing observed and modelled monthly mean precipitation on global maps, e.g. with contours incremented by factors of two. Such success is remarkable, considering that it is derived largely from first principle descriptions of physical processes. However, the GCMs must correctly simulate precipitation patterns on continental scales with better accuracies than factor-of-two for monthly means, if they are to be expected to achieve an adequate accuracy for modelling ET and runoff.

The success of climate models in reproducing precipitation patterns depends largely on their correctness in reproducing dynamically the large-scale patterns of atmospheric motions, including their many feedbacks to atmospheric and surface physical processes. It may not be possible to correct many of the unrealistic aspects of model precipitation patterns, except through the slow process of overall improvement of model resolution and of treatments of physical processes.

Overall biases in the model hydrology may, however, be less difficult to reduce. One such bias that now commonly occurs is an excess of precipitation over land. Averaged over all land, model precipitation may exceed that observed, typically by as much as 50%. This bias has sometimes been explained in terms of errors in the treatment of the model's atmospheric hydrological cycle, e.g. spurious horizontal or vertical diffusion connected with model orography or neglect of re-evaporation (within the atmosphere) of falling precipitation. However, this bias towards excess precipitation may be related in part to excess ET and oversensitivity of the models to moist convection. Excess ET would be promoted by excess surface radiation, or by excess water holding capacity, or by a tendency toward a relative deficit in runoff.

Betts *et al.*, (1993) have shown a close coupling between errors in a model's surface parameterisation and in its PBL, clouds, and moist convection. An excess of ET, by providing excess moisture to the PBL destabilises the PBL to moist convection and promotes excess precipitation. Precipitation at one location results in subsidence patterns that suppress precipitation elsewhere. Because of surface heterogeneities, substantial subgrid-scale variations in the instability of PBL air will occur, which may act to reduce the area covered by moist convection. Present convective parameterisations, which assume a uniform distribution of PBL air, may thus over-estimate overall precipitation volumes.

SUBGRID-SCALE PARAMETERISATION AT THE LAND SURFACE

One of the serious criticisms of present treatments of land processes in climate models is their failure to include many of the most essential aspects of subgrid-scale heterogeneity. Heterogeneity is manifested in the precipitation and radiative inputs, and in modelling the land processes themselves. The issue of precipitation heterogeneity has been addressed through a simple statistical model in several GCMs.

Thomas and Henderson-Sellers (1991) have reviewed these treatments. They assume precipitation to occur over some fraction μ of a model grid square: within μ the precipitation is assumed to have an exponential distribution of intensities. Simple runoff models with a nonlinear dependence on precipitation are integrated across the distribution of precipitation intensities to provide a grid-square runoff. This approach has been generalised to interception (Shuttleworth, 1988). Pitman *et al.* (1990) have demonstrated (Figure 2) that the partitioning between ET and runoff in the Biosphere-Atmosphere Transfer Scheme (BATS) can be very sensitive to μ for prescribed atmospheric forcing. On the other hand, Dolman and Gregory (1992), who allow for atmospheric feedbacks with a one-dimensional model, find very little sensitivity of average ET to assumptions about the rainfall distribution, but that the partitioning between interception versus evaporation and hence short time-scale rates of ET, can vary widely with assumptions about the precipitation distribution.

All the approaches reviewed above assume variability of precipitation but retain a homogeneous water storage. Spatial variability of storage can also have a substantial effect (Wood *et al.*, 1992). Entekhabi and Eagleson (1989) have developed separate statistical models for precipitation and soil moisture that have been tested in the Goddard Institute for Space Studies (GISS) GCM (Johnson *et al.*, 1993). They find a substantial variation in ET and runoff depending on the assumed model. Eltahir and Bras (1993) have generalised Shuttleworth's approach to a statistical description of interception. In particular, they assume a statistical distribution of leaf water stores and find for prescribed atmospheric forcing that interception changes substantially but that little change from homogeneous conditions would be realised if the leaf water stores were assumed to be uniform.

Another issue is the inclusion of heterogeneities in land surface cover (Avissar and Pielke, 1989) and hence inferred parameters such as roughness. At least three scales need to be considered. On a very fine scale, canopy air interacts between different surfaces, so that surface roughness is not necessarily an average inferred from any

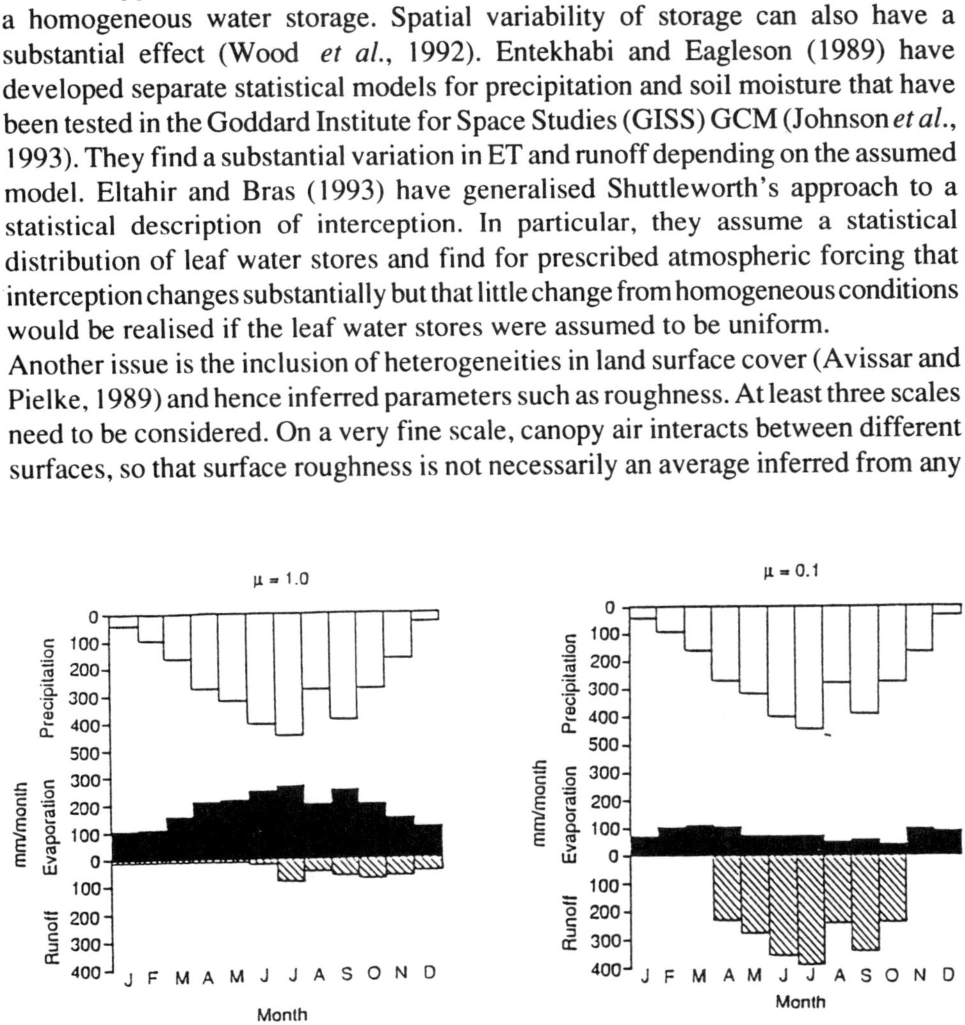

Figure 2 The seasonal variation in precipitation, runoff and evaporation for BATS with (a) $\mu=1.0$ and (b) $\mu=0.1$. Precipitation in each case is identical, and is plotted downward from the top abscissa. Runoff is plotted downward from the lower abscissa, evaporation is plotted upward from the lower abscissa and is shaded (after Pitman *et al.*, 1990).

single element. Koster and Suarez (1992) refer to this scale of heterogeneity as "mixture" and for a given roughness give a simple model for deriving a total canopy temperature and water vapour from individual elements. This type of heterogeneity needs to be provided as part of the overall land cover description.

On a coarser scale, surfaces independently interact with an overlying homogeneous atmosphere. Koster and Suarez (1992) refer to this as "mosaic," and Shuttleworth (1988) as "disordered" heterogeneity. This is the scale of typical agricultural fields or small stands of forest.

Finally, at scales of at least a few tens of km and certainly at the scales of GCM resolution, individual surfaces are overlaid by a PBL that is unaffected by individual neighbouring surfaces. Shuttleworth (1988) refers to these as "ordered" heterogeneity. In the presence of mesoscale heterogeneities, mesoscale circulations may add substantial additional vertical fluxes to those from small scale turbulence (e.g. Pielke *et al.*, 1991).

Another issue of heterogeneity is in the distribution of incident radiation. Subgrid-scale clouds, perhaps associated with precipitation, may be important. For example, if precipitation occurs over some fraction of a grid square and spatial variation of surface wetness is included, but radiation is assumed homogeneous, the estimated evaporation from the wet surfaces will be excessive. Besides clouds, surface slope can be a major cause of heterogeneity in the amounts of absorbed surface solar radiation (Avissar and Pielke, 1989). In sum, there are a wide range of subgrid-scale heterogeneities that may have sufficient effect on surface water and energy fluxes to indicate a need for their inclusion in climate models.

FEEDBACKS ON ATMOSPHERIC PROCESSES AND SENSITIVITY TO CHANGES IN LAND COVER

Processes at the land surface influence the atmosphere through fluxes of heat and moisture into the planetary boundary layer. These in turn affect atmospheric amounts and occurrence of precipitation and cloud-radiative effects. Sensitivity studies with GCMs, many looking at the question of the possible effect of Amazonian land cover modifications, have indicated major influences of the surface on the atmospheric hydrological processes. This should perhaps not be too surprising since variations in ocean fluxes driven by small changes in oceanic surface temperatures have long been known to have large climatic effects, e.g. the El Niño Southern Oscillation (ENSO) phenomena. However, the feedbacks of land surface processes to the atmosphere are more complex and still far from understood.

Moist convection occurs as a result of conditional instability, that is, if parcels of air, usually from the planetary boundary layer, can rise to a level of saturation, they became buoyant to further upward movement as a result of latent heat release. The probability of convection and intensity of resulting precipitation can be increased through additions of moist static energy (or equivalent potential temperature) to the planetary boundary layer. Fluxes of either latent or sensible heat add moist static energy, but the latter may be less effective as it also raises the level of saturation by warming the planetary boundary layer.

Studies of the response of the Amazon Basin climate to a change from forest to grassland have been published by Lean and Warrilow (1989), Lean and Rowntree (1993), Dickinson and Kennedy (1992), Nobre *et al.* (1991) and Henderson-Sellers *et al.* (1993b). A contrasting question is the effect of changing a semiarid grassland such as the Sahel in Africa to a desert (e.g. Xue and Shukla, 1993). Figures 3 and 4 from Dickinson and Kennedy (1992) are given as examples of an Amazonian modification study. Reduction in absorbed solar energy due to higher surface albedo

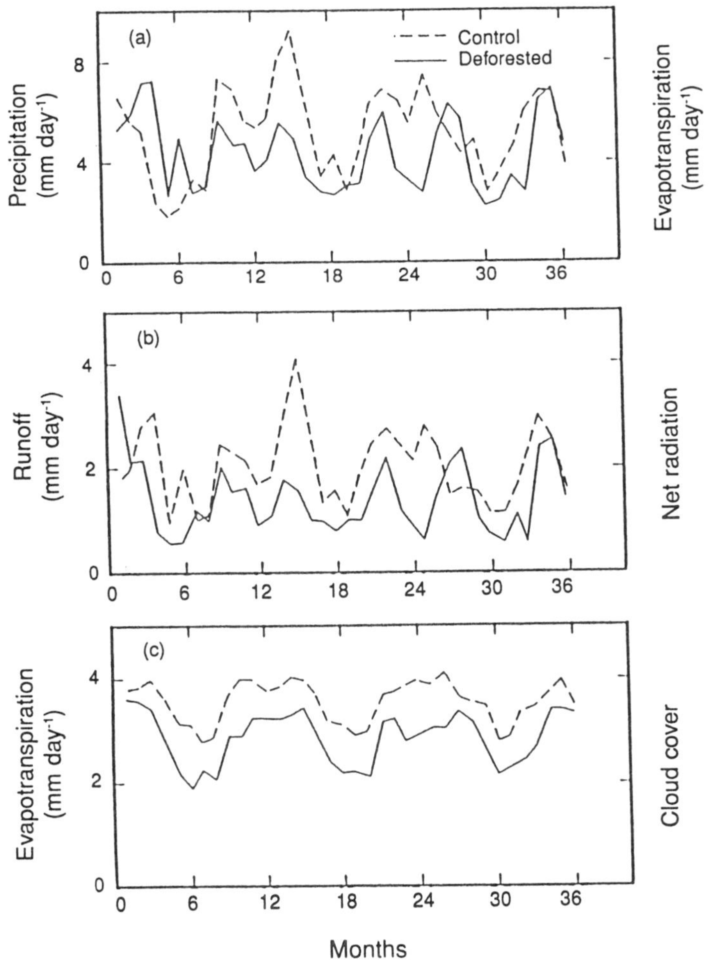

Figure 3 Modelling simulation of the climate response to complete conversion of Amazonian forest to pasture (Dickinson and Kennedy, 1992), showing averages over individual months of the control (forest) and deforested simulations over the three-year period of the integrations. Changes in evapotranspiration were regular from year to year, but changes in precipitation and run-off showed considerable differences between years.

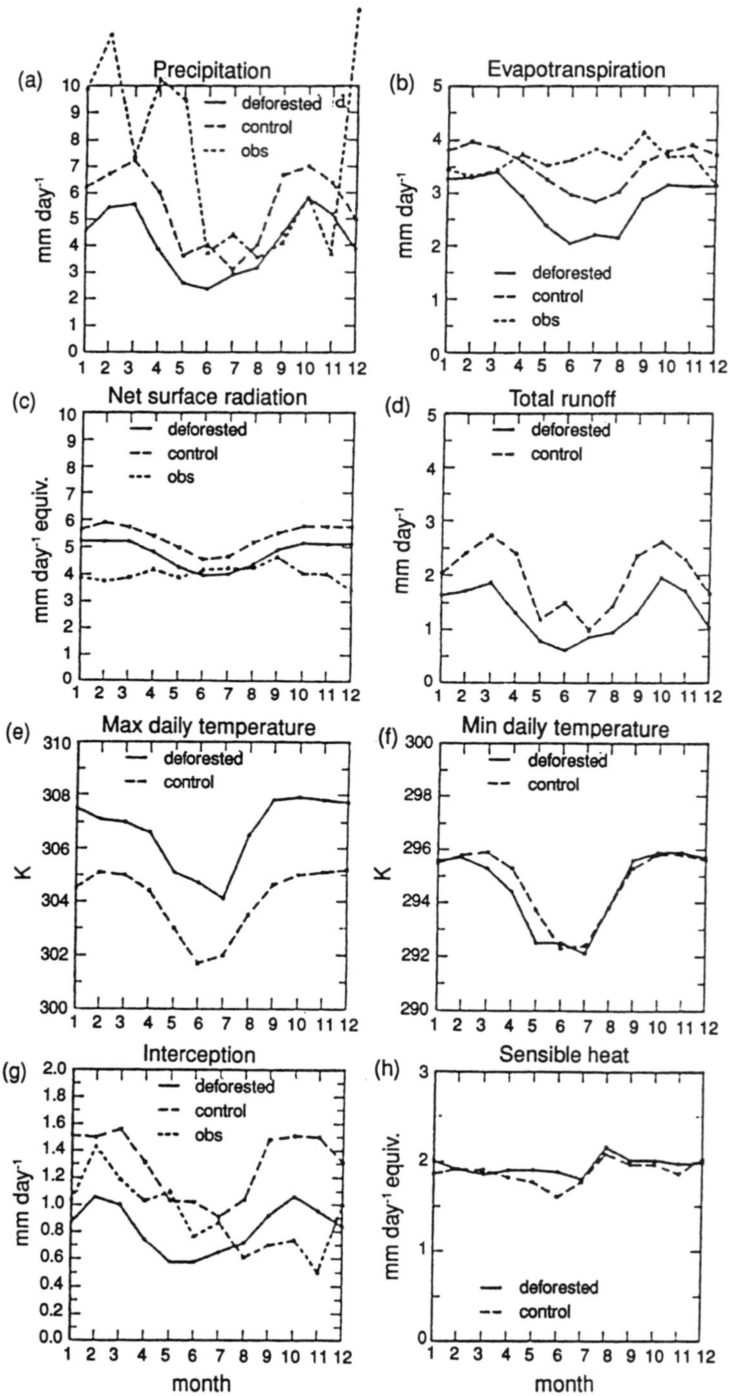

Figure 4 Seasonal cycle for indicated fields from monthly means averaged over the three-year integration and over the Amazon basin. Where indicated, fields reported by Shuttleworth (1988) for 2°S, 60°W are given as "obs" (Dickinson and Kennedy, 1992).

is one of the major drives. How this interacts with the moist static energy of the planetary boundary layer over the diurnal cycle variations of solar radiation is still not well understood. Reduction of moisture can readily be seen to reduce the probability of moist convection, but it is not so clear how a change in precipitation should be related to a change in ET. Many of the simulations have shown precipitation to reduce substantially more than ET, which implies a reduction in the convergence of moisture from the ocean. Perhaps further observational studies of water recycling over Amazonia can help provide better tests of how well the GCMs are treating the basic hydrological processes.

FIELD PROGRAMMES AND SATELLITE REMOTE SENSING

Besides ABRACOS, other international field programmes have recently been carried out to improve our understanding of the role of land processes for climate models, including the Hydrological Atmospheric Pilot Experiment (HAPEX), the First ISLSCP (International Satellite Land-Surface Climatology Project) Field Experiment (FIFE) and the Boreal Ecosystem-Atmosphere Study (BOREAS). FIFE was specifically designed to improve understanding of the role of biology in controlling the interactions between the atmosphere over a vegetated land surface and to investigate the use of satellite observations for inferring climatologically significant land surface parameters. It provided an opportunity to examine the issue of scaling from surface plot measurements to scales measurable by satellite (Figure 5; Sellers *et al.*, 1992). Betts *et al.* (1993) have effectively made use of FIFE data to help identify difficulties with the treatment of land processes in the European Centre for Medium-Range Weather Forecasting (ECMWF) model. In general, ABRACOS and the other recent field programmes have provided invaluable information about specific parameter values for the given systems studied and new and improved insights to the functioning of important processes.

The issue of global climate modelling also requires long period global data sets for model inputs and validation. For example, Running *et al.* (1994) propose a scheme for classification of global land cover. On a global scale, satellites offer the possibility of uniform descriptions of land cover properties, and perhaps more importantly, of their seasonal and interannual variations. In particular, for Amazonia, observational descriptions of year-to-year changes in forest cover along with improved climate models should allow estimates of the climatic effects of actual land cover changes rather than of idealised hypothetical situations.

DISCUSSION AND CONCLUSIONS

ABRACOS has been part of on-going international efforts to improve our understanding and prediction of human effects on the natural environment and, in particular, to improve the representations of the effects of land processes in climate

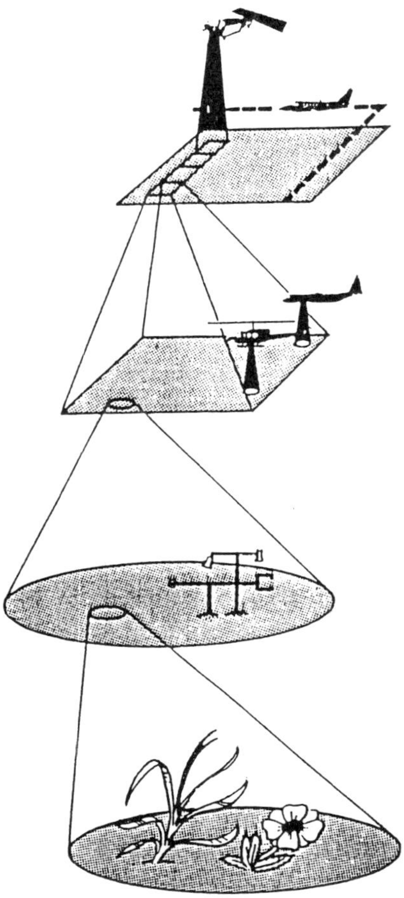

Figure 5 The FIFE experiment was designed to scale from individual plants to a much larger area through the use of surface measurements interpolated and extrapolated by aircraft measurements and satellite remote sensing.

models. Some of the latest modelling of land processes is recognising the close coupling between evapotranspiration and plant photosynthesis. This coupling between hydrological and carbon cycles helps unify and improve the treatments of both important problems.

Land surface parameterisations intended for models have recently proliferated. Having a variety of similar approaches and codes to choose from raises the question as to how these various schemes may differ when used as part of a climate model. Programmes for the intercomparison between models are helping to reduce the confusion that results from having so many models. Earlier model intercomparison studies by Cess *et al.* (1990) and Randall *et al.* (1992) have, respectively, shown a wide variety of differences between model cloud/radiation and surface energy flux responses to change in the prescribed ocean surface temperatures. The Project for Intercomparison of Land Process Schemes (PILPS) has been organised through the Global Energy and Water Cycle Experiment (GEWEX) (Henderson-Sellers *et al.*, 1993a) to focus directly on comparing different land schemes. The Atmospheric

Model Intercomparison Project (AMIP) is developing data sets from the simulations of many different atmospheric plus land climate models (Gates, 1992). Ocean surface temperatures are prescribed for a long period (greater than a decade).

REFERENCES

Avissar, R. and Pielke, R.A. 1989. A parameterization of heterogeneous land surfaces for atmospheric numerical models and its impact on regional meteorology. *Mon. Weath. Rev.* **117**, 2113-2136.

Betts, A.K., Ball, J.H. and Belijaars, A.C.M. 1993. Comparison between the land surface response of the ECMWF model and the FIFE-1987 data. *Q. J. R. Meteorol. Soc.* **199**, 975-1001.

Cess, R.D., Potter, G.L., Blanchet, J.P., Boer, G.J., Del Genio, A.D., Deque, M., Dymnikov, V., Galin, V., Gates, W.L., Ghan, S.J., Kiehl, J.T. Lacis, A.A. Le Treut, H., Li, Z.-X. Liang, X.-Z., McAvaney, B.J., Meleshko, V.P., Mitchell, J.F.B. Morcrette, J.-J.. Randall, D.A Rikus, L., Roeckner, E., Royer, J.F., Schlese, U., Sheinin, D.A., Slingo, A., Sokolov, A.P., Taylor, K.E. Washington, W.M., Wetherald, R.T., Yagi, I. and Zhang, M.-H. 1990. Intercomparison and interpretation of climate feedback processes in 19 atmospheric general circulation models. *J. Geophys. Res.* **97**, 3711-3724.

Darnell, W.L., W.F.Staylor, S.K.Gupta, N.A.Ritchey and A.C.Wilber, 1992. Seasonal variation of surface radiation budget derived from International Satellite Cloud Climatology Project C1 data. *J. Geophys. Res.* **97**, 15,741-15,760.

Dickinson, R.E. and Kennedy, P.J. 1991. Land surface hydrology in a General Circulation Model. 5. Global and regional fields needed for validation. *Surveys in Geophys.* **12**, 115-126.

Dickinson, R.E. and Kennedy, P.J. 1992. Impacts on regional climate of Amazon deforestation. *Geophys. Res. Lett.* **19**, 1947-1950.

Dickinson, R.E., 1995a. Land atmosphere interaction. *Rev. Geophys.* **33**, 917-922.

Dickinson, R.E., 1995b. Land processes in climate models. *Remote Sensing of Environment*, special issue, International Satellite Land Surface Climatology Project, **51**, 27-38.

Dolman, A.J. and Gregory, D. 1992. The parameterization of rainfall interception in GCMs. *Q. J. R. Meteorol. Soc.* **188**, 455-467.

Eltahir, E.A.B. and Bras, R.L. 1993. A description of rainfall interception over large areas. *J. Clim.* **6**, 1002-1008.

Entekhabi, D. and Eagleson, P.S. 1989. Land-surface hydrology parameterization for atmospheric general circulation models including subgrid-scale spatial variability. *J. Clim.*, **2**, 816-831.

Garratt, J.R., Krummel, P.B. and Kowalczyk, E.A. 1993. The surface energy balance at local and regional scales — A comparison of General Circulation Model results with observations. *J. Clim.* **6,** 1090-1109.

Garratt, J.R. 1994. Incoming short-wave fluxes at the surface — A comparison of GCM results with observations. *J. Clim.* **7**, 72-80.

Gates, W.L. 1992. AMIP. The atmospheric model intercomparison project. *Bull. Am. Met. Soc.* **73**, 1962-1970.

Hahmann, A.N., Ward, D.M. and Dickinson, R.E. 1995. Surface land temperature and radiative fluxes response of the NCAR CCM2/BATS land scheme to modifications in the optical properties of clouds. *J. Geophys. Res.* **100**, D11, 23239-23252.

Henderson-Sellers, A., Yang, Z.-L. and Dickinson, R.E. 1993a. The Project for Intercomparison of Land-surface Parameterization schemes. *Bull. Am. Met. Soc.* **74**, 1335-1349.

Henderson-Sellers, A., Dickinson, R.E., Durbidge, T.B., Kennedy, P.J., McGuffie K. and Pitman, A.J. 1993b. Tropical deforestation. Modelling local- to regional-scale climate change. *J. Geophys. Res.* **98**, 7289-7315.

Johnson, K.D., Entekhabi, D. and Eagleson, P.S. 1993. The implementation and validation of improved land-surface hydrology in an atmospheric General Circulation Model. *J. Clim.* **6**, 1009-1026.

Kiehl, J.T. 1994. Sensitivity of a GCM climate simulation to differences in continental versus maritime cloud drop size. *J. Geophys. Res.* **99**, D11, 23,107-23,116.

Koster, R.D. and Suarez, M.J. 1992. Modelling the land surface boundary in climate models as a composite of independent vegetation stands. *J. Geophys. Res.* **97**, 2697-2715.

Lean, J. and Warrilow, D.A. 1989. Simulation of the regional climatic impact of Amazon deforestation. *Nature* **342**, 411-413.

Lean, J. and Rowntree, P.R. 1993. A GCM simulation of the impact of Amazonian deforestation on climate using an improved canopy representation. *Q. J. R. Meteorol. Soc.* **199**, 509-530.

Nobre, C.A., Sellers P.J. and Shukla, J. 1991. Amazon deforestation and regional climate change. *J. Clim.* **4**, 957-988.

Pielke, R.A., Dalu, G.A., Snook, J.S., Lee, T.J. and Kittel, T.G.F. 1991. Nonlinear influence of mesoscale land use on weather and climate. *J. Clim.* **4**, 1053-1069.

Pinker, R.T. and Laszlo, K. 1992. Modelling surface solar irradiance for satellite applications on a global scale. *J Appl. Met.* **31**, 194-211.

Pitman, A.J., Henderson-Sellers, A. and Yang, Z.-L. 1990. Sensitivity of regional climates to localized precipitation in global models. *Nature* **346**, 734-737.

Randall, D.A., Cess, R.D., Blanchet, J.P., Boer, G.J., Dazlich, D.A., DelGenio, A.D., Deque, M., Dymnikov, V., Galin, V., Ghan, S.J., Lacis, A.A., .LeTreut, H, Li, Z.-Z., Liang, X.-Z., McAvaney, B.J., Meleshko, V.P., Mitchell, J.F.B., Morcrette, J.-J., Potter, G.L., Rikus, L., Roeckner, E., Royer, J.F., Schlese, U., Sheinin, D.A., Slingo, J., Sokolov, A.P., Taylor, K.E., Washington, W.M., Wetherald, R.T., Yagi, I. and Zhang, M.-H. 1992. Intercomparison and interpretation of surface energy fluxes in atmospheric General Circulation Models. *J. Geophys. Res.* **97**, 3711-3724.

Running, S.W., Loveland, T.R., Pierce L.L. and Hunt, E.R. Jr. 1994. A remote sensing based vegetation classification logic for global land cover analysis. *Remote Sensing Environ.* **51**, 39-48.

Sellers, P.J., Hall, F.G., Asrar, G., Strebel, D.E. and Murphy, R.E. 1992. An overview of the First International Satellite Land Surface Climatology Project (ISLSCP) Field Experiment (FIFE). *J. Geophys. Res.* **97**, D17, 18345-18371.

Shuttleworth, W.J. 1988. Evaporation from Amazonian rain forest. *Proc. R. Soc. Lond.* **B233**, 321-346.

Shuttleworth, W.J. and Dickinson, R.E. 1989. 'Comments on "Modelling tropical deforestation. A study of GCM land-surface parameterizations" by R.E. Dickinson and A. Henderson-Sellers'. *Q. J. R. Met. Soc.* **115**, 1177-1179.

Thomas, G. and Henderson-Sellers, A. 1991. An evaluation of proposed representations of subgrid hydrologic processes in climate models. *J. Clim.* **4**, 898-910.

Ward, D.M. 1995. Comparison of the surface solar radiation budget derived from satellite data with that simulated by the NCAR CCM2. Submitted to *J. Geophys. Res.*

Wood, E.F., Lettenmaier, D.P. and Zartarian, V.G. 1992. A land-surface hydrology parameterization with subgrid variability for general circulation models. *J. Geophys. Res.* **97**, 2717-2728.

Xue, Y. and Shukla, J. 1993. The influence of land-surface properties on Sahel climate. Part I. Desertification. *J. Clim.* **6**, 2232-2245.

RESUMO

O uso de parametrizações da superfície nos MCG (Modelos Climáticos de Circulação Geral) tem sido objeto de importantes estudos. Este artigo faz uma revisão destes estudos e aponta áreas onde novas pesquisas são necessárias. Esta revisão inclui: o tratamento conceitual dos processos de superfície, e sua evolução até algoritmos computacionais; a identificação da importante dependência das entradas de precipitação e radiação dependendo dos processos atmosféricos; o papel da variabilidade espacial em escalas muito pequenas para serem resolvidas pelos modelos climáticos globais atuais; uma melhor compreensão de como o clima depende das coberturas de superfície específicas, e como o clima poderia variar com uma alteração destas coberturas; projetos de campo bem sucedidos que, além da contribuição ao conhecimento; fornecem dados para validação de modelos; os avanços no entendimento da física da interação entre os fluxos de calor latente da superfície e os fluxos de dióxido de carbono utilizados pela vegetação na fotossíntese; o desenvolvimento e aplicação dos métodos para derivar a cobertura da superfície utilizada em MCG através de sensoriamento remoto global, e a implementação de parametrizações dos processos à superfície em MCG.

24 Observations of the atmospheric boundary layer in Rondônia

C.A. NOBRE[1], G. FISCH[2], H.R. da ROCHA[3], R.F. da F. LYRA[4], E.P. da ROCHA[5], A.C.L. da COSTA[5] and V.N. UBARANA[6]

[1] *Instituto Nacional de Pesquisas Espaciais, São José dos Campos, Brazil*
[2] *Centro Técnico Aeroespacial, São José dos Campos, Brazil*
[3] *Universidade de São Paulo, São Paulo, Brazil*
[4] *Universidade Federal de Alagoas, Maceió, Brazil*
[5] *Universidade Federal do Pará, Belém, Brazil*
[6] *Fund. Cearence de Meteorologia e Rec. Hídricos, Fortaleza, Brazil*

INTRODUCTION

Throughout the last decade, the Amazon region has received worldwide attention because of its mineral riches, its great biodiversity and the effects that deforestation may provoke on climate at regional and global scales. Determining the climatic effects of Amazonian deforestation has been the subject of many general circulation model (GCM) experiments over the past few years, for example Dickinson and Henderson-Sellers (1988), Lean and Warrilow (1989), Nobre *et al.* (1991), Lean and Rowntree (1993), Henderson-Sellers *et al.* (1993) and Manzi (1993). All these simulations have agreed that, following total deforestation of the Amazon basin, there would be an increase in air temperature (by 0.6 to 2.0 degrees C), a reduction in precipitation and evaporation (20 to 30%) and the dry season would become longer. Such modifications to the climate would have serious ecological consequences for the Amazon region.

The GCMs used for these numerical simulations contain complex parameterizations of land surface processes (e.g. Dickinson *et al.*, 1984; Sellers *et al.*, 1986; Noilhan and Planton, 1989) to make them as realistic as possible. Much effort is directed at calibrating these parameterizations accurately for different areas of the earth's surface (e.g. Wright *et al.*, 1996). However, the land surface processes and the atmosphere interact strongly to shape the weather and climate of a region. It is, therefore, also important to study the behaviour of the atmospheric boundary layer (ABL) in which these interactions take place, along with the land-surface processes, if GCMs are to be successful in predicting the climate.

The Rondônia Boundary Layer Experiment (RBLE-II) was conceived to collect data on the ABL over two representative surfaces in the Amazon region of Brazil; tropical forest and a deforested, pasture area. Previously, the influence of tropical vegetation on the thermodynamic structure and growth of the boundary layer has been

Amazonian deforestation and climate. Edited by J.H.C. Gash, C.A. Nobre, J.M. Roberts and R.L. Victoria.

addressed by Martin *et al.* (1988), for Manaus in central Amazon only and by Lyra *et al.* (1992) for equatorial Africa. The present paper deals with the observations of ABL growth and decay made during RBLE-II. A companion paper (Fisch *et al.*, 1996) describes some of the modelling aspects of the experiment.

DATA AND SITE

RBLE-II took place in the Ji-Paraná region of Rondônia in the southwest of the Amazonian tropical forest during the 1993 dry season (July). This area was chosen for the experiment because it has a distinct dry season and although it is suffering a high rate of deforestation, large areas of pasture and forest, suitable for boundary layer studies, coexist there (Gash *et al.*, 1996). The RBLE-II sites were the same sites as used for ABRACOS; the forest site was Reserva Jaru (10° 5' S, 61° 55' W, 120 m above sea level) and the pasture site was Fazenda Nossa Senhora da Aparecida (10° 45' S, 62° 22' W, 220 m above sea level). The forest canopy has an average height of 30 m, whilst the vegetation at the pasture site is mostly grass (*Brachiaria brizantha*), which was established 15 years ago and is typical of ranchland in the area. McWilliam *et al.* (1996) give more details about the experimental sites of ABRACOS; Gash *et al.* (1996) give an overview of ABRACOS and describe all the characteristics of vegetation at both sites.

ABL data were taken at both the forest and the pasture sites during the experiment although not simultaneously. At each site, radiosonde and tethered balloon profiles were made at 02.00, 05.00, 08.00, 11.00, 14.00, 17.00, 20.00 and 23.00 h local time. A Vaisala (Finland) radiosounding system was used along with sondes which measure the pressure, air temperature and humidity. The wind speed and direction were determined from the Omega network. These sondes are accurate to 0.1°C for air temperature, 0.1 hPa for atmospheric pressure, 1% for relative humidity, 0.5 m s^{-1} for windspeed and 1 degree for wind direction. The tethered instrument is manufactured by A.I.R. (USA) with a sonde measuring pressure, dry and wet bulb temperatures, windspeed and direction. Its accuracy is 0.1°C for temperature, 0.1 hPa for atmospheric pressure, 0.1 m s^{-1} for windspeed and 1 degree for wind direction. All the data were gathered and stored by a personal computer. To complement the ABL information, an automatic weather station (recording hourly values of air temperature, relative humidity, windspeed and direction, solar and net radiation fluxes and precipitation) and an eddy-correlation instrument (the Institute of Hydrology Hydra: Shuttleworth *et al.*, 1988) recording hourly values of momentum and energy fluxes were also run at both sites throughout the experiment.

As mentioned above, the ABL data were not collected simultaneously at the forest and the pasture sites. For seven days (3 to 10 July 1993) measurements were made at the forest site. This period will be referred to as Phase A. The equipment was then moved to the pasture site for a second seven-day measurement period, Phase B, which lasted from 18 to 25 July 1993. During Phase A, a cold air mass moved across the Ji-Paraná region from the south and modified the atmospheric structure. Such events, known locally as "friagens", are relatively common in the region. This particular event, and its influence on the atmospheric structure, is analysed in detail by Fisch (1995). In the current analysis, however, data gathered on days affected by the friagem are not considered.

To determine the main characteristics of the structure of the ABL over forest and pasture during the experiment, a composite profile was made for each site . For the pasture site, seven complete days were used in the analysis. Only four days were used in the forest case, however, as between 7 and 9 July the atmospheric structure was strongly influenced by the friagem.

CLIMATOLOGY OF THE JI-PARANÁ REGION

The climatology of the Ji-Paraná region will be discussed briefly, using conventional meteorological data measured at Ouro Prêto D'Oeste. This data-set has been kindly provided by Comissão Executiva da Lavoura do Cacau (CEPLAC). The CEPLAC station is about 20 km from the Fazenda Nossa Senhora site and about 80 km from Reserva Jaru. Figure 1 presents the climatological patterns of rainfall and air temperature for the Ji-Paraná area. Monthly average values are available from 1982 until 1992 and indicate strong seasonality: during the wet period (from November up to April) the monthly precipitation is higher than 200 mm per month. During the dry season (from June until August), the monthly total rarely reaches 20 mm. Usually, in July, the rainfall is less than 5 mm per month. However, during the dry season, there are some cold spells (between May until September there are, on average, seven cases a year, with two or three in July), which on some occasions have associated rainfall. The air temperature also shows seasonality with October (25.5°C) and July (22.6°C) the hottest and coldest months, respectively. October coincides with the end of dry season whilst the average temperature in July is low due to the influence of the friagem.

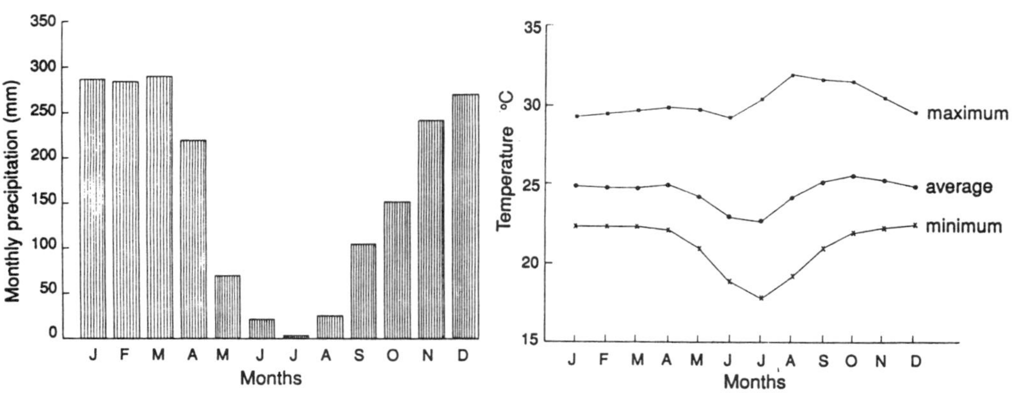

Figure 1 Climatological patterns of rainfall and temperature for the Ji-Paraná area, 1982-1992 (Source: CEPLAC)

MICROMETEOROLOGICAL ASPECTS OF RBLE-II

To characterise the weather patterns during the two phases, an average daily energy budget has been computed for the forest and pasture sites for both Phases A and B (Table 1). So that the average energy budget was calculated from the same number of days in each case, Phase A is considered for these purposes to run from 29 June until 5 July. Phase B has remained unchanged. During Phase A, there was moderate to strong convection with convective cloud cover. During Phase B, the weather in the region was dominated by the influence of a large-scale high pressure area and cloud convection was suppressed. The solar radiation at both sites increased from

Table 1 Average integrated values from fluxes of solar radiation (S), Net-radiation (NET), sensible (H) and latent (LE) heat fluxes, daily average of air temperature (TEMP) and integrated precipitable water (WP).

		S M.Jm^{-2} d^{-1}	NE MJ m^{-2} d^{-1}	H MJ m^{-2} d^{-1}	LE MJ m^{-2} d^{-1}	TEMP °C	WP g cm^{-2}
Forest	A	18.8	11.9	2.0	9.8	25.5	3.8
	B	20.6	12.1	2.5	9.6	24.4	
Pasture	A	17.7	10.1	2.7	7.0	25.5	
	B	20.1	11.2	2.2	8.7	24.0	2.6

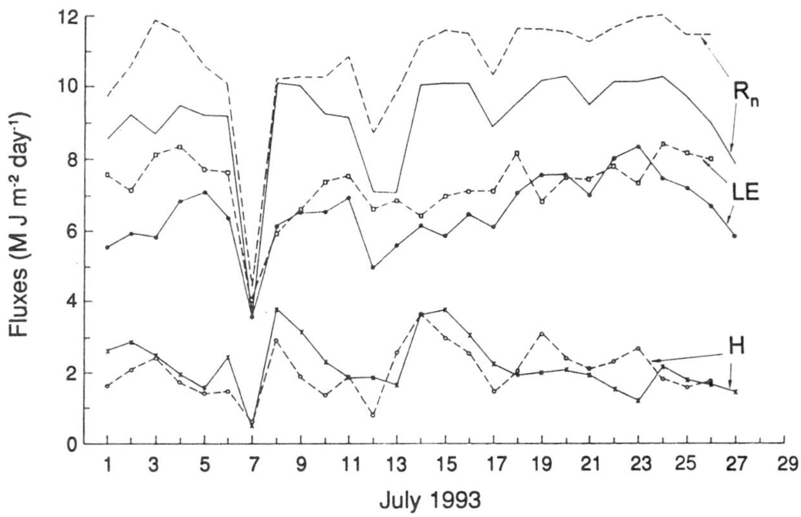

Figure 2 Time series of the components of the energy budget fluxes: net-radiation (Rn), sensible heat (H) and latent heat (LE), at the forest (---) and pasture (—) sites during RBLE-II

Phase A to Phase B, due to the reduced cloud cover. This difference was 10% for the forest and 13% for the pasture site and has led to a different partitioning of energy between the sensible and latent heat fluxes. Figure 2 presents a time series of the components of the energy budget. At the forest site, there was a small increase in net radiation, balanced by an increase in sensible heat and decrease in evapotranspiration. However, at the pasture site, there was a significant increase in net-radiation (11%) together with an increase in evaporation and soil heat flux and a decrease in the sensible heat flux. This increase in evaporation can be explained by the fact that there was precipitation in the preceding week which moistened the soil. There was a decrease in daily average temperature at both sites, but the diurnal curves show an increase in amplitude over the period (data not shown).

The precipitable water has been calculated for Phases A and B and is shown in Table 1. There was a decrease in precipitable water from 3.8 g cm^{-2} during Phase A to 2.6 g cm^{-2} during Phase B. However, as mentioned, this change may have been caused by the influence of the large-scale high pressure acting over the region.

The synoptic geopotential and meridional wind component charts obtained from the US National Meteorological Center for the two phases of RBLE-II are shown in Figure 3. During phase A when measurements were made at the forest site, Figure 3a shows that the height of the 1500 hPa geopotential surface was 1540 m, and Figure 3b indicates a weak meridional component of the wind of approximately 2 m s^{-1}. This synoptic pattern was very similar to the one observed during Phase B, when measurements were made at the pasture, although some influence of high pressure can be seen on 22 July (Figure 3c) and Figure 3d shows that there was a weak wind from the north on that day.

RESULTS

The analysis of the atmospheric boundary layer distinguishes between daytime and night-time cases, as different physical process are involved.

NOCTURNAL BOUNDARY LAYER (NBL)

For this analysis we use data collected by tethered balloon at 20.00, 23.00, 02.00 and 0500 h. Table 2 shows the main characteristics of the development of the nocturnal boundary layer at the forest and pasture sites. In the literature, there is no consensus about the best way to determine the height of the nocturnal boundary layer, but criteria commonly used are the height where the gradient of the virtual potential temperature (θ_v) is null, or the height where the turbulence has ceased (Stull, 1988). However, for the same temperature profile, Garratt (1992) has shown that, for different criteria, the height obtained will be different. In this study, we define the height of the nocturnal boundary layer as the height where the gradient of θ_v is null or less than 0.01 K m^{-1}. The strength of the inversion is defined as the difference between θ_v at the top of the nocturnal boundary layer and its surface value.

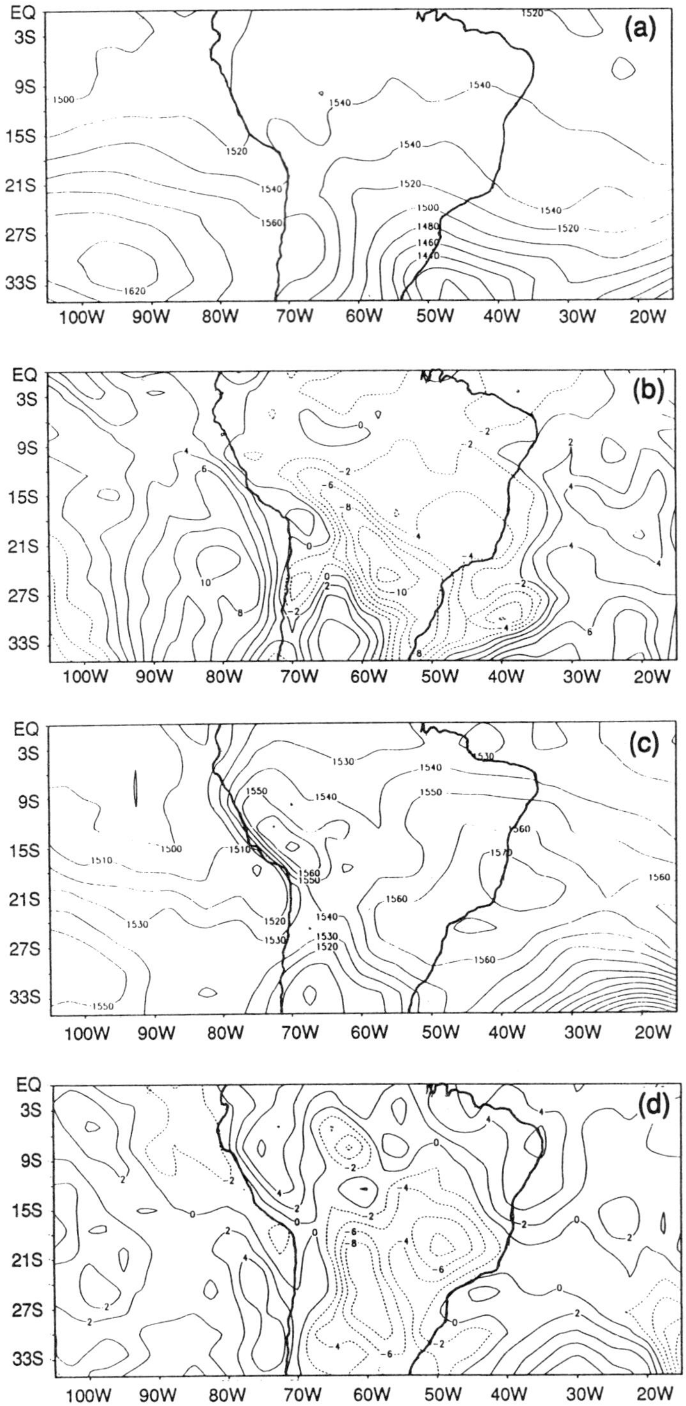

Figure 3 Synoptic charts of the height of the 1500 hPa geopotential surface and the meridional component of the wind during RBLE-II: (a) and (b) for the Phase A forest measurement period; (c) and (d) for Phase B, when measurements were made at the pasture.

In Figure 4, the nocturnal evolution of the virtual potential temperature profile is shown for the night of 4 July (forest) and 19 July (pasture). For the forest case, at 20.00 h this nocturnal layer is well developed, showing a height of 190 m with a thermal inversion of 6.4 K. As the nocturnal radiative cooling continues, the layer becomes deeper, reaching 210 m at 23.00 h, 260 m at 02.00 h and 385 m at 05.00 h. These values are slightly higher than observations from Martin *et al.* (1988) for Central Amazonia. During this evolution, the thermal inversion increased from 7.2 K at 23.00 h to 8.5 K at 05.00 h. The virtual potential temperature has decreased from 305.7 K (at 20.00 h) to 304.1 K, around the time of sunrise (05.00 h). The largest cooling rate observed was 1.1 K h^{-1} between 20.00 and 23.00 h. In the pasture, although there is the same general pattern, the influence of the surface is very strong, modifying the overall values: at 20.00 h the height of the nocturnal boundary layer is 140 m with a virtual potential temperature of 305.3 K. By 23.00, the height has increased to 150 m, 200 m (at 02.00) and finally 240 m (at 05.00 h). The strength of the nocturnal inversion increases from 8.5 K (at 20.00 h) to 11.6 K (at 05.00 h): meanwhile the average virtual potential temperature has dropped from 305.3 K to 303.9 K. The cooling rate was observed to be around 1.4 K h^{-1}.

CONVECTIVE ATMOSPHERIC BOUNDARY LAYER (CBL)

The data-set used in this analysis were radiosoundings collected at 08.00, 11.00, 14.00 and 17.00 h. The vertical resolution of radiosounding is not as good as that of the tethered balloon profiles, but the top of the CBL, marked by a discontinuity in the virtual potential temperature profile can usually be clearly seen.

Figure 5 shows typical profiles of virtual potential temperature over the forest on 5 July and over the pasture on 22 July at 17.00 h. The height of the CBL was determined from visual inspection of the virtual potential temperature profiles. The height of the CBL at the forest site has been estimated to be 200 m at 08.00 h, growing abruptly to 580 m at 11.00 h (the rate of growth being 127 m h^{-1}). At 14.00 h the height is 1150 m and reaches a final height (1250 m) at 17.00 h. During the afternoon the CBL grows very little. The average layer virtual potential temperature

Table 2 Principal characteristics of nocturnal boundary layer at the forest and pasture. height (h) of the top of boundary layer, virtual potential temperature (θ_v) at the top of boundary layer and the strength of nocturnal boundary layer ($\Delta\theta_v$)

	Forest			Pasture		
Local time h	Height m	θ_v K	$\Delta\theta_v$ K	Height m	θ_v K	$\Delta\theta_v$ K
20.00	190	305.7	6.4	140	305.1	8.5
23.00	210	304.6	7.2	150	303.9	9.2
02.00	260	303.9	7.8	200	303.8	10.6
05.00	385	304.1	8.5	240	303.9	11.6

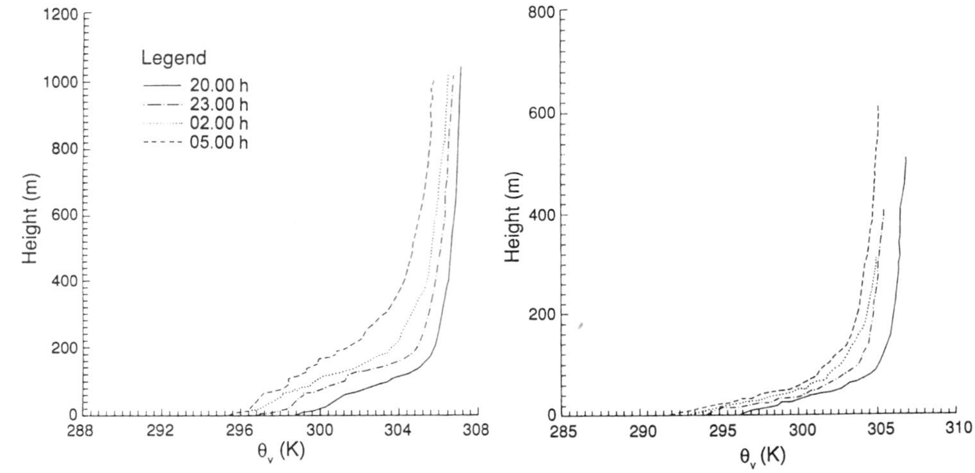

Figure 4 Nocturnal evolution of the virtual potential temperature profile (θ_v) on the nights of July 4 (left) and July 19 (right)

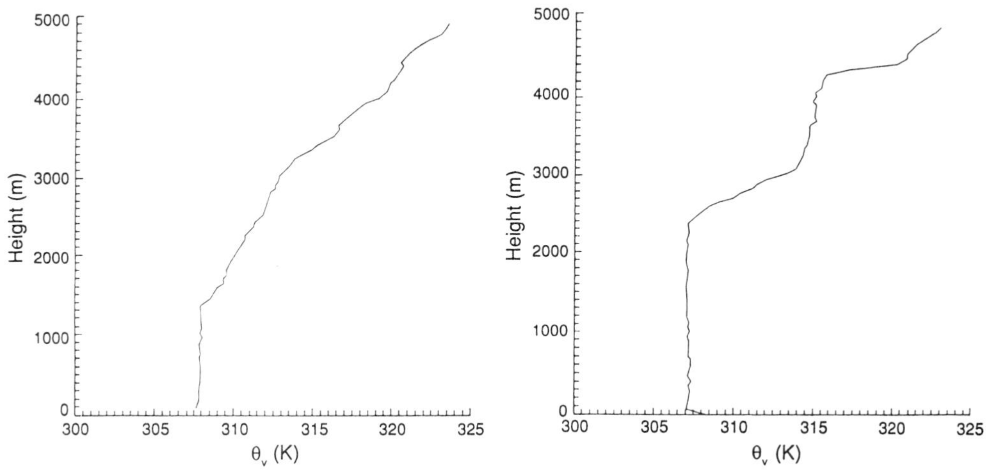

Figure 5 Typical late afternoon profiles of virtual potential temperature (θ_v) observed at the forest site (left) and the pasture site (right)

is 299.8 K during the early morning sounding (08.00 h) and reaches a final value of 307.2 K in the afternoon. The heating rate is high in the morning (1.6 K h^{-1}), when the nocturnal boundary layer is broken up by the surface heating. Later on, the heating rate is less (0.8 K h^{-1}) until 14.00 h, when the heating rate becomes stationary. In spite of the reduced surface heating in the late afternoon, the layer is still warmed by the entrainment flux of hot and dry air from above the CBL. Some authors (e.g. Tennekes, 1973; Driedonks, 1982) have considered the temperature gradient above the inversion as time-invariant, relating its change to large-scale variations. However, in the present case, this slope has a small time dependence, slowly increasing with time: at 8.00 h it is 18 K km^{-1} and it increases to 3.3 K km^{-1} at 17.00 h. These general observations of CBL height are similar to the observations reported by Martin *et al.* (1988).

Table 3 summarises the structure of the CBL for forest and pasture conditions. The pasture analysis shows that at 08.00 h the height of CBL is lower than for forest (110 m), although it has a fast growth rate (493 m h^{-1}) up to 11.00 h, when the height has been estimated to be 1590 m. The final height is around 2220 m at 17.00 h. The average layer virtual potential temperature increases from 298.6 K to 305.6 K at 11.00 h (the rate of increase is 2.3 K h^{-1}). The pasture dataset also shows a time variation of the gradient of virtual potential temperature above the CBL, from 3.6 K km^{-1} in the morning up to 8.0 K km^{-1} at 17.00 h. Figure 6 shows a schematic diagram of the atmospheric boundary layer structure summarising these observations.

Table 3 Principal characteristics of convective boundary layer at the forest and pasture: height of the top of convective boundary layer, temperature gradient above the top of boundary layer (S_o) and average virtual potential temperature of the layer (θ_v)

	Forest			Pasture		
Local time h	Height m	S_o K km^{-1}	$\Delta\theta_v$ K	Height m	S_o K km^{-1}	$\Delta\theta_v$ K
08.00	200	1.8	298.8	110	3.6	298.6
11.00	580	2.5	304.5	1590	5.0	305.6
14.00	1150	3.0	306.9	2100	6.3	307.1
17.00	1250	3.3	307.2	2220	8.0	307.4

CONCLUSIONS

Although the ABL measurements made in RBLE-II were not made simultaneously over the two different surface types, some insights can be gained from analysing and comparing their structure. The greater depth of the nocturnal boundary layer at the forest site may be due to the influence of mechanical turbulence. At night the forest's high value of aerodynamic roughness induces a vertical diffusion which can deepen the nocturnal layer. The pasture site is aerodynamically smoother and so the downward turbulent diffusion will be much less, resulting in a lower surface temperature. The strength of thermal inversion is, consequently, higher over the pasture (48.3 K km^{-1}) than over the forest (22.4 K km^{-1}). The development of the convective boundary layer is stronger over the pasture than over the forest. The influence of the surface (through the sensible heat flux) is important but may be not enough to explain the difference completely. The deforestation management policy in the Ji-Paraná area leads to the creation of areas with strips of pasture and forest

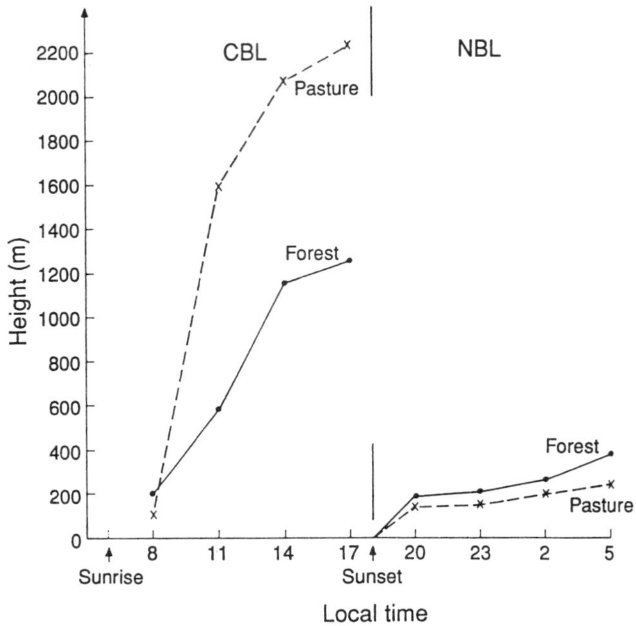

Figure 6 Schematic diagram of the ABL based on the composite observed data from RBLE-II

side-by-side and this contrast in vegetation (and the associated contrasting energy budgets) may lead to the generation of secondary thermal circulations. It seems that energy advection may occur from the wet and colder (forest) to the dry and warmer area (pasture), rapidly breaking up the nocturnal inversion. Such advection can explain the abrupt growth of the CBL at the pasture site during the early morning and is discussed further by Fisch *et al.* (1996).

ACKNOWLEDGEMENTS

The authors would like to express their thanks to all Brazilian and British participants of ABRACOS and RBLE-II who were involved in data collection at Ji-Paraná. RBLE-II was partially funded by Fundação de Amparo à Pesquisa do Estado de São Paulo (FAPESP) and Conselho Nacional Científico e Tecnológico (CNPq). One of the authors (G. Fisch) was funded by the Coordenadoria de Aperfeiçoamento de Pessoal de Nível Superior (CAPES) to carry out analysis at the Institute of Hydrology in the United Kingdom during 1993. Special thanks go to our colleagues J.H.C. Gash, J.M. Roberts, I.R. Wright and A.Culf (IH) and A. Beljaars (KNMI) for their help in preparing this paper. The authors would like to express their gratitude to the reviewers for their useful suggestions.

REFERENCES

Dickinson, R.E., Henderson-Sellers, A., Kennedy, P.J. and Wilson, M.F. 1984. Biosphere-Atmosphere Transfer scheme (BATS) for the NCAR community climate model, NCAR Technical Note TN-275+STR, 69 pp.

Dickinson, R.E. and Henderson-Sellers, A. 1988. Modelling tropical deforestation. a study of GCM land-surface parameterizations. *Q. J. Roy. Met. Soc.* **114**, 439-462.

Driedonks, A.C.M. 1982. Model and observations of the growth of the atmospheric boundary layer. *Bound.-Layer Met.* **23**, 283-306.

Fisch, G., 1995. Camada Limite Amazônica. aspectos observacionais e de modelagem. PhD Thesis, INPE, São José dos Campos, Brazil.

Fisch, G., Culf, A.D., and Nobre, C.A. 1996. Modelling convective boundary layer growth in Amazonia. This volume.

Garratt, J.R. 1992. *The atmospheric boundary layer.* Cambridge University Press, 316 pp.

Gash, J.H.C., Nobre, C.A., Roberts, J.M. and Victoria, R.L. 1996. An overview of ABRACOS. This volume.

Henderson-Sellers, A., Dickinson, R.E., Durbigde, T.B., Kennedy, P.J., McGuffie, K. and Pitman, A.J. 1993. Tropical deforestation modelling local to regional scale climate change. *J. Geophys. Res.* **98**, 7289-7315.

Lean, J. and Warrilow, D.A. 1989. Simulation of the regional climate impact of Amazon deforestation. *Nature* **342**, 411-413.

Lean, J. and Rowntree, P.R. 1993. A GCM simulation of the impact of Amazonian deforestation on climate using an improved canopy representation. *Q. J. Roy. Met. Soc.* **119**, 509-530.

Lyra, R., Druilhet, A., Benech, B. and Bouka Biorna, C. 1992. Dynamics above a dense equatorial rain forest to the free atmosphere. *J. Geophys. Res.* **97**, 12953- 12966.

Manzi, A.O. 1993. Introduction d'un schéma des tranferts sol-vegetation-atmosphère dans un modèle de circulation générale et application à la simulation de la deforestation Amazonienne. PhD Thesis, Paul Sabatier University, Toulouse (France), 230 pp.

Martin, C.M., Fitzjarrald, D.R., Garstang, M., Oliveira, A.P. de, Greco, S. and Browell, E. 1988. Structure and growth of the mixing layer over the Amazonian rain forest.*J. Geophys. Res.* **93**, 1361-1375.

McWilliam, A.-L.C., Cabral, O.M.R., Gomes, B.M., Esteves. J.L. and Roberts, J.M. 1996. Forest and pasture leaf-gas exchange in South-West Amazonia. This volume.

Nobre, C.A., Sellers, P.J. and Shukla, J. 1991. Amazonian deforestation and regional climate change. *J. Climate* **4**, 957-988.

Noilhan, J. and Planton, S. 1989. A simple parameterization of land surface process for meteorological models. *Mon. Weath. Rev.* **117**, 536-549.

Sellers, P.J., Mintz, Y., Sud, Y.C. and Dalcher, A. 1986. A simple biosphere model (SiB) for use within the General Circulation Models. *J. Amos. Sci.* **43**, 505-531.

Shuttleworth, W.J., Gash, J.H.C., Lloyd, C.R., McNeil, D.D., Moore, C.J. and Wallace, J.S. 1988. An integrated micrometeorological system for evaporation measurements. *Agric. For. Met.* **43**, 295-317.

Stull, R.B. 1988. *An introduction to boundary layer meteorology.* Kluwer Academic Publishers, Dordrecht, 666 pp.

Tennekes, H. 1973. A model for the dynamics of the inversion above a convective boundary layer. *J. Atmos. Sci.* **30**, 558-567.

Wright I.R., Nobre, C.A., Tomasella, J., da Rocha, H.R., Roberts, J.M., Vertamatti, E., Culf, A.D., Alvalá, R.C., Hodnett, M.G. and Ubarana, V. 1996. Towards a GCM surface parameterisation for Amazonia. This volume.

RESUMO

Observações do desenvolvimento de camadas limites atmosféricas sobre regiões de floresta tropical e pastagem são apresentadas. Estas medidas foram obtidas no Experimento da Camada Limite de Rondônia — Fase II, realizado na região de Ji-Paraná (sudoeste da Amazônia), durante o mês de julho de 1993. O conjunto de dados consiste em perfis atmosféricos medidos a partir de balão cativo, radiossondagem e de medidas de fluxos turbulentos à superfície. Os resultados mostram que a camada limite convectiva desenvolvida sobre a área de pastagem pode atingir alturas de até 2.200 m ao final da tarde, valores estes superiores aos observados sobre a floresta (1.250 m). A camada limite estável noturna é mais profunda sobre a floresta (385 m) do que sobre a pastagem (240 m). Os mecanismos atmosféricos que determinam estas características são discutidos.

25 Modelling convective boundary layer growth in Rondônia

G. FISCH[1], A.D. CULF[2] and C.A. NOBRE[3]
[1] *Centro Técnico Aeroespacial, São José dos Campos, Brazil*
[2] *Institute of Hydrology, Wallingford, UK.*
[3] *Instituto Nacional de Pesquisas Espaciais, Brazil*

INTRODUCTION

The dynamics of the convective boundary layer (CBL) play an important role in the land-surface atmosphere interaction by controlling the transfer of the turbulent fluxes from the surface to the free atmosphere. The depth of the CBL gradually increases during the day, not only due to direct heating from the surface, but also because the surface fluxes (thermal and mechanical) provide the energy necessary for air parcels to rise, overshoot the top of the inversion and return (due to their negative buoyancy in the air above the mixed layer) entraining warm and dry air from aloft into the layer. The CBL does not merely grow passively in response to this turbulence but has mechanisms of interaction in the opposite direction: the entrainment flux brings drier air into the layer, changing the saturation deficit and hence modifying the partition of energy in terms of sensible and latent heat fluxes at the surface.

As an active link between the surface and the free atmosphere, the CBL has to be well understood and accurately represented in General Circulation Models (GCMs) if they are to provide realistic simulations of the climate. It is also an essential component of pollutant dispersion models (Raynor and Watson, 1991). The parameterization of the CBL in GCMs has been reviewed by Garratt (1993) who summarized the most important components of a CBL scheme in such models as: the depth of the CBL for vertical distribution of momentum and energy; the values of the turbulent fluxes and other turbulent properties, and the role of clouds.

Deardoff (1972) has discussed two contrasting methods of representing the CBL in GCMs: the multi-layer schemes where the lowest levels of the GCM vertical grid are used to describe the CBL, and the bulk layer approach which considers the entire CBL as a single layer. Although schemes of the first type are more realistic, as the parameters are represented explicity, they can consume a great deal of computer time. In contrast, bulk layer representations are simple and permit one to obtain a prognostic equation for the height of the CBL. However, they have the disadvantage of being difficult to implement in GCMs as they require the model to have a variable vertical grid spacing to be able to represent the temporal variability of the CBL height.

The objective of the present work is to examine the ability of a slab model to

Amazonian deforestation and climate. Edited by J.H.C. Gash, C.A. Nobre, J.M. Roberts and R.L. Victoria.

simulate the evolution and the structure of the CBL, based on the theory developed by Tennekes (1973), using data collected over pasture and forest sites in the Amazon region of Brazil.

SITE AND INSTRUMENTATION

The dataset used in this paper was collected during the Rondônia Boundary Layer Experiment - II (RBLE II) which took place at the ABRACOS forest and pasture sites in Rondônia, Brazil in July and August 1993 and is described fully by Nobre *et al.* (1996). These two sites, Reserva Jaru and Fazenda Nossa Senhora, and the permanent ABRACOS instrumentation installed there, are described by Gash *et al.* (1996). At the forest site, the micrometeorological data and turbulent fluxes have been measured at the top of an aluminium tower 52 m high. The mean height of the canopy is 30 m, although some emergent trees reach up to 42 m. The pasture site, covered predominantly by *Brachiaria brizantha*, was deforested about 15 years ago and is typical ranchland for this part of Amazonia. During RBLE II, measurements were made by radiosondes, a tethered balloon system and surface micro-meteorological intrumentation. At the forest site the CBL measurements were made at the IBAMA field station, about 2 km from the ABRACOS tower. At the pasture site, all the CBL and surface measurements instrumentation was set up together, inside a circle of radius less than 500 m.

The radiosounding data were collected using radiosondes (Vaisala-Finland), during daytime hours at 08.00, 11.00, 14.00 and 17.00 h local time. The tethered balloon system (A.I.R. Inc., Colorado, USA) consists of a tethersonde measuring pressure, dry and wet bulb temperatures, windspeed and wind direction, attached to a 7.0 m^3 balloon, which can be operated at heights of up to 1000 m. The tethered

Table 1 Energy ($w'\theta_v'$) and momentum (proportional to u_*) surface fluxes observed at the forest and pasture sites and used as inputs for the CBL model

	Forest		Pasture	
Local time h	$w'\theta_v'$ m K s^{-1}	u_* m s^{-1}	$w'\theta_v'$ m K s^{-1}	u_* m s^{-1}
08.00	-0.04	0.04	0.009	0.06
09.00	0.016	0.16	0.06	0.14
10.00	0.059	0.25	0.136	0.26
11.00	0.08	0.26	0.182	0.31
12.00	0.092	0.3	0.19	0.3
13.00	0.133	0.38	0.21	0.3
14.00	0.135	0.43	0.194	0.28
15.00	0.098	0.41	0.14	0.28
16.00	0.075	0.38	0.094	0.27
17.00	0.021	0.29	0.029	0.23

balloon was operated during the night and in the early morning with the profile of temperature measured at 08.00 h being used as an initial value for the model simulations. Hourly values of surface fluxes were obtained from an eddy-correlation instrument (Hydra) described in detail by Shuttleworth *et al.* (1988). The average surface fluxes (momentum and energy) obtained from this instrument are used to prescribe the input of energy to the CBL model and are presented in Table 1. The data obtained over the forest were collected between 3 and 10 July 1993 and at the pasture site from 10 to 25 July 1993. However, during the forest data collection period, a cold front or "friagem" moved across the area, modifying the atmosphere structure. Surface fluxes measured during this event (7 to 8 July) have not been included in the calculation of average surface fluxes.

THE MODEL

The model of the CBL used here can be visualised as a single layer box. The box grows in height due the input of energy from the bottom (the surface fluxes) and from the top (the entrainment flux which includes contributions from both thermally and mechanically generated turbulence). Inside the box, the air is well mixed with a uniform profile of virtual temperature θ_v. This type of model has been studied under different atmospheric conditions and types of surface by several researchers (e.g. Tennekes, 1973; Driedonks, 1982; McNaughton and Spriggs, 1986; Batchavarova and Gryning, 1990; Culf, 1992). Following Tennekes (1973), Tennekes and Driedonks (1981) and Driedonks (1982), it is assumed that all transfer of energy occurs vertically (i.e. the model is one-dimensional) and that large-scale subsidence can be neglected. The conservation of energy of a homogeneous layer can then be written as:

$$\frac{d}{dt}(\theta_v) = \frac{1}{h}\left[\overline{(w'\theta_v')_o} - \overline{(w'\theta_v')_b}\right] \tag{1}$$

where θ_v is the average virtual potential temperature of the mixed layer, $w'\theta'_v$ represents virtual heat flux at the surface (subscript o) and at the inversion height (subscript b). h is the height of the mixed layer (see Figure 1).

Applying energy conservation at the step in θ_v at the top of the mixed layer gives:

$$\frac{d}{dt}(\Delta\theta_v) = S_o\left(\frac{dh}{dt}\right) - \frac{d\theta_v}{dt} \tag{2}$$

where $\Delta\theta_v$ is the size of the step in virtual potential temperature at the top of the mixed layer and S_o is the vertical gradient of θ_v above the CBL which is taken to be invariant with time. The entrainment flux is related to $\Delta\theta_v$ and the rate of boundary layer growth as (Lilly, 1968):

$$-\left(\overline{w'\theta_v'}\right)_b = \Delta\theta_v\,\frac{dh}{dt} \tag{3}$$

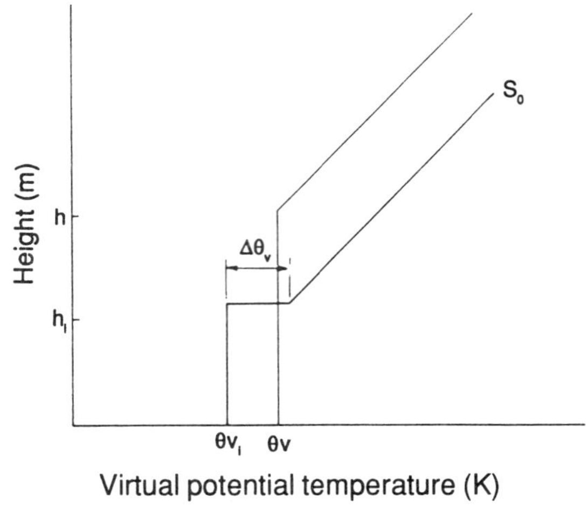

Figure 1 The vertical distribution of virtual potential temperature in and above a convective boundary layer (CBL).

Equations 1, 2 and 3 are the kernel of the CBL growth model, but it is not a closed system as it has four unknowns (θ_v, $\Delta\theta_v$, h, and $w'\theta'_v$). To close the system, Tennekes (1973) suggested that, based on the conservation of turbulence kinetic energy, the entrainment flux can be parameterized in terms of the other variables as:

$$-\left(\overline{w'\theta'_v}\right)_b = C_F \left(\overline{w'\theta'_v}\right)_o + A\,\frac{u_*^3\theta_v}{gh} \tag{4}$$

where C_F and A are empirical constants. The first term on the right-hand side is the contribution from thermally generated turbulence and the second term is the contribution from mechanically generated turbulence. We define the ratio of these two terms as G, a parameter which describes the relative importance of the two contributions to the entrainment flux. Substituting Equation 4 into Equation 3 yields, after some rearrangement:

$$\frac{dh}{dt} = 0.2\frac{\left(\overline{w'\theta'_v}\right)_o}{\Delta\theta_v} + 5\frac{u_*^3\theta_v}{gh\Delta\theta_v} \tag{5}$$

where C_F and A have been replaced with the usual values of 0.2 and 5 respectively (Driedonks, 1982; Driedonks and Tennekes, 1984) The system of Equations 1, 2 and 5 can now be solved numerically. Using a forward scheme for numerical integration the equations can be rewritten in finite difference form as:

$$\frac{h^{n+1}-h^n}{\Delta t} = 0.2\,\frac{\left(\overline{w'\theta'_v}\right)_o^{n+\frac{1}{2}}}{\Delta\theta_v^n} + \frac{5(u_*^3)^{n+\frac{1}{2}}\,\theta_v^n}{gh^n\Delta\theta_v^n} \tag{6}$$

$$\frac{\theta_v^{n+1} - \theta_v^n}{\Delta t} = \frac{1}{h^n}\left[1.2\overline{(w'\theta_v')}^{\,n+\frac{1}{2}} + \frac{5u_*^{3^{\,n+\frac{1}{2}}}\theta_v^n}{gh^n}\right] \tag{7}$$

$$\frac{\Delta\theta_v^{n+1} - \Delta\theta_v^n}{\Delta t} = S_o\left[\frac{h^{n+1} - h^n}{\Delta t}\right] - \frac{(\theta_v^{n+1} - \theta_v^n)}{\Delta t} \tag{8}$$

where the subscripts n and n+1 represent the variable at two sucessive time steps (a time interval of 60 s was used in the current model) and the subscript n+½ denotes the average value of a variable during the time interval. The average values were interpolated from the observed data given in Table 1. This forward scheme is potentially numerically unstable, but with a time step of 60 s instabilities do not appear. Tests with other explicit numerical schemes, such as Adam-Bashfort, Leap-Frog and Runge-Kutta (Mesinger and Arakawa, 1982), have given similar results that are not presented here.

RESULTS

The growth of the CBL over the forest and pasture sites has been simulated using Equations 6, 7 and 8, the boundary conditions given in Table 1, and the appropriate initial values given in Table 2. For each simulation, it is necessary to provide initial values for h, θ_v, $\Delta\theta_v$ and S_o measured at the time of the nocturnal boundary layer breakdown, or in the early morning soon afterwards. The first three of these values were obtained from a tethered balloon profile measured at 08.00 h whilst S_o was obtained from the simultaneous radiosonde ascent. The main results obtained from the simulations are given in Table 3. In Figure 2 the values of the height of the CBL and the virtual potential temperature are shown along with the observed values, extracted from Fisch (1995) and Nobre *et al.* (1996), for comparison. In all the analyses throughout this paper average values have been calculated and compared with average observational aspects. This approach (composite time evolution) has

Table 2 Initial conditions for boundary layer height, thermal gradient (S_o), virtual potential temperature (θ_v) and the discontinuity in virtual potential temperature at the top of the mixed layer ($\Delta\theta_v$) for the forest and the pasture sites.

	Height (m)	S_o (K km^{-1})	θ_v (K)	$\Delta\theta_v$ (K)
Forest	200	1.8	298.8	5.8
Pasture	110	3.6	298.6	5.3

Table 3 Results of the numerical simulation of the growth of the CBL over the forest and the pasture sites

	Forest					Pasture				
Local time h	height m	$\Delta\theta_v$ K	θ_v K	G	w_* ms^{-1}	height m	$\Delta\theta_v$ K	θ_v K	G	w_* ms^{-1}
08.00	200	5.80	298.8			110	5.3	298.6	0.17	0.32
09.00	201	5.66	298.9	0.99	0.47	117	3.95	300.0	0.30	0.61
10.00	211	4.77	299.8	0.96	0.74	175	0.77	303.4	0.57	0.92
11.00	234	3.26	301.4	0.72	0.85	645	0.38	305.4	0.20	1.56
12.00	281	1.69	303.1	0.81	0.94	975	0.55	306.5	0.11	1.81
13.00	461	0.40	304.7	0.70	1.25	1232	0.68	307.3	0.08	2.02
14.00	949	0.33	305.6	0.48	1.60	1445	0.79	307.9	0.06	2.07
15.00	1270	0.42	306.1	0.43	1.59	1600	0.87	308.4	0.08	1.93
16.00	1464	0.48	306.4	0.39	1.52	1702	0.92	308.7	0.10	1.72
17.00	1562	0.51	306.5	0.58	1.02	1754	0.96	308.9	0.19	1.17

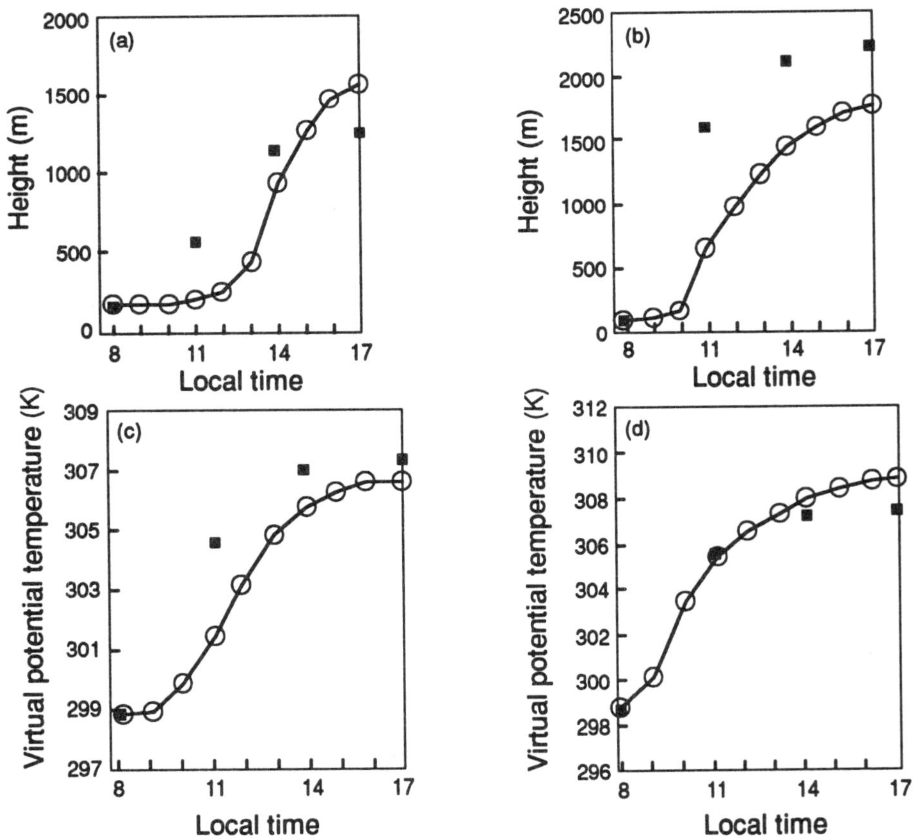

Figure 2 Comparisons of calculated average values (—o—) and observations (■) for the development of CBL: height of the CBL for forest (a) and pasture (b); average virtual potential temperature for forest (c) and pasture (d).

been chosen, instead of predicting values for individual days, to remove some of the observational noise and day-to-day variability from the results.

For the forest site, the model predicts that the CBL grows higher than observed: at 17.00 h the final height was computed as 1562 m against an observed value of 1250 m. The calculated value of θ_v was 0.7 K less than the observed value. The lower value of temperature computed by the model can be explained by the long time period used by the model to decrease the value $\Delta\theta_v$. Only after 13.00 h is the value of $\Delta\theta_v$ small enough to allow the entrainment flux to become significant. Because of this slow initial increase, the calculated value of θ_v is less than the observed value throughout the integration period. The results obtained from the simulation over the pasture site show a similar feature: the model is unable to reproduce the rapid development of the CBL during the early and mid morning. The height at 11.00 h was computed to be 645 m, compared with an observed value of 1590 m, resulting in a lower height at 17.00 h (1754 m compared with an observation of 2220 m). The value of θ_v calculated by the model is 1.5 K higher than observed, but at midday the model reproduces the observed temperature reasonably well.

The performanace of the model was investigated qualitatively by using an even simpler representation of the CBL. If the mechanical contribution to entrainment can be neglected, the height of the CBL can be considered to be dependent on square root of time (Driedonks, 1982) and can be computed as (Garratt, 1992):

$$h(t) = \left[2(1 + 2C_F) \int_{t8}^{t17} \frac{\overline{w'\theta_v'}}{S_o} \, dt \right]^{\frac{1}{2}} \tag{9}$$

Using typical data for forest and pasture (from Tables 1 and 2), Equation 9 gives a value of 1200 m for forest and 1150 m for pasture. For the forest site, this estimate of final CBL height is reasonable, as we have observed values of around 1250 m. However, for the pasture, the difference between the value from Equation 9 and the observations is very large (around 1200 m), indicating that an additional source of energy is missing from the model.

Further investigation into the cause of the poor representation of the CBL during the morning was carried out by performing a second numerical simulation of CBL growth for both the forest and the pasture sites. For these simulations, however, the time integration was started later in the day (11.00 h) when free convection is dominant. The results of this second pair of model runs are shown in Table 4 and Figure 3. For the forest site there is no improvement on the results of the first simulation, but for the pasture site, which previously had the worst results, the predictions of height and average virtual potential temperature of the CBL are improved. These results further identify the mid-morning before 11.00 h as the period poorly represented by the model for simulations of the CBL over the pasture site.

Table 4 Results of the numerical simulations of the growth of the CBL over the forest and pasture sites starting at 11.00 h when free convection is dominant

	Forest		Pasture	
Local Time	Height (m)	θ_v (K)	Height (m)	θ_v (K)
11.00	580	304.5	1590	305.6
12.00	771	305.1	1750	306.1
13.00	1026	305.7	1893	306.6
14.00	1281	306.2	2023	307.0
15.00	1471	306.6	2123	307.4
16.00	1596	306.8	2190	307.6
17.00	1663	307.0	2225	307.7

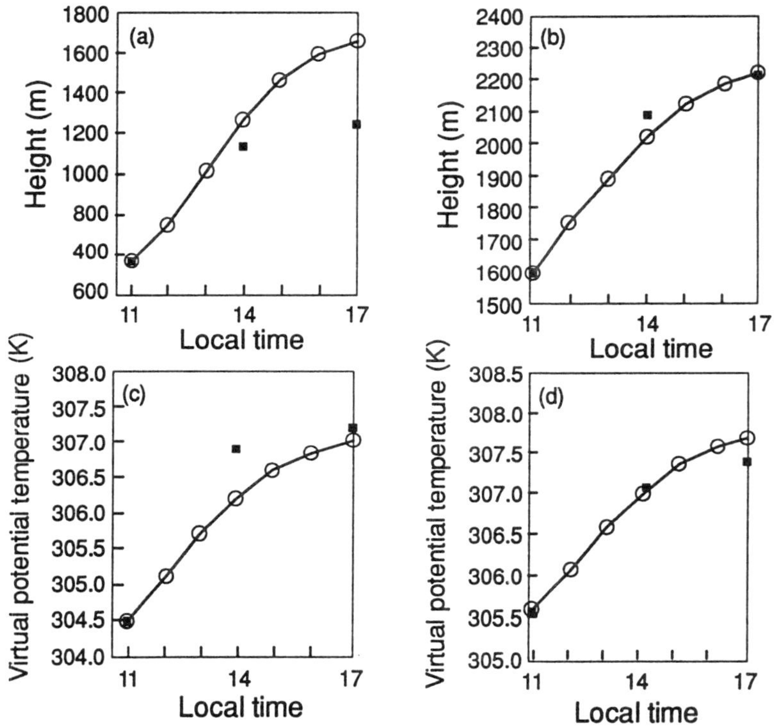

Figure 3 Comparisons of calculated average values (—⊖—) and observations (■) for the development of CBL after 11.00 h when free convection is predominant: height of the CBL for forest (a) and pasture (b); average virtual potential temperature for forest (c) and pasture (d).

DISCUSSION

In both the forest and the pasture cases, the model is unable to reproduce the observed rapid development of the CBL during the mid-morning period. The calculated values of the parameter G, defined above as the ratio of the thermal and mechanical contributions to the entrainment flux, and listed in Table 3, can help to explain this aspect of the model's performance. Table 3 shows that G is of the order of unity for both the forest and pasture simulations in the early morning when the mechanical contribution to the entrainment flux would be expected to be dominant. For example Culf (1992) showed that the contribution to the entrainment flux from mechanical turbulence can be ten times greater than the thermal contribution during the mid-morning period in the semi-arid Sahel. Despite these low values of G, the values of the vertical velocity scale, or mixing velocity, $w_* = g\ h\ w'\theta_v'/\theta_v$, which represents the thermally induced turbulence, are of the same order of magnitude as the values observed by Martin *et al.* (1988) in Manaus, with typical values of about 1.0 m s^{-1}. Since the values of w_* appear to be reasonable, the low values of G may be evidence of a significant underestimation of mechanically generated turbulence in the model.

The large deforested area around Ji-Paraná has strips of forest 2-4 km wide embedded in it (Skole and Tucker, 1993). These are typical of the type of surface features which may induce mesoscale thermal circulations (Segal and Arrit, 1992). The forest is typically wetter and colder than the pasture (see results from Nobre *et al.*, 1996) and it is possible that secondary thermal circulations develop in the morning leading to advection of energy and a more rapid breakdown of the nocturnal boundary layer than would otherwise be observed.

Another possible explanation for the apparent underestimation of the mechanically generated turbulence is that, on the area average scale relevant to CBL development, the nature of the landscape surrounding the pasture site may mean that the value of u_* is actually much larger than the values measured by micrometeorological instruments mounted at the surface and used for forcing data in these simulations. The use of an increased area average value of u_* in the model simulations would increase the mechanical contribution to the entrainment flux and increase the rate of CBL growth.

A third hypothesis is that strong wind shear develops over the pasture areas at night due to the decoupling of the nocturnal boundary layer flow from the free atmosphere above. Such a decoupling of the surface from the flow aloft has been shown to occur over a small pasture area near Manaus (Bastable *et al.*, 1993). The wind shear developed would tend to lower the Richardson number above the nocturnal boundary layer, increase the turbulence there and hasten the morning breakdown.

CONCLUSION

The model used to study the growth of CBL was able to reproduce some of the characteristics of the observed CBL, although there were discrepancies in the virtual potential temperature profiles and the CBL height. The simulation of the development of the CBL for the forest site is reasonable, with the final height only about 200 m higher than observed. For the pasture scenarios, the results of the numerical exercise showed poor agreement with the observations, especially in the mid-morning. The results of a second simulation, starting when the free convection is predominant, were better, indicating that the problem with the model simulation was occurring between sunrise and mid-morning. It is likely that the juxtaposition of patches of forest and pasture in the area of the experiment means that several processes can take place which all have the effect of increasing the turbulence in the early morning, leading to a more rapid breakdown of the nocturnal boundary layer than would be observed over a uniform surface.

ACKNOWLEDGEMENTS

The authors would like to express their thanks to all Brazilian and British participants of ABRACOS/RBLE II involved in the data collection at Ji-Paraná. RBLE II was partially funded by Fundação de Amparo à Pesquisa do Estado de São Paulo (FAPESP) and Conselho Nacional Científico e Tecnológico (CNPq). G.Fisch received a grant from the Coordenadoria de Aperfeiçoamento de Pessoal do Nível Superior (CAPES) to carry out analyses at the Institute of Hydrology (UK) during 1993. Special thanks go to J.H.C. Gash, A. Beljaars, J.M. Roberts and I.R. Wright for their encouragement to produce this work. The authors would also like to thank the anonymous referees for their useful comments and suggestions.

REFERENCES

Bastable, H.G., Shuttleworth, W.J., Dallarosa, R.L.G., Fisch, G. and Nobre, C.A. 1993. Observations of climate, albedo and surface radiation over cleared and undisturbed Amazonian Forest. *Int. J. Climatol.* **13**, 783-796.

Batchvarova, E. and Gryning, S.E. 1990. Applied model for the growth of the daytime mixed layer. *Bound.-Layer Met.* **56**, 261-274.

Culf, A.D. 1992. An application of simple models to Sahelian convective boundary layer growth. *Bound.-Layer Met.* **58**, 1-18.

Deardoff, J.W. 1972. Parameterization of the planetary boundary layer for use in general circulation models. *Mon. Weath. Rev.* **100**, 93 -106.

Driedonks, A.C.M. 1982. A model and observations of the growth of the atmospheric boundary layer. *Bound.-Layer Met.* **23**, 283-306.

Driedonks, A.C.M., and Tennekes, H. 1984. Entrainment effects in the well mixed atmospheric boundary layer. *Bound.-Layer Met.* **30**, 75-105.

Gash, J.H.C., Nobre, C.A., Roberts, J.M. and Victoria, R.L. 1996. An overview of ABRACOS. This volume.

Garratt, J.R. 1992. *The atmospheric boundary layer*. Cambridge Press University, London. 316p.

Garratt, J.R. 1993. Sensivitiy of climate simulations to land-surface and atmospheric boundary layer treatments - a review. *J. Climate* **6**, 419-449.

Fisch, G., 1995. Camada limite Amazonica: aspectos observacionais e de modelagem. PhD thesis, INPE, São José dos Campos, Brazil.

Lilly, D.K. 1968. Models of cloud topped mixed layers under a strong inversion. *Q. J. Roy. Met. Soc.* **94**, 292-309.

Martin, C.M., Fitzjarrald, D.R., Garstang, M., Oliveira, A.P. de, Greco, S. and Browell, E. 1988. Structure and growth of the mixing layer over the Amazonian rain forest. *J.Geophys. Res.* **93**, 1361-1375.

Mesinger, F. and Arakawa, A. 1982. Numerical methods used in atmospheric models. GARP Publications Series. **17**, 65pp.

McNaughton, K.G. and Spriggs, T.W. 1986. A mixed-layer model for regional evaporation. *Bound.-Layer Met.* **34**, 243-262.

Nobre, C.A., Fisch, G., Rocha, H.R. da, Lyra, R.F.F., Rocha, E.P., Costa, A.C.L.da and Ubarana, V.N. 1996. Observational aspects of atmospheric boundary layer in Rondonia. This volume.

Rayner, K.N. and Watson, I.D. 1991. Operational prediction of daytime mixed layer heights for dispersion modelling. *Atmos. Environ.* **25A**, 1427-1436.

Segal, M. and Arrit, R.W. 1992. Non-classical mesoscale circulations caused by surface sensible heat flux gradients. *Bull. Am. Met. Soc.* **73**, 1593-1604.

Shuttleworth, W.J., Gash, J.H.C., Lloyd, C.R., McNeil, D.D., Moore, C.J. and Wallace, J.S. 1988. An integrated micrometeorological system for evaporation measurements. *Agric. For. Met.* **43**, 295-317.

Skole, D. and Tucker, C. 1993. Tropical deforestation and habitat fragmentation in Amazonia: satellite data from 1978 to 1988. *Science* **260**, 1905-1910.

Tennekes, H. 1973. A model for the dynamics of the inversion above a convective boundary layer. *J. Atmos. Sci.* **30**, 558-567.

Tennekes, H. and Driedonks, A.C.M. 1981. Basic entraiment equations for the atmospheric boundary layer. *Bound.-Layer Met.* **20**, 515-531.

RESUMO

Um modelo uni-dimensional simples foi utilizado para estudar o crescimento da camada limite convectiva (CLC) desenvolvidas sobre superfícies de floresta e pastagem na região Amazônica do Brasil, durante o Experimento da Camada Limite de Rondônia. Este modelo foi baseado na teoria desenvolvida por Tennekes (1973) e é capaz de reproduzir algumas das características observadas da CLC, embora diferenças do valor final da temperatura potencial e virtual e altura da camada limite tenham sido observadas. Estas diferenças são maiores para a simulação realizada para a pastagem. Os resultados de uma nova simulação, na qual o modelo é integrado a partir do instante em que a convecção livre torna-se dominante, mostrou melhores resultados. Baseado nestas análises, sugere-se que possa existir uma fonte de turbulência extra no início da manhã na área de pastagem e que o modelo uni-dimensional não contabiliza. Esta turbulência é gerada pela existência de circulação térmica secundária. devida à justaposição de áreas de floresta e pastagem e é muito importante na erosão da camada limite noturna.

26 Modelling surface conductance for Amazonian pasture and forest

I.R. WRIGHT[1], J.H.C. GASH[1], H.R. da ROCHA[2] and J.M. ROBERTS[1]
[1]*Institute of Hydrology, Wallingford, UK*
[2]*Universidade do São Paulo, Brasil*

INTRODUCTION

The prediction of the stomatal control of transpiration by vegetation is commonly represented in GCM surface schemes (Sellers *et al.* 1986; Noilhan and Planton, 1989; Lean *et al.*, 1996) by a type of model, first proposed by Jarvis (1976), in which a theoretical maximum surface conductance is attenuated by various climatic influences. This paper presents the calibration of this type of model for two pairs of pasture and primary forest sites in Amazonia, and compares two methods of application. The comparison between similar vegetation types at different locations allows comment on the extent to which parameterisations are site specific and on the spatial validity of applying a single set of parameters across Amazonia. In particular, the optimised parameters from forest and pasture sites near Manaus (central Amazonia) are used to predict the transpiration from sites in Rondônia (south-west Amazonia), assuming that only hourly climate data and a minimum of other site details are available.

It is acknowledged that the Jarvis-type model has limitations associated with the method of calibration and is likely to be superseded by models which are more strongly based upon the leaf physiology of plants (Friend, 1991). This is especially the case as GCM surface schemes begin to require models of leaf gas exchange for processes such as carbon assimilation and large-scale vegetation growth. However, the robust performance of Jarvis-type models and the continuing demand for parameters offsets the limitations and justifies the calibration given here.

SITES AND INSTRUMENTATION

PASTURE SITES

Micrometeorological measurements were made over Amazonian pasture at the Fazenda Dimona site near Manaus (2° 19'S 60° 19'W) and at Fazenda Nossa Senhora near Ji-Paraná in Rondônia (10° 45'S 62° 22'W). The sites were chosen to

Amazonian deforestation and climate. Edited by J.H.C. Gash, C.A. Nobre, J.M. Roberts and R.L. Victoria.

be representative of post-deforestation cattle pasture in southern and eastern Amazonia (Uhl *et al.*, 1988; Eden *et al.*, 1990). The instrumentation was used in a similar configuration at both sites (Wright *et al.*, 1992) and measured sensible heat, water vapour and momentum fluxes together with the surface energy balance. Fluxes were derived from vertical profiles of temperature, humidity and windspeed together with independent measurements using a 'Hydra' eddy correlation device (Shuttleworth*et al.*, 1988). Surface radiation measurements were mostly derived from the permanently located automatic weather stations (AWSs) and soil moisture content was measured every 5 to 15 days using a neutron probe.

Principal differences between the two pasture installations were the height of the AWSs and the choice of reference height for height dependent variables, e.g., aerodynamic conductance. At Fazenda Nossa Senhora the chosen reference height was 5.5 m from the ground, the same height as the AWS. At Fazenda Dimona the AWS was installed at 1.5 m from the ground because of the very short grass at the site but the reference height was 3.5 m, the height of the 'Hydra' and the logarithmic mean height of the instrumentation on the micrometeorological profile tower.

Vegetation details for the pasture at Fazenda Dimona (spp.*Brachiaria decumbens* and *B. humidicola*) are given by Wright *et al.* (1992) and McWilliam *et al.* (1993), and for Fazenda Nossa Senhora (sp. *B. brizantha*) by McWilliam *et al.* (1996). The proportion of bare soil at each site was similar: 0.11 at Fazenda Dimona and 0.14 at Fazenda Nossa Senhora respectively, and both pastures were grazed by cattle during the intensive measurement periods. However, the grass had grown well at Fazenda Nossa Senhora, was not over-grazed and had reached a mean height of 0.6 m in April 1993 with total Leaf Area Index (including stems), L^*, of 3.9. In contrast, pasture growth on the clay soil at Fazenda Dimona was greatly restricted by water stress (Wright *et al*, 1992) and heavier grazing. In 1990 and 1991 the pasture height was typically 0.28 m with $L^* = 1.5$, but the grass species at this site are naturally shorter.

FOREST SITES

A description of the primary forest structure is given for Reserva Jaru, Ji-Paraná (10° 'S 61° 55'W) by McWilliam *et al.* (1996), and for Reserva Ducke, Manaus (2° 57'S 59°57'W) by Roberts *et al.* (1990). The climate and flux measurement instrumentation were mounted on a 53 m tower at Reserva Jaru and a 45 m tower at Reserva Ducke: 23 m and 10 m respectively above the mean canopy-top height.

No new flux measurements were recorded by the ABRACOS project over the forest at Reserva Ducke. Optimised parameters for this site, using similar field techniques and analytical methods, have been reported by Dolman *et al.* (1991).

During the 1993 measurement campaign 112 days of hourly evaporation, heat and momentum fluxes were recorded at Reserva Jaru using the 'Hydra' eddy correlation device. Comparison between these estimates and 40 days of coincident measurements by Grace *et al.* (1995) showed a satisfactory agreement between heat flux measurements, but the vapour flux measurements compared less well. In view of this uncertainty, both this study and Grace *et al.* (1995) estimated the hourly vapour flux

λE as a residual of the energy balance

$$\lambda E = R_n - G - H - S_{T.q} \tag{1}$$

where R_n is the net all-wave radiation, G is the soil heat flux, H is the sensible heat flux and $S_{T.q}$ is the hourly gain or loss of energy storage from the forest biomass and the canopy air space (Moore and Fisch, 1986). Following the method published by Moore and Fisch (1986) and based upon their calibration for Reserva Ducke, the storage term was calculated for Reserva Jaru using hourly AWS data and an estimate of the biomass. Hourly heat storage had a typical diurnal maximum of about 80 W m^{-2} at 1000-1100 h (local time). Most of the error in the estimate of δE originates from errors in H and $S_{T.q}$; it is estimated to be 10 - 20 W m^{-2} when vapour fluxes are small, rising to 15 - 30 W m^{-2} at typical midday maximum vapour fluxes of 500 W m^{-2}.

SOILS

The soils in the two areas, Manaus and Ji-Paraná, are predominantly clay and sand respectively, and are quite different in their water retention characteristics (Tomasella and Hodnett, 1996; Wright *et al.*, 1996). Forest rooting depths are discussed by Wright *et al.* (1996). However, because of the absence of any observed stress in the bulk transpiration measurements associated with soil moisture deficits, forest rooting depths are not relevant here. For the pasture sites the ABRACOS soil moisture measurements made over a two year period were used as a guide to evaluate an effective rooting depth for plant water extraction: 1.5 m at Fazenda Dimona and 2 m at Fazenda Nossa Senhora respectively. For these depths, the effective range of volumetric soil moisture available to plants, i.e. between wilting point and field capacity, was found to be 0.33 to 0.42 m^3 m^{-3} at Fazenda Dimona, and 0.15 to 0.32 m^3 m^{-3} at Fazenda Nossa Senhora.

DATA

Data for the calibration of the 'Jarvis-type' model of bulk surface conductance are taken from ABRACOS intensive measurement campaigns: 1990 and 1991 at Fazenda Dimona (Missions 1 and 2), and 1992 and 1993 over the forest at Reserva Jaru and the pasture at Fazenda Nossa Senhora (Missions 3 and 4/5). Details of the data periods for the five micrometeorological missions are given by Gash *et al.* (1996) and are summarised in Table 1 together with the data ranges over which the conductance models were calibrated.

Data were chosen for optimisation using only hours during which the evaporation was considered to be dominated by transpiration, that is with a minimum of contamination by evaporation from free water or the soil surface. The total hours available for optimisation at each site are shown in Table 1.

Table 1 Mission and optimisation data details

FAZENDA DIMONA, PASTURE			
Dates		*FROM*	*TO*
	Mission 1	18 Sep. 1990	2 Nov. 1990
	Optimisation period	9 Oct. 1990	1 Nov. 1990
	Total hours for optimisation	187	
	Mission 2	1 July 1991	10 Sep. 1991
	Optimisation period	19 July 1991	9 Sep. 1991
	Total hours for optimisation	153	
Data Ranges Mission 1 and 2	Spec. humidity deficit kg kg^{-1}	.0016	.0224
	Air Temperature °C	24.4	35.4
	Solar Radiation W m^{-2}	43	1003
	Soil Moisture m^3 m^{-3}	.37	.43
	Bulk conductance, g_s mm s^{-1}	.12	13.61
FAZENDA NOSSA SENHORA, PASTURE			
Dates	Mission 3	6 Aug. 1992	7 Oct. 1992
	Optimisation period	8 Aug. 1992	7 Oct. 1992
	Total hours for optimisation	159	
	Mission 4 and 5	1 Apr. 1993	28 July 1993
	Optimisation period	31 Mar. 1993	28 July 1993
	Total hours for optimisation	627	
Data Ranges Mission 3	Spec. humidity deficit kg kg^{-1}	.0004	.0116
	Air Temperature °C	23.9	33.8
	Solar Radiation W m^{-2}	111	994
	Soil Moisture m^3 m^{-3}	.207	.262
	Bulk conductance, g_s mm s^{-1}	.81	23.6
Mission 4 and 5	Spec. humidity deficit kg kg^{-1}	.0003	.0220
	Air Temperature °C	16.8	33.1
	Solar Radiation W m^{-2}	26	976
	Soil Moisture m^3 m^{-3}	.222	.335
	Bulk conductance, g_s mm s^{-1}	.05	23.1
RESERVA JARU, FOREST			
Dates	Mission 3	8 Aug. 1992	5 Oct. 1992
	Optimisation data	10 Aug. 1992	4 Oct. 1992
	Total hours for optimisation	311	
Mission 4 and 5	4 Apr. 1993	26 July 1993	
	Optimisation data	5 Apr. 1993	26 July 1993
	Total hours for optimisation	513	
Data Ranges Mission 3	Spec. humidity deficit kg kg^{-1}	.0017	.0188
	Temperature °C	20.8	32.8
	Solar Radiation W m^{-2}	164	987
	Bulk conductance, g_s mm s^{-1}	2.1	32.7
Data Ranges Mission 4 and 5	Spec. humidity deficit kg kg^{-1}	.0021	.0164
	Temperature °C	18.1	32.5
	Solar Radiation W m^{-2}	185	1053
	Bulk conductance, g_s mm s^{-1}	0.74	35.8

evaporation of dew. Any hours that were either one hour before, or up to 24 hours after, recorded rainfall were also excluded. Similar criteria were used for selecting the pasture data except that an additional restriction was necessary at Fazenda Nossa Senhora, where evaporation from the sandy soil surface was considered to be a significant component of the total evaporation for up to three days after rainfall. Therefore, at this site no data were used either from three days after rainfall or until 10 mm of measured evaporation had been accumulated since the last storm. It is recognised that a complete analysis of Fazenda Nossa Senhora data will require a 'two component' model which is able to partition the available energy between heat flux and evaporation processes at both the vegetation and soil surfaces, with full regard for the antecedent rainfall. Nevertheless, the data ranges that remain after the imposition of the restrictions to eliminate free water evaporation (Table 1) are considered more than adequate for a satisfactory calibration.

Figures 1 and 2 show the mean daily values of temperature and specific humidity deficit for pasture and forest, and daily totals of available energy, evaporation and rainfall. The period of measurement includes the transition from the wet season to dry

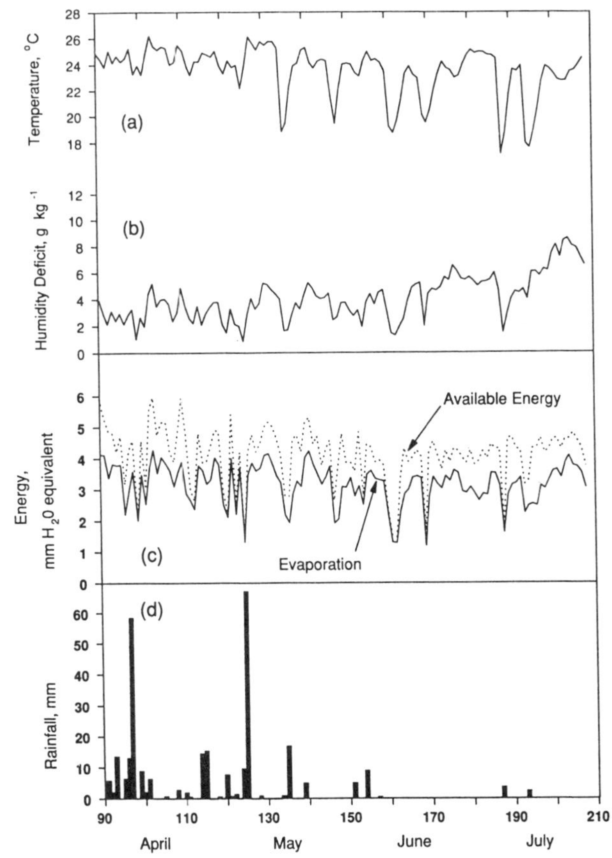

Figure 1 Daily mean temperature and humidity, and daily totals of evaporation, available energy and rainfall at Fazenda Nossa Senhora during the Mission 4/5 data period.

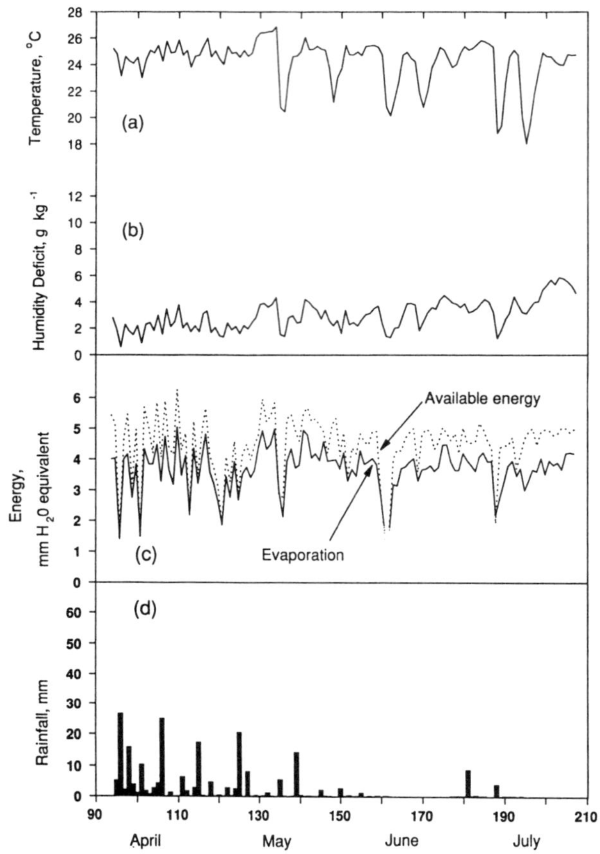

Figure 2 Daily mean temperature and humidity, and daily totals of evaporation, available energy and rainfall at Reserva Jaru during the Mission 4/5 data period.

season in 1993 in Rondônia. As the dry season progresses, the increase in humidity deficit can be seen at both sites as well as the fall-off in evaporation from the pasture. There is also a small reduction in available energy over both sites from increased cloudiness. The temperature at both sites clearly shows the regular occurrence of cold southerly winds, which have a marked effect on the energy balance on both pasture and forest.

Figure 3 shows the decrease in soil moisture in the top 3.6 m at the Ji-Paraná forest from April to July 1993 (see also Hodnett *et al.*, 1996). This decrease is in marked contrast to the absence of seasonality in forest evaporation shown in Figure 2c for the same period. Therefore, as with previous work (Dolman *et al.*, 1991), soil moisture has been excluded from the forest model of transpiration.

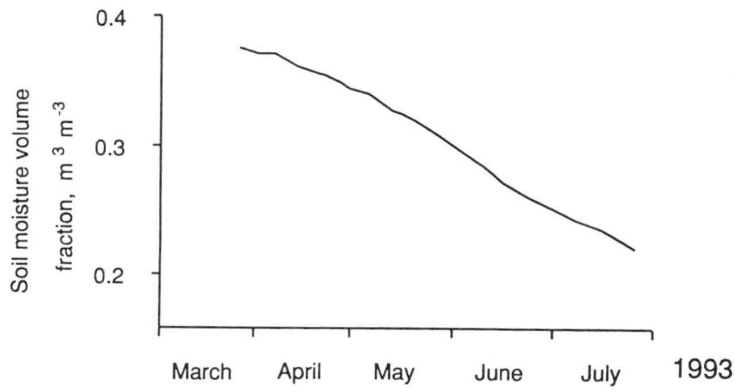

Figure 3 Observations of soil moisture content at the Reserva Jaru forest site during Mission 4/5

METHODS

The theoretical basis outlined in this section and the method of calibration are described fully by Wright *et al.*, 1995. Many models of the surface water balance are based on the Penman-Monteith evaporation equation (Monteith, 1965). When measurements of evaporation are available the equation may be rearranged to give the effective or bulk surface conductance per unit ground area, g_c - this parameter which represents the regulation of transpiration by the leaf stomata, as follows

$$\frac{1}{g_c} = \frac{\left(\frac{sH}{C_p E} - 1\right)}{g_{a,v}} + \frac{\rho D_{ref}}{E} \tag{2}$$

where E and H are the water mass and heat energy flux densities respectively, D_{ref} is the specific humidity deficit at the reference (measurement) height, $g_{a,v}$ is the aerodynamic conductance for water vapour transport, s is the rate of change of saturation specific humidity with temperature, ρ is the density of air and C_p is the specific heat of air at constant pressure. The aerodynamic conductance was calculated for this study using the surface roughness parameters (zero plane displacement and roughness length) derived for this site, including corrections for stability (Arya, 1988), and taking account of the difference between momentum sink and heat flux source heights (Verma, 1989; Wright *et al.*, 1992). It is important to note that the derived values of g_c will depend upon, and be sensitive to, the assumptions made in calculating g_a and the subsequent site calibrations given here for transpiration are only valid when g_a is calculated in a similar manner (Huntingford, 1995).

The model for the response of stomata to climatic variables is based on the original work of Jarvis (1976) and subsequently modified by Stewart (1988) to yield a relationship with parameters suitable for use in GCMs:

$$g_c = g_{s,Max} L^* f(D_0) f(T_0) f(S) f(\theta) \tag{3}$$

where $g_{s,Max}L^*$ is parameter a_1, the subscript '0' refers to canopy level climate and

$g_{s,Max}$ = maximum stomatal conductance per unit leaf area
L^* = leaf area index
T_0 = temperature
S = solar radiation
θ = soil moisture content
D_0 = specific humidity deficit, $= q^*(T_0) - q_0$ (4)

and where the control functions are

$$f(D_0) = \exp(-a_2 D_0) \tag{5}$$
$$f(T_0) = [(T_0-T_l)\times(T_h-T_0)^t]/[(a_3-T_l)\times(T_h-a_3)^t] \tag{6}$$
$$t = (T_h-a_3)/(a_3-T_l) \tag{7}$$
$$f(S) = [S/1000]\times[(1000+a_4)/(S+a_4] \tag{8}$$
$$f(\theta) = 0 \quad \text{for } \theta < \theta_w \tag{9}$$
$$f(\theta) = (\theta-\theta_w)/(a_5-\theta_w) \quad \text{for } \theta_w < \theta < a_5 \tag{10}$$
$$f(\theta) = 1 \quad \text{for } \theta > a_5 \tag{11}$$

T_l = lower temperature limit to transpiration
T_h = upper temperature limit to transpiration
θ_w = wilting point soil moisture content

It has been a common practice, especially for aerodynamically rough vegetation, to optimise the parameters for Equation 3 (a_1-a_5) using climatic variables measured above the surface, assuming that the efficient coupling of vegetation and atmosphere produces shallow gradients, and therefore the climate at reference height is an acceptable estimate of climate at canopy level. This assumption was used for the calibration of Reserva Ducke by Dolman *et al.* (1991), whose parameters are used here for comparison with those for Reserva Jaru. More recently, and especially for smoother surfaces, it has been shown that estimates of canopy climate, derived by downward extrapolation of reference height climate, yield a parameterization which is much less site specific and less height dependant, and has minimal loss of accuracy in estimating transpiration (Wright *et al.*, 1995). However, an additional complexity arises, described more fully by Wright *et al.* (1995), when applying this method to estimate evaporation, whereby prior knowledge of evaporation and heat flux is required to estimate the surface climate. Bearing in mind the computational economy required by GCM modellers, Wright *et al.*, 1995 showed that this problem can be overcome by limited and minimal iteration.

In this study we use field estimates of bulk surface conductance to calibrate a 'Jarvis-type' conductance model for the two sets of climate data: AWS measured at the reference height and canopy height climate calculated from flux gradient relationships derived for each site. Calibration is by non-linear least squares optimisation (Marquardt, 1963) of the predicted conductances calculated by Equation 3, with respect to the 'observed' conductances from Equation 2. Up to five parameters are evaluated (a_1-a_5) depending upon the need to include soil moisture or temperature. Wilting point soil moisture θ_w was estimated from the soil moisture measurements at each site, and the temperature limits to transpiration, T_l and T_h, were fixed at 0°C and 45°C respectively. Although potentially important, Wright *et al.* (1995) found that for pasture the upper limit to transpiration T_h could not be readily optimised and did not greatly affect the final optimum above a value of about 45°C. The value of 45°C is unfortunately rather low for completely satisfactory optimisation with respect to forest canopy climate — Dolman *et al.* (1991) used a value of 40°C — as there was a tendency in this study for the temperature parameter to approach this value. However, T_h was left at 45°C as there was little justification for raising the value on physiological grounds and little to be gained by adding T_h to the optimised parameters.

RESULTS

Field measurements of bulk surface conductance of the pasture at Fazenda Dimona during the 1990 and 1991 field campaigns have already been published (Wright *et al.*, 1992; Wright *et al.*, 1995). Similarly, Figure 4 shows typical diurnal trends in pasture and forest bulk surface conductances from wet and dry periods, ten weeks apart, during Mission 4/5. There is a noticeable reduction in conductance both for the

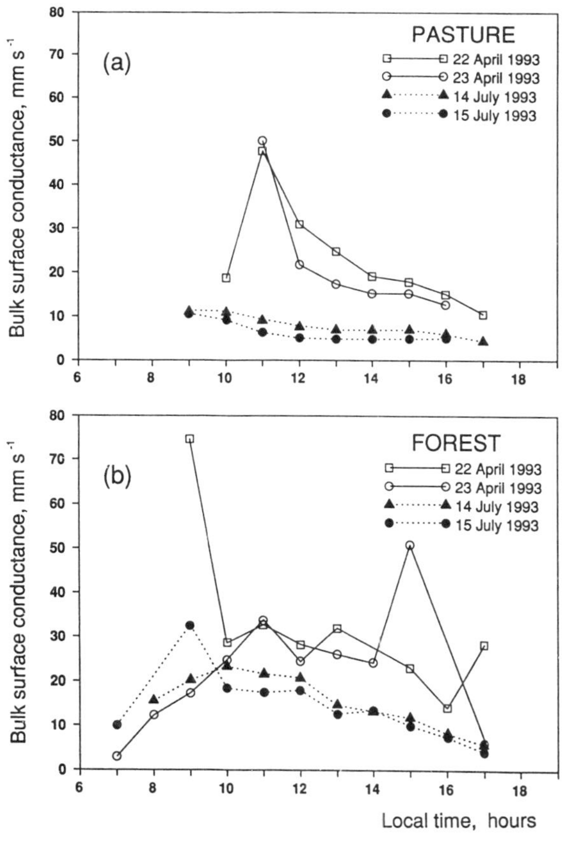

Figure 4 Examples of bulk surface conductances for two pairs of days representing the wet season (April 1993) and dry season (July 1993) at Ji-Paraná, at both (a) pasture and (b) forest sites

forest and the pasture, which can be attributed to the increase in specific humidity deficit over the period. The lowering of conductances in the pasture is enhanced by the additional stress imposed by a decrease in soil moisture.

PARAMETER OPTIMISATION

Table 2 gives the results of the non-linear optimisation of controlling parameters in Equation 3, (a_1-a_5), for pasture and forest sites and for various combinations of data and parameter set. Observed and predicted values of bulk surface conductance are shown in Figure 5 for the Ji-Paraná pasture and forest sites, both with and without the temperature dependence function. Whereas a small improvement in goodness of fit can be seen when the temperature function is excluded from the pasture model (Figure 5a and b), the forest model shows the opposite trend (Figure 5c and d). Also shown is an example of the reduced accuracy of fit associated with the use of reference height climate data (Figure 5e). The example shown is for pasture, where the difference between reference and canopy climate is large, but r^2 values in Table 2 show a similar result for the forest, although less severe. It should be noted that some of the improvement in goodness of fit, when using surface climate, may be related to use of the heat and water vapour fluxes in the calculation of both canopy climate and surface conductance. Nevertheless, the use of canopy climate retains the advantage of being more independent of the site specific aerodynamic characteristics.

For the pasture, the large and variable values for the solar radiation parameter in Table 2 are misleading, as values over 10,000 give an almost linear response to radiation and their influence upon the overall transpiration model is very similar. The specific humidity deficit parameters derived from surface climate at both the Manaus and Ji-Paraná sites are very similar. However, as humidity deficit is a dominant parameter in this model, the agreement might be expected, considering monocultures of similar grasses at both sites.

Examples of the functional form of the fitted lines for Mission 4/5 are shown for the Ji-Paraná pasture and forest (Figures 6 and 7), together with the hourly data on which the functions were optimised. The example for pasture (Figure 6 uses parameters derived from estimated canopy climate during Missions 3 and 4/5, and together with Table 2, includes the first published field values of an optimum temperature for grass transpiration. This has probably been possible because of the large range of temperatures encountered at this site: 16.8 - 33.1 °C (see Table 1). It is particularly useful to identify values which are consistent with those previously suggested (Wright *et al.*, 1995) with respect both to reference and to canopy climate. This permits the use of a universal model compatible with the requirements of forest transpiration modelling. However, the inclusion of the temperature function for grass pasture only has a small affect on the performance of the model and does not improve the optimised model fit. Kim and Verma (1991) also excluded temperature from their study of C_4 grassland in Kansas because of temperature insensitivity and the uncertainty in the optimum temperature for transpiration.

Although the fit for the forest site is fairly good (r^2=0.855), Figure 7 is a good

Table 2 Optimised parameters

Model	Mission	$g_{s,Max}L^*$ a_1	f(S) a_4	f(D) a_2	f(T) a_3	f(θ) a_5	r^2
Fazenda Dimona (Wright *et al.*, 1995)							
Surf (with T)	1 & 2	27.1	2922	.0670	28.5	.387	.909
Surf (no T)	▲1 & 2	27.1	2423	.0686	-	.391	.915
Ref (with T)	1 & 2	no solution					
Ref (no T)	1 & 2	30.7	470	.0369	-	.428	.817
Fazenda Nossa Senhora							
Surf (with T)	3	44.0	38528	.0869	41.1	.234	.852
	4 & 5	43.2	108934	.0592	23.5	.272	.664
	#3,4 & 5	43.3	19346	.0763	27.1	.254	.747
Surf (no T)	3	45.0	28713	.0882	-	.236	.897
	4 & 5	42.2	107808	.0757	-	.262	.820
	§3,4 & 5	43.0	17283	.0821	-	.252	.849
Ref (with T)	3	17.5	364	.0788	26.3	.240	.731
	4 & 5	36.4	768	.0846	18.9	.284	.454
	●3,4 & 5	32.0	654	.0922	22.9	.262	.559
Ref (no T)	3	17.8	367	.0780	-	.240	.772
	4 & 5	37.1	846	.1186	-	.250	.671
	3,4 & 5	33.1	671	.1127	-	.259	.562
Reserva Ducke (from Dolman *et al.*, 1991)							
Ref (with T)	» -	20.8	250	.064	30.2	-	
Reserva Jaru							
Surf (with T)	✱3,4 & 5	80.1	3916	.1248	43.1	-	.855
Surf (no T)	▫3,4 & 5	54.6	4500	.1141	-	-	.826
Ref (with T)	❖3,4 & 5	65.2	743	.1064	44.6	-	.704
Ref (no T)	3,4 & 5	39.4	765	.0956	-		.681

#	illustrated in Figures 5a, 6 and 8
§	illustrated in Figure 5b
✱	illustrated in Figures 5c and 7
▫	illustrated in Figure 5d
●	illustrated in Figure 5e
▲	illustrated in Figure 8
»	illustrated in Figure 9
❖	illustrated in Figure 9

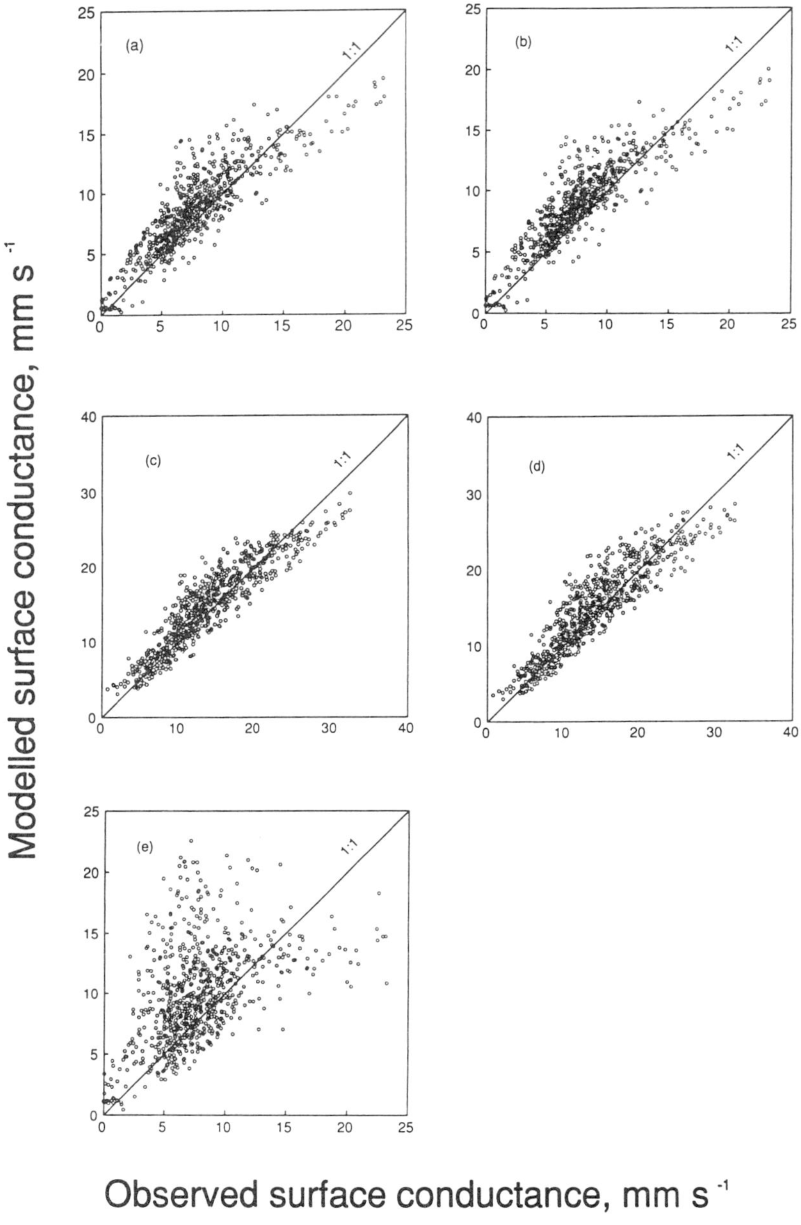

Figure 5 Observed and predicted bulk surface conductances for the Ji-Paraná pasture (a, b and e) and forest (c and d), using models which include a temperature dependence and use canopy climate (a and c), exclude a temperature dependence using canopy climate (b and d) and include a temperature dependence using reference height climate (e).

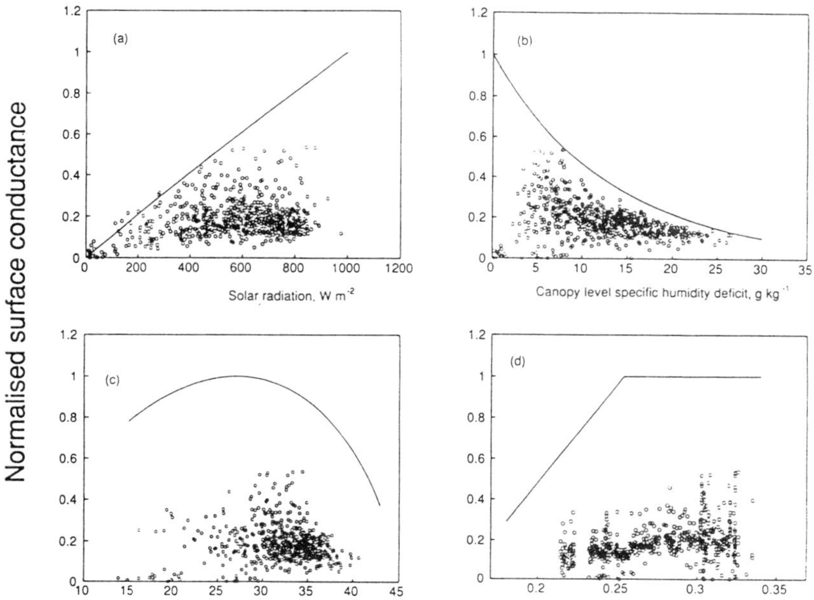

Figure 6 Ji-Paraná pasture: environmental control functions (lines), using optimised parameters a_1-a_5, in Eqs. 7-11. Also shown are the hourly surface conductance data from which the parameters were derived.

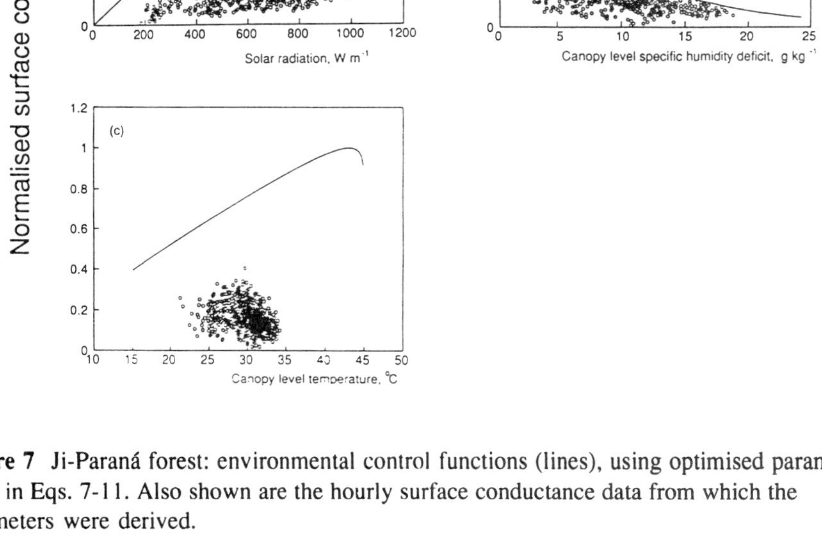

Figure 7 Ji-Paraná forest: environmental control functions (lines), using optimised parameters a_1-a_4, in Eqs. 7-11. Also shown are the hourly surface conductance data from which the parameters were derived.

example of how the application of a 'Jarvis-type' model may derive misleading parameters at some sites and how the method of application may have contributed to the seemingly high value of $g_{s,Max}L^*$. Compared to the pasture result, only a small number of data points approach the function line, with the maximum conductance parameter, $g_{s,Max}L^*$, over twice that of the highest measured value. A trial, in which $g_{s,Max}L^*$ was given various fixed values, showed that over the range of $g_{s,Max}L^*$ from 60 to 90 s m^{-1}, the remaining parameters could be fitted without a marked reduction in r^2. This is evidence of collinearity between maximum surface conductance and the other optimised parameters.

Figures 8 and 9 are a comparison of the shape of the climate functions from the pairs of pasture and forest sites respectively, and emphasises the similarity in the pasture functions and the dissimilarity for the forest. The comparison uses the calibration for pastures which omits temperature dependence, and the forest calibration, which uses reference height climate. These choices are necessary for compatibility with the previously published calibrations for the Manaus pasture and forest sites (Wright *et al.*, 1995 and Dolman *et al.*, 1991).

TRANSPIRATION PREDICTION WITH MINIMUM SITE INFORMATION

As a test of the performance of this model to represent Amazonia as a whole, the errors associated with predicting transpiration at an uncalibrated site were investigated. The parameters calibrated for the Manaus pasture and forest sites (Wright *et al.*, 1995 and Dolman *et al.*, 1991 respectively) were used to predict the evaporation at the Ji-Paraná sites, assuming that only hourly AWS data and basic site details for Ji-Paraná were available.

Hourly values of net long-wave radiation, soil heat flux, temperature, specific humidity deficit and wind speed were used from the AWS located at each of the Ji-Paraná sites. Aerodynamic conductance $g_{a,v}$ was calculated using the wind speed u and a simple relationship to relate canopy height (h_c) to zero plane displacement d and roughness length, z_0: $d=0.61h_c$ and $z_0=0.09h_c$ for the pasture (Wright *et al.*, 1992) and $d=0.86h_c$ and $z_0=0.06h_c$ for the forest (Shuttleworth, 1989). These values were then used in the momentum flux equation

$$\frac{1}{g_{a,v}} = \frac{\left[A^2 + A\left(\ln\left[\frac{z_0}{z_h}\right] + \Psi_m - \Psi_h\right)\right]}{k^2 u} \tag{12}$$

where

$$A = \ln\left[\frac{z-d}{z_0}\right] \tag{13}$$

and z_h is the roughness length for heat fluxes. The integrated stability parameters, Ψ_m and Ψ_h, were calculated by iteration of the stability correction formulae (Arya, 1988) and $\ln[z_0/z_h]$ is given a value of 2.0 for pasture and 1.5 for forest (Verma 1989).

All parameters were used unmodified except for the soil moisture parameter a_5.

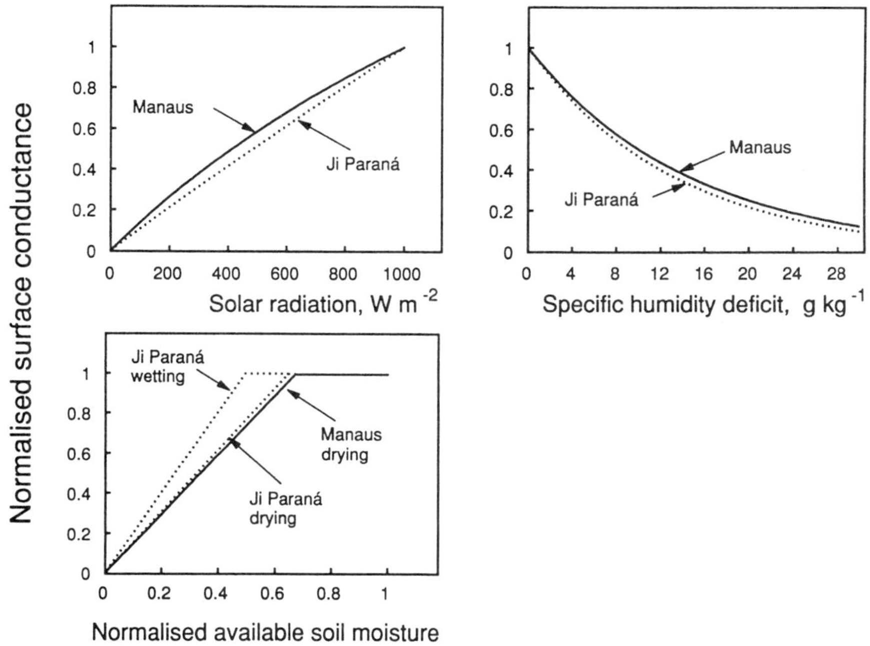

Figure 8 Optimised environmental control functions for the Ji-Paraná and Manaus pasture sites.

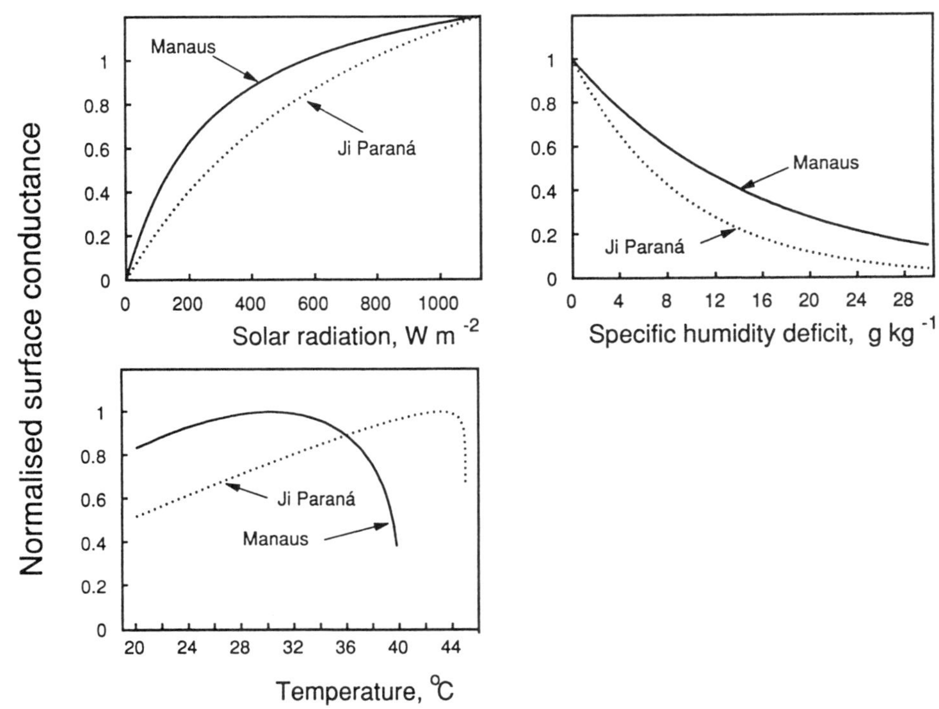

Figure 9 Optimised environmental control functions for the Ji-Paraná and Manaus forest sites.

Below this value of moisture content the surface conductance is linearly attenuated, reaching zero at the wilting point θ_w. To account for the considerable difference between the soils at the two sites, the soil moisture parameter for use at Fazenda Nossa Senhora was calculated to be at the same relative position with respect to the wilting point and field capacity moisture contents at Fazenda Dimona. That is, given the wilting point and field capacity at Fazenda Dimona of 0.33 and 0.42 respectively, the soil moisture parameter, $a_{5,\mathrm{Manaus}}=0.391$, is at 68% of available soil moisture content. The wilting point and field capacity at Fazenda Nossa Senhora are taken as 0.15 and 0.32 respectively, yielding a parameter value of $a_{5,\mathrm{Ji\text{-}Parana}}=0.265$ at 68% of available soil moisture. The leaf area index was left unmodified for reasons that are discussed below.

Figure 10 shows the predicted values of hourly transpiration plotted against the field measurements for the pasture and forest sites. Both applications of the Manaus parameters at Ji-Paraná show an underestimation of transpiration which is discussed below.

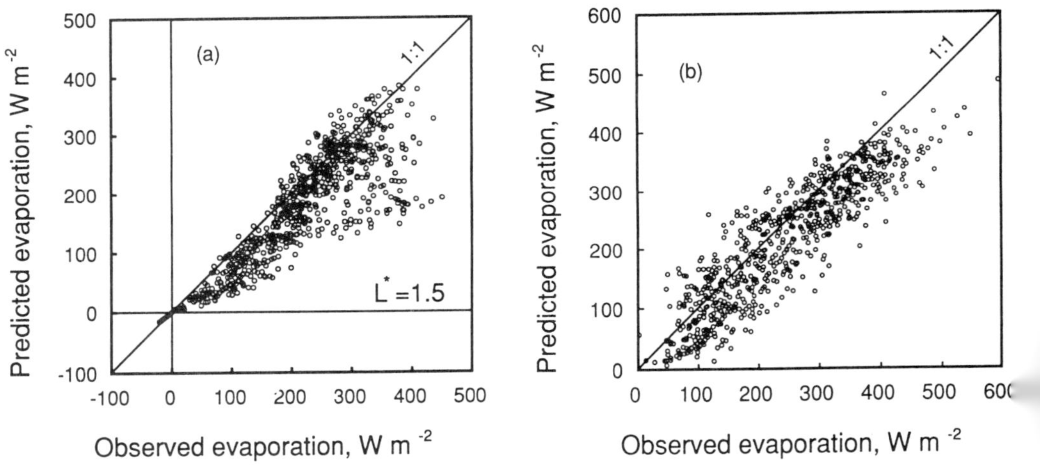

Figure 10 Transpiration at the Ji-Paraná sites predicted using Manaus site parameters and Ji-Paraná climate data, plotted against observations at the Ji-Paraná sites: (a) pasture and (b) forest

DISCUSSION

PASTURE CONDUCTANCE PARAMETERS

The parameter values derived for stomatal response for the grass at Fazenda Nossa Senhora highlight some important trends and support the tentative conclusions concerning the role of temperature and canopy climate suggested in previous work. The use of surface climate produces a stability of result which is not found in the parameters derived from reference height climate (see Table 2). Indeed the use of surface climate has a more profound effect on the result than the inclusion of a temperature component in the model. Also, in contrast to the forest results, the smoother surface of the pasture has smaller aerodynamic conductances, giving surface conductances that are less sensitive to the assumptions in the calculation of momentum flux and stability.

The soil moisture parameter is independent of the problems of diurnal interdependence of parameters (see below). However, this does not suggest that there is no diurnal soil moisture influence, but is a consequence of weekly data which can only relate to the seasonal trend in moisture stress to the plants. In all cases where Mission 3 results are separated from the Mission 4/5 values, the soil moisture parameter is typically 0.03 lower for Mission 3. This difference agrees with the hysteresis associated with seasonally rising (Mission 3) and falling (Mission 4/5) soil moisture. When expressed in relation to the wilting point and field capacity the difference is more marked (see Fig. 6c): upon wetting, the plants cease to experience stress at 50% of available soil moisture and upon drying become stressed at around 70%. In addition to the hysteresis in soil wetting and drying processes on the small scale, the model does not attempt any vertical distribution of moisture or roots within the specified root zone: wet soil over dry in Mission 3 (wetting) and dry soil over wet in Mission 4/5 (drying). The soil moisture parameter for Manaus pasture was derived during a drying sequence and agrees well with the Ji-Paraná Mission 4/5 (drying) parameter: expressed as a percentage of available soil moisture: 67.7% at Fazenda Dimona and 65.8% at Fazenda Nossa Senhora. Arguably the best overall value for the soil moisture parameter should be 58% of available soil moisture, being the mean of the wetting and drying parameters derived here.

FOREST CONDUCTANCE PARAMETERS

There is a clear improvement when fitting observed to predicted forest conductances with the use of canopy climate and, to a lesser extent, the inclusion of the temperature dependence function. However, it is important to note that the possibility of significant parameter interdependence requires that these sets of empirical parameters should remain as sets when used at uncalibrated sites. Clearly, it would be unwise to combine parameters from different calibrated sites in an attempt to describe vegetation that has a combination of similarities with those sites. Furthermore, in view of the rough forest canopy and the related sensitivity to atmospheric stability corrections, it is clearly advisable to calculate the momentum flux in a similar way to that used in the calibration analysis.

TRANSPIRATION PREDICTION WITH MINIMUM SITE INFORMATION

Pasture

Figure 10a shows that when the parameters for the Manaus pasture site are used to predict transpiration at the Ji-Paraná pasture, the overall effect was an underestimate when compared to the measured values. In this particular case, where all parameters

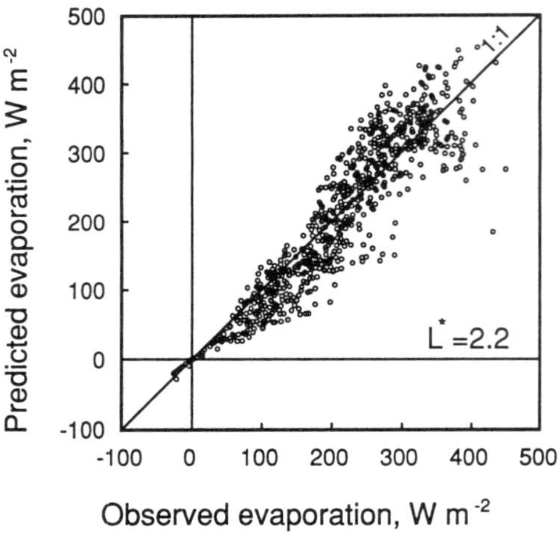

Figure 11 Transpiration for the Ji-Paraná pasture site predicted using Manaus site parameters, Ji-Paraná climate data and using a Leaf Area Index of 2.2 to achieve the best fit with observations.

for both sites are similar except for the maximum conductance, $g_{sMax}L^*$, it is reasonable to look to this parameter to explain the difference. Leaf area index was estimated as 1.6 for Mission 3 and 2.6-3.9 for Mission 4 and 5. Clearly the leaf area index at Manaus, 1.2, is too low to represent the Ji-Paraná pasture, yet values as high as 3.9 at the Ji-Paraná will produce a significant overestimate.

Although the total pasture evaporation will include a small contribution from the soil surface, when this component is ignored it is only necessary to raise L^* to 2.2 to effect an agreement between the observed and model transpiration, see Figure 11. This is consistent with the suggested proportionality between bulk surface conductance and L^* which is reported to be valid up to a value of about $L^* = 3$ (Schulze *et al.*, 1994; Rosenberg, *et al.*, 1983), at which value the structure of the canopy begins to feedback upon itself by self-shading, and conductance tends towards a maximum values or at least to one that is insensitive to L^* (Stewart and Verma, 1992). It is possible that in these pastures, where the grass takes the form of sparse but dense tufts, the value of L^* at which bulk surface conductance ceases to be a linear function of L^* is lower than three.

Forest

It is less easy to interpret the underestimation by the Manaus forest parameters when used for the Ji-Paraná forest because there are larger differences between several of the parameters at the two sites and greater natural differences between the two forests. Leaf area index at the Ji-Paraná forest is 4.7 (Roberts *et al.* 1996), 18% less than the 5.7 at Manaus (McWilliam *et al.*, 1996), and is not consistent with the maximum surface conductance at Ji-Paraná - three times that of Manaus. However, the large difference in maximum conductance is substantially compensated for by the difference in the humidity deficit response producing the fairly good prediction of evaporation when used at the non-calibration site. Although the multivariate optimisation may contribute to the interdependence between maximum conductance and the humidity parameter, there is evidence in the literature (Hall and Roberts, 1989) — albeit for a temperate forest — that tree species with a high maximum conductance are more reactive to humidity deficit. The species differences between the two tropical forests presented here could account for the different derived parameter values. The within-set compensation between empirical parameters shown here further emphasises the need to retain the parameters in sets.

LIMITATIONS OF JARVIS-TYPE MODELS

Comparison of the various parameters derived in this study with the results of other similar studies (Adams *et al.* 1991; Stewart and Verma, 1992; Dolman *et al.*, 1991; Wright *et al.*, 1995) clearly shows that most of the empirically derived parameters for Jarvis-type models have little meaning outside of their specific application. For example, whereas the soil moisture parameter has clear relevance with respect to measured soil moisture status, maximum surface conductance is an 'effective' parameter which is disassociated from canopy structure and stomatal physiology, and includes all hidden errors and processes that are proportional to measured conductance. Interdependence between parameter values cannot be ignored when optimisation is used, and its existence weakens the conclusions that may be drawn from the different solutions. For example, there are consistently higher values for the solar radiation parameter when surface climate is used rather than that at the reference height. This result is illogical, but is unavoidable when several of the climate elements, including those that vary with height, have a correlated diurnal pattern: these are solar radiation, specific humidity deficit and temperature.

Current new work to enhance GCM surface schemes with explicit models of plant growth and leaf gas exchange (Friend, 1991; Lloyd *et al.*, 1995; Monteith, 1995) are likely to remove some of the problems inherent in Jarvis-type models of surface conductance. These models are working towards specifying independent component processes, such as photosynthesis and light response, and will therefore remove empiricism and the consequent dependence upon multivariate optimisation.

CONCLUSIONS

A 'Jarvis-type' model of bulk surface conductance, representing stomatal response in conductance-based models of total transpiration, has been satisfactorily calibrated for two tropical vegetation types, pasture and forest, in Rondônia, south-west Amazonia. When these results are compared with similarly derived calibrations for pasture and forest in central Amazonia, it is possible to comment on the viability of using these calibrations to represent pasture and forest over the whole region.

This study has shown considerable coherence between calibrations for pasture and forest sites and has highlighted the key areas where care is needed when attempting to predict transpiration at uncalibrated sites. For pasture, the environmental control parameters agree particularly well, but more work is needed to understand the relationship between maximum bulk conductance and leaf area index. At the forest sites there appear to be compensating processes, creating large differences in individual parameters, but which combine to derive similar overall conductances. This means that good transpiration estimates may be derived for an uncalibrated forest site, but the model may be implying inappropriate component processes. Clearly, until more work is done to understand the physiological differences between types of tropical forest, it is particularly important to retain these groups of empirical parameters in sets.

Considering that there is generally good agreement between most conductance parameters so far derived for tropical pasture and forest, and that the main differences can be partially explained in terms of known structural and physiological site differences, it can be concluded that the parameters presented here can be used with some confidence to represent pasture and forest within Amazonia.

REFERENCES

Adams, R.S., Black, T.A. and Fleming, R.L. 1991. Evapotranspiration and surface conductance in a high elevation, grass-covered forest clearcut. *Agric. For. Meteorol.*, **56**, 173-193.

Arya, S.P., 1988. *Introduction to micrometeorology*. Academic Press (London). pp 307.

Dolman, A.J., Gash, J.H.C., Roberts, J. and Shuttleworth, W.J. 1991. Stomatal and surface conductance of tropical rainforest. *Agric. and Forest Meteorol.*, **54**, 303-318.

Eden, M.J., McGregor, D.F.M. and Vieira, N.A.Q. 1990. The Maracá rainforest project: III. Pasture development on cleared forest land in northern Amazonia. *Geograph. J.*, **156**, 283-296.

Friend, A.D. 1991. Use of a model of photosynthesis and leaf microenvironment to predict optimal stomatal conductance and leaf nitrogen partitioning. *Plant, Cell and Env.*, **14**, 895-905.

Gash, J.H.C., Nobre, C.A., Roberts, J.M. and Victoria, R. 1996. An overview of the ABRACOS project. This volume.

Grace, J., Lloyd, J., McIntyre, J., Miranda, A.C., Meir, P., Miranda, H.S., Moncrieff, J., Massheder, J., Wright, I.R. and Gash, J.H.C. 1995. Fluxes of carbon dioxide and water vapour over an undisturbed tropical rain forest in south-west Amazonia. *Global Climate Change*, **1**, 1-12.

Hall, R.L. and Roberts J.M. 1989. Hydrological aspects of new broad-leaf plantations. *SEESOIL*, 6, 2-38.

Hodnett, M.G., Tomasella, J. and Oyama, M. 1996. Comparisons of long-term soil water storage under pasture and forest in three areas of Amazonia. This volume.

Huntingford, C., 1995. Non-dimensionalisation of the Penman-Monteith model. *J. Hydrol.*, **170**, 215-232.

Jarvis, P.G. 1976. The interpretation of the variations in leaf water potential and stomatal conductances found in canopies in the field. *Phil. Trans. Roy. Soc., London, Ser. B*, **273**, 593-610.

Lean, J., Bunton, C.B., Nobre, C.A. and Rowntree, P.R. 1996. The simulated impact of Amazonian deforestation on climate using measured ABRACOS vegetation characteristics. This volume.

Lloyd, J., Grace J., Miranda, A.C., Meir, P., Wong, S.C., Miranda, H.S., Wright, I.R., Gash, J.H.C. and McIntyre, J. 1995. A simple calibrated (SiC) model of Amazon rainforest productivity based on leaf biochemical properties. *Plant, Cell and Env. In press.*

Marquardt, D.W. 1963. An algorithm for least squares estimation of nonlinear parameters. *J. Soc. Ind. Appl. Math.*, **11**, 431-441.

McWilliam, A-L. C., Roberts, J.M., Cabral, O.M.R., Leitao, M.V.B.R., de Costa, A.C.L., Maitelli, G.T. and Zamparoni, C.A.G.P. 1993. Leaf area index and above-ground biomass of *terra firme* rain forest and adjacent clearings in Amazonia. *Functional Ecology*, **7**, 310-317.

McWilliam, A-L. C., Cabral, O.M.R., Gomes, B.M., Esteves, J.L. and Roberts, J.M. 1996. Forest and pasture leaf-gas exchange in south-west Amazonia. This volume.

Monteith, J.L. 1965. Evaporation and environment. Symp. Soc. *Exp. Biol.*, 19, 205-224.

Monteith, J.L. 1995. A reinterpretation of stomatal responses to humidity. *Plant Cell and Env.*, **18**, 357-364.

Moore, C.J. and Fisch, G. 1986. Estimating heat storage in Amazonian tropical forests. *Agric. For. Meteorol.*, **38**, 147-169.

Noilhan, J. and Planton, S. 1989. A simple parameterization of land surface processes for meteorological models. *Mon. Weath. Rev.*, **117** 536-549.

Rosenberg, N.J., Blad, B.L. and Verma, S.B., 1983. *Microclimate: The biological environment.* Wiley, New York, 495 pp.

Roberts, J.M., Cabral, O.M.R. and de Aguiar, L.F. 1990. Stomatal and boundary-layer conductances in an Amazonian *terra firme* rain forest. *J. Appl. Ecol.*, **27**, 336-353.

Roberts, J.M., Cabral, O.M.R., da Costa, J.P., McWilliam, A-L.C. and Sa, T.D.A. 1996. Leaf area index, biomass and physiological responses in Amazonian forest and pasture: an overview. This volume.

Sellers, P.J., Mintz. Y, Sud, Y.C. and Dalcher, A. 1986. A simple biosphere model (SiB) for use within general circulation models. *J. Atmos. Sci.*, **43**: 505-531.

Schulze, E-D., Kelliher, F.M., Körner, C., Lloyd, J. and Leuning, R. 1994. Relationships among maximum stomatal conductance, ecosystem surface conductance, carbon assimilation rate and plant nitrogen nutrition - a global ecosystem scaling exercise. *Annual Review of Ecology and Systematics*, **25**.

Shuttleworth, W.J. 1989. Micrometeorology of temperate and tropical forest. *Phil. Trans. R. Soc. Lon. B* **324**, 299-334.

Shuttleworth, W.J., Gash, J.H.C., Lloyd, C.R., McNeil, D.D., Moore, C.J. and Wallace, J.S., 1988. An integrated micrometeorological system for evaporation measurement. *Agric. and Forest Meteorol.*, **43**, 295-317.

Stewart, J.B. 1988. Modelling surface conductance of pine forest. *Agric. For. Meteorol.*, **43**, 1 9 - 35.

Stewart, J.B. and Verma, S.B. 1992. Comparison of surface fluxes and conductivities at two contrasting sites within the FIFE area. *J. Geophys. Res.*, **97**, 18623-18628.

Tomasella, J. and Hodnett, M.G. 1996. Measurements and modelling of soil hydraulic properties of oxisols in central Amazonia. This volume.

Uhl, C., Buschbacher, R. and Serrão, E.A.S. 1988. Abandoned pastures in eastern Amazonia: 1. Patterns of succession. *J. Ecol.*, **76**, 663-681.

Verma, S.B. 1989. Aerodynamic resistances to transfers of heat, mass and momentum in estimation of areal evaporation. In: T.A.Black, D.L. Spittlehouse, M.D. Novak and D.T. Price (Eds.). *IAHS Publication 177*, IAHS Press, Wallingford, UK.

Wright, I.R., Gash, J.H.C., da Rocha, H.R., Shuttleworth, W.J., Nobre, C.A., Maitelli, G.T., Zamparoni, C.A.G.P. and Carvalho, P.R.A., 1992. Dry season micrometeorology of central Amazonian ranchland. Quart. J. R. Meteorol. Soc., 118: 1083-1099.

Wright, I.R., Manzi, A.O. and da Rocha, H.R. 1995. Surface conductance of Amazonian pasture: model application and calibration for canopy climate. *Agric. For. Meteorol.*, **75**, 51-70.

Wright, I.R., Nobre, C.A., Tomasella, J, da Rocha, H.R., Roberts, J.M., Vertamatti, E., Culf, A.D., Alvalá, R.C.S., Hodnett, M.G. and Ubarana, V. 1996. Towards a GCM land-surface parameterization for Amazonia. This volume.

RESUMO

São apresentados parâmetros otimizados, em um modelo de tipo "Jarvis", da resposta estomática "bulk" de dois sítios de pastagem e floresta na Amazônia brasileira. Condutâncias de superfícies "bulk" são derivadas a partir de mais de 800 horas de medidas selecionadas de evaporação e clima em cada sítio, durante as missões intensivas de 1992 e 1993. Nos sítios de pastagem, a máxima condutância de superfície, por unidade de área, foi consistente com uma relação linear com o índice de área foliar (IAF), até o valor estável de 43 mm s^{-1}, quando o índice de área foliar era maior que cerca de 2,0. Nos sítios de floresta, a máxima condutância foi de mais difícil interpretação, sendo 21 mm s^{-1} no sítio de Manaus, e entre 55 e 80 mm s^{-1} no sítio de Rondônia, dependendo da inclusão de uma função de dependência da temperatura. Limitações do modelo de tipo "Jarvis" são evidentes e são discutidos.

O efeito do uso de um nível de referência da Estação Meteorológica Automática é comparada com um método de base física mais rigoroso o qual usa o clima estimado do dossel. O primeiro método é simples de ser aplicado, mas somente se aplica para um sítio específico e é de uso limitado; o último facilita a comparação de parâmetros entre os sítios e permite estimativas mais confiáveis de evaporação nos sítios onde o modelo não tenha sido calibrado. Os parâmetros dos sítios de pastagem e floresta de Manaus fornecem resultados satisfatórios quando utilizados para estimar a transpiração nos sítios de Rondônia: em ambas, floresta e pastagem, a transpiração foi ligeiramente subestimada. Pode-se encontrar concordância para a pastagem ao se utilizar um ajuste para o índice de área foliar.

27 Calibration and use of the SiB2 model to estimate water vapour and carbon exchange at the ABRACOS forest sites

H.R. da ROCHA[1], P.J. SELLERS[2], G.J. COLLATZ[2], I.R. WRIGHT[3] and J. GRACE[4]
[1]*Instituto Nacional de Pesquisas Espaciais, Brazil*
present affiliation: Universidade de São Paulo, Brazil
[2]*Goddard Space Flight Center, USA*
[3]*Institute of Hydrology, Wallingford, UK*
[4]*University of Edinburgh, UK*

INTRODUCTION

Water vapour and CO_2 exchanges over vegetated surfaces are closely coupled mechanisms with strong dependences on interacting environmental factors and biological processes. Stomatal behaviour is affected by incident radiation, air temperature, humidity, CO_2 partial pressure, and soil moisture stress, all factors which affect hydroactive leaf cell processes of ion transport. The parameterization of latent heat and carbon dioxide flux in the SiB2 model (Sellers *et al.*, 1995) is an attempt to describe key canopy processes bearing on the energy partition at regional scales. SiB2 represents a significant improvement over SiB (Sellers *et al.*, 1986) in that the realism of surface energy, water and carbon fluxes was enhanced and the number of model parameters was reduced. The model employs the approach of Farquhar *et al.* (1980) modified by Collatz *et al.* (1991) for C3 plants: CO_2 assimilation by leaves is a rate-limited function of enzyme kinetics— specifically the amount and cycle time of the carboxylating enzyme rubisco— and electron transport, the latter dependent on photosynthetically active radiation (PAR) and its absorption by chlorophyll. Gross leaf assimilation rate is constrained by three important factors: the assimilation rates defined by the photosynthetic enzyme system, w_c; the amount of PAR absorbed by leaf chlorophyll, w_e; and the capacity of the leaf to export or utilize the products of photosynthesis, w_s. SiB was calibrated for Amazonian tropical forest by Sellers *et al.* (1989) and used in a number of numerical experiments coupled to atmospheric models, e.g. Sato *et al.* (1989), Nobre *et al.* (1991) and Rocha *et al.* (1996). The calibration of the SiB2 model for Amazonian forest should lead to further improvements in the simulation of land surface-atmospheric interaction in the tropics.

Amazonian deforestation and climate. Edited by J.H.C. Gash, C.A. Nobre, J.M. Roberts and R.L. Victoria.

Net ecosystem exchange (NEE) or total carbon flux is the balance between the gross primary carbon assimilation by the canopy, A, and respiration by the above ground vegetation, R_c, and the soil and root respiration, R_s,

$$NEE = R_s + R_c - A \qquad (1)$$

where the net canopy assimilation, A_n, takes account of gains and losses in the above ground vegetation by:

$$A_n = A - R_c \qquad (2)$$

so $$NEE = R_s - A_n \qquad (3)$$

Data collected by Grace *et al.* (1996) for a forest site in southern Amazonia during the 1992/1993 wet season showed a NEE of -0.27 kg C ha^{-1} h^{-1}, while Fan *et al.* (1990) reported comparable values of -0.25 kg C ha^{-1} h^{-1} in Central Amazonia during the wet season. Soil respiration measurements in the Amazonian forest indicate that soils contribute about 70 to 80% of the total carbon emission to the atmosphere, although the field measurements have shown a great deal of variability: Wofsy *et al.* (1988) reported soil respiration of 1.76 kg C ha^{-1} h^{-1} at the Reserva Ducke site, while Fan *et al.* (1990) has observed larger amounts of 2.22 kg C ha^{-1} h^{-1}.

Model studies of tropical forest evaporation since Shuttleworth (1988) agree on the evergreen character of Amazonian rainforests, a region defined to be where the dry season lasts no longer than three months on average. Culf *et al.* (1995) have more recently shown that a seasonal variation in albedo of between 2% and 4% over three forest sites in Amazonia is significantly correlated and in phase with soil moisture patterns, i.e. depletion in soil moisture takes place simultaneously with higher albedos. Deciduous species in the Amazonian rainforest tend to loose their leaves during the dry season, according to the reports of litter leaf area index for the forest from Roberts *et al.* (1996). We observe this type of phenology in the southern Amazonia region close to the forest boundary particularly with the 'cerrado' savannah type vegetation. The observations of Culf *et al.* (1995) that seasonal variations of albedo are linked with soil moisture suggest a biological control which should be represented in coupled biosphere-atmosphere models, instead of using a fixed albedo, especially in model studies that involve the prediction of drier climates in Amazonia (deforested or not).

This paper uses field eddy correlation data of water vapour, CO_2 and energy fluxes to calibrate the SiB2 model for an Amazonian forest site. The calibrated model was run with two years of data (September 1983 to August 1985) to produce estimates of monthly evapotranspiration which were then compared to those of Shuttleworth (1988). The model was also forced by data collected at the three forest sites of ABRACOS: Reserva Ducke, Reserva Jaru and Reserva Vale do Rio Doce in central, southern and eastern Amazonia respectively (see Gash *et al.*, 1996, and Culf *et al.*, 1996) for 1991 to 1994, thus providing monthly evapotranspiration and carbon fluxes, which allows us to explore the model performance for the three different climatic forcings.

MODEL CALIBRATION

The data used in this study are described by Grace *et al.* (1996) and consist of eddy correlation measurements of water vapour, sensible heat and carbon dioxide fluxes collected in the Reserva Jaru forest site during a period of 44 days in the wet season, 05 May to 22 June 1993.

To make use of the CO_2 flux data in the model calibration exercise, we formulated a model for soil respiration processes (not present in the version published by Sellers *et al.*, 1995). Most soil respiration models (Bonan, 1991; Norman *et al.*, 1992) account for the influence of leaf area index, soil moisture and soil temperature in the upper soil layers. An accurate calibration of the coefficients involved in such detailed formulations would require a long-term data set covering all the seasonal variability of soil respiration, soil moisture and temperature for a tropical forest. Meir *et al.*, (1996) have measured soil respiration, R_s (mol m^{-2} s^{-1}) at the same site and time as the carbon fluxes described above, which were used to fit a simple function on soil temperature, T_s (°C), given as

$$R_s = \exp\,[0.084(T_s - 0.23)] \tag{4}$$

Although simple, Equation (4) has a practical and useful application for the exercise carried out for this paper. The SiB2 model is run in its 'off-line' mode, i.e. forced by *in situ* observations of meteorological hourly average variables, namely: global solar radiation, net radiation, air temperature and humidity, air CO_2 concentration and precipitation. The calibration approach undertaken here involves initialising the model with a specified set of parameters, and assessing the quality of the simulation through the calculation of a specified function F_i, which represents the deviation between a calculated and an observed variable, the index i referring to the time period when the variables are averaged. The set of deviations for each period i is used to calculate the mean deviation error comprising all the simulation period and input to a least-squares minimisation program, ZXSSQ (IMSL, 1984), which calculates the partial derivatives of each deviation with respect to the optimized parameters. The procedure is repeated until the best set of parameters is found, i.e. until each deviation cannot be reduced further. This approach is based on the SiB model calibration procedure described by Sellers *et al.* (1989), and has also been used to calibrate the model for an Amazonian ranchland pasture by Rocha *et al.* (1996).

The calibration exercise was carried out in two stages: firstly, the function F was taken as the deviation between the measured and the calculated daily running averages of carbon fluxes, i.e. net ecosystem exchange, given as

$$F_i = (NEE_c - NEE_m)_i \tag{5}$$

The observed carbon fluxes exhibited diurnal cycles which are sensitive to the stability of the air in and above the canopy: during the night, large amounts of carbon dioxide are trapped within the canopy air space, a situation which is disturbed in the first daylight hours which then leads to oscillations in the observations. The use of daily averages in F_i smoothes these effects but still retains the natural bias of the data; that is, the net positive assimilation through the period. A data set of 44 days was used as input for the model (for details see Grace *et al.*, 1996).

In the first instance, the optimized parameter was the maximum rubisco capacity at canopy top, V_{max}, fixed initially as 60 μmol m^{-2} s^{-1}. The remaining parameters of the momentum and radiation transfer, rainfall interception and soil hydraulic diffusion modules were similar to those described by Sellers *et al.* (1989). After successive iterations there was a significant reduction in the total error (ΣF_i^2) converging to an optimized value of 81.8 μmol m^{-2} s^{-1}. (Table 1). The daily running averages of carbon flux for the two runs are shown in Figure 1 and compared with the measured data. While initial V_{max} did appear to underestimate canopy assimilation, thus showing positive fluxes of net ecosystem exchange, the optimized V_{max} converged to the observation of -0.27 kg C ha^{-1} h^{-1} by the end of the period.

Table 1 Initial and optimized values of the parameter V_{max} (_mol m^{-2} s^{-1}) and respective sum of square error of the optimization function F (kgC ha^{-1} h^{-1})

	initial	*optimized*
V_{max}	60	81.8
(ΣF_i^2)	13.58	0.05

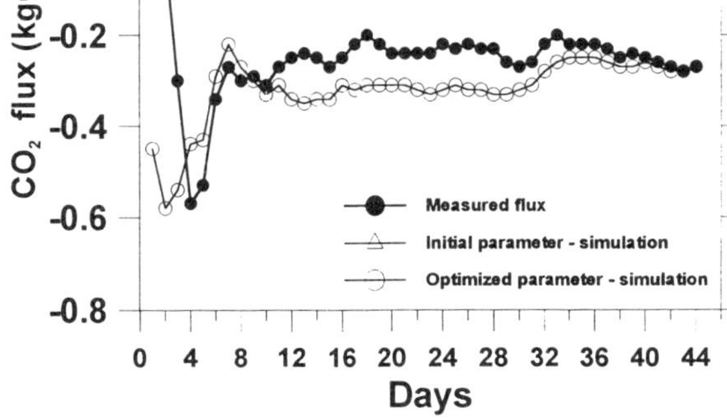

Figure 1 Daily running average of carbon flux calculated by SiB2 model with the initial (Δ) and optimized (O) parameter V_{max} and the associated measured carbon flux (●)in kgC ha^{-1} h^{-1}

Second, a sub-set of seven fine days from the original dataset was used in an attempt to optimize the parameters controlling the greenness factor parameter (fraction of the canopy with green leaves) *G* and the slope of stomatal conductance as a function of canopy assimilation *m* given by

$$g_s = m \frac{A_n}{C_s} H_s p + b \tag{6}$$

where g_s is the leaf stomatal conductance (mol m^{-2} s^{-1}), c_s is the CO_2 concentration at the leaf surface (Pa), h_s is the relative humidity at leaf surface, p is the atmospheric pressure (Pa) and b is an empirical coefficient from observations, equal to 0.01 for C_3 vegetation.

With V_{max} optimized previously, and *m* and *G* initially set to 9 and 0.7979 respectively, the optimization procedure computed the deviation between the calculated and the measured energy fraction, latent heat flux (λE) divided by the sum of sensible (H) plus latent heat flux, weighted by the absolute value of measured latent heat flux, at every time step, given as

$$F_i = [\lambda E_{calc} / (\lambda E + H)_{calc} - \lambda E_{meas} / (\lambda E + H)_{meas}]\, |\lambda E_{meas}| \tag{7}$$

The time step was fixed at one hour. Table 2 gives the initial and optimized values of *m* and *G*, with the respective sum of squared errors in each case. A significant improvement in fitting the observed latent heat flux was achieved with the optimized set, as shown in Figure 2.

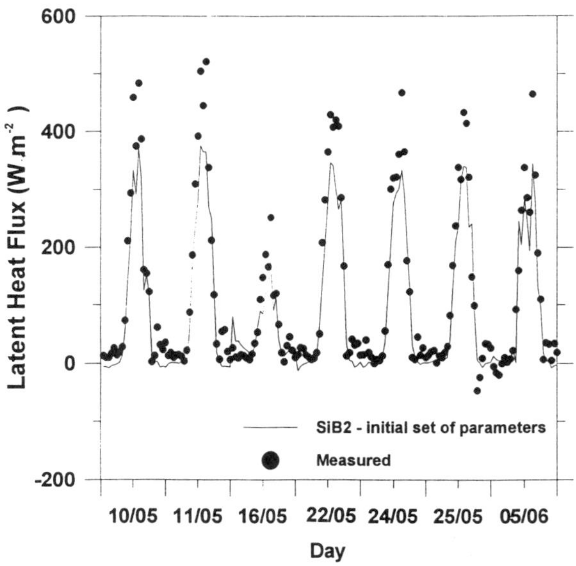

Figure 2 SiB2 model calculated (——) and measured (●) latent heat flux (W m^{-2}) for seven fine days over the Reserva Jaru forest site, using the initial set (above) and the optimised (overleaf) set of parameters *m* and *G*

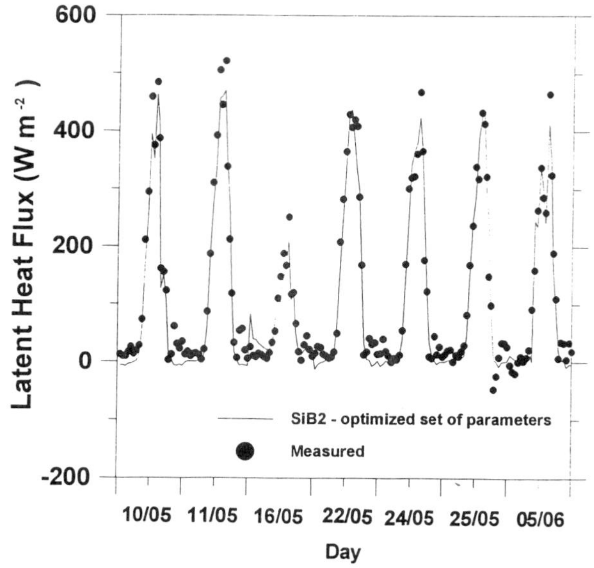

Figure 2 contd.

Table 2 Initial and optimized parameter *m* and *G* and respective sum of square error of the optimization function F (W m^{-2})

	initial	*optimized*
m	9	11.498
G	0.798	0.858
(Fi ^{2})	492	300

WATER VAPOUR AND CARBON DIOXIDE EXCHANGES

Following the optimizations, the calibrated model was run for the two-year period from September 1983 to August 1985, forced by hourly meteorological data collected at Reserva Ducke (Shuttleworth, 1988). The rainfall interception parameters derived by Sellers *et al.* (1989) were used throughout this analysis. The greenness factor, *G*, was varied monthly in the model, being interpolated between 0.89 in March (maximum value) to 0.80 in November (minimum value). A fixed value of air CO_2 concentration of 340 ppm was used (Wofsy *et al.*, 1988), and the time step was of 1 hour.

Figure 3 shows the calculated monthly average evapotranspiration. The integral average evapotranspiration from SiB2 was 3.65 mm day^{-1}, very similar to that of 3.59 mm day^{-1} of Shuttleworth (1988) for the same period. Calculated average daily sensible and soil heat fluxes were 13.9 and 0.1 W m^{-2}, respectively, leading to an

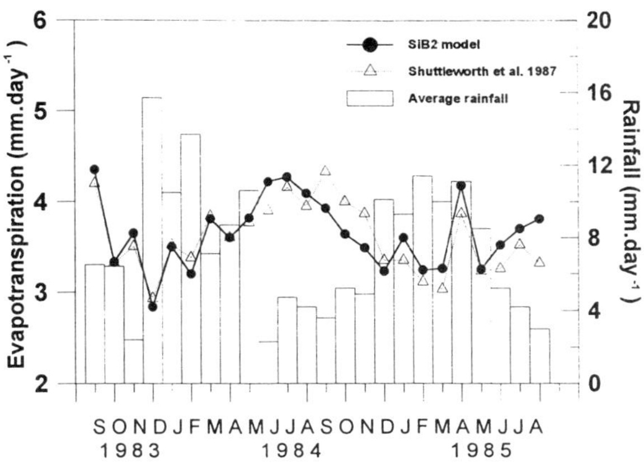

Figure 3 Calculated total evapotranspiration (mm day^{-1}) for the SiB2 model (●) and for Shuttleworth (1988) (Δ) for the period September 1983 to August 1985, with the associated average montly rainfall (bars).

average Bowen ratio of 0.13. Evaporation varied seasonally from 2.8 to 4.3 mm day^{-1}. Simulations closely followed data of Shuttleworth (1988), except during a short period of 3 months (September to November 1984) in which the SiB2 model estimate fell by around 0.5 mm day^{-1}. The forcing precipitation data (also shown in Figure 3) showed a dry period sufficient to cause soil moisture stress as calculated in the model. Concerning the carbon balance, shown in Figure 4, the average net canopy assimilation was -2.73 kg C ha^{-1} h^{-1}, consisting of a gross canopy photosynthetic rate of -3.23 kg C ha^{-1} h^{-1} and the canopy respiration of 0.505 kg C ha^{-1} h^{-1}. Soil respiration was 2.65 kg C ha^{-1} h^{-1}, giving a net ecosystem exchange of -0.074 kg C ha^{-1} h^{-1}. The

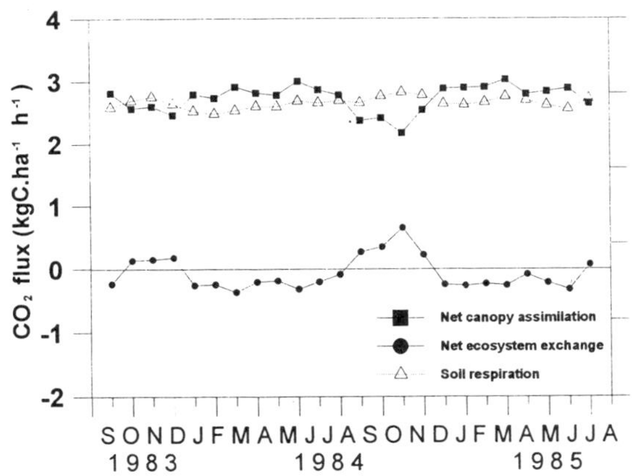

Figure 4 Carbon fluxes (kg C ha^{-1} h^{-1}) calculated by the SiB2 model for the period September 1983 to August 1985: net canopy assimilation (■), net ecosystem exchange (●) and soil respiration (Δ).

calculated net assimilation appeared to peak in the wet to dry season transition, in both June 84 and April 85. The relative minimum occurred in the dry period between September to November 1984, wherein the calculated soil moisture stress caused a small reduction in both transpiration and assimilation.

In a second simulation, the SiB2 model was used to calculate the carbon and surface energy fluxes of the three ABRACOS forest sites, for the periods: January 1991 to May 1994 (Reserva Ducke); September 1991 to September 1994 (Reserva Vale do Rio Doce); and November 1991 to December 1994 (Reserva Jaru). Missing values in the original weather station data used to force the model were discarded. The values of canopy height and leaf area index for Reserva Ducke, Reserva Jaru and Reserva Vale are, respectively, 35; 30 and 33 m (Wright *et al.* (1996a) and 5.7; 4.7 and 5.3 $m^2 m^{-2}$ (Roberts *et al.* 1996). The calibrated parameters and the annual variation in G were similar to the results of the first simulation.

The simulation for Reserva Ducke is shown in Figure 5. A seasonal pattern in evapotranspiration with values peaking in the drier months between August to October, and a minimum in the wet months around March can be seen in the time series. There appears to be a strong link with cloudiness and insolation. The sites all show seasonal variations. As shown in Table 3, the annual total evapotranspiration for Reserva Ducke is equal to that for Reserva Jaru although the latter shows a smaller intra-annual variability in evapotranspiration. The maximum evapotranspiration values are reduced at this site during the dry months mostly due to the large-scale air masses which often penetrate southern Amazonia during this time of year, an event not always observed at the other sites. It is noticeable that the

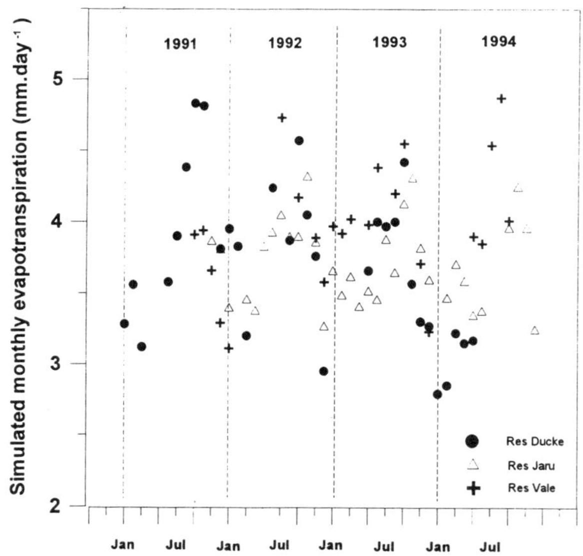

Figure 5 Average monthly evapotranspiration (mm day^{-1}) calculated by the SiB2 model for the three ABRACOS forest sites: Reserva Ducke (circles), Reserva Jaru (triangles) and Reserva Vale (crosses)

Table 3 Average values of evapotranspiration (λE); sensible heat flux (H), net canopy assimilation (A_n), soil respiration (R_s), and net ecosystem exchange (NEE), calculated by the SiB2 model, with the Reserva Jaru parameters, for the periods shown in Figures 5 and 6, for the three forest ABRACOS sites.

	Reserva Ducke	Reserva Jaru	Reserva Vale do Rio Doce
λE (mm day -1)	3.71 ± 0.55	3.71 ± 0.30	3.97 ± 0.45
H (W m^{-2})	24.00 ± 8.09	30.34 ± 6.07	17.63 ± 7.80
A_{n9} (kgC ha^{-1} h^{-1})	2.86 ± 0.13	2.84 ± 0.12	2.90 ± 0.16
R_s (kgC ha^{-1} h^{-1})	2.77 ± 0.09	2.71 ± 0.16	2.75 ± 0.07
NEE (kgC ha^{-1} h^{-1})	-0.09 ± 0.11	-0.11 ± 0.17	-0.15 ± 0.18

largest evapotranspiration rates occur at the Reserva Vale do Rio Doce, where they appear to peak earlier in the year than the other two sites. This lagging peak in the evapotranspiration must be related to the earlier climatological dry season in the eastern part of Amazonia.

The calculated monthly averages of carbon exchange for the three sites, shown in Figure 6, suggest initially a carbon accumulation for the whole period and for most of the months. Taking the integrals (Table 3), net assimilation is largest at Reserva Vale do Rio Doce (2.90±0.16 kgC ha^{-1} h^{-1}), as is evapotranspiration (3.97±0.15 mm): these values are not totally comparable to the other sites because there were many missing data during the wet months for this site that did not help in the overall calculation, particularly for 1992 and 1994. Net assimilation differences between Reserva Jaru and Reserva Ducke are small, of the order of 0.02.

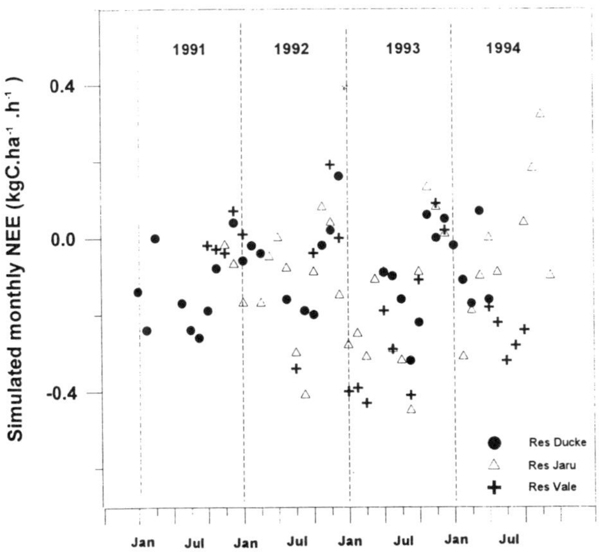

Figure 6 Average monthly net ecosystem exchange (kgC ha^{-1} h^{-1}) calculated by the SiB2 model for the three forest sites, symbols as in Figure 5.

A net uptake of carbon by the forests can be distinguished during the wet months for the three sites. During the dry months the soil respiration component reaches a maximum which makes the net ecosystem exchange more positive. However, the monthly sum of the total flux is negative for all three sites. Grace *et al.* (1996) calibrated their model with the same data set as used here, wherein a run from July 1992 to June 1993 has shown a carbon accumulation of -0.1 kg C ha^{-1} h^{-1}, a result comparable to those obtained here (Table 3).

Although it is not easy to evaluate the significance of inter-annual variability in the model calculations, during the period 1992/1993 the average temperature over the Amazon basin was likely to be reduced by 0.5°C, due to the volcanic emissions of Mount Pinatubo. According to Figure 5, 1993 appears to show relative minimum values of evapotranspiration in the dry months. In addition, the carbon exchange in Figure 6 shows a strong biospheric sink during the first months of 1993. This model result, an increase in the carbon accumulation during cooler years, agrees with the predictions of Grace *et al.* (1996).

CONCLUSIONS

The SiB2 model was calibrated for a tropical rain forest site in Amazonia (Reserva Jaru) using field measurements of eddy covariance surface energy and carbon dioxide fluxes collected during a period of 44 days in the wet season of 1993. An optimization of the maximum rubisco capacity of canopy top, V_{max}, was carried out employing the use of a non-linear method forced by daily averages of the total carbon flux. The daily averages were used to cope with noisy oscillations observed in the hourly data during the first daylight hours. The parameter relating stomatal conductance to canopy assimilation rates m was subsequently calibrated through the use of hourly water vapour flux data collected during seven fine days within the mission. Optimized model parameters controlling photosynthetic activity and stomatal conductance lead to a satisfactory fitting of the observations.

The monthly water and carbon balance were calculated for a tropical rain forest in Amazonia using the SiB2 model, optimized and measured parameters from ABRACOS. The model was then run for the three forest sites in ABRACOS for periods between October 1990 and December 1994 (see earlier). Simulations show that the forest behaves as a carbon sink throughout the year, accumulating carbon on the order of -0.1 kgC ha^{-1} h^{-1}, a result also reported by Grace *et al.* (1996) for a shorter period. The model shows a noticeable seasonal variability: net assimilation exceeds soil respiration during the wet months and in the early transition of the wet to dry season, while in the dry season the total flux might change sign, indicating a net release carbon to the atmosphere.

At least two major uncertainties exist in these calculations. The first involves using parameters calibrated for the site at other locations. Wright *et al.* (1996b) comment that their model underestimates fluxes for Reserva Jaru when using the calibration for Reserva Ducke, and that the lower surface conductance of the latter is not consistent with it having the largest leaf area index of the sites. They suggest that compensation occurs because of the humidity feedback on stomatal conductance. This produces a reasonable prediction of evapotranspiration. A sensitivity study using measurements from different sites during different periods is needed to separate the effects of the many environmental variables that affect the flux. That lies beyond the scope of this exploratory study.

The second uncertainty is associated with the use of parameters obtained for a specific period to describe fluxes for a different period, especially considering the lack of knowledge of seasonal variability of the surface energy fluxes and soil respiration. For example, the soil respiration model probably has smaller uncertainties during the wet season, i.e. the conditions when it was fitted, which allows us to say that the negative net ecosystem exchanges calculated for the wet periods are likely to be realistic. During the dry months the uncertainties are larger, as the model was not calibrated for these conditions and as it did not take account of changes in soil moisture and leaf area index. While soil moisture depletion is likely to decrease soil respiration, the increase in litter layer could compensate for this. For example, leaf area index data from Roberts *et al.* (1996) supports the idea of a peak in litter fall in the dry season, therefore increasing soil respiration, a mechanism also suggested by Wofsy *et al.* (1988). The lack of measurements in the dry season does not allow us to determine whether such compensating processes occur.

REFERENCES

Bonan, G.B. 1991. Atmosphere-biosphere exchange of carbon dioxide in boreal forests. *J. Geophys. Res.*, **96**(D4), 7301-7312

Collatz, G.J., Ball, J.T., Grivet, C. and Berry, J.A. 1991. Physiological and environmental regulation of stomatal conductance, photosynthesis and transpiration: a model that includes a laminar boundary layer. *Agric. For. Meteorol.*, **54**, 107-136.

Culf, A.D., Fisch, G. and Hodnett, M.G. 1995. The albedo of Amazonian forest and ranchland. *J. Climate*, **8**, *1544-1554*.

Culf, A.D., Esteves, J.L., Marques, A. de O., da Rocha, H.R. 1996. Radiation, temperature and humidity over forest and pasture in Amazonia. This volume.

Fan, S.M., Wofsy, S.C., Bakwin, P.S., Jacob, D.J. and Fitzjarrald, D.R. 1990. Atmosphere-biosphere exchange of CO_2 and O_3 in the central Amazon Forest. *J. Geophys. Res.*, **95**(D10):16851.

Farquhar, G.D., von Caemmerer, S. and Berry, J.A. 1980. A biochemical model of photosynthetic CO_2 fixation in leaves of C3 species, *Planta*, **149**,78-90.

Gash, J.H.C., Nobre, C.A., Roberts, J.M. and Victoria, R. L. 1996. An overview of ABRACOS. This volume.

Grace, J., Lloyd, J., McIntyre, J., Miranda, A.C., Meir, P., Miranda, H., Nobre, C.A., Moncrieff, J.,Massheder, J., Wright, I. and Gash, J.H.C. 1996. Net carbon dioxide uptake by an undisturbed tropical rain forest in South-West Amazonia during 1992/1993. This volume.

IMSL (International Mathematical and Statistical Library). 1984. Chapter Z IMSL - ZXSSQ-1 to ZXSSQ-7.

Meir, P., Grace, J., Miranda, A.C. and Lloyd, J. 1996. Soil respiration measurements in the Brazilian Tropical Forest and Cerrado vegetation during the wet season. This volume.

Nobre, C.A., Sellers, P.J. and Shukla, J. 1991. Amazonian deforestation and regional climate change. *J. Climate*, **4**, 957-987.

Tropical Forest and Cerrado vegetation during the wet season. This volume.
Nobre, C.A., Sellers, P.J. and Shukla, J. 1991. Amazonian deforestation and regional climate change. *J. Climate,* **4**, 957-987.
Norman, J.M., Garcia, R. and Verma, S.B. 1992. Soil surface CO_2 fluxes and the carbon budget of a grassland. *J. Geophys. Res.,* **97(D17)**, 18845-18853.
Roberts, J.M., Cabral, O.M.R., da Costa, J.P., McWilliam, A.L. and Sá, T.D.A. 1996. Leaf area index, biomass and physiological responses in Amazonian forest and pasture: an overview. This volume.
Rocha, H.R., Nobre, C.A., Bonatti, J.P., Wright, I.R. and Sellers, P.J. 1995. A vegetation-atmosphere interaction study for Amazonia deforestation using field data and a 'single column' model. *Q. J. Roy. Meteorol. Society*) In press.
Sato, N., Sellers, P.J., Randall, D.A., Schneider, E.K., Shukla, J., Kinter III, J.L., Hou, Y.T., and Albertazzi, E. 1989. Effects of implementing the Simple Biosphere model (SiB) in a general circulation model. *J. Atmosph. Sci.* **46**(18), 2757-2782.
Sellers, P.J., Shuttleworth, W.J., Dorman, L., Dalcher, A. and Roberts, J.M. 1986. Calibrating the Simple Biosphere Model for amazonian tropical forest using field and remote sensing data. Part I: average calibration with field data. *J. Appl. Meteorol.,* **28**, 727-759.
Sellers, P.J, Randall, D.A.,Collatz, C.J., Berry, J.A., Field, C.B., Dazlich, D.A., Zhang, C. and Collelo, G.D. 1995. A revised land surface parameterization (SiB2) for atmospheric GCMs. Part 1:Model formulation. *J. Climate.* In press.
Shuttleworth, W.J. 1988. Evaporation from Amazonian rain forest. *Proc. Roy. Soc. London, Ser. B,* **233**, 321-346.
Wofsy, S.C., Harriss, R.C. and Kaplan, W.A. 1988. Carbon dioxide in the atmosphere over the Amazon basin. *J. Geophys. Res.,* **93**(D2),1377-1387.
Wright, I.R., Nobre, C.A., Tomasella, J., da Rocha, H.R., Roberts, J.M., Vertamatti, E., Culf, A.D., Alvalá, R.C., Hodnett M.G. and Ubarana, V. 1996a. Towards a GCM surface parameterization for Amazonia This volume.
Wright, I.R., Gash, J.H.C. and da Rocha, H.R. 1996b. Modelling surface conductance for Amazonian pasture and forest. This volume.

RESUMO

O papel das florestas tropicais no balanço global de carbono ainda é impreciso. Pode-se entender melhor as interações de troca de carbono entre a atmosfera e a biosfera através da utilização de modelos biofísicos e de circulação da atmosfera. Os resultados destes modelos entretanto dependem muito de sua utilização e desempenho. Este trabalho descreve um método de calibração do modelo SiB2 utilizando medidas de fluxos de energia e de carbono, coletados através do método da covariância dos turbilhões na floresta tropical da Reserva Biológica do Jaru, sudoeste da Amazônia, durante a estação úmida de 1993. As simulações do modelo, utilizando-se os parâmetros otimizados de relacionamento da atividade fotossintética e condutância estomática, levaram a comparações satisfatórias com os dados observados. Os parâmetros otimizados foram utilizados em um exercício de simulação do modelo, forçado por dados meteorológicos de superfície coletados nos três sítios de floresta tropical amazônica do ABRACOS (Reserva Ducke, Reserva Biológica do Jaru e Reserva Vale do Rio Doce), de outubro de 1990 a Dezembro de 1994. A simulação representa uma tentativa de se avaliar os fluxos calculados por um modelo calibrado floresta tropical pode funcionar como um sorvedouro de carbono ao longo de toda a estação úmida.

28 Towards a GCM surface parameterization of Amazonia

I.R. WRIGHT[1], C.A. NOBRE[2], J. TOMASELLA[3], H.R. da ROCHA[4], J.M. ROBERTS[1], E. VERTAMATTI[5], A.D. CULF[1], R.C.S. ALVALÁ[2], M.G. HODNETT[1] AND V.N. UBARANA[6]

[1] *Institute of Hydrology, Wallingford, UK*

[2] *Instituto Nacional de Pesquisas Espaciais, São José dos Campos, Brazil*

[3] *Instituto de Pesquisas Hidraúlicas, Porto Alegre, Brazil*

[4] *Universidade de São Paulo, Brazil*

[5] *Instituto Tecnológico da Aeronaútica/ Centro Técnico Aerospacial, São José dos Campos, Brazil*

[6] *FUNCEME, Fortalesa, Brazil*

INTRODUCTION

Recent initiatives to improve models of soil, plant and atmospheric processes (Wilson and Henderson-Sellers, 1985; Lean and Rowntree, 1993; Manzi and Planton, 1994) and to calibrate model parameters accurately (Shuttleworth, 1988; Shuttleworth and Dickinson, 1989; Sellers *et al.*, 1989) have significantly improved the realism in models of land-surface processes (Sellers *et al.*, 1986; Warrilow *et al.*, 1986; Noilhan and Planton, 1989). These advances have made it possible to construct and compare plausibly modelled climate scenarios for forested and deforested Amazonia (Dickinson and Henderson-Sellers, 1988; Lean and Warrilow, 1989; Nobre *et al.*, 1991; Polcher and Laval, 1994; Manzi and Planton, 1996; Lean *et al.*, 1996). Generally, these simulations of deforestation have predicted increases in surface temperature and decreases in rainfall and evaporation when Amazonian forest is replaced with pasture.

Advances in climate modelling have identified that simulated regional and global climate can be very sensitive to the values of several key parameters describing the land surface (Mintz, 1984 and Garratt, 1993): particularly albedo (Charney, 1975; Sud and Fennessy, 1982; Lean *et al.*, 1996), aerodynamic roughness length (Sud and Smith, 1985; Dickinson and Henderson-Sellers, 1988; Sud *et al.*, 1988) and soil characteristics (Shukla and Mintz, 1982; Henderson-Sellers and Gornitz, 1984; Wilson *et al.*, 1987; Lean *et al.*, 1996). The importance of reliable surface parameterization for GCM evaporation estimates has been pointed out by several reviews (e.g. Mintz, 1984; Rowntree, 1991), including some with particular emphasis on the interpretation of hypothetical large-scale deforestation in tropical regions (Shuttleworth *et al.*, 1991; Henderson-Sellers, 1992), yet it is these remote areas

Amazonian deforestation and climate. Edited by J.H.C. Gash, C.A. Nobre, J.M. Roberts and R.L. Victoria.

which suffer most from a paucity of suitable information. Furthermore, Pitman (1993) has shown that not all studies agree on the relative importance of key parameters, and although this is partly due to the different models and methods of application, much uncertainty is removed with the use of accurate and representative parameters. It must partly be due to the enforced use of substitute parameters and the absence of validation data from remote places that so much effort has been spent on sensitivity studies.

Given this sensitivity of climate modelling to surface parameters, it is necessary that models should include the most accurate parameters if predictions are to be more plausible than in previous studies. Until now, Amazonian deforestation simulations have used parameters derived at a single location, Reserva Ducke, near Manaus, to describe the entire Amazonian forest, even though Amazonia encompasses regions with different soils, vegetations and climates. Furthermore, no data have been available to calibrate parameters for Amazonian pasture or to validate post-deforestation micro-climate. This is particularly unsatisfactory, for although a GCM may adequately describe the current forested climate, the modification of parameters, without the opportunity for model validation, cannot necessarily be expected to describe the deforested situation. Henderson-Sellers (1992) has demonstrated that compensating interactions between parameters within a particular GCM construction may create a fortuitously satisfactory current climate, yet could be quite unreliable once parameters have been changed: particularly albedo and aerodynamic roughness.

ABRACOS was established to satisfy these modelling requirements, by recording detailed measurements at both pasture and forest sites in three regions of Amazonia. Some of the results from ABRACOS fieldwork have been published, prior to this volume, on the following topics: pasture micrometeorology (Wright *et al.*, 1992), climate (Bastable *et al.*, 1993), biomass (McWilliam *et al.*, 1993); pasture surface conductance (Wright *et al.*, 1995) and albedo (Culf *et al.*, 1995), and parameters from these studies have already been used to improve GCM land surface sub-models (da Rocha *et al.*, 1996; Lean *et al.*, 1996; Manzi and Planton, 1996; Dias and Regnier, 1996). This paper aims to bring together the results obtained from the atmospheric, vegetation and soil disciplines of ABRACOS, enhanced where necessary by the results of other Amazonian research, to present a comprehensive tabulation of parameters for the GCM modelling community.

The parameters are divided into two groups under the following headings:

Vegetation parameters

a) Vegetation height and distribution
b) Rooting depth
c) Leaf area index
d) Albedo
e) Aerodynamic parameters
f) Bulk surface conductance
g) Forest rainfall interception

Soil parameters

a) Water release characteristics
b) Density and structure
c) Thermal properties
d) Spatial distribution of parameters

The vegetation parameters are assumed to be independent of soil type and need only be subdivided into pasture and forest subgroups. However, it is acknowledged that rooting depth is, at times, a function of soil type, especially in the transitional stage of pasture establishment after deforestation. No attempt is made to distribute or vary these parameters geographically.

The soil parameters relevant to four defined soil categories are geographically allocated to a 1°×1° grid map of Amazonia using various pedological sources. Where possible the parameters are presented for various depths to support the increasing number of Soil-Vegetation-Atmosphere Transfer Schemes (SVATS) that have multi-layer soil models. Only limited information is available concerning the modification of soil parameters when forest is replaced by pasture, and although some values and references are given in the text for this process, it is necessary to assume that all other soil parameters are independent of vegetation type.

FIELD SITES

The three ABRACOS field study areas, in which both forest and pasture sites were instrumented, were chosen to be representative of a range of typical Amazonian deforestation scenarios. The Manaus area, in central Amazonia, is predominantly forest whereas Marabá, in eastern Amazonia, is mostly pasture. The vegetation in the region of Ji-Paraná, south west Amazonia, comes between these extremes and is undergoing progressive deforestation (Gash *et al.*, 1996). The forest sites were all representative of *terra firme* forest which predominates in lowland Amazonia (Takeuchi, 1961; Pires, 1978). The pasture sites were considered representative of post-deforestation cattle ranches (Uhl *et al.*, 1988: Eden *et al.*, 1990), having been converted 10-15 years previously and sown with *Brachiaria* spp. and *Panicum* spp. pasture grasses (*B. decumbens, B. humidicola, B. brizantha, P. maximum*). Summaries of site and vegetation details are given by Gash *et al.* (1996) and Roberts *et al.* (1996) respectively. Further details of the vegetation at the various pasture and forest sites are given by Wright *et al.* (1992) and McWilliam *et al.* (1993) for Manaus, by McWilliam *et al.* (1996) for Ji-Paraná, and by Sá *et al.* (1996) and Salomão *et al.* (1991) for Marabá.

VEGETATION PARAMETERS

VEGETATION HEIGHT AND DISTRIBUTION

Pasture

Pasture heights were sampled at irregular time intervals throughout the year, by transects and randomly placed quadrats, and found to have an overall two year mean (and corresponding standard deviation) of 0.28 m (0.05 m), 0.58 m (0.12 m) and 0.76 m (0.24 m) for Fazenda Dimona (Manaus), Fazenda Nossa Senhora (Ji-Paraná) and Fazenda Boa Sorte (Marabá) respectively. Although the crop height surveys for Fazenda Dimona and Fazenda Nossa Senhora are biased towards the dry season months, grazing policy is an additional strong influence over the annual growth and senescence cycle. Crop height was very much less variable than green leaf area index (*q.v.*). At Fazenda Nossa Senhora the grass height at the end of the 1992 dry season had a mean value of 0.53 m (21 August 1992) yet had only reached a height of 0.60 m (9 April 1993) towards the end of the following wet season, while the green leaf area index was more than doubled.

The overall mean Amazonian pasture height is given in Table 1 as 0.53 m. However, in view of the relationship between pasture height and roughness length (see **Aerodynamic parameters**), and the importance of the latter in GCM estimates of surface fluxes (Sud and Smith, 1985; Dickinson and Henderson-Sellers, 1988), it would be clearly advantageous for GCMs to incorporate pasture height variability, assuming that the ability to make seasonal and geographic variations in pasture height becomes available in the models.

All pastures studied had significant areas of bare soil (see Figure 1), which were distributed on two spatial scales and occurred mostly for the following reasons. At a small scale, fertility and seasonal soil moisture constraints had prevented the grass from achieving a fully closed canopy since sowing. At Fazenda Nossa Senhora, even after 20 years, the inability of the grasses to spread, either by self-seeding or the subsurface development of rhizomes, left the original planting rows clearly visible (Figure 1b). At a larger scale, areas of open bare soil were the result of cattle tracks, termite activity and failed seed germination. At Fazenda Dimona overall canopy area was also limited by fallen tree trunks (3-5% of ground area) which had not been cleared or burnt since felling (Figure 1a). The areas of bare soil were 15%, 14% and 16% at Fazenda Dimona (including fallen trunks), Fazenda Nossa Senhora, and Fazenda Boa Sorte respectively, giving an overall mean value of 15%: this value appears in Table 1 as 85% canopy cover.

Forest

Mean canopy-top height of the forest at Reserva Ducke, Manaus, and Reserva Jaru, Ji-Paraná, was measured as 35 m and 30 m respectively. The canopy height at Reserva Vale do Rio Doce, Marabá, is more complex: the height of the closed canopy was typically only 25 m but the overall mean is increased by emergents, mainly Brazil nut trees (*Bertholletia excelsa*), typically 50 m tall (see Sá *et al.*, 1996). This is clearly illustrated in Figure 1c, and may be compared with the relatively uniform canopy-top at Reserva Ducke and Reserva Jaru shown in Figure 2a and 2b respectively. Therefore, although the mean canopy-top height at Reserva Vale do Rio Doce is about 35 m, it is questionable whether conventional near-surface aerodynamic theory or 'big leaf' models of radiation interception, energy partitioning and stomatal behaviour, based upon data from an AWS placed at 52 m, can be expected to model this particular type of forest structure successfully. Mean canopy top height appears in Table 1 as 33 m.

Canopy cover, for estimates of surface fluxes, is given in Table 1 as 100%.

However, for the purposes of rainfall interception modelling, the amount of rainfall reaching the ground without impacting the canopy is estimated as the fraction of sky seen from the ground. This fraction has been measured at the three forest sites and found to be 0.08 at Reserva Ducke (Lloyd *et al.*, 1988), and 0.031 and 0.044 at Reserva Jaru and Reserva Vale do Rio Doce respectively (Ubarana, 1996). The average value of 0.052 appears in Table 1 as the 'free throughfall' fraction, and also as 94.8% canopy cover.

ROOT DEPTH

There is mounting evidence that rooting depths have been underestimated in GCM simulations of forest and pasture root extraction of soil moisture. Furthermore, Wilson *et al.* (1987) showed a sensitivity to a doubling of total soil depth over only 10 days of GCM simulation. It is also important to consider the definition of rooting depth in relation to the intended application. A distinction should be made between (a) the depth of soil in which the vast majority of roots may lie, (b) the effective rooting depth for modelling (which may be deeper) and (c) the depth at which live roots may be discovered by digging (which is very much deeper still). Before ABRACOS, the deepest rooting depths so far used in GCM simulations have been 1.0 m for pasture and 2.0 m for forest (Dickinson and Henderson-Sellers, 1988).

Pasture

For pasture, neutron probe data (Hodnett *et al.*, 1995) show clear root extraction below 1.0 m depth at Fazenda Dimona, and Wright *et al.* (1995 and 1996) have calculated effective rooting depths of 1.5 m and 2.0 m for Fazenda Dimona (Oxisol, 65-80% clay) and Fazenda Nossa Senhora (Podsol, 10-35% clay) respectively. These were defined as the depths in which 95% of soil moisture changes had occurred over a two year period. Digging revealed live roots at 3 m depth at both of these ABRACOS pasture sites.

Forest

There is strong evidence for soil water extraction by deep roots which penetrate below depths of 3.5 m at all ABRACOS forest sites (e.g. Hodnett *et al.*, 1995). Similarly, during a dry season event in eastern Amazonia, Nepstad *et al.* (1994) observed that more than 75% of root water extraction occurred between 2 m and 8 m depth and estimated that 'most of the eastern and southern half of the Amazonian closed-canopy forest ... must rely on water uptake from deep soil'. Uhl (1988) has shown that forest root depths may reach 10 m and Nepstad *et al.* (1994) have observed roots at approximately 18 m. Micrometeorological studies have not shown any significant decline in forest transpiration during periods without rain (Wright *et al.*, 1996). Until soil-moisture-induced stress can be shown for tropical forests, it must be assumed that there is no quantifiable limit to the ability of the trees to obtain water for sustained transpiration, and therefore, the effective rooting depth for modelling must be represented by a sufficiently deep reservoir of available soil water, whose

Figure 1 Photographs of the ABRACOS pasture sites at (a) Fazenda Dimona, Manaus (b) Fazenda Nossa Senhora, Ji-Paraná (c) Fazenda Boa Sorte, Marabá

Figure 2 Photographs of the canopy top at (a) Reserva Ducke, Manaus (b) Reserva Jaru, Ji-Paraná (c) Reserva Vale do Rio Doce, Marabá

minimum depth is related to the water retention characteristics of the specified soil.

LEAF AREA INDEX

Pasture

Total green leaf area index (leaf and stem), L*, at the pasture sites was estimated by destructive sampling within randomly selected areas of 0.25 m^2 within the study site (Roberts *et al.*, 1996). Each estimate is the mean of between 10 and 12 samples. Values of L* for various months of the year are given in Table 2 of Roberts *et al.* (1996) for the three ABRACOS pasture sites. At Fazenda Dimona a single measurement was made during 1990 in the middle of a particularly dry period. The value, 1.22 ±0.61, is probably close to the lowest likely L* for this site. Fazenda Nossa Senhora has much higher values in both wet and dry seasons. At Fazenda Boa Sorte, L* was not high even though a considerable proportion of the leaf area comprises regenerating shrubs which had developed since the last burning event two years previously.

It is interesting to note that Fazenda Boa Sorte has lower L* values than either of the other ABRACOS pasture sites, yet it has the tallest grass (Figure 1c). This point is also referred to with respect to albedo (*q.v.*). With the exception of surface roughness, many of the pasture state variables discussed below are independent of grass height.

Clearly, representing Amazonian pasture with a single mean value of L* could obscure important seasonal variations and site differences. Also the limited spatial, annual and inter-annual sample presented here, weakens the justification for calculating such a value. However, the range of L* values from the sites with the largest and smallest values provides valuable information for GCM modelling: 0.49-1.64 at Fazenda Boa Sorte and 1.55-3.90 at Fazenda Nossa Senhora. If future GCMs include a spatial distribution of L*, the above ranges will enable simple modelling of a basic annual cycle. Such a cycle could be timed to correlate with the geographically variable dry season and be linked to models of plant growth and senescence, and changes in albedo (*q.v.*). From the two sites, Fazenda Nossa Senhora and Fazenda Boa Sorte, the average maximum and minimum values of L* are 2.7 and 1.0 respectively: the latter also being close to the minimum of L*=1.2 at Fazenda Dimona.

Great care is needed when using L* in models of bulk surface conductance, such as that described by Jarvis (1976). If L* is used as an independent multiplier to modify bulk conductance per unit leaf area (see **Bulk surface conductance**), the large range of pasture leaf areas shown here will generate a proportionally large range of conductances, and consequently have a marked effect on evaporation. Stewart and Verma (1992), and Shuttleworth *et al.* (1989) also found that a large range of L* values did not greatly affect bulk vapour fluxes from tall grass prairie in Kansas, USA. Until more is known about the negative feedback by self-shading of a canopy, effective bulk surface conductance (per unit ground area) should be used for pastures having a leaf area index greater than a specified threshold. Values of this threshold, ranging from 2-3, have been suggested by Rosenberg *et al.* (1983), Schulze *et al.* (1994) and supported by Wright *et al.* (1996), (see also **Bulk surface conductance**).

Forest

McWilliam *et al.* (1993) obtained a leaf area index of 5.7 ± 0.6 by destructive sampling of a 20 m x 20 m mature forest plot near Manaus, and Roberts *et al.* (1996, Table 1) provide two other estimates for this region. The first was derived from literature values of foliage fresh weight biomass and its vertical distribution (Roberts *et al.*, 1993; Klinge, 1973; Klinge *et al.*, 1975). This information was converted into a vertical profile of leaf area indices using assumptions about the specific leaf area and the ratio of leaf fresh weight to dry weight. The total L* was around 15 per cent greater than the measurement of McWilliam *et al.* (1993), but had the same vertical distribution. The second method (Roberts *et al.*, 1996) used the annual cumulative leaf litter area as an estimate of total canopy L*, giving an estimate of 6.1.

It still remains to be shown whether the estimation of L* from annual litter fall is a valid approach in tropical rain forest. Nevertheless, this method was used at Reserva Jaru and Reserva Vale do Rio Doce and gave values of 4.6 and 5.4 respectively. The estimate for Reserva Jaru is nearly identical to that independently estimated by Grace (Pers. Comm.) using a gap fraction approach. The mean L* from the three sites, based on the leaf litter and destructive sampling methods, is therefore 5.2 with a range of ±0.5. However, as discussed by Roberts *et al.* (1996), the very different leaf distributions and canopy complexities may require that a less simplistic approach be adopted: perhaps a more meaningful value of active leaf area should be defined which takes account of leaf distribution, light interception and canopy conductance.

ALBEDO

Culf *et al.* (1995) have presented between two and three years of albedo data from all ABRACOS forest and pasture sites. Overall mean albedos were found to be 0.180 and 0.134 for the pasture and forest respectively, numerically closer to each other than the average albedos used previously to represent these biomes: for example, 0.19 for pasture (Wilson and Henderson-sellers, 1985) and 0.123 for tropical forest (Shuttleworth *et al.*, 1984). Culf *et al.* (1996, Figure 3) illustrate the average seasonal trend in both forest and pasture, and these data are reproduced in Table 1.

For the forest sites, the seasonal variation was shown to be correlated with soil moisture. Care was taken to ensure that solar angle, cloudiness and the shadow or reflectance of the instrument tower were not influencing this result. It was concluded that seasonal changes in leaf reflectance were being detected, probably associated with leaf dehydration and changes in leaf angle.

At the pasture sites, Culf *et al.* (1995) concluded that, notwithstanding the seasonal changes in pasture growth and decay, there was no dependable cycle of monthly albedo on a seasonal basis. However, combining the specific monthly albedo data from Culf *et al.* (1995) with the leaf area index data taken from Roberts *et al.* (1996) revealed a relationship, shown in Figure 3, which appears to be reasonably site-independent and suggests several interesting points.

The albedo appears independent of grass height. Comparing Fazenda Dimona and Fazenda Boa Sorte: these sites are shown to have similar albedos (Figure 3) yet they have the shortest and tallest grass: 0.28 m and 0.78 m respectively (see Figure 1).

Low values of albedo, resulting from low L*, can be attributed to a higher proportion of dead leaves and greater soil exposure: assuming that the soil albedo is significantly lower than that of the green vegetation. However, wet soil albedo is very much lower than that of dry soil (Idso *et al.*, 1975; Allen *et al.*, 1994) and if soil moisture were influential its effect would be to lower the albedo at higher L* during the wet season. This is illustrated in Figure 4 where albedos after rainfall are shown for high and low L* at the two sites with contrasting leaf areas, Fazendas Nossa Senhora and Boa Sorte. When leaf areas are low at both sites during the dry season (Figures 4a and 4b) there is a small influence from preceding rainfall, yet the mean albedo at these times is lower than at high leaf areas (Figures 4c and 4d) and contrary to a notional influence from seasonal soil moisture status. When leaf area is high, the effect of rain and wet soil is undetectable, even though the high L* at Boa Sorte is similar to the low L* at Nossa Senhora. Therefore, it may be concluded that dead leaf material has a low albedo and is mostly responsible for the seasonal variations in pasture albedo. It should be noted in Figure 4, that some of the low albedos during the early morning and late afternoon are associated with overcast conditions.

Albedo at Fazenda Nossa Senhora shows an insensitivity to L*, which, when combined with data from the other sites, suggests no further increases in albedo above values of L* of about 2-3. This value is similar to the values of L* above which it has been suggested that canopy self-shading begins to occur and bulk surface conductance ceases to be linearly related to L* (Rosenberg *et al.*, 1983; Schulze *et al.*, 1994; Wright *et al.*, 1996).

In future studies, it may be possible to introduce a physically based relationship between composite albedo and L* into GCMs. Furthermore, because of the independence of L* from grass height, this study suggests that the relationship between L* and albedo, together with the maximum surface conductances associated with L* (see **Bulk surface conductance**) can be used independently from grass height and the aerodynamic parameters (*q.v.*). This is a particularly relevant result to GCM deforestation experiments and sensitivity studies, as it helps to reduce the uncertainty associated with the interdependence between L* and z_0: an interaction specifically identified by Henderson-Sellers (1992).

In addition to these natural trends, changes in albedo caused by pasture burning have been reported by Fisch *et al.* (1994). In a single burning event at Fazenda Boa Sorte, albedo was reduced from 0.19 to 0.10, and the subsequent recovery to a value of 0.19 took 11 weeks. This burning event is shown to constitute a small but significant perturbation in the seasonal energy balance.

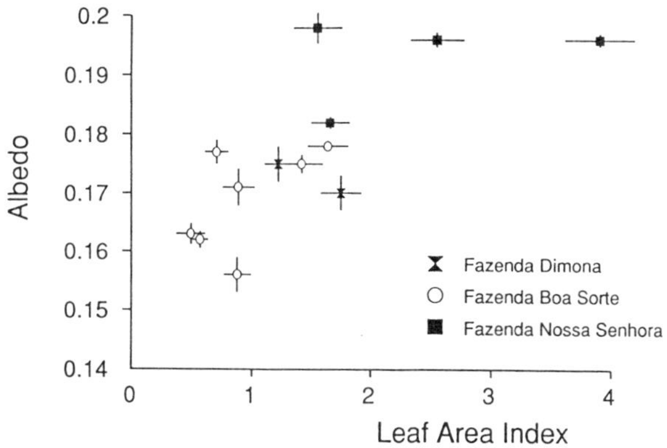

Figure 3 Monthly mean albedos from the three pasture sites plotted against the value of Leaf Area Index measured during that month.

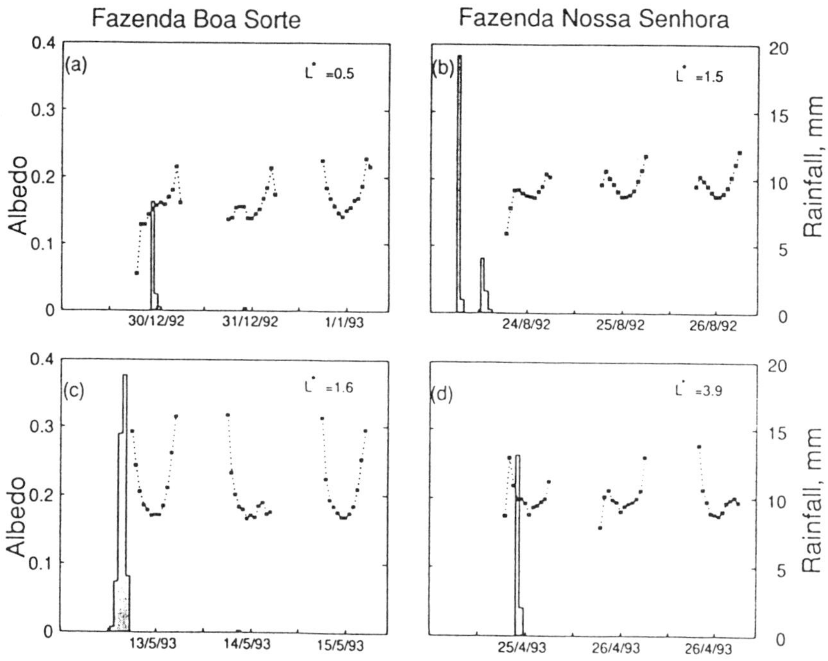

Figure 4 Diurnal changes in albedo (dotted lines) following rainfall (shaded bars) at the Marabá and Ji-Paraná pasture sites representing periods of high and low leaf area index at each site.

AERODYNAMIC PARAMETERS

Pasture

All values of roughness length, z_0, and zero plane displacement, d, have been derived from the momentum flux equation which states that under conditions of neutral buoyancy, wind-speed, u_z, is proportional to the logarithm of height, z, above the zero plane displacement,

$$\frac{ku_z}{u_*} = \ln\left[\frac{z-d}{z_0}\right] \qquad (1)$$

where u_* is the friction velocity and k is von Karman's constant, taken as 0.41.

Pasture roughness parameters for Fazenda Dimona and Fazenda Nossa Senhora were calculated from a 9 m high profile of anemometers and thermometers. Full details of the method are given by Wright *et al.* (1992) together with the derivation of d and z_0 for Fazenda Dimona during the 1990 field season: d = 0.17 ±0.03 m and z_0 = 0.026 ±0.003 m. During the 1991 field season the mean values of d and z_0 were not significantly different to those of 1990: d = 0.19 ±0.03 m and z_0 = 0.025 ±0.003 m.

Unlike Fazenda Dimona, where the wind came predominantly from one direction with an uninterrupted fetch of 900 m, at Fazenda Nossa Senhora there was no predominant wind direction and patches of 8 m high palm trees necessitated careful placement of the instruments. Of the hourly mean windspeeds greater than 1.0 m s^{-1}, 62% came from the northern and southerly sectors, totalling 130 degrees of arc: the remaining winds came mostly from an easterly direction (27% within 110 degrees of arc). The tower was placed to minimise the effect of these limitations and only 26% of winds over 1.0 m s^{-1} came from directions having a fetch that was interrupted by palms or the instrument tower. There were no palms closer than 150 m in any direction. The fetch in the northerly and southerly directions was level uniform pasture extending for at least 1000 m.

Using only hours in which the wind came from a fetch of uninterrupted pasture, the overall mean values of d and z_0 for Fazenda Nossa Senhora were found to be 0.38 ±0.09 m and 0.064 ±0.011 m respectively for the 1992 field season: 76% and 13% of the grass height. For the 1993 field season the overall mean value of d and z_0 were found to be 0.40 ±0.07 m and 0.064 ±0.008 m respectively: 67% and 11% of the grass height respectively. An estimate of z_0 from the wind sectors containing the scattered palm trees was 3% higher, but this difference is not significant, even at the 99% level of confidence.

At Fazenda Boa Sorte wind profile data were not available and it was necessary to use turbulent flux data recorded by the 'Hydra' eddy correlation device (Shuttleworth *et al.*, 1988). These data yield a value for u/u_* and hence $\ln[(z-d)/z_0]$ but do not allow evaluation of d or z_0 separately. However, when z is very much larger than d, the derivation of z_0 is insensitive to the value of d and a value of z_0 may be estimated within a reasonable margin of error by applying limits to d with respect to the vegetation height, h_c (Gash, 1986). Using d = 0.6 h_c, z_0 was estimated as 0.085 m and assigned an error of 0.02 m with regard to the various errors associated with the measurement of u/u_* and grass height.

Figure 5 shows the values of pasture roughness length from all sites, with error bars, plotted against the mean vegetation height. Also shown is the weighted mean relationship

$$z_0 = 0.101\ h_c \tag{2}$$

which is identical (to two decimal places) to the commonly accepted relationship of $0.10h_c$. Comparing Figure 5 with the albedos in Figure 3 there is no clear relationship between roughness length and either leaf area index or albedo. In particular, the pastures which represent the extremes of height and roughness, Fazenda Dimona and Fazenda Boa Sorte, have very similar albedos. As already discussed (under **Albedo**) this suggested independence is specifically relevant to GCM deforestation and sensitivity studies.

Forest

At the Reserva Ducke forest site no estimates of zero plane displacement or roughness length were made during the current project due to the previous extensive work by Molion and Moore (1983), Shuttleworth (1988) and Sellers *et al.* (1989). Here we take the values published by Shuttleworth (1989): d = 30.1 m ($0.86\ h_c$) and z_0 = 2.1 m ($0.06\ h_c$).

For the Reserva Vale do Rio Doce site, flux data were recorded for only a short period and there proved to be insufficient data to derive a well defined relationship between windspeed and friction velocity. Indeed, in view of the complex structure of the Reserva Vale forest canopy (Figure 2c) and the position of the instrument tower close to an emergent tree, it is likely that considerable data would be necessary to investigate the roughness characteristics of this site, and also likely that it might prove impossible to explain the observed momentum fluxes using conventional formulae.

For the best estimate of roughness for the Reserva Jaru forest, the 'Hydra' data were combined with independent data from a 'Solent' eddy flux system (see Grace *et al.*, 1996) operating concurrently on the same tower for carbon dioxide flux measurements. After filtering for wind direction, neutral stability and windspeeds greater than 0.5 ms^{-1}, 461 and 197 hours of momentum flux data were selected from

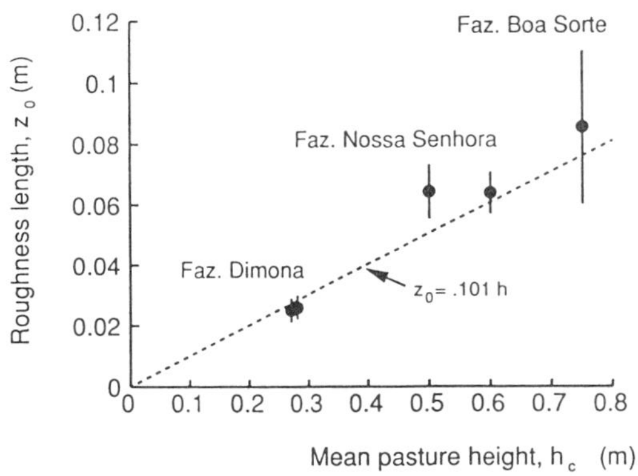

Figure 5 Zero plane displacement from all three pasture sites plotted against vegetation height.

the 1993 Hydra and Solent system data respectively. As d is of comparable size to the instrument height, it is not possible to evaluate d or z_0 separately. However, using the relationships $u/u_* = 0.173$ derived from the combined measurements and d = 0.86 h_c (25.8 m) from Shuttleworth (1989), and imposing an error of ±0.1 h_c on d, z_0 for the Reserva Jaru forest is estimated as 2.6 ±0.3 m. This value is higher than the 2.1 m observed at Reserva Ducke and raises the mean (two site) Amazonian roughness length to 2.35 m.

In terms of the commonly used height, z, and windspeed, u_z, dependent relationship for $g_{a,z}$, the aerodynamic conductance,

$$g_{a,z} = \frac{u_z}{f_z}, \qquad (3)$$

where

$$f_z = \left[\frac{1}{k}\ln\left(\frac{(z-d)}{z_0}\right)\right]^2 \qquad (4)$$

The value of f_z for use with AWS data at each site is 22.8 (z-d = 14.9 m) and 33.4 (z-d = 27.7 m) for Reserva Ducke and Reserva Jaru respectively. However, the difference between these two f_z values is misleading because of the different measurement heights, and masks the fairly similar roughness of the two sites. For a common reference height of, say, (z-d) = 25 m, f_z = 36.5 for Reserva Ducke and f_z = 0.5 at Reserva Jaru. When using Equation 3 to estimate aerodynamic conductance, it is important to calculate the correct value of f_z to match the windspeed reference height.

BULK SURFACE CONDUCTANCE

Evaporation measurements from both forest and pasture sites near Manaus and Ji-Paraná have been used to calibrate a Jarvis-type model (Jarvis, 1976) of surface conductance for each of the four sites. These results are described in detail by Wright *et al.* (1995 and 1996). Two calibrations are presented: the first is for use with reference (measurement) height data, and although site specific, gives the most accurate estimate of transpiration at the calibration site. The second calibration, which uses calculated canopy level climate, is much less site specific and may be used for comparison between sites and to derive a general calibration for Amazonia. Although both parameter sets are shown in Table 1, only the latter calibration should be used in GCMs. The reference level sets are provided for use at the sites at which they were derived, and have the advantage of avoiding the necessity to estimate the climate within the canopy.

The definition of the stomatal conductance parameters, a_1 - a_5, and the form of the equations in which they appear are given elsewhere in this volume by Wright *et al.*

(1996). However, it should be noted that there are different conventions for the parameterization of a_1, which in this study and Wright *et al.* (1996) is defined as the maximum bulk stomatal conductance *per unit ground area*, $L^*g_{s,Max}$. This definition is necessitated by the non-linearity between a_1 and L^*, especially when pasture $L^* > 2$. Many studies, including Wright *et al.* (1995), exclude L^* from a_1, equating a_1 to the maximum bulk stomatal conductance *per unit leaf area*, and also use leaf area index without the inclusion of green stem area. This study, and Wright *et al.* 1996, use L^* equal to the total green leaf (and stem) area index as given by Roberts *et al.* (1996, Table 2).

Pasture

The similarity between conductance parameters a_2 - a_5 for the two pasture sites (see Wright *et al.*, 1996) is sufficiently good for the calibration from one of the sites, Fazenda Nossa Senhora, to be placed into Table 1 to represent Amazonia. Maximum stomatal conductance per unit ground area, a_1, has been given a constant value of 43.0 mm s^{-1}: representing pastures having a leaf area index L^* greater than about 2. The calibrations at Fazenda Nossa Senhora suggest that there is little change in a_1 above $L^*=2$. However, for L^* values lower than 2, $g_{s,Max}$ can be made equal to 21.5 mm s^{-1} as a constant of proportionality (i.e. $a_1 = 21.5\ L^*$). This would give a value of $a_1=32.3$ mm s^{-1} for Fazenda Dimona, which is fairly consistent with the 27.1 mm s^{-1} optimised for that site (Wright *et al.*, 1995). New theoretical work on the relationship between bulk surface conductance, L^* and $g_{s,Max}$ is becoming available (Kelliher *et al.*, 1995; Schulze *et al.*, 1994), however much more field data is needed to calibrate this work. The relationship between L^* and $g_{s,Max}$ is clearly an important component in the modelling of seasonal pasture development and evaporation flux.

For a global value of the soil moisture parameter, a_5, a normalised value is derived, $a_{5,G}$, with respect to the available soil moisture of the different soils, thus

$$a_{5,G} = \frac{a_5 - \theta_r}{\theta_{sat} - \theta_r} \tag{5}$$

where θ_r and θ_{sat} are the residual and saturation soil moisture contents respectively. The normalised parameter was found to be consistent between the two Amazonian pasture sites, especially during similar seasonal conditions (Wright *et al.*, 1996). When soil moisture was decreasing during the dry season, $a_{5,G}$ was estimated as 0.68 and 0.66 at Fazendas Dimona and Nossa Senhora respectively. During a period of increasing soil moisture at Fazenda Nossa Senhora, this critical soil moisture parameter was lower: $a_{5,G}= 0.50$. Wright *et al.* (1996) suggest that this hysteresis can be expected when representing the soil as a single layer without regard for the root distribution, and recommend a mean value of $a_{5,G}= 0.58$. Winkworth (1970) suggested a value of $a_{5,G}= 0.50$ for a grassland site in Australia (Lat. 23°S) where the soil was a Red Earth (Soil type 3 in **Spatial distribution of parameters** below).

Forest

For forest conductances, Wright *et al.* (1996) showed that, when using the site calibration from Reserva Ducke to predict evaporation from Reserva Jaru, the evaporation estimate compared moderately well with observations even though some of the parameters were very different. The parameters associated with solar radiation and temperature were similar, but those for humidity deficit and maximum conductance were different in magnitude yet mutually compensating when their influences are combined. This result may be caused either by interdependence between parameters or physiological differences between forests producing similar vapour fluxes from a different combination of climatic interactions. Although some parameter interdependence is unavoidable with the calibration method used by Wright *et al.* (1996), there is evidence from leaf conductance measurements at the Reserva Ducke and Reserva Jaru (Roberts *et al.* 1996), and for temperate forest by Hall and Roberts (1989), that forests having a wide species diversity can also have a wide range of reactions to climate variables. Wright *et al.* (1996) show a similar relationship to Roberts *et al.* (1996) and Hall and Roberts (1989) where species with a high maximum conductance have a rapid response to humidity deficit and *vice versa*. Furthermore, these studies show that the different, but compensating, responses to humidity at each site converge to give similar conductances at or around the predominant humidity deficits measured above the forest. Clearly, as larger vapour fluxes coincide with larger humidity deficits, it is not surprising that the calibrations are almost interchangeable. It is important to note that it is the calibrations (parameters sets) that are interchangeable and not the individual parameters, especially in view of the likely sensitivity of GCM representations of Amazon forest to the humidity deficit parameter (Sellers *et al.*, 1989).

Table 1 contains the forest calibration representative of all available data for Reserva Jaru. This set is chosen to represent tropical forest as it is the most comprehensive calibration based on canopy level climate. The only justifiable adjustment to this calibration would be for forests of a different leaf area index. However, this is not to suggest that the influence of L^* is necessarily linear. The reader is referred to Dolman *et al.* (1991) and Wright *et al.* (1996) for details of other calibrations.

FOREST RAINFALL INTERCEPTION

Measurements of throughfall and stemflow at the Reserva Jaru and Reserva Vale do Rio Doce forest sites have been used to derive the canopy and trunk storage capacities, and the proportions of rainfall diverted to the trunks and reaching the ground without impacting the canopy (the 'free throughfall' parameter). This work is described fully by Ubarana (1996). The experimental design was based on the earlier work of Lloyd *et al.* (1988) at the Reserva Ducke forest site and assumes a Rutter-type model of canopy water dynamics (Rutter *et al.*, 1971). Combining the results of Ubarana (1996) and Lloyd *et al.* (1988), mean interception parameters have been calculated as follows (values in parentheses are for Reservas Ducke, Jaru and Vale do Rio Doce, respectively): canopy capacity, 1.01 mm (0.74 mm, 1.03 mm, 1.25 mm); 'free throughfall' fraction, 0.052 (0.080, 0.031, 0.044); trunk storage capacity, 0.11 mm (0.15 mm, 0.09 mm, 0.10 mm) and the proportion of rainfall that is diverted to the trunks as stemflow, 0.023 (0.036, 0.010, 0.023). The mean values appear in Table 1 as representative of Amazonian *terra firme* forest.

No parameters have been derived for the process of evaporation of intercepted rainfall from the pasture sites.

SOIL PARAMETERS

WATER RELEASE CHARACTERISTICS

Table 1 Amazonian vegetation parameters
Parameters that are not derived in this paper are accompanied by the relevant reference

	Pasture		Forest
Vegetation height (m)			
Faz. Dimona (Wright *et al.*, 1992)	0.28	Res. Ducke (Shuttleworth *et al.*, 1984)	35
Faz. Nossa Senhora (McWilliam *et al.*, 1996)	0.58	Res. Jaru (Roberts *et al.*, 1996)	30
Faz. Boa Sorte (Sá *et al.*, 1996)	0.76	Res. Vale do Rio Doce (Sá *et al.*, 1996)	20-50[1]
Mean vegetation height	0.53		33
Canopy cover (%)	85		100 (94.8[2])
Rooting depth (m)	1.5-2.0 (Wright *et al.*, 1995,1996)		>4.0[3]
Green leaf area index	1.0-2.7[4] (See also Roberts *et al.*, 1996)		5.2 (McWilliam *et al.*, 1993; Roberts *et al.* 1996)

Notes

1 Complex canopy structure, see **Vegetation height and distribution: *Forest***
2 94.8% of rainfall impacts the canopy for interception modelling
3 No soil/plant stress has yet been identified under forest, see **Root depth: *Forest***
4 Mean annual minimum and maximum
5 h_c = mean canopy height
6 Maximum conductance per unit ground area, $L^*g_{s,Max}$
7 When pasture $L^* < 2$ then $a_1 = 21.5L^*$, see **Bulk surface conductance: *Pasture***
8 a_5 = Site specific critical soil moisture content
9 $a_{5,G}$ = Normalised critical soil moisture content, $(a_5 - \theta_r)/(\theta_{sat} - \theta_r)$

Table 1 Amazonian vegetation parameters (continued)

	Pasture		Forest	
Albedo *Month* (Culf *et al.*, 1995, 1996)				
January	0.175		0.126	
February	0.171		0.123	
March	0.181		0.121	
April	0.184		0.124	
May	0.187		0.132	
June	0.186		0.139	
July	0.192		0.141	
August	0.187		0.144	
September	0.171		0.142	
October	0.172		0.143	
November	0.177		0.141	
December	0.180		0.134	
Mean albedo	0.180		0.134	
Zero plane displacement[(5)] **(m)**	$0.66h_c$ (See also Wright *et al.*, 1992)		$0.86h_c$ (Shuttleworth, 1989)	
Roughness length[(5)] **(m)**	$0.10h_c$		2.35	
Surface conductance *For use with reference height climate*				
	Site		**Site**	
	Faz. Dimona	Faz. N. Senhora	Res. Ducke	Res. Jaru
	(Wright *et al.*, 1995 & 1996)		(Dolman *et al.*, 1991)	(Wright *et al.*, 1996)
a_1[(6)] mm s^{-1}	30.7	33.1	20.8	65.2
a_2 kg g^{-1}	0.0369	0.1127	0.064	0.1064
a_3 °C	-	-	30.2	44.6
a_4 W m^{-2}	470	671	250	743
a_5 m^3 m^{-3}	0.428[(8)]	0.259[(8)]	-	-
For use with canopy climate (Wright *et al.*, 1996)				
a_1[(6)] mm s^{-1}	43.0[(7)]		80.1	
a_2 kg g^{-1}	0.0821		0.1248	
a_3 °C	-		44.2	
a_4 W m^{-2}	17280		3916	
$a_{5,G}$ m^3 m^{-3}	0.58[(9)]		-	
Forest interception (See also Ubarana, 1996)				
Canopy capacity (mm)	-		1.01	
Free throughfall fraction	-		0.052	
Trunk storage (mm)	-		0.11	
Fraction of rain to trunks	-		0.023	

Notes — see previous page

Water release model

The model of soil water release characteristics (van Genuchten, 1980), for which parameters have been derived for the ABRACOS soils, has the principal advantages of being a continuous function with clearly defined limits. For this reason the model has been adopted for use in many GCM descriptions of soil water movement. Soil water content, S_θ, is related to the soil matric potential by

$$S_\theta = [\,1 + (\alpha\Psi)^n]^{(1-n)/n} \quad (6)$$

and hydraulic conductivity by

$$K(S_\theta) = K_{sat} S_\theta^{\,L}\,(1-[1-S_\theta^{\,n/(n-1)}]^{(n-1)/n})^{0.5} \quad (7)$$

where

$$S_\theta = (\theta_r - \theta)/(\theta_{sat} - \theta_r) \quad (8)$$

θ_r = residual soil moisture content

θ_{sat} = saturation soil moisture content

K_{sat} = saturation hydraulic conductivity

Ψ = matric potential

and, n and L are curvature parameters. Tomasella and Hodnett (1996) derived parameter values from routine neutron probe and tensiometer measurements at Fazenda Dimona, and enhanced by intensive field measurement campaigns employing permeameters and the 'instantaneous profile method' of soil water release (Hillel *et al.*, 1972). Full details of the optimisation procedure, application to the Fazenda Dimona data, and the role and relative importance of the various parameters are discussed by Tomasella and Hodnett (1996). Table 2 gives the optimised parameter values derived for four ABRACOS sites: forest and pasture at Manaus and Ji-Paraná. Also shown in Table 2 are the results of similar optimisation studies van Lier (Pers. Comm., van Lier; Moraes, 1991; van Lier and Neto, 1993) in which van Genuchten parameters were derived for a Brazilian 'Structured Red Earth' or 'Terra Roxa Estuturada'.

Parameter interpretation and vegetation influences

Tomasella and Hodnett (1996) recorded data for the Manaus soil under forest close (1500 m) to the Fazenda Dimona pasture site (rather than at Reserva Ducke), and give a good comparison between forest and pasture on the same soil. The soil at Fazenda Dimona is a yellow latosol (Oxisol or Haplic Acrorthox), with a high clay content of typically 65-80%, but has weathered to give high moisture conductivities when close to saturation. This is a particular characteristic of Amazonian latosols (Sanchez, 1976; Hodnett *et al.*, 1995) and is not easily described by single parameter models of soil matric potential and conductivity. This soil has one of the lowest capacities of available soil moisture in Amazônia.

The 'van Genuchten' parameters for latosol shown in Table 2a and 2b vary considerably with depth, yet are remarkably similar under both pasture and forest. Figure 6 shows the shape of the function using parameters for a depth of 0.5 m. This similarity suggests that, even after 15 years, the conversion to pasture has not greatly affected the soil structure at depths below 0.2 m. However, Hodnett *et al.* (1995) observed impeded infiltration and a small amount of runoff at the pasture site, K_{sat} = 50 mm h^{-1}, whereas runoff was never observed in the forest: Medina and Leite (1985) give infiltration rates for undisturbed forest near Manaus as 223 mm h^{-1}. Although very poor infiltration at pastures sites can be caused by the method of forest clearance, particularly when heavy machinery is used (Dias and Nortcliff, 1985; Medina, 1985), this is not likely to be the case at Fazenda Dimona where forest clearance was by 'slash and burn'. At this site the relatively mild reduction in infiltration is consistent with the effects of compaction by cattle (Reategui *et al.*, 1990; Grimaldi *et al.*, 1993).

The podsol at Ji-Paraná (Arenosol or Paleudult) has a particularly high sand content, especially near the surface (85%), and has contrasting water release characteristics to that of the Manaus latosol. The optimised parameters given in Table 2c and 2d were derived in the same way as those for the Manaus soil (Tomasella and Hodnett, 1996) and are published here for the first time. Optimised water release curves for pasture and forest at 0.4 m depth are shown in Figure 6.

Compared to the Manaus clay, there is less similarity between the forest and pasture podsol parameters. However, at most depths there is consistent variation in θ_{sat} and θ_r with depth, indicating the increased weathering and sand content closer to the surface at both sites. Surface compaction in the pasture is evident in the reduced surface conductivity and there is also a suggestion that in the 15 years since conversion there has been a change in the soil characteristics at 0.2 m when compared to the forest. Although the water release parameters at 0.2 m are very

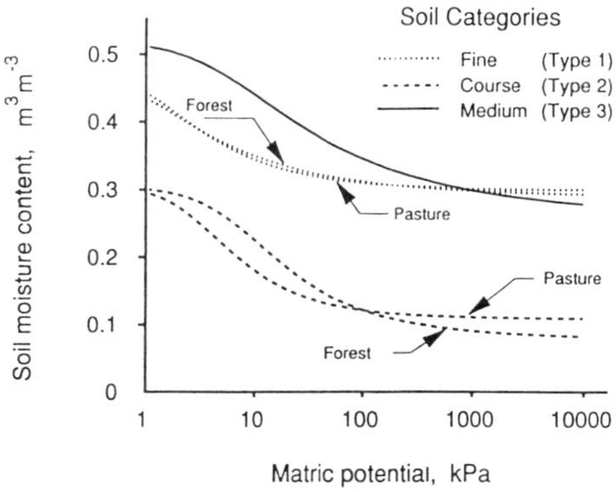

Figure 6 Water release curves derived using the van Genuchten equation and optimised parameters for fine, medium and coarse soils.

different, they describe similar water release characteristics at lower moisture contents: the principal difference is that the pasture has poorer water retention properties closer to saturation. Unlike the Manaus sites, these two soils are less easily compared as they are about 90 km apart, and the seasonally shallow water table at the forest site will account for the change in parameter values at 1.5 m.

The soils of Manaus and Ji-Paraná are close to the textural extremes of fine and coarse soils in Amazonia. It is, therefore, fortunate that a calibration is available for a Structured Red Earth (Alfisol or Kanhapludulf), which represents a third Amazonian soil type and which has contrasting water release characteristics to the ABRACOS soils. The water release parameters for this soil are given in Table 2e and were derived by van Lier (Pers. Comm.) using similar criteria to those used for the ABRACOS soil calibrations (van Lier and Neto, 1993; Moraes, 1991). Figure 6 shows the shape of water release curve for this soil at 0.45 m, and clearly illustrates the contrasting water retention characteristics of the Red Earth when compared to the clay latosol near Manaus and the podsol under pasture at Ji-Paraná. The release characteristic for available soil moisture is fairly similar to that under the Ji-Paraná forest, however, the higher water capacity and, in the other cases, greater range of available soil moisture are clearly shown. Although the data for a Red Earth are from an area south of Amazônia, this soil type and similar well structured soils occur extensively in Amazônia, being derived from the same podsolic pedogenesis. This Red Earth has been extensively studied at the University of São Paulo Agricultural Faculty, Piracicaba. Apart from the work already cited, Table 2e has been enhanced using surface infiltration data from Reichardt *et al.* (1978), and further validated using conductivity data from Saunders *et al.* (1987) and soil texture information from Vieira and Santos (1987).

Density and structure

Bulk densities and particulate content of the three soil categories are given for various depths in Table 2. Values for the ABRACOS sites were derived from laboratory analysis of field samples and those for the Structured Red Earth are taken from Van Lier (Pers. Comm.) and Vieira and Santos (1987). Soil particle density at the ABRACOS sites was 2.6 ± 0.1 Mg m^{-3} and did not vary with depth or between sites. Therefore, this value can be used to infer porosities from the tabulated bulk densities. The clay content profiles are very typical for the soil type and are similar to those published by Ranzani (1980). The reduced bulk density near the surface of the forest soils compared to the rest of the profile, and the weakening of this effect after deforestation, is consistent with the results presented for Amazonian latosols by Martins *et al.* (1991).

Thermal properties

Alvalá *et al.* (1996) derived thermal diffusivity from a 0.40 m profile of soil thermistors, operated in the field during the micrometeorological missions, and recorded temperatures at four levels every 10 minutes. Diffusivity was calculated using a numerical finite difference method, and two analytical methods which consider the change in either the phase or the amplitude of diurnal soil temperature at different depths. These methods gave various estimates for diffusivity at different depths and at different moisture contents, however the methods did not always agree, probably as a result of the vertical heterogeneity of the soils. For this reason a single best estimate of diffusivity is given for each of the two ABRACOS soil types, without regard for depth or moisture content: $0.15\text{-}0.45 \times 10^{-6}\ m^{-2}\ s^{-1}$ for the Manaus (fine) soils and $1.45 \times 10^{-6}\ m^{-2}\ s^{-1}$ for the Ji-Paraná (coarse) soils.

Spatial distribution of soil parameters

Many previous studies have needed to consider the water release characteristics of the soils of the Amazon basin. Most of these have been GCM studies of hypothetical deforestation scenarios, which in the absence of representative water release measurements, have either placed a single best estimate soil type over the whole of Amazonia (e.g. Nobre *et al.*, 1991) or used land-surface classifications (e.g. Dickinson, 1984; Wilson and Henderson-Sellers, 1985) to obtain $1° \times 1°$ soil texture information. This textural information is then used to infer water release characteristics (e.g. Dickinson and Henderson-Sellers, 1988; Lean and Warrilow, 1989). Clearly, both of these methods are unsatisfactory although they were the best that could be done at the time. Also, bearing in mind that Tomasella and Hodnett (1996) and Hodnett *et al.* (1995) have demonstrated that it is unwise to 'import' empirical water release functions to represent soils of the Amazon basin, the pedological and textural detail that is currently available in global data bases could be grossly misleading. For example, for the predominant fine soils of Amazonia, 'imported' estimates of available soil moisture would be typically 50% greater than those suggested from Table 2. This difference would radically affect the predicted hydrology of the shallow rooted pastures by delaying the onset of stress in the grass at the beginning of the dry season.

Many Amazonian soils have been extensively studied, but mainly from an agricultural point of view, and only very limited data have been published concerning water retention. From the limited number of studies that have derived parameters for Amazonian soils that are relevant to contemporary GCM modelling (ABRACOS, and Van Lier and Neto, 1993) four categories of soil type have been identified:

1. *Fine soils* (Table 2a and 2b)

   The Manaus clay latosol is probably the most common and most studied Amazonian soil, (Correa, 1984; Tomasella and Hodnett, 1996; Hodnett *et al.*, 1995) and is used here to represent most of the fine soils of the area. Although this category covers large areas of latosol (commonly fine) and podsol (commonly medium-coarse), Ranzani (1980) has shown that these pedological units do not have a unique texture and their clay content can vary widely. With the limited parameters available these soils are considered to be sufficiently well represented by a clay-like soil with high conductivity close to saturation. However, there may be some bias in using a soil with a particularly high clay content as given here.

2. *Coarse soils* (Table 2c and 2d)

   Although the soil around Ji-Paraná is a podsol it has a high sand content for that pedological type and is considered representative of coarse soils in Amazonia, i.e., Arenosols, Lithosols and hydromorphic podsols.

3. *Medium texture soils* (Table 2e)
   This category is based on the work of van Lier (Pers. Comm.) on a Structured Red Earth from the south of Amazonas (see also Moraes, 1991; van Lier and Neto, 1993). Soil types represented by this category are Red Earths, Vertisols and some oxidised and well structured latosols.
4. *Plinthitic soils* (Table 2b - c)
   Plinthitic soils, which cover a large and relatively undisturbed area of Amazonia, could have a considerable impact upon regional hydrology if disturbed by large-scale deforestation. These soils have been found to oxidise rapidly when exposed by machinery or aerated by a lowering of the water table. When disturbed by machinery, this soil has been observed to reduce, in less than 12 months, from a heavy clay podsol to a concreted and coarse structured material with the water holding characteristics of a sandy soil. Although this disturbance represents rather extreme circumstances, it is not unreasonable to suggest that the characteristics of this soil could change from 'fine' to 'coarse' over a period of less than, say, five years. During this transition, the poorly structured soil could not necessarily be represented by the medium textured soil type (3 above) because it is unlikely that the water holding capacity would develop to any great extent.

For GCM experiments investigating the effects of deforestation in Amazônia, these soils should be represented by the 'fine' soil category while they remain undisturbed. Until more is known about the oxidisation of these soils, it is recommended that these areas should be represented by 'coarse' soil parameters when describing the deforested state. In multi-layer soil models it would be relevant to modify only the upper soil layers in a way that is consistent with the estimated depth and rate of oxidation and the type of land use after deforestation.

Figure 7 shows the 1°×1° allocation of soil categories recommended by this study to represent Amazonia. The figure is not intended to be an accurate soils map. The distribution of categories is based on pedological information from various sources (RADAM, 1980; Wilson and Henderson-Sellers, 1985; Vieira and Santos, 1987; Kineman and Ohrenschall, 1992; Webb and Rosenzweig, 1993), and the personal experience of one of the authors. However, particular emphasis has been placed on maintaining the relative proportion of coarse (16%), medium (5%) and fine (79%, including plinthitic) textured soils based on the proportion suggested by the various sources. It should be noted that there is a great disparity between published datasets for classifying Amazonian soil textures. Apart from datasets being superseded by legitimate improvements, this disparity is probably due to, and made more confusing by, the differing pedological classifications to which the Amazonian soils have been required to conform (e.g. FAO, Brazilian, North American).

In summary, although the mapping of soil parameters is greatly simplified in this study, further detail is not relevant until more Amazonian soils have been parameterized for water retention characteristics. More work on mapping actual soil texture, rather than inferred texture, is needed to complement the existing wealth of detailed

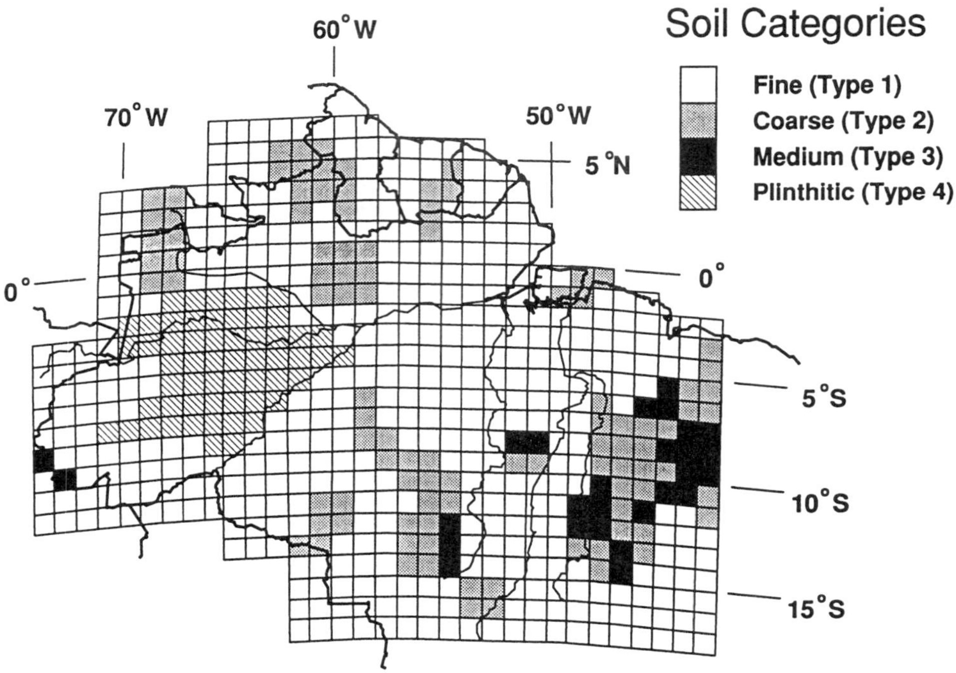

Figure 7 A 1°×1° distribution of the three parameterized soil types over the Amazon region of Brazil, including the area of plinthitic soils where soil characteristics may change after deforestation.

pedological information. However, this study provides sufficient information to investigate the sensitivity of climate and hydrology to simple but well calibrated differences in Amazonian soil type: an important first step to indicate the direction of future work. For example, assuming that forested areas are insensitive to soil type because of their unlimited access to deep water, it is not clear whether pasture on coarse soils will have a significant impact on modelled hydrology or change the severity of dry seasons. For although the available soil moisture is greater in coarse soils, they only cover up to 16% of the total area in this study: 26% if plinthitic soils are included.

CONCLUSIONS

The parameters presented in this paper are a summary of research in many areas of environmental science, resulting in a pair of tables that are designed to be used and interpreted by the GCM modelling community. Vegetation, soil and surface flux related parameters have been derived from data recorded at three contrasting areas within Amazonia, including typical pasture and forest of each area, and encompassing representative Amazonian soils and vegetation structures. Although these six study sites are a very small sample of the vast Amazon basin, this study should result in

Table 2 Soil parameters

Depth m	n -	α kPa⁻¹	1[1] -	θ_{sat} m³ m⁻³	θ_r m³ m⁻³	K_{sat} mm h⁻¹	Bulk density Mg m⁻³	Clay %	Sand %
a) PASTURE - FINE SOIL (Fazenda Dimona)									
0.0	-	-	-	-	-	50	1.06	65	20
0.3	1.50	0.745	5.40	0.448	0.305	546	1.07	75	15
0.5	1.62	0.967	1.98	0.486	0.299	676	0.94	80	10
0.75	1.33	4.62	-1.77	0.573	0.304	1096	1.10	80	10
1.05	1.20	6.64	-4.91	0.565	0.355	422	1.12	80	10
1.35	1.37	0.200	3.65	0.526	0.420	18	-	-	-
b) FOREST - FINE SOIL (Fazenda Dimona)									
0.0	-	-	-	-	-	223[2]	0.93	65	20
0.3	1.47	0.704	5.29	0.447	0.304	560	1.04	75	15
0.5	1.44	1.49	1.38	0.489	0.290	1256	1.04	80	10
0.75	1.28	5.45	-2.01	0.574	0.289	1495	1.15	80	10
1.05	1.18	6.35	-6.26	0.565	0.359	626	1.15	80	10
1.35	1.31	0.190	5.32	0.511	0. 382	23	-	-	-
c) PASTURE - COARSE SOIL (Fazenda Nossa Senhora)									
0.0	-	-	-	-	-	6[3]	1.50	7	85
0.2	1.92	0.202	0.5	0.259	0.046	20-60[3]	1.50	12	78
0.4	1.77	0.359	0.5	0.309	0.109	-	1.30	16	72
0.6	1.16	0.730	0.5	0.389	0.131	-	1.30	33	58
0.8	1.35	0.293	0.5	0.418	0.257	-	1.30	33	58
1.0	1.56	0.251	0.5	0.465	0.298	-	1.24	36	53
1.2	1.59	0.206	0.5	0.425	0.275	-	1.24	36	53
1.5	1.25	1.103	0.5	0.383	0.191	-	1.24	36	53
d) FOREST - COARSE SOIL (Reserva Jaru)									
0.0	-	-	-	-	-	-	1.38	4	85
0.2	1.34	2.209	0.5	0.483	0.025	63	1.55	4	82
0.4	1.60	0.164	0.5	0.305	0.079	66	1.52	6	77
0.6	1.73	0.304	0.5	0.343	0.155	10	1.49	18	63
0.8	1.46	0.209	0.5	0.397	0.212	(10)	1.49	35	58
1.0	1.39	0.212	0.5	0.410	0.231	-	-	35	58
1.2	1.40	0.252	0.5	0.408	0.207	-	-	36	53
1.5	1.57	0.213	0.5	0.418	0.189	-	-	36	53
e) MEDIUM TEXTURE SOIL (parameters derived from field data by van Lier (Pers. Comm.)									
0.0	-	-	-	-	-	6[4]	-	38	34[5]
0.15	1.28	1.896	(0.5)	0.493	0.243	-	1.54	49	32
0.3	1.68	0.131	(0.5)	0.527	0.294	2.8	1.43	61	24
0.45	1.36	0.226	(0.5)	0.520	0.262	5.5	1.40	64	21
0.6	1.20	1.167	(0.5)	0.516	0.241	46	1.38	63	22
0.75	1.42	0.629	(0.5)	0.502	0.273	108	1.25	62	22
0.9	1.53	0.390	(0.5)	0.533	0.259	87	1.23	62	22
1.05	1.60	0.478	(0.5)	0.535	0.255	186	1.20	59	24
1.20	1.49	0.516	(0.5)	0.531	0.239	168	1.23	58	24
1.35	1.66	0.263	(0.5)	0.550	0.239	136	1.21	58	24
1.50	1.64	0.365	(0.5)	0.558	0.240	-	1.21	56	25

Notes

1 When insufficient data are available l = 0.5, when not in parentheses l = 0.5 was used in the optimisation of the van Genuchten parameters.
2 Not optimised - measured by Medina and Leite (1985)
3 Not optimised - measured by ABRACOS
4 Reichardt *et al.* (1978)
5 Vieira and Santos (1987)

a marked improvement to the validity of GCM deforestation experiments and will contribute to identifying important areas of future research. Vegetation parameters, such as albedo, canopy structure and aerodynamic roughness, have shown a level of coherence between sites that validates some of the necessary generalities that have to be made in representing the Amazon basin in a GCM. This work has also provided some surface seasonality to complement the modelled GCM annual cycle.

Bulk stomatal conductance and its control on transpiration is an area of empiricism that is still poorly understood, especially for tropical forest. Although the total transpiration from the forest appears to be similar between sites, the complex physiological and structural diversity at each forest site produces different, yet compensating, parameters and creates an element of uncertainty when applying evaporation models to new forest sites. No soil-induced reduction in transpiration has yet been observed from tropical forest: a result which must be considered when setting effective forest rooting depths.

For pasture, the grazing of cattle and the combined influences of soil fertility, dry season severity and species composition, has produced at each pasture site a grass cover whose height is only weakly related to its leaf area index. As the dry season develops and green leaf area declines, the canopy structure remains largely the same. Between sites there is a consistent relationship between height and roughness length, and to a weaker extent between leaf area index and albedo. This means that seasonal variations in leaf area and albedo can be legitimately investigated without necessarily varying crop height. When pasture leaf area index is below about 2, bulk stomatal conductance and albedo are reduced: above this value both of these parameters have a suggested maximum. Although there is insufficient information to derive a functional relationship, future work may lead to a more mechanistic model of plant structure with fewer and more meaningful parameters.

Soil research within ABRACOS has highlighted the paucity of well calibrated parameters to describe the Amazonian surface and sub-surface hydrology. However, a framework for essential future work has been identified. Water release and moisture movement in Amazonian soils cannot be readily represented with parameters 'imported' from soils outside of Amazonia. There may yet be soils which can be treated as more typical of global soils, but until more Amazonian soils are studied with respect to the needs of GCM modelling, it is necessary to exercise some caution.

With the advances in the land surface parameterization of Amazonia, it is now possible to embark upon more focused sensitivity studies: identifying the most relevant areas of future work. The design of future experiments can also be influenced by identifying the relative importance of individual parameters or areas of research. The ABRACOS results show that both soil parameterization and the understanding of vegetation processes require further work. Within the current GCM land-surface schemes there is clearly a mutually dependant relationship between the sub-models of rainfall, evaporation and soil water release. Therefore, for a reliable model of wet season runoff and dry season stress, it is necessary to continue with accurate calibrations to improve realism in GCMs and reduce the empiricism in modelling the biosphere.

ACKNOWLEDGEMENT

In addition to the acknowledgement of support from the very many Brazilian scientific institutions mentioned elsewhere in this volume, the authors would especially like to acknowledge the valuable contribution of 'Terra Roxa' parameters supplied by Q. Jong van Lier of São Paulo University. Also the advice and additional soils information supplied by Walter Chomentowski of the University of New Hampshire, USA, and Dr. Antonio Nobre of the Instituto Nacional de Pesquisas da Amazonia, Manaus.

REFERENCES

Allen, S.J., Wallace, J.S., Gash, J.H.C. and Sivakumar, M.V.K. 1994. Measurements of albedo variation over natural vegetation in the Sahel. *Int. J. Climatol.* **14**, 625-636.

Alvalá, R.C.S., Gielow, R., Wright, I.R. and Hodnett, M.G. 1996. Thermal diffusivities of Amazonian soils. This volume.

Bastable, H.G., Shuttleworth, W.J., Dallarosa, R.L.G., Fisch, G. and Nobre, C.A. 1993. Observations of climate, albedo and surface radiation over cleared and undisturbed Amazonian forest. *Int. J. Climatol.* **13**, 783-796.

Charney, J.G. 1975. Dynamics of deserts and drought in the Sahel. *Q. J. R. Met. Soc.* **101**, *193-202*.

Correa, J.C. 1984. Características físico-hídricas dos solos latossolo amarelo, podzólico vermelio-amarelo e podzol hidromórfico do estado do Amazônas. *Pesquisa Agropecuaria Brasileira* **20**, 1317-1322.

Culf, A.D., Fisch, G. and Hodnett, M.G. 1995. The albedo of Amazonian Forest and ranchland. *J. Climate* **8**, 1544-1554.

Culf, A.D., Esteves, J.L., Marques Filho, A.O. and da Rocha, H.R 1996. Observations of radiation, temperature and humidity over forest and pasture in Amazonia. This volume.

Dias, A.C.C.P. and Nortcliff, S. 1985. Effects of two land clearing methods on the physical properties of an oxisol in the Brazilian Amazon. *Tropical Agriculture (Trinidad)* **62**, 202-212.

Dias, A.C.C.P. and Regnier, P. 1996. Simulations of mesoscale circulations in a deforested area of Rondonia in the dry season. This volume.

Dickinson, R.E. 1984. Modelling evapotranspiration from three-dimensional global climate models. In: *Climate Processes and Climate Sensitivity*. Eds. J.E. Hanson and T. Takahashi. Amer. Geophys. Union. Washington.

Dickinson, R.E. and Henderson-Sellers, A. 1988. Modelling tropical deforestation: A study of GCM land-surface parametrizations. *Q. J. R. Met. Soc.* **114**, 439-462.

Dolman, A.J., Gash, J.H.C., Roberts, J. and Shuttleworth, W.J. 1991. Stomatal and surface conductance of tropical rainforest. *Agric. For. Met.* **54**, 303-318.

Eden, M.J., McGregor, D.F.M. and Vieira, N.A.Q. 1990. 111. Pasture development on cleared forest land in northern Amazonia. *Geogr. J.* **156**, 283-296.

Fisch, G. and Wright, I.R. and Bastable, H.G. 1994. Albedo of Tropical Grass: A case study of pre- and post burning. *Int. J. Climatol.* **14**, 103-107.

Garratt 1993. Sensitivity of climate simulations to land-surface and atmospheric boundary-layer treatments - a review. *J. Climate* **6**, 419-449.

Gash, J.H.C. 1986. Observations of turbulence downwind of a forest-heath interface. *Bound.- Layer Met.* **36**, 227-237.

Gash, J.H.C., Nobre, C.A., Roberts, J.M. and Victoria, R. L. 1996. An overview of the ABRACOS project. This volume

Grace, J., Lloyd, J., McIntyre, J., Miranda, A., Meir, P. and Miranda, H. 1996. Carbon fluxes to Amazon forest in Rondônia. This volume.

Grimaldi, M., Sarrazin, M., Chauval, A., Luizao, F., Nunes, N., Rodriguez, M.R.L., Amblard, P., Tessier, D. 1993. Effets de la déforestation et des cultures sur la structure des sols argileux d'Amazonie Brésilienne. *Cahiers agricultures* **2**, 36-47.

Hall, R.L. and Roberts J.M. 1989. Hydrological aspects of new broad-leaf plantations. *SEESOIL* **6**, 2-38.

Henderson-Sellers, A. 1992. Assessing the sensitivity of a land-surface scheme to parameters used in tropical deforestation experiments. *Q. J. R. Met. Soc.* **118**, 1101-1116.

Henderson-Sellers and Gornitz, V. 1984. Possible climatic impacts of land cover transformations with particular emphasis on tropical deforestation. *Climate Change* **6**, 231-258.

Hillel, D., Krentos, J.D., and Stylianou, Y. 1972. Procedure and test of an internal drainage method for measuring soil hydraulic characteristics in situ. *Soil Sci.* **114**, 395-400.

Hodnett, M.G., da Silva, L.P., da Rocha, H.R. and Cruz Senna, R.C. 1995. Seasonal soil water storage changes beneath central Amazonian rainforest and pasture. *J. Hydrol.* **170**, 233-254.

Idso, S.B., Jackson, R.D., Reginato, R.J. Kimball, B.A. and Nakayama, F.S. 1975. The dependence of bare soil albedo on soil water content. *J. Appl. Met.* **14**, 109-113.

Jarvis, P.G. 1976. The interpretation of the variations in leaf water potential and stomatal conductances found in canopies in the field. *Philos. Trans. Roy. Soc. Lond.* ***B*** **273**, 593-610.

Kelliher, F.M., Leuning, R., Raupach, M.R. and Schulze, E-D. 1995. Maximum conductance for evaporation of global vegetation types. *Agric. For. Met.* **73**, 1-16.

Kineman, J.J. and Ohrenschall, M.A. 1992. Global ecosystems database. US Dept of Commerce, Boulder, Colorado.

Klinge, H. 1973. Struktur und Artenreichtum des zentral amazonischen Regenwaldes. *Amazoniana* **4**, 283-292.

Klinge, H., Rodriques, W.A., Brunig, E.F. and Fittkau, E.J. 1975. Biomass and structure in a central Amazonian Rain Forest. In: Golley, F.B. and Medina, E. (Eds.), *Tropical Ecological Systems. Trends in Terrestrial and Aquatic Research*. Springer Verlag, Berlin, 115-122.

Lean, J., Bunton, C.B., Nobre, C.A. and Rowntree, P.R. 1996. The simulated impact of Amazonian deforestation on climate using measured ABRACOS vegetation characteristics. This volume.

Lean, J. and Rowntree, P.R. 1993. A GCM simulation of the impact of Amazonian deforestation on climate using an improved canopy representation. *Q. J. R. Met. Soc.* **119**, 509-530.

Lean, J. and Warrilow, D.A. 1989. Climatic impact of Amazon deforestation. *Nature* **342**, 411-413.

Lloyd, C.R., Gash, J.H.C., Shuttleworth, W.J. and Marques, A.O. 1988. The measurement and modelling of rainfall interception by Amazonian rain forest. *Agric For. Met.* **43**, 277-294.

Manzi, A.O. and Planton, S. 1994. Implementation of the ISBA parameterization scheme for land surface processes in a GCM - an annual cycle experiment. *J. Hydrol.* **155**, 353-387.

Manzi, A.O. and Planton, S. 1996. A simulation of Amazonian deforestation using a GCM calibrated with ABRACOS and ARME data. This volume.

McWilliam, A-L.C., Roberts, J.M., Cabral, O.M.R., Leitao, M.V.B.R., de Costa, A.C.L. Maitelli, G.T. and Zamparoni, C.A.G.P. 1993. Leaf area index and above-ground biomass of terra firme rain forest and adjacent clearings in Amazonia. *Functional Ecol.* **7**, 310-317.

McWilliam, A-L.C., Cabral, O.M.R., Gomes, B.M., Esteves, J.L., Roberts, J.M. 1996. Forest and pasture leaf-gas exchange in south-west Amazonia. This volume.

Martins, P.F.S., Cerri, C.C., Volkoff, B., Andreux, F. and Chauvel, A. 1991. Consequences of clearing and tillage on the soil of a natural Amazonian ecosystem. *For. Ecol. Manage.* **38**, 273-282.

Medina, B.F. 1985. Influência de dois métodos de preparo de área na compactação de um latossolo amarelo no Estudo do Amazônas. In: *Anais do II Simposio do trópico úmido, Belém,* EMBRAPA-CPATU, Brazil, 446-455.

Medina, B.F. and Leite, J.A. 1985. Influência de Trés sistemas de manejo e duas coberturas vegetais na infiltração de água em um latassolo amarelo em Manaus AM. *Pesq. Agropec. Bras.*, Brasilia,

20, 1323-1331.

Mintz, Y. 1984. The sensitivity of numerically simulated climates to land surface boundary conditions. In: Houghton, J.T. (Ed.) *The Global Climate*. Cambridge University Press. 79-105.

Molion, L.C.B. and Moore, C.J. 1983. Estimating the zero-plane displacement for tall vegetation using a mass conservation method. *Bound.-Layer Met.* **26**, 115-125.

Moraes, S.O. 1991. Heterogeneidade hidráulica de uma Terra Roxa Estruturada. Thesis. Escola Superior de Agricultura 'Luiz de Queiroz'. Universidade São Paulo, Piracicaba, Brasil.

Nepstad, D.C., de Carvalho, C.R., Davidson, E.A., Jipp, P.H., Lefebvre, P.A., Negreiros, G.H., da Silva, E.D., Stone, T.A., Trumbore, S.E. and Vieira, S. 1994. The role of deep roots in the hydrological and carbon cycles of Amazonian forests and pastures. *Nature* **372**, 666-669.

Nobre, C.A., Sellers P.J. and Shukla J. 1991. Amazonian deforestation and regional climate change. *J. Climate* **4**, 957-988.

Noilhan, J. and Planton, S. 1989. A simple parameterization of land surface processes for meteorological models. *Mon. Weath. Rev.* **117**, 536-549.

Pires, J.M. 1978. The forest ecosystems of the Brazilian Amazon: Description, functioning and research needs. *Tropical Forest Ecosystems*, 607-627. UNESCO, Paris.

Pitman, A.J. 1993. The sensitivity of the land surface to the parameter values: a reassessment. In: Mesoscale Modelling of the Hydrosphere. *IAHS Publ. No. 214*. 125-132.

Polcher, J. and Laval, K. 1994. The impact of African and Amazonian deforestation on tropical climate. *J. Hydrol.* **155**, 389-405.

RADAM 1980. Projeto RADAMBRASIL, Ministério de Minas e Energia, Rio de Janeiro, Brazil.

Ranzani, G. 1980. Identificação e caracterização de alguns solos da Estação Experimental de Silvicultura Tropical do INPA. *Acta Amazonica* **10**, 7-41.

Reategui, K., Ruiz, R., Cantera, G. and Lascano, C. 1990. Persistencia de pasturas asociados con diferentes manejos del pastores en un Ultisol arcilloso de Puerto Bermudez, Peru. *Pasturas Tropicales* **12**, 16-24.

Reichardt, K., Libardi, P.L. and do Nascimento, V.F. 1978. Condutividade hidraúlica saturada de um perfil de Terra Roxa Estruturada (alfisol). *R. Bras. Ci. Solo* **2**, 21-24.

Roberts, J., Cabral, O.M.R., Fisch, G., Molion, L.C.B., Moore, C.J. and Shuttleworth, W.J. (1993). Transpiration from an Amazonian rainforest calculated from stomatal conductance measurements. *Agric. For. Met.* **65**, 175-196.

Roberts, J.M., Cabral, O.M.R., da Costa, J.P., McWilliam, A-L.C. and Sa, T.D.A 1996. Leaf area index, biomass and physiological responses in Amazonian forest and pasture: an overview. This volume.

da Rocha, H.R., Nobre, C.A., Bonatti, J.P., Wright, I.R. and Sellers, P.J. 1996. A Vegetation-Atmosphere Interaction Study for Amazonian deforestation using field data and a single column model. *Q. J. R. Met. Soc.* In Press.

Rosenberg, N.J., Blad, B.L. and Verma, S.B. 1983. *Microclimate: The biological environment.* Wiley, New York, 495 pp.

Rowntree, P.R. 1991. Atmospheric parametrization schemes for evaporation over land: basic concepts and climate modelling aspects. In: T.J. Schmugge and J-C. Andre (Eds) *Land Surface Evaporation, Measurement and Parametrization*, Springer-Verlag, New York, USA. 5-29.

Rutter, A.J., Kershaw, K.A., Robins, P.C. and Morton, A.J. 1971. A predictive model of rainfall interception in forest. 1. Derivation of the model from observations in a plantation of Corsican pine. *Agric. Met.* **9**, 367-389.

Sá, T.D.A., Costa, P.C. and Roberts, J.M. 1996. Forest and pasture conductances in southern Pará, Amazonia. This volume.

Salomão, R. de P. 1991. Uso de parcelas permanentes para estudos da vegetação da floresta tropical úmida. 1. Município de Marabá, Pará. *Bol. Mus. Para. Emilio Goeldi, sér. Bot.* **7**, 543-604.

Sanchez., P.A. 1976. *Properties and management of soils in the tropics.* John Wiley, New York.

Saunders, L.C.U., Libardi, P.L. and Reichardt, K. 1987. Condutividade hidraulica da terra roxa estrutarada em condiçoes de campo. *R. Bras. Ci. Solo* **2**, 164-167.

Sellers, P.J., Mintz. Y, Sud, Y.C. and Dalcher, A. 1986. A simple biosphere model (SiB) for use within general circulation models. *J. Atmos. Sci.* **43**, 505-531.

Sellers, P.J., Shuttleworth, W.J., Dorman, J.L., Dalcher, A. and Roberts, J.M. 1989. Calibrating the Simple Biosphere Model (SiB) for Amazonian data; part 1, average calibration with field data. *J. Clim. Appl. Met.* **28**, 727-759.

Schulze, E-D., Kelliher, F.M., Korner, C., Lloyd, J. and Leuning, R. 1994. Relationships between maximum stomatal conductance, ecosystem surface conductance, carbon assimilation rate and plant nitrogen nutrition: a global ecology scaling exercise. *Oecologia* **95**, 153-163.

Shukla, J. and Mintz, Y. 1982. Influence of land-surface evapotranspiration on the earth's climate. *Science* **215**, 1498-1501.

Shuttleworth, W.J. 1988. Evaporation from Amazonian rainforest. *Proc. R. Soc. Lond.* ***B*** **233**, 321-346.

Shuttleworth, W.J. 1989. Micrometeorology of temperate and tropical forest. *Phil. Trans. R. Soc. Lond.* ***B*** **324**, 299-334.

Shuttleworth, W.J., Gash, J.H.C., Lloyd, C.R., Moore, C.J., Roberts, J.M., Marques Filho, A. de O., Fisch, G., Silva Filho, V. de P., Ribeiro, M. de N.G., Molion, L.C.B., de Abreu Sa, L.D., Nobre, C.A., Cabral, O.M.R., Patel, S.R. and de Moraes, J.C. 1984. Observations of radiation exchange above and below Amazonian forest. *Q. J. R. Met. Soc.* **110**, 1143-1162.

Shuttleworth, W.J., Gash, J.H.C., Lloyd, C.R., McNeil, D.D., Moore, C.J. and Wallace, J.S. 1988. An integrated micrometeorological system for evaporation measurement. *Agric. For. Met.* **43**, 295-317.

Shuttleworth, W.J., Gurney, R.J., Hsu, A.Y. and Ormsby, J.P. 1989. FIFE: the variation in energy partition at surface flux sites. *IAHS. Publ. No. 186*, 67-74.

Shuttleworth, W.J. and Dickinson, R.E. 1989. Comments on Dickinson and Henderson-Sellers (1988). *Q. J. R. Met. Soc.* **115**, 1177-1179.

Shuttleworth, W.J., Gash, J.H.C., Roberts, J.M., Nobre, C.A., Molion, L.C.B. and Ribeiro, M.N.G. 1991. Post-deforestation Amazonian climate: Anglo-Brazilian research to improve prediction. *J. Hydrol.* **129**, 71-85.

Stewart, J.B. and Verma, S.B. 1992. Comparison of surface fluxes and conductivities at two contrasting sites within the FIFE area. *J. Geophys. Res.* **97**, 18623-18628.

Sud, Y.C. and Fennessy, M.J. 1982. A study of the influence of surface albedo on July circulation in semi-arid regions using the GLAS GCM. *J. Climatol.* **2**, 105-125.

Sud, Y.C. and Smith, W.E. 1985. The influence of surface roughness of deserts on July circulation - a numerical study. *Bound.-Layer Met.* **33**, 15-49.

Sud, Y.C., Shukla, J. and Mintz, Y. 1988. Influence of land surface roughness on atmospheric circulation and precipitation: a sensitivity study with a GCM. *J. Appl. Met.* **27**, 1036-1054.

Takeuchi, M. 1961. The Structure of the Amazonian vegetation 2. Tropical forest. Journal of the Faculty of Science, University of Tokyo, Section 3, Botany, 1-26.

Tomasella, J. and Hodnett, M.G. 1996. Soil hydraulic properties of an oxisol under pasture in central Amazonia. This Volume.

Ubarana, V. 1996. Observation and modelling of rainfall interception in two experimental sites in Amazonian forest. This volume.

Uhl, C. 1988. Restoration of degraded lands in the Amazon basin. In Wilson, E.O. (ed.) *Biodiversity*. National Academy Press. Washington. 326-332.

Uhl, C., Buschbacher, R. and Serrao, A.S. 1988. Abandoned pastures in eastern Amazonia. 1. Patterns of plant succession. J. Ecol. **76**, 663-681.

Van Genuchten, M. Th. 1980. A closed formulation for predicting hydraulic conductivity in unsaturated soils. *Soil Sci. Soc. Am. J.* **44**, 892-898.

Van Lier, Q. J. and Neto, D.D. 1993. Valores extremos de umidade do solo referentes ao modelo de Van Genuchten. *R. Bras. Ci. Solo. Campinas* 17, 325-329.

Vieira, L.S. and Santos, P.C.T.C. 1987. Amazônia: seus solos e outros recursos naturais. Editora

Agronômica Ceres, São Paulo, Brasil.
Warrilow, D.A., Sangster, A.B. and Slingo A. 1986. Modelling of land surface processes and their influence on European climate. Met O 20 DCTN 38, Meteorological Office, Bracknell, UK.
Webb, R.S. and Rosensweig, C.E. 1993. Specifying land surface characteristics in General Circulation Models: Soil profile data set and derived water-holding capacities. *Global Biogeochem. Cycles* **7**, 97-108.
Wilson, M.F. and Henderson-Sellers, A. 1985. A global archive of land cover and soils data for use in general circulation climate models. *J. Climatol.* **5**, 119-143.
Wilson, M.F., Henderson-Sellers, A., Dickinson, R.E. and Kennedy, P.J. 1987. Sensitivity of the Biosphere-Atmosphere Transfer Scheme (BATS) to the inclusion of variable soil characteristics. *J. Clim. Appl. Met.* **26**, 341-362.
Winkworth, R.E. 1970. The soil water regime of an arid grassland (*Eragrostis eriopoda* Benth.) community in Central Australia. *Agric. Met.* **7**, 387-399.
Wright, I.R., Gash, J.H.C., da Rocha, H.R., Shuttleworth, W.J., Nobre, C.A., Maitelli, G.T., Zamparoni, C.A.G.P. and Carvalho, P.R.A. 1992. Dry season micrometeorology of central Amazonian ranchland. *Q. J. R. Met. Soc.* **118**, 1083-1099.
Wright, I.R., Manzi, A.O. and da Rocha, H.R. 1995. Surface conductance of Amazonian pasture: model application and calibration for canopy climate. *Agric. For. Met.* **75**, 51-70.
Wright, I.R., Gash, J.H.C., da Rocha, H.R. and Roberts J.M. 1996. Modelling surface conductance for Amazonian pasture and forest. This volume.

RESUMO

Parâmetros de superfície, para vegetação de pastagem e floresta na Amazônia, são apresentados para a utilização em experimentos de desmatamento realizados por modelos de circulação geral (MCG). Os valores dos parâmetros são baseados, predominsntemente, nas medições registrada peolo projeto ABRACOS, e complementados, quando necessário, pelos resultados de outras pesquisas na Amazônia. Os parâmetros são independentes dos diversos esquemas de transferência Solo-Vegetação-Atmosfera (ETSVA) atualmente em uso, entretanto, quando um particular submodelo é utilizado para calibrar um parâmetro, a estrutura do submodelo é apresentada. A variação sazonal dos parâmetros de vegetação é apresentada quando possível.

Na floresta, sob a influência da umidade de solo, o albedo médio mensal variou de 0,121 a 0,144, ao passo que, na pastagem, sob uma variação no índice de área foliar entre 0,5 e 4,0, o albedo variou de 0,155 a 0,20. A rugosidade aerodinâmica, z_0, nos sítios de pastagem, foi consistentemente 10% da altura da vegetação, variando de $z_0 = 0{,}025$ m a $z_0 = 0{,}08$ m, e foi encontrada como sendo independente do índice de área foliar. A média da rugosidade de dois sítios de floresta foi de $z_0 = 2{,}35$ m. Parâmetros obtidos em três diferentes tipos de solo na Amazônia (Latossol, Podssol e Terra Vermelha) são dados com uma distribuição geográfica simples, consistente com a limitada informação disponível para uma modelagem representativa do sistema solo-água. Os parâmetros de retenção e relativos à dinâmica da água no solo são apresentados para várias profundidades, entretanto, a existência de raízes profundas nas florestas, consequência da ausência de sazonalidade na transpiração, requer cuidado na escolha, pelos modelos, da profundidade do sistema radicular.

Para a modelagem da transpiração, a condutância estomática "bulk" continua como o resultado empírico dos métodos de otimização: a falta de independência para com os outros parâmetros dificulta a compreensão da diferença entre os sítios. Também a influência da estrutura das plantas no albedo, transpiração e evaporação do solo não está ainda bem acoplada em muitos ETSVA. Para modelagem hidrológica, o detalhado mapeamento pedológico da Amazônia é insuficiente e pouco completo com relação aos parâmetros do sistema solo-água. Estudos de sensibilidade utilizando os parâmetros publicados pelo estudo do ABRACOS são agora necessários para investigar a importância relativa de cada uma das áreas de pesquisa em trabalhos futuros.

29 A simulation of Amazonian deforestation using a GCM calibrated with ABRACOS and ARME data

A.O. MANZI[1] and S. PLANTON[2]
[1]Instituto Nacional de Pesquisas Espaciais, São José dos Campos, Brazil
[2]Météo-France, Centre National de Recherches Météorologiques, Toulouse, France

INTRODUCTION

The climatic effects of large scale deforestation are caused by complex interactions between the soil-vegetation system, clouds, radiation and the general circulation of the atmosphere. GCMs allow these to interact freely, making them uniquely powerful tools for simulating different climate change scenarios. A new land surface scheme ISBA (Intéractions Sol-Biosphère et Atmosphère) has been developed at CNRM (Centre National de Recherches Météorologiques), the research institute of Météo-France, for inclusion in the forecast, climate and mesoscale models. This scheme (Noilhan and Planton, 1989), despite its simplicity, represents all the principle physical processes present in more complex hydrological and ecological models of the soil-vegetation interface. Water and heat transfer through the soil depend on soil texture and moisture content; vegetation dependent processes such as the interception of rainfall and dew are treated explicitly, and transpiration is restricted by a surface resistance.

A first step in validating any soil-vegetation atmosphere transfer scheme consists of validating its ability to reproduce local scale observations of the water and energy balance. ISBA was tested and validated in this way using data from a variety of vegetation types and bare soil, drawn from the HAPEX-MOBILHY, FIFE and EFEDA datasets (Noilhan *et al.*, 1992). In this paper, these tests are enhanced by data from ARME (Amazon Region Micrometeorological Experiment, 1983 - 1985; Shuttleworth *et al.*, 1984a) and from ABRACOS (Gash *et al.*, 1996) to improve the ability of ISBA to represent tropical forest and the pasture which normally replaces it after deforestation. Further description of this work is given by Manzi (1993). The added realism gained by incorporating the ISBA scheme into EMERAUDE has allowed it to be used to simulate the impact of tropical deforestation.

The sensitivity of Amazonian climate to deforestation has been studied previously. Recent papers include those by Lean and Warrilow (1989), Nobre *et al.* (1991), Dickinson and Kennedy (1992), Dirmeyer (1992), Lean and Rowntree (1993), and

Amazonian deforestation and climate. Edited by J.H.C. Gash, C.A. Nobre, J.M. Roberts and R.L. Victoria.

Polcher and Laval (1994a and 1994b). The results of these modelling studies agree with regard to the sign of the resultant climatic changes but not always in their magnitude. The principal points on which they agree are the likely increase of the surface temperature and the reduction in evaporation and rainfall following deforestation. The motivation for carrying out another deforestation simulation is that the main climatic impacts are dependent on the individual models' representation of the physical processes, such as surface hydrology, convection and cloudiness, which are still poorly described and differ between models.

CALIBRATION OF ISBA WITH ARME AND ABRACOS DATA

A one-dimensional version of the ISBA surface scheme (Noilhan and Planton, 1989) was tested against observational data, both to verify the realism of the representation of the physical processes and to calibrate those parameters which could not be derived directly from the observations. In addition to the tests which were carried out using the data from HAPEX-MOBILHY, FIFE and EFEDA (Noilhan *et al.*, 1992) a one dimensional version of the model was tested against observational data from the ARME (Shuttleworth *et al.*, 1984a,b; Shuttleworth, 1988) and from ABRACOS (Gash *et al.*, 1996). These tests allowed verification of the ability of the model to reproduce the energy and water balances of equatorial rainforest and the tropical pasture which normally replaces it after deforestation.

AMAZONIAN FOREST

The ARME forest data were collected in the Reserva Ducke between September 1983 and September 1985, at the same site later used by ABRACOS (see Gash *et al.*, 1996). The site was considered representative of the *terra firme* forest (not seasonally flooded) of central Amazonia.

The tests carried out with the ARME data showed that some specific details of the parameterization needed to be improved. Firstly, because of the height of the vegetation it was necessary to introduce a zero plane displacement into the calculation of the turbulent exchange coefficients. Secondly it was necessary to introduce a term to account for the storage of heat in the vegetation, which can amount to up to 10 per cent of the daily net radiation. Adjusting the thermal coefficient of the vegetation, C_{veg}, to a value of $2\ 10^{-5}\ m^2KJ^{-1}$ gave correct simulation of heat storage in the forest soil-vegetation system, which agreed with the observations of Moore and Fisch (1986). This also improved the calculation of surface temperature, which was previously too low - particularly during the night. Without this calibration, *i.e.*, using the value of C_{veg} originally proposed by Noilhan and Planton (1989), ISBA calculates a flux from storage heat which is almost zero, producing night time surface temperatures which are too low.

Evaporation measurements after rainfall have shown that during the daytime water intercepted by forest canopies evaporates within a few hours. The ISBA scheme previously evaporated this water too rapidly and has therefore been modified to reduce the rate of evaporation of water intercepted by dense vegetation and giving a lower rate of evaporation for rainfall intercepted by the lower levels of the canopy.

Some parameters have been extracted directly from the data. Roughness length, z_0, has been taken as 2 m, and zero plane displacement, *d*, as 30 m. These values are consistent with those derived by other authors using the same dataset. Following the observations of McWilliam *et al.* (1993), for a similar site to Reserva Ducke, leaf area index, L^*, has been taken as 6.0. Taking into account the low values of radiation observed at the forest floor by Shuttleworth *et al.* (1984b) vegetation fraction, *veg*, has been taken as 0.99; for the albedo, α, the value of 0.12 reported in the same analysis has been used.

The calibration of the scheme was carried out on a selection of days on which the maximum amount of data were available. For these periods the thickness of the layer of hydrologically active soil (medium texture) was taken as 1 m.

Finally, a value of 250 sm^{-1} was derived for the minimum stomatal resistance per unit leaf area, R_{smin}, by fitting the model to the measured Bowen ratios.

The value of the maximum canopy interception capacity, $W_{rmax} = 0.1\ veg\ L^*$, was derived by running the scheme for a period of 25 months, forced by the observed atmospheric variables, and comparing the predicted interception loss with the measurements of Lloyd *et al.* (1988). Using the same approach a parameter sensitivity analysis was also carried out - the results can be summarised as:

(1) Reduction in z_0 modified the energy balance giving a reduction in evaporation (particularly evaporation directly from the soil and evaporation of intercepted water), and an increase in sensible heat flux and the amplitude of the daily surface temperature cycle.

(2) The effect of the initial value of the soil water reservoir on the energy balance is limited to the first month of the integration, the high rate of precipitation quickly produces an equilibrium situation with the soil moisture level close to saturation.

(3) Soil texture has very little effect on the energy or water balance.

(4) The effect of increasing the depth of the active soil layer on the energy balance is very small; however increasing the capacity of the soil water reservoir did cause an increase in runoff and reduced the seasonal variation in soil water content.

(5) The energy balance is highly sensitive to the vegetation fraction, *veg*. The relative contributions of evaporation directly from the soil and transpiration to the total evapotranspiration are strongly affected by the value of *veg*.

evapotranspiration are strongly affected by the value of *veg*.

AMAZONIAN PASTURE

Nineteen days of consecutive data from the first ABRACOS intensive observation period have been used. These were collected during the 1990 dry season (Wright *et al.*, 1992) at Fazenda Dimona (see Gash *et al.*, 1996) which is 80 km north of Manaus and about 80 km from the Reserva Ducke. Fazenda Dimona is an open clearing where, about 15 years previously, the forest had been replaced with cattle pasture by cutting and burning the primary forest, then sowing grass on the clay soil.

Most of the parameters needed to describe the ABRACOS site are available from the published results: a clayey soil covered with grass of leaf area index 1.2; a vegetation fraction of 0.85; a roughness length for momentum of 0.026 m; a zero plane displacement of 0.17 m; and an albedo of 0.163 (Wright *et al.*, 1992; Bastable *et al.*, 1993; and McWilliam *et al.*, 1993).

Using these parameter values in the calibration tests for the soil heat flux has allowed the mean thermal coefficient for the soil-vegetation system C_T to be evaluated as 3 10^{-5} m^2 kJ^{-1} and the mean thermal coefficient of the vegetation C_{veg} as 5 10^{-5} m^2 kJ^{-1}. This adjustment to the previously assigned values gave much improved prediction of the surface temperature, which was otherwise too low during the night. ISBA has been calibrated against the ABRACOS data in three stages. In the first stage the roughness length for vapour and sensible heat transfer has been taken as being the same as that for momentum transfer. This is the assumption made in most surface schemes and which was made in the version of ISBA incorporated into the EMERAUDE GCM (Manzi and Planton, 1994).

An optimised value of the minimum surface resistance of 144 s m^{-1} gave very satisfactory agreement between the observed energy balance and that calculated by ISBA - both for the daily trend and for the cumulative values of the fluxes.

The cumulative evapotranspiration predicted by ISBA for the whole period was 47 mm, which can be compared with the observed value of 46 mm. Evaporation directly from the soil, transpiration, and evaporation of intercepted water amounted to 33, 64 and 3 per cent respectively of the total evapotranspiration, however this result cannot be validated as separate measurements of the components were not made. The loss of water as evapotranspiration and the gains by rainfall and dew are balanced by the observed variation in soil moisture content down to a depth of approximately 2 m.

The second stage of the calibration took account of the differences in roughness length between vapour and sensible heat transport, and that for momentum. These tests allow comparison between the surface temperature, T_s, predicted by ISBA with those estimated by Wright *et al.* (1992) who took $ln(z_{0m}/z_{0h}) = 2$, as suggested by Brutsaert (1982). Good agreement between T_s, as calculated by ISBA and as by Wright *et al.* is obtained with $ln(z_{0m}/z_{0h}) = 2.3$. The amplitude of the cycle of T_s is increased by 2.5 °C compared to the case when $ln(z_{0m}/z_{0h}) = 2$. This increase is dominated by the maximum temperature being raised by 2 °C, which causes an calibrated to reproduce the surface resistance estimated from the Penman-Monteith model. With the assumption that the vegetation totally covers the ground ($veg = 1$) and $z_{0m} = 10\ z_{0h}$, ISBA correctly reproduced the observed surface resistance and energy balance. However, the observed vegetation cover is not unity and sensitivity studies have shown that ISBA is very sensitive to vegetation cover, particularly the estimates of direct evaporation from the soil. (Jacquemin and Noilhan, 1990; Manzi, 1993). Therefore, when there is incomplete surface cover the surface resistance calculated by ISBA is not comparable to that calculated by the Penman-Monteith equation.

It can be concluded that ISBA allows several combinations of parameters for tropical pasture, all of which are capable of satisfactorily reproducing the observed reduction in evaporation in a drying soil. The only real improvement to the predictions that can be obtained by taking into account the difference between z_{0m} and z_{0h}, appears to be the slightly higher surface temperature of approximately 2 °C. This, together with the fact that after the calibration, tests with $z_{0m} = z_{0h}$ gave results that were just as good those with z_{0m} z_{0h} justifies not complicating ISBA, either by introducing procedures to account for the differences in transport of vapour and sensible heat compared to momentum, or by introducing different coefficients to describe the controls on surface resistance.

THE EMERAUDE MODEL

The GCM EMERAUDE is the spectral model used by Météo-France for short range operational forecasting between 1985 and 1991 (Ernie, 1985; Coiffier *et al.*, 1987; Geleyn *et al.*, 1988). The version used for the experiments reported here includes the modifications by Royer *et al.* (1990) and Planton *et al.* (1991) to improve the model's ability to make climate simulations. The horizontal resolution corresponds to a T42 truncation, so that the physical processes are resolved on a Gaussian grid of approximately 2.8° by 2.8°. There are 20 levels in the vertical, with five levels in the troposphere below 2000 m. The physical parameters include: radiation with interactive cloudiness, deep convection with a Kuo-type scheme, large scale precipitation, and orographic gravity waves. Boundary layer processes and turbulence are represented by the parameterization scheme of Louis *et al.* (1981).

METHOD

Two model runs each of 50.5 months were carried out with EMERAUDE coupled to the ISBA surface scheme: a control run where the vegetation has been classified according to Wilson and Henderson-Sellers (1985), and a deforested run in which the original humid tropical forest was replaced by grassland. The control run is identical to that described as the "ISBA experiment" by Manzi and Planton (1994), except that the integration time has been increased by 12 months. The only difference between the control run and the deforested run is the change made to the Amazonian vegetation.

The model runs were initialised using climatological snow and ice cover. Sea surface temperature was specified throughout the experiment using AMIP monthly data fields given by COLA/CAC (Center for Ocean-Land-Atmosphere Studies and NOAA Climate Analysis Center) averaged over the ten year period 1979-1988 (Gates, 1992). The initial state of the atmosphere was that on 15 December 1983, interpolated from the ECMWF (European Centre for Medium-Range Weather Forecasting) analysis. The initial soil moisture content was the climatological value given by Mintz and Serafini (1989) using the formula

$$w_s = w_2 = H_{ser} (w_{fc} - \text{veg } w_{fl} - 0.02) + \text{veg } w_{fl} + 0.02 \quad (1)$$

where w_s and w_2 are the volumetric water content of the surface and lower reservoirs respectively; H_{ser} is the Mintz and Serafini (1989) climatological relative humidity, which varies between 0 and 1; w_{fc} and w_f are volumetric soil water content at field capacity and the wilting point respectively; and 0.02 is the minimum value of soil water content obtained by Manzi and Planton (1992) during a previous annual integration of EMERAUDE with ISBA. This equation gives the maximum value equivalent to field capacity when H_{ser} is equal to unity, and when H_{ser} is zero gives values slightly above wilting point for vegetation with complete canopy cover and very low values for bare soil. For both simulations the deep soil temperature is reset to a climatological value derived from the ECMWF analysis every twenty days. The soil water content is not reset.

Table 1 shows the values of the parameters used in ISBA to describe tropical forest and pasture. For the forest case these parameters have been calibrated with the ARME data as described in Section 2. For the pasture case the parameters were prescribed in a similar way to other recent deforestation experiments (for example Nobre *et al.*, 1991). The characteristics of this hypothetical grassland are similar to the parameters calibrated against the ABRACOS data collected over actual Amazonian pasture, as described in Section 2. The actual values derived in Section 2 could not be used as the data were not available until after the model experiments were completed.

The extent of the modelled deforestation can be seen in Figure 1 from the difference between the roughness lengths of the original forest and the modelled pasture, which

Table 1 The ISBA parameters representing Amazonian forest and the grassland which replaces it in the deforestation experiment

	albedo	leaf area index (L*)	R_{smin}/L (sm^{-1})	z_0 (m)	d (m)	veg.	depth of soil (m)
forest	0.13	6	42	2	28	0.99	4
grass	0.20	2	75	.06	.20	0.85	2

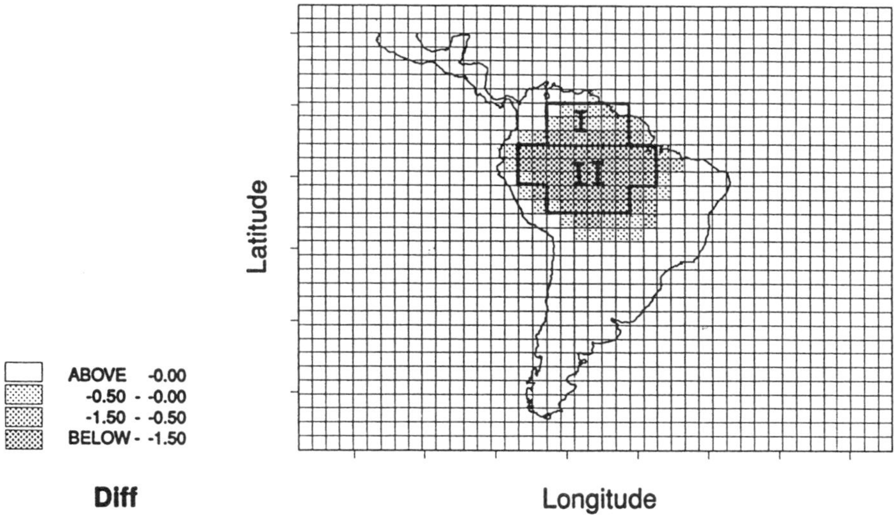

Figure 1 The extent of deforestation represented by the surface roughness length anomaly (grass minus forest). The areas chosen to represent northern and southern Amazonia are shown as I and II respectively.

is reduced from 2.0 m for the forest to 0.06 m for the pasture. This reduction in aerodynamic roughness modifies the wind field in the lower layers of the model and interacts strongly with the dynamics of the model.

The albedo for diffuse solar radiation is increased with deforestation from 13 to 20 per cent, which causes a decrease in surface net radiation and a reduction in convection and humidity convergence. The other important difference between the two vegetation types is the decrease, from 4 to 2 m, in the depth of the hydrologically active soil layer. This reduction in soil water holding capacity can create increased runoff and also reduction in evapotranspiration during the driest periods.

The increase in minimum surface resistance ($R_{sm} = R_{smin}/L^*$) for the grass, compared to the forest, reduces the transpiration, whilst the reduction in leaf area index, coupled with the reduction in roughness length, reduces the interception loss. On the other hand the increase in the fraction of bare soil (1 - veg) can contribute to increasing evaporation directly from the soil, which in part compensates for the reduction in transpiration and interception.

RESULTS AND DISCUSSION

The results have been averaged over three years of integration (from the 2 March to 5 February). The first 14.5 months, from initialisation on 15 December to the end of February 14.5 months later, have been ignored. This allows the model to reach a stage of quasi-equilibrium, with only a weak dependence on the initial soil water content.

Figure 2 shows the anomalies (deforestation minus control) for the fields of

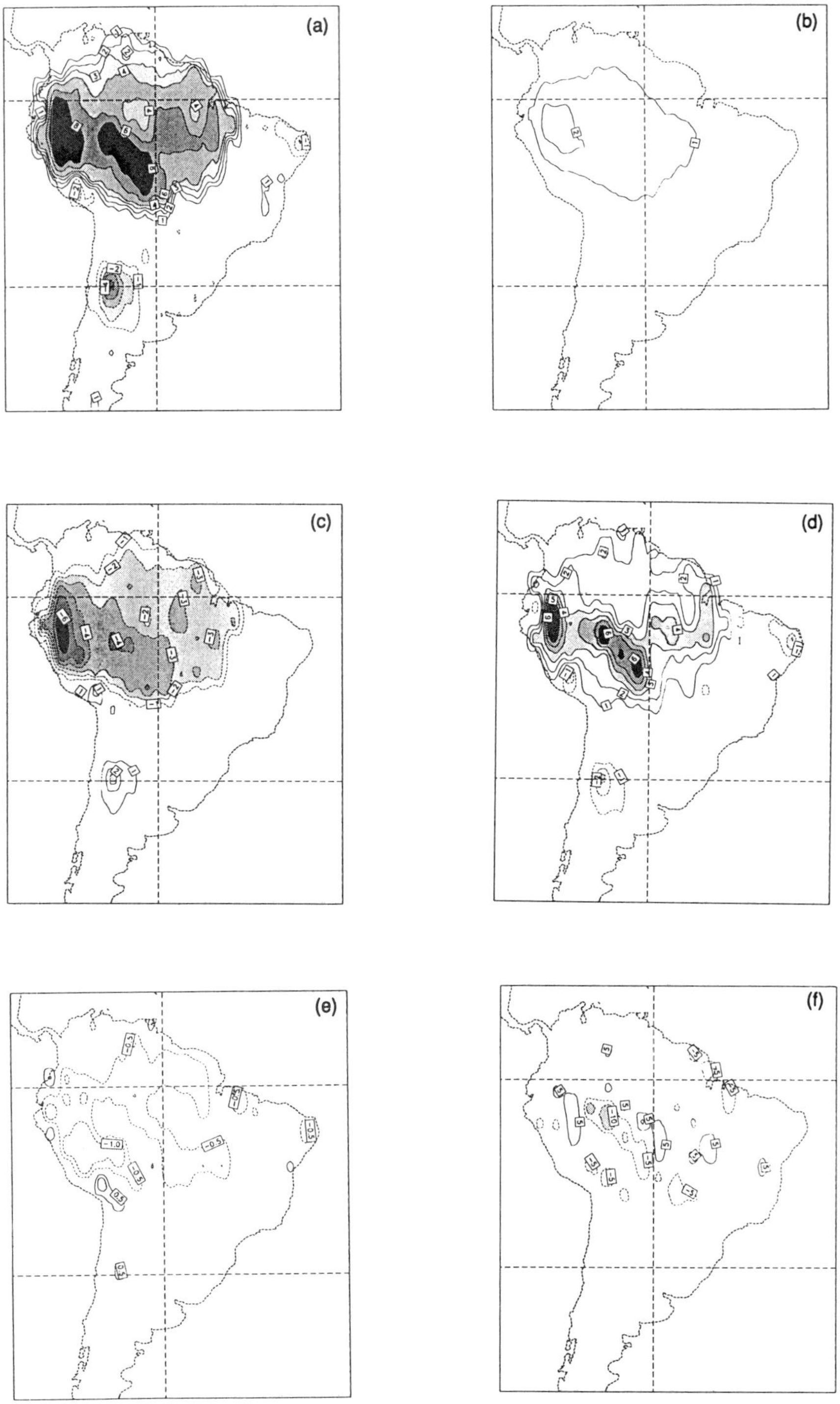
(a)
(b)
(c)
(d)
(e)
(f)

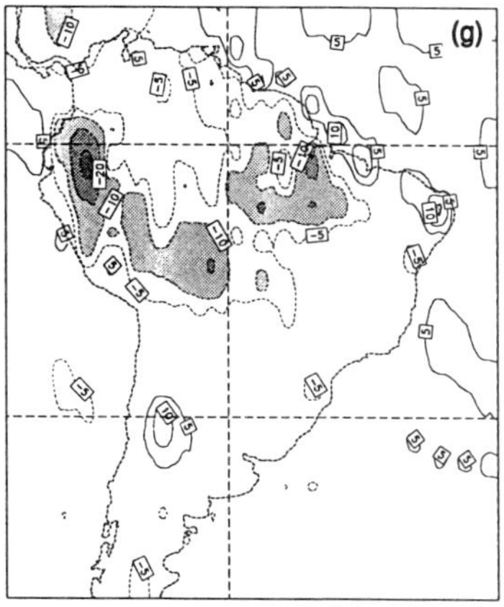

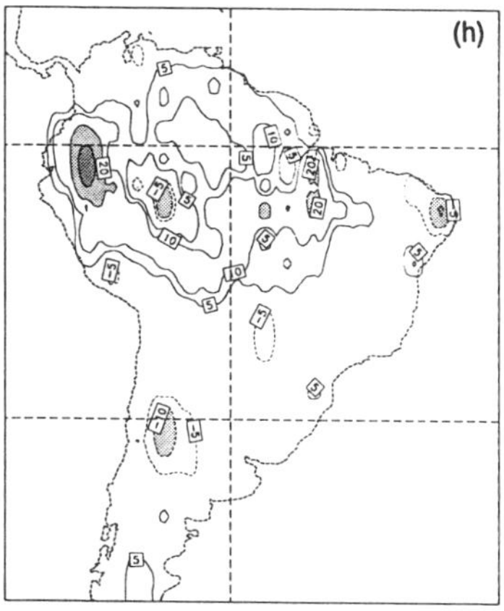

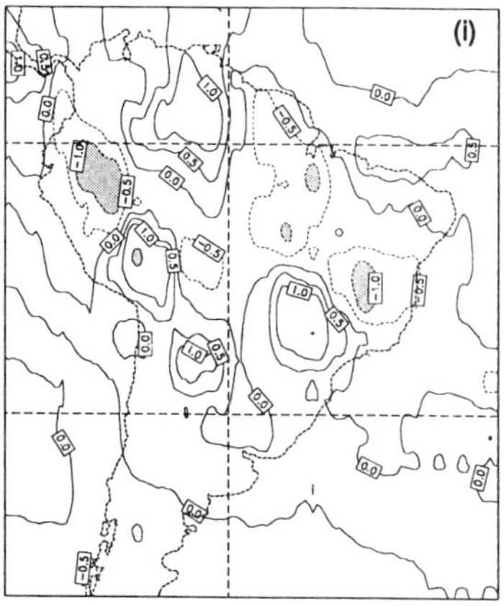

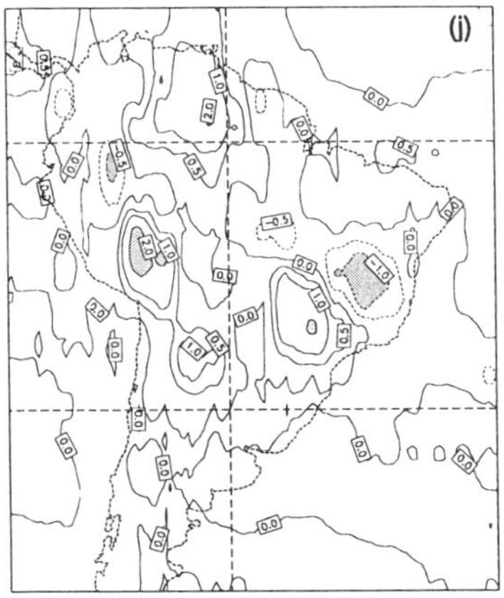

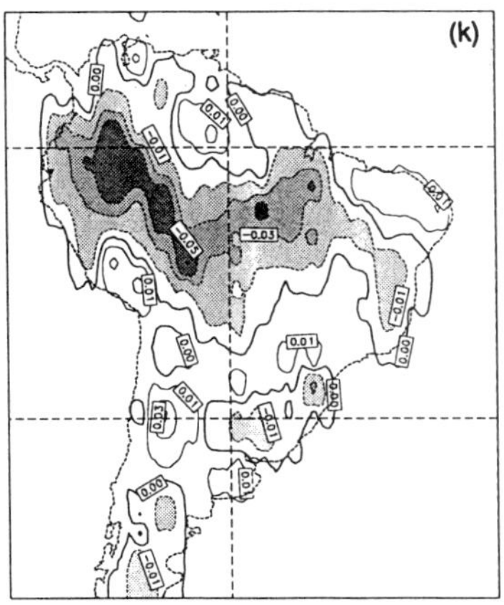

Figure 2 The anomalies (deforestation minus control) of the fields of meteorological variables over South America averaged over 36 months: (**a**) amplitude of the daily temperature cycle (°C); (**b**) air temperature (°C); (**c**) daily minimum surface temperature (°C); (**d**) daily maximum surface temperature (°C) (**e**) average surface temperature (°C); (**f**) net all-wave radiation (Wm^{-2}); (**g**) latent heat flux (Wm^{-2}); (**h**) sensible heat flux (Wm^{-2}); (**i**) rainfall (mm per day); (**j**) runoff (mm per day); and (**k**) volumetric soil moisture content (m^3/m^3).

meteorological variables averaged over the 36 months discussed above. The analysis has been limited to the region which has been deforested. The control run has been analysed by Manzi and Planton (1994) for the first 38.5 months and no significant changes have been detected in the extra 12 months. The values of these variables, averaged over the 36 months, are also presented in the pairs of Tables 4 and 5, 6 and 7, and 2 and 3 respectively for Region I (north), Region II (south) and Region I + II (total) as defined in Figure 1. Monthly values averaged over Region I + II, and over the last three years of the integration (both deforested and control) are given in Figure 3. In the north of Amazonia (shown as Region I in the figure) the wet season occurs during the northern hemisphere summer, whereas the wet season further south (Region II) occurs in the first months of the year (Figueroa and Nobre, 1990).

Figures 3c and 3j show the average monthly radiation and rainfall respectively measured at the Reserva Ducke site, near Manaus, during the 25 months of the ARME (Shuttleworth, 1988; Lloyd *et al.*, 1988; Manzi, 1993). The estimates of evapotranspiration and interception produced by ISBA (Noilhan*et al.*, 1992; Manzi, 1993) are shown in Figures 3h and 3i, respectively. The ARME results are presented as a separate line since they represent the particularly humid conditions of Manaus, whereas the modelling experiments are averaged over the whole of Amazonia, with its varied climatic conditions. However, the model predictions are in good general agreement with the observations, with the exception of the dry season, which is predicted to be too short and two months early in comparison with the observations. The model also predicts a second less-rainy period which does not occur in Manaus.

Following the modelled deforestation, the amplitude of the daily surface temperature cycle (Figure 2a) increased by an average of 6 °C. This increase in amplitude is the result of an increase of 3 °C in the daily maximum temperature (Figures 2d and 3e) and a decrease of 3 °C in the daily minimum. The average surface temperature is slightly decreased by about 0.5 °C (Figures 2e and 3d, and Table 2), but the temperature of the air close to the surface increased by an average of 1.3 °C (Figure 2b and Table 2). The increase in average air temperature, but decrease in the average surface temperature, is explained by the fact that the turbulent transfer is more efficient during the day when the surface temperature is higher over the grass than the forest.

The increase in the amplitude of the daily temperature cycle has also been found in other deforestation simulations (for example: Lean and Warrilow, 1989; Nobre*et al.*, 1991; Dickinson and Kennedy, 1992). They report an increase in maximum surface temperature, which produces an increase in average surface and air temperature. In the simulation reported by Polcher and Laval (1994b) the average surface temperature and the amplitude of the daily cycle are both slightly decreased by deforestation. The ABRACOS observations reported by Bastable *et al.* (1993) showed that the amplitude of the daily temperature cycle was greater over grass than over nearby forest, principally during the dry season, but that the average temperature was the same for the two vegetation types. In agreement with our results this was found for both dry and wet seasons.

The impact of deforestation on the temperature field is more important in Region

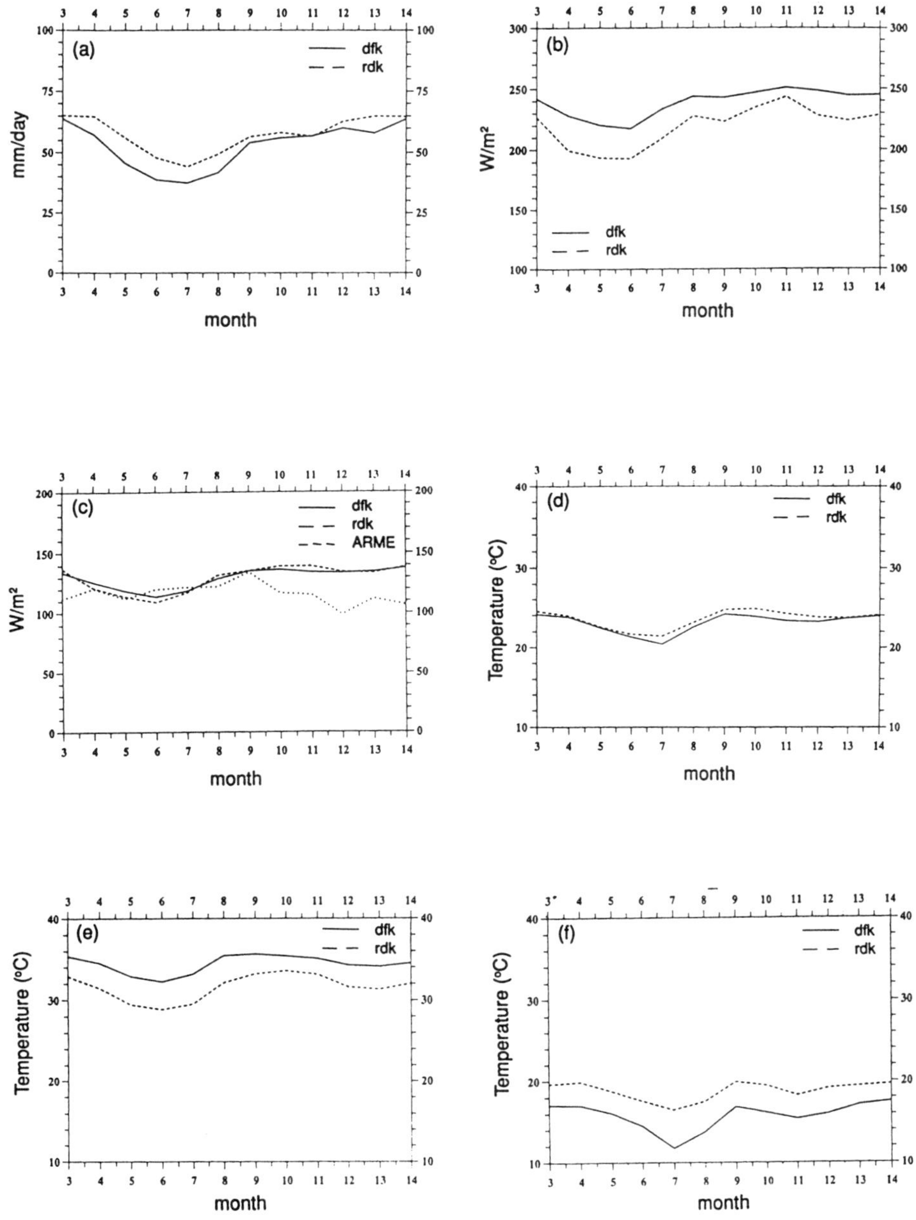

Figure 3 Monthly averages, over the 36 months for the whole of Amazonia (Regions I and II) for the deforestation experiment (solid line) and the control run (dashed line) and the ARME observations (dotted line): (**a**) cloudiness (%); (**b**) incident solar radiation (Wm^{-2}); (**c**) net all-wave radiation (Wm^{-2}); (**d**) average surface temperature (°C); (**e**) daily maximum surface temperature (°C); (**f**) daily minimum surface temperature (°C).

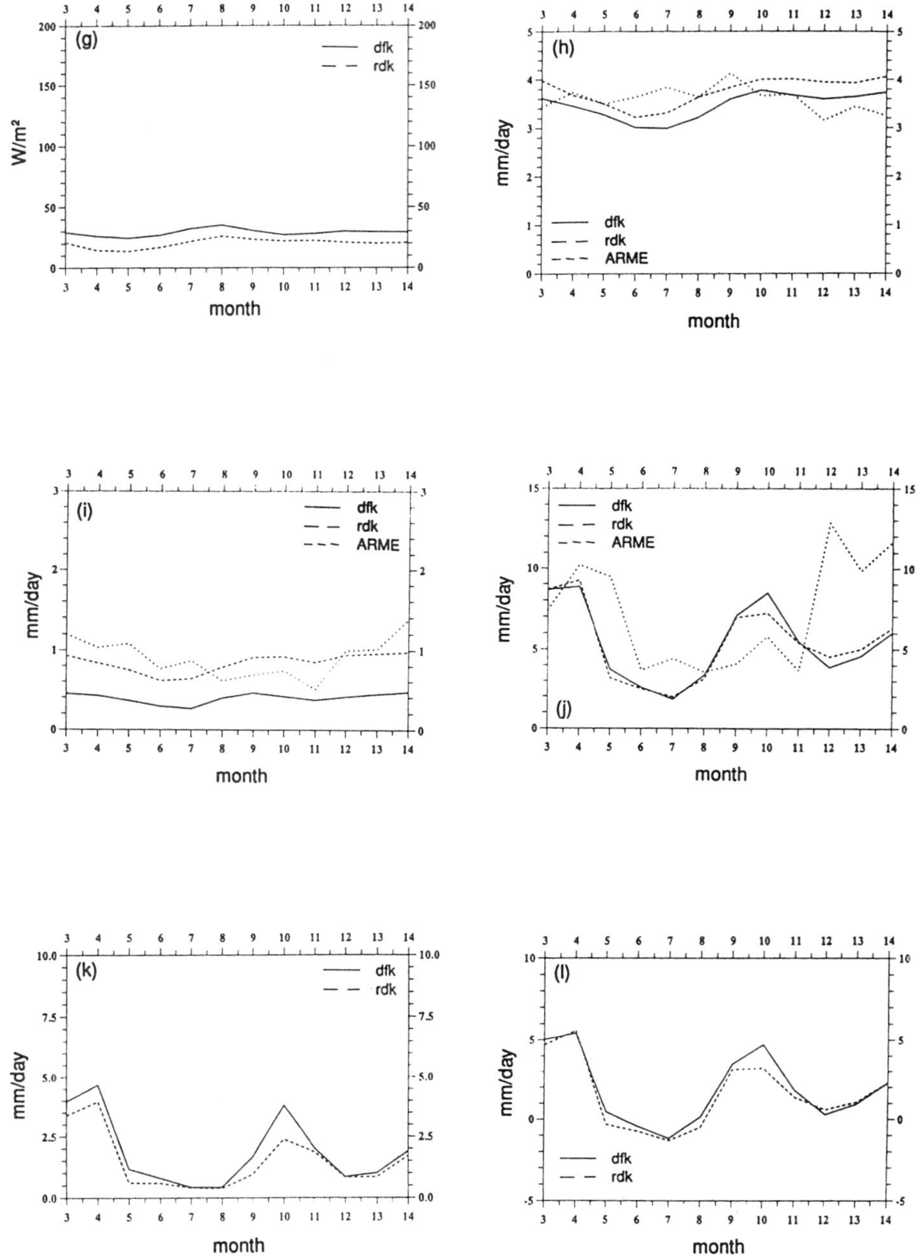

Figure 3 (contd.) (**g**) sensible heat flux (Wm^{-2}); (**h**) evapotranspiration (mm per day); (**i**) interception (mm per day); (**j**) rainfall (mm per day) (**k**) runoff (mm per day); and **l**) rainfall minus runoff (mm per day).

Table 2 The energy balance of Amazonia (Regions I and II) over the 36 months of the deforestation and control runs. S_D is the incident solar radiation; $(1 - \alpha)S_D$ is the net solar radiation; L_D is the downward longwave radiation; L_n is the net longwave radiation; R_n is the net all-wave radiation; LE is latent heat flux; H is sensible heat flux T_s(mean), T_s(max) and T_s(min) are the mean, daily maximum and daily minimum surface temperatures; T_a is air temperature at the lowest model level; N(tot) and N(low) are the percentages of total and low cloud cover; and U_a is horizontal windspeed at the lowest model level. Fluxes are expressed in Wm^{-2}, temperatures in °C and windspeed in ms^{-1}.

	S_D	$(1-\alpha)$ S_D	L_D	L_n	R_n	LE	H	T_s (mean)	T_s (max)	T_s (min)	T_a	N (tot)	N (low)	U_a
control	218	191	376	62	129	109	19	23.5	31.4	18.8	24.4	58	38	2.2
deforested	238	196	370	66	130	100	29	23.0	34.3	15.8	25.7	53	30	4.0
deforested – control	20	5	-6	4	1	-9	10	-0.5	2.9	-3.0	1.3	-5	-8	1.8

II (Table 3), to the south of the Equator, than in Region I (Table 4) to the north. This result is also found for the meteorological variables. On average there is no change to the net radiation, R_n (Figures 2f and 3c). Figure 3a shows that total cloudiness is reduced by about 5 per cent for the deforested region, but this is not uniform in the vertical. There was in fact a decrease of more than 8 per cent in low cloud amount, but this was partly compensated for by an increase in high cloud. The reduction in

Table 3 The energy balance of Region II of Amazonia for the 36 months of the deforestation and control run

	S_D	$(1-\alpha)$ S_D	L_D	L_n	R_n	LE	H	T_s (mean)	T_s (max)	T_s (min)	T_a	N (tot)	N (low)	U_a
control	221	193	374	65	128	106	23	23.7	32.1	18.6	24.7	56	35	2.1
deforested	244	199	367	71	128	95	35	23.2	35.7	15.1	26.2	50	26	4.3
deforested – control	23	6	-7	6	0	-11	12	-0.5	3.6	-3.5	1.5	-6	-9	2.2

Table 4 The energy balance of Region I of Amazonia for the 36 months of the deforestation and control run.

	S_D	$(1-\alpha)$ S_D	L_D	L_n	R_n	LE	H	T_s (mean)	T_s (max)	T_s (min)	T_a	N (tot)	N (low)	U_a
control	212	185	380	55	130	117	9	23.0	29.9	19.1	23.7	62	46	2.4
deforested	226	188	376	56	131	113	15	22.5	31.1	17.1	24.5	59	40	3.3
deforested – control	14	3	-4	1	1	-4	6	-0.5	1.2	-2.0	0.8	-3	-6	0.9

result is also found for the meteorological variables. On average there is no change to the net radiation, R_n (Figures 2f and 3c). Figure 3a shows that total cloudiness is reduced by about 5 per cent for the deforested region, but this is not uniform in the vertical. There was in fact a decrease of more than 8 per cent in low cloud amount, but this was partly compensated for by an increase in high cloud. The reduction in cloud cover leads to an average 20 Wm^{-2} increase in solar radiation reaching the surface. Because the albedo of the grass is higher than that of the forest some of this excess radiation is reflected by the surface and the net solar radiation over the grass is increased by only 5 Wm^{-2} compared to the forest. The downward longwave radiation is also reduced by about 6 Wm^{-2} as a result of the decreased cloud cover. However, the average surface temperature of the grass is less than the forest and it emits an average of about 2 Wm^{-2} less than the forest. The net longwave loss is thus increased by an average of about 4.5 Wm^{-2} and the net all-wave radiation increased by less than 1 Wm^{-2} (1.4 Wm^{-2} north of the Equator and 0.2 Wm^{-2} to the south). This increase is in contrast to other simulations of Amazonian deforestation. For example Lean and Warrilow (1989) reported a decrease in R_n of 21 Wm^{-2} as a result of the increase in surface albedo. In this case the longwave radiation did contribute, since with an increase in albedo of 5.2 per cent, in their simulation, it would have been necessary for the average incident solar radiation to have been above 400 Wm^{-2} to reduce R_n by 21 Wm^{-2} (whereas it did not exceed 250 Wm^{-2}). Nobre *et al.* (1991) used a fixed climatological cloudiness and found a reduction of 26 Wm^{-2} in R_n, which resulted in an increase of 8 Wm^{-2} in longwave loss (caused by an increase in surface temperature) and a decrease of 18 Wm^{-2} in net solar radiation (caused by an increase in the albedo). Dickinson and Kennedy (1992) also reported a decrease of 18 Wm^{-2} in R_n, dominated by an increase in longwave loss of 15 Wm^{-2}. This result was attributed to a reduction in the incident longwave radiation, caused by 7 per cent less cloud cover - the increase in incident solar radiation was compensated by the higher albedo of the grass. The disagreement between these simulations shows the strong dependence of the results of model experiments on the parameterization of clouds and radiative transfer. In fact, for Lean and Warrilow (1989) and Nobre *et al.* (1991) the decrease in net radiation at the surface is dominated by the effects of albedo, whilst for Dickinson and Kennedy (1992) it is dominated by changes to the longwave balance. In the simulation described in this paper the increase in longwave loss is compensated by an increase in net incident solar radiation.

Following deforestation the average latent heat flux (LE) decreased by 9 Wm^{-2} and the average sensible heat flux (H) increased by the same amount. The Bowen ratio, H/LE, therefore increased from 0.17 over the forest to 0.26 over the grass (Table 2). These changes do not occur uniformly over the whole area of deforestation - they are greater to the south of the Equator (Table 6) than to the north (Table 4). However, monthly mean evapotranspiration for the grass averaged over the whole region is always less than that for the forest (Figure 3h) and the sensible heat flux is always greater (Figure 3g). This comparatively lower latent heat flux from the grass is caused by a strong reduction in transpiration and the evaporation of intercepted water, which is only partly compensated by increased evaporation directly from the soil (Table 5,

Table 5 The water balance of Amazonia (Regions I and II) over the 36 months of the deforestation and control runs. P is rainfall; E evapotranspiration; R runoff; E_{tr} transpiration; E_g evaporation from the soil; I is interception (all expressed in mm per day). Ptt is precipitable water (mm) and HU_{soil} is relative soil moisture (%).

	P	E	R	E_{tr}	E_g	I	Ptt	HU_{soil}
control	5.33	3.78	1.51	2.44	0.51	0.83	37.43	82.3
deforested	5.29	3.47	1.84	1.74	1.33	0.39	37.04	76.2
deforested – control	-0.4	-0.31	0.33	-0.69	0.82	-0.44	-0.38	-0.61

Table 6 The water balance of Region I of Amazonia for the 36 months of the deforestation and control run

	P	E	R	E_{tr}	E_g	I	Ptt	HU_{soil}
control	6.47	4.04	2.43	2.45	0.97	0.62	37.33	100.0
deforested	6.97	3.90	3.15	2.02	1.49	0.39	37.81	100.0
deforested – control	0.50	-0.14	0.72	-0.43	0.52	-0.23	0.48	0.0

Table 7 The water balance of Region I of Amazonia for the 36 months of the deforestation and control run

	P	E	R	E_{tr}	E_g	I	Ptt	HU_{soil}
control	4.83	3.67	1.12	2.43	0.31	0.92	37.47	74.7
deforested	4.56	3.28	1.27	1.63	1.26	0.40	36.72	65.9
deforested – control	-0.27	-0.39	01.6	-0.81	0.95	-0.53	-0.75	-8.8

6 and 7). The reduction in transpiration is due to the reduced canopy cover and increased minimum surface resistance. The reduction in evaporation of intercepted water is caused both by the smaller canopy capacity of the grass, 0.3 mm, compared to the forest, 0.6 mm, and by the reduction in roughness length from 2 m for the forest to 0.06 m for the grass. The increase in direct evaporation from the soil follows from the increased proportion of exposed soil in the grassland and the high rainfall climate.

The reduction in evapotranspiration of about 8 per cent (0.31 mm per day) is less than that found in the other experiments discussed previously: 27 per cent (0.85 mm per day) by Lean and Warrilow, 30 per cent (1.36 mm per day) by Nobre *et al.*, and 20 per cent (0.7 mm per day) by Dickinson and Kennedy (1992). However the reduction is in good agreement with that found by Polcher and Laval (1994b), 11 per cent (0.35 mm per day). The different response of these various models is mainly associated with the performance of their surface schemes, but also with the model calculations of available energy and rainfall. For Lean and Warrilow (1989) the

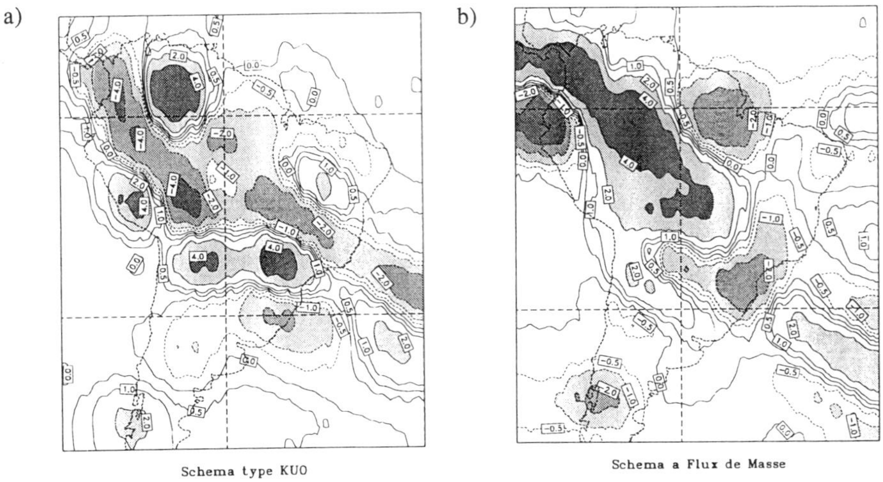

Figure 4 Anomalies in the rainfall fields (mm per day) for the deforestation minus control run for a) the Kuo-type convection scheme and b) the mass flux convection scheme averaged over January to March inclusive. The dotted lines represent negative values.

reduction in evapotranspiration (which was dominated by the interception loss) was attributed to the reduction in roughness and is the same size as the reduction in R_n. Dickinson and Kennedy (1992) reported a similar conclusion from their experiment. However Nobre *et al.* (1991) found that transpiration and evaporation directly from the soil accounted for 70 per cent of the reduction in evapotranspiration. In the deforestation simulation reported here the reduction in evapotranspiration is mainly caused by reduced evaporation of intercepted water, as shown in Figure 3i, since the reduction in transpiration is compensated by the increased evaporation from the soil surface.

It is important to note that the deforested areas chosen by the various authors are not necessarily coincident. However, with the exception of Polcher and Laval (1994b), they all represent the greater part of Amazonia. Polcher and Laval averaged over an area which was smaller than the area used here.

Rainfall is reduced in some places, but is increased in others (Figure 2i). It increases particularly in the north, south-west and south-east of Amazonia. This clearly implies that the deforestation produces a change in atmospheric circulation over Amazonia. The mean monthly rainfall over Amazonia is shown in Figure 3j, which shows that there is no stable tendency in the monthly regional rainfall anomaly during the year after deforestation. This seems to result from the seasonal cycle of moisture convergence (taken as the difference between rainfall and evapotranspiration) (Figure 3l). However the runoff is always higher from the grass compared to the forest (Figure 3k). The anomalies in the runoff field (Figure 2j) follow those in the rainfall field (Figure 2i).

Figures 3 and 5 show that the water balance is intensified to the north of the Equator (with the exception of a slight decrease in the evapotranspiration) with an increase in rainfall. This increase is likely to be caused by the combination of higher windspeeds close to the ground surface and the more pronounced topography in that region. The hydrological cycle is certainly less intense south of the Equator, with a reduction in rainfall of 0.27 mm per day (5.5 per cent) and of evapotranspiration of 0.39 mm per day (10.5 per cent) for the grassland compared to the control run. Nevertheless the runoff and the moisture convergence has been slightly increased. The relative moisture content of the soil (HU_{soil} in Tables 5, 6 and 7, defined here as the difference between actual soil moisture content and that at the wilting point, normalised by the difference between soil moisture content at field capacity and at the wilting point) is less for the grassland to the south of the Equator, but is not changed to the north – where soil moisture content never falls below field capacity for either vegetation type.

The response to the deforestation can be most clearly seen in its effect on soil moisture content. This is shown in Figure 2k, where it can be seen that those areas with reduced soil moisture content correspond exactly to those areas where the rainfall has been reduced (Figure 2i), which are also the areas with the strongest reduction in evapotranspiration (Figure 2i), greatest increase in maximum surface temperature (Figure 2e), and greatest increase in sensible heat flux (Figure 2h).

It is difficult to conclude whether or not the modification to the rainfall field in the grassland is a direct result of the reduction in evapotranspiration. As is shown in the next section, it appears that the reduction in evapotranspiration, and thus the change in the atmospheric circulation over the deforested region is brought about by the effect of surface roughness. It can also be seen from Tables 2, 4 and 6 that there is a strong increase, of a factor two, in the surface windspeed.

Contrary to the results of the other deforestation experiments discussed here Tables 5, 6, and 7 show that the moisture convergence has increased over the region after deforestation, partly compensating for the reduction in evapotranspiration. It can also be seen from these tables that the precipitable water vapour content (the specific humidity integrated over a complete vertical column of the atmosphere) has not changed significantly.

SENSITIVITY TO THE ALBEDO AND SURFACE ROUGHNESS

As has been shown in numerous numerical climate simulations (see the review by Rowntree, 1991) anomalies in albedo and surface roughness can produce significant changes in the atmospheric circulation. Deforestation modifies both of these parameters, and therefore, to separate the effects of these changes two additional short simulations were carried out, each of three months duration and covering the period December to February inclusive. The initial conditions for these two simulations are taken from the predictions of the control run at the end of November on the first year of integration – i.e. 11.5 months after the start of the integration. In the first experiment all the parameters in ISBA representing the forest have been retained, apart from the

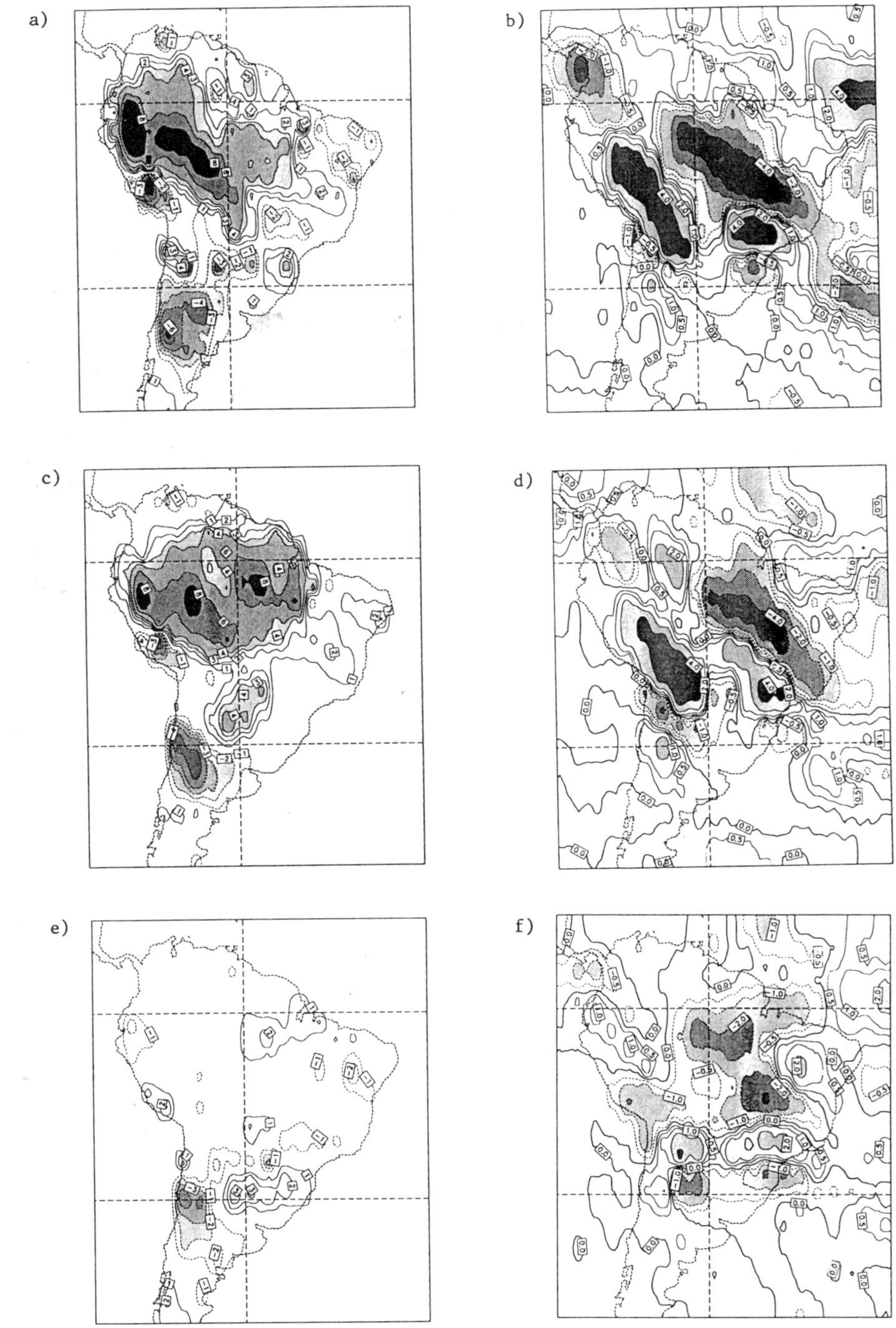

Figure 5 Anomalies in the fields of the amplitude of the daily surface temperature cycle (°C) and rainfall (mm per day) for the following experiments respectively: (a) and (b) complete deforestation minus control; (c) and (d) change in roughness length minus control; (e) and (f) change in albedo minus control.

roughness length which has been changed to the post-deforestation value. In the second experiment all the forest parameters were retained and the albedo was changed. These two simulations were then compared with control and deforested model runs for the same three month period.

The energy balance parameters and hydrological averages for Region II (see Figure 1) are given in Table 8. The analysis has been limited to Region II because it is entirely covered in tropical forest during the control run, and the impact of deforestation is currently stronger in that part of Amazonia. The map showing the anomalies in the fields of the amplitude of the daily surface temperature cycle and rainfall over South America are shown in Figure 5, for the three cases: deforestation minus control, roughness change minus control and albedo change minus control.

The albedo increase changes the radiation balance by reducing the net solar radiation and thus the net all-wave radiation. This reduction of net all-wave radiation of 9 Wm^{-2} produces a decrease in the daily average surface temperature of 0.4 °C, but without as much change in the amplitude in the daily cycle (Figure 5e). The energy balance has been changed with a reduction of 9 Wm^{-2} in the latent heat flux, but with no change in the sensible heat flux. All the components of evapotranspiration have been reduced: by 10 per cent for evaporation from the soil surface and the interception, and by 7 per cent for the transpiration. The different response of the direct evaporation and the transpiration is the result of a decrease of 13 per cent in the rainfall over the region (Figure 5f). The fact that latent heat flux changed in response to the reduction in net radiation whilst the sensible heat flux did not is associated with the strong reduction in rainfall. This is a positive feedback process where reduction in evapotranspiration goes on to produce a reduction in rainfall. It may be that the surface cooling which follows the increase in albedo leads to a weakening of the moisture convergence over the region and thus also contributes to the reduction in rainfall. These results are in good agreement with those of Milne and Rowntree (1992), who carried out a similar simulation with the UK Meteorological Office GCM (see their Table IVa), and with that of Dirmeyer (1992), who carried out a similar sensitivity analysis with the COLA GCM (see his Table 3.7 and 3.8).

The increase in albedo over Amazonia produces a weakening of the energy and water exchange over the region. Evapotranspiration decreases by 0.3 mm per day, rainfall by 0.7 mm per day, runoff by 0.34 mm per day, and moisture convergence by approximately 0.4 mm per day. The drier atmosphere produces a slight decrease in cloudiness at all levels in model. In practice this reduction in cloud cover produces an increase in incident solar radiation of 7 Wm^{-2} and a reduction of downward longwave radiation of 2 Wm^{-2}, this latter effect being compensated by reduction in surface temperature. These consequences of increased albedo are characteristic of the Charney effect (Charney, 1975; Charney *et al.*, 1977).

Analysing the impact of reducing the roughness length from 2 m to 0.06 m is not so straightforward as analysing the effect of increased albedo as the model responds indirectly. There is a marked increase in average windspeed at the first model level from 2.3 ms^{-1} in the control run, to 5.1 ms^{-1} for the reduced roughness experiment. Total cloudiness is reduced from an average of 67.5 to 61.2 per cent, and low cloud

Table 8 The energy and water balance of Region II of Amazonia for the period January to March inclusive of the deforestation, control, albedo change and roughness length change runs. S_D is the incident solar radiation; $(1 - \alpha) S_D$ is the net solar radiation; L_D is the downward longwave radiation; L_n is the net longwave radiation; R_n is the net all-wave radiation; LE is latent heat flux; H is sensible heat flux; T_s(mean), T_s(max) and T_s(min) are the mean, daily maximum and daily minimum surface temperatures; T_a is air temperature at the lowest model level; N(tot) is the percentage of total cloud cover; P is rainfall, and I is interception. Fluxes are in W m^{-2}, temperatures in °C, and rainfall and interception in mm per day.

	S_D	$(1-\alpha)$ S_D	L_D	L_n	R_n	LE	H	T_s (mean)	T_s (max)	T_s (min)	T_a	N (tot)	P	I
Control	232	203	380	65	138	112	27	24.7	32.8	19.8	25.3	68	6.2	1.12
Deforested	255	209	376	70	139	102	38	24.6	36.2	17.4	26.3	62	5.5	0.48
Albedo change	239	194	378	65	129	103	27	24.3	32.5	19.3	25.1	65	5.4	1.01
Roughness change	258	226	375	78	148	101	46	25.8	37.8	18.5	27.4	61	5.9	0.66

from 36.2 to 24.6 per cent, making a substantial change to the radiation balance (Table 8). In comparison with the control run the incident solar radiation is increased by 26 Wm^{-2} (11 per cent). There is a decrease of 5 Wm^{-2} in the downward longwave radiation, but as the surface temperature is on average 1.1°C warmer together these give an increase in longwave loss of 13 Wm^{-2}, which leads to an increase of net all-wave radiation of 10 Wm^{-2}.

The Bowen ratio changes from 0.24 in the control run to 0.46 in the roughness change experiment. This is partly a result of the less evapotranspiration, mainly caused by less interception loss, but also a result of the increased net all-wave radiation leading to a higher sensible heat flux.

The average air temperature near the surface rises by 2.1 °C in response to an increased maximum surface temperature of 5 °C. On average the decrease in roughness alone produced an increase in the daily surface temperature cycle of 6.3 °C, compared to 5.8 °C for the deforestation experiment. To strengthen the hypothesis that the increase in daily temperature amplitude found in the deforestation experiment is the result of the low roughness length of the grass, Figures 4a and 4c compare the fields of this variable for the deforestation and roughness change experiments respectively.

These sensitivity analyses are too short to produce stable rainfall fields. However, comparing Figures 4b and 4d, for the deforestation and roughness change experiments respectively, it appears that the changes to the atmospheric circulation over Amazonia after deforestation which lead to a change in rainfall are principally caused by the reduction in roughness length. This result is in agreement with that of Sud *et al.* (1988) who predicted a reduction in rainfall over Amazonia following a large reduction in roughness length, even though the energy balance had not been modified.

The results presented here are in good agreement with those of Lean and Warrilow (1989). In their simulation of the effects of reducing roughness over Amazonia they also found reduced evapotranspiration, moisture convergence and rainfall.

The sensitivity analyses presented in this section show that the effects of albedo and surface roughness play an important role in the determination of the radiation and energy balance, and in modifying the rainfall field when the forest is replaced by grass.

SENSITIVITY TO THE CONVECTION SCHEME

In introducing the ISBA scheme into the GCM EMERAUDE, Manzi and Planton (1994) emphasised the importance of considering the interactions between the surface and the other parameterizations in the model. This section looks at the interaction between the surface and convection schemes. This is achieved with two further deforestation experiments.

In the first experiment, which is the same as that described previously, the GCM is coupled to a Kuo-type cloud scheme; in the second experiment the GCM is coupled with a mass flux scheme (Bougeault, 1985). The initial conditions and the prescribed perturbations are identical for both experiments.

The results are presented as averages for the whole of Amazonia (Regions I and II), for the period January to March inclusive. The first 17 days of integration have been neglected. The model is certainly not in equilibrium at the beginning of January, however that will not affect this analysis which compares scheme response, rather than predicting the effects of deforestation.

Table 9 compares the predictions with the Kuo-type scheme. After deforestation net all-wave radiation is slightly decreased by 3 Wm^{-2}, evapotranspiration is decreased by 14 per cent, and Bowen ratio is increased from 0.16 to 0.30. The average surface temperature remains the same at 24.9 °C. The anomaly in the rainfall field (deforested minus control), presented in Figure 5a, shows a variable change in rainfall. The rainfall is reduced over the greater part of Amazonia, but it is increased in the north,

Table 9 The energy and water balance of Amazonia (Region I and II) averaged over the period January to March inclusive with EMERAUDE/ISBA coupled with a Kuo-type convection scheme. T_s(mean) is mean surface temperature (°C); E evapotranspiration; (mm per day); R_n is the net all-wave radiation (Wm^{-2}); H is sensible heat flux (Wm^{-2}); P is rainfall (mm per day); and P - E is moisture convergence (mm per day).

	T_s(mean)	E	R_n	H	P	P – E
control	24.9	4.2	142	20	7.3	3.1
deforested	24.9	3.6	139	32	6.8	3.2
deforested – control	0	-0.6	-0.3	12	-0.5	0.1

Table 10 The energy and water balance of Amazonia (Region I and II) averaged over the period January to March inclusive with EMERAUDE/ISBA coupled with a mass flux convection scheme. T_s(mean) is mean surface temperature (°C); E evapotranspiration; (mm per day); R_n is the net all-wave radiation (Wm^{-2}); H is sensible heat flux (Wm^{-2}); P is rainfall (mm per day); and P - E is moisture convergence (mm per day).

	T_s(mean)	E	R_n	H	P	P – E
control	25.6	4.0	140	22	4.7	0.7
deforested	26.6	3.6	146	41	6.7	3.1
deforested – control	1.0	-0.4	6	19	2	2.4

south and south-west.

The predictions for the mass flux scheme are compared in Table 10. In contrast to the results from the Kuo scheme, the net all-wave radiation and the average surface temperature for the grass are increased by 6 Wm^{-2} and 1 °C respectively compared to the control. The evapotranspiration is decreased by 10 per cent and the Bowen ratio changes from 0.19 to 0.39. However, it is for the moisture convergence and the rainfall that the two schemes differ most. For the mass flux scheme the rainfall for the grass increases by more than 40 per cent compared to the control, with a consequent increase of 2.4 mm per day in moisture convergence. In fact the moisture convergence over Amazonia is too weak for the forested control run, whilst it appears correct for the grassland simulation. The spatial distribution of the rainfall anomaly (deforested minus control) shows that, following deforestation, rainfall is increased over the whole of Amazonia (Figure 4b). Again showing a result in contrast to the Kuo-type scheme.

The reasons for this poor functioning of the mass flux scheme over Amazonia have not yet been established. But this analysis clearly makes the point that the predictions of climate change simulations made by GCMs must be treated with caution, because they are strongly dependent on the model formulation.

CONCLUSION

The model experiment described in this paper predicts that the impact of deforestation will be regional, with a greater impact south of the Equator. Deforestation is predicted to lead to a reduction in evapotranspiration, largely as a result of the reduced evaporation of intercepted rainfall. Rainfall is predicted to decrease in some parts of Amazonia, but increases in the north, south-west and south-east of Amazonia. In contrast to the results of other recent simulations (Lean and Warrilow, 1989; Nobre *et al.*, 1991; Dickinson and Kennedy, 1992; Polcher and Laval, 1994b) the moisture convergence and runoff are, on average, increased for the deforested case. However this effect is less apparent where the reduction in rainfall is less.

The sensitivity analyses have shown that the changes in surface roughness and albedo play an important role in determining the overall effect of deforestation, but that the type of convection scheme used in the model can also play a critical part in determining the overall conclusions that can be drawn from the simulation. The principal difference between the results reported here and those reported previously by other authors, appears to be in the model dependent schemes for the surface energy, convection, cloud, and radiative transfer. The reduction in cloud cover and the modification to the radiation balance in this simulation was also reported by Dickinson and Kennedy (1992), but not found by Lean and Warrilow (1989) or Nobre *et al.* (1991). Moisture convergence increased for this simulation, but decreased in the other studies. It can be concluded that Amazonian deforestation will change the regional climate, but more reliable estimates require improvements to made to be the GCMs' parameterisation of cloudiness and convection.

ACKNOWLEDGEMENTS

We thank Mrs Sophie Tyteca (Météo-France/CNRM) for help in conducting these simulation experiments and processing model outputs.

A.O. Manzi was partly supported by the Brazilian Conselho Nacional de Desenvolvimento Científico e Tecnológico (CNPq/RHAE). This work was also supported by the European Centre for Medium Range Weather Forecasting as part of the Special Project SPFRVESO, by the Programme National d'Etude de la Dynamique du Climat (PNEDC) and by the Commission of the European Communities (Contract EPOC-0003-C-MB).

We are grateful to John Gash for translating this paper.

REFERENCES

Bastable, H.G., Shuttleworth, W.J., Dallarosa, R.L.G., Fisch, G. and Nobre, C.A. 1993. Observations of climate, albedo and surface radiation over cleared and undisturbed Amazonian forest. *Int. J. Climatol.*, **13**, 783-796.

Bougeault, Ph. 1985. A simple parameterization of the large-scale effects of cumulus convection. *Mon. Weather. Rev.*, **113**, 2108-2121.

Brutsaert, W.H. 1982. *Evaporation into the atmosphere: theory, history and applications.* D. Reidel Publishing Company, Dordrecht.

Charney, J.G. 1975. Dynamics of deserts and drought in the Sahel. *Q. J. Roy. Meteorol. Soc.*, **101**, 193-202.

Charney, J.G., Quirks, W.J., Chow, S.H., and Kornfield, J. 1977. A comparative study of the effects of albedo change on drought in semi-arid regions. *J. Atmos. Sci.*, **34**, 1366-1385.

Coiffier, J., Ernie, Y., Geleyn, J.F., Clochard, J., Hoffman, J., and Dupont, F. 1987. The operational hemispheric model at the French Meteorological Service. *J. Meteorol. Soc. Japan, Special NWP Symposium*, 337-345.

Dickinson, R.E., and Kennedy, P. 1992. Impacts on regional climate of Amazon deforestation. *Geophysical Research Letters*, **19**, 1947-1950.

Dirmeyer, P.A. 1992. GCM studies of the influence of vegetation on the general circulation: the role of albedo in modulating climate change. *PhD thesis*, University of Maryland.

Ernie, Y. 1985. Experiments with the French spectral model. In Proceedings of 7th American Meteorological Society Conference on Numerical Weather Prediction, Montreal, 486-489.

Figueroa, S.N. and Nobre, C.A. 1990. Precipitation distribution over central and western tropical South America. *Climánalise*, **5**, 36-45.

Gash, J.H.C., Nobre, C.A., Roberts, J.M. and Victoria, R.L. 1996. An overview of ABRACOS. 1996. This volume.

Gates, W.L. 1992. AMIP: The atmospheric model intercomparison project. *Bull. Amer. Meteor. Soc.*, **73**, 36-45.

Geleyn, J.F., Bougeault, P., Rochas, M., Cariolle, D., Lafore, J.P., Royer, J.F. and André, J.C. 1988. The evolution of numerical weather prediction and atmospheric modelling at the French Weather Service. *J. Theoretical Appl. Mechanics, Suppl. 2 to vol 7*, 87-110.

Jacquemin, B. and Noilhan, J. 1990. Sensitivity study and validation of a land surface parameterization using the HAPEX-MOBILHY data set. *Bound.-Layer Meteorol.*, **52**, 93-134.

Lean, J. and Rowntree, P.R., 1993. A GCM simulation of the impact of Amazonian deforestation on climate using an improved canopy representation. *Q. J. R. Meteorol. Soc.*, **119**, 509-530.

Lean, J. and Warrilow, D. 1989. Simulation of the regional climatic impact of Amazon deforestation. *Nature*, **342**, 411-413.

Lloyd, C.R., Gash, J.H.C., Shuttleworth, W.J. and Marques, A. de O. 1988. The measurement and modelling of rainfall interception by Amazonian rain forest. *Agric. For. Meteorol.*, **43**, 277-294.

Louis, J.F., Tiedke, M. and Geleyn, J.F. 1981. A short history of the operational PBL parameterization of the ECMWF. In Workshop on Planetary Boundary Layer, ECMWF, 59-79.

Manzi, A.O. and Planton, S. 1992. Implementation of the ISBA parameterization scheme for land surface processes in a GCM. An annual cycle experiment. Notes de Centre 11, Météo-France/ CNRM/GMGEC.

Manzi, O. and Planton, S. 1994. Implementation of the ISBA scheme for land surface processes in a GCM - an annual cycle experiment. *J. Hydrol.*, **155**, 353-387.

Manzi,O. 1993. Introduction d'un schéma des transfert sol-vegetation-atmosphere dans un modèle de circulation general et application a la deforestation Amazonienne. PhD thesis, Université Paul Sabatier, Toulouse.

McWilliam, A.L.C., Roberts, J.M., Cabral, O.M.R., Leitão, M.V.B.R., de Costa, A.C.L., Maitelli, G.T. and Zamporoni, C.A.G.P. 1993. Leaf area index and above-ground biomass of *terra firme* rain forest and adjacent clearings in Amazonia. *Functional Ecol.*, **7**, 310-317.

Milne, M.F. and Rowntree, P.R. 1992. Modelling the effects of albedo change associated with tropical deforestation. *Climate Change*, **21**, 317-343.

Mintz, Y. and Serafini, Y.V. 1989. Global monthly climatology of soil moisture and water balance. *UNESCO Studies and Reports in Hydrology.*

Moore, C.J. and Fisch, G. 1986. Estimating heat storage in Amazonian tropical forest. *Agric. and Forest Meteorol.*, **38**, 147-169.

Nobre, C.A., Sellers, P.J. and Shukla, J. 1991. Amazonian deforestation and regional climate change. *J. Climate*, **4**, 957-988.

Noilhan, J. and Planton, S. 1989. A simple parameterization of land surface processes for meteorological models. *Mon. Wea. Rev.*, **117**, 536-549.

Noilhan, J., Mahfouf, J.F., Manzi, A. and Planton, S. 1992. Validation of land-surface parameterizations: Developments and experiments at the French weather service. In: *Proceedings of the workshop in validation of models over Europe*, Reading (England), ECMWF.

Planton, S., Déqué, M. and Bellevaux, C. 1991. Validation of an annual cycle simulation with a T42-L20 GCM. *Climate Dynamics*, **5**, 189-200.

Polcher, J. and Laval, K. 1994a. The impact of African and Amazonian deforestation on tropical climate. *J. Hydrol.*, **155**, 389-405.

Polcher, J. and Laval, K. 1994b. A statistical study of the regional impact of deforestation on climate in the LMD GCM. *Climate Dynamics*, **10**, 205-219.

Rowntree, P.R. 1991. Atmospheric parameterization schemes for evaporation over land: basic concepts and climate modelling aspects. In: *Land Surface Evaporation - Measurement and Parameterization*, (Eds. T.J. Schmugge and J.C. André). Springer-Verlag. 5-29.

Royer, J.F., Planton, S. and Déqué, M. 1990. A sensitivity experiment for the removal of Arctic Sea ice with the French Spectral General Circulation Model. *Climate Dynamics*, **5**, 1-17.

Shuttleworth, W.J. 1988. Evaporation from Amazonian rain forest. *Proc. Roy. Soc.(Lond.) B*, **233**: 321-346.

Shuttleworth, W.J., Gash, J.H.C., Lloyd, C.R., Moore, C.J., Roberts, J., Marques, A. de O., Fisch, G., Silva, V. de P., Ribeiro, M.N.G., Molion, L.C.B., de Abreu Sa, L.D., Nobre, C.A., Cabral, O.M.R., Patel, S.R. and de Moraes, J.C. 1984a. Eddy correlation measurements of energy partition for Amazonian forest. *Q.J. Roy. Meteorol. Soc.*, **110**, 1143-1162.

Shuttleworth, W.J., Gash, J.H.C., Lloyd, C.R., Moore, C.J., Roberts, J., Marques, A. de O., Fisch, G., Silva, V. de P., Ribeiro, M.N.G., Molion, L.C.B., de Abreu Sa, L.D., Nobre, C.A., Cabral, O.M.R., Patel, S.R. and de Moraes, J.C. 1984b. Observations of radiation exchange above and below Amazonian forest. *Q.J. Roy. Meteorol. Soc.*, **110**, 1163-1169.

Sud, Y.C., Shukla, J. and Mintz, Y. 1988. Influence of land surface roughness on atmospheric circulation and rainfall: a sensitivity study with a General Circulation Model. *J. Appl. Meteor.*, **27**, 1036-1054.

Wilson, M.F. and Henderson-Sellers, A. 1985. A global archive of land cover and soils data sets for use in General Circulation Climate Models. *J. Climatol.*, **5**, 119-143.

Wright, I.R., Gash, J.H.C., da Rocha, H.R., Shuttleworth, W.J., Nobre, C.A., Maitelli, G.T.M., Zamparoni, C.A.G.P. and Carvalho, P.R.A. 1992. Dry season micrometeorology of central Amazonian ranchland. *Q.J. Roy. Meteorol. Soc.*, **118**, 1083-1099.

RESUMO

As parametrizações dos processos de superfície usadas nos MCGA (Modelos de Circulação Geral da Atmosfera) foram bastante aprimoradas na última década, com a inclusão de uma representação explícita da vegetação e dos processos físicos relacionados a ela. No CNRM (Météo-France) um novo esquema, chamado de esquema ISBA (Interações entre Solo, Biosfera e Atmosfera) desenvolvido por Noilhan e Planton (1989), foi implementado no AGCM Espectral EMERAUDE (Manzi e Planton, 1994). Este acoplamento ISBA/AGCM é uma ferramenta poderosa para se investigar mudanças de origem natural e antropogênica nas superfícies continentais, como os processos de desertificação ou desmatamento. Neste artigo são apresentados resultados de uma simulação de desmatamento de 4 anos sobre a Amazônia, onde a floresta tropical chuvosa natural e as savanas de uma extensa área da América do Sul foram substituídas por pastagens degradadas. O esquema ISBA foi testado cuidadosamente usando dados dos experimentos observacionais ARME (Experimento Micrometeorológico da Região Amazônica) e ABRACOS, respectivamente, para floresta e pastagem. A simulação com a Amazônia desmatada mostrou um enfraquecimento do ciclo hidrológico e uma amplificação do ciclo diurno da temperatura de superfície quando comparada com a simulação de controle. Um estudo de sensitividade mostrou a grande importância do albedo de superfície e do comportamento da rugosidade na resposta ao desmatamento.

30 Simulation of mesoscale circulations in a deforested area of Rondônia in the dry season

M.A.F.S. DIAS[1] **and P. REGNIER**[2]
[1]*Instituto Astronômico e Geofísico, Universidade de São Paulo*
[2]*Université de Paris VI, France*

INTRODUCTION

Local circulations induced by differences in vegetation coverage are included in the category of non-classical mesoscale circulations, i.e. those induced by differences in surface sensible heat fluxes other than the classical type (sea/land breeze, mountain/valley circulations). However, most of the time, nature presents the added effects of classical and non-classical circulations, showing enhancements and reductions when these circulations accelerate the winds in the same direction or in the opposite direction, respectively. Pielke *et al.* (1991) demonstrated that the influence of mesoscale landscape spatial variability is in the form of fluxes of heat and moisture which are of the same order as the turbulent fluxes, but differ in vertical structure. They suggested that the mesoscale fluxes due to horizontal variations in surface characteristics should be properly included in parametrizations of sub-grid scale processes used in large scale and general circulation models of the atmosphere.

In the Amazon region, the tropical rain forest is interrupted by patches of land that have been cleared by burning, cutting and bulldozing, and then replaced by pasture. This provides a good opportunity for measuring the differences in the surface energy budget in these two types of vegetation in the tropics, which has been one of the objectives of ABRACOS.

The objective of this paper is to use the observed components of the surface energy budget to validate surface fluxes simulated with a mesoscale numerical model and to perform a realistic simulation of the local circulation in Rondônia in the dry season. This is intended as a contribution to understanding the impact of deforested areas in the Amazon region on the larger scale.

NUMERICAL MODEL AND DATA

Our simulations are performed with the latest, "3a", version of the Regional Atmospheric Modelling System (RAMS) which was developed at Colorado State

Amazonian deforestation and climate. Edited by J.H.C. Gash, C.A. Nobre, J.M. Roberts and R.L. Victoria.

University (Tripoli and Cotton, 1982, Pielke *et al.* 1992). This numerical tool is a multipurpose prediction model that is designed to simulate weather systems from the hemisphere scale down to large eddy simulations of the planetary boundary layer (PBL). RAMS includes numerous physical parametrizations appropriate for simulating local circulations induced by varying surface land cover and topography. Precipitation processes have been treated with the Kuo convective parameterization (Kuo, 1974) and the radiation scheme used is the method developed by Chen and Cotton (1983) which takes explicit account of the radiative properties of the modelled clouds with respect to short and long wave radiation. We used the Mellor and Yamada (1982) parameterization for the turbulence in the PBL and a fully compressible non- hydrostatic formulation.

The Cartesian grid used has a horizontal grid separation of 20 km in both x and y directions and covers an 800 km by 1000 km area in the southern Amazonian forest basin, namely, in Rondônia. The vertical grid has 25 levels with an increasing resolution by a stretching factor of 1.2 starting from 80 m above ground level up to a resolution of 2000 m and then constant up to model top located at about 24 km. The parameterization of land-surface processes in RAMS has been modified according to a new scheme developed at the University of São Paulo (Regnier, 1995) for the computation of the surface turbulent heat flux. A brief overview of this scheme is given in the next section.

The micrometeorological measurements obtained between 22 and 25 July 1993 during the ABRACOS campaign have been used, namely, surface turbulent heat fluxes measured above the tropical forest and above the pasture. The horizontal extent of the deforested area has been obtained from LANDSAT satellite images analysed at INPE.

We shall simulate successively the situations with and without deforestation. In the Control Simulation (CS), all the grid points of the numerical domain are considered as covered by tropical forest. In the Deforested Simulation (DFS), the 84 points corresponding to the deforested area as defined by the INPE satellite image were considered to be covered by pasture (Figure 1).

The initialization of the simulations used the Isentropic Analysis Package (ISAN), available in RAMS. This package provides the meteorological fields at 0 UTC of each day between 22 and 26 July for all points of the grid, by spatial interpolation of the 2.5° resolution data set of the European Centre for Medium Range Weather Forecasting - ECMWF. The 22 July file was used to initialize the atmospheric fields at the beginning of the simulation (Figure 2). The five daily files at 0 UTC were used to update the boundary conditions in a band having a thickness of four grid points, thus representing the impact of synoptic trends over the four-day period of the simulation. The synoptic trend, however, is quite weak so that the local circulations are similar over the whole period of the simulation.

FLUX SIMULATIONS

Figure 3 shows the Bowen ratios for the forest and for the pasture for the hourly

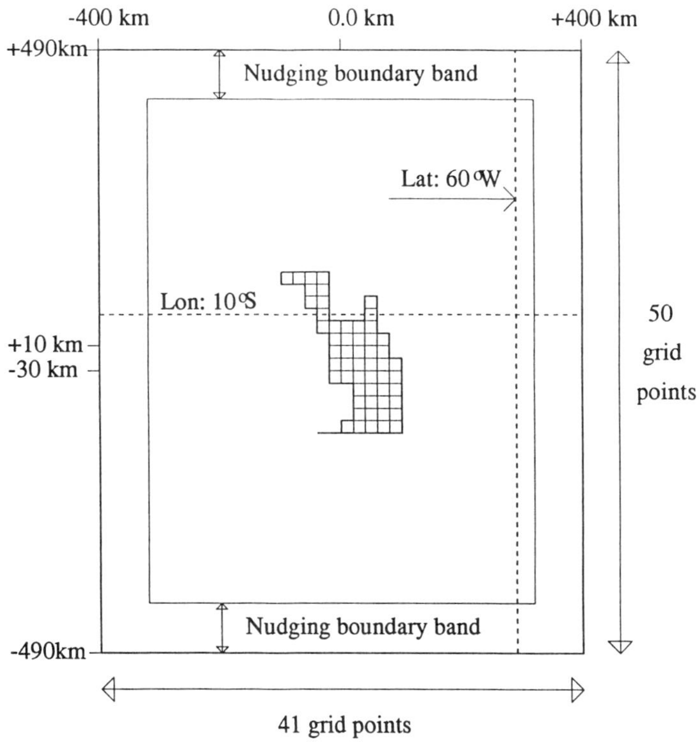

Figure 1 Simulation area. The gridded area corresponds to deforestation.

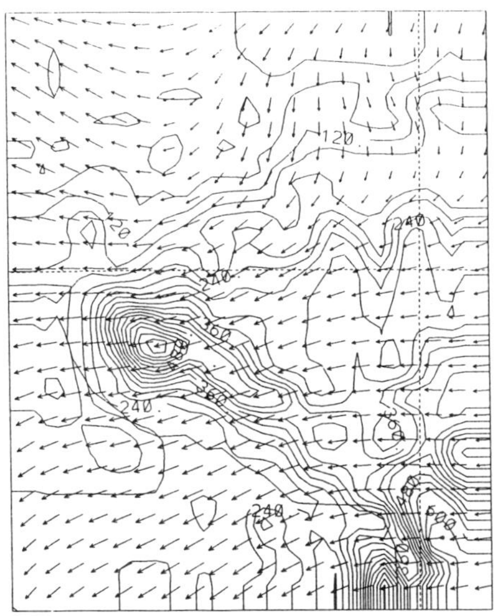

Figure 2 Initial condition at z = 38.2 m for the horizontal wind vector field at 0 UTC, 22 July 1993. The size of the wind vectors is proportional to the wind velocity, with maximum speed of 5.9 m s^{-1}. The height contours are shown with spacing of 30 m.

daytime observations for the whole month of July 1993 during the ABRACOS field campaign. The Bowen ratio is slightly more elevated above the pasture. Than above the tropical forest. One would have expected a larger difference, but although the period corresponds to the end of the dry season in Rondônia, a few rainy events took place during this month, maintaining a high soil moisture content and thus reducing the contrasts in the fluxes.

To simulate the impact of the different vegetation coverage, we needed a reasonably simple scheme in the RAMS to represent the differences between tropical forest and pasture. A new vegetation model has been implemented based on

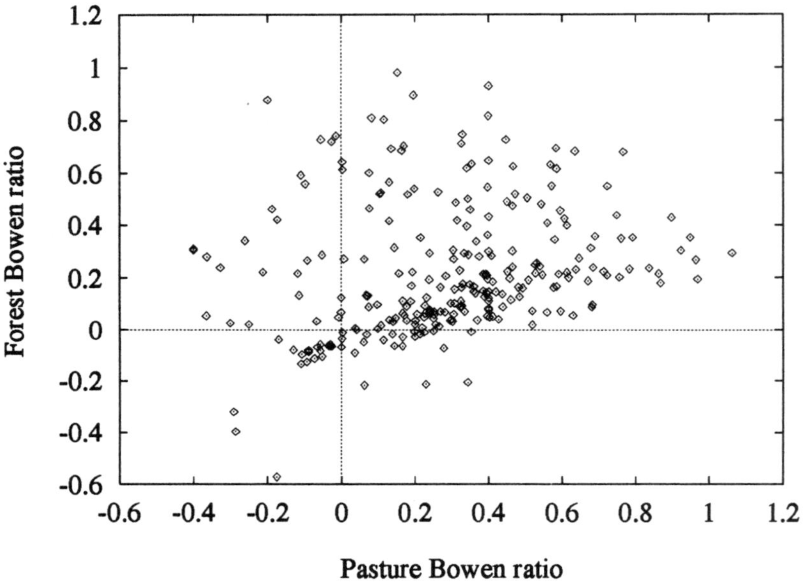

Figure 3 Forest versus pasture Bowen ratio for the month of July 1993 during ABRACOS: diurnal values only are plotted.

the "Big Leaf model" principle (Deardorff, 1978) and its recent developments: BATS (Dickinson *et al.* 1986), SECHIBA, (Ducoudré, 1993) and ISBA (Noilhan and Planton, 1989). This Model of Interaction of the Surface with the Atmosphere (MISA) takes into account the vegetation stress factors (soil moisture, solar radiation, vapour pressure deficit and leaf temperature) for the calculation of transpiration through a stomatal conductance as suggested by Avissar and Mahrer (1988). The parameterization of heat and humidity diffusion in the soil is taken from McCumber and Pielke (1981) using four layers between 0 and 2 metres below the ground surface. Both temperatures of the 3 cm upper soil layer and of the vegetation are used for the turbulent flux calculations. The vegetation type parameters are taken from the results of Wright *et al.* (1992) and Rocha (1992) for the pasture and from Sellers *et al.* (1989) and Manzi (1993) for the tropical forest (Table 1). Sensitivity studies and validation of the turbulent heat fluxes generated by the MISA using both types of vegetation cover have been performed with an off-line version of the model, forced by measured meteorological fields of the 1990 ABRACOS campaign at "Fazenda Dimona" (Wright *et al.* 1992) and the 1983 ARME campaign at "Reserva Ducke" (Shuttleworth, 1984). These will not be shown here.

The measurements of soil moisture were not available at the time of the analysis so the soil moisture values have been initialized arbitrarily to obtain a satisfactory agreement between the measured and simulated turbulent heat fluxes. Table 2 shows the values used for each soil layer. Since the simulation took place at the end of the dry season, the soil moisture is nowhere near saturation. The numerical effects of the initialization by the ECMWF fields, or the model adjustment to the initial data for the 22-26 July period, becomes undetectable after about 24 hours. We shall, therefore, analyse the results of the last three days.

The forest has a deep rooting system so that this relatively low soil moisture initialization does not induce any reduction in transpiration through stomata control as it is compensated for by the extraction of deep soil water. As shown in Figure 4c, the agreement between the measured and simulated latent heat fluxes for this is quite satisfactory. The lower values of the measured sensible heat fluxes (Figure 4a) during the last two days was caused by advection of drier air which increased the transpiration of the vegetation, and a slightly stronger windspeed which increased the aerodynamic coefficient. This trend is not reproduced by the simulation.

The transpiration of the pasture is merely reduced by the soil moisture stress as we can see from Figure 4d. Despite the obvious disagreement between the sensible heat fluxes simulated and measured above the pasture as seen in Figure 4b, we chose to conserve this overestimated dry soil moisture initialization in order to strengthen the amplitude of the local circulation induced by the horizontal turbulent heat flux gradient. Sensible heat fluxes of this magnitude were observed in other periods of July 1993. The difference in surface sensible heat flux between forest and pasture

Table 1 Parameters for each type of vegetation

Parameter	*Tropical Forest*	*Pasture*
Leaf Area Index (LAI)	5.1	1.22
Vegetation Fraction (σ_v)	0.98	0.84
Albedo (α)	0.12	0.15
Displacement height (d)	25.0 m	0.26 m
Roughness length (z_0)	2.0 m	0.03 m
Canopy interception capacity (W_i)	0.70 mm	0.122 mm

Table 2 Soil moisture initialisation

Layer	1	2	3	4
Relative soil moisture	0.70	0.68	0.65	0.65

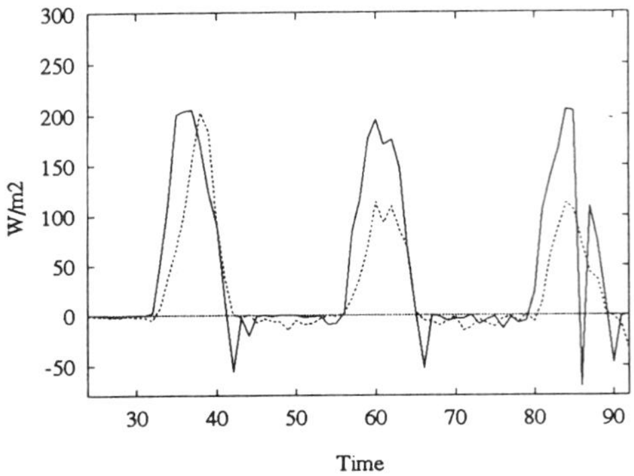

(a) Forest sensible heat fluxes.

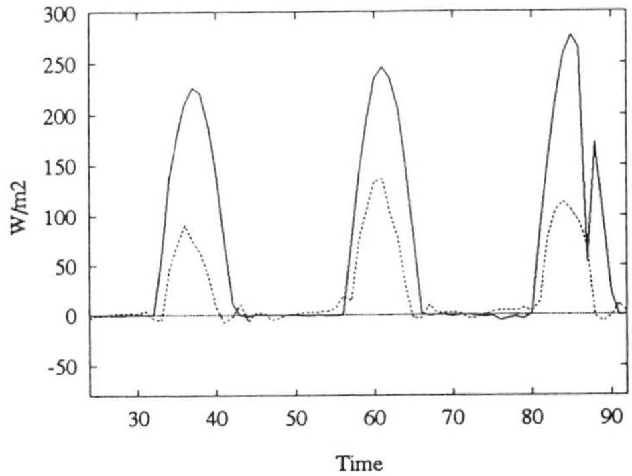

(b) Pasture sensible heat fluxes.

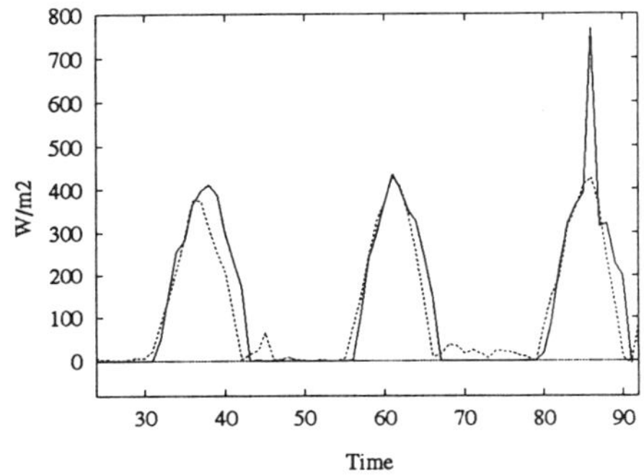

(c) Forest latent heat fluxes.

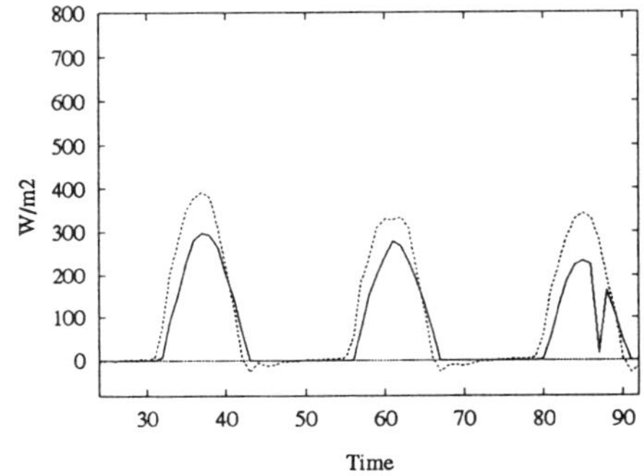

(d) Pasture latent heat fluxes.

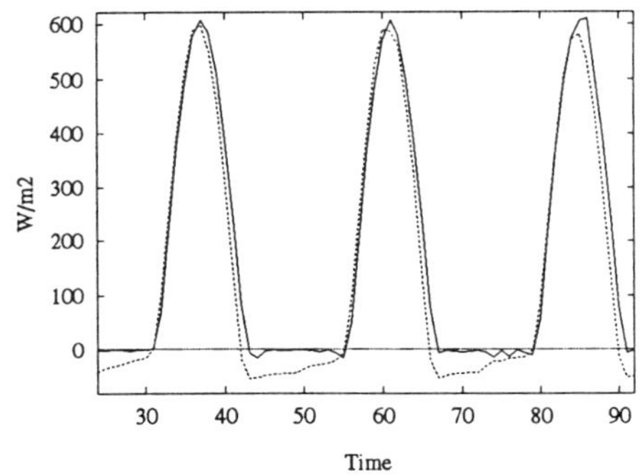

(e) Forest net radiation fluxes.

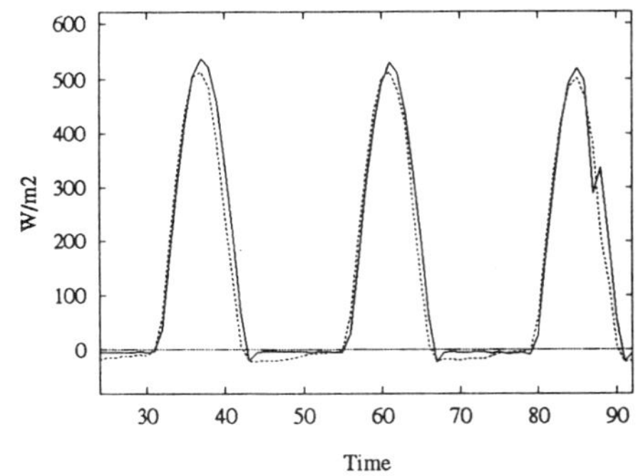

(f) Pasture net radiation fluxes.

Figure 4 Forest (left) and pasture (right) sensible heat (top), latent heat (middle) and net radiation fluxes (bottom). The full line represents measured values and the dotted line corresponds to simulated values.

on the third day (between 48 and 72 hours) reaches about 30 W m^{-2}, for the observations and about 60 W m^{-2}, for the simulation. The simulation may be regarded as a more energetic version of the real case, in which we observe enhanced circulations that would take place in a case of low soil moisture.

As we may observe in Figures 4e and 4f, respectively, our vegetation scheme failed to reproduce the night time upward long wave radiation flux of about 40 W m^{-2} for the tropical forest and about 25 W m^{-2} for the pasture. This is the consequence of our low vegetation calorific capacity value which does not represent the total biomass capacity but only that of the canopy foliage. According to the study on heat storage by tropical forest.by Moore and Fisch (1986), 25 to 50% of the nocturnal upward long wave radiation is balanced by the heat stored by the vegetation during daytime, the other 50 to 75% could come from the downward sensible heat flux, although there is no observational evidence.

The computation of the aerodynamic coefficient uses a function proposed by Louis (1979) dependent on the Richardson number. This function quickly tends to zero in the stable case and thus, the turbulent heat fluxes calculated by the MISA at night are very close to zero. This could be improved by reducing the slope of the aerodynamic coefficient dependency function on the Richardson number.

MESOSCALE CIRCULATIONS

The discussion of the numerical simulation of the mesoscale circulations will focus on the 48 -72 hours of simulation and on the three-dimensional evolution of the lower troposphere including the PBL. Upper level structure should be important in the moist season with the development of deep convection. In the case of this dry season simulation, there was a short lived convective development, in the CS and in the DFS, but this feature will not be addressed here. Throughout the daily evolution of the meteorological fields close to the surface and in the mixed layer we will point out, in the next paragraphs, that there is a delicate balance between the large scale forcing and the topographical influence in the CS and those two features plus the vegetation variability impact in the DFS. Local times are UTC minus four hours.

The daily evolution of the low level circulation (height of 38.2 m above ground level) in the CS and in the DFS may be seen in Figure 5 for night and early morning (6 UTC and 12 UTC) and in Figure 6 for afternoon and early evening (18 UTC and 0 UTC). During the daytime this structure is similar to the one observed in other levels within the mixed layer. We have chosen this level closer to the surface so that the nightime evolution in the surface inversion layer may be followed.

In the CS, the night situation in Figure 5a shows basically the influence of topography on the low level flow, since the stable nocturnal layer decouples this flow from the larger scale influence above. A downslope flow is seen to dominate all slopes. The presence of the pasture in the centre of the domain (*cf.* Figure 1) has an impact on the wind speed that may be observed in Figure 5b. Since the simulated fluxes of sensible heat and latent heat are almost zero at this time (Figure 4) the

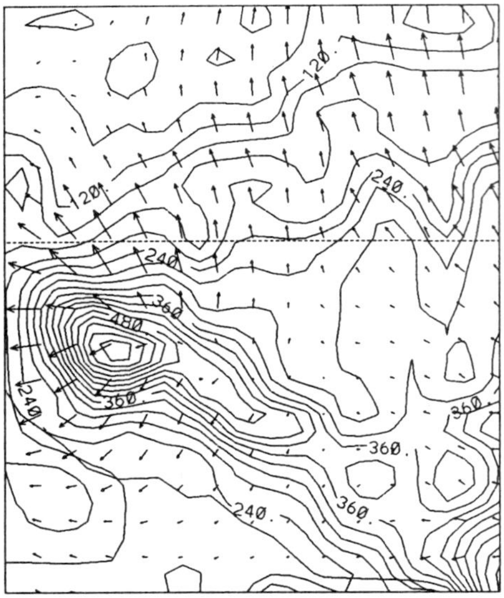

(a) Control Case, 6 UTC, maximum wind speed 2.4 $m.s^{-1}$.

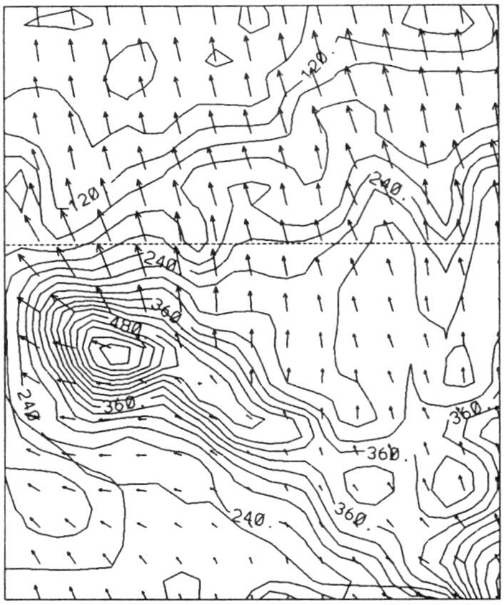

(b) Deforestation Case, 6 UTC, maximum wind speed 3.2 $m.s^{-1}$.

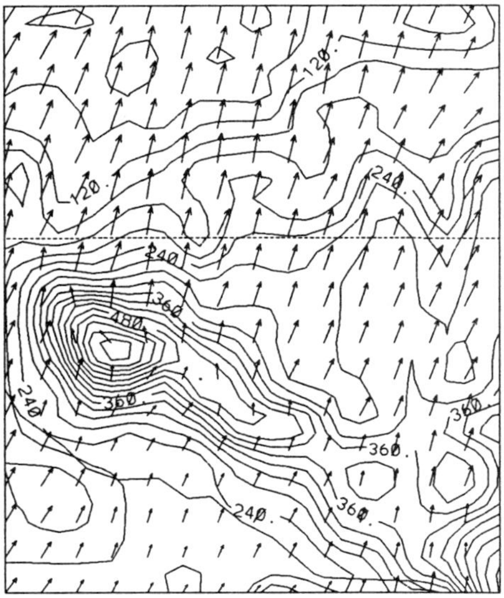

(c) Control Case, 12 UTC, maximum wind speed 1.6 $m.s^{-1}$.

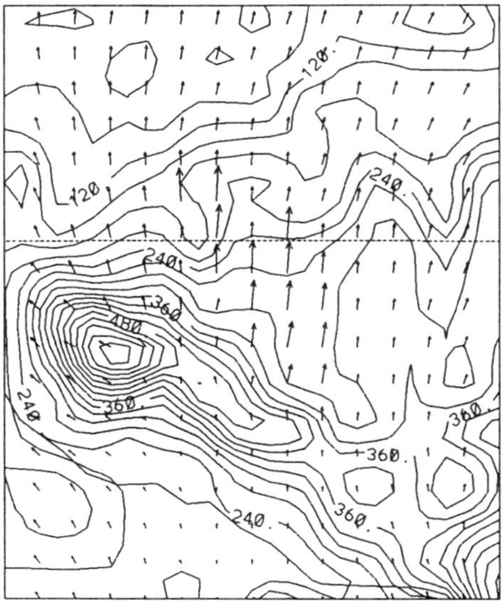

(d) Deforestation Case, 12 UTC, maximum wind speed 3.1 $m.s^{-1}$.

Figure 5 Simulated wind field at 38.2 m above ground level. (a) CS, 6 UTC, maximum speed 2.37 m s^{-1}; (b) DFS, 6 UTC, maximum speed 3.22 m s^{-1}; (c) CS, 12 UTC, maximum speed 1.58 m s^{-1}; (d) DFS, 12 UTC, maximum speed 3.1 m s^{-1}. Contours and arrows as in Figure 2.

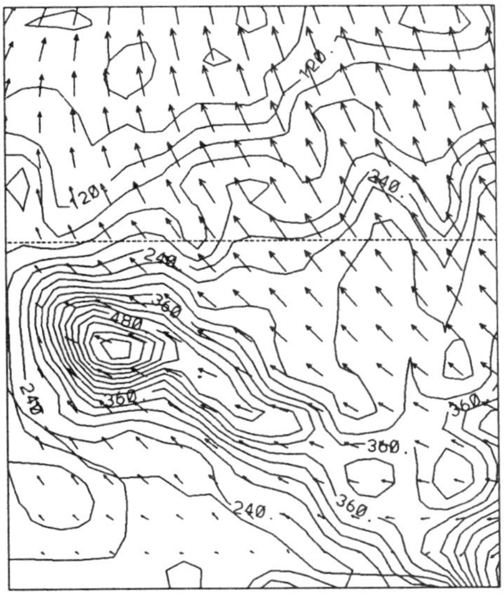

(a) Control Case, 18 UTC, maximum wind speed 2.6 $m.s^{-1}$.

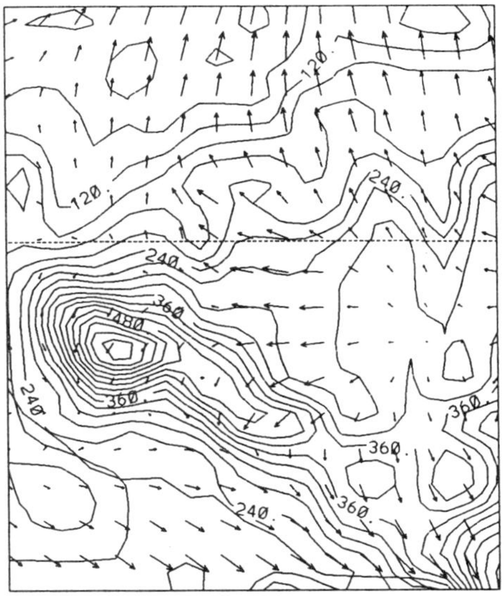

(b) Deforestation Case, 18 UTC, maximum wind speed 1.8 $m.s^{-1}$.

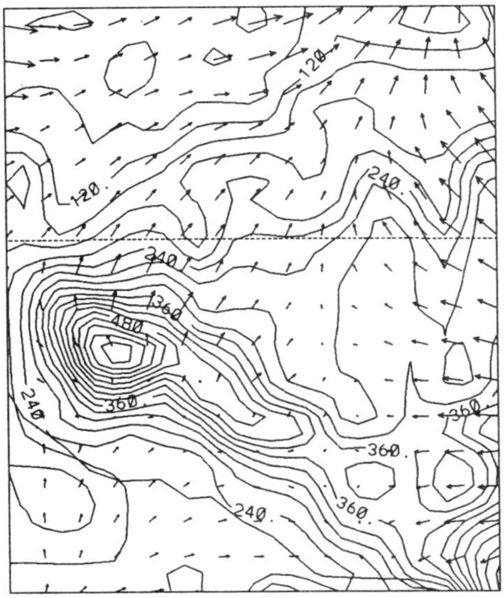

(c) Control Case, 0 UTC, maximum wind speed 2.5 $m.s^{-1}$.

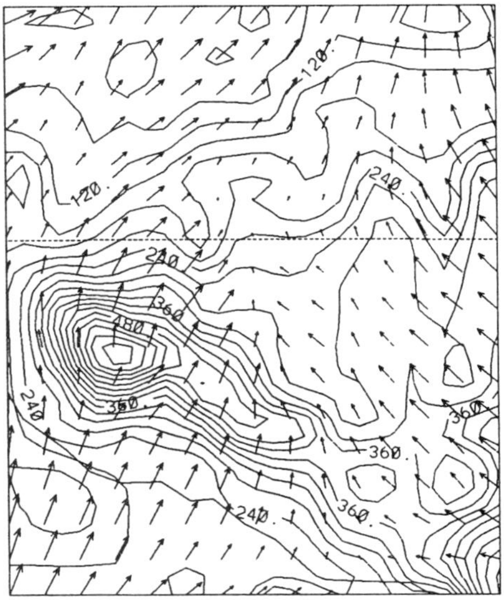

(d) Deforestation Case, 0 UTC, maximum wind speed 2.9 $m.s^{-1}$.

Figure 6 Simulated wind field at 38.2 m above ground level. (a) CS, 18 UTC, maximum speed 2.6 m s^{-1}; (b) DFS, 18 UTC, maximum speed 1.75 m s^{-1}; (c) CS, 0 UTC, maximum speed 2.45 m s^{-1}; (d) DFS, 0 UTC, maximum speed 2.92 m s^{-1}. Contours and arrows as in Figure 2.

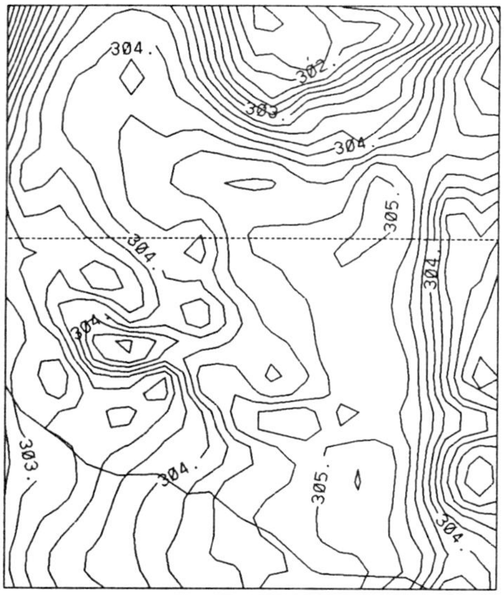

(a) Control Case, 0 UTC, maximum potential temperature 305.2 K.

(b) Deforestation Case, 0 UTC, maximum potential temperature 305.4 K.

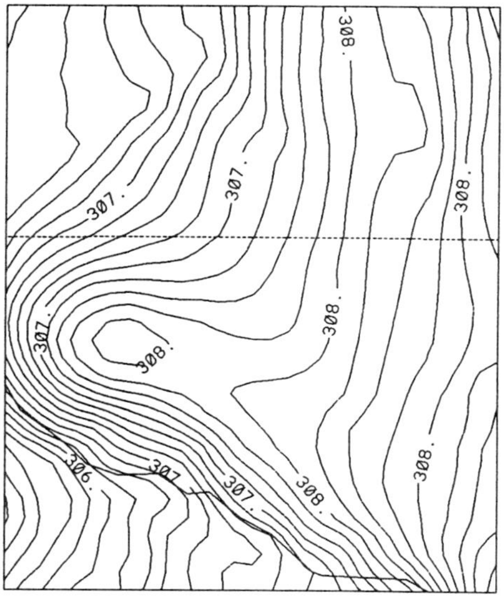

(c) Control Case, 18 UTC, maximum potential temperature 308.4 K.

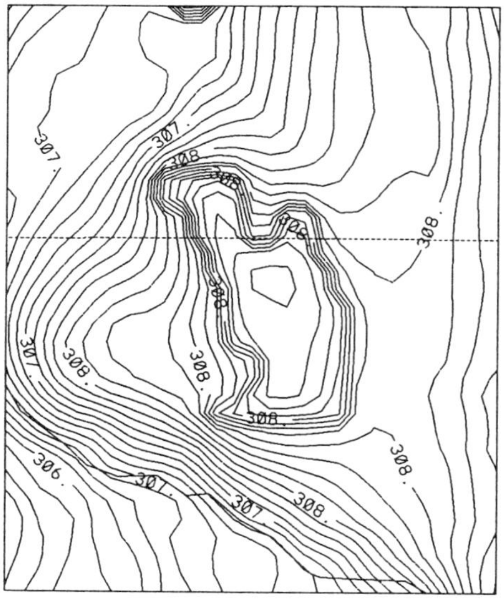

(d) Deforestation Case, 18 UTC, maximum potential temperature 308.8 K.

Figure 7 Potential temperature at 38.2 m above ground level, contour interval 0.25 K at 0 UTC. (a) CS 0UTC; (b) DFS 0 UTC; (c) CS 18 UTC; (d) DFS 18 UTC stronger winds over the pasture in the DFS case are a combined result of the smaller roughness length implying less stable layer turbulence and a stronger horizontal temperature gradient. The potential temperature at 6 UTC has a gradient between the pasture area and the mountain crest to the south west of about 1.5 K/200 km in the CS and 2.5 K/200 km in the DFS (not shown). The higher temperature at the pasture at 6 UTC is a residual of the daytime maximum temperature observed in the afternoon in the DFS in Figure 7d.

The early morning situation (12 UTC) shows that the erosion of the nocturnal stable layer has already reached the level of 38.2 m so that the wind field depicted in Figure 5c is a representation of the larger scale situation with winds predominantly from the south-southwest. The DFS has stronger winds still prevailing over the pasture with velocity almost double the CS for the same location over the pasture. As the daily heating acts over the area the wind field in the centre of the domain responds primarily to the stronger heating in the elevated terrain to the west. In the CS (Figure 6a) the resulting wind is a result of the large scale component plus the topographical forcing with the resulting southeasterly wind at 18 UTC. In the DFS, the afternoon wind field in Figure 6b shows the clear impact of the pasture area with strong velocity gradients responding to the local temperature gradient seen in Figure 7d. The temperature difference at this level (38.2 m) is of almost 1 K in the interface between the two types of vegetation. This is associated with a gradient of water vapour mixing ratio of about 2 g kg^{-1} in the eastern interface as seen in Figure 8b due to the different rates of evaporation between the pasture and the forest (*cf.* Figures 4 c and d).

The wind coming from the east is accelerated as it meets the forest to pasture

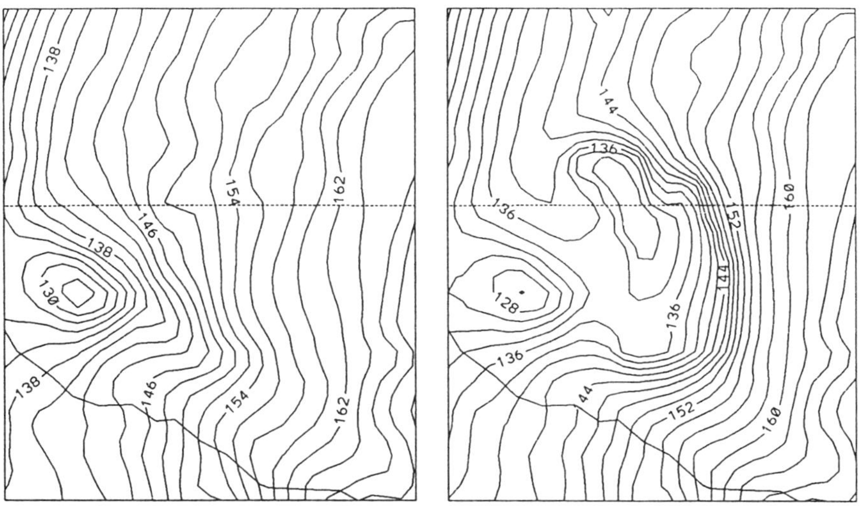

(a) Control Case, 18 UTC, maximum mixing ratio 17.0 $g.kg^{-1}$.

(b) Deforestation Case, 18 UTC, maximum mixing ratio 16.8 $g.kg^{-1}$.

Figure 8 Water vapour mixing ratio at 38.2 m above ground level at 18 UTC, with contour interval of 0.2 g kg^{-1}; (a) CS; (b) DFS interface and then decelerated as it meets the pasture to forest interface. These horizontal gradients in speed give rise to vertical motion and, thus, to a vertical circulation cell. However, the southwestern side of the pasture is already at the slopes of the mountain range, and so an interaction with the upslope circulation is expected.

The vertical cross-section of vertical velocity for 18 UTC and 0 UTC may be seen for an east-west cross-section along y = 10 km in Figure 9 and for a north-south cross section along x = 40 km in Figure 10. Most of the area is dominated by large scale subsidence but the difference in the vertical velocity between the CS and the DFS in low levels, shows the clear impact of the pasture in the middle of the forest. Figures 9b and 9d have the pasture between x = 0 and x = 100 km. At 18 UTC the upward vertical motion on the western side of the pasture area, associated with the deceleration of the low level horizontal flow, has a speed of about 6.10^{-3} m s^{-1} and is more localized than at 0 UTC when it reaches 14.10^{-3} m s^{-1}. At 18 UTC the downward motion associated with the acceleration of the horizontal winds in the eastern side of the pasture has a value of -18.10^{-3} m s^{-1}. At the same time, to the west of the upward motion, a weaker circulation cell may be seen associated with the slope.

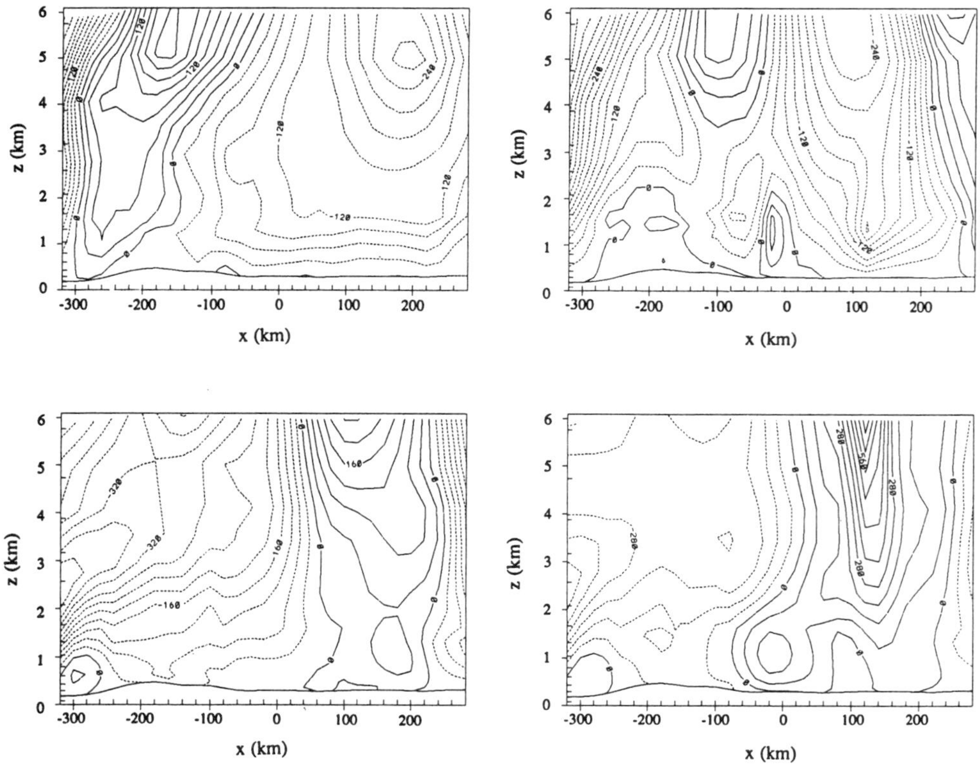

Figure 9 Vertical motion in an east-west cross-section along y = 10 km. Pasture is between x = 0 m and x = 10 km. (a) CS, 18 UTC, contour interval 3. 10^{-3} m s^{-1}; (b) DFS, 18 UTC, contour interval 3. 10^{-3} m s^{-1}; (c) CS, 0 UTC, contour interval 4. 10^{-3} m s^{-1}; (d) DFS, 0 UTC, contour interval 7. 10^{-3} m s^{-1}.

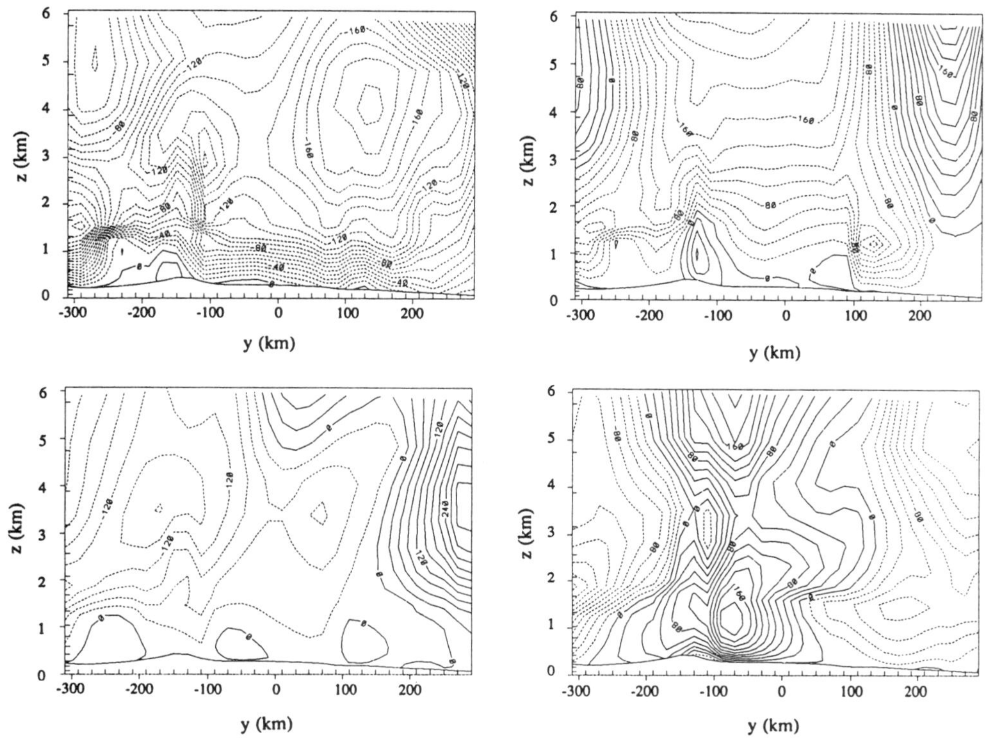

Figure 10 Vertical motion in a north-south cross-section along x = 0 km. Pasture is between y = -1100 m and y = 90 km. (a) CS, 18 UTC, contour interval 1. 10^{-3} m s^{-1}; (b) DFS, 18 UTC, contour interval 2. 10^{-3} m s^{-1}; (c) CS, 0 UTC, contour interval 3. 10^{-3} m s^{-1}; (d) DFS, 0 UTC, contour interval 2. 10^{-3} m s^{-1}.

In both the slope cell and in the forest pasture cell, the downward speed is larger than the upward speed since the downward motion is enhanced by larger scale subsidence.

The CS shows that at 18 UTC (Figure 9a) the slope circulation is only beginning to act with a speed of less than 3.10^{-3} m s^{-1} and at 0 UTC it has already dissipated. In the DFS the upward vertical motion at 18 UTC associated with the forest/pasture western interface reaches 2 km above sea level and in both simulations the upslope circulation is confined below 1 km.

In the north-south cross-section, Figure 10a and 10c show that in the absence of the vegetation difference, the upward vertical motion associated with the mountain range is associated with upslope flow at 18 UTC (Figure 6a) and with convergence of two downslope flows at 0 UTC. In the DFS, the pasture is between y = -110 km and y =90 km so that the southern edge almost reaches the mountain crest. Figure 10b shows that the upward motion at 18 UTC is again very close to the southern side of the pasture, coincidentally very close to the mountain top. The northern edge of the pasture area has the maximum downward velocity. By 0 UTC, Figure 10d shows that the upward and downward branches have increased in size and intensity. The convergence seen in Figure 6c for the CS is enhanced in Figure 6d for the DFS since the pasture area is still warmer than the forest (Figure 7b) and keeps accelerating the windspeed. There is triggering of convection at this time and the upward velocity reaches the upper troposphere.

Besides the effect on the wind field, the difference in vegetation coverage has an impact on the mixed layer structure and height as may be seen in Figures 11 and 12. Figures 11a and 11b show a vertical east-west cross-section of potential temperature and Figure 12a and 12b shows a north-south vertical cross-section of water vapour mixing ratio. In both cases the presence of the pasture in the middle of the forest increases the mixed layer height by about 300 m and changes the horizontal gradients over the mixed layer depth.

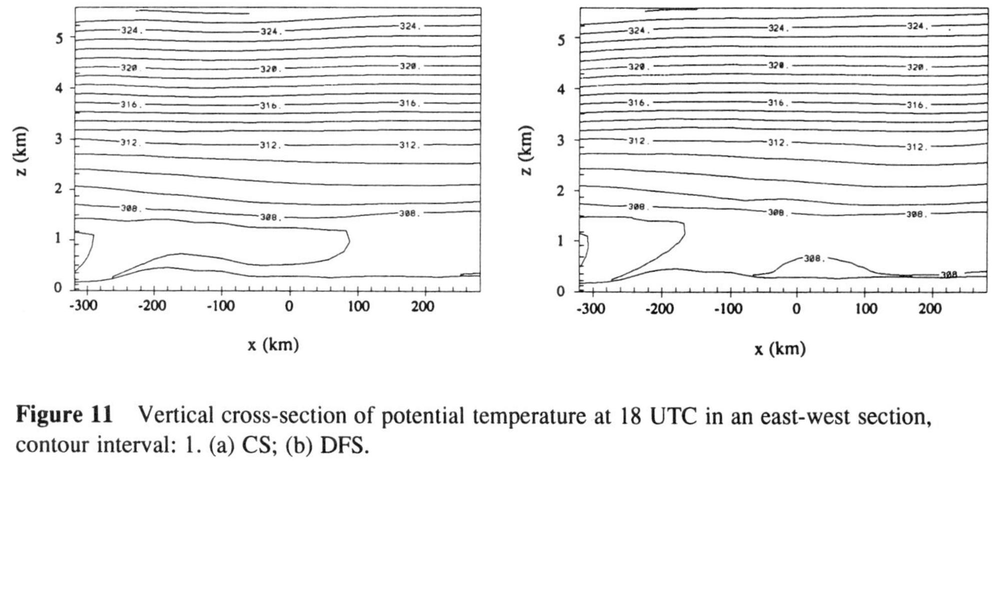

Figure 11 Vertical cross-section of potential temperature at 18 UTC in an east-west section, contour interval: 1. (a) CS; (b) DFS.

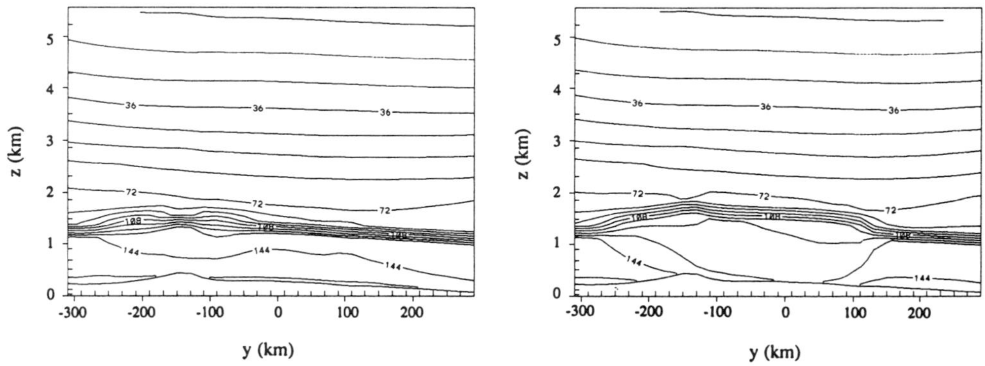

Figure 12 Vertical cross-section of water vapour mixing ratio at 18 UTC in a north-south section, contour interval: 9. 10^{-4} g kg^{-1} (a) CS; (b) DFS.

SUMMARY AND CONCLUSION

In the preceding sections we have shown that the difference in surface fluxes between a forest area and a deforested area, where pasture replaces the native vegetation, produces a complex interaction between mesoscale circulations produced by the vegetation coverage with the circulation produced by the topography and with the larger scale flow. The combination of these effects leads to an enhancement of the magnitude and depth of the upward vertical motion, to horizontal gradients of wind speed, temperature and moisture and to a deeper mixed layer. Given enough moisture and the proper thermodynamic stability, convection may be triggered by these features and the whole troposphere may be affected by the convective transports of momentum, heat and moisture. Even without the convective development, these features indicate a possible serious impact on the vertical transport of momentum, heat and moisture that is not a function of local turbulence alone, but depends on the horizontal gradients of vegetation and of topography. Pielke *et al.*(1991) suggest that the sub-grid scale vegetation variability has to be taken into account in a large scale or in a general circulation model to properly include its effect on the generation of vertical motion. We may argue that taking into account the vegetation variability*per se* is not enough since the particular distribution of vegetation with respect to the topography may completely alter the magnitude of the vertical motion, and thus, of the vertical transports.

ACKNOWLEDGMENTS

This work has been supported by grants from FAPESP and CNPq. The CSU RAMS and the NCAR-Graphics have been used with license from the respective owners. P. R. acknowledges the scholarship that made possible his visit to INPE and the University of São Paulo and the participation in the ABRACOS field campaign. We are grateful to Dr Carlos Nobre for encouragement and to Dr José Eduardo Prates for help in earlier phases of this work. Prof Eliana Kalil from INPE provided the LANDSAT information.

REFERENCES

Avissar, R. and Mahrer, Y. 1988. Mapping frost-sensitive areas with a three dimensional local scale model. Part I: Physical and numerical aspects. *J. Appl. Meteor.* **27**, 400-413.

Chen, C. and Cotton, W. R. 1983. A one-dimensional simulation of thee stratocumulus capped mixed layer. *Boundary-Layer Meteor.* **25**, 289-321.

Deardorff, J. W. 1978. Efficient prediction of ground surface temperature and moisture with the inclusion of a layer of vegetation. *J. Geophys. Res.* **83**, 1889-1903.

Dickinson, R. E., Hendeson-Sellers, A., Kennedy, P. J. and Wilson, M. F. 1986. Biosphere-Atmosphere Transfer Scheme (BATS) for the NCAR community climate model.*NCAR Technical Note,* NCAR/TN-275+STR, 69.

Ducoudré, N.I., Laval, K. and Perrier, A. 1993. SECHIBA, A new set of parametrizations of the hydrology exchanges at the land-atmosphere interface within the LMD atmospheric general circulation model. *J. Climate* **6**, 248-273.

Kuo, H.L. 1974. Further studies of the parameterization if the influence of cumulus convection on large-scale flow. *J. Atmos. Sci.*, **31**, 1232-1240.

Louis, J.F. 1979. A parametric model of vertical eddy fluxes in the atmosphere. *Boundary-Layer Meteorol.* **17**, 187-202.

McCumber, M.C. and Pielke, R.A. 1981. Simulation of the effects of surface fluxes of heat and moisture in a mesoscale numerical model. *J. Geophys. Res.* **86**, 9929-9938.

Manzi, A.O., 1993. Introduction d'un schéma des transferts sol-végétation-atmosphère dans un modèle de circulation générale et application à la simulation de la déforestation amazonienne. *Thèse de doctorat.* Université Paul Sabatier, Toulouse.

Mellor, G.L. and Yamada, T. 1982. Development of a turbulence closure model for geophysical fluid problems. *Rev. of Geophys. and Space Phys.* **20**(D4), 851-875.

Moore, C.J. and Fisch, G. 1986. Estimating heat storage in Amazonian tropical forest. *Agri. For. Meteorol.* **38**, 147-169.

Noilhan, J. and Planton, S. 1989. A simple parameterization of land surface processes for meteorological models. *Mon. Wea. Rev.* **117**, 536-549.

Pielke, R.A., Cotton, W. R., Walko, R. L., Tremback, C. J., Lyons, W. A., Grasso, L., Nicholls, M. E., Moran, M. D., Wesley, D. A., Lee, T. J. and Copeland, J. H. 1992. A comprehensive Meteorological Modelling System - RAMS. *Meteorol. and Atmos. Phys.* **49**, 69-9.1

Pielke, R.A., Dalu, G.A., Snook, J.S., Lee, T.J. and Kittel, T.G.F. 1991. Nonlinear influence of mesoscale land use on weather and climate. *J. Climate* **4**, 1053-1069.

Regnier, P. 1995. Etude de l'interaction de la forèt tropicale et des paturages post-déforestation avec le climat amazonien. *Thèse de doctorat.* Université de Paris VI.

Rocha, H.R. da, 1992. Modelagem da interaçao biosfera-atmosfera para areas de floresta e de pastagem na Amazonia. *M.S. thesis,* INPE, Sao Jose dos Campos.

Sellers, P.J., Shuttleworth, W.J., Dorman, J.L., Dalcher, A. and Roberts, J. M. 1988. Calibrating the simple biosphere model for Amazonian tropical forest using field and remote sensing data: Part 1. Average calibration with field data. J. *Appl. Meteorol.* **28**, 727-759.

Shuttleworth, W.J., Gash, J.H.C., Lloyd, C.R., Moore, C.J., Roberts, J., Marques, A., Fisch, G., Silva, V., Ribeiro, M.N.G., Molion, L.C.B., Abreu Sa, L., Nobre, C.A., Cabral, O.M.R., Patel, S. R., and de Moraes, J.C. 1984. Eddy correlation measurements of energy partition for Amazonian forest. *Q. J. R. Meteorol. Soc.* **110**, 1143-1162.

Tripoli, G.J. and Cotton, W.R. 1982: The Colorado State University three-dimensional cloud/ mesoscale model - 1982 Part I: general theoretical framework and sensitivity experiments.*J. Rech. Atmos.* **16**, 185-219

Wright , I.R., Gash, J.H.C., Rocha, H.R. da, Shuttleworth, W. J., Nobre, C. A., Maitelli, G. T., Zamparoni, C.A.G.P. and Carvalho, P.R.A. 1992. Dry season micrometeorology of central Amazonian ranchland. *Q. J. R. Meteorol. Soc.* **118**, 1083-1099.

RESUMO

As circulações de mesoscala numa area desmatada em Rondônia durante a estação seca são simuladas através de um modelo tridimensional de mesoescala inicializado com os campos de grande escala do dia 22 de julho de 1993 e levando em conta as tendências sinóticas através de uma banda nas condições de contorno laterais que tende a seguir as condições de grande escala. Os fluxos de superfície simulados são comparados com os fluxos medidos durante a campanha do ABRACOS de julho de

1993. Mostra-se que a diferença nos fluxos de superfície entre uma área com floresta e uma área desmatada, onde uma pastagem cresceu no lugar da vegetação nativa, produz uma interação complexa com as circulações produzidas pela topografia e pelo escoamento de grande escala. A combinação desses efeitos leva ao aumento da magnitude e da profundidade dos movimentos ascendentes, a gradientes horizontais de velocidade do vento, de temperatura e de umidade e a uma camada de mistura mais profunda. Essas características indicam um possível sério impacto no transporte vertical de momentum, calor e umidade que não é uma função apenas da turbulência local mas depende dos gradientes horizontais associados à vegetação e sua distribuição sobre a topografia.

31 The simulated impact of Amazonian deforestation on climate using measured ABRACOS vegetation characteristics

J. LEAN[1], C.B. BUNTON[1], C.A. NOBRE[2] and P.R. ROWNTREE[1]
[1]*Hadley Centre for Climate Prediction and Research, Meteorological Office, UK*
[2]*CPTEC-INPE, São José dos Campos, Brazil*

INTRODUCTION

In recent years a number of general circulation model (GCM) experiments have been conducted to assess the impact of complete removal of the South American tropical rainforest on climate. A summary of the most recent experiments is presented in Table 1 (an updated version from Pitman *et al.*, 1993). Differences in the specification of two important vegetation characteristics, albedo and roughness are given; the imposed increases in albedo ranges from 5% to 8% and the decreases in roughness from 0.76 to 2.57 m. Results vary widely between experiments: increases in surface temperature range from 0.1 to 2.3°C, decreases in precipitation from 15 to 640 mm year^{-1} and decreases in evaporation from 25 to 500 mm year^{-1}. This study will address four key issues raised by an assessment of these experiments.

The prescription of the land surface parameters in the forested and deforested scenarios — the wide disparity in results may be partially explained by the large differences in the specification of vegetation and soil characteristics between experiments. The availability of extensive measurements taken during the ABRACOS field campaigns for paired sites of forest and cleared areas provides a major opportunity to introduce some consensus across the breadth of experiments. In this study measured ABRACOS values of vegetation characteristics are used where possible, to provide a more representative and accurate description of the forest and pasture within Amazonia in our model.

The response to particular aspects of the deforestation scenario — previous experiments have indicated that changes in albedo and roughness are important in determining the final deforested climate (Lean and Warrilow, 1989). Two companion experiments that include separate changes to albedo and roughness, from the forest to pasture values, allow clearer identification of the mechanisms involved in the

Amazonian deforestation and climate. Edited by J.H.C. Gash, C.A. Nobre, J.M. Roberts and R.L. Victoria.

response to complete removal of the forest.

Sensitivity to the change in maximum infiltration capacity of the soil — previous experiments have made changes in the maximum infiltration rate to represent the observed compaction of soil and consequent reduction in infiltration following deforestation (Wright *et al.*, 1996). This quantity has not been clearly defined from measurements and so an experiment in which all vegetation characteristics except the maximum infiltration capacity are changed from the forest to the pasture values is used to assess the impact of this change on the total deforested climate.

Response to partial removal of the forest - most previous Amazonian deforestation experiments have evaluated the response to total removal of the forest, a further experiment in which deforestation is confined to a more realistic reduced area is presented here.

MODEL DESCRIPTION

The Hadley Centre model, the climate version of the Meteorological Office Unified Forecast/Climate GCM, used in these experiments, has been described in some detail by Jones *et al.* (1995). The atmospheric model has 19 layers, spaced unevenly in the vertical, to give optimum resolution in terms of pressure in the upper troposphere and stratosphere and in the boundary layer, in which there are five layers below 750 h Pa. The horizontal mesh is 2.5° latitude by 3.75° longitude, the same as used by Lean and Rowntree (1993) (called LR in this paper) in an earlier deforestation experiment.

The dynamics use a two-step Heun scheme for horizontal advection with fourth order accuracy, formulated in terms of total water (vapour plus liquid and frozen cloud water) and liquid water temperature (temperature adjusted to allow for the latent heat of cloud water and ice).

The physics of the model is similar to that of the model used by LR. The main differences may be summarised as:

- A four-layer soil hydrology scheme based on Richards's equation is used, with the same vertical layer depths as used by Warrilow *et al.* (1986) for thermal processes.
- Over the South American tropical forest region, stomatal resistance is computed as a function of atmospheric variables (solar radiation, vapour pressure deficit below saturation and temperature) and soil moisture as described by Dolman *et al.* (1991) for forest and Wright *et al.* (1995) for pasture.
- Cloud optical properties are computed as a function of the liquid and ice cloud water contents, with cloud assumed to be liquid for temperatures above 0°C, and ice below -15°C with mixed phase cloud in between.
- The convection scheme has been extended to include representation of a downdraught (Gregory and Allen, 1991).

Table 1 Comparison of some recent GCM Amazonian deforestation experiments adapted from Pitman *et al*. (1993)

	Dickinson and Kennedy (1992)	Henderson-Sellers *et al*.(1993)	Lean and Rowntree (1993)	Nobre *et al*. (1991)	Polcher and Laval (1994a)	Polcher and Laval (1994b)	Manzi (1993)	This expt.
	CCM1	CCM1-0Z	UKMO	NMC	LMD	LMD	EMERAUDE	UKMO
Resolution	4.5°x7.5°	4.5°x7.5°	2.5°x3.75°	1.8°x2.8°	2.0°x5.6°	2.0°x5.6°	2.8°x2.8°	2.5°x3.75°
Surface	BATS	BATS	Warrilow (1986)	SIB	SECHIBA	SECHIBA	ISBA	Warrilow (1986) + interactive resistance + multilayer hydrology
Ocean	mixed layer model	mixed layer model	climatological SSTs	perpetual December SST	observed SSTs 1978-1988	climatological SSTs	climatological SSTs	climatological SSTs
Length	3 years	6 years	3 years	1 year	11 years	1.1 years	4.2 years	10 years
roughness length (m)	2.0 / 0.05	2.0 / 0.2	0.8 / 0.04	2.65 / 0.08	2.3 / 0.06	no change	2.0 / 0.06	2.1 /0.026
albedo	0.12 / 0.19	0.12 / 0.19	0.14 / 0.19	0.13 / 0.20	0.14 / 0.22	0.1 / 0.18	0.13 / 0.20	0.13 /0.18
ΔT (°C)	+0.6	+0.6	+2.1	+2.0	+0.1	+3.8	+1.3	+2.3
ΔP (mm)	-511.0	-588.0	-295.7	-640.0	-186.2	+394.0	-15.0	-157.0
ΔE (mm)	-25.5	-232.0	-198.0	-500.0 -	127.8	-985.0	-113.0	-295.7
Change in moisture convergence	decreased	decreased	decreased	decreased	decreased	increased	increased	increased

- The atmospheric boundary layer occupies up to five layers (about 2.5 km), depending on the position of the lowest statically stable layer. In unstable conditions, the increased vertical extent of turbulent eddies is accounted for by allowing non-local mixing (Smith, 1993).

EXPERIMENT DESIGN

The following set of experiments was designed to give a more realistic portrayal of Amazonian deforestation and to provide a clearer indication of the mechanisms involved.

(1) FOREST - the control experiment - a ten-year, three-month experiment with vegetation parameter values for all points in South America defined as tropical rainforest in the model (see Figure 1) taken mostly from ABRACOS measurements of tropical rainforest.

(2) PASTURE - the deforested experiment - a ten-year, three-month experiment with vegetation parameter values for all points in South America to be deforested (see Figure 1) taken mostly from ABRACOS measurements of pasture.

(3) REDRGH - a five-year, three-month experiment as FOREST, except with roughness reduced from the FOREST value to the PASTURE value and all other vegetation characteristics unchanged.

(4) INCALB - a five-year, three-month experiment as FOREST, except with surface albedo increased from the FOREST value to the PASTURE value and all other vegetation characteristics unchanged.

(5) FIXINF - a five-year, three-month experiment as PASTURE, except the maximum infiltration capacity which was held constant at the value given for FOREST.

(6) HALF - a 10 year three month experiment as PASTURE except that only approximately half of the points were assumed to be deforested

Figure 1 shows the points on the 2.5° x 3.75° grid over South America that were defined as tropical forest. These points are based on the interpolation of the 1° x 1° grid of vegetation types as defined in the SiB model (Sellers *et al.*, 1986) with some alterations based on more recent observations. Those points that were deforested in the HALF experiment are shown. The reduced deforested area was selected on the basis of (i) the assessment of Brazilian deforestation by Skole and Tucker (1993) and (ii) Myers' (1991) national estimates of deforestation fractions for other South American countries. The resulting area includes most of the forests of Colombia, Ecuador and northern Peru in the northwest, and, in the south and east, those of southeastern Peru, northern Bolivia, and southern and eastern parts of the Brazilian Amazonia. Of the total of 66 forested points, 34, just over half, were assumed to be deforested. Figure 1 also shows the two regions that were chosen for analysis. Region 1 essentially encompasses the southern Amazon basin and corresponds most closely with the area used for analysis in LR. Region 2 represents northern Amazonia covering all points north of 1.25°N. Experiments were started at the beginning of

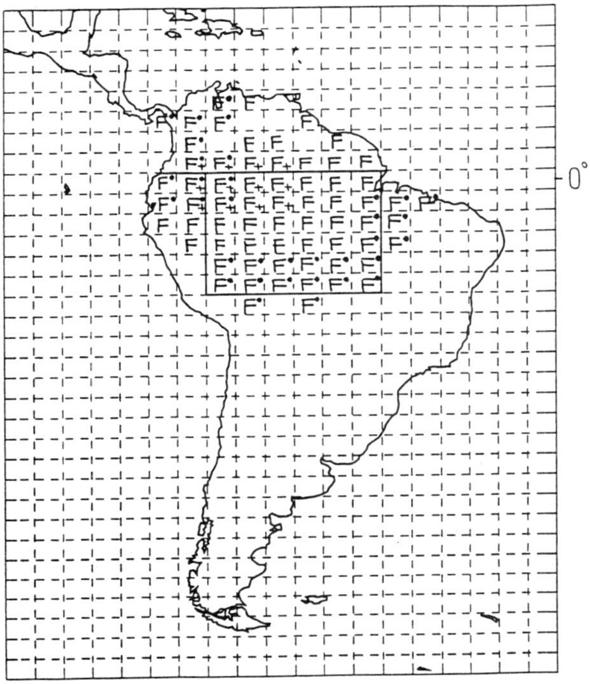

Figure 1 Model grid for South America showing those points that are defined as forest (F) and replaced by pasture in the deforested scenario. Points marked with a solid circle are forest points that are replaced by pasture in the HALF deforestation experiment. Those points marked with a + are defined as having a constant albedo value all year. The box encompassed by solid black lines shows the southern area (Region 1) and all points north of 1.25°N the northern area (Region 2).

December from initial fields generated by a data assimilation scheme for that time of year. All results presented are averaged over the ten years of the experiment excluding the first three months.

Table 2 shows the vegetation and soil parameters for the FOREST and PASTURE simulations. The measured values of monthly-varying albedo as summarised by Culf *et al.* (1995) were used for those points that have their dry season (and highest albedos) in southern winter or spring and the seasonal cycle was reversed for points where the observed rainfall cycle is approximately opposite, namely most points north of the equator. An area of western Amazonia centred near the equator which effectively has no dry season (a dry season was arbitrarily defined as having less than 300 mm per three month season) was located, using the seasonal rainfall maps of Figueroa and Nobre (1990). For this region the lowest values of albedo (0.12 for FOREST and 0.17 for PASTURE) were assigned throughout the year. All vegetation parameters are for the forest and pasture site at Manaus except values for albedo which are averages over the forest and pasture sites at Ji-Paraná, Marabá and Manaus. Although field measurements have shown that root extraction occurs at depths greater than 3.5 m at all ABRACOS sites (Hodnett *et al.*, 1995), the rooting depth was limited to the maximum depth of soil (about 2.3 m) in the model.

At present only a few point measurements of soil parameters are available from the ABRACOS field campaigns and as variations in soil are apparent on much smaller scales than even vegetation characteristics it was decided that it was inappropriate to assume that these measurements could be taken as representative of the whole Amazon basin. Soil parameters were therefore taken from values given in our standard model; these parameters vary across the basin with average values as given in Table 2. Most soil parameters were kept constant between the FOREST and PASTURE experiments except the ratio of saturated hydraulic conductivity at the surface relative to the subsurface soil. This was altered to correspond with observations of reduced infiltration due to compaction of the soil following deforestation (Wright *et al.* 1996).

Table 2 Vegetation and soil characteristics used in the FOREST and PASTURE experiments

Parameter	**FOREST**			**PASTURE**		
	Month (south of equator)		Month (north of equator)	Month (south of equator)		Month (north of equator)
Albedo	Jan	12.6	Jul	Jan	17.5	Jul
	Feb	12.3	Aug	Feb	17.1	Aug
	Mar	12.1	Sep	Mar	18.1	Sep
	Apr	12.4	Oct	Apr	18.4	Oct
	May	13.2	Nov	May	18.7	Nov
	Jun	13.9	Dec	Jun	18.6	Dec
	Jul	14.1	Jan	Jul	19.2	Jan
	Aug	14.4	Feb	Aug	18.7	Feb
	Sep	14.2	Mar	Sep	17.1	Mar
	Oct	14.3	Apr	Oct	17.1	Apr
	Nov	14.1	May	Nov	17.8	May
	Dec	13.4	Jun	Dec	18.0	Jun
	Mean	13.4		Mean	18.0	
Surface resistance (s m^{-1})	Interactive			Interactive		
Rooting depth (m)	2.26			1.5		
Roughness (m)	2.1			0.026		
Canopy capacity (mm)	0.74			0.5		
Vegetation fraction (%)	100			85		
Soil moisture conc. at wilting point	0.047			0.047		
Soil moisture conc. at critical point	0.113			0.113		
Soil moisture conc. at saturation point	0.343			0.343		
Saturated hydraulic conductivity (K_s) (kg m^{-2} s^{-1})	0.0026			0.0026		
B exponent used in hydraulic conductivity dependence on soil moisture	3.98			3.98		
Heat capacity (J K^{-1} m^{-3})	2.371×10^6			2.371×10^6		
Heat conductivity (W m^{-1} K^{-1})	0.531			0.531		
Soil water suction (m)	0.50			0.50		
Surface K_s / subsurface K_s	1.5			0.3		

MODEL VALIDATION

The following section assesses the performance of the model's simulations of the climates of northern South America associated with both forest and pasture covered locations. Until now it has only been possible to compare with climatological and point measurements for undisturbed tropical rainforest (Shuttleworth, 1988) but if we want to improve the confidence we have in the modelled climate scenarios then it is vital to validate the climate associated with deforested sites as well. Measurements taken during the ABRACOS campaign described by Wright *et al.* (1992) and Bastable *et al.* (1993) provide realistic descriptions of the climate typical of cleared areas within the Amazonian rainforest. All model results are averaged over ten years with the first three months of the experiments excluded.

Figure 2 compares modelled rainfall with extensive climatological observations taken over a number of years (Figueroa and Nobre, 1990) for the southern hemisphere summer (December-February) and winter (June-August). Rainfall coverage and amounts over northern South America are well simulated by the model

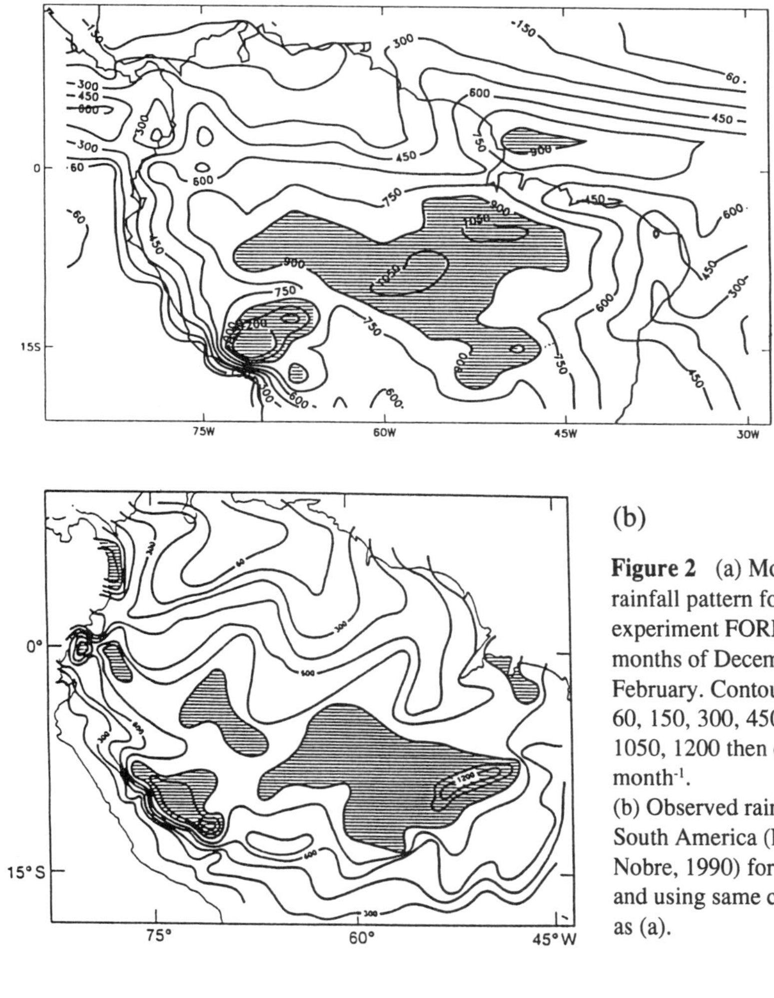

(a)

(b)

Figure 2 (a) Modelled mean rainfall pattern for the control experiment FOREST for the months of December, January and February. Contours are at 0, 30, 60, 150, 300, 450, 600, 750, 900, 1050, 1200 then every 300mm month^{-1}.
(b) Observed rainfall pattern over South America (Figueroa and Nobre, 1990) for the same period and using same contour interval as (a).

(c)

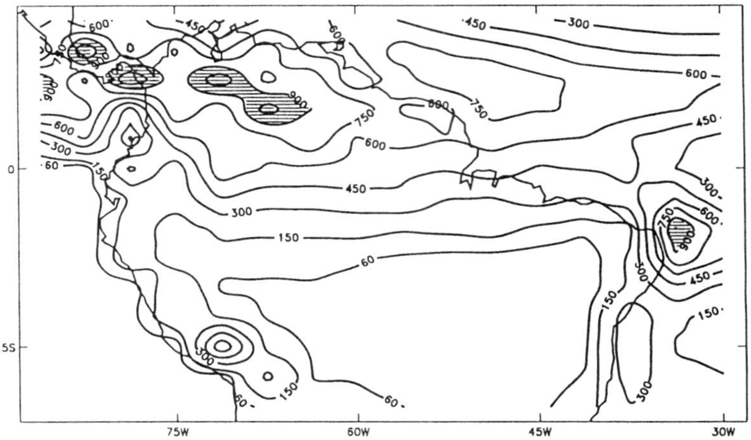

(d)

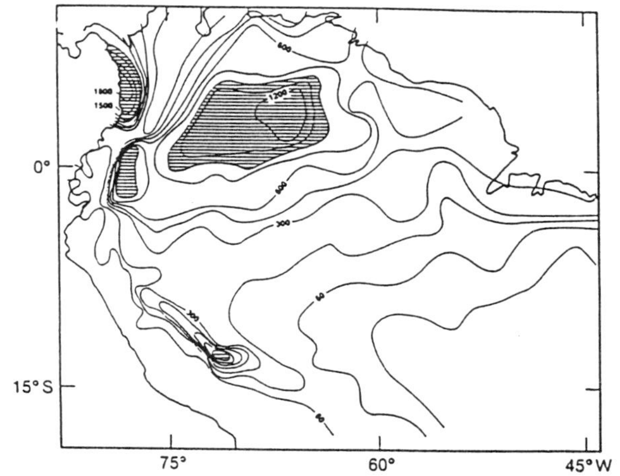

Figure 2 contd.
(c) As (a) but for June, July, August .
(d) As (b) but for June, July, August.

in both these seasons. The most notable deficiencies are that the model fails to emulate some of the high rainfall totals recorded in the more mountainous regions of Colombia, Ecuador and Peru. The simulations for the other two seasons (not shown) are equally realistic.

Modelled net radiation at the surface is overestimated by about 25%, in common with a number of other GCMs (Lean and Rowntree, 1993). As a result of this deficiency evaporation rates are overestimated and so to compare with observations it is more useful to normalise the evaporation rates by available net radiation at the surface. Figure 3 compares normalised evaporation, for both FOREST and the control experiment used by LR, with observations for Manaus. The new control shows a marked improvement in modelling evaporation in all months but especially so from July to October when the model previously suffered from an excessive limitation for dry periods. This more realistic simulation can be attributed to a combination of the new formulation of stomatal resistance and the increased rooting depth (previously a rooting depth of 1.27 m was used for forest) together with the

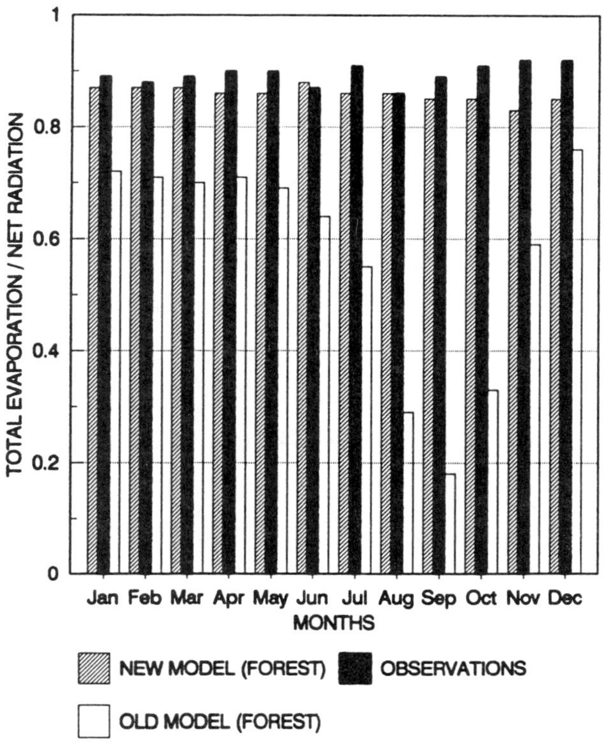

Figure 3 Modelled monthly mean evaporation as a fraction of net radiation for a single gridpoint centred at 2.5°N 60°W, close to Manaus, for the current model and averaged over four gridpoints centred near Manaus for the old version of the model used by Lean and Rowntree (1993) compared with observations for Manaus (Shuttleworth, 1988).

implementation of the new multilayer hydrology scheme. Modelled interception accounts for 30% of the total evaporation over the year which agrees reasonably well with the value of 25% quoted by Shuttleworth (1988).

Wright *et al.* (1992) reported that during periods when there was negligible soil moisture limitation the observed evaporation rate for a pasture area was more comparable with the transpiration rate from the forest, but during dry periods evaporation rates were much lower. The model is able to reproduce this behaviour. The average reduction in evaporation from 4.0 to 2.8 mm day^{-1} during the relatively dry months of July through to September agrees well with the magnitude of the change reported by Wright *et al.* from 3.8 to 2.1 mm day^{-1} during the 1990 dry season.

Bastable *et al.* (1993) showed the much greater daily variation in climatic variables over clearings compared with those observed over forest. Figure 4 shows the daily cycle in modelled surface temperature for the gridpoint representative of the Manaus site, both with forest and with pasture cover. These values can be compared with measurements of screen height temperature to judge performance although strictly

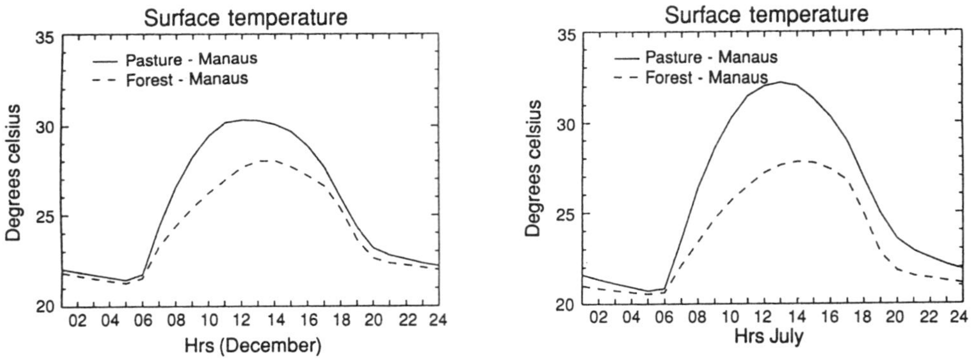

Figure 4 Modelled hourly surface temperature over forest and pasture for December (left) and July (right).

these variables are not the same. Two months are illustrated: the first for December, typical of a wet month (Figure 4 - left), shows a difference in the diurnal range of 9.1°C for PASTURE and 6.9°C for FOREST which may be compared with the measured values for screen height temperature of 7.1°C and 5.3°C. In contrast, during the relatively dry month of July (Figure 4 - right) there is a much larger difference in diurnal range, 11.4°C for PASTURE and 7.3°C for FOREST compared with the measured values for screen height temperature of 12.7°C and 6.5°C. However, Bastable *et al.* showed that there was an increase in temperature during the day but a decrease in temperature at night over the clearing relative to the forest, whereas the model shows an increase during the day and night.

Bastable *et al.* also present an analysis of the surface radiation balance for the forest and cleared sites. Table 3 shows the modelled surface radiation components averaged over Region 1 for PASTURE and FOREST experiments.

Table 3 Average daily radiation components at the surface for the FOREST and PASTURE experiments and the difference (PASTURE–FOREST) averaged over Region 1.

Radiation component	FOREST ($W m^{-2}$)	PASTURE ($W m^{-2}$)	PASTURE–FOREST ($W m^{-2}$)
Solar downward	223.9	233.7	+9.8
Solar upward	29.6	41.8	+12.2
Net solar	194.3	191.9	-2.4
Longwave downward	404.2	407.5	+3.3
Longwave upward	441.5	456.3	+14.8
Net longwave	-37.3	-48.7	-11.4
Net surface	157.0	143.2	-13.8

In line with the results presented by Bastable *et al.* the difference in the longwave radiation balance makes a large contribution to the difference in the surface net radiation. The longwave radiation balance is influenced by changes in surface temperature and cloudiness. Higher surface temperatures dominate the change in the longwave balance leading to increased longwave cooling from the surface. Increased albedo results in a large increase in the amount of solar radiation that is reflected at the surface. This is largely offset in the model by increased incoming solar radiation as a result of reduced cloudiness; this cloudiness change would not be expected in the observations from the paired sites described by Bastable *et al.*, 1993. because of the very different scale of the forest clearance at their site.

We have shown that the model is able to reproduce the important features of the climate of Amazonia associated with both forested and cleared sites for time scales which range from diurnal to the seasonal.

RESULTS

This presents the results from ten-year averages of the FOREST, PASTURE and HALF experiments and five-year averages of INCALB, REDRGH and FIXINF which exclude the first three months when the initial conditions are likely to be influencing results. These relatively long simulations (see Table 1) allow a more confident assessment of the statistical significance of the results. The null hypothesis that is put forward is that there has been no change in, for example, rainfall due to deforestation. Student's t statistics for the difference between the annual mean rainfalls for the PASTURE and FOREST experiments are then calculated, using the rainfall data averaged over each year from each experiment. For the ten-year (five-year) experiments the number of degrees of freedom is 18 (8) and hence for a 95% confidence level requiring 2.5 % in each tail, a test for t less than -2.101 (-2.306) or greater than +2.101 (+2.306) is performed. If this criterion is met then the null hypothesis is rejected and it can be concluded that deforestation is responsible for the change in rainfall at the 95% level of confidence. Regional maps are shown with changes significant at the 95% and 99% level of confidence delineated.

TOTAL DEFORESTATION EXPERIMENT

Table 4 summarises the main changes to the climate of Region 1 brought about by clearance of the forest.

In broad terms results are similar to previous experiments reported by LR with reductions in evaporation and rainfall and an increase in surface temperature. The most notable difference is that the reduction in evaporation of 0.81 mm day^{-1} is partially compensated by increased moisture convergence, resulting in a smaller reduction in rainfall of 0.43 mm day^{-1}. This is contrary to results from most previous deforestation experiments but agrees with Polcher and Laval (1994b) and Manzi (1993) — see Table 1.

Table 4 Comparison of Climate Variables for the FOREST and PASTURE experiments and the difference (PASTURE–FOREST) averaged over Region 1 for this experiment and from Lean and Rowntree (1993) (LRPAS–LRFOR).

Variable	FOREST	PASTURE	PASTURE–FOREST	LRPAS–LRFOR
Total evaporation (mm day^{-1})	4.32	3.51	-0.81 (-19%)	-0.55 (-24%)
Interception (mm day^{-1})	1.56	1.07	-0.49 (-31%)	-0.08 (-10%)
Rainfall (mm day^{-1})	6.56	6.13	-0.43 (-7%)	-0.81 (-14%)
Moisture convergence (mm day^{-1})	2.24	2.62	+0.38 (+17%)	-0.26 (-8%)
Soil moisture (mm)	346.0	150.4	-195.60 (-57%)	-87.9 (-65%)
Total runoff (mm day^{-1})	2.21	2.60	+0.39 (+18%)	-0.2 (-6%)
Surface runoff (mm day^{-1})	0.52	2.22	+1.70 (+327%)	+0.47 (+27%)
Net radiation (W m^{-2})	157.0	143.2	-13.80 (-9%)	-18.5 (-13%)
Sensible heat (W m^{-2})	31.9	41.6	+9.70 (+30%)	-2.9 (-4%)
Latent heat (W m^{-2})	125.0	101.6	-23.40 (-19%)	-15.8 (-24%)
Velocity at level 1 (ms^{-1})	1.03	2.28	+1.25 (+120%)	+1.03 (+44%)
Surface temperature (°C)	23.8	26.1	+2.3	+2.3

Figure 5 shows monthly means for FOREST and PASTURE for southern Amazonia, Region 1 and northern Amazonia, Region 2 (see Figure 1). Evaporation shows substantial decreases throughout the year with the largest shortfalls occurring when soil moisture restricts evaporation. This is apparent from the large reduction in combined evapotranspiration and soil evaporation (shown as evapotranspiration on graph) at these times of year. Interception is reduced in all months. For southern Amazonia rainfall has decreased for all months but for northern Amazonia rainfall shows small increases as well as decreases; these are associated with regional variations which are discussed later. Increases in surface temperature are due partly to reduced evaporative cooling and partly to reduced surface roughness. Surface runoff is much higher in all months as a result of the imposed reduction in maximum infiltration rate at the surface. The largest increases coincide with times when rainfall is at a maximum and so the reduction in maximum infiltration rate can have the greatest impact. Subsurface runoff (defined as drainage from the bottom of the root zone) is less in all months because of the drier soil resulting from the increased surface

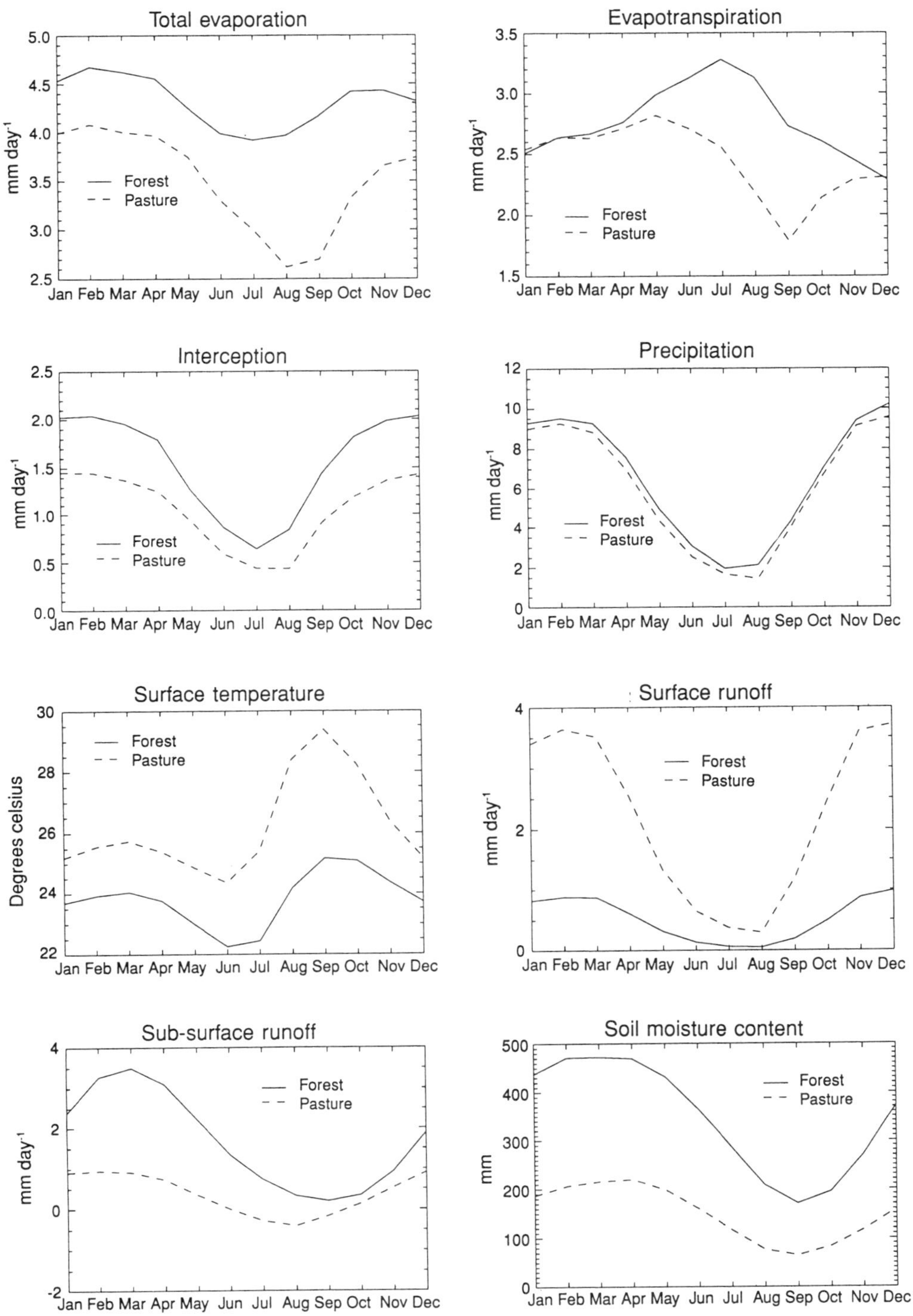

Figure 5 Modelled monthly mean climate variables for southern Amazonia (Region 1) and for northern Amazonia (Region 2) (overleaf).

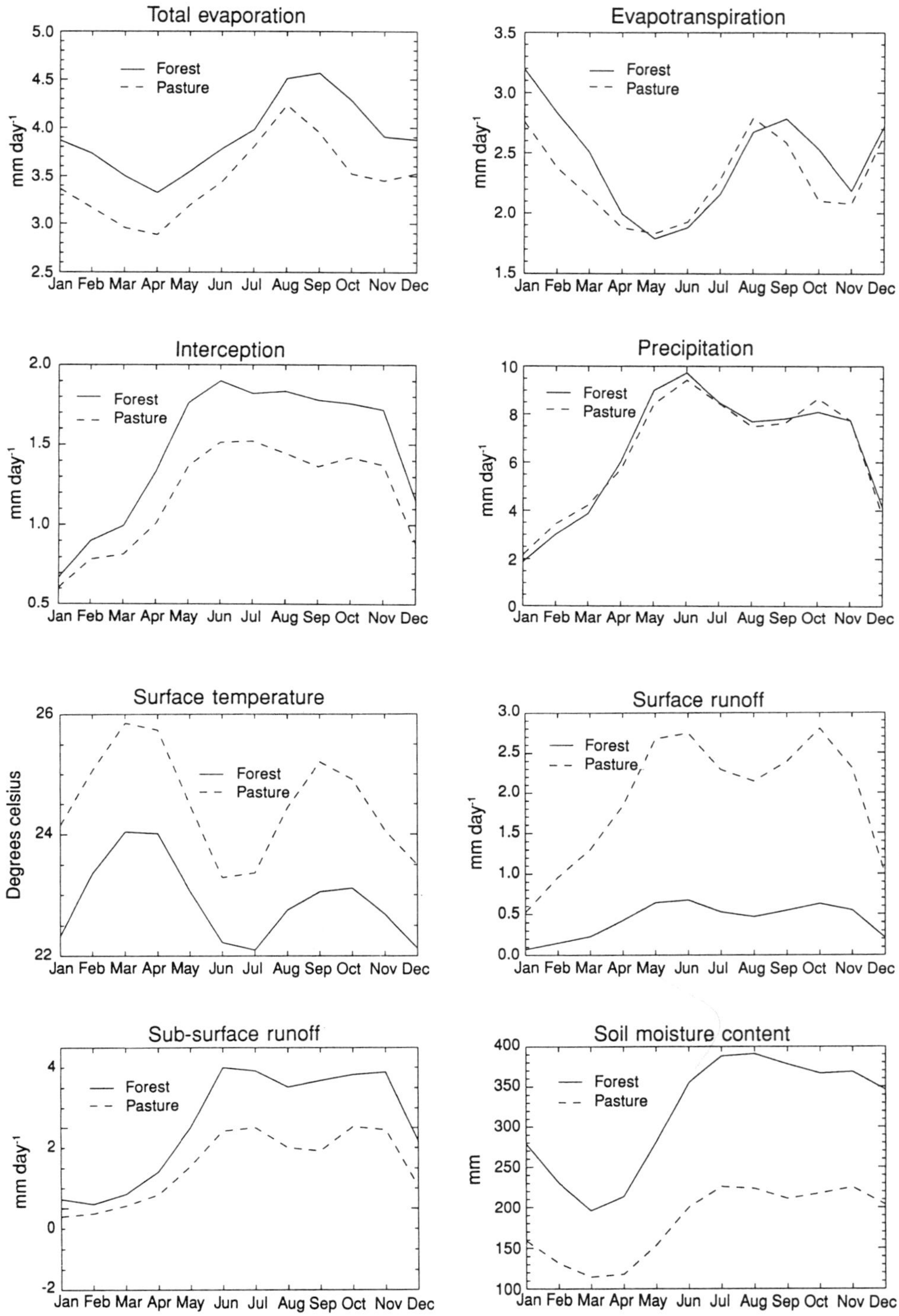

Figure 5 (contd.)

runoff. During dry months soil moisture is drawn up from below the root zone as low rainfall and evapotranspiration at these times deplete the soil moisture within the root zone.

Figure 6 shows the annual spatial distribution of changes in rainfall, evaporation

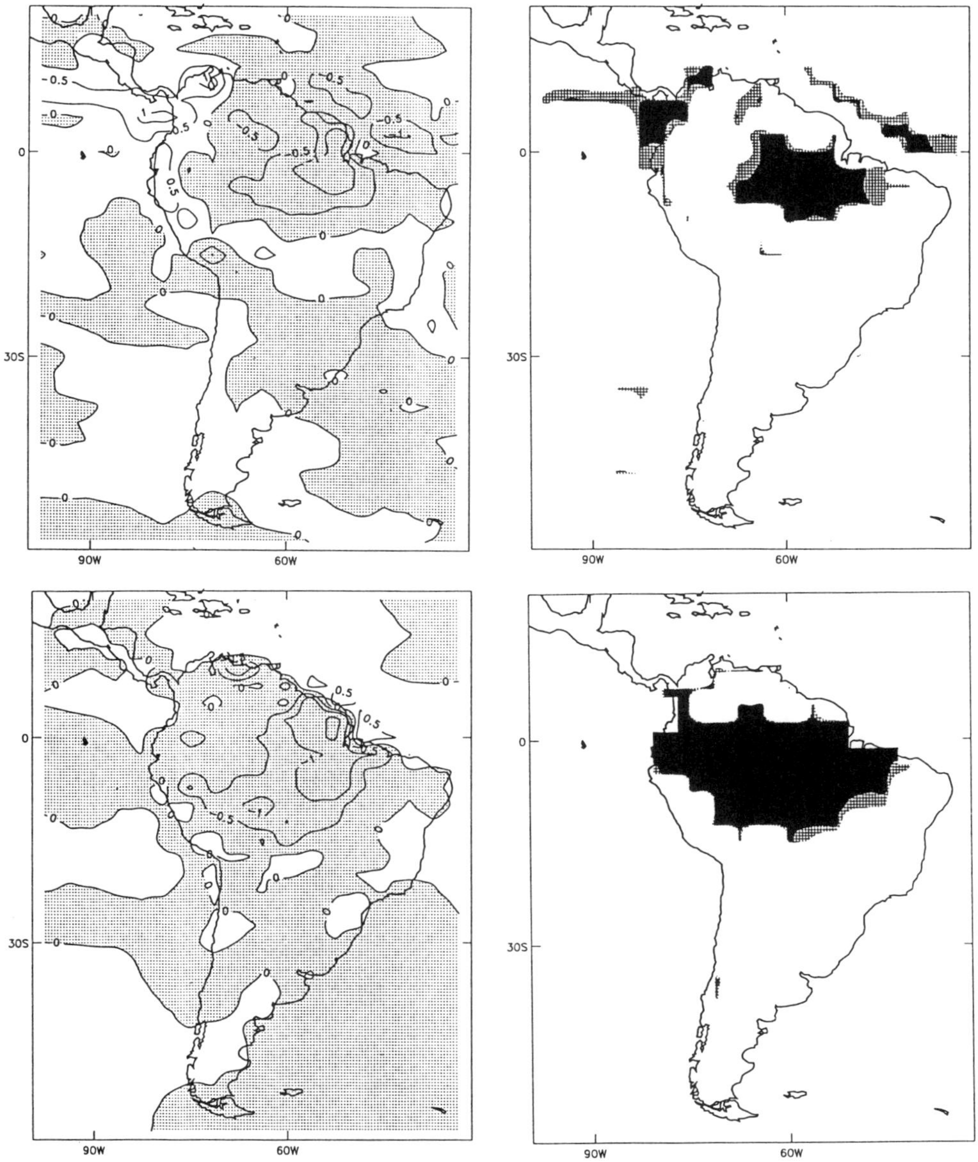

Figure 6 (a) Modelled annual mean difference (PASTURE–FOREST) in rainfall. Contours are at 0.0, ±0.5, 1.0 and then every 1.0 mm day^{-1} with areas of decrease shaded. (b) Student's t fields showing changes in rainfall that are locally significant at the 95% (cross-hatched shaded areas) and 99% (solid shaded areas) confidence levels. (c) As for (a) except for evaporation. Contours are at 0.0, ±0.5, 1.0 and then every 1.0 mm day^{-1} with areas of decrease shaded.
(d) As for (b) except for evaporation.

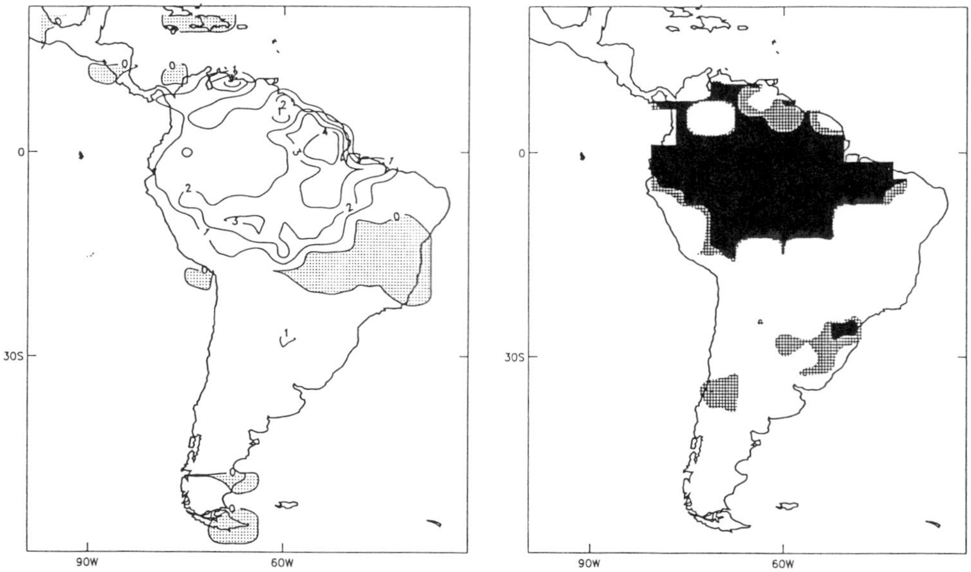

Figure 6 (contd.) (e) As for (a) except for surface temperature. Contours are at 0.0, ±0.5, 1.0 then every 1°C with areas of decrease shaded. (f) As for (b) except for surface temperature.

and surface temperature together with the statistical significance. Decreases in rainfall cover most of the deforested area. Maximum decreases of over 1 mm day^{-1} or up to about 20% of the FOREST rainfall are located near the mouth of the Amazon and show a high level of significance. This is in contrast to results shown by LR where the maximum decreases coincided with the areas of maximum rainfall. There are significant increases over the coastal regions of northwestern South America including some deforested gridboxes. Decreases in evaporation cover most of northern South America with statistically significant decreases over most of the deforested area. Reductions reach 1 mm day^{-1} in the southern part of the basin which is most strongly influenced by the low rainfall in June, July and August allowing the soil to dry and so restrict evapotranspiration. Statistically significant increases in surface temperature cover almost the whole of the deforested area with maximum increases of over 3°C. These maxima broadly coincide with the maximum reductions in evaporation.

As already stated, the notable difference from most previous deforestation scenarios is the overall increase in annual moisture convergence so this merits closer examination. Figure 7 shows the average monthly mean differences (PASTURE–FOREST) in rainfall, evaporation and the moisture convergence (represented as rainfall minus evaporation) for Region 1. This shows that moisture convergence increases through most of the year with the largest increases taking place from July through to November. This indicates that moisture is flowing into the basin in response to the increased drying during this period. The associated changes in the horizontal velocity potential and divergent wind field at about 200 m above the

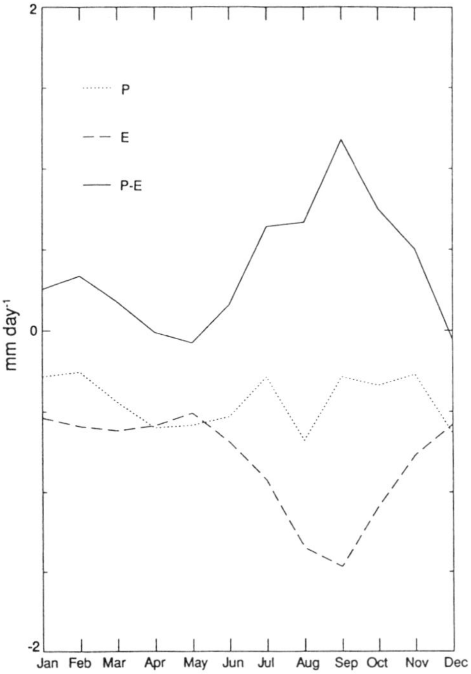

Figure 7 Modelled monthly mean difference (PASTURE–FOREST) in rainfall (P), evaporation (E) and moisture convergence (P–E).

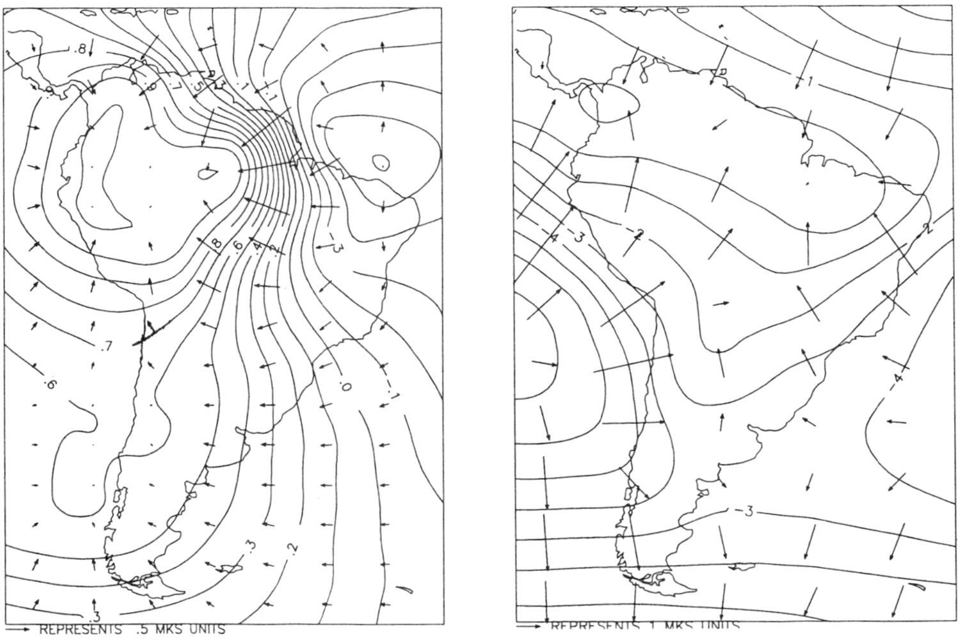

Figure 8 (a) Modelled annual mean difference (PASTURE–FOREST) in velocity potential and divergent wind field at about 200 m (model level 2) over South America. Contours are every $0.1 \times 10^6\ m^2\ s^{-1}$. (b) Modelled annual mean FOREST velocity potential and divergent wind field at about 200 m (model level 2) over South America. Contours are every $1.0 \times 10^6\ m^2\ s^{-1}$.

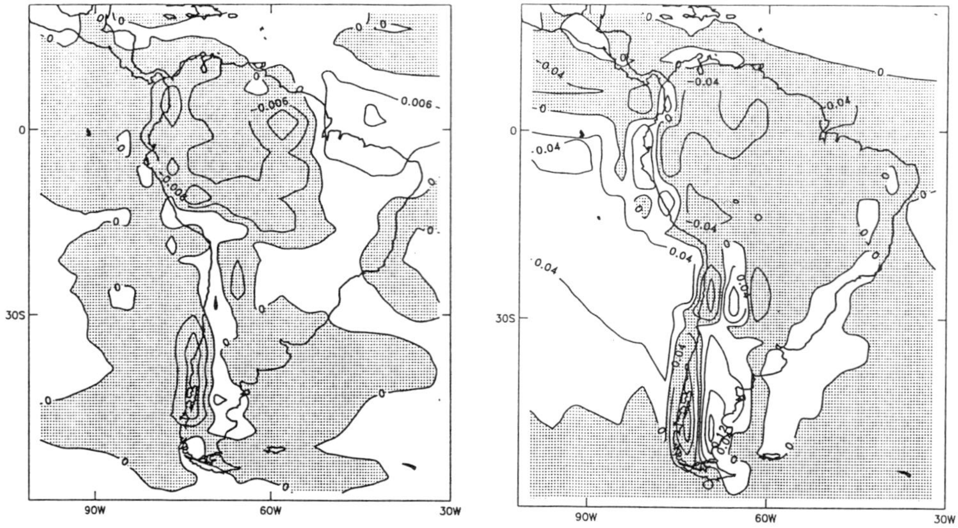

Figure 9 (a) Modelled annual mean difference (PASTURE - FOREST) in vertical velocity at 700 hPa. Contours every 0.006 hPa s^{-1}. Negative areas are shaded and indicate increased ascent. (b) Modelled annual mean FOREST vertical velocity at 700mb. Contours at 0.0, ± 0.04, 0.12 and then every 0.08 hPa s^{-1}. Negative areas are shaded and indicate areas with ascent.

surface are shown in Figure 8. This clearly shows an increase in the convergence over western and central Amazonia, which extends up through the boundary layer (not shown). Figure 9 demonstrates that there is a corresponding increase in vertical velocity at 700 h Pa, well above the boundary layer. The impact on moisture convergence, velocity potential and vertical velocity is in agreement with the results shown by Polcher and Laval (1994b).

COMPANION EXPERIMENTS

In this section results are presented from the set of companion experiments. Area means are averages over all deforested points to give a representative picture of the total changes within both northern and southern Amazonia.

INCREASED ALBEDO EXPERIMENT

The increased albedo experiment INCALB has retained all the vegetation parameters as defined for FOREST except the surface albedo which has been increased to that of the PASTURE experiment - the average change over the year is +0.046. Table 5 compares some climate variables for the difference between INCALB and FOREST.

Comparing the difference (INCALB - FOREST) with the difference (PASTURE - FOREST), evaporation has decreased but by a much smaller amount. The change

Table 5 Comparison of some climate variables for the FOREST experiment and the differences (PASTURE–FOREST), (INCALB–FOREST), (REDRGH–FOREST) and (FIXINF–FOREST) averaged over all deforested points.

Variable	FOREST	PASTURE –FOREST	INCALB –FOREST	REDRGH –FOREST	FIXINF –FOREST
Total evaporation (mm day^{-1})	4.12	-0.76 (-18%)	-0.14 (-3%)	-0.11 (-3%)	-0.3 (-7%)
Evapotranspiration (mm day^{-1})	2.61	-0.31 (-12%)	-0.09 (-3%)	+0.02 (+1%)	+0.13 (+5%)
Interception (mm day^{-1})	1.51	-0.45 (-30%)	-0.05 (-3%)	-0.14 (-9%)	-0.43 (-28%)
Rainfall (mm day^{-1})	6.40	-0.27 (-4%)	-0.46 (-7%)	+0.46 (+7%)	-0.3 (-5%)
Moisture convergence (mm day^{-1})	2.28	+0.49 (+22%)	-0.32 (-14%)	+0.57 (+25%)	0.0
Soil moisture (mm)	351.4	-186.4 (-53%)	-21.9 (-6%)	+23.6 (+7%)	-103.5(-29%)
Total runoff (mm day^{-1})	2.24	+0.51 (+23%)	-0.44 (-20%)	+0.45 (+20%)	+0.0 3(+1%)
Subsurface runoff (mm day^{-1})	1.76	-1.17 (-66.5%)	-0.38 (-22%)	+0.48 (+27%)	+0.15 (+9%)
Surface runoff (mm day^{-1})	0.48	+1.68 (+350%)	-0.06 (-13%)	-0.03 (-6%)	-0.13 (-27%)
Net radiation (W m^{-2})	153.0	-12.7 (-8%)	-5.35 (-4%)	-3.91 (-3%)	-10.2 (-7%)
Sensible heat (W m^{-2})	33.77	+9.31 (+28%)	-1.42 (-4%)	-0.70 (+2%)	-1.51(+5%)
Latent heat (W m^{-2})	119.21	-22.0 (-18%)	-3.87 (-3%)	-3.20 (-3%)	-8.68 (-8%)
Horizontal velocity at level 1(m s^{-1})	1.17	+1.23 (+105%)	+0.01(+1%)	+1.14 (+97%)	+1.17 (+100%)
Surface temperature (°K)	296.29	+2.27	-0.09	+1.30	+1.38

in evaporation is driven by the decrease in available energy at the surface. This decrease is much reduced compared with the total deforestation experiment as there is little change in surface temperature, and in the longwave energy balance. The reductions in rainfall and moisture convergence are substantial. There is a small mean decrease in surface temperature.

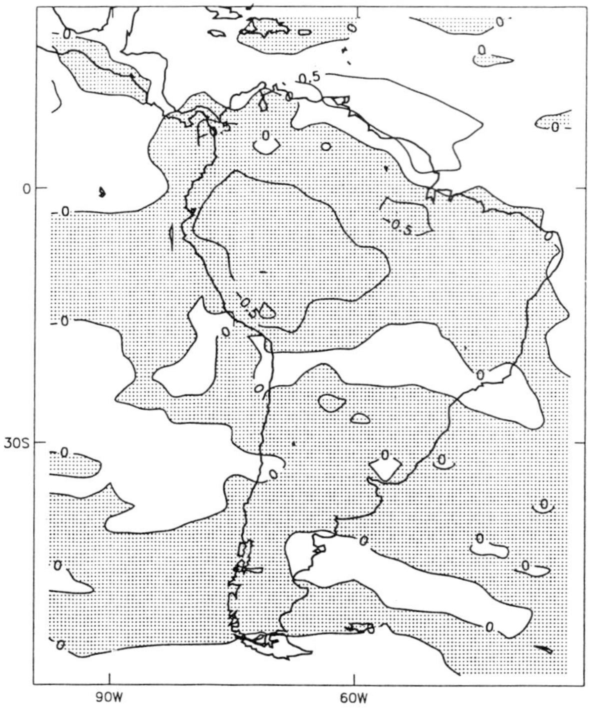

Figure 10 Modelled annual mean difference (INCALB–FOREST) in rainfall. Contours are at 0.0, ±0.5, 1.0 and then every 1.0 mm day^{-1} with areas of decrease shaded.

Figure 10 shows the annual mean change in rainfall and is to be compared with changes in the total deforestation experiment (Figure 6). Decreases are more widespread than in the total deforestation experiment although maximum reductions are less. Areas of maximum reduction tend to coincide with the location of rainfall maxima. Figure 11 shows the reduction in convergence at about 200 m above the surface. This reduction extends well up through the atmospheric column.

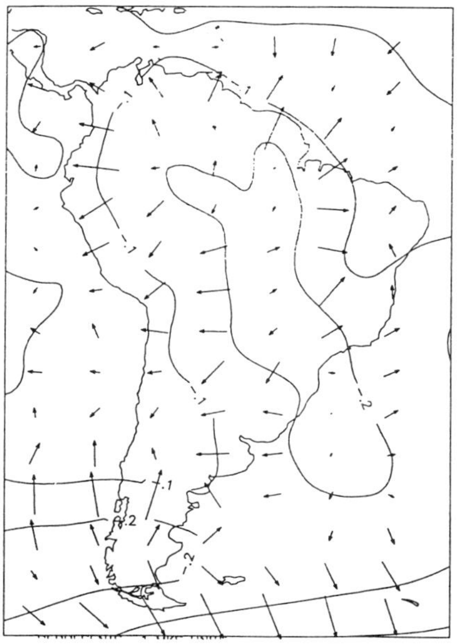

Figure 11 Modelled annual mean difference (INCALB–FOREST) in velocity potential and divergent wind field at about 200 m (model level 2) over South America. Contours every 0.1×10^6 m^2 s^{-1}. See Figure 8(b) for control.

REDUCED ROUGHNESS EXPERIMENT

The reduced roughness experiment REDRGH has retained all the vegetation parameters as defined for FOREST except the roughness length which has been reduced from the FOREST value of 2.1 m to the PASTURE value of 0.026 m. Table 5 compares some climate variables for the difference between REDRGH and FOREST.

Comparing the difference (REDRGH - FOREST) with the difference (PASTURE - FOREST), evaporation is reduced much less. Rainfall has increased, in contrast to the decrease in the total deforestation experiment and there is an even larger increase in moisture convergence. Surface temperature has risen but by a smaller amount; these increases are a result of the higher surface temperatures required to maintain the surface fluxes due to the reduced efficiency of the turbulent exchange of energy at the surface.

Figure 12(a) shows the annual change in rainfall and is to be compared with changes in the total deforestation experiment (Figure 6). There is a broad band of increase through central southern and western Amazonia with increases that are as large as the decreases in the total deforestation experiment. This NW-SE band of increase extends well outside the deforested area into southeastern Brazil. Both these increases and decreases to the north east of eastern Brazil are statistically significant (Figure 12(b)). The increased convergence at about 200m is shown in Figure 13. The increased moisture convergence resulting in increased rainfall can partially be explained by the stronger boundary layer winds with weaker frictional stress. Figure 14 shows that there is an increase of about 40% in wind speeds.

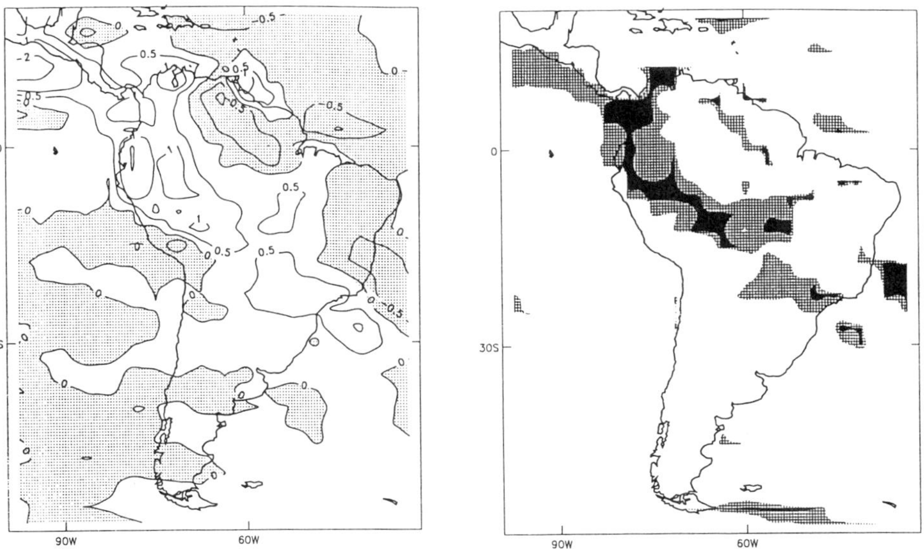

Figure 12 (a) Modelled annual mean difference (REDRGH–FOREST) in rainfall. Contours are at 0.0, ±0.5, 1.0 and then every 1.0 mm day^{-1} with areas of decrease shaded. (b) Student's t fields showing changes in rainfall that are locally significant at the 95% (cross hatched shaded areas) and 99% (solid shaded areas) confidence levels.

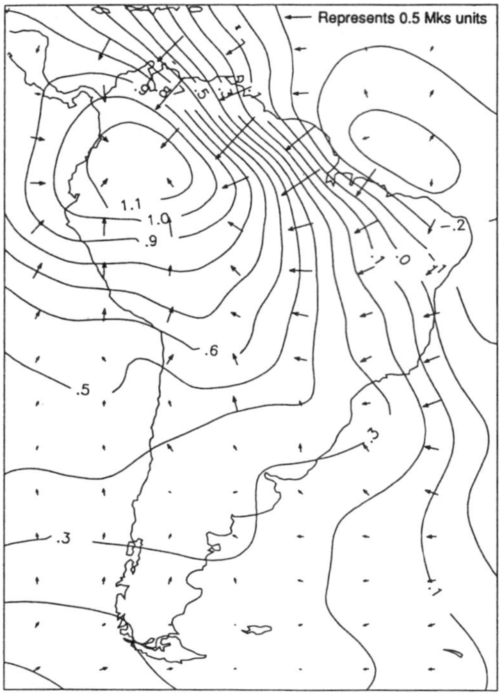

Figure 13 Modelled annual mean difference (REDRGH–FOREST) in velocity potential and divergent wind field at about 200 m (model level 2) over South America. Contours every $0.1 \times 10^6\ m^2\ s^{-1}$. See Figure 8(b) for control.

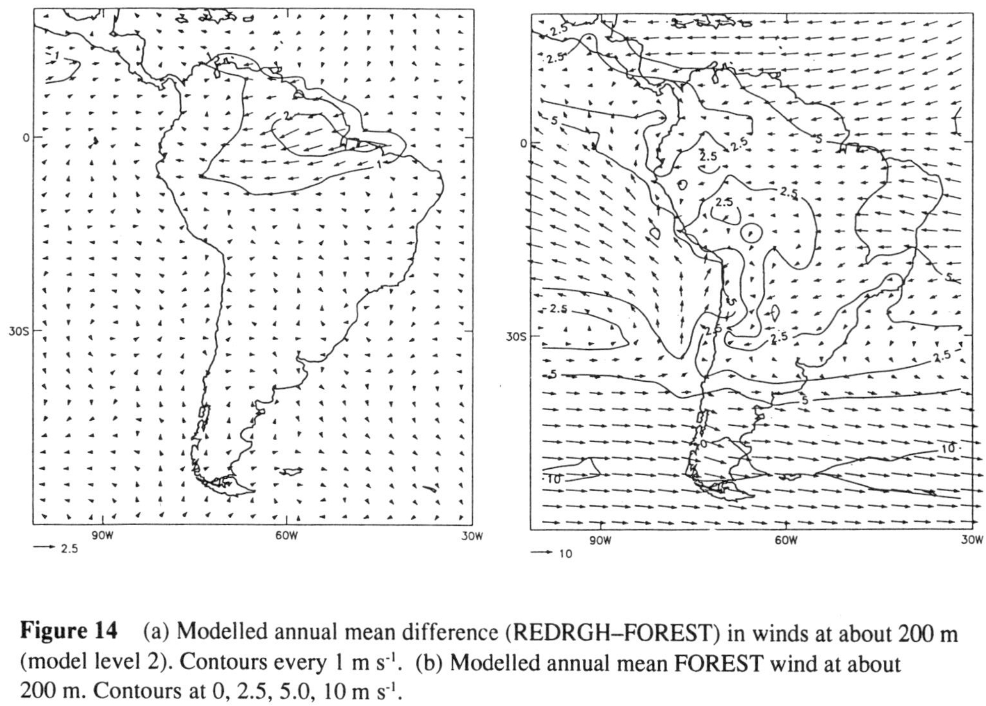

Figure 14 (a) Modelled annual mean difference (REDRGH–FOREST) in winds at about 200 m (model level 2). Contours every 1 m s^{-1}. (b) Modelled annual mean FOREST wind at about 200 m. Contours at 0, 2.5, 5.0, 10 m s^{-1}.

FIXED INFILTRATION RATE EXPERIMENT

The fixed maximum infiltration rate experiment FIXINF has all surface parameters modified to the PASTURE values except the ratio of saturated hydraulic conductivity at the surface relative to the subsurface soil which has retained the FOREST value of 1.5. Table 5 shows differences between FIXINF and FOREST.

Comparing the difference (FIXINF - FOREST) with the difference (PASTURE - FOREST), allowing more water to infiltrate into the soil results in a decrease in surface runoff compared with the very large increase in surface runoff in the total deforestation experiment. As more water infiltrates into the soil the reduction in soil moisture total is much less than in the total deforestation experiment (approximately half). Drainage is increased relative to FOREST as a result of the reduced rooting depth. The increases in total runoff and hence the loss of water from the system are less than in (PASTURE - FOREST). The most significant impact of this smaller reduction in moisture within the system is shown by the increase in evapotranspiration. Interception is reduced by similar amounts to those in the total deforestation experiment. Overall the reduction in total evaporation is less than half that in the total deforestation experiment. Surface temperature has increased but by a smaller amount, consistent with the smaller reduction in evaporation. Changes in rainfall are more closely related to changes in evaporation with no compensating increase in moisture convergence as in the total deforestation experiment. The increase in moisture convergence in the total deforestation experiment was largest during the dry seasons in response to the large reductions in soil moisture at these times. There is still a small increase in moisture convergence during these seasons in (FIXINF - FOREST) associated with the relative drying of the soil (not shown).

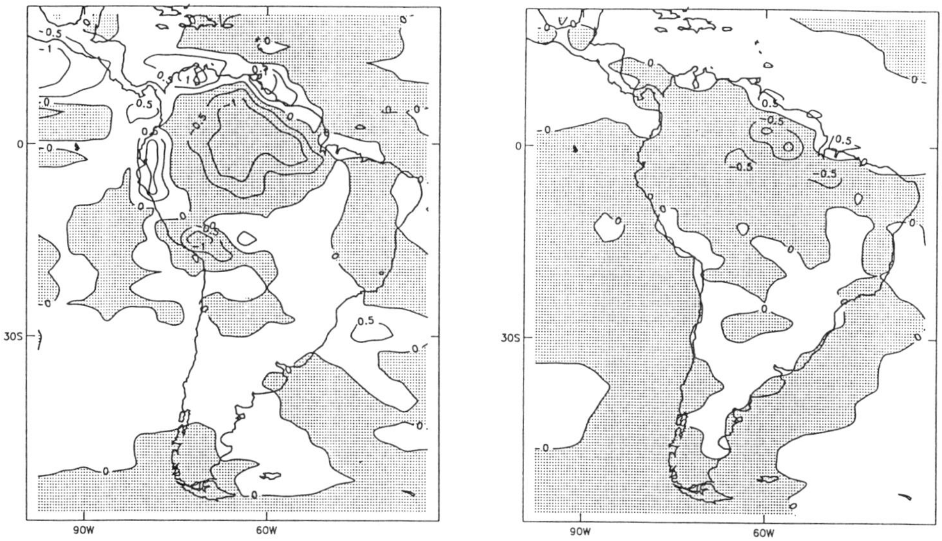

Figure 15 (a) Modelled annual mean difference (FIXINF–FOREST) in rainfall. Contours are at 0.0, ±0.5, 1.0 and then every 1.0 mm day^{-1} with areas of decrease shaded. (b) As for (a) except for evaporation.

Figure 15 shows the annual mean changes in rainfall and evaporation and is to be compared with changes in the total deforestation experiment (Figure 6). The area of maximum decrease in rainfall of more than 1 mm day^{-1} is shifted towards the west and covers a much larger area. Reductions in evaporation cover most of northern South America but the large decreases in the southern part of the deforested area present in the total deforestation experiment are absent due to the retention of higher soil moisture levels during the winter months.

HALF DEFORESTATION EXPERIMENT

In this experiment, the full set of deforestation changes including the reduction in maximum infiltration rate was applied to a reduced deforested area.

Figure 16 shows that the annual mean rainfall changes are similar in some respects to those from the total deforestation experiment. There are increases over western and southern regions of the original forested region, and decreases over most of the rest of northern South America. However, the largest decreases are not over the deforested region but over part of the remnant forest of central Amazonia. Further analysis will explain these patterns, one obvious feature is that the model's response is on a larger scale than in the relatively small deforested regions. It is interesting that the region of increases over Colombia and Ecuador coincides with that of statistically significant increases in observed rainfall deduced from trends in outgoing longwave radiance for 1974-90 by Chu *et al.* (1994). The evaporation changes (not shown) were much as expected on the basis of the total deforestation experiment, with widespread reductions in parts of the deforested region where there is a pronounced dry season.

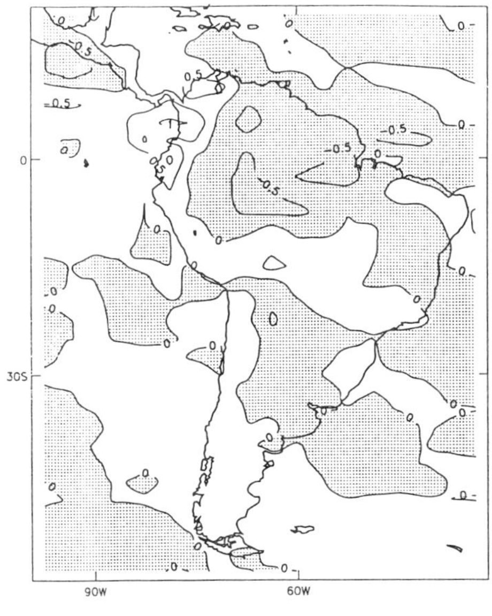

Figure 16 Modelled annual mean difference (HALF–FOREST) in rainfall. Contours are at 0.0, ±0.5, 1.0 and then every 1.0 mm day^{-1} with areas of decrease shaded.

DISCUSSION AND CONCLUSIONS

The aim of these experiments was to find the climatic response to a more realistic portrayal of Amazonian deforestation by using the most up-to-date measurements of vegetation characteristics for forest and replacement pasture locations. These relatively long ten-year simulations have allowed more confidence to be placed in the results because of their higher statistical significance. The series of single component change experiments have provided a clearer understanding of the mechanisms involved in the response to total deforestation.

The effect of increased albedo broadly agreed with the Charney (1975) hypothesis that increased albedo reduces the energy available to the column and so forces descent in order to maintain a temperature similar to the adjacent regions. In the Amazon basin this will tend to suppress ascent and rainfall and, by implication, the moisture convergence into the region. Results from the experiment in which albedo alone was increased from the forest to the pasture value shows that rainfall was reduced in a rather uniform manner across northern South America with decreases of 0.5 to 1.0 mm day^{-1} or about 10% in the annual mean. Results from the experiment in which roughness was reduced from the forest to the pasture value show a marked increase in moisture convergence leading to increased rainfall; there is broad band of increase in rainfall through central southern and western Amazonia with maximum increases of over 1 mm day^{-1}. One explanation of this response is that the boundary layer winds are much stronger with weaker frictional stresses. These results can be compared with a similar but shorter eight-month experiment carried out by Lean and Warrilow (1989). They showed a reduction in moisture convergence and reduced rainfall. It is difficult to attribute the difference in response to any one cause although a number of differences existed between the models and the experiments. Two of the most likely reasons to explain the differences are the prescribed change in roughness (in this experiment a change of 2.1 m was imposed compared with the 0.8 m reduction in Lean and Warrilow) and the deficiencies in the earlier climate simulation, notably the overestimation of interception and the re-evaporation of rainfall by the canopy. Sud *et al.* (1988) obtained decreased convergence and rainfall over northern South America in July with reduced roughness. This disparity in results may be explained by the lower Andes barrier in their model (typically about half the height used in our model); a mountain barrier such as the Andes tends to force convergence to be contained within the deforested area, whereas in the absence of this barrier all extra downwind convergence would occur outside of this region. The response to changes in roughness is therefore rather complex and dependent on a number of factors including the model and its climate simulation, the magnitude of the change in roughness and the height of the mountain barrier to the west.

The experiment in which all vegetation parameters were modified except the maximum infiltration rate was originally designed because of the notably large impact that the reduction in infiltration and reduced evaporation were having together with the uncertainty in defining the magnitude of the change in this parameter. Results from this experiment were the most unexpected. The reduction in average annual evaporation of 0.35mm day^{-1} (8%) was substantially less than the 0.81mm day^{-1} (19%) decrease in the total deforestation experiment. This could be attributed mainly to the retention of more soil moisture during the dry season. Changes in rainfall were more closely related to changes in evaporation with no compensating increase in moisture convergence as in the total deforestation experiment.

If it is assumed that the changes in roughness and albedo dominate the response in the total deforestation experiment to the exclusion of changes in all other vegetation parameters, then by adding the changes together from these two individual experiments one might expect to get some way towards the experiment in which all parameters have been modified except the reduced maximum infiltration rate. However there are substantial discrepancies in rainfall and evaporation shown by the difference patterns in Figure 17 - this is formed by the difference in fixed maximum infiltration rate experiment from the sum of the changes due to albedo and roughness. Although there are only small differences in evaporation which can probably be attributed to the stomatal resistance and rooting depth changes, discrepancies in rainfall are substantial. Over eastern Amazonia, there are no differences above 0.5 mm day^{-1} but within western Amazonia there are discrepancies of over 1 mm day^{-1}. This implies that there are a number of non-linear interactions taking place.

Results have been shown to be highly sensitive to the specification of a number of key parameters - the most surprising result was the response to fixing maximum infiltration rate at the surface which reduced the area-averaged evaporation decrease from 0.81mmday^{-1} (19%) to 0.35 mmday^{-1} (8%). If GCMs are to produce credible

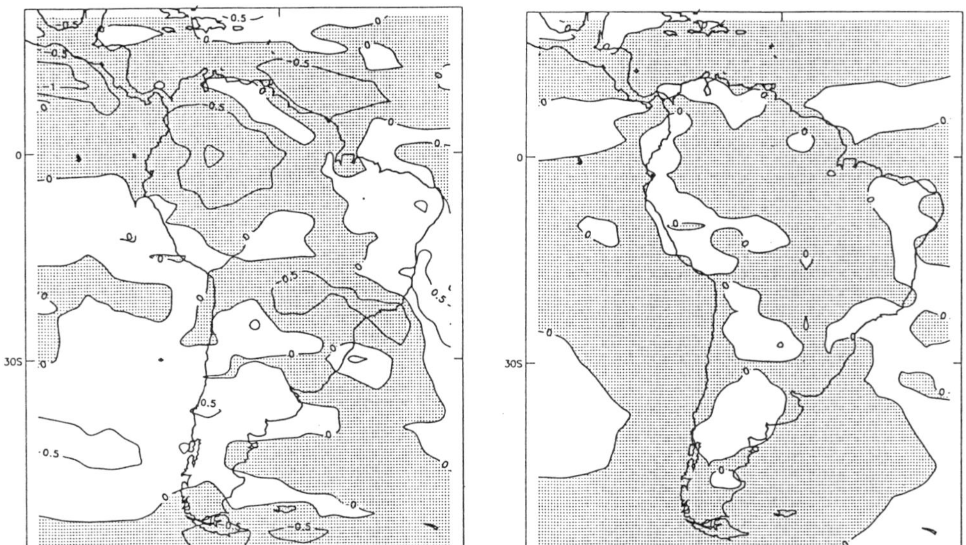

Figure 17 (a) Modelled annual mean difference (FIXINF–FOREST) – { (INCALB–FOREST) + (REDRGH–FOREST) } in rainfall. Contours are at 0.0, ±0.5, 1.0 and then every 1.0 mm day^{-1} with areas of decrease shaded. (b) As (a) except for evaporation.

predictions of regional changes due to deforestation it is necessary to use realistic and accurate specification of these parameters. Use of ABRACOS measurements of vegetation and soil characteristics in future Amazonian GCM deforestation experiments will signify an important evolution in adding more realism to such simulations.

REFERENCES

Bastable, H.G., Shuttleworth, W.J., Dallarosa, R.L.G, Fisch, G., and Nobre, C.A., 1993. Observations of climate, albedo, and surface radiation over cleared and undisturbed Amazonian forest. *Int. J. Climatol.* **13**, 783-796.

Charney, J.G. 1975. Dynamics of deserts and drought in the Sahel. *Q.J.R. Meteorol. Soc.* **101**, 193-202.

Chu, Pao-Shin, Yu, Zhi-Ping and Hastenrath, S. 1994. Detecting climate change concurrent with deforestation in the Amazon basin: which way has it gone? *Bull. Amer Meteorol. Soc.* **75**, 579-583.

Culf, A.D., Fisch, G. and Hodnett, M.G., 1995. The albedo of Amazonian forest and ranchland. *J. Climate* **8**, 1544-1554.

Dickinson, R.E. and Kennedy, P. 1992. Impacts on regional climate of Amazon deforestation. *Geophys. Res. Letters* **19**, 1947-1950.

Dolman, A.J., Gash, J.H.C., Roberts, J. and Shuttleworth, W.J. 1991. Stomatal and surface conductance of tropical rainforest. *Agric. Forest Meteorol.* **54**, 303-318.

Figueroa S.N. and Nobre, C., 1990. Rainfall distribution over central and western tropical South America. *Climánalise* **5**, 36-45.

Gregory, D. and Allen, S. 1991. The effect of convective scale downdraughts upon NWP and climate simulations. *Ninth Conf. on Numerical Weather Prediction, Denver, Col., Amer. Met. Soc.*, 122-123.

Henderson-Sellers, A.,Dickinson, R.E., Durbridge, T.B., Kennedy, P.J., McGuffie, K. and Pitman, A.J. 1993. Tropical deforestation:Modelling local to regional-scale climate change *J Geophys. Res.* **98**, 7289-7315.

Hodnett, M.G., Oyama, M.D. and Tomasella, J. 1995. Comparison of long term soil water storage behaviour under pasture and forest in three areas of Amazonia. this volume.

Jones, R.J., Murphy, J.M. and Noguer, M. 1995. Simulation of climate change over Europe using a nested regional climate model. Part I. Assessment of control climate including sensitivity to location of lateral boundaries. *Q.J.R. Meteorol. Soc.*, 121, 1413-1449.

Lean, J. and Rowntree, P.R. 1993. A GCM simulation of the impact of Amazonian deforestation on climate using an improved canopy representation. *Q.J.R. Meteorol. Soc.* **119**, 509-530.

Lean, J., and Warrilow, D.A. 1989. Simulation of the regional impact of Amazon deforestation. *Nature* **342**, 411-413.

Manzi, A.O. 1993. Introduction d'un schema des transferts sol-vegetation-atmosphere dans un modele de circulation generale et application a la simulation de la deforestation Amazonienne. *These de doctorat de l'universite Paul Sabatier. Meteo-France/CNRM/GMGEC/UDC, 42 Av. G. Coriolis - 31057 Toulouse Cedex.*

Myers, N. 1991. Tropical forests: present status and future outlook. *Climatic Change* **19**, 3-32.

Nobre, C.A., Sellers, P.J. and Shukla, J. 1991. Amazonian deforestation and regional climate change. *J. of Climate* **4**, 957-988.

Pitman, A.J., Durbidge, T.B. and Henderson-Sellers, A. 1993. Assessing climate model sensitivity to prescribed deforested landscapes. *Int. J. Climatol.* **13**, 879-898.

Polcher, J. and Laval, K. 1994a. A statistical study of regional impact of deforestation on climate in the LMD-GCM. *Climate Dynamics* **10**, 205-219.

Polcher, J. and Laval, K. 1994b. The impact of African and Amazonian deforestation on tropical climate. *J. Hydrol.* **155**, 389-405.

Sellers, P.J., Mintz, Y., Sud, Y.C. and Dalcher, A. 1986. A simple biosphere model (SIB) for use within general circulation models. *J. Atmos. Sci.* **43**, 505-531.

Shuttleworth, W.J. 1988. Evaporation from Amazonian rainforest. *Proc. R. Soc. Lond. B* **233**, 321-346.

Skole, D. and Tucker, C. 1993. Tropical deforestation and habitat fragmentation in the Amazon : satellite data from 1978 to 1988. *Science* **260**, 1905-1910.

Smith, R.N.B. 1993. Subsurface, surface and boundary layer processes. *Unified Model Documentation Paper Number 24*. UK Meteorological Office, Bracknell, England.

Sud, Y.C., Shukla, J. and Mintz, Y. 1988. Influence of land surface roughness on atmospheric circulation and rainfall: a sensitivity study with a general circulation model. *J. Appl. Meteor.* **27**, 1036-1054.

Warrilow, D.A., Sangster, A.B. and Slingo, A. 1986. Modelling of land surface processes and their influence on European climate. *Met O 20 DCTN 38,* UK Meteorological Office Bracknell, England.

Wright, I.R., Manzi, A.O. and da Rocha, H.R. 1995. Surface conductance of Amazonian pasture: Model application and calibration for canopy climate. *Agric. For. Met.* 75, 51-70.

Wright, I.R., Nobre, C.A., Tomasella, J., Da Rocha, H.R., Roberts, J.M., Vertamatti, E., Culf, A.D., Alvala, R.C., Hodnett, M.G. and Ubarana, V. 1996. Towards a GCM surface parameterization for Amazonia. This volume.

Wright, I.R., Gash, J.H.C., Da Rocha, H.R., Shuttleworth, W.J., Nobre, C.A., Maitelli, G.T., Zamparoni, C.A.G.P. and Carvalho, P.R.A. 1992. Dry season micrometeorology of central Amazonian ranchland. *Q.J.R. Met. Soc.* **118**, 1083-1099.

RESUMO

Experimentos de desmatamento da Amazônia foram conduzidos utilizando-se o MCG do Hadley Centre. As caracterísitcas da floresta e pastagem foram obtidas a partir dos dados do Projeto ABRACOS. A completa retirada da floresta Amazônica produziu as seguintes modificações médias (para toda Amazônia): redução na evaporação em 0,81 mm d^{-1} (19%) e na precipitação em 0,43 mm d^{-1} (7%), e aumento na temperatura da superfície em 2,3°C. Estes resultados assemelham-se aos obtidos em experimentos anteriores. Entretanto, as diferenças relativas em magnitude indicam que o aumento na convergência de umidade compensa, parcialmente, a redução de evaporação, em contraste com a maioria dos experimentos anteriores. Os resultados também mostram grandes diferenças regionais na variação da precipitação média anual sobre a América do Sul, com redução sobre a maior parte da área desmatada e aumento próximo aos Andes. Um experimento de desmatamento, confinado a uma área reduzida mais realista, mostrou que as maiores reduções de precipitação se localizaram sobre a floresta remanescente na Amazônia Central, ao invés de sobre a área desmatada. Em outra série de experimentos, variando-se separadamente a rugosidade, albedo e infiltração máxima do solo, foi examinada a resposta do clima em face a essas modificações. Estes experimentos indicaram que o clima final, gerado pelo modelo, é muito dependente dessas variações discretas nos parâmetros e é fortemente influenciado pela interação não linear destas variações.

32 Conclusions from ABRACOS

C.A. NOBRE[1], J.H.C. GASH[2], J.M. ROBERTS[2] and R.L. VICTORIA[3]
[1] *Instituto Nacional de Pesquisas Espaciais, São José dos Campos, Brazil*
[2] *Institute of Hydrology, Wallingford, Oxfordshire, UK*
[3] *Centro de Energia Nuclear na Agricultura, Universidade de São Paulo, Brazil*

INTRODUCTION

The experience from projects whose primary motivation is data collection is that often the full benefits of the work are not gained until some years after completion of the data collection phase. The results described in this book should therefore be regarded as only the start of the analysis and application of the ABRACOS data. However, many new insights into the soil-vegetation-atmosphere interactions for Amazonian forest and pasture have already been derived. This paper presents a summary of the main results so far and the conclusions to which they lead.

CLIMATOLOGY

On average the forest sites were found to absorb about 11 per cent more radiation than the pasture (Culf *et al.*, 1996). This is because the forest reflects less solar radiation and emits less longwave radiation. The mean forest albedo was found to be 0.134, slightly higher than the value generally used in deforestation simulations, while the mean pasture albedo, 0.180, is slightly lower than the value usually used. Surprisingly the pasture's grasses failed to show any strong seasonality in albedo, whereas the forest albedos have a well defined seasonal variation which is not due to the effects of the changing solar elevation angle or to variations in cloudiness, but is correlated with the soil moisture (Culf *et al.*, 1995, 1996). Although the albedo of the pasture sites has no clearly defined seasonal trend, there is some month to month variation which has been shown to be linked to the leaf area index (Wright *et al.*, 1996b). At Ji-Paraná systematic differences in the incoming solar radiation between the forest and pasture sites were also observed during the dry season (Culf *et al.*, 1996). These differences may be due to increased cloud cover over the pasture at that time of year. If this is the case, it is an important result as it indicates a direct effect of the change in surface cover on a mesoscale atmospheric phenomenon.

Pastures showed higher daytime maximum temperature and higher diurnal amplitude of temperature (Culf *et al.*, 1996). Generally the minimum temperature was lower at

Amazonian deforestation and climate. Edited by J.H.C. Gash, C.A. Nobre, J.M. Roberts and R.L. Victoria.

night for the pasture. This result is likely to be associated with the low windspeeds near the surface for pasture at night, which can reduce vertical mixing and increase stability. The daytime temperature in urban Manaus was higher than at the two rural sites, but the night-time city temperature was similar to that above the forest (Maitelli and Wright, 1996). There is little seasonal variation of temperature at Manaus or Marabá, but there is a pronounced cooling during the dry season at Ji-Paraná. Both Marabá and Ji-Paraná have a marked annual cycle of humidity which is not observed in Manaus.

SOIL MOISTURE AND SOIL PHYSICAL PROPERTIES

Representing the spatial variability of soil hydraulic properties in large scale models remains a major challenge to the soil physics and GCM modelling communities. Soil properties are highly variable, yet the parameterizations of large scale models require values which represent large areas. The ABRACOS observations of soil properties were necessarily limited, but by adopting a pragmatic simplification to the soil map of Amazônia, and with at least some measurements of the properties of three of the major soil types, Wright *et al.* (1996b) have derived a far more realistic representation of soils than was previously available for use in large-scale models of Amazônia.

Tomasella and Hodnett (1996) found the hydraulic conductivity of the forest soil at the surface to be so high that it was impossible to measure. This is a result of the rooting system and macropores. However, the measurements in the Manaus pasture showed a reduction in conductivity at the soil surface, which was probably a result of trampling by cattle. Pasture saturated hydraulic conductivity varied strongly with depth, from more than 100 mm h^{-1} at the surface to 17 mm h^{-1} at 1.3 m. Zones of high conductivity had a substantial number of meso- and macropores. Most of the impact of deforestation occurs in the top 20 cm of soil, below which the properties of soil under forest and under pasture were quite similar. However this conclusion is dependent on the management practice of using little or no heavy machinery and maintaining very low cattle stocking densities. Under these conditions compaction or low values of top soil hydraulic conductivity were not observed for any sites. Water release and unsaturated hydraulic conductivity values were generally well fitted by the van Genuchten model, but were very different to published values for clay soils measured in temperate regions. Soil hydraulic properties of these soils should not be estimated from texture using relationships derived from temperate soil; these relationships would overestimate water availability and underestimate unsaturated conductivity if applied to the Manaus pasture soil.

Thermal conductivity was found to be less variable than the hydraulic properties, ranging from 1.2 to 2.9 W m^{-1} $°C^{-1}$ (Alvalá *et al.*, 1996). For pasture, de Souza *et al.* (1996) found that the heat flux was downward during the dry season and upward during the rainy season, whereas for the forest site it was found that there was a small but consistently upward flux. On average, for the Marabá site, the difference in top soil temperature between pasture and forest was 2.4°C. These differences in flux and temperature can be explained in terms of the different variations in moisture content and different soil-surface energy balances of the two vegetation types.

Hodnett *et al.* (1996a) show that during the dry season there is more extraction of water from soil below the forest resulting in a drier profile, within the range of measurement levels, at the end of the dry season. There are clear indications, substantiated by other studies (Chauvel *et al.*, 1992; Nepstad *et al.*, 1994), that the forest is abstracting water from levels below the 3.6 m depth of the deepest tubes. These indications are supported by the measurements of maximum recorded storage changes for the three sites. Consistent with a longer dry season, the maximum storage changes to 3.6 m depth for Marabá were 742 mm and 376 mm for the forest and pasture respectively. Both storage changes and the difference between forest and pasture were much smaller for Manaus where the dry season is less severe, however the low water holding capacity of the Manaus soil still resulted in water being extracted from depth. Interestingly, the soil moisture profile did not completely recharge in the wet season following the maximum dry season depletion below the Marabá forest, but was 349 mm below its value at the same time the previous year. The storage term in the hydrological balance was not zero over the annual cycle and if these results were extrapolated to the regional scale the difference between rainfall and river runoff would not have given large scale annual evaporation. Although this may be an uncommon occurrence, it has obvious implications for the use of hydrological measurements as a check on estimates of evaporation in large scale models. The overall picture for soil moisture shows that there are large differences in the seasonal changes in soil moisture content, both between forest and pasture, and between sites. These differences are a result of the different rainfall climates, combined with the different soil properties, water table behaviour and rooting depths of forest and grass.

Models were developed for forest and pasture to predict soil water storage in the top 2 m of soil using only daily rainfall as input data (Hodnett *et al.*, 1996b). The models were developed and calibrated using rainfall, soil water content and micrometeorological data from the ABRACOS study period. Evaporation and deep drainage were related to the soil water storage through linear and exponential functions respectively, allowing soil water storage to be determined from a simple water balance. Soil water storage was then predicted from the 27 year rainfall record from the Reserva Ducke site, near Manaus, assuming that the forest evaporation was not affected by soil water stress. For the forest, the model predicted uptake from below 2 m depth in 20 out of the 27 years. Mean predicted uptake from below 2 m was 60 mm, with a maximum of 231 mm in 1967. Because of the low water holding capacity of the soils near Manaus this uptake must be provided by a very large depth of soil, and deep rooting is essential.

There is still much to be learnt about the properties and behaviour of Amazonian forest soils. Measurements of soil hydraulic properties are needed for the remainder of the major Amazonian soil types. Both soil hydraulic properties and soil moisture content need to be measured down to depths in excess of 4 m to give a complete picture of the water cycle beneath Amazonian forest.

FOREST RAINFALL INTERCEPTION

Rainfall interception data were collected by Ubarana (1996) at the Reserva Jaru and Reserva Vale do Rio Doce sites. This work followed the same sampling methodology as used previously by Lloyd and Marques (1988) in Reserva Ducke and therefore provide directly comparable measurements. A higher percentage of the rainfall was evaporated directly back into the atmosphere at the two new sites: 12 and 13 per cent for Reservas Jaru and Vale do Rio Doce respectively, compared with 9 per cent found at Reserva Ducke by Lloyd *et al.* (1988). The tropical rainfall climate makes the interception loss a relatively small difference between rainfall and throughfall, and the latter, because of the structure of the vegetation, is highly variable. These factors make measurements of interception loss in a tropical forest subject to large errors. The differences between the measurements from the different sites are of comparable size to the measurement errors and it is therefore not possible to prove that the interception losses at the two new sites were greater than at the original site at Ducke. However the differences were consistent with the differences found in the values of the forest structure parameters controlling the evaporation: higher canopy capacities and more closed canopies were found at Jaru and Vale do Rio Doce than at Ducke. Within the limits of the measurements, the evaporation from all the sites can be successfully estimated with the Rutter model.

STOMATAL BEHAVIOUR

A key parameter in global circulation models is the surface conductance to water vapour transport from vegetated surfaces. Surface conductance integrates the influence of individual stomatal conductances over the canopy. While bulk surface conductance can be measured directly using micrometeorological methods a more detailed understanding of its fluctuations requires information about stomatal conductance at the leaf level and its responses to factors which change diurnally and seasonally, as well as the magnitude and fluctuations in leaf area index.

At the pasture sites at Ji-Paraná and Manaus, McWilliam *et al.* (1996) observed strong relationships between stomatal conductance and soil moisture deficit, but at Marabá stomatal conductance was less affected by increasing soil moisture deficit (Sá *et al.*, 1996). Total transpiration is controlled by the combination of stomatal conductance and leaf area. However, leaf area index at the pasture sites also responds to soil moisture and varies by up to 50 per cent over a year. The consequent seasonal variation in the transpiration is qualitatively consistent with that observed in the soil moisture data and by the micrometeorological measurements of surface fluxes.

Despite the differences in botanical composition between and within the three forest sites the physiological responses of the different species were similar (see Roberts*et al.*, 1996). There was generally no response to soil moisture, and similar responses to atmospheric humidity deficit. Stomatal conductance of all species depended on height, declining downwards through the canopy. This lack of variation between species is a useful result because it means that the considerable literature on the stomatal behaviour of tropical forest species can be exploited to derive generally applicable results.

The above ground biomass and leaf area index of one site was measured directly by destructive sampling (McWilliam *et al.*, 1993), and leaf area indices of the other forest sites were estimated from measurements of leaf litter fall (Roberts *et al.*, 1996). The leaf area index is estimated to be 5.7, 5.4 and 4.4, at Manaus, Marabá and Ji-Paraná respectively, with a seasonal variation of probably 15 per cent. These figures depend on the assumption that the lifetime of leaves is one year; this assumption remains to be proved, but appears to be valid.

The climate and radiation levels change through the depth of the forest canopy, but empirical functions of the variables driving transpiration have been derived by Cabral *et al.* (1996) so that the total forest transpiration can be calculated using above canopy weather station measurements with a multi-layer canopy model and a knowledge of how the leaf area and stomatal conductance vary with height and season. Application of the model revealed that the lower part of the canopy contributes relatively little to the total transpiration. This is despite the high proportion of the canopy at lower levels, and is because in this part of the canopy lower stomatal conductances and radiation levels are coupled with lower vapour pressure deficits and windspeeds. Studies of the isotopic composition of water vapour in the canopy space by Ribeiro *et al.* (1996) show that the lower part of the canopy is effectively decoupled from the middle part of the canopy which contributes most to the total transpiration. Measurements of the variation in the carbon isotope composition suggest that the physiological behaviour of vegetation in the lower canopy (Kruijt *et al.*, 1996) may also be different to that in the upper canopy. Thus for the complex canopy of the rainforest the concept of a simple, single value of leaf area index, which is used in most predictive models, is likely to be a poor representation of the actual situation.

MICROMETEOROLOGY

Over the four years of ABRACOS field measurements there were seven intensive monitoring campaigns. The objective of these field campaigns was to evaluate the physical parameters that describe the micrometeorology of the sites, and to provide estimates of all components of the energy balance, including evaporation, which may be used to calibrate land surface models. It was found that the aerodynamic roughness characteristics of the sites conformed to conventional relationships based on vegetation structure (Wright *et al.*, 1996b). During the wet seasons, the partitioning of energy to evaporation was similar in pasture and forest, but the total evaporation from pasture during the wet season was typically 10% - 15% lower than that of the forest due to the reduced available energy at pasture sites and smoother aerodynamic roughness (Wright *et al.*, 1992).

During the dry seasons the shallow rooted pastures were all affected by depletion of the soil water reserves, albeit at different intensities depending upon the soil type and rainfall. On the clay soil near Manaus, transpiration from the pasture declined rapidly after only 10 days without rain (Wright *et al.*, 1992). In contrast, no significant attenuation in transpiration was observed at any of the forest sites during dry periods, including the prolonged dry seasons at Marabá and Ji-Paraná (Wright *et al.*, 1996a). Current GCM parameterizations of Amazonian forest tend to underestimate forest rooting depth and erroneously reduce evaporation when soil water in the upper profile is depleted.

Evaporation flux data from the first five campaigns have been used to calibrate models which derive bulk stomatal conductance of pasture and forest canopies from the climate near the surface (Wright *et al.*, 1995 and 1996a). This is an important calibration for GCMs as it determines the diurnal and seasonal partitioning of energy into heat for convection and cloud formation, and the amount of water vapour returned to the atmosphere from the soil (da Rocha *et al.*, 1996b). The similarity between pasture conductance parameters from Manaus and Ji-Paraná and the relative dissimilarity in the individual forest parameters, has given insight into the accuracy and suitability of the model for predicting the evaporation from Amazonia as a whole. Using the sets of parameters from the Manaus sites to predict the transpiration from Ji-Paraná forest and pasture produced encouraging results (Wright *et al.*, 1996a), but emphasised the importance of only applying optimised parameters as complete sets.

CARBON DIOXIDE

Eddy correlation measurements over the Reserva Jaru forest have shown that CO_2 flux can be measured reliably and continuously over long periods. The measurements of Grace *et al.* (1995a, 1995b, 1996) showed that the forest at Jaru was accumulating carbon. Grace *et al.* applied the model developed by Lloyd *et al.* (1995) and da Rocha *et al.* (1996a) applied the model developed by Sellers *et al.* (1995) to estimate the annual carbon balance from the climate data collected above the forest with the automatic weather station. In both cases this modelling predicted that the forest was a significant sink for carbon. However, the net flux of carbon is a small difference between photosynthesis and respiration: Meir (1996) found that the net flux is only 6 per cent of the soil respiration and because soil respiration shows high sensitivity to temperature the net flux is also highly sensitive to temperature. The forest may change from being a source to a sink for an average change of a degree or less. If scaled up to the whole of Amazonia, these estimates imply Amazonia would be a net sink of about 0.5 G tonnes of carbon per year. This is highly significant in terms of the so-called missing sink of carbon of about 1 G tonne of carbon per year, but given the short period of observations and the model limitations, these figures must be regarded as provisional and interpreted with caution. There is an urgent need to collect more carbon flux data. These data must be collected over sufficiently long periods to sample the year-to-year variation, and measurements are needed from all the major ecosystems in Amazonia.

Measurements over cerrado vegetation by Miranda *et al.* (1996) demonstrate the very different behaviour found in areas with a long dry season. There is a strong seasonal variation, with accumulation of carbon by the system during the wet season which becomes a loss by the end of the dry season. At the site studied, annual accumulation is estimated to be about 2 to 3 tonnes per hectare per year. However, as cerrado vegetation is highly variable and dependent on its previous history of fire, these first results must be regarded as site specific.

Estimates of biomass in secondary forest by Honzak *et al.* (1996) confirm the importance of regrowth as a sink for carbon. The extent and type of young secondary forest can be determined by remote sensing (Lucas *et al.*, 1996), thus allowing the possibility of scaling up plot estimates of carbon accumulation to the whole Amazon basin.

RONDÔNIA BOUNDARY LAYER STUDIES

The atmospheric boundary layer during day time was found to be 700 to 1000 m higher over areas with substantial clearing than over adjacent forest (Nobre *et al.*, 1996). Fisch *et al.* (1996) applied a simple 1-D model of boundary layer growth (forced by sensible heating at the bottom and entrainment at the top) and found that it provided an adequate representation of boundary layer growth over the forest throughout the day, and over pasture from mid-morning onward. However it failed to simulate the rapid boundary layer growth over the pasture in the morning before 11.00 h local time. A possible explanation is that the particular spatial pattern of land cover change in the region (strips of forest and grass) is not adequately represented in the simple model used to explain boundary layer growth. Dias and Regnier (1996) used a complex 3-D atmospheric model (RAMS) to study the impact on the wind, temperature and moisture distributions of the observed land cover in the Ji-Paraná region: a large deforested area surrounded by forests; the main conclusion was that the vertical transports of momentum, heat and moisture by mesoscale circulations are significant and are not accounted for by the local turbulence parameterizations in GCMs.

CLIMATE MODELLING

The ABRACOS data from the 1990 and 1991 campaigns have been used to calibrate the UK Hadley Centre climate model (Lean *et al.*, 1996), the French meteorological service climate model (Manzi and Planton, 1996), and also SiB (da Rocha *et al.*, 1996b), the land surface model which is used both at the Brazilian climate centre (Centro de Previsão de Tempo e Estudos Climáticos, CPTEC), at NASA and at the Center for Ocean-Land-Atmosphere Studies (COLA) in the United States. GCM deforestation experiments have been carried out with all these models.

The most recent model results (Lean *et al.*, 1996) indicate a substantial response to deforestation which varies regionally: annual rainfall changes are predicted to be up to 20 per cent, with a tendency for NW — SE orientation of rainfall changes extending SE of the deforested area (especially in December — February); in this numerical experiment, response is sensitive to (at least) albedo, roughness length and surface maximum infiltration, but the sensitivity analysis with the French GCM (Manzi and Planton, 1996) also revealed considerable sensitivity to the type of convection scheme used. The results of an experiment by Lean*et al.* (1996) in which only part of Amazônia was deforested, a more realistic scenario, could not easily be deduced from the full deforestation experiment. These differences highlight the fact that all GCM predictions of the climate are very sensitive to the representation of the land surface characteristics.

CONCLUSION

At the start of ABRACOS the climate modelling community had advanced their models to the point where progress was limited by lack of data rather than by model development. The ABRACOS data changes this situation. The priority now is to apply the wealth of data which has been collected under ABRACOS to these models. The data have already been used to derive some of the required parameters and the data are available for all to use to develop or validate their models. A summary of the parameters generated by the ABRACOS studies has been compiled into a table by Wright *et al.* (1996b). The use of the information in this table would make an immediate improvement to the realism of climate modelling studies of Amazônia. More than that, the improvement in the understanding our how the processes in the soil, plants and atmosphere change as forest is replaced by pasture means that we can now have more confidence in the interpretation of the model predictions.

REFERENCES

Alvalá, R.C., Gielow, R., Wright, I.R. and Hodnett, M.G. 1996. Thermal diffusivities of Amazonian soils. This volume.

Cabral, O.M.R., McWilliam, A.L.C. and Roberts, J.M. 1996. In-canopy microclimate of Amazonian forest and estimates of transpiration. This volume.

Chauvel, A. Vital, A.R.T., Lucas, Y., Desjardins, T., Franken, W.K., Luizão, F.J., Araguás, L.A., Rozanski, K. and Bedmar, A.P. 1992. O papel das raízes no ciclo hidrológico da florestaAmazônica. Anais do VII Congresso Brasileiro de Meteorologia, São Paulo, Brazil, 298-302.

Culf, A.D., Fisch, G. and Hodnett, M.G. 1995. The albedo of Amazonian forest and ranchland. *J. Climate* **8**, 1544-1554.

Culf, A.D., Esteves, J.L., Marques, A.de O. and da Rocha, H.R. 1996. Radiation, temperature and humidity over forest and pasture in Amazonia. This volume.

Cutrim, E., Martin, D. and Rabin, R. 1995. Enhancements of cumulus clouds over deforested lands in Amazonia. Bull. Amer. Met. Soc. **76**, 1801-1805.

da Rocha, H.R., Nobre, C.A., Bonatti, J.P., Wright, I.R. and Sellers, P.J. 1996a. A vegetation-

atmosphere interaction study for Amazonian deforestation using field data and a single-column model. *Q. J. R. Meteorol. Soc.*, in press.
da Rocha, H.R., Sellers, P.J., Collatz, G.J., Wright, I.R. and Grace, J., 1996b. Calibration and use of the SiB2 model to estimate water vapour and carbon exchange at the ABRACOS sites. This volume.
de Souza, J.R.S., Pinheiro, F.M.A., de Araujo, R.L.C., Pinheiro, H.S. and Hodnett, M.G. 1996. Temperature and moisture content profiles in the soil beneath forest and pasture areas in eastern Amazonia. This volume.
Dias, M.A.F.S. and Regnier, P. 1996. Simulation of meso-scale circulations in a deforested area of Rondonia in the dry season. This volume
Fisch, G., Culf, A.D. and Nobre, C.A. 1996. Modelling convective boundary layer dynamics in Amazonia. This volume.
Grace, J., Lloyd, J., McIntyre, J., Miranda, A., Meir, P., Miranda, H., Moncrieff, J., Massheder, J., Wright, I. and Gash, J. 1995a. Fluxes of carbon dioxide and water vapour over an undisturbed tropical forest in south-west Amazonia. *Global Change Biology* **1**, 1-12.
Grace, J., Lloyd, J., McIntyre, J, Miranda, A., Meir, P., Miranda, H., Nobre, C., Moncrieff, J., Massheder, J., Mahli, Y., Wright, I. and Gash, J. 1995b. Carbon dioxide uptake by an undisturbed tropical rain forest in South-West Amazonia, 1992-1993. *Science*, **270**, 778-780.
Grace, J., Lloyd, J., McIntyre, J., Miranda, A., Meir, P. and Miranda, H. 1996. Carbon flux over Amazon forest in Rondônia. This volume.
Hodnett, M.G., Oyama, M.D., Tomasella, J. and Marques, A. de O. 1996a. Comparisons of long-term soil water storage behaviour under pasture and forest in three areas of Amazonia. This volume.
Hodnett, M.G., Tomasella, J., Marques, A. de O., and Oyama, M.D. 1996b. Deep soil water uptake by forest in central Amazonia: predictions from long term daily rainfall data using a simple water balance model. This volume.
Honzak, M., Foody, G.M., Lucas, R.M., Curran, P.J. Amaral, I. and Amaral, S. 1996. Estimation of the leaf area index and total biomass of tropical regenerating forest: a comparison of methodologies. This volume.
Kruijt, B., Lloyd, J., Grace, J., McIntyre, J., Farquhar, G., Miranda, A.C. and Cracken, P. 1996. Sources and sinks of CO_2 in Rondonian tropical rainforest, inferred from concentrations and turbulence along a vertical gradient. This volume.
Lean. J, Bunton, C.B., Nobre, C.A. and Rowntree, P.R. 1996. The simulated impact of Amazonian deforestation on climate using measured ABRACOS vegetation characteristics. This volume.
Lloyd, C.R., Gash, J.H.C., Shuttleworth, W.J. and Marques, A. de O. 1988. The measurement and modelling of rainfall interception by Amazonian rain forest. *Agric. and Forest Meteorol.* **43**, 277-294.
Lloyd, C.R. and Marques, A.de O, 1988. Spatial variability of throughfall and stemflowmeasurements in Amazonian rainforest. *Agric. and Forest Meteorol.* **42**, 63-73.
Lloyd, J., Grace, J., Miranda, A.C., Meir, P., Wong, S.C., Miranda, H., Wright, I., Gash, J.H.C. and McIntyre, J. 1995. A simple calibrated model of Amazon rainforest productivity based on leaf biochemical properties. *Plant Cell and Environment* **18**, 1129-1145.
Lucas, R.M., Curran, P.J., Honzak, M., Foody, G.M., Amaral, I. and Amaral, S. 1996. Disturbance and recovery of tropical forests: balancing the carbon account. This volume.
Maitelli, G. and Wright, I.R. 1996. On the climate of a riverside city in the Amazon Basin: urban-rural differences in temperature and humidity. This volume.
Manzi, O. and Planton. S. 1996. Calibration of a GCM using ABRACOS and ARME data and a simulation of Amazonian deforestation. This volume.
McWilliam, A.L.C., Roberts, J.M., Cabral, O.M.R., Leitão, M.V.B.R., de Costa, A.C.L. Maitelli, G.T. and Zamporoni, C.A.G.P. 1993. Leaf area index and above-ground biomass of *terra firme* rain forest and adjacent clearings in Amazonia. *Functional Ecol.* **7**, 310-317.
McWilliam, A.L.C.,Cabral, O.M.R., Gomes, B.M. Esteves, J.L. and Roberts, J. 1996. Forest and pasture leaf-gas exchange in south-west Amazonia. This volume.
Meir, P., Grace, J., Miranda, A. and Lloyd, J. 1996. Soil respiration in a rainforest in Amazonia, and

in cerrado in central Brazil. This volume.
Miranda, A.C., Miranda, H.S., Lloyd, J., Grace, J., McIntyre, J.A., Meir, P., Riggan, P., Lockwood, R. and Brass, J. 1996. Fluxes of CO_2 over a cerrado *sensu stricto* in Central Brazil. This volume.
Nepstad, D.C., de Carvalho, C.R., Davidson, E.A., Jipp, P.H., Lefebvre, P.A., Negreiros, G.H., da Silva, E.D., Stone, T.A., Trumbore, S.E. and Vieira, S. 1994. The role of deep roots in the hydrological and carbon cycles of Amazonian forests and pastures. *Nature* **372**: 666-669.
Nobre, C.A., Fisch, G., da Rocha, H.R., Lyra, R.F.F., da Rocha, E.P. and Ubarana,V.N. 1996. Observations of the atmospheric boundary layer in Rondônia. This volume.
Ribeiro, A., Victoria, R.L., Martinelli, L.A. Moreira, M.Z., and Roberts, J. 1996. The isotopic composition of the atmospheric water vapour inside a canopy in the Amazon forest: vertical and diurnal variation. This volume.
Roberts, J., Cabral, O.M.R, de Costa, J.P., McWilliam, A.L.C. and Sá. T.D. de A. 1996. An overview of the leaf area index and physiological measurements during ABRACOS. This volume.
Sá, T.D. de A., da Costa, J. de P.R., Roberts, J.M. 1996. Forest and pasture conductances in southern Pará, Amazonia. This volume.
Sellers, P.J., Randall, D.A., Collatz, C.J., Berry, J.A., Field, C.J., Dazlich, D.A., Zhang, C. and Collelo, 1996. Calibrating the Simple Biosphere Model for Amazonian tropical forest using field and remote sensing data. Part I: average calibration with field data. *J. Climate*, in press.
Tomasella, J. and Hodnett, M.G. 1996. Soil hydraulic properties and van Genuchten parameters for an oxisol under pasture in central Amazonia. This volume.
Ubarana, V.N. 1996. Observation and modelling of rainfall interception loss in two experimental sites in Amazonian forest. This volume.
Wright, I.R., Gash, J.H.C., da Rocha, H.R., Shuttleworth, W.J., Nobre, C.A., Maitelli, G.T.M., Zamparoni, C.A.G.P. and Carvalho, P.R.A. 1992. Dry season micrometeorology of central Amazonian ranchland. *Q. J. R. Meteorol. Soc.*, **118**, 1083-1099.
Wright. I.R., Manzi, A.O. and da Rocha, H.R. 1995. Canopy surface conductance of Amazonian pasture: model application and calibration for canopy climate. *Agric. and Forest Meteorol.* **75**: 51-70.
Wright, I.R., Gash. J.H.C., da Rocha, H.R. and Roberts, J.M. 1996a. Modelling surface conductance for Amazonian pasture and forest. This volume.
Wright, I.R., Nobre, C.A., Tomasella, J., da Rocha, H.R., Roberts, J.M., Vertamatti, E., Culf, A.D., Alvalá, R.C., Hodnett, M.G. and Ubarana, V.N. 1996b. Towards a GCM surface parameterisation for Amazonia. This volume.

Conclusões do projeto ABRACOS

INTRODUÇÃO

A experiência de projetos cuja motivação primária é a coleta de dados mostra que freqüentemente os benefícios plenos do trabalho somente são obtidos alguns anos após o término da fase de coleta de dados. Os resultados descritos nesse livro devem ser considerados somente como o início da análise e aplicação dos dados do Projeto

ABRACOS. No entanto, foram revelados uma série de novos conhecimentos sobre as interações solo-vegetação-atmosfera para florestas e pastagens na Amazônia. Este capítulo apresenta um sumário dos resultados principais obtidos até o momento e as conclusões a que levam esses resultados.

CLIMATOLOGIA

Em média os sítios experimentais de florestas absorveram 11% mais radiação do que as pastagens (Culf et al., 1996). Isto advém do fato de a floresta refletir menos radiação solar e emitir menos radiação de onda longa. O albedo médio da floresta foi de 0,134, ligeiramente mais alto do que o valor usualmente utilizado em simulações numéricas de desmatamento, enquanto o albedo médio da pastagem, 0,180, foi ligeiramente menor que os valores habitualmente utilizados. Surpreendentemente as gramíneas das pastagens não apresentaram uma forte sazonalidade no albedo, ao passo que o albedo da floresta mostrou uma variação sazonal bem definida que não ocorre devido aos efeitos de variações do ângulo de elevação solar ou a variações de nebulosidade, mas está correlacionada umidade do solo (Culf et al., 1995, 1996). Embora o albedo dos sítios de pastagem não tenha mostrado uma clara tendência sazonal, variações de mês a mês foram observadas, estando associadas ao índice de área foliar (Wright et al., 1996b). Em Ji-Paraná, diferenças sistemáticas na radiação solar incidente entre os sítios experimentais de floresta e pastagem foram observadas durante a estação seca (Culf et al., 1996). Estas diferenças podem estar relacionadas com o aumento de nebulosidade sobre a pastagem durante aquela época do ano, fato este evidenciado também no trabalho de Cutrim et al. (1995). Em se confirmando essas observações, é um resultado importante na medida que indica um efeito direto da mudança de cobertura vegetal num fenômeno atmosférico de mesoescala.

As pastagens apresentaram temperaturas máximas durante o dia mais altas e amplitudes de temperatura (temperatura máxima menos temperatura mínima) também mais altas (Culf et al., 1996). Geralmente o mínimo de temperatura foi menor à noite para a pastagem. Este resultado está associado provavelmente às baixas velocidades do vento prôximo à superfície para as pastagens à noite, o que pode levar à redução dos processos turbulentos de mistura na vertical e maior estabilidade. A temperatura durante o dia na área urbana de Manaus foi sistematicamente mais alta que aquela nas duas áreas rurais (Fazenda Dimona e Reserva Ducke), mas as temperaturas na cidade à noite foram similares àquelas sobre a floresta (Maitelli and Wright, 1996). Há pequena variação sazonal de temperatura em Manaus ou Marabá, porém há um resfriamento considerável durante a estação seca em Ji-Paraná, associado à advecção de ar frio de latitudes extratropicais do Hemisfério Sul devido à passagem de sistemas frontais na América do Sul subtropical. Um marcante ciclo anual de umidade foi observado em Ji-Paraná e Marabá, mas não foi observado em Manaus. Valores mais baixos de umidade durante a estação seca estão associados a subsidência de grande escala, que é predominante próximo às fronteiras do domínio florestal (como nas regiões de Marabá e Ji-Paraná no sudeste e sudoeste da Amazônia, respectivamente).

UMIDADE E PROPRIEDADES FÍSICAS DO SOLO

A representação da variabilidade espacial das propriedades hidráulicas do solo nos modelos de grande escala permanece como um grande desafio para as comunidades de física do solo e de modelagem de MCG. As propriedades do solo são altamente variáveis, entretanto as parametrizações de modelos de grande escala requerem valores médios representativos de áreas muito grandes. As observações de propriedades do solo do Projeto ABRACOS foram necessariamente limitadas, mas ao adotar uma simplificação pragmática dos mapas de solos da Amazônia, e com pelo menos algumas medições das propriedades de três tipos principais de solos, Wright *et al.* (1996) derivaram uma representação de solos para utilização em modelos de grande escala da Amazônia que é, atualmente, muito mais realística do que as representações que existiam anteriormente.

Tomasella e Hodnett (1996) acharam que a condutividade hidráulica da floresta à superfície é tão alta que torna-se impossível medi-la. Isso é devido ao sistema de raízes e aos macroporos. No entanto, as medidas na pastagem de Manaus mostraram uma redução na condutividade do solo à superfície, o que foi provavelmente devido ao pisoteio do gado. A condutividade hidráulica saturada na pastagem variou enormemente com a profundidade, de mais de 100 mm h^{-1} à superfície até 17 mm h^{-1} a 1,3 m de profundidade. Zonas de alta condutividade apresentaram um número substancial de meso e macroporos. A maior parte do impacto do desmatamento ocorre nos primeiros 20 cm do solo, abaixo disso as propriedades observadas dos solos sob floresta e sob pastagem foram bastante similares. No entanto, esta conclusão é dependente da prática de manejo de utilizar pouca ou nenhuma máquina pesada e de manter baixas densidades de cabeças de gado. Sob estas condições, não foram encontrados compactação ou baixos valores de condutividade hidráulica na parte superior do solo para qualquer um dos sítios. A curva de retenção hídrica e a condutividade hidráulica não saturada puderam ser bem ajustados através do modelo de van Genuchten, mas os parâmetros foram muito diferentes de valores publicados para solos argilosos de regiões temperadas. As propriedades hidráulicas desses solos não deveriam ser estimadas a partir de textura utilizando relações derivadas para solos de regiões temperadas; essas relações implicariam em superestimar a disponibilidade hídrica e subestimar a condutividade não saturada se aplicadas ao solo de pastagem de Manaus.

A condutividade térmica apresentou menor variabilidade quando comparada às propriedades hidráulicas, variando de 1,2 a 2,9 W m^{-1} C^{-1} (Alvalá *et al.*, 1996). Para a pastagem Souza *et al.* (1996) encontrou que o fluxo de calor direciona-se para baixo durante a estação seca e para cima durante a estação chuvosa, enquanto que, para a área de floresta, foi pequeno mas consistentemente para cima. Em média, para o sítio experimental de Marabá, a diferença de temperatura da porção superior do solo entre pastagem e floresta foi de 2,4 °C. Estas diferenças em fluxo e temperatura podem ser explicadas em termos das diferentes variações de conteúdo de água e diferenças nos balanços de energia à superfície do solo para os dois tipos de vegetação.

Hodnett *et al.* (1996) mostrou que durante a estação seca houve sistematicamente maior extração de água de perfis de água no solo sob a floresta, resultando em perfis mais secos, dentro da faixa de mediçoes, ao final da estação seca. Há claras indicações provenientes de outros estudos (Chauvel *et al.*, 1992; Nepstad *et al.*, 1994) que confirmam e consubstanciam esse resultado de que a floresta está extraindo água de profundidades maiores que 3,6 m (profundidade máxima dos tubos de acesso no ABRACOS). Estas indicações encontram suporte nas medidas de variações máximas de armazenamento registradas para cada um dossítios experimentais. Consistente com uma estação seca mais longa, as máximas variações de armazenamento até a profundidade de 3,6 m em Marabá foram de 742 mm e 376 mm para floresta e pastagem, respectivamente. Tanto as variações de armazenamento quanto as diferenças de armazenamento entre floresta e pastagem foram bem menores para Manaus onde a estação seca é menos severa, no entanto a baixa capacidade de retenção d'água para os solos de Manaus acabou resultando que água do solo foi extraída de profundidades maiores. É interessante observar que não houve recarga completa do perfil de água no solo durante a estação chuvosa que se seguiu à máxima depleção observada durante a estação seca sob a floresta em Marabá, sendo 349 mm abaixo de seu valor na mesma época do ano anterior. O termo de armazenamento do balanço de água não foi nulo ao final do ciclo anual e, se esses resultados fossem extrapolados para a escala regional, a diferença entre precipitação e vazão fluvial não resultariam na evaporação média anual de grande escala. Embora isso possa ser uma ocorrência incomum, tem óbvias implicações na utilização de medições hidrológicas para verificar e validar estimativas de evaporação obtidas em modelos de grande escala. O quadro geral para água no solo mostra que existem grande diferenças nas variações sazonais de conteúdo de água no solo, tanto entre floresta e pastagem como entre os sítios experimentais. Essas diferenças ocorrem como resultado dos diferentes regimes de precipitação, combinados com as diferenças de propriedades do solo, comportamento do lençol freático e profundidade das raízes das florestas e das gramíneas.

Modelos foram desenvolvidos para calcular o armazenamento de água nos 2 m superiores do solo para floresta e pastagem usando somente precipitação diária como dado de entrada (Hodnett *et al.*, 1996b). Os modelos foram desenvolvidos e calibrados utilizando-se precipitação, conteúdo de água no solo e dados micrometeorológicos do período de estudo do Projeto ABRACOS. Evaporação e drenagem profunda foram relacionados ao armazenamento de água no solo através de funções lineares e exponenciais, respectivamente, possibilitando determinar o armazenamento de água no solo a partir de um cálculo simples de balanço hídrico. O armazenamento de água no solo foi, então, calculado a partir da série histórica de 27 anos de dados de precipitação diária da Reserva Ducke, próxima à Manaus, assumindo que a evaporação da floresta não é afetada pelo estresse hídrico no solo.

Para floresta o modelo previu extração de profundidades abaixo de 2 m em 20 dos 27 anos da série histórica. A extração média de profundidades abaixo de 2 m foi de 60 mm, com um máximo de 231 mm em 1967. Pelo fato de que a capacidade de retenção d'água dos solos da região de Manaus é baixa, essa extração deve ser oriunda de uma grande faixa de solos profundos e, assim, enraizamento profundo é essencial.

Ainda há muito a ser aprendido sobre as propriedades e comportamento dos solos da floresta Amazônica. Medições das propriedades hidráulicas de solos são necessárias para o restante dos tipos de solos majoritários da Amazônia. Tanto as propriedades hidráulicas dos solos como o conteúdo de água no solo precisam ser medidos a profundidades maiores que 4 m de modo a fornecer um quadro completo do ciclo de água abaixo da floresta.

INTERCEPTAÇÃO DE PRECIPITAÇÃO PELA FLORESTA

Dados de interceptação de chuva foram coletados por Ubarana (1996) nos sítios experimentais na Reserva Jaru e na Reserva Vale do Rio Doce. Este trabalho seguiu a mesma metodologia utilizada previamente por Lloyd e Marques (1988) na Reserva Ducke e, portanto, forneceu resultados diretamente comparáveis. Uma fração mais alta da chuva evaporou diretamente de volta à atmosfera nos dois novos sítios: 12 e 13 porcento nas Reservas Jaru e Vale do Rio Doce, respectivamente, comparado a 9 porcento na Reserva Ducke (Lloyd *et al.*, 1988). O regime tropical de precipitação pluviométrica faz com que asperdas por interceptação sejam uma diferença relativamente pequena entre precipitação sobre o dossel e precipitação sob o dossel ("throughfall"), e esta última, devido a estrutura morfológica da vegetação, é muito variável. Estes fatores tornam as medidas de perdas por interceptação sujeitas a grandes erros. As diferenças entre as medidas dos diferentes sítios experimentais são comparáveis aos valores dos erros de mensuração e, portanto, não é possível provar conclusivamente que as perdas por interceptação nos dois novos sítios foram sistematicamente maiores do que as medidas no sítio original na Reserva Ducke. Por outro lado, as diferenças foram consistentes com as diferenças encontradas nos parâmetros da estrutura da floresta que controlam a evaporação: dossel mais fechado e maiores capacidades de armazenamento deste foram observadas nas Reservas Jaru e Vale do Rio Doce em comparação à Reserva Ducke. Dentro dos limites das medições, a evaporação de todos os sítios pode ser estimada com sucesso com o modelo de Rutter.

COMPORTAMENTO ESTOMÁTICO

Um parâmetro crucial nos modelos de circulação geral é a condutância de superfície para o transporte da vapor d'água de superfícies vegetadas. A condutância de superfície integra a influência de condutâncias estomáticas individuais no volume do dossel. Enquanto a condutância de superfície "bulk" pode ser medida diretamente utilizando métodos micrometeorológicos, um entendimento mais completo de suas flutuações requer informação sobre a condutância estomática ao nível da folha e sua resposta a fatores que mudam diuturna e sazonalmente, como a magnitude e variações no índice de área foliar.

Nos sítios de pastos em Ji-Paraná e Manaus, McWilliam *et al.* (1996) observou estreitos relacionamentos entre condutância estomática e déficit de água no solo, mas em Marabá a condutância estomática foi menos afetada pelo aumento do déficit de água no solo (Sá *et al.*, 1996). A transpiração total é controlada pela combinação de condutância estomática e área foliar. No entanto, o índice de área foliar nos sítios de pastagem também foi afetado pela água no solo e variou até 50 porcento durante o curso de um ano, A variação sazonal resultante é qualitativamente consistente com a variação observada nos dados de água no solo e medidas micrometeorológicas dos fluxos de superfície.

Apesar das diferenças em composição botânica entre e dentro dos três sítios de florestas, as respostas fisiológicas das diferentes espécies foram bastante similares (Roberts *et al.*, 1996). Não havia em geral resposta à água no solo e foram observadas respostas similares ao déficit de umidade do ar. A condutância estomática de todas as espécies depende da altura — diminuindo de cima para baixo no dossel. Esta falta de variação entre as espécies é um resultado útil porque significa que a considerável quantidade de trabalhos na literatura científica sobre o comportamento estomático de espécies da floresta tropical pode ser explorada para derivar resultado de aplicação mais geral.

A biomassa acima do solo e o índice de área foliar de um sítio foram medidos diretamente através de amostragem destrutiva (McWilliam *et al.*, 1993), e índices de área foliar de outros sítios de floresta foram estimados a partir de medidas de folhas caídas na liteira (Roberts *et al.*, 1996). O índice de área foliar foi estimado em 5,7, 5,4 e 4,4 em Manaus, Marabá e Ji-Paraná, respectivamente, com variação sazonal de somente cerca de 15 porcento. Estas estimativas dependem da suposição de que a vida média das folhas é de um ano; esta suposição ainda necessita ser comprovada, mas parece ter validade.

O microclima e os níveis de radiação são diferentes para cada profundidade do dossel, mas funções empíricas das variáveis responsáveis pela transpiração foram derivadas por Cabral *et al.* (1996) de tal modo que a transpiração total da floresta pode ser calculada usando medições de estagues meteorológicas acima do dossel, um modelo do dossel de múltiplas camadas e oconhecimento de como a área foliar e a condutância estomática variam com a altura e sazonalmente. A aplicação desse modelo revelou que a parte inferior do dossel contribui relativamente pouco à transpiração total. Isso ocorre apesar do fato de que alta proporção do dossel encontra-se nos baixos níveis e é explicado porque, nessa parte do dossel, condutâncias estomáticas mais baixas e níveis de radiação estão acoplados a baixos valores dos déficit de pressão de vapor e baixas velocidades do vento. Estudos da composição isotópica do vapor d' água no dossel por Ribeiro *et al.* (1996) mostraram que a parte inferior do dossel esta efetivamente desacoplada da parte intermediária, a qual contribui o máximo para a transpiração total. Medidas da variação da composição isotópica de carbono sugerem que o comportamento fisiológico da vegetação na parte inferior do dossel (Kruijt *et al.*, 1996) pode também ser diferente daquela da parte superior. Portanto, para o dossel complexo da floresta tropical, o conceito de um simples e único valor para o índice de área foliar, o qual é utilizado na maioria dos modelos preditivos, provavelmente é uma representação inadequada na situação real.

MICROMETEOROLOGIA

Nos quatro anos de medidas de campo do Projeto ABRACOS houve sete campanhas intensivas de monitoramento. O objetivo dessas campanhas de campo foi o de avaliar os parâmetros físicos que descrevem a micrometeorologia dos sítios experimentais, e fornecer estimativas de todas as componentes do balanço de energia, incluindo evaporação, a qual pode ser utilizada para calibrar modelos da superfície vegetada. Encontrou-se que as características da rugosidade aerodinâmica dos sítios conformam-se a relacionamentos convencionais baseados na estrutura da vegetação (Wright *et al.*, 1996b). Durante a estação chuvosa, a partição de energia para evaporação foi similar para floresta e pastagem, mas a evaporação total da pastagem nessa estação foi tipicamente 10 a 15 porcento menor em comparação com a floresta devido a reduzida energia disponível na pastagem e a rugosidades aerodinâmicas mais suavizadas (Wright *et al.*, 1992).

Durante a estação seca, as pastagens, que têm raízes mais rasas, foram todas afetadas pela diminuição das reservas de água no solo, apesar que com intensidades variadas dependendo do tipo de solo e da precipitação. Nos solos argilosos próximos à Manaus, a transpiração das pastagens declinou rapidamente depois de somente 10 dias sem chuva (Wright*et al.*, 1992). Em contraste, nenhuma atenuação significativa em transpiração foi observada em qualquer dos sítios de floresta durante os períodos secos, inclusive durante as estações secas mais longas em Marabá e Ji-Paraná (Wright *et al.*, 1996a). Parametrizações atuais de MCG para floresta Amazônica tendem a subestimar a profundidade da zona de raízes e, portanto, erroneamente reduzem a evaporação quando a água no solo na parte superior do solo é reduzida.

Dados de fluxos evaporativos das primeiras cinco campanhas foram utilizadas para calibrar modelos os quais calculam condutâncias estomáticas "bulk" para dosséis de florestas e pastagens a partir de dados climáticos acima do dossel (Wright *et al.*, 1995 e 1996a). Isso é uma importante calibração para os MCG na medida que determina a partição diurna e sazonal de energia em calor para convecção e formação de nuvens, e a quantidade de vapor d'água retornada do solo à atmosfera (Rocha*et al.*, 1996b). A similaridade entre os parâmetros de condutância de Manaus e Ji-Paraná e a relativa falta de similaridade nos parâmetros individuais para cada sítio de floresta indicam a acurácia e adequação desse modelo para calcular a evaporação para toda a Amazônia. A utilização dos conjuntos de parâmetros dos sítios de Manaus para calcular a transpiração de Ji-Paraná na floresta e pastagem produziu resultados encorajadores (Wright *et al.*, 1996a), porém ressaltaram a importância de somenteaplicar parâmetros otimizados como conjuntos completos.

DIÓXIDO DE CARBONO

Medições com a técnica de correlação de vórtice turbulentos ("eddy correlation") sobre a floresta da Reserva Jaru mostraram que o fluxo de CO_2 pode ser medido confiável e continuamente durante prolongados permodos. As medições de Grace *et al.* (1995a, 1995b, 1996) mostraram que a floresta em Jaru estava acumulando carbono. Grace *et al.* (1996) aplicou o modelo desenvolvido por Lloyd *et al.* (1995), e Rocha *et al.* (1996a) aplicou o modelo desenvolvido por Sellers *et al.* (1995), para estimar o balanço de carbono a partir de dados climáticos coletados acima da floresta na estação meteorológica automática. Em ambos os casos, esta modelagem calculou que a floresta era um sorvedouro significativo para carbono. No entanto, o fluxo líquido de carbono é uma pequena diferença entre fotossíntese e respiração: Meir (1996) mostrou que o fluxo líquido é somente 6 porcento da respiração do solo e, porque a respiração do solo apresenta alta sensibilidade à temperatura, o fluxo lmquido é também altamente sensível a uma variação de um grau ou menos. Se for extrapolado para toda a Amazônia, estas estimativas significam que a Amazônia seria um sorvedouro de aproximadamente 0,5 G tons de carbono por ano. Este valor é altamente significativo em relação ao chamado "missing carbon" de cerca de 1 G ton de carbono por ano, porém considerando o curto período de observações, as limitações dos modelos e a variabilidade espacial, na Amazônia, das forçantes climáticas de entrada dos modelos, estas estimativas devem ser tomadas com muita cautela. Há uma urgente necessidade de coletar mais dados de fluxos de carbono. Estes dados de fluxos dever ser coletados por períodos longos para amostrar a variabilidade interanual e medidas são necessárias em todos os principais ecossistemas da Amazônia.

Medidas de fluxos de carbono sobre a vegetação dos cerrados por Miranda *et al.* (1996) mostram um comportamento bem diferente para áreas com estações secas muito longas. Há uma variação sazonal marcante, com acumulação de carbono pelo sistema durante a estação chuvosa e perda de carbono no final da estação seca. No sítio experimental estudado, a acumulação anual é estimada situar-se entre 2 a 3 tons de carbono por hectare por ano. No entanto, como a vegetação do cerrado é altamente variável e dependente da história privia de fogo do local, estes resultados iniciais devem ser considerados como específicos àquele sítio experimental.

Estimativas de biomassa em floresta secundária por Honzack *et al.* (1996) confirmam a importância da rebrota (crescimento secundário) como um sorvedouro de carbono. A extensco e tipo de florestas secundárias em crescimento podem ser determinados por sensoriamento remoto (Lucas *et al.*, 1996), permitindo, desse modo, a extrapolação das estimativas de acumulação de carbono em pequenas áreas experimentais para toda a Bacia Amazônica.

ESTUDOS DA CAMADA LIMITE ATMOSFÉRICA

A camada limite atmosférica durante o dia atingiu alturas 700 a 1000 m mais altas sobre áreas com desmatamentos substanciais do que áreas adjacentes com florestas na região de Ji-Paraná (Nobre *et al.*, 1996), o que mostrou-se qualitativamente consistente com as observações de aumento do fluxo de calor sensível à superfície e diminuição da evapotranspiração sobre a pastagem em comparação com a floresta. Nobre *et al.* (1996) também observaram uma pequena diminuição na quantidade total de vapor d' água sobre a pastagem em relação à floresta. Fisch *et al.* (1996) aplicou um modelo simples, unidimensional de crescimento dacamada limite (forgado por calor sensível à superfície e entranhamento no topo) e concluiu que este modelo fornece uma adequada representação do crescimento da camada limite diurna sobre a floresta durante todo o dia e, para a pastagem, da metade da manhã para a frente. No entanto, este modelo falhou em representar adequadamente o rápido crescimento da camada limite sobre a pastagem durante a manhã antes das 11 horas (hora local). Uma explicação possível é que o padrão espacial de mudanças na cobertura vegetal naquela região de Rondônia (faixas de florestas e pastagens, padrão este conhecido como "espinha de peixe") não é adequadamente representado nesse modelo de crescimento da camada limite. Dias e Regnier (1996) utilizaram um modelo tridimensional complexo (RAMS) para estudar o impacto nos campos de vento, temperatura e umidade da distribuição observada de cobertura de vegetação e topografia na região de Ji-Paraná: uma grande área desmatada cercada por florestas; a principal conclusão foi a de que o transporte vertical de momentum, calor e vapor d'água por circulações de mesoescala é significativo e não pode ser explicado pelas parametrizações de turbulência local em MCG.

MODELAGEM CLIMÁTICA

O dados do Projeto ABRACOS das campanhas de 1990 e 1991 foram utilizados para calibrar o modelo climático do Hadley Centre da Grã-Bretanha (Lean *et al.*, 1996), o modelo climático do serviço de meteorologia da França (Manzi e Planton, 1996), e também o SiB (Rocha *et al.*, 1996b), sendo este o modelo da superfície vegetada utilizado no centro brasileiro de modelagem (Centro de Previsão de Tempo e Estudos Climáticos - CPTEC), e no GSFC/NASA e Center for Ocean-Land-Atmosphere Studies (COLA), ambos nos EUA. Experimentos de desmatamentos com MCG foram realizados com todos esses modelos.

Os resultados mais recentes desses experimentos numéricos de desmatamento (Lean *et al.*, 1996) indicam uma resposta substancial ao desmatamento que varia regionalmente: alterações do total anual de precipitação podem chegar a até 20 porcento (diminuição para o caso da Amazônia coberta por pastagens), com uma tendência de orientação NW-SE das modificações de precipitação, estendendo-se a SE da área desmatada; neste experimento numérico em particular, a resposta é sensível a (pelo menos) albedo, rugosidade e infiltração máxima à superfície, mas a analise de sensibilidade do MCG francês (Manzi e Planton, 1996) também revelou considerável sensibilidade ao tipo de esquema de convecção utilizado. Os resultados de um experimento por Lean *et al.* (1996), no qual somente parte da Amazônia foi desmatada, isto é, um cenário mais realístico, não poderia ser deduzido facilmente do experimento de desmatamento total. Estas diferenças enfatizam o fato de que todas as previsões dos MCG do clima são muito sensíveis à representação das características da superfície vegetada.

É interessante notar que as modificações climáticas previstas nas distribuições de temperatura e umidade como resultado do desmatamento da Amazônia nesses estudos de simulação são da mesma magnitude daquelas previstas para a Amazônia ao se dobrar a concentração atmosférica de CO_2, fazendo o impacto regional de modificações da cobertura da vegetação na Amazônia (pelo menos) tão importante como os potenciais efeitos regionais do aquecimento global.

COMENTARIO FINAL

No início do Projeto ABRACOS a comunidade de modelagem climática estudiosa da questão dos impactos climáticos dos desmatamentos da Amazônia havia obtido avanços em seus modelos a um ponto onde progressos adicionais estavam mais limitados pela falta de dadosdo que por desenvolvimentos de modelagem. Os dados coletados pelo Projeto ABRACOS modificam esse quadro. A prioridade agora é aplicar, nos modelos, essa enorme quantidade e variedade de dados coletados sob os auspícios do Projeto ABRACOS. Os dados já foram utilizados para derivar alguns dos parâmetros requeridos e econtram-se amplamente disponíveis para todos que queiram utiliza-los para desenvolver, calibrar ou validar seus modelos. Um sumario dos parâmetros gerados pelos estudos do ABRACOS é apresentado em forma tabular por Wright *et al.* (1996). A utilização da informação nessa tabela já representaria uma melhora imediata ao realismo de estudos de modelagem climática da Amazônia. Mais do que isso, o aumento no entendimento de como os processos, no solo, plantas e atmosfera, modificam-se quando a floresta é substituída por pastagem significa que podemos ter mais confiança nas interpretações das previsões dos modelos.

Index

Index compiled by Colin Will